T0180897

Lecture Notes in Computer Science   12757

Advanced Research in Computing and Software Science
Subline of Lecture Notes in Computer Science

Paola Flocchini · Lucia Moura (Eds.)

Combinatorial Algorithms

32nd International Workshop, IWOCA 2021
Ottawa, ON, Canada, July 5–7, 2021
Proceedings

 Springer

Editors
Paola Flocchini ⓘ
University of Ottawa
Ottawa, ON, Canada

Lucia Moura ⓘ
University of Ottawa
Ottawa, ON, Canada

ISSN 0302-9743   ISSN 1611-3349 (electronic)
Lecture Notes in Computer Science
ISBN 978-3-030-79986-1   ISBN 978-3-030-79987-8 (eBook)
https://doi.org/10.1007/978-3-030-79987-8

LNCS Sublibrary: SL1 – Theoretical Computer Science and General Issues

This Springer imprint is published by the registered company Springer Nature Switzerland AG
The registered company address is: Gewerbestrasse 11, 6330 Cham, Switzerland

Preface

The 32nd International Workshop on Combinatorial Algorithms (IWOCA 2021) was originally scheduled to take place at the University of Ottawa, Canada. Due to the COVID-19 pandemic, it was decided, a few months before the paper submission deadline, that the conference would move online. The conference took place during July 5–7, 2021, organized and coordinated from Ottawa by the IWOCA 2021 Organizing Committee, with online logistics support by The Fields Institute for Research in Mathematical Sciences.

IWOCA is an annual conference series that started in 1989 as AWOCA (Australasian Workshop on Combinatorial Algorithms) and became an international conference in 2007, having been held in Australia, Canada, Czech Republic, Finland, France, Indonesia, India, Italy, Japan, Singapore, South Korea, UK, and USA. The conference brings together researchers on diverse topics related to combinatorial algorithms, such as algorithms and data structures; algorithmic game theory; approximation algorithms; complexity theory; combinatorics and graph theory; combinatorial generation and enumeration; combinatorial optimization; combinatorics of words and strings; computational geometry; computational biology; cryptography and information security; graph algorithms; graph drawing and labelling; decompositions and combinatorial designs; distributed and network algorithms; dynamic and evolving networks; mobile agents; new paradigms of computation; online algorithms; parallel algorithms; parameterized and exact algorithms; probabilistic and randomized algorithms; and streaming algorithms.

The Program Committee (PC) of IWOCA 2021 received 107 submissions. Each submission was reviewed by at least three PC members and some trusted external referees, and evaluated on its quality, originality, and relevance to the conference. The PC selected 38 papers for presentation at the conference and inclusion in the proceedings.

Four invited talks were given at IWOCA 2021, by *Maria Chudnovsky* (Princeton University, USA), *Anna Lubiw* (University of Waterloo, Canada), *David Peleg* (Weizmann Institute of Science, Israel), and *Alfred Wassermann* (University of Bayreuth, Germany). This volume contains the abstracts of the four invited talks and the paper versions of two of those invited talks.

The PC selected two contributions for the best paper and the best student paper awards. The best paper award was given to *Benjamin Merlin Bumpus and Kitty Meeks* for their paper "Edge Exploration of Temporal Graphs". The best student paper award was given to *Stefan Lendl, Gerhard J. Woeginger, and Lasse Wulf* for their paper "Non-preemptive Tree Packing".

IWOCA 2021 was held more than a year after a global pandemic was declared in March 2020, a period that has been a trying time for everyone around the world. During this period, computer scientists and mathematicians have quickly learned how to teach, research, and collaborate remotely. In the midst of these uncertain times, the response to our call for papers was a record number of submissions, for which we are very grateful. While we long for a time when we will be sharing knowledge in the same geographical location, we are also very thankful to many individuals and organizations who worked together to build a successful online conference.

We would like to thank all the authors who responded to the call for papers, the invited speakers, the members of the PC, the external referees, and the members of the Organizing Committee.

We also thank Springer for publishing the proceedings of IWOCA 2021 in their ARCoSS/LNCS series and for their financial support towards the best paper and the best student paper awards.

We gratefully acknowledge the financial and logistics support from the following institutions: The Fields Institute for Research in Mathematical Sciences, the Faculty of Engineering at uOttawa, and the University of Ottawa.

Finally, we thank the Steering Committee for giving us the opportunity to serve as program chairs of IWOCA 2021 and for their continuous support.

July 2021   Paola Flocchini
                                                                                                 Lucia Moura

Organization

Steering Committee

Maria Chudnovsky	Princeton University, USA
Charles Colbourn	Arizona State University, USA
Costas Iliopoulos	King's College London, UK
Ralf Klasing	CNRS and University of Bordeaux, France
Wing-Kin (Ken) Sung	National University of Singapore, Singapore

Program Committee

Hans L. Bodlaender	Utrecht University, Netherlands
Ljiljana Brankovic	University of Newcastle, UK
Charles Colbourn	Arizona State University, USA
Alessio Conte	University of Pisa, Italy
Celina M. H. de Figueiredo	Federal University of Rio de Janeiro, Brazil
Giuseppe Antonio Di Luna	University of Rome La Sapienza, Italy
Vida Dujmović	University of Ottawa, Canada
Cristina G. Fernandes	University of São Paulo, Brazil
Henning Fernau	University of Trier, Germany
Gabriele Fici	University of Palermo, Italy
Paola Flocchini (Co-chair)	University of Ottawa, Canada
Florent Foucaud	Clermont Auvergne University, France
Dalibor Froncek	University of Minnesota Duluth, USA
Travis Gagie	Dalhousie University, Canada
Luisa Gargano	University of Salerno, Italy
Leszek Gąsieniec	University of Liverpool, UK
Konstantinos Georgiou	Ryerson University, Canada
Sylvie Hamel	University of Montreal, Canada
Giuseppe F. Italiano	LUISS Guido Carli University, Italy
Taisuke Izumi	Osaka University, Japan
Ralf Klasing	CNRS and University of Bordeaux, France
Rastislav Královič	Comenius University, Slovakia
Thierry Lecroq	University of Rouen Normandy, France
Zsuzsanna Lipták	University of Verona, Italy
Bernard Mans	Macquarie University, Australia
George Mertzios	Durham University, UK
Othon Michail	University of Liverpool, UK
Matúš Mihalák	Maastricht University, Netherlands
Lucia Moura (Co-chair)	University of Ottawa, Canada
Maura Paterson	Birkbeck University of London, UK
Nicola Prezza	Ca' Foscari University of Venice, Italy

Tomasz Radzik	King's College London, UK
Joe Sawada	University of Guelph, Canada
Michiel Smid	Carleton University, Canada
Wing-Kin Sung	National University of Singapore, Singapore
Jukka Suomela	Aalto University, Finland
Koichi Wada	Hosei University, Japan

Organizing Committee

Saman Bazargani	University of Ottawa, Canada
Jean-Lou De Carufel	University of Ottawa, Canada
Vida Dujmović (Co-chair)	University of Ottawa, Canada
Paola Flocchini (Co-chair)	University of Ottawa, Canada
Lucia Moura (Co-chair)	University of Ottawa, Canada
Kanstantsin Pashkovich	University of Waterloo, Canada

External Reviewers

Abdeddaim, Said
Akrida, Eleni C.
Almethen, Abdullah
Asahiro, Yuichi
Babu, Jasine
Bampas, Evangelos
Bazgan, Cristina
Boeckenhauer, Hans-Joachim
Bonnet, François
Bramas, Quentin
Bulteau, Laurent
Cameron, Ben
Cao, Yixin
Casel, Katrin
Centeno, Carmen Cecilia
Cenzato, Davide
Chakraborty, Dibyayan
Chaudhary, Juhi
Chekuri, Chandra
Choudhary, Pratibha
Cicalese, Ferdinando
Colini Baldeschi, Riccardo
Connor, Matthew
Cordasco, Gennaro
Cotumaccio, Nicola
Cunha, Luis
Dailly, Antoine

Danziger, Peter
de Castro Mendes Gomes, Guilherme
de Freitas, Rosiane
de Lima, Paloma
Dev, Subhadeep
Dey, Sanjana
Dirk, Arnold
dos Santos, Vinicius F.
Eiben, Eduard
Escoffier, Bruno
Fages, François
Fernández-Baca, David
Frei, Fabian
Gajjar, Kshitij
Ganian, Robert
Gavenčiak, Tomáš
Gibney, Daniel
Giuliani, Sara
Gkikas, Angelos
Gomez, Renzo
Gourdel, Garance
Goyal, Pooja
Groenland, Carla
Grzesik, Andrzej
Guinand, Frédéric
Gómez Brandón, Adrián
Hahn, Gena

Hamm, Thekla
Hasunuma, Toru
Hoang, Chinh
Huang, Shenwei
Italiano, Davide
Jiménez, Andrea
Kamali, Shahin
Katayama, Yoshiaki
Kesselheim, Thomas
Kindermann, Philipp
Klobas, Nina
Kobayashi, Yusuke
Kolay, Sudeshna
Komusiewicz, Christian
Labourel, Arnaud
Lafond, Manuel
Lee, Edward
Lee, Euiwoong
Lefebvre, Arnaud
Leupold, Peter
Liedloff, Mathieu
Lintzmayer, Carla Negri
Manabe, Yoshifumi
Manlove, David
Martins, Simone
Menasché, Daniel Sadoc
Mercas, Robert
Milanič, Martin
Miltzow, Till
Mishra, Suchismita
Mizuki, Takaaki
Molter, Hendrik
Mondal, Debajyoti
Moura, Phablo
Mynhardt, Kieka
Ono, Hirotaka
Ooshita, Fukuhito

Paixao, Joao
Paschos, Vangelis
Pedrosa, Lehilton L. C.
Pelantova, Edita
Picouleau, Christophe
Pouly, Amaury
Prabhakaran, Veena
Punzi, Giulia
Rescigno, Adele
Richomme, Gwenaël
Rizzo, Riccardo
Romana, Giuseppe
Rossi, Massimiliano
Rote, Günter
Rubei, Elena
Rucci, Davide
Saona Urmeneta, Raimundo
Schouery, Rafael
Shalom, Mordechai
Skretas, George
Sopena, Eric
Stipulanti, Manon
Sudo, Yuichi
Takaoka, Asahi
Theofilatos, Michail
Unger, Walter
Vaccaro, Ugo
Venturini, Rossano
Vialette, Stéphane
Wilczynski, Anaëlle
Williams, Aaron
Wolf, Petra
Wong, Dennis
Yanagisawa, Nayuta
Zhu, Binhai
Zschoche, Philipp

Sponsors

Abstracts of Invited Talks

Induced Subgraphs and Tree Decompositions

Maria Chudnovsky

Princeton University, USA
mchudnov@math.princeton.edu

Tree decompositions are a powerful tool in structural graph theory, that is traditionally used in the context of forbidden graph minors. Connecting tree decompositions and forbidden induced subgraphs has so far remained out of reach. Recently we obtained several results in this direction: the talk will be a survey of these results.

Token Swapping

Anna Lubiw

University of Waterloo, Canada
alubiw@uwaterloo.ca

Given a graph where every vertex has exactly one labeled token, a *swap* exchanges the tokens at the two endpoints of an edge. The situation can be modelled as an exponentially large Cayley graph with a vertex for each permutation of the labels and an edge for each swap.

It is easy to see that the Cayley graph is connected (every permutation of labels can be realized by a sequence of swaps). Of interest are the diameter of the Cayley graph (the worst case length of a sequence of swaps) and the complexity of computing the minimum length sequence of swaps to realize a given permutation.

These token swapping problems have been studied by disparate groups of researchers in discrete mathematics, theoretical computer science, robot motion planning, game theory, and engineering.

I will survey this work and talk about hardness and approximation algorithms for token swapping on trees.

New Directions in Network Realization

David Peleg

Weizmann Institute of Science, Israel
david.peleg@weizmann.ac.il

Network realization problems concern situations where given a specification S, detailing the desired values of a certain network parameter, it is required to construct a network adhering to S, or decide that no such network exists. A variety of network realization problems have been studied over the past 70 years, focusing mainly on the parameters of vertex degrees and inter-vertex distances. The talk will present some recent developments in the area of network realization.

Search for Combinatorial Objects Using Lattice Algorithms - Revisited

Alfred Wassermann

University of Bayreuth, Germany
Alfred.Wassermann@uni-bayreuth.de

In 1986, Kreher and Radziszowski first used the famous LLL algorithm to construct combinatorial designs. Subsequently, lattice algorithms have been applied to construct a large variety of objects in design theory, coding theory and finite geometry. Unfortunately, the use of lattice algorithms in combinatorial search is still not well established. Recently, a new enumeration strategy based on "limited discrepancy search" was used to further improve the power of this method. In this talk, we will describe the search strategy based on lattice basis reduction, compare it to widely used backtracking algorithms and integer linear programming algorithms, and will outline the recent progress.

Contents

Invited Papers

Relaxed and Approximate Graph Realizations

Amotz Bar-Noy[1], Toni Böhnlein[2], David Peleg[2(✉)], Mor Perry[2], and Dror Rawitz[3]

[1] City University of New York (CUNY), New York, USA
amotz@sci.brooklyn.cuny.edu
[2] Weizmann Institute of Science, Rehovot, Israel
{toni.bohnlein,david.peleg,mor.perry}@weizmann.ac.il
[3] Bar Ilan University, Ramat-Gan, Israel
dror.rawitz@biu.ac.il

Abstract. A network realization problem involves a given specification π for some network parameters (such as vertex degrees or inter-vertex distances), and requires constructing a network G that satisfies π, if possible. In many settings, it may be difficult or impossible to come up with a precise realization (e.g., the specification data might be inaccurate, or the reconstruction problem might be computationally infeasible). In this expository paper, we review various alternative approaches for coping with these difficulties by relaxing the requirements, discuss the resulting problems and illustrate some (precise or approximate) solutions.

1 Introduction

1.1 Background: Precise (pure) Network Realization Problems

Network realization problems are fundamental questions pertaining to the ability to construct a network conforming to pre-defined requirements. Given a *specification* (or *information profile*) detailing constraints on some network parameters, such as the vertex degrees, distances or connectivity, it is required to construct a network abiding by the specified profile, i.e., satisfying the requirements, or to determine that no such network exists.

Realization problems may arise in two general types of contexts. In scientific contexts, the information profile may consist of the outcomes of some measurements obtained from observing some real world network (e.g., social networks and information networks) whose full structure is unknown. In such a setting, our goal is to construct a model that may possibly explain the empirical observations. Many of the studies in the field of phylogenetics and evolutionary trees (see, e.g., [16,23,41,42,51,63,68,84,88]) as well as in the field of discrete tomography and microscopic image reconstruction (see, e.g., [3–5,11,15,46,47,52,64,77]) belong to this class.

Supported in part by a US-Israel BSF grant (2018043).

P. Flocchini and L. Moura (Eds.): IWOCA 2021, LNCS 12757, pp. 3–19, 2021.
https://doi.org/10.1007/978-3-030-79987-8_1

A second, engineering-related context where realization problems come up is network design. Here, the profile may be defined based on a specification dictated by the future users of the network, and the goal is to construct a network that obeys the specification. For example, the profile may specify the required connectivity, flow capacities, or distances between vertex pairs in the network. In particular, network realization techniques may be useful in the area of *software defined networks* (SDN). For example, in service chain placement, the specification can define a directed acyclic graph (DAG) of *virtual network functions* (VNF), and the realization must determine the placement of one of the paths of the DAG in the physical network [38, 39, 72].

Two of the most well-studied families of realization problems concern *vertex degrees* and *inter-vertex distances*. The following is a brief review of the literature on these two problems.

Degree Profile. The *degree sequence* of a simple (no parallel edges or self-loops) and undirected graph $G = (V, E)$ with the vertex set $V = \{1, \ldots, n\}$ is an integer sequence $\mathrm{DEG}(G) = (d_1, \ldots, d_n)$, where $d_i = \deg_{G,i}$ is the degree of vertex i in G. The degree sequence occurs as a central and natural parameter in many network applications, and provides information on the significance, centrality, connectedness, and influence of each vertex in the network, contributing to the understanding of the network structure and some of its important properties. Given a sequence D of n non-negative integers, the *degree realization* problem asks to decide whether there exists a graph G whose degree sequence $\mathrm{DEG}(G)$ equals D. A sequence admitting such a realization is called *graphic* (or *graphical*).

The two key questions studied extensively in the past concern identifying characterizations (or, necessary and sufficient conditions) for a sequence to be graphic, and developing effective and efficient algorithms for finding a realizing graph for a given sequence if exists. A necessary and sufficient condition for a given sequence of integers to be graphic (also implying an $O(n)$ decision algorithm) was presented by Erdös and Gallai in [36]. (For alternative proofs see [2, 27, 32, 90–92].) Havel [60] and Hakimi [55] (independently) described an algorithm that given a sequence of integers computes in $O(m)$ time an m-edge graph realizing it, or proves that the given sequence is not graphic.

A number of related questions were considered in the literature, including the following: (a) Given a degree sequence D, find all the (non-isomorphic) graphs that realize it. (b) Given D, count all its (non-isomorphic) realizing graphs. (c) Given D, sample a random realization as uniformly as possible. (d) Determine the conditions under which a given D defines a unique realizing graph (a.k.a. the *graph reconstruction* problem). These questions are well-studied, cf. [27, 36, 55, 60, 67, 78, 85, 90, 94, 99, 101], and have found several interesting applications, most notably in network design, randomized algorithms, and recently in the study of social networks [17, 30, 35, 75] and chemical networks [89]. Degree realization with given constraints on some vertex was studied in [69]. For surveys on graphic sequences see [95–97].

The *subgraph realization* problem adds the restriction that the realizing graph must be a subgraph (or *factor*) of some fixed input graph. Subgraph realization problems are generally harder. For instance, it is easy to construct an n-vertex connected graph whose degrees are all 2, but the same problem for subgraph-realization is NP-hard (being essentially the Hamiltonian cycle problem). For more on this interesting line of work see, e.g., [8,53,61,74,93]. Here we focus on the case where the host graph is the complete graph.

Graphic sequences were studied also on specific graph families. The problem is straightforward on trees (see [54] for an elegant proof) but of interest on more complex classes. For example, characterizations were given to sequence pairs that can be realized as the degree sequences of a *bipartite* graph [21,49,82,102]. The family of *planar* graphs was studied to some extent, but the degree realization problem in this setting is still far from being resolved, and existing results are restricted to characterizing planar graphic *k-sequences* (in which the difference between $\max d_i$ and $\min d_i$ is at most k) for $k = 0, 1, 2$ [1,83]. *Split* graphs and *difference* graphs are considered in [58,98] and in [57], respectively. *Chordal, interval,* and *perfect* graphs were studies in [25]. Degree realization in *directed* graphs was studied in [7,24,37,48,71], and the NP-hardness of degree realization by directed acyclic graphs was proved in [59].

Distance Profile. In a graph G, define the *distance* $\mathrm{dist}_G(u, v)$ between two vertices u and v as the length of the shortest path connecting them in G. In the *distance realization* problem, the input DIST-profile consists of a matrix $D \in (\mathbb{N} \cup \{\infty\})^{n \times n}$, which specifies the required distance between every two vertices $i \neq j$ in the graph. $D_{i,j} = \infty$ represents the case where i and j are in different disconnected components. The goal is to compute a realizing graph G (if exists), such that any two vertices i, j in G satisfy $\mathrm{dist}_G(i, j) = D_{i,j}$.

In the *unweighted distance realization* problem it is assumed that each edge is of length 1. It follows that the graph is fully determined by D: there is an edge (i, j) in the graph if and only if $D_{i,j} = 1$. It follows that there is only one graph that may serve as a realization. This was observed by Hakimi and Yau [56], who provided a characterization for distance specifications realizable by unweighted graphs, implying also a polynomial-time algorithm for distance realization.

In the *weighted distance realization* problem the edges of the realizing graph may have arbitrary integral weights. (We assume that the minimum edge weight is 1.) Hakimi and Yau [56] also studied the weighted problem. They proved that the necessary and sufficient condition for the realizability of a given martix D is that D is a metric. Furthermore, they gave a polynomial-time algorithm that given a distance specification D, which is a metric, computes a realization. More specifically, their algorithm constructs a unique realizing graph which contains edges that are necessary in every possible realization of D.

Patrinos and Hakimi [81] considered the case where weights can be negative. They showed that any symmetric matrix with zero diagonal is a distance matrix of some graph G. They gave necessary and sufficient conditions for realizing such a matrix as a tree, and they showed that the tree realization is unique.

Precise distance realization by *weighted trees* were considered in [9], which presented a characterization for realizability. (For unweighted trees, there is a straightforward realization algorithm, based on the algorithm of [56] for general unweighted graphs, and on the fact that the realization, if exists, is unique.) Precise distance realization restricted to bipartite graphs was studied in [14], where it was observed that it is sufficient to check the unique realization in the unweighted case or the minimal realization in the weighted case.

1.2 Limitations of Precise Realizations

Most of the research activity on realizations dealt with precise realizations in which the specification must be met *exactly*, in terms of the graph size and the attributes of each vertex in the realizing graph. We refer to such realizations as pure realizations because they forbid any deviation or relaxation from the specifications. Unfortunately, pure realizations suffer from certain limitations, making them hard to utilize, and motivating the use of *relaxed* versions of network realization. We now discuss several of these limitations.

Hardness of Pure Realization. Our ability to solve a specific realization problem, i.e., construct networks satisfying a given specification, depends heavily on the attributes considered. While some profile types are handled easily, for other certain profile types it turns out to be very hard to derive characterizations for realizability or construction algorithms for pure realizations. For example, the realizability of vertex connectivity profiles is to date poorly understood. Moreover, for certain types of information profiles, it may be feasible to find characterizations or construction algorithms for general graphs, but hard to do so for some specific graph classes. For example, degree realizability is well understood on general graphs, but not on planar graphs. In such cases, it makes sense to resort to various relaxations, or to realizations that well-approximate the given specification. Specifically, efficient approximation algorithms for various hard to realize profiles may turn out to have practical significance in the context of network design, as it may provide designers with tools for constructing networks that obey a detailed pre-specified behavior in terms of the attributes most relevant to the purposes for which the network is designed.

Flexibility of Imprecise Realizations. When solutions are restricted to precise realizations, it could be the case that the space of all possible realizations may be too small or even empty. As a result, there is very little freedom in selecting a realization that optimizes some objective functions. By relaxing the specifications the space of all feasible realizations might become large enough to produce optimal or near optimal realizations in regard to predefined optimization goal. Consider for example the precise degree realization problem. By allowing the degrees of vertices to deviate by a constant number from the precise degree, one could find a realizing graph with a smaller diameter and/or higher connectivity. Now consider the precise distance realization problem. By allowing the actual distances to be larger by some factor, one could find a realizing graph with fewer edges and smaller degrees.

Imperfect Data. In certain cases, the input specification data is imprecise (e.g., describing the results of inaccurate measurements on a real-life network with unknown parameters), or ill-chosen (e.g., based on the unrealistic expectations of clients describing their "dream network"). In such cases, it might well happen that the specification is infeasible with no network conforming to it. As a result, characterization tests and construction algorithms will simply return "non-realizable" and halt. This might be an unsatisfactory outcome in the practical world. As an alternative, we could aim for a more positive outcome, in the form of an actual realizing graph, if only approximate.

Hardness of Composition. In some settings we may be given a *composed* profile, combining two or more profiles that must be realized simultaneously (see [13] in these proceedings). As one might suspect, handling profile compositions may sometimes be significantly harder, especially when insisting on pure realizations. Hence for such composed profiles, finding relaxed or approximate realizations might often be our only recourse.

In short, the above discussion makes it evident that in many situations one may be interested in considering some *alternative* approaches as a substitute for pure realizations. In this expository paper we review and illustrate various alternative approaches to network realization, based on relaxed or approximate network realizations. Specifically, Sect. 2 gives an overview of relaxed realizations, Sect. 3 illustrates the notion of minimum deviation distance realizations and some related results, and Sect. 4 discusses *multigraph* realizations of degree profiles with minimum multiplicity.

2   Relaxed and Approximate Realizations

We next overview a number of variations studied in the literature, which introduce flexibility in one way or another. For simplicity, we focus only on attributes of vertices or vertex pairs.

Bounding Specifications. The specifications considered so far were *precise*, i.e., they specified the exact value of the attribute required for every vertex (or vertex pair). In contrast, a *lower-bounding* specification ϕ for an information profile f (f-profile) provides lower bounds on the required attributes. Such ϕ is realizable if there is a graph G whose f-profile satisfies $f(G) \geq \phi$. More explicitly, when ϕ and f concern *vertex* attributes of n-vertex graphs, G satisfies ϕ if $f_i \geq \phi_i$ for every $1 \leq i \leq n$ (a similar definition applies for attributes of vertex pairs). An analogue notion can be defined for *upper-bounding* specifications, as well as *range* specifications, specifying both lower and upper bounds on each attribute. Clearly, bounding specifications are more flexible than precise ones; whereas precise specifications sometimes (although certainly not always) force a unique realization, bounding specifications are more likely to admit many realizing graphs.

As an example, a natural generalization of the graphic sequence problem is the *degree range* profile. In this variant we are given a *range sequence* consisting of n ranges, $\mathcal{S} = ([a_1, b_1], \ldots, [a_n, b_n])$ such that $0 \leq a_i \leq b_i \leq n - 1$ for every i, which is said to be realizable if there exists a graphic sequence $D = (d_1, \ldots, d_n)$ falling within the specified ranges, namely, such that $a_i \leq d_i \leq b_i$ for $1 \leq i \leq n$. (cf. [12,22,50,53] and the references therein). Two natural questions are to find an efficient algorithm for verifying the realizability of a given range sequence $\mathcal{S}$, and to compute a certificate (that is, a suitable graphic sequence D) for a given realizable range sequence $\mathcal{S}$. (Given D, one can readily construct a graph whose degrees realize D, as discussed earlier.) An easy to verify characterization for realizable range sequences is provided in [22], crucially using the (g, f)-Factor Theorem of [74]. A constructive proof of this characterisation is given in [50]. Another algorithm for computing a graph whose degrees realize a given range sequence (if exists), based on [55, 60], is presented in [62].

As another example, range variants of the distance realization problem, where we are given a range $[D_{i,j}^-, D_{i,j}^+]$ for every i, j and the realizing G must satisfy $D_{i,j}^- \leq \text{dist}_G(i, j) \leq D_{i,j}^+$, were studied in [14], and the resulting problems were classified as polynomial or NP hard.

Handling Unrealizable Specifications. When faced with a given *unrealizable* specification ϕ for some f-profile, a natural question that presents itself is to find a graph G that "resembles the specification most," namely, whose f-profile $f(G)$ "minimally deviates" from ϕ. This requirement is rather broad, and may lead to different optimization problems, depending on the setting. For our canonical example of degree realizations in general graphs, an interesting instantiation of this problem is the *minimum total discrepancy* degree realization problem [62, 75], a.k.a. the *graphic deviation* problem [19]. It requires one to find, for a given non-graphic sequence D, a graph G whose degree sequence is closest to D.

Several measurements for closeness or the quality of an approximation are possible. As an example consider the unrealizable sequence $D_0 = (4, 4, 1, 1, 1)$. We may look for a realizing graph G that minimizes the *sum* of the deviations at all the vertices $\sum_i |\deg_{G,i} - d_i|$, i.e., with respect to the L_1-norm. The realizable sequence $(4, 4, 2, 2, 2)$ approximates D_0 with minimum total deviation 3.

Naturally, other norms may be of interest as well. In particular, the problem can be solved for the L_∞ norm (see Sect. 4) where the deviation measure to be minimized is $\max_i |\deg_{G,i} - d_i|$. A solution for D_0 is the sequence $(3, 3, 2, 1, 1)$, which is graphic and has maximum deviation 1. Beyond being a natural extension of the degree sequence problem, a practical motivation for this problem is raised and studied in [18, 20], in the context of probabilistic and game-theoretic analyses of population models.

Another approach is to minimize the *number* of vertices that do not conform to their specification (are mismatches), that is, to find a realizing graph G minimizing the number of vertices i such that $\deg_{G,i} \neq d_i$. A possible solution for the example D_0 is the sequence $(4, 1, 1, 1, 1)$, which is graphic (the realizing graph being the 4-star) and has one mismatch.

Returning to the example of degree range profiles, for a non-realizable range sequence $\mathcal{S}$, the algorithm of [62] computes a graph whose deviation with respect to L_1-norm is minimum. The time complexity of the algorithm is $O(\sum_{i=1}^n b_i)$, where b_i is the upper limit of the ith range, which can be as high as $\Theta(n^2)$. In [12] we presented an $O(n \log n)$ time algorithm for computing a graphic certificate (if exists) for any given range sequence. In addition, given a range sequence $\mathcal{S}$, our algorithm can obtain (in the same time) a degree-certificate corresponding to graphs with minimum (resp. maximum) possible edges. We also gave an $O(n \log n)$ time algorithm for efficiently computing graphic sequences having the least possible L_1-deviation when the input range sequence is non-realizable. Again, the problem can be solved also for the L_∞ norm, i.e., where the deviation measure to be minimized is $\max_i\{0, \deg_{G,i} - b_i, a_i - \deg_{G,i}\}$. These tools allow also studying other interesting applications, such as computing a minimum extension of non-graphic sequences to graphic ones.

Super-Realizations. It is sometimes acceptable to allow the realizing graph G to be *larger* than indicated by the specification (i.e., have more than n vertices). We refer to such a realizing graph as an *super-realization* of the given specification. The flexibility of adding new vertices is useful in tackling problems that are otherwise intractable. Consequently, super-realizations were extensively studied for various profile types.

As an example let us consider distance profiles. The *optimal distance realization* problem was introduced in [56]. In this problem, a distance matrix D is given over a set S of n *terminal* vertices, and the goal is to find a graph G including S, with possibly additional vertices, that realizes the given distance matrix for S. Necessary and sufficient conditions are given for a symmetric matrix with nonnegative entries to be realizable by a weighted or an unweighted graph. It is also shown that if G is an n-vertex realization of D without redundant edges (i.e., no edge can be removed without violating the distance matrix), then G is unique. It is shown in [34] that an optimal realization can have at most n^4 vertices (recall that the number of terminal vertices is n), and therefore, there is a finite (but exponential) time algorithm to find an optimal realization. In [6] it is shown that finding optimal realizations of distance matrices with integral entries is NP-complete, and evidence to the difficulties in approximating the optimal realization is provided in [28]. Over the years, various heuristics for optimal realizations are discussed in many papers [76, 86, 87, 100].

Since optimal realization seems hard even to approximate, special cases and other variations have been studied. In [28], a *weak* realization is defined, where the distance matrix is a bounding specification that sets *lower bounds* on the required distances. it is shown that this weak realization problem is also NP-complete, but its optimum solution can be 2-approximated. In [40], an optimizing variant is defined to be the one with minimum *number of edges*. It is shown therein that if additional vertices are not allowed, then an edge-minimal graph can be found in polynomial time. On the other hand, in a setting allowing additional vertices, if the distance matrix has to be realized by an unweighted graph, then the problem is essentially as hard to approximate as graph coloring and maximum clique. In addition, polynomial approximation algorithms are presented for specific cases.

Special attention has been given to the optimal distance realization problem where the realizing graph is a tree. In [56], a procedure is given for finding a tree realization of D if exists. It is also shown therein that a tree realization, if exists, is unique and is the optimum realization of D. Necessary and sufficient conditions for a distance matrix to be realizable by a tree were given in [10, 33, 87]. Finally, an $O(n^2)$ time algorithm for optimal tree-realization is given in [31].

For degree realizations, we may look for a realizing graph G having a subset of n vertices whose parameters conform precisely to the given profile. The typical goal would be to minimize the number of additional vertices, i.e., $|V(G)| - n$. This type of relaxation may be useful, e.g., when each component f_i of the specification f represents a vertex that *must* exist in the constructed network, and cannot be omitted, whereas extra vertices imply extra cost but are not prohibitive. For the above example of $D = (4, 4, 1, 1, 1)$, one can obtain a super-realization with $n' = 7 = n + 2$ by realizing the sequence $D' = (4, 4, 2, 1, 1, 1, 1)$, which is graphic. In general graphs, this type of approximate degree realization problem can be solved efficiently, using algorithms for degree range profiles.

Another option is to allow a realizing graph G that is *smaller* than indicated by the specification, namely, has fewer than n vertices. We call such a graph an *sub-realization* of the given specification. For example, for the profile DEG, the specification $D = (4, 3, 1, 1, 1)$ cannot be realized, but by removing one vertex we obtain $D' = (3, 1, 1, 1)$ that can be realized using a star. The natural goal in such a setting is to compute a sub-realization with maximum number of vertices, i.e., with minimum number of vertex deletions. This type of relaxation may arise, e.g., when external constraints or physical resource bounds limit the size of the constructed network to no more than n vertices, and it is desired to utilize the available resources to the extent possible.

Yet another interesting variant of super-realizations is obtained if we allow only *splitting* vertices. For example, $D = (4, 3, 1, 1, 1)$ cannot be realized, but $(3, 2, 2, 1, 1, 1)$, obtained by splitting the requirement 4 into 2 and 2, is realizable.

Optimizing Realizations. Optimization goals may be of interest also for *realizable* profiles. It is often the case that the given profile has many possible realizing graphs, and it may be of interest to seek a realizing graph that also optimizes some desirable quality measure. This type of optimization problem often arises in contexts where the specification is bounding rather than precise, since the flexibility of bounding specifications, resulting in the larger variety of realizing graphs, makes it attractive to select the most suitable realizing graph for the quality measure at hand.

For example, in the context of connectivity realization, a natural problem to consider is *minimum edge connectivity threshold realization*. The input for this problem is an $n \times n$ connectivity threshold matrix CT^e serving as a bounding, rather than precise, specification, namely, specifying the required *minimum* edge connectivity $CT^e(i, j)$ between every two vertices $i \neq j$ in the graph. Hence a graph G satisfies the specification if e-conn$_G(i, j) \geq CT^e(i, j)$ for any two vertices i, j. Note that for any connectivity threshold matrix CT^e, the complete graph on V is always a legal realization (provided that $CT^e(i, j) \leq n - 1$ for every i and j). However, the minimum edge connectivity threshold realization problem imposes an additional desirable goal, i.e., to find the *sparsest possible* realizing graph G. In [45], this problem was studied in the context of survivable network design, and a 2-approximate solution is provided for this problem, guaranteeing that the number of edges is at most twice the minimum possible. See also [26, 43, 44, 66] for related studies.

Another approach to relax the the degree realization problem is to ask for a multigraph (with parallel edges and self-loops) instead of a simple graph. Then it is natural to try to maximize (or minimize) the number of edges in the underlying graph, i.e., to try to minimize the number of mismatches which are parallel edges or self-loops in this case. We take a closer look at these variants in Sect. 4.

We note that all the above deviations also have two one-sized (i.e., bounding) variants. For example, in the case of minimizing the number of mismatches, a subclass of possible bounding realizations arises when we allow only $\deg_{G,i} \geq d_i$, or only $\deg_{G,i} \leq d_i$, depending on the application at hand.

3   Minimum Deviation Distance Realization

If a given distance matrix D is unrealizable, one may want to find a graph G whose distance matrix is closest to D, say, with respect to the L_1-norm, i.e., such that the sum of deviations of all matrix entries is minimized, or with respect to the L_∞-norm, i.e., the maximum deviation of a matrix entry is minimized. Note that here we allow both *downwards deviation*, where the actual distance satisfies $\mathrm{dist}_G(i, j) < D_{i,j}$, and *upwards deviation*, where $\mathrm{dist}_G(i, j) > D_{i,j}$. Other deviation functions can be considered, for example, the number of matrix entries for which there is a deviation. For each of the above deviation functions (which allow both downwards and upwards deviation) we may consider also two more variants: one allowing only downward deviations, and one allowing only upward deviations.

In this section we focus on the version of only downward deviations, which is particularly interesting for distance realizations, since in system design, we would like to get as close as possible to the specification matrix, but never exceed the specified distances.

There are several variants of the distance realization problem, depending on whether the distance matrix specifies exact values or ranges at each entry, and whether the realizing graph is required to be unweighted or weighted. For all three deviation functions mentioned above, it turns out that finding a graph G whose distance matrix is closest to D is NP-hard when G must be unweighted, and is polynomial for weighted graphs. This holds for both types of distance matrix, namely, precise distances and distance range.

As a first illustration we show a polynomial algorithm for the weighted case.

Theorem 1 ([14]). *For every deviation function and every distance matrix D, there is a polynomial time algorithm that finds a weighted graph G with minimum downward deviation from D.*

The theorem is established by presenting a polynomial time algorithm for the most general case of a distance-range matrix (a precise-distance matrix is a special case, for which our algorithm applies as well). Given a distance-range matrix D, construct a weighted clique graph G by assigning the weight

$\omega(i,j) = D_{i,j}^+$ for every edge (i,j). We now sketch the argument showing that G is a minimum downward deviation graph. First, observe that by construction, $\text{dist}_G(i,j) \leq D_{i,j}^+$ for every i and j, since there exists an edge of weight $D_{i,j}^+$ connecting i and j, so there are no upward deviations in G. Next, we argue that for every pair (i,j), and for every graph whose distance matrix deviates only downwards from D, the deviation of (i,j) is at least $D_{i,j}^- - \text{dist}_G(i,j)$, i.e., G is a minimum deviation graph. To see this, consider a path $i = v_0, \ldots, v_\ell = j$ in G for which the total weight is $\text{dist}_G(i,j)$. This path implies a set of requirements $D_{v_k, v_{k+1}}$, for $k = 0, \ldots, \ell - 1$, such that the distance between v_k and v_{k+1} is at most $D_{v_k, v_{k+1}}^+ = \omega(v_k, v_{k+1})$. Since every realizing graph G' is not allowed to deviate upwards, the distance between i and j in G' is at most $\text{dist}_G(i,j)$, i.e., $D_{i,j}^- - \text{dist}_{G'}(i,j) \geq D_{i,j}^- - \text{dist}_G(i,j)$.

Our second complementary example is an NP-hardness result for the unweighted case.

Theorem 2 ([14]). *The problem of finding an unweighted graph G with minimum sum of downward deviations from D, is NP-hard.*

The hardness of this problem is established for the case of a precise-distance matrix, implying also the hardness of the distance-range case. Hardness is established by reducing the *diameter-2 augmentation* problem, which is known to be NP-hard [73], to our deviation problem. Given a graph $G = (V, E)$, input to the diameter-2 augmentation problem, the goal is to find the minimal set of edges E' such that the diameter of the graph $G' = (V, E \cup E')$ is at most 2. This instance is reduced to an instance D of the *minimum sum of downward deviations* problem by setting $D_{i,j}$ to be 0 if $i = j$, 1 if $(i,j) \in E(G)$, and 2 otherwise, yielding a distance matrix for n vertices. It then remains to verify that D has a realization with minimum sum of downwards deviations k if and only if G has a diameter-2 augmentation of k edges, establishing the theorem.

The same reduction actually proves also the hardness of *minimum number of downward deviations* problem. Moreover, one can prove that the problem *min-max downward deviation* is NP-hard by a reduction from the k-coloring problem. A systematic study of these types of minimum-deviation realizations for distance profiles, and their classification into polynomial and NP-hard cases, can be found in [14], where we also consider other deviation functions, such as multiplicative deviation, two-sided deviation and only upward deviation.

4   Optimizing Multigraph Realizations of Degree Profiles

In this section, we consider degree profiles where the realization must satisfy the given specification, but it is allowed to use *parallel edges*. Namely, we allow realizations by *loopless multigraphs* rather than by only simple graphs. The goal is to find a realization which is as "close" to a simple graph as possible.

Let $H = (V, E)$ be a (loopless) multigraph. That is, E is a multiset. Denote by $E_H(v, w)$ the multiset of edges connecting v and w in H, and let $\#_H(v, w) = |E_H(v, w)|$. Given a vertex u, let $N(u)$ be the neighborhood of u, namely $N(u) = \{v : (u,v) \in E\}$. Also, let $E(u) = \{(u,v) \in E : v \in V\}$ be the multiset containing edges that are adjacent to u. Observe that $\deg(u) = |E(u)|$, while it could be that $\deg(u) > |N(u)|$. Finally, let $\mathcal{P}(H) = \{(v,w) \mid E(v,w) \neq \emptyset\}$. Observe that $(V, \mathcal{P}(H))$ is the underlying simple graph of the multigraph H.

Define the *total multiplicity* as the number of parallel edges $|E| - |\mathcal{P}(H)|$. It follows that

$$\mathsf{TotMult}(H) = \sum_{(v,w)\in\mathcal{P}(H)}(\#_H(v,w) - 1).$$

Observe that in a simple graph $\mathsf{TotMult}(H) = 0$.

A realization minimizing the total multiplicity can be computed efficiently (see [70, 79, 80]). In [79] this problem is also solved when only loops and loops and parallel edges are allowed. Interestingly, finding a multigraph realization that maximizes the number of parallel edges is NP-hard [65].

Define the *maximum multiplicity* as

$$\mathsf{MaxMult}(H) = \max_{(v,w)\in\mathcal{P}(H)}(\#_H(v,w)).$$

Observe that in a simple graph $\mathsf{MaxMult}(H) = 1$.

Theorem 3 (Chungphaisan [29]). *Consider a sequence* $d = (d_1,\dots,d_n)$ *where* $\sum_{i=1}^{n} d_i$ *is even and an integer* $r \geq 1$. *There is a multigraph* H *where* $\mathsf{MaxMult}(H) \leq r$ *if and only if for* $k = 1,\dots,n$ *we have*

$$\sum_{i=1}^{k} d_i \leq rk(k-1) + \sum_{i=k+1}^{n} \min\{rk, d_i\}\ .$$

We conclude the discussion by illustrating how the generalized variant of the Havel-Hakimi algorithm outlined in [29] can be used to compute a multigraph H where $\mathsf{MaxMult}(H) \leq r$. The algorithm receives as input a sequence d and a parameter r and looks for a realization of d which uses at most r copies of each edge. An explicit description of the algorithm is given below. Each recursive call of this algorithm connects an arbitrary vertex ℓ to at most d_ℓ vertices with the highest degrees (not including itself) while using at most r copies per edge. The initial call is $\mathrm{MaxMult}(d,r)$.

Algorithm 1: $\mathrm{MaxMult}(d,r)$

---

1 **if** $d = 0$ **then return** $\emptyset$
2 Let ℓ be an arbitrary vertex such that $d_\ell > 0$
3 $V_\ell \leftarrow V \setminus \{\ell\}$; $E_\ell \leftarrow \emptyset$
4 **while** $d_\ell > 0$ **do**
5 $\quad$ $j \leftarrow \mathrm{argmax}_{q\in V_\ell} d_q$
6 $\quad$ **if** $d_j \leq 0$ **then** ABORT $\qquad\qquad\qquad\qquad$ ▷ d is not r-graphic
7 $\quad$ Add (ℓ,j) to E_ℓ
8 $\quad$ $d_\ell \leftarrow d_\ell - 1$; $d_j \leftarrow d_j - 1$
9 $\quad$ **if** $\#_{E_\ell}(j,\ell) = r$ **then** $V_\ell \leftarrow V_\ell \setminus \{j\}$
10 **return** $E_\ell \cup \mathrm{MaxMult}(d,r)$

For completeness, we include an analysis of the algorithm, starting with the running time. There are $O(n)$ recursive calls, and each can be implemented to run in $O(n)$ time, yielding a total of $O(n^2)$ time. A more careful implementation yields a running time of $O(\sum_i d_i)$.

For correctness, the next lemma proves that if the algorithm terminates successfully, then the computed edge multiset induces a realization.

Lemma 1. *Let d be a nonincreasing sequence and r be a positive integer. If the Algorithm MaxMult terminates without aborting, then the computed multiset induces a realization of d that contains at most r copies per edge.*

Proof. By induction on the number of recursive calls. In the base case, $d = 0$, and thus $E = \emptyset$ is a realization. For the induction step, let d' denote the value of d in Line 10, and let E' be the multiset of edges returned by the recursive call in Step 10. By the inductive hypothesis, E' induces a realization of d' that contains at most r copies per edge. The while loop realizes $d - d'$ and Line 9 makes sure that the number of copies per edge is bounded by r. Also, observe that E' does not contain edges adjacent to ℓ, since $d'_\ell = 0$ Hence $E' \cup E_\ell$ induces a realization of d that contains at most r copies per edge. □

The following lemma shows that if d admits a realization, then the algorithm will terminate successfully.

Lemma 2. *Let d be a nonincreasing sequence and r be a positive integer. If d has a realization that contains at most r copies per edge, then the algorithm will terminate successfully.*

Proof. By induction on the number of recursive calls. In the base case, $d = 0$, and in this case an $H = (V, \emptyset)$ is the only realization. For the induction step, let H be a realization of d that contains at most r copies per edge. First, observe that since there is a realization of d, the while loop would terminate successfully. If E_ℓ is contained in H, then we are done. Otherwise, we show that there is another realization H' of d that contains E_ℓ. Transform H into H' using edge-swaps. Let $v \in V$ such that $\#_H(\ell, v) < \#_{E_\ell}(\ell, v)$, where $\#_{E_\ell}(\ell, v)$ is the multiplicity of the edge (ℓ, v) in E_ℓ. It follows that there is a vertex u such that $\#_H(\ell, u) > \#_{E_\ell}(\ell, u)$. Since the last copy of (ℓ, v) was chosen over (ℓ, u), it must be that

$$d_v - \#_H(\ell, v) \geq d_v - (\#_{E_\ell}(\ell, v) - 1) \geq d_u - \#_{E_\ell}(\ell, u) > d_u - \#_H(\ell, u) \; .$$

Consequently, there must be a vertex $w \neq \ell, v, u$ such that $\#_H(v, w) > 0$ and $\#_H(u, w) < r$. Modify H by applying the following edge swap: remove the edges $(\ell, u), (v, w)$ and add the edges $(\ell, v), (u, w)$. It is not hard to verify that the resulting multigraph is a realization d that contains at most r copies per edge. Moreover, $\#_H(\ell, v)$ is increased by 1. Continue with edge swaps in this manner until H contains E_ℓ. □

The best r can be obtained with Algorithm 1 using binary search on r.

Lemma 3. *There exist a polynomial time algorithm that, given a sequence d, computes a realization H of d such that $\mathsf{MaxMult}(H) = \mathsf{MaxMult}(d)$.*

References

1. Adams, P., Nikolayevsky, Y.: Planar bipartite biregular degree sequences. Discr. Math. **342**, 433–440 (2019)
2. Aigner, M., Triesch, E.: Realizability and uniqueness in graphs. Discr. Math. **136**, 3–20 (1994)
3. Alpers, A., Gritzmann, P.: Reconstructing binary matrices under window constraints from their row and column sums. Fundamenta Informaticae **155**(4), 321–340 (2017)
4. Alpers, A., Gritzmann, P.: Dynamic discrete tomography. Inverse Probl. **34**(3), 034003 (2018)
5. Alpers, A., Gritzmann, P.: On double-resolution imaging and discrete tomography. SIAM J. Discr. Math. **32**, 1369–1399 (2018)
6. Althöfer, I.: On optimal realizations of finite metric spaces by graphs. Discret. Comput. Geom. **3**, 103–122 (1988)
7. Anstee, R.: Properties of a class of (0,1)-matrices covering a given matrix. Canad. J. Math. **34**, 438–453 (1982)
8. Anstee, R.P.: An algorithmic proof of Tutte's f-factor theorem. J. Algorithms **6**(1), 112–131 (1985)
9. Baldisserri, A.: Buneman's theorem for trees with exactly n vertices. CoRR (2014)
10. Bandelt, H.: Recognition of tree metrics. SIAM J. Discr. Math. **3**, 1–6 (1990)
11. Bar-Noy, A., Böhnlein, T., Lotker, Z., Peleg, D., Rawitz, D.: The generalized microscopic image reconstruction problem. In: 30th ISAAC, volume 149 of LIPIcs, pp. 42:1–42:15 (2019)
12. Bar-Noy, A., Choudhary, K., Peleg, D., Rawitz, D.: Efficiently realizing interval sequences. SIAM J. Discr. Math. **34**(4), 2318–2337 (2020)
13. Bar-Noy, A., Peleg, D., Perry, M., Rawitz, D.: Composed degree-distance realizations of graphs. In: 32nd IWOCA (2021)
14. Bar-Noy, A., Peleg, D., Perry, M., Rawitz, D., Schwartz, N. L.: Distance realization approximations. In: preparation (2021)
15. Battaglino, D., Frosini, A., Rinaldi, S.: A decomposition theorem for homogeneous sets with respect to diamond probes. Comput. Vis. Image Underst. **117**, 319–325 (2013)
16. Baum, D.A., Smith, S.D.: Tree Thinking: an Introduction to Phylogenetic Biology. Roberts and Company, Greenwood Village, CO (2013)
17. Blitzstein, J.K., Diaconis, P.: A sequential importance sampling algorithm for generating random graphs with prescribed degrees. Internet Math. **6**(4), 489–522 (2011)
18. Broom, M., Cannings, C.: A dynamic network population model with strategic link formation governed by individual preferences. J. Theoret. Biol. **335**, 160–168 (2013)
19. Broom, M., Cannings, C.: Graphic deviation. Discr. Math. **338**, 701–711 (2015)
20. Broom, M., Cannings, C.: Game theoretical modelling of a dynamically evolving network i: general target sequences. J. Dyn. Games **335**, 285–318 (2017)
21. Burstein, D., Rubin, J.: Sufficient conditions for graphicality of bidegree sequences. SIAM J. Discr. Math. **31**, 50–62 (2017)
22. Cai, M.-C., Deng, X., Zang, W.: Solution to a problem on degree sequences of graphs. Discr. Math. **219**(1–3), 253–257 (2000)
23. Camin, J.H., Sokal, R.R.: A method for deducing branching sequences in phylogeny. Evolution **19**, 311–326 (1965)

24. Chen, W.-K.: On the realization of a (p, s)-digraph with prescribed degrees. J. Franklin Inst. **281**(5), 406–422 (1966)
25. Chernyak, A.A., Chernyak, Z.A., Tyshkevich, R.I.: On forcibly hereditary p-graphical sequences. Discr. Math. **64**, 111–128 (1987)
26. W. Chou and H. Frank. Survivable communication networks and the terminal capacity matrix. IEEE Trans. Circ. Theory, CT-17, 192–197 (1970)
27. Choudum, S.A.: A simple proof of the Erdös-Gallai theorem on graph sequences. Bull. Austral. Math. Soc. **33**(1), 67–70 (1991)
28. Chung, F.R.K., Garrett, M.W., Graham, R.L., Shallcross, D.: Distance realization problems with applications to internet tomography. J. Comput. Syst. Sci. **63**, 432–448 (2001)
29. Chungphaisan, V.: Conditions for sequences to be r-graphic. Discr. Math. **7**(1–2), 31–39 (1974)
30. Cloteaux, B.: Fast sequential creation of random realizations of degree sequences. Internet Math. **12**(3), 205–219 (2016)
31. Culberson, J.C., Rudnicki, P.: A fast algorithm for constructing trees from distance matrices. Inf. Process. Lett. **30**(4), 215–220 (1989)
32. Dahl, G., Flatberg, T.: A remark concerning graphical sequences. Discr. Math. **304**(1–3), 62–64 (2005)
33. Dahlhaus, E.: Fast parallel recognition of ultrametrics and tree metrics. SIAM J. Discr. Math. **6**(4), 523–532 (1993)
34. Dress, A.W.M.: Trees, tight extensions of metric spaces, and the cohomological dimension of certain groups: a note on combinatorial properties of metric spaces. Adv. Math. **53**, 321–402 (1984)
35. Erdös, D., Gemulla, R., Terzi, E.: Reconstructing graphs from neighborhood data. ACM Trans. Knowl. Discov. Data **8**(4), 23:1–23:22 (2014)
36. Erdös, P., Gallai, T.: Graphs with prescribed degrees of vertices [Hungarian]. Matematikai Lapok **11**, 264–274 (1960)
37. Erdös, P.L., Miklós, I., Toroczkai, Z.: A simple Havel-Hakimi type algorithm to realize graphical degree sequences of directed graphs. Electr. J. Comb. **17**(1) (2010)
38. Even, G., Medina, M., Patt-Shamir, B.: On-line path computation and function placement in SDNs. Theory Comput. Syst. **63**(2), 306–325 (2019)
39. Even, G., Rost, M., Schmid, S.: An approximation algorithm for path computation and function placement in SDNs. In: Suomela, J. (ed.) SIROCCO 2016. LNCS, vol. 9988, pp. 374–390. Springer, Cham (2016). https://doi.org/10.1007/978-3-319-48314-6_24
40. Feder, T., Meyerson, A., Motwani, R., O'Callaghan, L., Panigrahy, R.: Representing graph metrics with fewest edges. In: Alt, H., Habib, M. (eds.) STACS 2003. LNCS, vol. 2607, pp. 355–366. Springer, Heidelberg (2003). https://doi.org/10.1007/3-540-36494-3_32
41. Felsenstein, J.: Evolutionary trees from DNA sequences: a maximum likelihood approach. J. Mol. Evol. **17**, 368–376 (1981)
42. Fitch, W.M.: Toward defining the course of evolution: Minimum change for a specific tree topology. Syst. Biol. **20**, 406–416 (1971)
43. Frank, A.: Augmenting graphs to meet edge-connectivity requirements. SIAM J. Discr. Math. **5**, 25–43 (1992)
44. Frank, A.: Connectivity augmentation problems in network design. In: Mathematical programming: state of the art, pp. 34–63. Univ. Michigan (1994)
45. Frank, H., Chou, W.: Connectivity considerations in the design of survivable networks. IEEE Trans. Circuit Theory, CT-17, 486–490 (1970)

46. Frosini, A., Nivat, M.: Binary matrices under the microscope: a tomographical problem. Theor. Comput. Sci. **370**(1–3), 201–217 (2007)
47. Frosini, A., Nivat, M., Rinaldi, S.: Scanning integer matrices by means of two rectangular windows. Theor. Comput. Sci. **406**(1–2), 90–96 (2008)
48. Fulkerson, D.: Zero-one matrices with zero trace. Pacific J. Math. **12**, 831–836 (1960)
49. Gale, D.: A theorem on flows in networks. Pacific J. Math. **7**, 1073–1082 (1957)
50. Garg, A., Goel, A., Tripathi, A.: Constructive extensions of two results on graphic sequences. Discr. Appl. Math. **159**(17), 2170–2174 (2011)
51. Gontier, N.: Depicting the tree of life: the philosophical and historical roots of evolutionary tree diagrams. Evol. Educ. Outreach **4**, 515–538 (2011)
52. Gritzmann, P., Langfeld, B., Wiegelmann, M.: Uniqueness in discrete tomography: three remarks and a corollary. SIAM J. Discr. Math. **25**, 1589–1599 (2011)
53. Guo, J., Yin, J.: A variant of Niessen's problem on degree sequences of graphs. Discr. Math. Theor. Comput. Sci. **16**, 287–292 (2014)
54. Gupta, G., Joshi, P., Tripathi, A.: Graphic sequences of trees and a problem of Frobenius. Czechoslovak Math. J. **57**, 49–52 (2007)
55. Hakimi, S.L.: On realizability of a set of integers as degrees of the vertices of a linear graph -I. SIAM J. Appl. Math. **10**(3), 496–506 (1962)
56. Hakimi, S.L., Yau, S.S.: Distance matrix of a graph and its realizability. Quart. Appl. Math. **22**, 305–317 (1965)
57. Hammer, P.L., Ibaraki, T., Simeone, B.: Threshold sequences. SIAM J. Algebra. Discr. **2**(1), 39–49 (1981)
58. Hammer, P.L., Simeone, B.: The splittance of a graph. Combinatorica **1**, 275–284 (1981)
59. Hartung, S., Nichterlein, A.: Np-hardness and fixed-parameter tractability of realizing degree sequences with directed acyclic graphs. SIAM J. Discr. Math. **29**, 1931–1960 (2015)
60. Havel, V.: A remark on the existence of finite graphs [in Czech]. Casopis Pest. Mat. **80**, 477–480 (1955)
61. Heinrich, K., Hell, P., Kirkpatrick, D.G., Liu, G.: A simple existence criterion for $(g < f)$-factors. Discr. Math. **85**, 313–317 (1990)
62. Hell, P., Kirkpatrick, D.: Linear-time certifying algorithms for near-graphical sequences. Discr. Math. **309**(18), 5703–5713 (2009)
63. Hennig, W.: Phylogenetic Systematics. Illinois University Press, Champaign (1966)
64. Herman, G.T., Kuba, A.: Discrete Tomography: Foundations, Algorithms, and Applications. Springer Science & Business Media (2012)
65. Hulett, H., Will, T.G., Woeginger, G.J.: Multigraph realizations of degree sequences: maximization is easy, minimization is hard. Oper. Res. Lett. **36**(5), 594–596 (2008)
66. Jayadev, S.P., Narasimhan, S., Bhatt, N.: Learning conserved networks from flows. Technical report, CoRR (2019). http://arxiv.org/abs/1905.08716
67. Kelly, P.: A congruence theorem for trees. Pacific J. Math. **7**, 961–968 (1957)
68. Kidd, K.K., Sgaramella-Zonta, L.: Phylogenetic analysis: Concepts and methods. American J. Hum. Gen. **23**, 235–252 (1971)
69. Kim, H., Toroczkai, Z., Erdos, P.L., Miklos, I., Szekely, L.A.: Degree-based graph construction. J. Phys. Math. Theor. **42**, 1–10 (2009)
70. Kleitman, D.J.: Minimal number of multiple edges in realization of an incidence sequence without loops. SIAM J. Appl. Math. **18**(1), 25–28 (1970)

71. Kleitman, D.J., Wang, D.L.: Algorithms for constructing graphs and digraphs with given valences and factors. Discr. Math. **6**, 79–88 (1973)
72. Kutiel, G., Rawitz, D.: Service chain placement in SDNs. Discr. Appl. Math. **270**, 168–180 (2019)
73. Li, C., McCormick, S., Simchi-Levi, D.: On the minimum-cardinality-bounded-diameter and the bounded-cardinality-minimum-diameter edge addition problems. Oper. Res. Lett. **11**, 303–308 (1992)
74. Lovász, L.: Subgraphs with prescribed valencies. J. Comb. Theory **8**, 391–416 (1970)
75. Mihail, M., Vishnoi, N.: On generating graphs with prescribed degree sequences for complex network modeling applications. In: 3rd ARACNE (2002)
76. Nieminen, J.: Realizing the distance matrix of a graph. J. Inf. Process. Cybern. **12**(1/2), 29–31 (1976)
77. Nivat, M.: Sous-ensembles homogénes de z2 et pavages du plan. Comptes Rendus Mathematique **335**(1), 83–86 (2002)
78. O'Neil, P.V.: Ulam's conjecture and graph reconstructions. Am. Math. Mon. **77**, 35–43 (1970)
79. Owens, A., Trent, H.: On determining minimal singularities for the realizations of an incidence sequence. SIAM J. Appl. Math. **15**(2), 406–418 (1967)
80. Owens, A.B.: On determining the minimum number of multiple edges for an incidence sequence. SIAM J. Appl. Math. **18**(1), 238–240 (1970)
81. Patrinos, A.N., Hakimi, S.L.: The distance matrix of a graph n and its tree realizability. Q. Appl. Math. **30**(3), 255 (1972)
82. Ryser, H.: Combinatorial properties of matrices of zeros and ones. Canad. J. Math. **9**, 371–377 (1957)
83. Schmeichel, E.F., Hakimi, S.L.: On planar graphical degree sequences. SIAM J. Appl. Math. **32**, 598–609 (1977)
84. Schuh, R.T., Brower, A.V.: Biological Systematics: Principles and Applications. Cornell University Press (2009)
85. Sierksma, G., Hoogeveen, H.: Seven criteria for integer sequences being graphic. J. Graph Theory **15**(2), 223–231 (1991)
86. Simões-Pereira, J.M.S.: An optimality criterion for graph embeddings of metrics. SIAM J. Discr. Math. **1**(2), 223–229 (1988)
87. Simões-Pereira, J.M.S.: An algorithm and its role in the study of optimal graph realizations of distance matrices. Discr. Math. **79**(3), 299–312 (1990)
88. Swofford, D.L., Olsen, G.J.: Phylogeny reconstruction. Mol. Syst. 411–501 (1990)
89. Tatsuya, A., Nagamochi, H.: Comparison and enumeration of chemical graphs. Comput. Struct. Biotechnol. **5**(6), e201302004 (2013)
90. Tripathi, A., Tyagi, H.: A simple criterion on degree sequences of graphs. Discr. Appl. Math. **156**(18), 3513–3517 (2008)
91. Tripathi, A., Venugopalan, S., West, D.B.: A short constructive proof of the Erdös-Gallai characterization of graphic lists. Discr. Math. **310**(4), 843–844 (2010)
92. Tripathi, A., Vijay, S.: A note on a theorem of Erdös & Gallai. Discr. Math. **265**(1–3), 417–420 (2003)
93. Tutte, W.T.: Graph factors. Combinatorica **1**, 79–97 (1981)
94. Tyshkevich, R.: Decomposition of graphical sequences and unigraphs. Discr. Math. **220**, 201–238 (2000)
95. Tyshkevich, R.I., Chernyak, A.A., Chernyak, Z.A.: Graphs and degree sequences: a survey I. Cybernetics **23**, 734–745 (1987)
96. Tyshkevich, R.I., Chernyak, A.A., Chernyak, Z.A.: Graphs and degree sequences: a survey II. Cybernetics **24**, 137–152 (1988)

97. Tyshkevich, R.I., Chernyak, A.A., Chernyak, Z.A.: Graphs and degree sequences: a survey III. Cybernetics **24**, 539–548 (1988)
98. Tyshkevich, R.I., Mel'nikov, O.I., Kotov, V.M.: Graphs and degree sequences: canonical decomposition. Cybernetics **17**, 722–727 (1981)
99. Ulam, S.: A Collection of Mathematical Problems. Wiley, Hoboken (1960)
100. Varone, S.C.: A constructive algorithm for realizing a distance matrix. Eur. J. Oper. Res. **174**(1), 102–111 (2006)
101. Wormald, N.: Models of random regular graphs. Surveys Combin. **267**, 239–298 (1999)
102. Zverovich, I.E., Zverovich, V.E.: Contributions to the theory of graphic sequences. Discr. Math. **105**(1–3), 293–303 (1992)

Search for Combinatorial Objects Using Lattice Algorithms – Revisited

Alfred Wassermann$^{(\boxtimes)}$ (iD)

Department of Mathematics, University of Bayreuth, 95440 Bayreuth, Germany
`alfred.wassermann@uni-bayreuth.de`

Abstract. In 1986, Kreher and Radziszowski were the first who used the famous LLL algorithm to construct combinatorial designs. Subsequently, lattice algorithms have been applied to construct a large variety of objects in design theory, coding theory and finite geometry. Unfortunately, the use of lattice algorithms in combinatorial search is still not well established. Here, we provide a list of problems which could be tackled with this approach and give an overview on exhaustive search using lattice basis reduction. Finally, we describe a different enumeration strategy which might improve the power of this method even further.

Keywords: Lattice enumeration · Combinatorial search

1 Introduction

In 1982, Lenstra, Lenstra and Lovász published the groundbreaking LLL algorithm, which finds in polynomial time "short" vectors in a lattice. As soon as in 1983, Lagarias and Odlyzko [44] applied the LLL algorithm successfully to break certain cryptosystems based on the subset sum problem.

It seems that in the field of constructive combinatorics, Kreher and Radziszowski were the first who used the LLL algorithm. In [40,41] they used it to construct a 6-$(14, 7, 4)$ design. Subsequently, lattice basis reduction was used by the author and collaborators to find the first combinatorial designs for $t = 7$, 8, and 9 with small parameters and other variants of designs, see e.g. [2–4,14,42,43,45] (see [16] for a comprehensive overview on combinatorial design theory). The same program has been used successfully in the search for large sets of designs [5,46,47], in coding theory (linear codes, codes over rings, two-weight codes, covering codes) [6,12,32,34,37,54,62], subspace designs and their variants [8–11,13,33], as well as in finite geometry [7] and other problems.

All of these combinatorial search problems can be reduced to the solution of a Diophantine linear system which is a generalization of the subset sum problem studied by Lagarias and Odlyzko [44] and has the following form.

Let $A \in \mathbb{Z}^{m \times n}$, $\mathbf{d} \in \mathbb{Z}^m$, and $\mathbf{l}, \mathbf{r} \in \mathbb{Z}^n$. Determine all vectors $\mathbf{x} \in \mathbb{Z}^n$ such that

$$A \cdot \mathbf{x} = \mathbf{d} \text{ and } \mathbf{l} \leq \mathbf{x} \leq \mathbf{r}, \tag{1}$$

where $\mathbf{l} \leq \mathbf{r}$ for vectors $\mathbf{l}, \mathbf{r} \in \mathbb{Z}^n$ is defined as $l_i \leq r_i$ for all $0 \leq i < n$.

© Springer Nature Switzerland AG 2021
P. Flocchini and L. Moura (Eds.): IWOCA 2021, LNCS 12757, pp. 20–33, 2021.
https://doi.org/10.1007/978-3-030-79987-8_2

With the substitution $\mathbf{x} := \mathbf{x} - \mathbf{l}$, $\mathbf{d} := \mathbf{d} - A \cdot \mathbf{l}$ and $\mathbf{r} := \mathbf{r} - \mathbf{l}$, it suffices to consider $\mathbf{l} = \mathbf{0}$ as a lower bound on the variables.

Kramer and Mesner [39] reduced the search for combinatorial designs with prescribed automorphism group to such a problem.

Solving equation (1) is a special instance of the *multi-dimensional subset sum problem* which is known to be NP-complete [22]. Since problem (1) can be reduced to many other NP-hard problems it is no surprise that there are many solving algorithms available. See [23, 31, 49] for a survey.

In case the right hand side vector $\mathbf{d}$ in (1) is the all-one vector, the problem is also called *exact cover problem* and the fastest algorithm seems to be the *dancing links* algorithm[1] by Knuth [35] or – in special cases – *maximum clique search*, see e.g. [53]. In case there is a "$\leq$" instead of "$=$" in (1), it seems that typical integer linear programming algorithms [24, 27] are most promising.

However, in the special case that there is "$=$" in (1), and $\mathbf{d}$ is much larger than the all-one vector, and $\mathbf{r}$ is the all-one vector (i.e. solution vectors $\mathbf{x}$ are $\{0, 1\}$ vectors), reduction of the problem to a lattice point enumeration problem has been very successful as shown in the above references. The algorithm has been described in detail in [60, 61] but unfortunately there are not many other implementations if any. This may be due to the widespread misconception that lattice basis reduction is only able to find random solutions which was the case in the implementation of [41]. It has been overlooked that lattice basis reduction can be followed by an exhaustive enumeration of all solutions of (1) with a backtracking algorithm.

In the sequel we will give an overview to exhaustive enumeration of all solutions of (1) using lattice point enumeration and also hint to a different enumeration scheme (limited discrepancy search) which allows to find the first solutions sometimes much more quickly.

2   Lattices

Let $\mathbb{R}^n$ denote the real Euclidean n-dimensional space. Its elements $\mathbf{v} \in \mathbb{R}^n$ are written as column vectors $\mathbf{v} = (v_0, v_1, \ldots, v_{n-1})^\top$. Let $\langle \mathbf{v}, \mathbf{w} \rangle = \sum_{i \in n} v_i \cdot w_i$ be the standard bilinear form for $\mathbf{v}, \mathbf{w} \in \mathbb{R}^n$. For $q \in \mathbb{R}$, $q \geq 1$, the ℓ_q-*norm* is defined by

$$\|-\|_q : \mathbb{R}^n \to \mathbb{R} : \mathbf{v} \mapsto \left(\sum_{i \in n} |v_i|^q \right)^{1/q},$$

and the ℓ_∞-*norm* is defined as:

$$\|-\|_\infty : \mathbb{R}^n \to \mathbb{R} : \mathbf{v} \mapsto \max_{i \in n} |v_i|.$$

Let $\mathbf{b}^{(0)}, \mathbf{b}^{(1)}, \ldots, \mathbf{b}^{(m-1)}$ be m linearly independent vectors in $\mathbb{R}^n$. The *discrete additive subgroup* of $\mathbb{R}^n$

---

[1] Updated versions available at https://www-cs-faculty.stanford.edu/~knuth/progr ams.html.

$$\mathcal{L}(\mathbf{b}^{(0)}, \mathbf{b}^{(1)}, \ldots, \mathbf{b}^{(m-1)}) = \{\sum_{i=0}^{m-1} u_i \cdot \mathbf{b}^{(i)} \mid u_i \in \mathbb{Z}, \, i \in m \,\} \subset \mathbb{R}^n$$

is called *lattice* with *basis* $\mathbf{b}^{(0)}, \mathbf{b}^{(1)}, \ldots, \mathbf{b}^{(m-1)}$.

The *rank* m of a lattice $\mathcal{L}$ with basis $\mathbf{b}^{(0)}, \mathbf{b}^{(1)}, \ldots, \mathbf{b}^{(m-1)}$ is the dimension of the $\mathbb{R}$-subspace $\langle \mathbf{b}^{(0)}, \mathbf{b}^{(1)}, \ldots, \mathbf{b}^{(m-1)} \rangle$ which is spanned by the basis. The corresponding $n \times m$-matrix

$$B = \left(\mathbf{b}^{(0)} \mid \ldots \mid \mathbf{b}^{(m-1)} \right)$$

is called a *generator matrix* of $\mathcal{L}$ if $\mathcal{L} = \mathcal{L}(\mathbf{b}^{(0)}, \mathbf{b}^{(1)}, \ldots, \mathbf{b}^{(m-1)})$.

For a lattice $\mathcal{L} \subset \mathbb{R}^n$, the most important algorithmic problems are:

- *Shortest vector problem* (SVP): Find an ℓ_q-shortest vector in L, i.e. find an element $\mathbf{w}$ in $\mathcal{L}$ such that

$$\|\mathbf{w}\|_q = \min\{\|\mathbf{w}'\|_q \mid \mathbf{w}' \in \mathcal{L} \setminus \{0\}\}.$$

This question is most interesting for the Euclidean norm ℓ_2, the ℓ_1-norm, and the ℓ_∞-norm.
- *Closest vector problem* (CVP): Given a vector $\mathbf{v} \in \mathbb{R}^n$, find a lattice vector $\mathbf{w} \in \mathcal{L}$ which is closest to $\mathbf{v}$ in the ℓ_q-norm, i.e. such that

$$\|\mathbf{v} - \mathbf{w}\|_q = \min\{\|\mathbf{v} - \mathbf{w}'\|_q \mid \mathbf{w}' \in \mathcal{L}\}.$$

- *Lattice basis reduction*: Given a basis $\mathbf{b}^{(0)}, \mathbf{b}^{(1)}, \ldots, \mathbf{b}^{(m-1)}$ of the lattice $\mathcal{L}$ compute a new basis $\mathbf{b}'^{(0)}, \mathbf{b}'^{(1)}, \ldots, \mathbf{b}'^{(m-1)}$ of L consisting of "shortest" vectors. Here, the meaning of short will have to be made precise, compare Fig. 1.

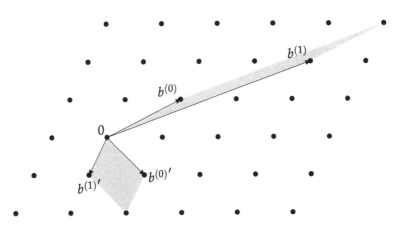

Fig. 1. Two different bases for $\mathbf{b}^{(0)}, \mathbf{b}^{(1)}$ and $\mathbf{b}^{(0)'}, \mathbf{b}^{(1)'}$ of the same lattice

For an overview on the algorithmic complexity of the above problems we refer e.g. to [50] and the literature cited there.

Concerning the last of the mentioned problems, we remark that the problem of finding a basis consisting of shortest vectors is not exactly defined provided the dimension is at least three. In fact, many different versions of the concept of a shortest basis exist. Two classical concepts are the reduced bases in the sense of Minkowski [51] and the reduced quadratic forms in the sense of Korkine and Zolotarev [38] which rely on the computation of shortest lattice vectors in sublattices and related lattices. Therefore, the problem of computing a reduced lattice basis in the sense of Minkowski or Korkine and Zolotarev is at least as hard as the shortest vector problem.

Let $B = (\mathbf{b}^{(0)} \mid \ldots \mid \mathbf{b}^{(m-1)})$ be a generator matrix of a lattice $\mathcal{L}$. The matrix $G(B) = (\langle \mathbf{b}^{(i)}, \mathbf{b}^{(j)} \rangle)_{i,j \in m} \in \mathbb{R}^{m \times m}$ is called *Gram matrix* of the lattice basis. The volume of the lattice $\mathcal{L}$ is defined as $\mathrm{Vol}(\mathcal{L}) = \det(\mathcal{L}) = \sqrt{\det(G(B))}$, it does not depend on the choice of the basis. Further invariants of a lattice – independent from the choice of the basis – are the *successive minima* of Minkowski [51].

Let $\mathcal{L} \subset \mathbb{R}^n$ be a lattice of rank m. For an integer $i \in m$ let $\lambda_i(\mathcal{L})$ be the least positive real number for which there exist $i+1$ linearly independent lattice vectors $\mathbf{v} \in \mathcal{L} \setminus \{0\}$ with $\|\mathbf{v}\|_2 \leq \lambda_i(L)$. The numbers $\lambda_0(\mathcal{L}), \lambda_1(\mathcal{L}), \ldots, \lambda_{m-1}(\mathcal{L})$ are the *successive minima* of the lattice $\mathcal{L}$. From the definition it follows that $\lambda_0(\mathcal{L}) \leq \lambda_1(\mathcal{L}) \leq \ldots \leq \lambda_{m-1}(\mathcal{L})$. A classical result due to Hermite [26] gives an upper bound for the ℓ_2-shortest vector of a lattice $\mathcal{L} \subset \mathbb{Z}^n$, namely $\mathcal{L}$ contains a nonzero vector $\mathbf{v}$ such that

$$\|\mathbf{v}\|_2^2 \leq (4/3)^{(m-1)/2} \cdot \det(\mathcal{L})^{2/m}\,.$$

3   Lattice Basis Reduction

Let $\mathbf{b}^{(0)}, \mathbf{b}^{(1)}, \ldots, \mathbf{b}^{(m-1)}$ be a set of linearly independent vectors $\in \mathbb{R}^n$.

Gram–Schmidt orthogonalization (GSO) is the orthogonal family defined for $0 \leq i < m$ by

$$\hat{\mathbf{b}}^{(i)} = \mathbf{b}^{(i)} - \sum_{j=0}^{i-1} \mu_{ij} \cdot \hat{\mathbf{b}}^{(j)}\,,$$

where

$$\mu_{ij} = \frac{\langle \mathbf{b}^{(i)}, \hat{\mathbf{b}}^{(j)} \rangle}{\langle \hat{\mathbf{b}}^{(j)}, \hat{\mathbf{b}}^{(j)} \rangle}\,. \tag{2}$$

For $0 \leq t < m$ and $\mathbf{v} \in \mathbb{R}^n$ the *orthogonal projection* $\pi_t(\mathbf{v})$ is defined by

$$\pi_t : \mathbb{R}^n \to \langle \mathbf{b}^{(0)}, \mathbf{b}^{(1)}, \ldots, \mathbf{b}^{(t-1)} \rangle^{\perp}, \quad \mathbf{v} \mapsto \sum_{j=t}^{m-1} \frac{\langle \mathbf{v}, \hat{\mathbf{b}}^{(j)} \rangle}{\langle \hat{\mathbf{b}}^{(j)}, \hat{\mathbf{b}}^{(j)} \rangle} \cdot \hat{\mathbf{b}}^{(j)}$$

and $\hat{\mathbf{b}}^{(t)} = \pi_t(\mathbf{b}^{(t)})$. The orthogonal projection of a lattice $\mathcal{L}$ is the lattice $\mathcal{L}_t$ defined by

$$\mathcal{L}_t = \mathcal{L}(\pi_t(\mathbf{b}^{(t)}), \pi_t(\mathbf{b}^{(t+1)}), \ldots, \pi_t(\mathbf{b}^{(m-1)}))\,.$$

A basis $\mathbf{b}^{(0)}, \mathbf{b}^{(1)}, \ldots, \mathbf{b}^{(m-1)}$ of a lattice $\mathcal{L} \subset \mathbb{R}^n$ is *reduced in the sense of Korkine and Zolotarev* [38] if

1. $\mathbf{b}^{(0)}$ is an ℓ_2-shortest vector in $\mathcal{L}$ and
2. for $0 \le t < m$, $\hat{\mathbf{b}}^{(t)}$ is an ℓ_2-shortest vector in the lattice $\mathcal{L}_t(\mathbf{b}^{(t)}, \ldots, \mathbf{b}^{(m-1)})$.

However, no polynomial time algorithm to compute a Korkine–Zolotarev-reduced basis is known. A major breakthrough was achieved by Lenstra, Lenstra, and Lovász in their seminal work [48]. They compute a different type of reduced basis, which is now called an *LLL-reduced basis*, see the original paper [48] or textbooks like [15,52].

The *LLL algorithm* computes an LLL-reduced basis. The input is a basis $\mathbf{b}^{(0)}, \ldots, \mathbf{b}^{(m-1)}$ of the lattice $\mathcal{L}$ of rank m.

(1) Let $\delta \in \mathbb{R}$ with $\frac{1}{4} < \delta < 1$.
(2) **Set** $k := 0$.
(3) **do**
(4)   1. **for** $j = 0, \ldots, k-1$
(5)   **replace** $\mathbf{b}^{(k)}$ **by** $\mathbf{b}^{(k)} - \lfloor \mu_{kj} \rceil \mathbf{b}^{(j)}$,
(6)   where μ_{kj} is the Gram-Schmidt coefficient (2).
(7)   2. **if** $\delta \| \pi_k(\mathbf{b}^{(k)}) \|^2 > \| \pi_k(\mathbf{b}^{(k+1)}) \|^2$ **then**
(8)   **swap** $\mathbf{b}^{(k+1)}$ and $\mathbf{b}^{(k)}$
(9)   **update** $\hat{\mathbf{b}}^{(k+1)}$, $\hat{\mathbf{b}}^{(k)}$ and μ
(10)   **set** $k := \max(k-1, 0)$
(11)   **else**
(12)   **set** $k := k+1$
(13) **until** $k = m-1$.   □

The output $\mathbf{b}^{(0)}, \mathbf{b}^{(1)}, \ldots, \mathbf{b}^{(m-1)}$ of the LLL-algorithm with $\frac{1}{4} < \delta < 1$ is called δ-reduced basis of the lattice $\mathcal{L}$.

The LLL algorithm runs in polynomial time in m, n, and the size of the entries of the basis vectors. In [52, Chapters 4 and 5] recent developments are described, e.g. how to approximate the LLL algorithm by using floating point numbers.

The LLL algorithms can not be expected to compute shortest vectors in a lattice. Let $\mathbf{b}^{(0)}, \mathbf{b}^{(1)}, \ldots, \mathbf{b}^{(m-1)}$ be a δ-reduced basis of the lattice $\mathcal{L} \subset \mathbb{Q}^n$. Then, the following bounds can be proved [48].

$$\| \mathbf{b}^{(j)} \|^2 \le \left(\frac{4}{4\delta - 1} \right)^i \cdot \| \hat{\mathbf{b}}^{(i)} \|^2 \text{ for } 0 \le j \le i < m. \tag{3}$$

$$\det(\mathcal{L}) \le \prod_{i=0}^{m-1} \| \mathbf{b}^{(i)} \| \le \left(\frac{4}{4\delta - 1} \right)^{m(m-1)/4} \cdot \det(\mathcal{L}). \tag{4}$$

$$\| \mathbf{b}^{(0)} \| \le \left(\frac{4}{4\delta - 1} \right)^{(m-1)/4} \cdot \det(\mathcal{L})^{1/m}. \tag{5}$$

The fascinating mystery behind the LLL algorithm is that in many cases it produces a much better approximation to the shortest vector of the lattice than the proven bounds guarantee.

Nevertheless, a full reduction in the sense of Korkine and Zolotarev would require exponential complexity. In [56,57] *blockwise Korkine–Zolotarev reduction* (BKZ) was introduced which restricts enumeration in the sense of Korkine and Zolotarev to blocks of a fixed size β of basis vectors, i.e. searches by exhaustive enumeration for nontrivial integer linear combinations

$$u_k \mathbf{b}^{(k)} + u_{k+1}\mathbf{b}^{(k+1)} + \ldots + u_{k+\beta-1}\mathbf{b}^{(k+\beta-1)}$$

which minimize the Euclidean length of

$$\pi_k\left(u_k\mathbf{b}^{(k)} + u_{k+1}\mathbf{b}^{(k+1)} + \ldots + u_{k+\beta-1}\mathbf{b}^{(k+\beta-1)}\right).$$

The original LLL algorithm can be interpreted as blockwise Korkine–Zolotarev reduction for $\beta = 2$. For a further description of improved practical versions and recent developments, e.g. sieving methods, we refer to [52,57,58]. In a blockwise Korkine–Zolotarev-reduced basis of a lattice of rank m the factor $(\frac{4}{4\delta-1})^{(m-1)/2}$ can be replaced by $(1+\epsilon)^m$ for any fixed $\epsilon > 0$. Of course, the time complexity increases exponentially as ϵ approaches 0.

4 Lattice Embedding of Diophantine Linear Systems

In [44], Lagarias and Odlyzko described the reduction of problem (1) for $\{0,1\}$ vectors $\mathbf{x}$, i.e. $\mathbf{r} = \mathbf{1}$. In [17,18] their embedding of (1) into a lattice problem was be improved. In turn, the following generalization to arbitrary upper bounds $\mathbf{r}$ has been given in [61].

The basis of the lattice to which the LLL algorithm is applied consists of the columns of the following generator matrix of size $(m + n + 1) \times (n + 1)$:

$$
\begin{pmatrix}
-N \cdot d & N \cdot A \\
-r_{\max} & 2c_1 & 0 & \cdots & 0 \\
-r_{\max} & 0 & 2c_2 & \cdots & 0 \\
\vdots & \vdots & & \ddots & \vdots \\
-r_{\max} & 0 & \cdots & \cdots & 2c_n \\
r_{\max} & 0 & \cdots & \cdots & 0
\end{pmatrix}
\tag{6}
$$

where $N \in \mathbb{Z}_{>0}$ is a large constant and

$$r_{\max} = \mathrm{lcm}\{r_1, \ldots, r_n\} \quad \text{and} \quad c_i = \frac{r_{\max}}{r_i}, \quad 1 \le i \le n.$$

If N is large enough, see [1], the reduced basis will consist of $n - m + 1$ vectors with only zeroes in the first m rows and m vectors which contain at least one nonzero entry in the first m rows. The latter vectors can be removed from the basis. From the remaining $n - m + 1$ vectors we can delete the first m rows which contain only zeroes. This gives a basis $\mathbf{b}^{(0)}, \mathbf{b}^{(1)}, \ldots, \mathbf{b}^{(n-m)} \in \mathbb{Z}^{n+1}$.

In the second step of the algorithm, see Sect. 5, all integer linear combinations of the basis vectors $\mathbf{b}^{(0)}, \mathbf{b}^{(1)}, \ldots, \mathbf{b}^{(n-m)} \in \mathbb{Z}^{n+1}$ are enumerated which correspond to solutions of (1).

Theorem 1 ([61]). *Let*

$$\mathbf{w} = u_0 \cdot \mathbf{b}^{(0)} + u_1 \cdot \mathbf{b}^{(1)} + \ldots + u_{n-m} \cdot \mathbf{b}^{(n-m)} \tag{7}$$

be an integer linear combination of the basis vectors with $w_0 = r_{\max}$. $\mathbf{w}$ is a solution of the system (1) if and only if

$$\mathbf{w} \in \mathbb{Z}^{n+1} \text{ where } -r_{\max} \leq w_i \leq r_{\max}, 1 \leq i \leq n.$$

5   Lattice Point Enumeration

Usually, we are interested in finding all solutions to problem (1), or to conclude that there are none. In terms of the associated lattice (6), this mean that we wish to enumerate all lattice points which are subject to a certain set of constraints. Such an approach has first been described by Ritter [55] for $\{0, 1\}$ problems. Here we solve the general problem with arbitrary bounds on the variables.

A priori, a lattice $\mathcal{L} = \{\sum_{i \in m} u_i \mathbf{b}^{(i)} \mid u_i \in \mathbb{Z}\}$ of rank m contains infinitely many elements. However, it will turn out that there are bounds on the integers $|u_i|$, $i \in m$, which depend solely on the lattice basis $\mathbf{b}^{(0)}, \mathbf{b}^{(1)}, \ldots, \mathbf{b}^{(m-1)}$. These bounds reduce the problem of finding solution vectors to a finite set of lattice vectors. Each solution vector $\mathbf{v}$ has the upper bounds

$$\|\mathbf{v}\|_2^2 \leq (n+1) \cdot r_{\max}^2 \quad \text{and} \quad \|\mathbf{v}\|_\infty \leq r_{\max}.$$

The exhaustive enumeration is arranged as backtracking algorithm. Starting from $u_{n-m} \in \mathbb{Z}$, successively all possible $u_t \in \mathbb{Z}$ for $t = n - m, n - m - 1, \ldots, 1, 0$ are tested. The enumeration can be pruned at stage t if certain conditions are violated. These pruning tests have quite a long history and are based on the work of [19–21, 28, 29, 36, 55].

In each level t of the backtracking algorithm, $\mathbf{w}^{(t)} = \pi_t(\sum_{j=t}^{n-m} u_j \mathbf{b}^{(j)})$ is the projection of the linear combination of the already fixed variables u_t, u_{t+1}, $\ldots$, u_{n-m} into the subspace of $\mathbb{R}^{n+1}$ which is orthogonal to the linear span $\langle b_0, \ldots, b_{t-1} \rangle$.

Starting from $\mathbf{w}^{(n-m+1)} = \mathbf{0}$, $\mathbf{w}^{(t)}$ can be computed iteratively from $\mathbf{w}^{(t+1)}$ by

$$\mathbf{w}^{(t)} = \left(\sum_{i=t}^{n-m} u_i \mu_{it} \right) \hat{\mathbf{b}}^{(t)} + \mathbf{w}^{(t+1)}$$

with Gram-Schmidt coefficients μ_{it}. In each level t, $n - m \geq t \geq 0$, all possible integer values for the variable u_t are tested. The following two main tests allow to restrict the possible values of u_t.

First pruning condition. For all $j \leq t$ the vectors $\hat{\mathbf{b}}^{(j)}$ are orthogonal to $\mathbf{w}^{(t+1)}$ and therefore

$$\|\mathbf{w}^{(t)}\|_2^2 = \left(\sum_{i=t}^{n-m} u_i \mu_{it} \right)^2 \|\hat{\mathbf{b}}^{(t)}\|_2^2 + \|\mathbf{w}^{(t+1)}\|_2^2.$$

We notice that $\mathbf{w}^{(0)} = \sum_{j=0}^{n-m} u_j \mathbf{b}^{(j)}$. Using $\|\mathbf{w}^{(j)}\|_2 \geq \|\mathbf{w}^{(t)}\|_2$ for $j \leq t$ we can backtrack as soon as

$$\|\mathbf{w}^{(t)}\|_2^2 > c := (n+1) \cdot r_{\max}^2 \, .$$

For fixed $u_{t+1}, \ldots, u_{n-m}$, this gives a bound for u_t:

$$\left(u_t + \sum_{i=t+1}^{n-m} u_i \mu_{it}\right)^2 \leq \frac{c - \|\mathbf{w}^{(t+1)}\|_2^2}{\|\hat{\mathbf{b}}^{(t)}\|_2^2} \, .$$

Second pruning condition. The second test is an adaption to the special situation that we are searching for an integer linear combination of the basis vectors which consists solely of components whose absolute value is bounded by $r_{\max}$. It is based on the following theorem by Ritter [55].

Theorem 2 ([55]). *If the given sequence of integers $u_t, u_{t+1}, \ldots, u_{n-m} \in \mathbb{Z}$ can be extended to $u_0, \ldots, u_t, \ldots, u_{n-m} \in \mathbb{Z}$ such that $\sum_{i=0}^{n-m} u_i \mathbf{b}^{(i)}$ is a solution of (1), then for all $y_t, y_{t+1}, \ldots, y_{n-m} \in \mathbb{R}$:*

$$\left| \sum_{i=t}^{n-m} y_i \|\mathbf{w}^{(i)}\|_2^2 \right| \leq r_{\max} \cdot \left\| \sum_{i=t}^{n-m} y_i \mathbf{w}^{(i)} \right\|_1 \, .$$

We use this theorem in the enumeration algorithm in the following way. Taking $(y_t, y_{t+1}, \ldots, y_{n-m}) = (1, 0, \ldots, 0)$ results in the test

$$\|\mathbf{w}^{(t)}\|_2^2 \leq r_{\max} \|\mathbf{w}^{(t)}\|_1 \, .$$

If this inequality is violated for some vector $\mathbf{w}^{(t)} = x\hat{\mathbf{b}}^{(t)} + \mathbf{w}^{(t+1)}$, then it will also fail for all vectors of the form $(x + r)\hat{\mathbf{b}}^{(t)} + \mathbf{w}^{(t+1)}$ with $r \in \mathbb{Z}$ and $xr > 0$.

Summarizing, a high level description of the algorithm to solve (1) is as follows.

Lattice point enumeration
Given the generator matrix (6) of the lattice $\mathcal{L} \subset \mathbb{R}^{m+n+1}$ of rank $n + 1$ of problem (1) all nonzero vectors $\mathbf{v} \in \mathcal{L}$ such that $\|\mathbf{v}\|_\infty \leq r_{\max}$ are determined.

- Compute an LLL/BKZ-reduced basis $\mathbf{b}^{(0)}, \mathbf{b}^{(1)}, \ldots, \mathbf{b}^{(n)}$ of the lattice $\mathcal{L}$.
- Delete the unnecessary columns and rows of the generator matrix. The remaining basis $\mathbf{b}^{(0)}, \mathbf{b}^{(1)}, \ldots, \mathbf{b}^{(n-m)} \subset \mathbb{R}^{n+1}$ has rank $n - m + 1$.
- Compute the Gram–Schmidt vectors $\hat{\mathbf{b}}^{(0)}, \hat{\mathbf{b}}^{(1)}, \ldots, \hat{\mathbf{b}}^{(n-m)}$ together with the Gram–Schmidt coefficients μ_{ij}.
- Set $R := (n+1) \cdot r_{\max}^2$.
- The recursive backtracking algorithm enum() is initiated with the call of enum$(n - m, \mathbf{0})$.

(1)   **function** enum$(t, \mathbf{w}')$
(2)   **begin**
(3)   onedirection := false

(4) $y_t := \sum_{i=t+1}^{n-m} u_i \mu_{it}$

(5) $u_t := \lfloor -y_t \rceil$

(6)   **while** true

(7) $\mathbf{w} := (\sum_{i=t}^{n-m} u_i \mu_{it}) \hat{\mathbf{b}}^{(t)} + \mathbf{w}'$

(8)   **if** $\|\mathbf{w}\|_2^2 > R$ **then** **return**   /* step back */

(9)   **if** $t > 0$ **then**

(10)   **if** $\|\mathbf{w}\|_2^2 > r_{\max} \cdot \|\mathbf{w}\|_1$ **then**

(11)   **if** onedirection **then** **return**   /* step back */

(12)   **else**

(13)   next(u_t)

(14)   onedirection := true

(15)   **goto** line (7)

(16)   **end if**

(17)   **else**

(18)   enum($t - 1$, $\mathbf{w}$)   /* step forward */

(19)   **else**   /* $t = 0$ → solution */

(20)   output solution w

(21)   next(u_t)

(22)   **end while**

(23) **end**

The procedure next() in lines (13) and (21) determines the next possible integer value of the variable u_t. Initially, when entering a new level t, in line (5) u_t is set to be the closest integer value of $-y_t := -\sum_{i=t+1}^{n-m} u_i \mu_{it}$, say u_t^1. The next value u_t^2 of u_t is the second closest integer to $-y_t$ followed by u_t^3 and so forth. In other words, the values of u_t alternate with increasing distance around $-y_t$.

If the condition in line (10) is true then we do one more regular call of the procedure next() in line (13), i.e. u_t is set to be the next closest integer to $-y_t$. In Fig. 2 this happens while u_t^4 is determined. After that, the enumeration proceeds only in this remaining direction, see the computation of u_t^5 in Fig. 2. Finally, the second time when the condition in line (10) is true, the algorithm steps back and increases the enumeration level, see line (11).

6   Limited Discrepancy Search

For some problems of the form (1) it might be not interesting or may be impossible to enumerate all solutions. But nevertheless one is interested to find at least one solution as quick as possible. It turns out that in this situation, the enumeration algorithm in the previous section might not be optimal. In the following we will try to motivate a different enumeration algorithm.

The enumeration algorithm in Sect. 5 performs depth first search. In particular, when entering enumeration level t, u_t is chosen for $\mathbf{w} := (\sum_{i=t}^{n-m} u_i \mu_{it}) \hat{\mathbf{b}}^{(t)} + \mathbf{w}'$ in line (7) such that $\|\mathbf{w}\|_2$ is minimal among all choices for u_t. In other words, the depth first search is organized using the heuristic that choosing in each level

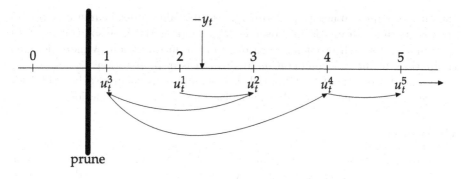

Fig. 2. Enumeration in level t and pruning after u_t^3

the vector $\mathbf{w}$ such that $\|\mathbf{w}\|_2$ is minimal will most probably lead to a solution vector. However, it may be that this choice for u_t in one of the first levels might lead to no solution, but nevertheless the algorithm will enumerate a huge search tree below u_t.

This is a general problem of depth first search algorithm. In 1995, Harvey and Ginsberg [25] described a simple, novel enumeration scheme called *limited discrepancy search* which aims to overcome this weakness of depth first search.

Assume that a backtrack algorithm has to examine a search tree. Each level corresponds to a variable and the algorithm has to assign a value to that variable, followed by a test if this assignment might lead to a solution. If yes, we can proceed to the next level, otherwise we have to assign a different value. If we have tested all values, we have to step back to the previous level. If values could be assigned to all variables, a solution has been found.

We assume that variable ordering is fixed and in each level of the backtracking there exists a heuristic which determines the order in which the values are assigned to the variable corresponding to that enumeration level. A *discrepancy* is defined as an deviation from the heuristic.

Harvey and Ginsberg suggest to enumerate the search tree in increasing number of discrepancies. In the first step, only the optimal choice in each level of enumeration in Sect. 5 is assigned to the variables until there is a contradiction or a solution is found. In the next step, all possible paths in the search tree with exactly one deviation (i.e. discrepancy) from the heuristic are examined. After that, all paths in the search tree with two deviations from the optimal choice are enumerated, and so forth.

In [25], the algorithm is given for binary search trees. In [30], the algorithm is described for general search trees, also a stop condition is given which allows to use the algorithm for exhaustive enumeration. The latter is mostly interesting to show the non-existence of solutions. There are other variants, see e.g. [59] for an overview.

Limited discrepancy search requires higher book keeping efforts than depth first search. Therefore, enumerating the whole search tree with limited discrepancy search will always be slower than with depth first search. But first tests with the lattice point enumeration algorithm and its value order heuristic in Sect. 5 show sometimes dramatic improvements for finding the first solution in hard combinatorial search problems mentioned in the introduction. A more detailed comparison of the two enumeration algorithms is in preparation.

It may be remarked that limited discrepancy search can also be useful for the enumeration algorithm in blockwise Korkine–Zolotarev reduction.

References

1. Aardal, K., Hurkens, C., Lenstra, A.K.: Solving a linear diophantine equation with lower and upper bounds on the variables. In: Bixby, R.E., Boyd, E.A., Ríos-Mercado, R.Z. (eds.) IPCO 1998. LNCS, vol. 1412, pp. 229–242. Springer, Heidelberg (1998). https://doi.org/10.1007/3-540-69346-7_18

2. Betten, A., Kerber, A., Laue, R., Wassermann, A.: Simple 8-designs with small parameters. Des. Codes Crypt. **15**, 5–27 (1998)

3. Betten, A., Kerber, A., Kohnert, A., Laue, R., Wassermann, A.: The discovery of simple 7-designs with automorphism group $P\Gamma L(2, 32)$. In: Cohen, G., Giusti, M., Mora, T. (eds.) AAECC 1995. LNCS, vol. 948, pp. 131–145. Springer, Heidelberg (1995). https://doi.org/10.1007/3-540-60114-7_10

4. Betten, A., Klin, M., Laue, R., Wassermann, A.: Graphical t-designs. Discrete Math. **197**(198), 111–121 (1999)

5. Betten, A., Laue, R., Wassermann, A.: New t-designs and large sets of t-designs. Discrete Math. **197**(198), 83–109 (1999)

6. Bouyukliev, I., Bouyuklieva, S., Kurz, S.: Computer classification of linear codes. CoRR abs/2002.07826 (2020). https://arxiv.org/abs/2002.07826

7. Braun, M., Kohnert, A., Wassermann, A.: Construction of (n, r)-arcs in PG$(2, q)$. Innovations Incidence Geom. **1**, 133–141 (2005)

8. Braun, M., Kerber, A., Laue, R.: Systematic construction of q-analogs of t-(v, k, λ)-designs. Des. Codes Crypt. **34**(1), 55–70 (2005). https://doi.org/10.1007/s10623-003-4194-z

9. Braun, M., Kiermaier, M., Kohnert, A., Laue, R.: Large sets of subspace designs. J. Comb. Theory Ser. A **147**, 155–185 (2017). https://doi.org/10.1016/j.jcta.2016.11.004

10. Braun, M., Kiermaier, M., Wassermann, A.: Computational methods in subspace designs. In: Greferath, M., Pavčević, M.O., Silberstein, N., Vázquez-Castro, M.Á. (eds.) Network Coding and Subspace Designs. SCT, pp. 213–244. Springer, Cham (2018). https://doi.org/10.1007/978-3-319-70293-3_9

11. Braun, M., Kohnert, A., Östergård, P.R.J., Wassermann, A.: Large sets of t-designs over finite fields. J. Comb. Theory A **124**, 195–202 (2014)

12. Braun, M., Kohnert, A., Wassermann, A.: Optimal linear codes from matrix groups. IEEE Trans. Inform. Theory **51**(12), 4247–4251 (2005). https://doi.org/10.1109/TIT.2005.859291

13. Buratti, M., Kiermaier, M., Kurz, S., Nakić, A., Wassermann, A.: q-analogs of group divisible designs. In: Pseudorandomness and Finite Fields, Radon Series on Computational and Applied Mathematics, vol. 23. DeGruyter (2019)

14. Buratti, M., Wassermann, A.: On decomposability of cyclic triple systems. Australas. J. Comb. **71**(2), 184–195 (2018)

15. Cohen, H.: A Course in Computational Algebraic Number Theory. Graduate Texts in Mathematics, vol. 138. Springer, Berlin (1993). https://doi.org/10.1007/978-3-662-02945-9

16. Colbourn, C.J., Dinitz, J.H.: Handbook of Combinatorial Designs, Second Edition (Discrete Mathematics and Its Applications). Chapman & Hall/CRC, Boca Raton (2006)
17. Coster, M., Joux, A., LaMacchia, B., Odlyzko, A., Schnorr, C., Stern, J.: Improved low-density subset sum algorithms. Comput. Complex. **2**, 111–128 (1992)
18. Coster, M.J., LaMacchia, B.A., Odlyzko, A.M., Schnorr, C.P.: An improved low-density subset sum algorithm. In: Davies, D.W. (ed.) EUROCRYPT 1991. LNCS, vol. 547, pp. 54–67. Springer, Heidelberg (1991). https://doi.org/10.1007/3-540-46416-6_4
19. Coveyou, R., MacPherson, R.: Fourier analysis of uniform random number generators. J. ACM **14**, 100–119 (1967)
20. Dieter, U.: How to calculate shortest vectors in a lattice. Math. Comput. **29**(131), 827–833 (1975)
21. Fincke, U., Pohst, M.: Improved methods for calculating vectors of short length in a lattice, including a complexity analysis. Math. Comput. **44**, 463–471 (1985)
22. Garey, M.R., Johnson, D.S.: Computers and Intractability: A Guide to the Theory of NP-Completeness. W. H. Freeman and Company, New York (1979)
23. Gibbons, P.B., Östergård, P.R.J.: Computational methods in design theory. In: Colbourn, C.J., Dinitz, J.H. (eds.) Handbook of Combinatorial Designs, chap. VII.6, 2 edn, pp. 755–783. Chapman & Hall/CRC, Boca Raton (2007)
24. Gurobi Optimization: Gurobi optimizer reference manual (2016). http://www.gurobi.com
25. Harvey, W.D., Ginsberg, M.L.: Limited discrepancy search. In: Proceedings of the 14th International Joint Conference on Artificial Intelligence, IJCAI 1995, vol. 1. pp. 607–613. Morgan Kaufmann Publishers Inc., San Francisco (1995)
26. Hermite, C.: Extraits de lettres de M.Ch. Hermite à M. Jacobi sur différents objets de la théorie des nombres. J. reine angew. Math. **40**, 279–290 (1850)
27. IBM: ILOG CPLEX Optimizer (2010)
28. Kaib, M., Ritter, H.: Block reduction for arbitrary norms. Preprint, Universität Frankfurt (1995)
29. Kannan, R.: Minkowski's convex body theorem and integer programming. Math. Oper. Res. **12**, 415–440 (1987)
30. Karoui, W., Huguet, M.-J., Lopez, P., Naanaa, W.: YIELDS: a yet improved limited discrepancy search for CSPs. In: Van Hentenryck, P., Wolsey, L. (eds.) CPAIOR 2007. LNCS, vol. 4510, pp. 99–111. Springer, Heidelberg (2007). https://doi.org/10.1007/978-3-540-72397-4_8
31. Kaski, P., Östergård, P.R.: Classification Algorithms for Codes and Designs. Springer, Heidelberg (2006). https://doi.org/10.1007/3-540-28991-7
32. Kiermaier, M., Kurz, S., Solé, P., Stoll, M., Wassermann, A.: On strongly walk regular graphs, triple sum sets and their codes. ArXiv e-prints, **abs/1502.02711** (2020)
33. Kiermaier, M., Laue, R., Wassermann, A.: A new series of large sets of subspace designs over the binary field. Des. Codes Crypt. **86**(2), 251–268 (2018). https://doi.org/10.1007/s10623-017-0349-1
34. Kiermaier, M., Wassermann, A., Zwanzger, J.: New upper bounds on binary linear codes and a $\mathbb{Z}_4$-code with a better-than-linear Gray image. IEEE Trans. Inf. Theory **62**(12), 6768–6771 (2016). https://doi.org/10.1109/TIT.2016.2612654
35. Knuth, D.E.: Dancing links. In: Davies, J., Roscoe, B., Woodcock, J. (eds.) Millennial Perspectives in Computer Science: Proceedings of the 1999 Oxford-Microsoft Symposium in Honour of Sir Tony Hoare. Palgrave (2000)

36. Knuth, D.: The Art of Computer Programming, Vol. 2: Seminumerical Algorithms. Addison-Wesley, Reading (1969)
37. Kohnert, A.: Constructing two-weight codes with prescribed groups of automorphisms. Discret. Appl. Math. **155**(11), 1451–1457 (2007). https://doi.org/10.1016/j.dam.2007.03.006
38. Korkine, A., Zolotareff, G.: Sur les formes quadratiques. Math. Ann. **6**, 366–389 (1873)
39. Kramer, E.S., Mesner, D.M.: t-designs on hypergraphs. Discret. Math. **15**(3), 263–296 (1976). https://doi.org/10.1016/0012-365X(76)90030-3
40. Kreher, D.L., Radziszowski, S.P.: The existence of simple 6-$(14, 7, 4)$ designs. J. Comb. Theory Ser. A **43**, 237–243 (1986)
41. Kreher, D.L., Radziszowski, S.P.: Finding simple t-designs by using basis reduction. Congr. Numer. **55**, 235–244 (1986)
42. Krčadinac, V.: Some new designs with prescribed automorphism groups. J. Comb. Des. **26**(4), 193–200 (2018). https://doi.org/10.1002/jcd.21587
43. Krčadinac, V., Pavčević, M.O.: New small 4-designs with nonabelian automorphism groups. In: Blömer, J., Kotsireas, I.S., Kutsia, T., Simos, D.E. (eds.) MACIS 2017. LNCS, vol. 10693, pp. 289–294. Springer, Cham (2017). https://doi.org/10.1007/978-3-319-72453-9_23
44. Lagarias, J., Odlyzko, A.: Solving low-density subset sum problems. J. Assoc. Comp. Mach. **32**, 229–246 (1985). Appeared already in Proceedings of 24th IEEE Symposium on Foundations of Computer Science, pp. 1–10 (1983)
45. Laue, R.: Constructing objects up to isomorphism, simple 9-designs with small parameters. In: Betten, A., Kohnert, A., Laue, R., Wassermann, A. (eds.) Algebraic Combinatorics and Applications, pp. 232–260. Springer, Heidelberg (2001). https://doi.org/10.1007/978-3-642-59448-9_16
46. Laue, R., Magliveras, S., Wassermann, A.: New large sets of t-designs. J. Comb. Des. **9**, 40–59 (2001)
47. Laue, R., Omidi, G.R., Tayfeh-Rezaie, B., Wassermann, A.: New large sets of t-designs with prescribed groups of automorphisms. J. Combin. Des. **15**(3), 210–220 (2007). https://doi.org/10.1002/jcd.20128
48. Lenstra, A., Lenstra Jr., H., Lovász, L.: Factoring polynomials with rational coefficients. Math. Ann. **261**, 515–534 (1982)
49. Mathon, R.: Computational methods in design theory. In: Keedwell, A.D. (ed.) Surveys in Combinatorics, Proc. 13th Br. Comb. Conf., Guildford/UK 1991, vol. 166, pp. 101–117. London Mathematical Society Lecture Note (1991)
50. Micciancio, D., Goldwasser, S.: Complexity of Lattice Problems. Kluwer Academic Publishers (2002)
51. Minkowski, H.: Geometrie der Zahlen. Teubner, Leipzig (1896)
52. Nguyen, P.Q., Vallée, B.: The LLL Algorithm: Survey and Applications, 1st edn. Springer, Heidelberg (2009). https://doi.org/10.1007/978-3-642-02295-1
53. Niskanen, S., Östergård, P.R.J.: Cliquer user's guide, version 1.0. Technical report T48, Helsinki University of Technology (2003)
54. Östergård, P.R.J., Quistorff, J., Wassermann, A.: New results on codes with covering radius 1 and minimum distance 2. Des. Codes Crypt. **35**, 241–250 (2005)
55. Ritter, H.: Aufzählung von kurzen Gittervektoren in allgemeiner Norm. Ph.D. thesis, Universität Frankfurt (1997)
56. Schnorr, C.: A hierachy of polynomial time lattice basis reduction algorithms. Theoret. Comput. Sci. **53**, 201–224 (1987)

57. Schnorr, C.P., Euchner, M.: Lattice basis reduction: improved practical algorithms and solving subset sum problems. In: Budach, L. (ed.) FCT 1991. LNCS, vol. 529, pp. 68–85. Springer, Heidelberg (1991). https://doi.org/10.1007/3-540-54458-5_51

58. Schnorr, C.P., Hörner, H.H.: Attacking the Chor-Rivest cryptosystem by improved lattice reduction. In: Guillou, L.C., Quisquater, J.-J. (eds.) EUROCRYPT 1995. LNCS, vol. 921, pp. 1–12. Springer, Heidelberg (1995). https://doi.org/10.1007/3-540-49264-X_1

59. van Beek, P.: Backtracking search algorithms. In: Rossi, F., van Beek, P., Walsh, T. (eds.) Handbook of Constraint Programming, Foundations of Artificial Intelligence, vol. 2, pp. 85–134. Elsevier (2006). https://doi.org/10.1016/S1574-6526(06)80008-8

60. Wassermann, A.: Finding simple t-designs with enumeration techniques. J. Comb. Des. **6**(2), 79–90 (1998)

61. Wassermann, A.: Attacking the market split problem with lattice point enumeration. J. Comb. Optim. **6**, 5–16 (2002)

62. Wassermann, A.: Computing the minimum distance of linear codes. In: Eighth International Workshop Algebraic and Combinatorial Coding Theory (ACCT VIII), Tsarskoe Selo, Russia, pp. 254–257 (2002)

Contributed Papers

Linear Algorithms for Red and Blue Domination in Convex Bipartite Graphs

Nesrine Abbas[(✉)] [iD]

Department of Computer Science, MacEwan University,
Edmonton, AB T5J 4S2, Canada
abbasn3@macewan.ca

Abstract. In the k red (blue) domination problem for a bipartite graph $G = (X, Y, E)$, we seek a subset $D \subseteq X$ (respectively $D \subseteq Y$) of cardinality at most k that dominates vertices of Y (respectively X). The decision version of this problem is NP-complete for perfect elimination bipartite graphs but solvable in polynomial time for chordal bipartite graphs. We present a linear time algorithm to solve the minimum cardinality red domination problem for convex bipartite graphs. The algorithm presented is faster and simpler than that in the literature. Due to the asymmetry in convex bipartite graphs, the algorithm does not extend to k blue domination. We present a linear time algorithm to solve the minimum cardinality blue domination problem for convex bipartite graphs.

Keywords: Convex bipartite graph · Red dominating set · Blue dominating set

1 Introduction

A k red (blue) dominating set in a bipartite graph $G = (X, Y, E)$ is a subset $D \subseteq X$ (respectively $D \subseteq Y$) of cardinality at most k that dominates vertices of Y (respectively X). The name originates from the view of a graph as having partitions *Red* and *Blue* (or X and Y as in this paper). The problem is NP-complete for perfect elimination bipartite graphs [1], but solvable in polynomial time for chordal bipartite graphs [2] which are a proper subset of perfect elimination bipartite graphs. In [2], the authors present an $\mathcal{O}(n \cdot \min(|E| \cdot \log n, n^2))$ delay enumeration algorithm for minimum cardinality red dominating sets in chordal bipartite graphs. In [1], a linear space linear delay algorithm is presented for enumerating minimum cardinality red dominating sets in convex bipartite graphs, a proper subset of chordal bipartite graphs. The first output of that algorithm needs $\mathcal{O}(n + |E|)$ time. Because convex bipartite graphs are not symmetric, the algorithms in [1] do not extend to blue domination. The aim of this paper is to further study the red and blue domination problem for convex bipartite graphs. We present linear time algorithms for both blue and red domination in convex bipartite graphs. The algorithm presented for red domination is faster and simpler than that presented in [1].

© Springer Nature Switzerland AG 2021
P. Flocchini and L. Moura (Eds.): IWOCA 2021, LNCS 12757, pp. 37–48, 2021.
https://doi.org/10.1007/978-3-030-79987-8_3

To the best of the author's knowledge red and blue domination has not been studied for convex bipartite graphs, apart from the work in [1] and in [2]. The problem has been studied for general graphs in [3] where the authors develop an exact exponential time algorithm for connected red dominating set. Other domination problems have been studied for convex bipartite graphs and classes of bipartite graphs that include them. In [4], the authors show that various domination problems are solvable in polynomial time for convex bipartite graphs. A linear time algorithm for paired domination in convex bipartite graphs is developed in [5]. Minimum paired domination is studied for chordal bipartite and perfect elimination bipartite graphs in [6]. Conditions for the tractability of independent domination for tree convex bipartite graphs are presented in [7]. The minimum dominating set problem for generalizations of convex bipartite graphs is studied in [8].

The next section contains definitions that will be used throughout the paper, and preliminary propositions that will be needed by all algorithms. Other definitions will be introduced as needed. Following that, we present the algorithm for computing a minimum cardinality blue dominating set in convex bipartite graphs. We then present the algorithm for computing a minimum cardinality red dominating set in convex bipartite graphs. We then conclude the paper.

2   Definitions and Preliminaries

A graph $G = (V, E)$ in this paper is finite, connected, undirected, with no loops and no parallel edges. We will use n to denote $|V|$. The *neighbourhood set of vertex* v is $N(v) = \{u : uv \in E\}$. The *degree* of v is $deg(v) = |N(v)|$. The graph induced by a subset $D \subseteq V$ is denoted by $G[D]$. $G[D]$ (or D) is an *independent set* if it induces a graph that has no edges. A set of vertices $D \subseteq V$ *dominates* another set $U \subseteq V$ if $\forall u \in U$, there is a $v \in D$ such that $u = v$ or $uv \in E$. When U is a single vertex u, we say that D dominates u. A set $D \subseteq V$ is *dominating* if it dominates V. A *path* of length k, or a k-path, in G is a sequence of distinct vertices $(u_1, \ldots, u_{k+1})$ such that $u_j u_{j+1} \in E$, $\forall j = 1, \ldots, k$. A *connected graph* is one that has a path between each pair in its vertex set. A *cycle* of length k in G is a sequence of distinct vertices $(u_1, \ldots, u_k)$ such that $u_j u_{j+1} \in E$ $\forall j = 1, \ldots, k-1$ and $u_1 u_k \in E$. A *chord* in a cycle is an edge joining non-consecutive vertices.

A *bipartite graph* $G = (X, Y, E)$ is one whose vertex set $X \cup Y$ can be partitioned into two independent sets X and Y. We will denote $|X|$ by n_X and $|Y|$ by n_Y. A *complete bipartite graph* contains all possible edges between its two partitions. An edge xy is *bisimplicial* if $N(x) \cup N(y)$ induces a complete bipartite graph. Let $\sigma = (e_1, \ldots, e_p)$, where $e_i = x_i y_i$, be an ordering of pairwise disjoint edges of G, $S_j = \{x_1, x_2, \ldots, x_j\} \cup \{y_1, y_2, \ldots, y_j\}$, and $S_0 = \emptyset$. σ is said to be a *perfect edge elimination scheme* if e_{j+1} is bisimplicial in $G[(X \cup Y) \setminus S_j]$, for $j = 0, \ldots, p-1$, and $G[(X \cup Y) \setminus S_p]$ has no edges. A bipartite graph that admits a perfect edge elimination scheme is called a *perfect elimination bipartite graph*. A bipartite graph $G = (X, Y, E)$ is said to be *chordal bipartite* if each cycle of length greater than 4 has a chord.

A bipartite graph $G = (X, Y, E)$ is said to be *Y-convex* if vertices of Y can be ordered so that for each $x \in X$ neighbours of x appear consecutively in Y. We will simply refer to that graph as *convex bipartite*. Such an ordering is called a *convex ordering* and can be computed in linear time [9]. Figure 1 shows a convex bipartite graph. Convex bipartite graphs are a proper subset of chordal bipartite graphs [10] which in turn are properly contained in perfect elimination bipartite graphs [11].

A convex bipartite graph $G = (X, Y, E)$ is *proper* if there is no vertex pair $x, x' \in X$ where $N(x) \subseteq N(x')$, i.e., its X vertices have no neighbourhood containment.

Let the neighbourhood set of some vertex $x \in X$ in the convex bipartite graph $G = (X, Y, E)$ be $N(x) = \{y_a, y_{a+1}, \ldots, y_{a+b}\}, y_a < \cdots < y_{a+b}$. Then $\text{left}(x) = y_a$ is the leftmost neighbour of x, and $\text{right}(x) = y_{a+b}$ is the rightmost neighbour of x. The *neighbourhood array* $\mathcal{N}$ stores the two values $\text{left}(x)$ and $\text{right}(x)$ for each vertex $x \in X$, i.e., $\mathcal{N}[i] = (\text{left}(x_i), \text{right}(x_i))$. The next proposition shows that we can populate $\mathcal{N}$ in linear time. We will use $\mathcal{N}$ as the data structure to represent a convex bipartite graph in our algorithms.

Proposition 1. *Let $G = (X, Y, E)$ be a convex bipartite graph whose vertices are in convex ordering. Entries in the array $\mathcal{N}$ can be calculated in $\mathcal{O}(n + |E|)$ time.*

Proof. A simple procedure that starts at $y = y_1$, moves sequentially in Y, and assigns the smallest value to $\text{left}(x)$ for all $x \in N(y)$ can be used to calculate $\text{left}(x)$. A similar procedure can be used for $\text{right}(x)$. Each such procedure needs $\mathcal{O}(n_X)$ time to initialize the values of $\text{left}(x)$ or $\text{right}(x)$, and $\mathcal{O}(|E|)$ time to go through neighbours of vertices in Y.   □

A *lexicographic convex ordering (lex-convex)* is a convex ordering for the vertices of G, and for x_i, x_j, $i < j$ if $\text{left}(x_i) < \text{left}(x_j)$, or $\text{left}(x_i) = \text{left}(x_j)$ and $\text{right}(x_i) < \text{right}(x_j)$. Cases where $\text{left}(x_i) = \text{left}(x_j)$ and $\text{right}(x_i) = \text{right}(x_j)$, i.e., $N(x_i) = N(x_j)$, can be numbered sequentially without disturbing other vertices' order. The ordering of the vertices in the graph in Fig. 1 is lex-convex. The proposition below shows that such an ordering can be computed in linear time. It is worth noting that lex-convex ordering has been used in other papers such as [5]. We assume that vertices are in lex-convex ordering for all convex bipartite graphs in this paper.

Proposition 2. *Let $G = (X, Y, E)$ be a convex bipartite graph whose vertices are in convex ordering. Given the array $\mathcal{N}$, a lex-convex ordering of the vertices of X can be calculated in $\mathcal{O}(n)$ time.*

Proof. We may assign the label $\text{left}(x)\,\text{right}(x)$ to each vertex $x \in X$. We may then use the set of those labels as keys for radix sort. The first pass sorts vertices based on right, and the second pass sorts vertices based on left. Since there are at most n values for each of right and left, the keys to be sorted can be regarded as base-n integers that are 2 digits long. Therefore the sorting may be completed in $\mathcal{O}(n)$ time.   □

The reader is referred to [12] and [11] for any missing graph theory definitions or notations.

We formally define the k *BLUE DOMINATION* and the k *RED DOMINATION* decision problems and the minimum cardinality versions.

Definition 1. k *BLUE DOMINATION: Given a bipartite graph $G = (X, Y, E)$ and integer k, $1 \leq k \leq |Y|$, is there a subset $D \subseteq Y$ of cardinality at most k that dominates X, i.e., $\forall x \in X$ does there exist some $y \in D$, s.t. $xy \in E$?*

Definition 2. *MCBD (Minimum Cardinality Blue Domination): Given a bipartite graph $G = (X, Y, E)$, find a minimum cardinality blue dominating set in G.*

The problems k RED DOMINATION and MCRD are analogously defined.

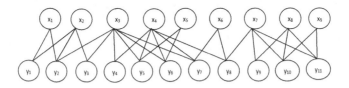

Fig. 1. A convex bipartite graph.

3   MCBD for Convex Bipartite Graphs

In this section we present a linear time algorithm to compute an MCBD set in a convex bipartite graph. The next two lemmas present an algorithmic property of particular MCBD sets that we will use in the algorithm.

Lemma 1. *Let $G = (X, Y, E)$ be a convex bipartite graph. There is an MCBD set for G, $D = \{v_1, \ldots, v_k\} \subseteq Y$, $v_1 < v_2 < \ldots < v_k$, such that for all i, $1 \leq i \leq k$, $v_i = \mathrm{right}(x)$, for some $x \in X$.*

Proof. Let $G = (X, Y, E)$ be a convex bipartite graph and D be an MCBD set for G, $D = \{v_1, \ldots, v_k\} \subseteq Y$, $v_1 < v_2 < \cdots < v_k$. Suppose for some $v_a \in D$ there is no $x \in X$ such that $v_a = \mathrm{right}(x)$. Let $u_a \in N(v_a)$ be the vertex with the smallest right among all neighbours of v_a, i.e., $v_a < \mathrm{right}(u_a) \leq \mathrm{right}(u_j)$, for all $u_j \in N(v_a)$. Since $\mathrm{left}(u_j) \leq v_a$ for all $u_j \in N(v_a)$, therefore $\mathrm{left}(u_j) < \mathrm{right}(u_a) \leq \mathrm{right}(u_j)$ for all $u_j \in N(v_a)$, and $\mathrm{right}(u_a)$ is adjacent to all neighbours of v_a and may replace v_a in D. This can be repeated for all vertices in D that are not the right of some vertex in X to generate an MCBD that satisfies the lemma. $\square$

Lemma 2. *Let $G = (X, Y, E)$ be a convex bipartite graph and $D = \{v_1, \ldots, v_k\} \subseteq Y$, $v_1 < v_2 < \ldots < v_k$, be an MCBD set for G that satisfies Lemma 1, i.e., for all v_i, $v_i = \mathrm{right}(x)$, for some $x \in X$. Let $u_i = x : v_i = \mathrm{right}(x)$, $1 \leq i \leq k$. Then there is such an MCBD where $u_1 < u_2 < \ldots < u_k$.*

Proof. Let $G = (X, Y, E)$ be a convex bipartite graph and D be an MCBD set for G that satisfies Lemma 1, $D = \{v_1, \ldots, v_k\} \subseteq Y$, $v_1 < v_2 < \ldots < v_k$. Suppose for some pair $v_a, v_{a+1}, u_a > u_{a+1}, u_i < u_{i+1}, \forall i, 1 \leq i < a$. Let D have the largest such index a among all MCBD sets that satisfy Lemma 1. Then by the lex-convex ordering left$(u_{a+1}) <$ left(u_a), and by our supposition right$(u_a) = v_a <$ right$(u_{a+1}) = v_{a+1}$. Let $N(v_{a+1}) = \{w_1, \ldots, u_{a+1}, \ldots, w_b\}$. v_a is adjacent to all $w_i \leq u_a$ because right$(w) \geq v_{a+1} > v_a$ for all $w \in N(v_{a+1})$, and left$(w) \leq$ left(u_a) for all $w \in N(v_{a+1}), w \leq u_a$. Among all vertices $w \in N(v_{a+1}), w > u_a$, let w' be the vertex with the smallest right, i.e., $w' \in N(v_{a+1})$, $w' > u_a$, right$(w') \leq$ right(w) for all $w \in N(v_{a+1}), w > u_a$. right$(w')$ is adjacent to all neighbours of v_{a+1}, w, $w > u_a$ and may replace v_{a+1} in D. This will result in an MCBD set that satisfies Lemma 1 and that has a longer sequence of vertices u_i where $u_1 < u_2 < \ldots < u_a < w' < \ldots < u_k$, a contradiction to our choice of D. $\qquad\square$

Definition 3. *Proper MCBD set:* $D = \{v_1, \ldots, v_k\}, v_1 < v_1 < \ldots < v_k$ *is a proper MCBD set for the bipartite convex graph* $G = (X, Y, E)$ *if D is an MCBD set for G, $v_i =$ right(x) for some $x \in X$, and if $u_i = x : v_i =$ right(x) then $u_i < u_j, \forall 1 \leq i < j \leq k$.*

The lemmas above say that we can choose the innermost right in the MCBD set and we can traverse vertices of X sequentially. We apply this result in the algorithm we present next. The algorithm computes a proper MCBD set for a convex bipartite graph in linear time. We assume that $|X| > 1$.

Applying the algorithm to the graph in Fig. 1 results in the MCBD set $\{y_2, y_6, y_8, y_{11}\}$.

Theorem 1. *Algorithm 1 outputs a proper MCBD set and runs in $\mathcal{O}(n_X)$ time.*

Proof. We first prove the correctness of the algorithm. Let $D = \{v_1, \ldots, v_k\}, v_1 < v_2 < \ldots < v_k$ be the output of the algorithm. We start by showing that the following observations about the output of the algorithm are true.

(a) Each $v_i =$ right(x) for some $x \in X$: To see that we observe that line 15 is the only line where vertices are added to D.
(b) Let $u_i = x : v_i =$ right$(x), 1 \leq i \leq k$. $u_1 < u_2 < \ldots < u_k$: This is clear because x_p goes in non-decreasing order from one iteration of the outer loop to the next.

Define u_i as in observation (b). Define w_i to equal the value of x_p set in line 16 immediately after v_i is added to the set D, $1 \leq i \leq k$. Let $w_0 = x_1$ and assume it is set to that value before the first iteration of the outer loop. Let $D_i = \{v_1, \ldots, v_i\}$. We use induction to prove that D_i is an MCBD set in the subgraph H_i induced by $\{x : x < w_i\} \cup \{y : y <$ left$(w_i)\}$. Then we show that $H_k = G$ thereby proving the correctness of the algorithm. We start with the following claim.

Claim. v_i is adjacent to all vertices $\{x : w_{i-1} \leq x < w_i\}, 1 \leq i \leq k$.

Algorithm 1. MCBD set for convex bipartite graph.

Input: The neighbourhood array $\mathcal{N}$ for convex bipartite graph $G = (X, Y, E)$
Output: MCBD set D for G

1: $D \leftarrow \emptyset$;
2: $x_p \leftarrow x_1$;
3: $x \leftarrow x_p + 1$;
4: **while** $x \leq x_{n_X}$ **do**
5:    // the next loop skips vertices x dominated by right(x_p) and whose neighbourhood set is not contained in $N(x_p)$
6:    **while** $x \leq x_{n_X}$ and left$(x_p) \leq$ left$(x) \leq$ right$(x_p) <$ right(x) **do**
7:       $x \leftarrow x + 1$;
8:    **end while**
9:    // the next loop updates x_p so that it is the largest indexed vertex that has the innermost right, i.e., right$(x_p) \leq$ right(x) for $N(x_p) \subseteq N(x)$
10:    **while** $x \leq x_{n_X}$ and left$(x_p) \leq$ left$(x) \leq$ right$(x) \leq$ right(x_p) **do**
11:       $x_p \leftarrow x$;
12:       $x \leftarrow x_p + 1$;
13:    **end while**
      // check if we have reached an x that is not dominated by right(x_p) or the end of the graph
14:    **if** left$(x) >$ right(x_p) or $x > x_{n_X}$ **then**
15:       $D \leftarrow D \cup \{\text{right}(x_p)\}$;
16:       $x_p \leftarrow x$;
17:       $x \leftarrow x_p + 1$;
18:    **end if**
19: **end while**

Proof. $x_p = w_{i-1}$ at the start of the loop in line 4 immediately after w_{i-1} was set in the if statement or when $i - 1 = 0$. The value of x_p does not change until the loop in line 10. Vertices in $\{x : w_{i-1} \leq x < w_i\}$ are the vertices skipped in both inner loops from when w_{i-1} is set and until the condition of the if statement becomes true. u_i is the value of x_p used in line 15. For vertices x skipped in the loop in line 6, either $x < u_i$ (i.e., x is skipped then x_p is updated in the loop in line 10) and left$(x) \leq$ left$(u_i) \leq$ right$(u_i) \leq$ right(x), or $x > u_i$ (i.e., x is skipped and x_p is not updated in the loop in line 10) and left$(u_i) \leq$ left$(x) \leq$ right$(u_i) <$ right(x). In either case, right$(u_i) = v_i$ is adjacent to each skipped vertex x in that loop. Each vertex x skipped in the loop in line 10, satisfies $N(u_i) \subseteq N(x)$, $x < u_i$. Therefore left$(x) \leq$ right$(u_i) \leq$ right(x) for all those vertices, and right$(u_i) = v_i$ is adjacent to all those vertices as well. This proves the claim. □

By the above claim, v_1 is adjacent to all x, where $w_0 = x_1 \leq x < w_1$. $D_1 = \{v_1\}$ is a blue dominating set in H_1 and it has minimum cardinality. Therefore, the hypothesis is true for $i = 1$.

Assume the hypothesis is true for all $i < a \leq 1$. $D_a = D_{a-1} \cup \{v_a\}$. D_{a-1} is an MCBD set in H_{a-1} by the induction hypothesis. By the above claim, v_a is adjacent to all x, $w_{a-1} \leq x < w_a$. Therefore, D_a is a blue dominating set in H_a. We must now show that it has minimum cardinality. Since left$(w_{a-1}) > v_{a-1} =$

right(u_{a-1}) by the condition of the if statement, then w_{a-1} has no neighbours in D_{a-1}. Since as well D_{a-1} is an MCBD set in H_{a-1} by the induction hypothesis, therefore an MCBD set in H_a must have cardinality $\geq a$. Therefore, D_a is an MCBD set in H_a.

By the principle of mathematical induction, the output of the algorithm $D = D_k$ is an MCBD set in H_k. D is a proper MCBD set by observations (a) and (b).

To see that $H_k = G$, we note that the last iteration of the outer while loop will take place when the if statement condition becomes true because $x > x_{n_X}$. Thus, $H_k = G$ and $D_k = D$ is a proper MCBD set in G.

To prove the time cost, we note that the operations inside the loops take constant time, and the order of the inner loops ensures that no vertex in X is visited more than once. Hence the number of iterations of the outer while loop is $\mathcal{O}(n_X)$. $\square$

4   MCRD for Convex Bipartite Graphs

In this section we present a linear time algorithm to compute an MCRD set. We show that it suffices to consider a particular maximal subgraph of a convex bipartite graph to obtain an MCRD set. The following lemma was proved in [1]. We include the proof here for completeness.

Lemma 3. *Let $D = \{u_1, u_2, \ldots, u_k\}$, $u_1 < u_2 < \ldots < u_k$, be an MCRD set of cardinality k for the convex bipartite graph $G = (X, Y, E)$. The following is true.*

- *For distinct vertices $u, v \in D$, it cannot be the case that $\text{left}(u) \leq \text{left}(v) \leq \text{right}(v) \leq \text{right}(u)$, i.e., $N(v) \not\subseteq N(u), \forall u, v \in D, u \neq v$.*
- *For all consecutive pairs $u, v, u < v$ in D, $\text{left}(u) < \text{left}(v) \leq \text{right}(u) + 1 \leq \text{right}(v)$, i.e., $N(u) \cup N(v)$ consists of consecutive vertices in Y.*
- *For each vertex $u \in D$, D cannot contain more than one vertex $v > u$ such that $\text{left}(u) < \text{left}(v) \leq \text{right}(u) + 1 \leq \text{right}(v)$.*

Proof. Let D be an MCRD set of cardinality k for the convex bipartite graph $G = (X, Y, E)$. $N(v) \subseteq N(u)$ for some pair $u, v \in D$, i.e. $\text{left}(u) \leq \text{left}(v) \leq \text{right}(v) \leq \text{right}(u)$, implies $D - \{v\}$ is a red dominating set of cardinality $k - 1$, which contradicts the minimality of D.

Note that this implies that for all vertices $u, v \in D$, $u \neq v$, $\text{left}(u) \neq \text{left}(v)$ and $\text{right}(u) \neq \text{right}(v)$.

The union of the neighbourhood sets, $N(u) \cup N(v)$, of two consecutive vertices $u, v, u < v$ in a red dominating set in a convex bipartite graph whose vertices are in lex-convex ordering must consist of consecutive vertices in Y, i.e., $\text{right}(u) + 1 \in N(v)$, for otherwise some vertices in Y would not be dominated. Therefore $\text{left}(u) < \text{left}(v) \leq \text{right}(u) + 1 \leq \text{right}(v)$ by the above paragraph.

Suppose there are vertices $u < v < w \in D$ such that $\text{left}(u) < \text{left}(v) \leq \text{right}(u) + 1 \leq \text{right}(v)$ and $\text{left}(u) < \text{left}(w) \leq \text{right}(u) + 1 \leq \text{right}(w)$. Then $N(u) \cup N(v) \subseteq N(u) \cup N(w)$ if $\text{right}(v) < \text{right}(w)$, and $N(u) \cup N(w) \subseteq N(u) \cup N(v)$ otherwise. In either case one of v or w may be removed from D to produce a $k - 1$ red dominating set, contradicting the minimality of D. $\square$

Definition 4. *A maximal proper subgraph:* $H(X_H, Y, E_H)$ *is a maximal proper subgraph of the convex bipartite graph* $G = (X, Y, E)$ *if* H *is a proper convex graph,* $H = G[X_H]$, *and for all* $x \in X - X_H, \exists x' \in X_H, s.t. N(x) \subseteq N(x')$.

If we remove $\{x_1, x_5, x_6, x_8, x_9\}$ from the vertex set of the graph in Fig. 1, we obtain its maximal proper subgraph. We can obtain such a subgraph in linear time as shown next.

Lemma 4. *Given an arbitrary convex bipartite graph* $G = (X, Y, E)$, *a maximal proper subgraph of* G *can be obtained in* $\mathcal{O}(n_X)$ *time.*

Proof. We present Algorithm 2 that takes as input the neighbourhood array $\mathcal{N}$ for an arbitrary convex bipartite graph $G = (X, Y, E)$ whose vertices are in lex-convex ordering and outputs a subset X_p of X that induces a maximal proper subgraph of G. An invariant of the algorithm is that the set X_p contains only vertices that belong to a maximal proper subgraph of G. The value of i increases from one iteration to the next which ensures that each vertex is visited once. The operations inside each loop take constant time. Therefore, the algorithm runs in $\mathcal{O}(n_X)$ time. $\square$

Algorithm 2. Maximal Proper Subgraph.

Input: The neighbourhood array $\mathcal{N}$ for the convex bipartite graph $G = (X, Y, E)$ whose vertices are in lex-convex ordering

Output: $X_p \subseteq X$, X_p induces a maximal proper convex bipartite subgraph of G

1: $X_p \leftarrow \emptyset$;
2: $i \leftarrow 1$;
3: $j \leftarrow 2$;
4: **while** $x_i \leq x_{n_X}$ **do**
5:     // the next loop finds the last vertex x_i adjacent to left(x_i)
6:     // each skipped vertex x_i has $N(x_i) \subseteq N(x_j)$
7:     **while** $x_j \leq x_{n_X}$ and left(x_j) = left(x_i) and right(x_j) $\geq$ right(x_i) **do**
8:         $i \leftarrow j$;
9:         $j \leftarrow j + 1$;
10:    **end while**
11:    $X_p \leftarrow X_p \cup \{x_i\}$;
       // the next loop skips vertices x_j whose $N(x_j)$ is contained in $N(x_i)$
12:    **while** $x_j \leq x_{n_X}$ and right(x_j) $\leq$ right(x_i) **do**
13:        $j \leftarrow j + 1$;
14:    **end while**
15:    // at this point left(x_j) > $left(x_i)$ and right(x_j) > right(x_i), i.e., $N(x_i) \not\subseteq N(x_j)$ and $N(x_j) \not\subseteq N(x_i)$
16:    $i \leftarrow j$;
17: **end while**

Corollary 1. *Let $G = (X, Y, E)$ be an arbitrary convex bipartite graph and D be a subset of X. D is an MCRD set in G if and only if D is an MCRD in its maximal proper subgraph.*

Proof. The proof follows directly from Lemma 3 and the definition of a maximal proper subgraph.   □

Proposition 3. *Let $D = \{u_1, u_2, \ldots, u_k\}$, $u_1 < u_2 < \ldots < u_k$ be a k red dominating set for the proper convex bipartite graph $G = (X, Y, E)$, where $X = \{x_1, x_2, \ldots, x_{n_X}\}$. Both x_1 and x_{n_X} are in D, i.e., $u_1 = x_1$ and $u_k = x_{n_X}$.*

Proof. Because G is proper and by the lex-convex ordering, x_1 is the only vertex in X adjacent to y_1, and x_{n_X} is the only vertex in X adjacent to y_{n_Y}. Therefore both x_1 and x_{n_X} must be in any red dominating set.   □

Corollary 1 allows us to focus on a maximal proper subgraph of an arbitrary convex bipartite graph. This will be done in a preprocessing step that takes linear time as was shown in Lemma 4. We will next see that there is an MCRD set whose vertices are the largest indexed neighbour of vertices in Y.

Definition 5. *$n_R(y)$: The neighbour of $y, y \in Y$ with the most right-reach $n_R(y)$ is defined as $n_R(y) = x_j : x_j \in N(y), \text{right}(x_j) \geq \text{right}(x_{j'}) \ \forall x_{j'} \in N(y)$.*

Proposition 4. *If $G = (X, Y, E)$ is a proper convex bipartite graph then $n_R(y) = x_j : x_j \geq x_{j'}, \forall x_j, x_{j'} \in N(y)$, i.e., it is the largest-indexed vertex among all neighbours of y.*

Proof. Suppose not, let $n_R(y) = x_j$, such that $\exists x_{j'} \in N(y), x_{j'} > x_j$. By the definition of $n_R(y)$, $\text{right}(x_j) \geq \text{right}(x_{j'})$. By the lex-convex ordering, $\text{left}(x_{j'}) > \text{left}(x_j)$ or $\text{left}(x_{j'}) = left(x_j)$ and $\text{right}(x_{j'}) \geq \text{right}(x_j)$. The first case implies $N(x_{j'}) \subset N(x_j)$. The second case implies $N(x_j) = N(x_{j'})$. Both cases contradict that G is proper.   □

To achieve linear time for the algorithm that computes MCRD, we calculate $n_R(y)$ for all $y \in Y$ beforehand.

Lemma 5. *Given the neighbourhood array $\mathcal{N}$ of a proper convex bipartite graph $G = (X, Y, E)$, $n_R(y)$ for all $y \in Y$ can be calculated in $\mathcal{O}(n_Y)$ time.*

Proof. We present Algorithm 3 to calculate $n_R(y)$ for all $y \in Y$ given the array $\mathcal{N}$ of a proper convex bipartite graph $G = (X, Y, E)$.

Algorithm 3. $n_R(y)$ for all $y \in Y$ for the proper convex bipartite graph $G = (X, Y, E)$.

Input: $\mathcal{N}$ for proper convex bipartite graph $G = (X, Y, E)$
Output: $n_R(y)$ for all $y \in Y$

1: $right \leftarrow y_{n_Y}$;
2: $i \leftarrow n_X$;
3: **while** $right \geq y_1$ **do**
4:    // note that because G is connected and proper, right is adjacent to x_i
5:    **for all** y such that $\text{left}(x_i) \leq y \leq right$ **do**
6:       $n_R(y) \leftarrow x_i$;
7:    **end for**
8:    $right \leftarrow \text{left}(x_i) - 1$;
9:    $i \leftarrow i - 1$;
10: **end while**

The loop in line 3 stops once all vertices of Y are assigned a value for n_R visiting each exactly once. Therefore Algorithm 3 calculates $n_R(y)$ for all $y \in Y$ in $\mathcal{O}(n_Y)$ time.   □

The algorithm that computes an MCRD set for a proper convex bipartite graph $G = (X, Y, E)$ follows. It starts with an empty set D. In each iteration of the loop, it adds the vertex x that is adjacent to vertices of Y not dominated thus far and that has the most right-reach, i.e., D will consist of select vertices $n_R(y)$. Applying the algorithm to the maximal proper subgraph of the graph in Fig. 1 results in the set $D = \{x_2, x_4, x_7\}$.

Algorithm 4. MCRD set for proper convex bipartite graph.

Input: The neighbourhood array $\mathcal{N}$ for the convex bipartite graph $G = (X, Y, E)$, $n_R(y)$ for all $y \in Y$
Output: MCRD set D for G

1: $D \leftarrow \emptyset$;
2: $y \leftarrow y_1$;
3: **while** $y \leq y_{n_Y}$ **do**
4:    $x \leftarrow n_R(y)$;
5:    $D \leftarrow D \cup \{x\}$;
6:    $y \leftarrow \text{right}(x) + 1$;// y is the first thus-far un-dominated vertex in Y
7: **end while**

Theorem 2. *Algorithm 4 correctly computes an MCRD set for a proper convex bipartite graph $G = (X, Y, E)$ in $\mathcal{O}(n_Y)$ time.*

Proof. Let $G = (X, Y, E)$ be a proper convex bipartite graph and D be the output of Algorithm 4. It is easy to see that D is a red dominating set since, by Lemma 3, the loop does not skip any un-dominated vertices in Y. We will prove that the algorithm computes a minimum cardinality red dominating set by contradiction. Suppose to the contrary that there is a smaller cardinality red dominating set for G, T, $|T| = p < k = |D|$. We assume that elements in T are in lex-convex ordering. By Proposition 3, $T = \{x_1, t_1, \ldots, t_{p-1}, x_n\}$. Note that x_1 will be the first vertex added by the algorithm to D because G is proper. Let $D = \{x_1, d_1, \ldots, d_{k-1}, d_k\}$. Suppose t_a, $a > 1$, is the first vertex in T that is not in D, i.e., $d_j = t_j, 1 \leq j < a$. Assume T is chosen so that no other MCRD set agrees more with D on the number of vertices after x_1, i.e., a is the largest such index. Let y_b be the last vertex in Y to be dominated by the vertices $\{x_1, \ldots, t_{a-1}\} = \{x_1, \ldots, d_{a-1}\}$. By our choice of d_a in the algorithm, $d_a = n_R(y_{b+1})$. Therefore $\{d_a, t_a\} \subseteq N(y_{b+1})$ and $\mathrm{right}(d_a) \geq \mathrm{right}(t_a)$. This implies that d_a is adjacent to all neighbours of t_a that are larger than or equal to y_b, and d_a may replace t_a in T. This will result in an MCRD set $T - \{t_a\} \cup \{d_a\}$ that agrees more with D, which contradicts our choice of T.

Since the loop goes n_Y iterations, the algorithm runs in $\mathcal{O}(n_Y)$ time.   □

Corollary 2. *MCRD is solvable in $\mathcal{O}(\max(n_X, n_Y))$ time for an arbitrary convex bipartite graph $G = (X, Y, E)$.*

Proof. To prove the time cost we notice that obtaining $\mathrm{right}(x)$ for any $x \in X$, for the purpose of Algorithm 4, can be done in constant time given the array $\mathcal{N}$. The corollary then follows from Lemma 5, Lemma 4, and Theorem 2.   □

5   Conclusion

We have presented an algorithm to find a minimum cardinality red dominating set in convex bipartite graphs in $\mathcal{O}(n)$ time. This algorithm is faster than the known algorithm for this problem. Because convex graphs are not symmetric, the algorithm cannot be used to find a minimum cardinality blue dominating set. We have presented an algorithm to find a minimum cardinality blue dominating set in convex bipartite graphs in $\mathcal{O}(n)$ time. The enumeration of minimum cardinality red dominating sets in convex bipartite graphs has been studied in the literature and it has been shown that it can be done in linear delay and linear space. It would be interesting to study the enumeration of minimum cardinality blue dominating sets in convex bipartite graphs and find if similar efficiency can be achieved. One of the challenges in this case is proving properties that apply to all minimum cardinality blue dominating sets, as opposed to proving properties for particular minimum cardinality blue dominating sets.

References

1. Abbas, N.: Red Domination in Perfect Elimination Bipartite Graphs. Submitted for publication (2021)
2. Golovach, P.A., Heggernes, P., Kanté, M.M., Kratsch, D., Villanger, Y.: Enumerating minimal dominating sets in chordal bipartite graphs. Discret. Appl. Math. **199**, 30–36 (2016)

3. Abu-Khzam, F.N., Mouawad, A.E., Liedloff, M.: An exact algorithm for connected red-blue dominating set. J. Discrete Algorithms **9**, 252–262 (2011)
4. Damaschke, P., Müller, H., Kratsch, D.: Domination in convex and chordal bipartite graphs. Inf. Process. Lett. **36**, 231–236 (1990)
5. Panda, B.S., Pradhan, D.: A linear time algorithm for computing a minimum paired-dominating set of a convex bipartite graph. Discret. Appl. Math. **161**, 1776–1783 (2013)
6. Panda, B.S., Pradhan, D.: Minimum paired-dominating set in chordal bipartite graphs and perfect elimination bipartite graphs. J. Comb. Optim. **26**(4), 770–785 (2012). https://doi.org/10.1007/s10878-012-9483-x
7. Song, Yu., Liu, T., Xu, K.: Independent domination on tree convex bipartite graphs. In: Snoeyink, J., Lu, P., Su, K., Wang, L. (eds.) AAIM/FAW -2012. LNCS, vol. 7285, pp. 129–138. Springer, Heidelberg (2012). https://doi.org/10.1007/978-3-642-29700-7_12
8. Pandey, A., Panda, B.S.: Domination in some subclasses of bipartite graphs. In: Ganguly, S., Krishnamurti, R. (eds.) CALDAM 2015. LNCS, vol. 8959, pp. 169–180. Springer, Cham (2015). https://doi.org/10.1007/978-3-319-14974-5_17
9. Booth, K.S., Lueker, G.S.: Testing for the consecutive ones property, interval graphs, and graph planarity using PQ-tree algorithms. J. Comput. Syst. Sci. **13**(3), 335–379 (1976)
10. Branstädt, A. and Le, V. B. and Spinrad, J. P.: Graph Classes: A Survey. SIAM (1999)
11. Golumbic, M.C.: Graphs. North-Holland, Amsterdam, Academic Press, New York (1980)
12. Berge, C.: Graphs. North-Holland, Amsterdam (1985)

Combinatorics and Algorithms
for Quasi-chain Graphs

Bogdan Alecu[1], Aistis Atminas[2], Vadim Lozin[1(✉)],
and Dmitriy Malyshev[3,4]

[1] Mathematics Institute, University of Warwick, Coventry CV4 7AL, UK
{B.Alecu,V.Lozin}@warwick.ac.uk
[2] Department of Mathematical Sciences, Xi'an Jiaotong-Liverpool University,
111 Ren'ai Road, Suzhou 215123, China
Aistis.Atminas@xjtlu.edu.cn
[3] Laboratory of Algorithms and Technologies for Networks Analysis,
National Research University Higher School of Economics,
136 Rodionova Str., 603093 Nizhny Novgorod, Russia
[4] National Research Lobachevsky State University of Nizhny Novgorod,
23 Gagarina Ave., Nizhny Novgorod 603950, Russia

Abstract. The class of quasi-chain graphs is an extension of the well-studied class of chain graphs. The latter class enjoys many nice and important properties, such as bounded clique-width, implicit representation, well-quasi-ordering by induced subgraphs, etc. The class of quasi-chain graphs is substantially more complex. In particular, this class is not well-quasi-ordered by induced subgraphs, and the clique-width is not bounded in it. In the present paper, we show that the universe of quasi-chain graphs is at least as complex as the universe of permutations by establishing a bijection between the class of all permutations and a subclass of quasi-chain graphs. This implies, in particular, that the induced subgraph isomorphism problem is NP-complete for quasi-chain graphs. On the other hand, we propose a decomposition theorem for quasi-chain graphs that implies an implicit representation for graphs in this class and efficient solutions for some algorithmic problems that are generally intractable.

Keywords: Bipartite chain graphs · Implicit representation ·
Polynomial-time algorithm

1 Introduction

A *chain graph* is a bipartite graph such that the neighbourhoods of the vertices in each part of its bipartition form a chain with respect to the inclusion relation. The class of chain graphs appeared in the literature under various names such

This work was supported by the Russian Science Foundation Grant No. 21-11-00194.

P. Flocchini and L. Moura (Eds.): IWOCA 2021, LNCS 12757, pp. 49–62, 2021.
https://doi.org/10.1007/978-3-030-79987-8_4

as difference graphs [8] or half-graphs [5]. In model theory, half-graphs appear as an instance of the order property [17]. The class of chain graphs is closely related to one more well-studied class of graphs, known as threshold graphs, and together they share many nice and important properties. In particular,

- chain graphs have bounded clique-width (and even linear clique-width), which implies polynomial-time solutions for a variety of algorithmic problems that are generally NP-hard;
- chain graphs are well- (and even better-) quasi-ordered under induced subgraphs. This is because another important parameter, graph lettericity, is bounded for chain graphs [22];
- chain graphs admit an implicit representation, which in turn implies a small induced-universal graph for the class. More specifically, there is a chain graph with $2n$ vertices containing all n-vertex chain graphs as induced subgraphs [14].

In the terminology of forbidden induced subgraphs, the class of chain graphs is precisely the class of $2P_2$-free bipartite graphs, i.e., bipartite graphs that do not contain the disjoint union of two copies of P_2 as an induced subgraph (P_n denotes the chordless path on n vertices).

In the present paper, we study a class of bipartite graphs that forms an extension of chain graphs defined by relaxing the chain property of the neighbourhoods in the following way. We say that a linear ordering $(a_1, \ldots, a_\ell)$ of vertices is *good* if for all $i < j$, the neighbourhood of a_j contains at most 1 non-neighbour of a_i. We call a bipartite graph G a *quasi-chain* graph if the vertices in each part of its bipartition admit a good ordering. Alternatively, quasi-chain graphs are bipartite graphs that do not contain an "unbalanced" induced copy of $2P_3$. To explain what we mean by this, we observe that $2P_3$ admits two bipartitions: one with parts of equal size (balanced) and the other with parts of different sizes (unbalanced). In the unbalanced bipartition, one of the parts does not admit a good ordering and hence quasi-chain graphs are free of unbalanced $2P_3$. On the other hand, if a bipartite graph G does not contain an unbalanced induced copy of $2P_3$, then by ordering the vertices in each part in a non-increasing order of their degrees we obtain a good ordering, i.e., G is a quasi-chain graph.

The class of quasi-chain graphs is substantially richer and more complex than the class of chain graphs. In particular, it is not well-quasi-ordered by induced subgraphs [13] and the clique-width is not bounded in this class [15]. To emphasize the complex nature of this class, in Sect. 2 we establish a bijection f between the class of all permutations and a subclass of quasi-chain graphs such that a permutation π contains a permutation ρ as a pattern if and only if the graph $f(\pi)$ contains the graph $f(\rho)$ as an induced subgraph. Together with the NP-completeness of the PATTERN MATCHING problem for permutations this implies the NP-completeness of the INDUCED SUBGRAPH ISOMORPHISM problem for quasi-chain graphs.

In spite of the more complex structure, the quasi-chain graphs inherit some attractive properties of chain graphs. To show this, in Sect. 3 we propose a structural characterisation that describes any quasi-chain graph as the symmetric difference of two graphs Z and H, where Z is a chain graph and H is a graph of vertex degree at most 2. This characterisation allows us to prove that quasi-chain graphs admit an implicit representation (Sect. 4) and that some algorithmic problems that are NP-complete for general bipartite graphs admit polynomial-time solutions when restricted to quasi-chain graphs (Sect. 5).

All graphs in this paper are simple, i.e., undirected, with neither loops nor multiple edges. The vertex set and the edge set of a graph G are denoted $V(G)$ and $E(G)$, respectively. The *neighbourhood* of a vertex $v \in V(G)$ is the set of vertices adjacent to v. We denote the neighbourhood of v in the graph G by $N_G(v)$ and omit the subscript if it is clear from the context. The subgraph of G induced by a set $U \subseteq V(G)$ is denoted $G[U]$.

A bipartite graph $G = (V, E)$ given together with a bipartition $V = A \cup B$ is denoted $G = (A, B, E)$. Once such a bipartition has been fixed, we may define the *bipartite complement* $\widetilde{G} = (A, B, E')$ of G, in which two vertices $a \in A$ and $b \in B$ are adjacent if and only if they are not adjacent in G (that is, $E' = (A \times B) - E$).

2 Quasi-chain Graphs and Permutations

Given two permutations $\pi = (\pi(1), \ldots, \pi(n))$ and $\rho = (\rho(1), \ldots, \rho(m))$, we will write $\pi \subseteq \rho$ to indicate that π is contained in ρ as a pattern, i.e., there is an order-preserving injection $e : \{1, 2, \ldots, n\} \to \{1, 2, \ldots, m\}$ such that $\pi(i) < \pi(j)$ if and only if $\rho(e(i)) < \rho(e(j))$ for all $1 \le i < j \le n$. The pattern containment relation on permutations is the subject of a vast literature, see, e.g., the book [12] and the references therein. By mapping each permutation to its permutation graph, we transform the pattern containment relation on permutations into the induced subgraph relation on graphs. This mapping, however, is not injective, as it can map different permutations to the same (up to an isomorphism) graph. In the present section we propose an alternative mapping from permutation to graphs: we map permutations to quasi-chain graphs, in such a way that two permutations are comparable if and only if their images are comparable. To make this mapping injective, we require the quasi-chain graphs to be coloured. That is, we will assume that every quasi-chain graph is given together with a partition of its vertex set into an independent set A of white vertices and an independent set B of black vertices and we will write $G \subseteq H$ to indicate that G is a coloured induced subgraph of H, i.e., there is an induced subgraph embedding of G into H that respects the colours. The distinction between coloured and uncoloured graphs matters, for instance, in the assignment problem.

We denote our mapping from permutations to graphs by f and define it as follows. If $\pi = (\pi(1), \pi(2), \ldots, \pi(n))$ is an n-entry permutation, then $f(\pi)$ is a bipartite graph with parts $A = \{a_1, a_2, \ldots, a_{2n}\}$ and $B = \{b_1, b_2, \ldots, b_{2n}\}$ and the following edges:

(i) For any $1 \le i \le j \le 2n$, we have $a_i b_j \in E(G)$.
(ii) For any $1 \le i \le n$, we have $a_{n+i} b_{\pi(i)} \in E(G)$.

We write $G_\pi := f(\pi)$ and say that G_π is the *quasi-permutation graph* of π. Any graph G isomorphic to G_π for some π will be called a quasi-permutation graph. It follows easily from the definition that f is order-preserving, in that $\pi \subseteq \rho$ implies $f(\pi) \subseteq f(\rho)$.

Claim. Any quasi-permutation graph G is a quasi-chain graph.

Proof. We observe that the edges of type (i) define a chain subgraph of G in which $N(a_j) \subseteq N(a_i)$ for all $1 \leq i < j \leq 2n$. The edges of type (ii) form a matching and therefore in the graph G we have $|N(a_j) - N(a_i)| \leq 1$ for all $1 \leq i < j \leq 2n$. Similarly, $|N(b_i) - N(b_j)| \leq 1$ for all $1 \leq i < j \leq 2n$ in G. This shows that A and B have good orderings, and so any quasi-permutation graph G is a quasi-chain graph. □

Claim. f is a bijection from the class of all permutations to the (non-hereditary) class of quasi-permutation graphs.

Proof. f is surjective by the definition of quasi-permutation graphs. Now notice that in the graph $f(\pi)$ the degree sequence of vertices in both A and B is $(2, 3, 4, \ldots, n+1, n+1, n+2, \ldots, 2n)$. In particular, $f(\pi)$ uniquely determines the size of π.

The unique vertex of A with degree 2 is adjacent to vertices b_{2n} and $b_{\pi(n)}$ in part B. Vertex b_{2n} has degree $2n$ and vertex $b_{\pi(n)}$ has degree k, for some $k \leq n+1$. Inspecting the value of k allows us to determine the value of $\pi(n)$, which is $k - 1$. Similarly, the unique vertex of degree 3 has three neighbours: b_{2n}, b_{2n-1} and $b_{\pi(n-1)}$, which allows us to determine the value of $\pi(n-1)$. In this way, we see that $f(\pi)$ uniquely determines $\pi(i)$ for all $2 \leq i \leq n$. But two permutations with the same number of elements cannot disagree in exactly one entry, hence the graph $f(\pi)$ uniquely determines the permutation π. Therefore f is injective. □

Lemma 1. *Let π and ρ be two permutations with n and m entries, respectively, with $n \leq m$ and $\pi(1) \neq n$. If $f(\pi) \subseteq f(\rho)$, then $\pi \subseteq \rho$.*

Proof. Assume $f(\pi) \subseteq f(\rho)$. We denote the vertices of $f(\rho)$ as $A = \{a_1, \ldots, a_{2m}\}$ and $B = \{b_1, \ldots, b_{2m}\}$ and edges $a_i b_j$ if either $1 \leq i \leq j \leq 2m$ or $m+1 \leq i \leq 2m$ and $j = \rho(i - m)$. Also, we denote the vertices of $f(\pi)$ as $A' = (a'_1, a'_2, \ldots, a'_{2n})$, and $B' = (b'_1, b'_2, \ldots, b'_{2n})$ with edges $a'_i b'_j$ if either $1 \leq i \leq j \leq 2n$ or $n+1 \leq i \leq 2n$ and $j = \pi(i - n)$. The mapping that embeds $f(\pi)$ into $f(\rho)$ as an induced subgraph will be denoted by $a'_i \mapsto a_{e(i)}$, $b'_i \mapsto b_{w(i)}$.

Firstly, observe that all but at most one entry from the set $\{w(1), \ldots, w(n)\}$ are less than or equal to m. Indeed, the vertices $b'_1, b'_2, \ldots, b'_n$ have pairwise incomparable neighbourhoods, and this must also be the case for their images; however, if $i, j > m$, the neighbourhoods of b_i and b_j are comparable. Moreover, since b'_{i+1} has two private neighbours with respect to b'_i for any $i \leq n - 1$, we must have $w(i) < w(i+1)$ for any $i \leq n - 1$, and hence we must have $w(1) < w(2) < \ldots < w(n-1) \leq m$ and $w(n-1) < w(n)$. Similarly, we can deduce that $m + 1 \leq e(n+2) < e(n+3) < \ldots < e(2n)$ with $e(n+1) < e(n+2)$.

Now, $a'_1, a'_2, \ldots, a'_{n-2}$ are adjacent to two vertices b'_{n-2}, b'_{n-1} with $w(n-2) < w(n-1) \leq m$. Therefore, we conclude that $\{e(1), e(2), \ldots, e(n-2)\}$ must all be smaller than or equal to m. As $a_1, a_2, \ldots, a_m$ form a chain graph together with the vertices in B, in order to have $N(a_{e(i)}) \supsetneq N(a_{e(j)})$ for $1 \leq i < j \leq n-2$, we conclude that we must have $1 \leq e(1) < e(2) < \ldots < e(n-2) \leq m$. To preserve correct adjacencies between $\{a'_1, \ldots, a'_{n-2}\}$ and $\{b'_1, \ldots, b'_{n-1}\}$ we must have

$$e(1) \leq w(1) < e(2) \leq w(2) < \ldots < e(n-2) \leq w(n-2) < w(n-1) \leq m.$$

Now $b_{w(n-1)}$ is already adjacent to $a_{e(1)}, a_{e(2)}, \ldots, a_{e(n-2)}$, but it has to be adjacent to two more vertices, $a_{e(n-1)}$ and $a_{e(n+\pi^{-1}(n-1))}$. Clearly, at least one of $e(n + \pi^{-1}(n-1))$ and $e(n-1)$ must be at most $w(n-1)$. Hence there are two cases: either both $e(n + \pi^{-1}(n-1))$ and $e(n-1)$ are at most $w(n-1)$, or one of them is at most $w(n-1)$ and the other is at least $m+1$, in which case $e(n-1)$ is the one that is at most $w(n-1)$, as a'_{n-1} has a private neighbour with respect to $a'_{n+\pi^{-1}(n-1)}$. In either case, we must have $e(n-1) \leq w(n-1)$. As a'_{n-1} is non-adjacent to b'_{n-2}, we must also have $w(n-2) < e(n-1)$, implying that

$$e(1) \leq w(1) < e(2) \leq w(2) < \ldots < e(n-2) \leq w(n-2) < e(n-1) \leq w(n-1) \leq m.$$

By symmetry, we derive that

$$m + 1 \leq e(n+2) \leq w(n+2) < e(n+3) \leq \ldots < e(2n) \leq w(2n).$$

We are only left with determining the location of the embeddings of the four vertices $a'_n, b'_n, a'_{n+1}, b'_{n+1}$. Since $\pi(1) \neq n$, we have that a'_{n+1} is not connected to b'_n, but connected to $b'_{\pi(1)}$ (with $\pi(1) < n$). It follows that $e(n+1) \geq m+1$. Clearly, for a'_{n+1} to have two private neighbours with respect to a'_{n+2} we must also have $e(n+1) < e(n+2)$. The two private neighbours of a'_{n+1} are $b'_{\pi(1)}$ and b'_{n+1}; since $a_{e(n+1)}$ only has one neighbour b_i with $i < e(n+1)$ (namely $b_{\pi(1)}$), the embedding of b'_{n+1} must satisfy $e(n+1) \leq w(n+1) < e(n+2)$. Now b'_n, which is not adjacent to a'_{n+1} but adjacent to $a'_{n+\pi^{-1}(n)}$ (note $e(n+\pi^{-1}(n)) \geq m+1$ since $\pi^{-1}(n) > 1$) must therefore satisfy $w(n) \leq m$. As b'_n has two private neighbours with respect to b'_{n-1}, we must have $w(n-1) < w(n)$, and as above, the private neighbour a'_n of b'_n must satisfy $w(n-1) < e(n) \leq w(n)$.

Summarizing, we conclude that

$$e(1) \leq w(1) < \ldots < e(n) \leq w(n) \leq m < m + 1$$
$$\leq e(n+1) \leq w(n+1) < \ldots < e(2n) \leq w(2n).$$

We may now alter this embedding of $f(\pi)$ into $f(\rho)$ if necessary to guarantee that $e(i) = w(i)$ for all $i = 1, 2, \ldots, 2n$. Indeed, it follows from the above inequalities that, for $1 \leq i \leq n$, $a_{e(i)}$ and $a_{w(i)}$ have the same set of neighbours among the embedded b-vertices, and similarly, for $n+1 \leq i \leq 2n$, $b_{w(i)}$ and $b_{e(i)}$ have the same set of neighbours among the embedded a-vertices. We may thus keep the embeddings of $b'_1, \ldots, b'_n, a'_{n+1}, \ldots, a'_{2n}$ where they are, and move the embeddings of the remaining vertices as appropriate to ensure $e(i) = w(i)$ for $1 \leq i \leq 2n$.

From this altered embedding, it is easy to see that $\pi \subseteq \rho$ as claimed (for instance, interpret the matching between $b_1, \ldots, b_m$ and $a_{m+1}, \ldots, a_{2m}$ as a line segment intersection model for ρ, and note that the intersection of this matching with the embedded graph $f(\pi)$ gives a line segment intersection model for π). □

Lemma 1 cannot, in general, be extended to permutations π with $\pi(1) = n$ (except trivially, when $n = 1$ or $m = n$). For example, if $\pi = (2, 1)$ and $\rho = (1, 2, 3, 4)$, then one can easily see that $f(\rho) \supseteq f(\pi)$, but ρ does not contain π. One underlying reason for this phenomenon is that whenever $\pi(1) = n$, the vertices a_n and a_{n+1} have exactly the same neighbourhoods, which makes it possible for the graphs to be embedded with more flexibility, not necessarily forcing embedding of permutations. For this reason, we introduce a slight modification of the embedding, which allows us to always avoid the case $\pi(1) = n$.

Definition 1. *Given a permutation* $\pi = (\pi(1), \pi(2), \ldots, \pi(n))$, *define* $\pi^* = (1, \pi(1) + 1, \pi(2) + 1, \ldots, \pi(n) + 1)$. *Define* $f^*(\pi) = f(\pi^*)$, *where* f *is the map from permutations to quasi-permutation graphs.*

Theorem 1. f^* *is an injection from the class of permutations to the class of quasi-permutation graphs such that for any two permutations* π *and* ρ *we have* $f^*(\pi) \subseteq f^*(\rho)$ *if and only if* $\pi \subseteq \rho$.

Proof. f^* is a composition of two injective maps $\pi \mapsto \pi^*$ and $\pi^* \mapsto f(\pi^*)$, with the image of the second map being a quasi-permutation graph. Therefore, f^* is an injection from the class of permutations to the class of quasi-permutation graphs. Further, $f^*(\pi) \subseteq f^*(\rho)$ means, by definition, that $f(\pi^*) \subseteq f(\rho^*)$, which happens if and only if $\pi^* \subseteq \rho^*$ (this follows from Lemma 1 as $\pi^*(1) = 1 \neq n$). Finally, it is easy to see that $\pi^* \subseteq \rho^*$ if and only if $\pi \subseteq \rho$, from which the second part of the theorem follows. □

3   The Structure of Quasi-chain Graphs

For two graphs $G_1 = (V, E_1)$ and $G_2 = (V, E_2)$ on the same vertex set we denote by $G_1 \otimes G_2$ the graph $G = (V, E_1 \otimes E_2)$, where $\otimes$ denotes the symmetric difference of two sets. The main result in this section is the following theorem.

Theorem 2. *If a bipartite graph* $G = (A, B, E)$ *is a quasi-chain graph, then* $G = Z \otimes H$ *for a chain graph* Z *and a graph* H *of vertex degree at most two such that* $E(H) \cap E(Z)$ *and* $E(H) - E(Z)$ *are matchings. Such a decomposition* $G = Z \otimes H$ *can be obtained in polynomial time.*

In the proof of this result, we use a word representation for our graphs, which builds on a special case of *letter graph representations*, introduced in [22]. The starting point is as follows: there is a bijective, order-preserving mapping between words over the alphabet $\{a, b\}$ (under the subword relation) and coloured chain graphs (under the coloured induced subgraph relation). This mapping sends a word w to the graph whose vertices are the entries of w, and we have edges

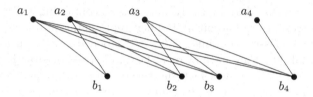

Fig. 1. The graph corresponding to the word $w = aababbab$

between each a and each b appearing after it in w. See Fig. 1 for an example (the indices of the letters indicate the order of their appearance in w).

We would like to extend this representation to graphs with the structure claimed in Theorem 2. To do so, we enhance the letter representation described above by allowing bottom edges between pairs a, b with the a appearing before the b in w and top edges between pairs a, b with the a appearing after the b in w. We require, in addition, that the set of top edges forms a matching and the set of bottom edges forms a matching, and interpret the bottom edges as an instruction to remove the corresponding matching from the chain graph represented by w, and the top edges as an instruction to add the corresponding matching. We call such a word an *enhanced word*. For instance, $w' = \underline{aab}ab\overline{bab}$ is an enhanced word obtained from $w = aababbab$ by adding the bottom edge connecting the first a to the first b and the top edge connecting the second b to the last a.

If G is the graph described by an enhanced word w, we say that w is an *enhanced letter representation* for G. In particular, $w' = \underline{aab}ab\overline{bab}$ is an enhanced letter representation of the graph obtained from the graph in Fig. 1 by removing the edge a_1b_1 and adding the edge b_2a_4. It is immediate from our discussion that Theorem 2 can be restated as follows.

Theorem 3. *Any quasi-chain graph admits an enhanced letter representation that can be found in polynomial time.*

The proof is by induction on the number of vertices of a quasi-chain graph G, noting that either G or $\widetilde{G}$ has a vertex of degree at most 1. By considering various cases, we show that such a vertex can always be added to an enhanced letter representation obtained inductively. The case analysis can be made algorithmic, yielding a polynomial bound.

4   Implicit Representation of Quasi-chain Graphs

The idea of implicit representation of graphs (also known in the literature as adjacency labelling scheme) was introduced in [11] and can be described as follows. A representation of an n-vertex graph G is said to be *implicit* if it assigns to each vertex of G a binary code of length $O(\log n)$ so that the adjacency of two vertices is a function of their codes.

Not every class of graphs admits an implicit representation, since a bound on the length of a vertex code implies a bound on the number of graphs admitting such a representation. More precisely, only classes containing $2^{O(n \log n)}$ labelled graphs with n vertices can admit an implicit representation. In the terminology of [3], hereditary classes containing $2^{O(n \log n)}$ labelled graphs on n vertices are at most factorial, i.e., they have at most factorial speed of growth. Whether all hereditary classes with at most factorial speed admit an implicit representation is a big open question known as the *implicit representation conjecture* [23]. The conjecture holds for a variety of factorial classes such as interval graphs, permutation graphs (which include chain graphs), line graphs, planar graphs, etc. It also holds for all graph classes of bounded vertex degree, of bounded clique-width, of bounded arboricity (including all proper minor-closed classes), etc.; see [2] and the references therein for more information on this topic.

The class of quasi-chain graphs is known to be factorial, which was shown in [1]. However, the question whether this class admits an implicit representation remains open. In this section, we answer this question in the affirmative. To this end, we introduce the following general tool.

For a graph $G = (V, E)$, let A_G denote the adjacency matrix of G, and for two vertices $x, y \in V$, let $A_G(x, y)$ be the element of the matrix corresponding to x and y. Given a Boolean function f of k variables and graphs $H_1 = (V, E_1), \ldots, H_k = (V, E_k)$, we will write $G = f(H_1, \ldots, H_k)$ if

$$A_G(x, y) = f(A_{H_1}(x, y), \ldots, A_{H_k}(x, y))$$

for all distinct vertices $x, y \in V$. If $G = f(H_1, \ldots, H_k)$, we say that G is an f-function of $H_1, \ldots, H_k$.

Theorem 4. *Let X be a class of graphs, k a natural number, f a Boolean function of k variables, and $Y_1, \ldots, Y_k$ classes of graphs admitting an implicit representation. If every graph in X is an f-function of graphs $H_1 \in Y_1, \ldots, H_k \in Y_k$, then X also admits an implicit representation.*

Proof. To represent a graph $G = f(H_1, \ldots, H_k)$ in X implicitly, we assign to each vertex of G k labels, each of which represents this vertex in one of the graphs $H_1, \ldots, H_k$. Given the labels of two vertices $x, y \in V(G)$, we can compute the adjacency of these vertices in each of the k graphs and hence, using the function f (which we may encode in each label with a constant number of bits), we can compute the adjacency of x and y in the graph G. □

According to Theorem 2, any quasi-chain graph is a $\oplus$-function of a chain graph and a graph of vertex degree at most 2, where $\oplus$ is addition modulo 2. As we mentioned earlier, chain graphs and graphs of vertex degree at most 2 admit an implicit representation. Together with Theorem 4 this implies the following conclusion.

Corollary 1. *The class of quasi-chain graphs admits an implicit representation.*

The same conclusion can be derived in an alternative way, which is of independent interest, because it deals with a parameter motivated by some biological applications. This parameter was introduced in [7] under the name contiguity and it can be defined as follows.

Graphs of contiguity 1 are graphs that admit a linear order of the vertices in which the neighbourhood of each vertex forms an interval. Not every graph admits such an ordering, in which case one can relax this requirement by looking for an ordering in which the neighbourhood of each vertex can be split into at most k intervals. The minimum value of k which allows a graph G to be represented in this way is the *contiguity* of G.

Theorem 5. *Contiguity of quasi-chain graphs is at most 3.*

Proof. It is not difficult to see that chain graphs have contiguity 1. Let G be a quasi-chain graph, and use Theorem 2 to obtain a decomposition $G = Z \otimes H$. Consider a linear order of the vertices of G such that their neighbourhoods in Z are intervals. Z can be transformed into G by adding at most one edge and at most one non-edge incident to each vertex. By adding a non-edge, we split the interval of neighbours of v into at most two intervals, and by adding a neighbour to v, its neighbourhood spans at most one additional interval consisting of a single vertex. $\qquad\square$

It is routine to check that graphs of bounded contiguity admit an implicit representation. Therefore, Corollary 1 follows from Theorem 5 as well.

5 Optimisation in Quasi-chain Graphs

Many algorithmic problems that are NP-complete for general graphs remain computationally intractable for bipartite graphs, which is the case, for instance, for HAMILTONIAN CYCLE [20], MAXIMUM INDUCED MATCHING [16], ALTERNATING CYCLE-FREE MATCHING [19], BALANCED BICLIQUE [10], MAXIMUM EDGE BICLIQUE [21], DOMINATING SET, STEINER TREE [18], INDEPENDENT DOMINATION [6], INDUCED SUBGRAPH ISOMORPHISM [9].

The simple structure of chain graphs implies bounded clique-width and therefore polynomial-time solvability of all these and many other problems. However, in quasi-chain graphs the clique-width is unbounded and hence no solution comes for free in this class. Moreover, INDUCED SUBGRAPH ISOMORPHISM remains intractable, as we show in Sect. 5.1 based on the relationship between quasi-chain graphs and permutations revealed in Theorem 1.

On the other hand, the structure of quasi-chain graphs revealed in Theorem 2 allows us to prove polynomial-time solvability of three problems in the above list, which we do in Sect. 5.2.

5.1 NP-completeness of INDUCED SUBGRAPH ISOMORPHISM in Quasi-chain Graphs

The INDUCED SUBGRAPH ISOMORPHISM problem can be stated as follows: given two graphs H and G, decide whether H is an induced subgraph of G or not.

This problem is known to be NP-complete even when both graphs are bipartite permutation graphs [9]. A related problem on permutations is known as PATTERN MATCHING: given two permutations π and ρ, it asks whether π contains ρ as a pattern. This problem is also NP-complete [4]. Together with Theorem 1 this immediately implies that *coloured* INDUCED SUBGRAPH ISOMORPHISM is NP-complete for quasi-chain graphs. Below we extend this conclusion to uncoloured graphs.

Theorem 6. *The* INDUCED SUBGRAPH ISOMORPHISM *problem is NP-complete for quasi-chain graphs.*

Proof. Let H and G be two coloured connected quasi-chain graphs. The NP-completeness of PATTERN MATCHING together with Theorem 1 imply that determining whether there is an embedding of H into G as an induced subgraph that respects the colours is an NP-complete problem. To reduce the problem to uncoloured graphs, we modify the instance of the problem as follows.

Let p be a natural number greater than the maximum vertex degree in G, and let $K_{1,p}$ be a star with the center x. We add this star to G, connect x to all the black vertices of G and denote the resulting graph by G^*. Similarly, we add this star to H, connect x to all the black vertices of H and denote the resulting graph by H^*. Clearly, G^* and H^* are quasi-chain graphs.

Now we ignore the colours and ask whether G^* contains H^* as an induced subgraph. If G^* contains H^*, then vertex x in H^* must map to vertex x in G^* (due to the degree condition), and the vertices of H in H^* are mapped to the vertices of G in G^* in a colour-preserving way (due to the connectedness of G and H). Therefore, G contains H as a *coloured* induced subgraph if and only if G^* contains H^* as an induced subgraph. Since G^* and H^* are quasi-chain graphs and these graphs can be obtained from G and H in polynomial time, we conclude that INDUCED SUBGRAPH ISOMORPHISM is NP-complete for quasi-chain graphs. $\square$

5.2   Polynomial-Time Algorithms for Quasi-chain Graphs

In this section, we use Theorem 2 to prove polynomial-time solvability of the following problems in quasi-chain graphs: BALANCED BICLIQUE, MAXIMUM EDGE BICLIQUE, and INDEPENDENT DOMINATION. We emphasize that Theorem 2 not only provides a structural characterisation of quasi-chain graphs, it also proves that a quasi-chain graph can be transformed into a chain graph by removing a matching and adding a matching in polynomial time, which is an important ingredient in all three solutions. We start with an auxiliary lemma.

Lemma 2. *A quasi-chain graph G with n vertices contains a collection $\mathcal{I}$ of $O(n)$ subsets of vertices that can be found in polynomial time such that every subset $I \in \mathcal{I}$ induces a graph of vertex degree at most 1, and every independent set in G is contained in one of these subsets.*

Proof. First, we observe that there are $O(n)$ inclusion-wise maximal independent sets in a chain graph, and that all of them can be found in polynomial time.

Now let $G = Z \otimes H$ be a quasi-chain graph and let S be an independent set in G. Then in the graph Z, the vertices of S either form an independent set, or induce some bottom edges, i.e., some edges of $E(H) \cap E(Z)$. Since bottom edges form a matching and Z is $2P_2$-free, we conclude that S contains at most one bottom edge in the graph Z.

If S is an independent set in Z, then it is contained in a maximal independent set I in Z. For each maximal independent set I in the graph Z, the vertices of I induce in G a subgraph $G[I]$ of vertex degree at most 1, because all edges of $G[I]$ are top edges and therefore they form a matching.

Assume now that S contains an edge $a_i b_j$ in the graph Z. We denote the set of non-neighbours of a_i in G by A_i and the set of non-neighbours of b_j in G by B_j, and let $I = A_i \cup B_j$. In particular, $S \subseteq I$. In Z, the vertices of I induce a subgraph $Z[I]$ containing exactly one edge $a_i b_j$. Indeed, no edge $e \neq a_i b_j$ in $Z[I]$ can be incident to a_i or b_j, because otherwise both e and $a_i b_j$ are bottom edges, which is impossible, and if e is not incident to a_i and b_j, then e and $a_i b_j$ create an induced $2P_2$ in Z, which is not possible either. Since $a_i b_j$ is the only edge in $Z[I]$ and this edge is not present in $G[I]$, we conclude that all edges of $G[I]$ are top edges and hence $G[I]$ is a graph of vertex degree at most one.

Putting everything together, our collection $\mathcal{I}$ consists of two types of sets: the maximal independent sets from Z, and the sets constructed as above from each of the bottom edges. This collection thus has $O(n)$ sets, and can be found in polynomial time as claimed.   □

Bicliques in Quasi-chain Bipartite Graphs. A *biclique* is a complete bipartite graph $K_{p,q}$ for some p and q. In a bipartite graph, the problem of finding a biclique with the maximum number of vertices can be solved in polynomial time. However, the problem of finding a biclique with the maximum number of edges, known as the MAXIMUM EDGE BICLIQUE problem, is NP-complete for bipartite graphs [21]. Additionally, the problem of finding a biclique $K_{p,p}$ with the maximum value of p, known as the BALANCED BICLIQUE problem, is NP-complete for bipartite graphs [10]. We show that both problems can be solved in polynomial time when restricted to quasi-chain graphs.

Theorem 7. *The* MAXIMUM EDGE BICLIQUE *and* BALANCED BICLIQUE *problems can be solved in polynomial time for quasi-chain graphs.*

Proof. Let $G = (A, B, E)$ be a quasi-chain graph. A biclique in G becomes an independent set in the bipartite complement $\widetilde{G}$ of G. Since an unbalanced $2P_3$ is self-complementary in the bipartite sense, we note that $\widetilde{G}$ is a quasi-chain graph too.

Let $\mathcal{I}$ be as in Lemma 2 for $\widetilde{G}$. Every independent set in $\widetilde{G}$ is contained in a maximal independent set, which in turn is contained in one of the subsets of $\mathcal{I}$. In G, those subsets induce almost complete bipartite graphs, i.e., graphs in which every vertex has at most one non-neighbour in the opposite part. Therefore, to solve both problems for G, it suffices to solve them for this collection of $O(n)$ almost complete bipartite graphs.

But those problems are both easy for almost complete bipartite graphs: suppose a graph is obtained from $K_{s,t}$ by deleting a matching of size $m \leq s \leq t$. It is not difficult to see that the number of edges in a maximum edge biclique in this graph equals $\max_{0 \leq i \leq m} (t - m + i) \cdot (s - i)$. As for the BALANCED BICLIQUE problem, the optimal solution is given by $p = s$ if $t - s \geq m$, and by $\lfloor \frac{t-m+s}{2} \rfloor$ if $t - s < m$. $\qquad\square$

Independent Domination in Quasi-chain Graphs. The INDEPENDENT DOMINATING SET problem asks to find in a graph G an inclusion-wise maximal independent set of minimum cardinality. This problem is NP-complete for general graphs and remains intractable in many restricted graph families. In particular, it is NP-complete both for $2P_3$-free graphs [24] and for bipartite graphs [6]. In the following theorem, we prove polynomial-time solvability of the problem for quasi-chain graphs.

Theorem 8. *The* INDEPENDENT DOMINATING SET *problem can be solved for quasi-chain graphs in polynomial time.*

Proof. Let $G = (A, B, E)$ be a quasi-chain graph and S an optimal solution to the problem in G, and let $\mathcal{I}$ be as in Lemma 2. Note that S is contained in at least one of the elements of $\mathcal{I}$. Moreover, crucially, for any $I \in \mathcal{I}$, all maximal independent sets in $G[I]$ have the same size. This suggests the following way of finding an optimal solution:

1. For each $I \in \mathcal{I}$, determine if I contains an independent set that dominates G, and if yes, find such a set.
2. Among the sets we found, pick one with minimum size.

We claim that this produces an optimal solution to the problem. Indeed, this procedure is guaranteed to produce a set S, since any optimal solution to the problem dominates G and is contained in some $I \in \mathcal{I}$. Moreover, since all maximal independent sets in $G[I]$ have the same size (and S dominates G, so it is maximal in both G and $G[I]$), S must be an optimal solution.

It thus suffices to show that Step 1 can be done efficiently. To do this, let $I \in \mathcal{I}$. Let $I' \subseteq I$ be the subset of I of vertices that have degree 1 in $G[I]$, and put $I'' := I - I'$. We note that any independent subset of I dominating G must contain all vertices of I'', and exactly one vertex from each edge of $G[I']$. Let A'' and B'' be the sets of vertices in A, respectively B that have at least one neighbour in I''. We also denote $I'_A := I' \cap A$ and $I'_B := I' \cap B$, and let A' and B' be the sets of vertices in $A - (A'' \cup I'_A)$, respectively $B - (B'' \cup I'_B)$ that have at least one neighbour in I'.

If I does not dominate G, then no subset of I dominates G; we may thus assume I dominates G, that is, $A - I = A' \cup A''$ and $B - I = B' \cup B''$. Since G does not contain an unbalanced $2P_3$, the graphs $G[I'_A \cup B']$ and $G[I'_B \cup A']$

are $2P_2$-free, i.e., chain graphs. It follows that I'_A and I'_B each have vertices that dominate B' and A' respectively. If there exists such a pair $x \in I'_A$ and $y \in I'_B$ that is *non-adjacent*, then we are done: we pick x and y in their respective edges, and arbitrarily choose vertices from each other edge of I' to complete our independent dominating set. Otherwise, the unique vertices $x \in I'_A$ and $y \in I'_B$ that dominate B' and A' respectively belong to the same edge of I'. In this case, no independent set of I dominates G, since vertices A' and B' have no neighbours in I'' by construction, and (using $2P_2$-freeness) $I'_A - \{x\}$ does not dominate A', and $I'_B - \{y\}$ does not dominate B'. This proves the theorem. $\square$

References

1. Allen, P.: Forbidden induced bipartite graphs. J. Graph Theory **60**, 219–241 (2009)
2. Atminas, A., Collins, A., Lozin, V., Zamaraev, V.: Implicit representations and factorial properties of graphs. Discrete Math. **338**, 164–179 (2015)
3. Balogh, J., Bollobás, B., Weinreich, D.: The speed of hereditary properties of graphs. J. Combin. Theory Ser. B **79**, 131–156 (2000)
4. Bose, P., Buss, J.F., Lubiw, A.: Pattern matching for permutations. Inf. Process. Lett. **65**, 277–283 (1998)
5. Conlon, D., Fox, J.: Bounds for graph regularity and removal lemmas. Geom. Funct. Anal. **22**, 1191–1256 (2012)
6. Damaschke, P., Müller, H., Kratsch, D.: Domination in convex and chordal bipartite graphs. Inform. Process. Lett. **36**, 231–236 (1990)
7. Goldberg, P., Golumbic, M., Kaplan, H., Shamir, R.: Four strikes against physical mapping of DNA. J. Comput. Biol. **2**, 139–152 (1995)
8. Hammer, P., Peled, U.N., Sun, X.: Difference graphs. Discrete Appl. Math. **28**, 35–44 (1990)
9. Heggernes, P., van 't Hof, P., Meister, D., Villanger, Y.: Induced subgraph isomorphism on proper interval and bipartite permutation graphs. Theor. Comput. Sci. **562**, 252–269 (2015)
10. Johnson, D.S.: The NP-completeness column: an ongoing guide. J. Algorithms **8**, 438–448 (1987)
11. Kannan, S., Naor, M., Rudich, S.: Implicit representation of graphs. SIAM J. Discrete Math. **5**, 596–603 (1992)
12. Kitaev, S.: Patterns in Permutations and Words. Monographs in Theoretical Computer Science. An EATCS Series. Springer, Heidelberg (2011). https://doi.org/10.1007/978-3-642-17333-2
13. Korpelainen, N., Lozin, V.: Bipartite induced subgraphs and well-quasi-ordering. J. Graph Theory **67**, 235–249 (2011)
14. Lozin, V., Rudolf, G.: Minimal universal bipartite graphs. Ars Combin. **84**, 345–356 (2007)
15. Lozin, V., Volz, J.: The clique-width of bipartite graphs in monogenic classes. Int. J. Found. Comput. Sci. **19**, 477–494 (2008)
16. Lozin, V.: On maximum induced matchings in bipartite graphs. Inform. Process. Lett. **81**, 7–11 (2002)
17. Malliaris, M., Shelah, S.: Regularity lemmas for stable graphs. Trans. Am. Math. Soc. **366**, 1551–1585 (2014)

18. Müller, H., Brandstädt, A.: The NP-completeness of STEINER TREE and DOM-INATING SET for chordal bipartite graphs. Theoret. Comput. Sci. **53**, 257–265 (1987)
19. Müller, H.: Alternating cycle-free matchings. Order **7**, 11–21 (1990)
20. Müller, H.: Hamiltonian circuits in chordal bipartite graphs. Discrete Math. **156**, 291–298 (1996)
21. Peeters, R.: The maximum edge biclique problem is NP-complete. Discrete Appl. Math. **131**, 651–654 (2003)
22. Petkovšek, M.: Letter graphs and well-quasi-order by induced subgraphs. Discrete Math. **244**, 375–388 (2002)
23. Spinrad, J. P.: Efficient graph representations. Fields Institute Monographs, 19. American Mathematical Society, Providence, RI (2003)
24. Zverovich, I.E.: Satgraphs and independent domination. Part 1. Theor. Comput. Sci. **352**, 47–56 (2006)

Composed Degree-Distance Realizations
of Graphs

Amotz Bar-Noy[1], David Peleg[2(✉)], Mor Perry[2], and Dror Rawitz[3]

[1] City University of New York (CUNY), New York City, USA
amotz@sci.brooklyn.cuny.edu
[2] Weizmann Institute of Science, Rehovot, Israel
{david.peleg,mor.perry}@weizmann.ac.il
[3] Bar Ilan University, Ramat-Gan, Israel
dror.rawitz@biu.ac.il

Abstract. Network realization problems require, given a specification π for some network parameter (such as degrees, distances or connectivity), to construct a network G conforming to π, or to determine that no such network exists. In this paper we study *composed profile* realization, where the given instance consists of *two* or more profile specifications that need to be realized simultaneously. To gain some understanding of the problem, we focus on two classical profile types, namely, *degrees* and *distances*, which were (separately) studied extensively in the past. We investigate a wide spectrum of variants of the composed distance & degree realization problem. For each variant we either give a polynomial-time realization algorithm or establish NP hardness. In particular:

- We consider both *precise* specifications and *range* specifications, which specify a range of permissible values for each entry of the profile.
- We consider realizations by both weighted and unweighted graphs.
- We also study settings where the realizing graph is restricted to specific graph classes, including trees and bipartite graphs.

1 Introduction

This paper considers the family of *network realization* problems. A Π-*realization* problem concerns some type of network parameter Π on networks, such as the vertex degrees, inter-vertex distances, centrality, connectivity, and so on. With every network G one can associate its Π-profile, $\Pi(G)$, giving[1] the values of Π on G. An instance of the Π-realization problem consists of a *specification* π, detailing the requirements on Π. Given such a specification π, it is necessary to construct a network G conforming to it, i.e., satisfying $\Pi(G) = \pi$, or to determine that no such network exists. The motivation for network realization problems stems from both "exploratory" contexts where one attempts to reconstruct an

---

Supported in part by a US-Israel BSF grant (2018043).
[1] We consider profile types for which $\Pi(G)$ is polynomial-time computable given G.

© Springer Nature Switzerland AG 2021
P. Flocchini and L. Moura (Eds.): IWOCA 2021, LNCS 12757, pp. 63–77, 2021.
https://doi.org/10.1007/978-3-030-79987-8_5

existing network of unknown structure based on the outcomes of experimental measurements, and engineering contexts related to network design.

Two cannonical examples of profile types are vertex *degrees* and *distances*.

Degree Realization. The most well-studied family of realization problems concerns vertex degrees. The *degree profile* of a (simple undirected) graph $G = (V, E)$ with vertex set $V = \{v_1, \ldots, v_n\}$ is an integer sequence $\text{DEG}(G) = (\delta_1, \ldots, \delta_n)$, where $\delta_i = \deg_{G,i}$ is the degree of vertex i in G. The *degree realization* problem asks to decide, given a sequence of n non-negative integers $\bar{\delta} = (\delta_1, \ldots, \delta_n)$, whether there exists a graph G whose degree sequence $\text{DEG}(G)$ equals $\bar{\delta}$. A sequence that admits such a realization is called *graphic*. The main questions studied in the past concerned characterizations for a sequence to be graphic and algorithms for finding a realizing graph for a given sequence if exists. For a brief review and some references to previous work on degree realization see, e.g., [6] in these proceedings. Graphic sequences were studied also on specific graph families, such as trees and bipartite graphs [14, 24, 26, 34, 43].

Rather than *precise* degree requirements, some studies concerned *degree-range* specifications, which define a range of allowable degrees for each vertex. An entry in the specification $\bar{\delta}$ consists of a pair $[\delta_i^-, \delta_i^+]$, and the realizing graph G must satisfy $\delta_i^- \leq \deg_G(i) \leq \delta_i^+$. Again see [6] for references to prior work.

Distance Realization. In a graph G, the *distance* $dist_G(u, v)$ between two nodes u and v is the length of the shortest path connecting them in G. (The length of a path is the sum of its edge weights; in an unweighted graph, the weight of each edge is taken to be 1.) The *distance profile* of an n-vertex graph $G = (V, E)$ with vertex set $V = \{v_1, \ldots, v_n\}$ consists of an $n \times n$ matrix $\text{DIST}(G) = D$, where $D_{i,j} \in \mathbb{N} \cup \{\infty\}$ for every $1 \leq i < j \leq n$, which specifies the required distance between every two nodes $i \neq j$ in the graph ($D_{i,j} = \infty$ when i and j are in different disconnected components).

The *unweighted distance realization* problem is defined as follows. An instance consists of an $n \times n$ matrix D, where $D_{i,j}$ is a nonnegative integer or ∞, for every $1 \leq i < j \leq n$. The goal is to compute an n-vertex unweighted undirected graph $G = (V, E)$ realizing D, i.e., such that $V = \{1, \ldots, n\}$ and $dist_G(i, j) = D_{i,j}$ for every $1 \leq i < j \leq n$, or to decide that no such realizing graph exists. In the *weighted* distance realization problem the edges of the realizing graph may have arbitrary integral weights. (We assume that the minimum edge weight is 1.)

Distance realization problems were introduced and studied in a seminal paper of Hakimi and Yau [27] (see [6] for a review). Precise distance realizations by trees and bipartite graphs were considered as well [4, 11]. The distance realization problem was studied also for *distance ranges*, i.e., where an entry in the given distance matrix D consists of a pair $[D_{i,j}^-, D_{i,j}^+]$, and the realizing graph G must satisfy $D_{i,j}^- \leq dist_G(i, j) \leq D_{i,j}^+$, for every $1 \leq i < j \leq n$ [33, 39].

Profile Composition. In reality, it is often required to address multiple network parameters simultaneously. In particular, one may be given specifications for two or more different profile types, and be requested to find a realizing graph conforming to all of these specifications simultaneously. We refer to the input as a "composed" profile specification. *Profile composition* is one of the fundamental, yet little understood, aspects of the realization problem. Specifically, we are interested in the following setting. Consider two different profile types, Π^A and Π^B. Given two profile specifications π^A and π^B corresponding to these profile types, the goal is to solve the realization problem of the composed profile specification $\pi^A \wedge \pi^B$, namely, construct a graph G that realizes both specifications simultaneously, if exists, or decide that this is impossible.

As a first concrete example, we focus in this paper on the composition of degree and distance profiles. In the composed *degree and distance (D&D) realization* problem, we are given both a degree vector $\bar{\delta}$ and a distance matrix D, and the goal is to decide whether there is an n-vertex (unweighted/weighted) undirected graph $G = (V, E)$ realizing $\bar{\delta}$ and D simultaneously, i.e., such that $V = \{1, \ldots, n\}$, $\deg_G(i) = \delta_i$ and $dist_G(i, j) = D_{i,j}$ for every $1 \leq i < j \leq n$.

We study such compositions in a variety of settings, including *precise* and *range* specifications, and realizations by *weighted* and *unweighted* graphs. In addition, we consider realizations by more restricted graph classes, such as (weighted or unweighted) *trees* and *bipartite graphs*. A somewhat different type of restriction that we consider is where the input to the problem consists also of a given graph G^+ (the *supergraph*), and the realizing graph G must be a *subgraph* of G^+, or in other words, G must belong to the class of subgraphs[2] of G^+.

Terminology. The above exposition outlines a wide collection of problems, classified by a number of characteristics, including the following:

- the *degree specification types* $\tilde{\delta} \in \{P, [\,]\}$:
  indicates whether the input specifies exact degrees or degree-ranges,
- the *distance specification types* $\tilde{D} \in \{P, [\,]\}$:
  indicates whether the input specifies exact distances or distance-ranges,
- the *class of graphs* $\tilde{g} \in \{U, W, UT, WT, UB, WB, US, WS\}$:
  indicates whether the realizing graph should be a (unweighted or weighted) general graph, tree, bipartite graph, or subgraph of a given graph.

Accordingly, the studied variants of the realization problem are:

- DEG($\tilde{\delta}, \tilde{g}$):   degree realization problems[3],
- DIST($\tilde{D}, \tilde{g}$):   distance realization problems,
- D&D($\tilde{\delta}, \tilde{D}, \tilde{g}$):   composed D&D realization problems.

Our Contributions. In Sect. 2 we present some basic properties of profile composition, concerning the relations between the properties of the two profile types Π^A and Π^B and the properties of their composition, $\Pi^A \wedge \Pi^B$. In Sect. 3 we focus

---

[2] Such a problem may arise naturally in a setting where it is known a priori that certain connections are impossible, infeasible or disallowed, due to the environment, the user specified requirements, or other reasons.

[3] For the pure degree-realization problem, there is no distinction between weighted and unweighted graphs.

on the properties of profile composition when one of the composed profiles is distances. Then, in Sect. 4, we turn to studying composed D&D realization problems. Most of the literature on realization problems concerns either efficient realization algorithms or characterizations (i.e., necessary or sufficient conditions for realizability), and negative results are scarce. Here, we are interested in classifying the resulting problems according to their complexity. We present, for each variant of this problem, either a polynomial-time algorithm or a proof that it is NP-hard. Specifically, we show that when the distance matrix specifies precise values, there are polynomial-time algorithms for all variants (precise degrees or degree-ranges realizations by weighted or unweighted arbitrary graphs, trees, bipartite graphs, and subgraphs of a given graph). However, when the distance matrix specifies distance-ranges, all variants are NP-hard. Along the way, we also fill in the gaps for some distance realization problems whose status was unresolved so far.

Related Work. A concrete type of composition that has received considerable attention concerns degree sequences that satisfy an additional graph property P. Let P be an invariant property of graphs. A graphic degree sequence $\bar{\delta}$ is said to be *potentially P-graphic* if there exists at least one graph G conforming with $\bar{\delta}$ that has the property P. The sequence $\bar{\delta}$ is *forcibly P-graphic* if every graph G conforming with $\bar{\delta}$ enjoys P. For example, consider the property that the graph is k-edge-connected. The composed profile specification consists of a degree specification and a k-edge-connectivity specification. An algorithm for this composed profile was given in [3]. Conditions for the existence of k-edge connected realizations of degree sequences, for $k \geq 1$, are also known [19]. Necessary and sufficient conditions for the realization of k-vertex connected graphs where $k \geq 2$, as well as a realization algorithm, were presented in [41]. For a survey on potentially and forcibly P-graphic sequences, see [32].

The *optimal distance realization* problem was also introduced in [27] and studied further in [1,5,13,15–18,20,28,30,31,33,35–38,40,42]. In this problem, a distance matrix D is given over a set S of n terminal vertices, and the goal is to find a *minimum-weight* graph G containing S, with possibly additional *auxiliary* vertices, that realizes the given D for S. In contrast, our "pure" distance realization problem requires the realizing graph to have exactly n vertices, and does not allow adding auxiliary ones. Hence, the results obtained for the optimal distance realization problem do not seem to carry over easily to our setting.

Realization questions were studied in the past for other types of network information *profiles*, including eccentricites [12,29], connectivity and flow [21–23,25] and maximum or minimum neighborhood degrees [7,10]. Several other realization problems are surveyed in [8,9].

2   Profile Composition

Consider a Π-realization problem for some network parameter Π. An instance π of the Π-realization problem is *realizable* by a graph of the class $\mathcal{GC}$ if there exists a network $G \in \mathcal{GC}$ satisfying $\Pi(G) = \pi$. We assume that an instance π (of any profile type) always specifies also the network size n (i.e., the number of vertices). Let $\mathcal{RG}(\pi, \mathcal{GC})$ denote the set of graphs $G \in \mathcal{GC}$ realizing π.

We are interested in several properties of profile types. Consider a profile type Π and a graph class $\mathcal{GC}$.

- Π is *enumerable* over $\mathcal{GC}$ if for each of its instances π, the set $\mathcal{RG}(\pi, \mathcal{GC})$ has size polynomial in n, and moreover, there is a polynomial-time procedure that given π generates all the graphs of $\mathcal{RG}(\pi, \mathcal{GC})$ successively.
- Π is *unique* over $\mathcal{GC}$ if each instance π has $|\mathcal{RG}(\pi, \mathcal{GC})| \leq \infty$, and moreover, there is a polynomial-time procedure that given π generates the unique graph of $\mathcal{RG}(\pi, \mathcal{GC})$ or declares that $\mathcal{RG}(\pi, \mathcal{GC}) = \emptyset$.
- Π is *verifiable* over $\mathcal{GC}$ if there is a polynomial-time procedure that given a specification π and $G \in \mathcal{GC}$ checks whether G realizes π (or, $\Pi(G) = \pi$).
- Π is *realizable* over $\mathcal{GC}$ if there is a polynomial-time procedure that given a specification π finds a graph $G \in \mathcal{GC}$ such that $\Pi(G) = \pi$, if π is realizable, and otherwise indicates that no such graph exists[4].
- Π is *super-realizable* over $\mathcal{GC}$ if there is a polynomial-time procedure that given a specification π and $G = (V, E)$ finds a supergraph[5] $G' = (V, E') \in \mathcal{GC}$ of G such that $\Pi(G) = \pi$, if π is realizable by a supergraph of G, and otherwise indicates that no such graph exists.

Towards gaining an understanding of profile composition, a central goal is to identify relationships between the realizability of individual profile types and the realizability of their composition. For example, one might hope to establish some connections between the following two conditions:

$$\text{``both } \Pi^A \text{ and } \Pi^B \text{ are realizable.''} \tag{C1}$$

$$\text{``}\Pi^A \wedge \Pi^B \text{ is realizable.''} \tag{C2}$$

Unfortunally, profile composition turns out to be more intricate. To illustrate this, let us consider the following examples.

Example 1. Let Π^A be the *full information* profile type, i.e., an instance of it is of the form $\pi^A = \langle G_0 \rangle$, a complete description of some n-vertex graph G_0. This profile type is realizable, and has a unique realization for every instance, namely, G_0 itself. Let Π^B be the *clique* profile type, i.e., an instance for it consists of two integers, $\pi^B = \langle k_0, n \rangle$, where $k_0 \leq n$, and a realizing n-vertex graph is required to have a clique of size k_0. This profile type is also realizable, simply by taking the complete n-vertex graph K_n. Hence both profile types are realizable in a trivial way. However, for the composed profile type $\Pi^A \wedge \Pi^B$, the realization problem is NP-hard, since in order to decide if a given instance $\pi^A \wedge \pi^B = \langle G_0 \rangle \wedge \langle k_0, n \rangle$ is realizable, we must determine if G_0 has a k_0-clique. $\qquad\square$

Hence, (C1) is not a sufficient condition for (C2), unless NP=P.

---

[4] Note the difference between a realizable specification π ("π has a realization") and a realizable profile type Π ("there is a polynomial-time algorithm deciding, for every specification π of Π, if it is realizable").

[5] satisfying $E \subseteq E'$.

Example 2. Let Π^A be the *2-coloring* profile type, i.e., an instance of it is of the form $\pi^A = \langle n \rangle$, and a realizing n-vertex graph is required to have a legal 2-vertex coloring. This profile is realizable, simply by taking an n-vertex path. Let Π^B be the *graph 3-coloring* profile type, i.e., an instance of it is of the form $\pi^B = \langle G_0 \rangle$, a complete description of some n-vertex graph G_0 and the realizing graph is also required to have a legal 3-vertex coloring. This realization problem is NP-hard, since in order to decide if it is realizable, we must determine if G_0 has a legal 3-vertex coloring. Hence profile type Π^A is realizable in a trivial way, but profile type Π^B is not realizable, unless NP = P. However, the composed profile type $\Pi^A \wedge \Pi^B$, is realizable, since in order to decide if a given instance $\pi^A \wedge \pi^B = \langle n \rangle \wedge \langle G_0 \rangle$ is realizable, we only need to determine if G_0 has a legal 2-vertex coloring, which is known to have a polynomial algorithm.   □

Hence, if P $\neq$ NP, then (C1) is not a necessary condition for (C2).

We establish the following sufficient condition.

Lemma 1. *For any profile types Π^A and Π^B and graph class $\mathcal{GC}$, if Π^A is enumerable over $\mathcal{GC}$ and Π^B is verifiable over $\mathcal{GC}$, then the composed profile type $\Pi^A \wedge \Pi^B$ is realizable over $\mathcal{GC}$.*

Proof. Let Π^A and Π^B be as in the lemma. The following algorithm is a polynomial realization for the composed profile type $\Pi^A \wedge \Pi^B$ over $\mathcal{GC}$. Let π^A and π^B be specifications of Π^A and Π^B, respectively. Succesively generate the graphs of $\mathcal{RG}(\pi^A, \mathcal{GC})$, and for each generated graph G, check whether it realizes π^B, and if so, return G and halt. If every $G \in \mathcal{RG}(\pi^A, \mathcal{GC})$ does not realize π^B, return "Impossible". The correctness of this algorithm is immediate. Since Π^A is enumerable over $\mathcal{GC}$, there is a polynomial-time procedure that generates all $G \in \mathcal{RG}(\pi^A, \mathcal{GC})$, and since π^B is verifiable over $\mathcal{GC}$, checking whether G realizes π^B is polynomial. Overall, the running time of this algorithm is polynomial.   □

Interestingly, note that the verifiability of both profiles Π^A and Π^B is not always a *necessary* condition for the verifiability of the composed profile $\Pi^A \wedge \Pi^B$. To see this, consider the following example.

Example 3. Consider the following two profile types. Profile type Π^A requires that all degrees in G are 2. Profile type Π^B requires that G contains a Hamiltonian cycle. Consequently, the composed profile type $\Pi^A \wedge \Pi^B$ necessitates that G is a cycle. Note that over general graphs, Π^A is both always realizable and verifiable. Π^B is always realizable too, but verifying it is NP-hard. Yet their composition $\Pi^A \wedge \Pi^B$ is still both always realizable and verifiable.   □

Lemma 2. *Let Π^A be a unique profile type over the graph class $\mathcal{GC}$. Then for any profile type Π^B, the composed $\Pi^A \wedge \Pi^B$ is realizable over $\mathcal{GC}$ if and only if Π^B is verifiable over $\mathcal{GC}$.*

Proof. Fix $\mathcal{GC}$, let Π^A be a unique profile type, and consider some profile type Π^B. Let ALG_A be the realization procedure of Π^A over $\mathcal{GC}$. We first show that if Π^B is verifiable over $\mathcal{GC}$ then $\Pi^A \wedge \Pi^B$ is realizable over $\mathcal{GC}$. Suppose Π^B

is verifiable over $\mathcal{GC}$, and let ALG_B be a verification algorithm for it. Then the following algorithm $ALG_{A \wedge B}$ solves the $\Pi^A \wedge \Pi^B$-realization problem over $\mathcal{GC}$. Given an instance specification $\pi^A \wedge \pi^B$ of $\Pi^A \wedge \Pi^B$, invoke procedure ALG_A on π^A. If procedure ALG_A returns "impossible", then return "impossible". If it returns the (unique) graph $G \in \mathcal{GC}$ realizing π^A, then invoke Alg_B on G and return its response. It is clear that algorithm $ALG_{A \wedge B}$ is correct.

In the opposite direction, we need to show that if $\Pi^A \wedge \Pi^B$ is realizable over $\mathcal{GC}$ then Π^B is verifiable over $\mathcal{GC}$. Suppose $\Pi^A \wedge \Pi^B$ is realizable over $\mathcal{GC}$, and let $ALG_{A \wedge B}$ be an algorithm for the $\Pi^A \wedge \Pi^B$-realization problem. Then the following procedure ALG_B verifies Π^B over $\mathcal{GC}$. Given an instance specification π^B of Π^B and a graph $G \in \mathcal{GC}$, do the following. Set $\pi^A = \Pi^A(G)$. Then $\pi^A \wedge \pi^B$ is an instance of the composed profile type $\Pi^A \wedge \Pi^B$. Invoke algorithm $ALG_{A \wedge B}$ to $\pi^A \wedge \pi^B$. If its response is a graph G', then return "yes", and if the response is "impossible", then return "no". Indeed, in the former case, the returned graph G' must equal the given G, implying that G satisfies π^B, and in the latter case, necessarily G (which is the only possible realization of π^A over $\mathcal{GC}$) has $\Pi(G) \neq \pi^B$, i.e., it fails to satisfy π^B.   □

3   Composing the Distance Profile

We next focus on compositions where one of the profile types is distances. The following algorithm for precise distance realization by unweighted graphs has been presented as part of the proof of [27, Theorem 7].

Algorithm 1

1: **for** every pair (i, j) **do**   ▷ Construction
2:     **if** $D_{i,j} = 1$, **then** add an edge (i, j) to G.
3: **end for**
4: **for** every pair (i, j) **do**   ▷ Verification
5:     **if** $dist_G(i, j) \neq D_{i,j}$ **then return** "Impossible" and halt.
6: **end for**
7: **return** G.

Hereafter, we refer to the constructed graph as the *base graph* of D, and denote it by G_{base}. The next lemma is implicit in the proof of Thm. 7 of [27].

Lemma 3 [27]. *If D is realizable, then G_{base} is its unique realization.*

Note that in particular, the lemma implies the following.

Corollary 1. *The precise distance profile is unique over unweighted graphs.*

Hence by Lemma 2, we immediately have the following for the precise distance profile DIST.

Corollary 2. *For every profile Π, the composed profile DIST $\wedge$ Π is realizable by unweighted graphs if and only if Π is verifiable over unweighted graphs.*

We next consider profile compositions where one of the two profiles is the distance profile and the realization is by weighted graphs. The following algorithm solves the distance realization problem in the weighted case [27].

Algorithm 2

1: Initially set $V = \{1, \ldots, n\}$ and $E = \emptyset$.   ▷ At this stage $dist_G(i,j) = \infty$ for every i, j.
2: Sort the vertex pairs (i,j) for $1 \leq i < j \leq n$ by nondecreasing distances $D_{i,j}$.
3: Go over the pairs in this order.
4: **for** every pair (i,j) **do**   ▷ Construction
5:    **if** $dist_G(i,j) > D_{i,j}$ then add an edge (i,j) of weight $D_{i,j}$ to E.
6: **end for**
7: **for** every pair (i,j) **do**   ▷ Verification
8:    **if** $dist_G(i,j) < D_{i,j}$ then return "Impossible" and halt.
9: **end for**
10: Return G.

The algorithm's correctness follows from the next lemma, also due to [27].

Lemma 4 [27]. *If the precise distance specification D is realizable by a weighted graph, then the output G of Algorithm 2 is a minimal realization of D, i.e., every realization of D must contain the edges of G.*

Lemma 5. *For every profile type Π on unweighted graphs, the composed profile type $DIST \wedge \Pi$ is realizable on weighted graphs if and only if Π is super-realizable.*

Proof. Suppose Π is super-realizable on unweighted graphs and let ALG_π be the realization algorithm. We have to show that the profile type $DIST \wedge \Pi$ is realizable on weighted graphs. The following algorithm $ALG_{D \wedge \pi}$ solves the $DIST \wedge \Pi$-realization problem. Consider a specification $D \wedge \pi$. The algorithm first applies Algorithm 2 to the precise distance specification D. If the response is "Impossible", then return "Impossible". If the response is a weighted graph $G_{\min}(D) = (V, E_{\min})$ realizing D, then invoke Algorithm ALG_π on the graph $G_{\min}(D)$ (ignoring the weights). If the response of ALG_π is "Impossible", then return "Impossible". If the response is an unweighted graph G_π then return the weighted graph $G_{\pi,D}$ which is $G_\pi = (V, E_\pi)$ where the weight of every edge $(i,j) \in E_\pi$ is $D_{i,j}$. Note that if the algorithm returns "Impossible" after the execution of Algorithm 2, the specification $D \wedge \pi$ is indeed unrealizable (since D is unrealizable). Now consider the case where D is realizable. By Lemma 4, all of the edges of the returned graph $G_{\min}(D)$ are necessary in any realization of D. Therefore, any realization of $D \wedge \pi$ must be a supergraph of $G_{\min}(D)$. If an unweighted supergraph G_π of $G_{\min}(D)$ realizing π exists, then it will be found by Algorithm ALG_π (and if not, both ALG_π and $ALG_{D \wedge \pi}$ return "Impossible"). Note that by definition of a supergraph $E_{\min} \subseteq E_\pi$. Since The weights assigned to the edges of $G_{\pi,D}$ are all from the specification matrix D, if $G_{\min}(D)$ is a weighted realization of D, then so is $G_{\pi,D}$. It follows that $ALG_{D \wedge \pi}$ returns the correct answer.

Conversely, suppose DIST $\wedge$ Π is realizable on weighted graphs, and let $ALG_{D\wedge\pi}$ be the realization algorithm. We have to show that Π is super-realizable on unweighted graphs. The following algorithm ALG_π solves the Π-realization problem on supergraphs of a given unweighted graph. Consider a specification π of Π, and let G be the given unweighted graph. Let $D_G = \mathrm{DIST}(G)$, the distance matrix of G. Note that $G_{\min}(D_G) = G$, since G is unweighted, so $G_{\min}(D_G)$ consists only of the distance-1 entries of D_G, which are exactly the edges of G. Create the composed profile specification $D_G \wedge \pi$, apply Algorithm $ALG_{D\wedge\pi}$ to this specification, and return its response. Note that any super graph of $G_{\min}(D_G)$ can be a realization of D_G where the weight of every edge is exactly its corresponding distance in D_G. Therefore, there exists a realization of the composed profile specification $D_G \wedge \pi$ on weighted graphs if and only if π is realizable by a supergraph of G. Hence, the output of $ALG_{D\wedge\pi}$ is the correct response for ALG_π. □

Note that these lemmas hold unconditionally, i.e., their proof does not rely on $NP \neq P$.

4 D&D Realizations

In this section, we illustrate profile composition by presenting our results concerning the concrete example of composing *degree and distance* profiles.

4.1 Precise Distance D&D Realizations

We begin with the problems that involve *precise* distance profiles. These variants turn out to be easy in general graphs and in specific graph classes.

Theorem 1. *Given a precise distance matrix (i.e., $\tilde{D} = P$), for both weighted and unweighted realizing graphs (i.e., $\tilde{g} \in \{U, W\}$), and for both precise degrees and degree-ranges (i.e., $\tilde{\delta} \in \{P, [\,]\}$), the composed D&D realization problem $D\&D(\tilde{\delta}, P, \tilde{g})$ is solvable in polynomial time.*

Proof. We present our algorithms for degree-range sequences, which in particular solve also instances of precise degrees. In other words, our algorithms assume instances where the degree sequence $\bar{\delta} = ([\delta_1^-, \delta_1^+], \ldots, [\delta_n^-, \delta_n^+])$ is given with ranges and the distance matrix D is given with exact values. We start with the unweighted problem: D&D($[\,], P, U$). A degree sequence $\bar{\delta}$ can be verified on a specific graph in linear time. Hence, by Corollary 2, the composed D&D($[\,], P, U$) problem is realizable in polynomial time.

We now turn to show a polynomial-time algorithm for the weighted problem: D&D($[\,], P, W$). According to Lemma 5, we need to show that finding a realization of a degree sequence which is lower bounded by a graph G can be done efficiently. Let $\bar{\delta}^0 = (\delta_1^0, \ldots, \delta_n^0)$ be the degree sequence of G, and let $f = (f_1, \ldots, f_n)$, where $f_i = \delta_i^+ - \delta_i^0$ for every $1 \leq i \leq n$, is the *degree shortage* vector. There are two cases to consider. If $f_i < 0$ for some $1 \leq i \leq n$, then there is no realizing graph for $\bar{\delta}$ which is lower bounded by G. Else, $f_i \geq 0$ for all $1 \leq i \leq n$. In this case, let $\bar{G} = K_n \backslash G$, where K_n is the complete graph on n vertices. Edges of $\bar{G}$ can now be added to G in order to increase some of the degrees. Let $g = (g_1, \ldots, g_n)$, where $g_i = \max\{0, \delta_i^- - \delta_i^0\}$. The problem now reduces to that of finding a (g, f)-factor in $\bar{G}$, which is known to be solvable in polynomial time [2]. □

(Hereafter, most proofs are omitted due to lack of space.)

Theorem 2. *Given a precise distance matrix (i.e., $\tilde{D} = P$), for weighted or unweighted tree or bipartite realizing graphs (i.e., $\tilde{g} \in \{UT, WT, UB, WB\}$), as well as for weighted or unweighted subgraphs of a given graph (i.e., $\tilde{g} \in \{US, WS\}$), and for both precise degrees and degree-ranges (i.e., $\tilde{\delta} \in \{P, [\,]\}$), the following realization problems are all solvable in polynomial time:*

- *The distance realization problems $DIST(P, \tilde{g})$,*
- *The composed D&D realization problems $D\&D(\tilde{\delta}, P, \tilde{g})$.*

4.2 Distance-Range D&D Realizations

We next consider problems involving distance-range profiles. As mentioned earlier, a polynomial-time algorithm for distance-range realization (not composed with degrees) by a weighted graph was given in [33,39]. However, no algorithm was known for the unweighted version. We now show that this is no coincidence: testing distance-range realizability by an unweighted graph is NP-hard. The proof method serves also (with small variations) for deriving hardness results for some of the D&D realization problems discussed later on.

Theorem 3. *The unweighted distance-range realization problem $DIST([\,], U)$ is NP-hard.*

Proof. We prove that the problem $DIST([\,], U)$ is NP-hard by a reduction from the coloring problem. Consider an instance (G, k) of the coloring problem, namely, an unweighted undirected graph G and an integer k, where it is required to decide whether G can be legally colored with k or fewer colors. We reduce this instance to an instance D of $DIST([\,], U)$ defined as follows. D is a distance matrix for $n + k + 1$ vertices, $\{1, \ldots, n + k + 1\}$. Intuitively, we think of the first n vertices of D, $V_{orig} = \{1, \ldots, n\}$, as representing the *original* n vertices of the given graph G, and of additional k vertices of D, $V_{col} = \{n + 1, \ldots, n + k\}$, as representing the *colors*, and the last vertex $n + k + 1$ represents a *coordinator*.

First, we impose the following requirements on the color vertices and the coordinator. Let $D_{n+\ell, n+k+1} = 1$ for every $1 \leq \ell \leq k$, and $D_{n+\ell, n+t} = 2$ for every $1 \leq \ell < t \leq k$. This allows only one possible realization for the subgraph induced by the vertices $V_{col} \cup \{n + k + 1\}$, namely, a star rooted at $n + k + 1$ (with no edges between the leaves).

Next, for the connections between the original vertices and the star structure, define the distance constraints as follows. For every $1 \leq i \leq n$, let $D_{i, n+k+1} = 2$

and let $D_{i,n+\ell} = [1,3]$ for every $1 \leq \ell \leq k$. This forces each of the original vertices to be connected to one (or more) of the color vertices, but not to the coordinator.

Finally, for every two original vertices $1 \leq i < j \leq n$, let

$$D_{i,j} = \begin{cases} 4, & (i,j) \in E(G), \\ [1,4], & \text{otherwise.} \end{cases}$$

It remains to show that the input G is k-colorable if and only if the distance matrix D is realizable.

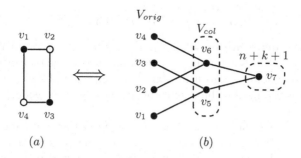

(a)   (b)

Fig. 1. An example of the reduction in the proof of Theorem 3, for $n = 4$ and $k = 2$. (a) The graph G is 2-colorable. (b) A realization of D which is the result of the reduction from the 2-coloring problem on G.

($\Rightarrow$): Suppose G is k colorable. Identify the colors as $1,\ldots,k$ and let $\varphi : V \mapsto \{1,\ldots,k\}$ be the coloring function. For the matrix D defined from G, construct a realizing graph G as follows. Start with a star rooted at $n+k+1$ with the k color vertices $n+1,\ldots,n+k$ as leaves. Connect each original vertex i to the color vertex $\varphi(i)$. It is easy to verify that G is a valid realization for D (see Fig. 1 for an example).

($\Leftarrow$): Suppose D has a valid realization G. Note that the restrictions of D force the color vertices to form a star rooted at $n+k+1$. Note that every original vertex i must be connected to at least one of the leaves of that star. Define a coloring function for G as follows. For every original vertex i, let ℓ be some color vertex connected to i, and let $\varphi(i) = \ell$. The distance constraints defined for the original vertices specify that if two original vertices i and j are connected by an edge in G, then their distance must be 4. This ensures that none of the color vertices are connected to both i and j (as this would make their distance 2). It follows that if $(i,j) \in E$ then i and j are assigned different colors.

This establishes the correctness of the reduction and shows that $\text{DIST}([\,],U)$ is NP-hard.   □

As shown next, all variants of the D&D realization problem are hard both in general graphs and in specific graph classes.

Theorem 4. *Given a distance matrix with distance-ranges (i.e., $\tilde{D} = [\ ]$), for both weighted and unweighted realizing graphs (i.e., $\tilde{g} \in \{U, W\}$), and for both precise degrees and degree-ranges (i.e., $\tilde{\delta} \in \{P, [\ ]\}$), the composed D&D realization problem $D\&D(\tilde{\delta}, [\ ], \tilde{g})$ is NP-hard.*

Theorem 5. *Given a distance matrix with distance-ranges (i.e., $\tilde{D} = [\ ]$), for weighted and unweighted general, tree and bipartite realizing graphs (i.e., $\tilde{g} \in \{UT, WT, UB, WB\}$), as well as for weighted and unweighted subgraphs of a given graph (i.e., $\tilde{g} \in \{US, WS\}$), and for both precise degrees and degree-ranges (i.e., $\tilde{\delta} \in \{P, [\ ]\}$), the following distance realization problems are NP-hard:*

- *The distance realization problems $DIST([\ ], \tilde{g})$,*
- *the composed D&D realization problems $D\&D(\tilde{\delta}, [\ ], \tilde{g})$.*

Note that the difference between our results for arbitrary realizing graphs and specific graph classes is for the weighted distance-range realization problem. Specifically, a polynomial algorithm exists in the general case, but when the realizing graph is required to be one of the structures we consider, this problem becomes NP-hard.

One of our results for trees is that the weighted distance-range realization problem, $DIST([\ ], WT)$, is NP-hard. It is interesting to note that in fact, this problem is NP-hard even if the realizing graph is required to be a simple path (denoted $DIST([\ ], Wpath)$), but when the realizing graph is required to be a star (denoted $DIST([\ ], Wstar)$), the problem becomes polynomial, as shown in the following two theorems.

Theorem 6. *The distance-range realization problem on weighted paths, $DIST([\ ], Wpath)$, is NP-hard.*

Theorem 7. *There exists a polynomial-time algorithm for the distance-range realization problem on weighted stars, $DIST([\ ], Wstar)$.*

5   Conclusions and Open Problems

This paper initiates the study of composed profiles and their realization, so naturally, many interesting research questions are ignored or touched upon only cursorily. Let us briefly mention some of those. We focused on the questions of deciding the realizability of a given specification, and generating a realizing graph if one exists. Some equally important questions, studied in the literature for various profile types, involve determining the *number* of different realizations, efficiently generating all realizations, establishing conditions for the existence of a unique realization, and so on. Our study of profile composition assumed that the two given specifications are *aligned*, i.e., vertex i in π^A is the same as vertex i in π^B. A different set of problems arises when we decouple the two specifications. In the context of D&D degree and distance profile composition, we may consider a graph G over $V = \{1, \ldots, n\}$ as an acceptable realization for the two given specifications if there exists a permutation π such that $dist_G(i, j) = D_{i,j}$ for every $1 \le i < j \le n$ and $\deg_G(\pi(i)) = \delta_i$ for every $1 \le i \le n$.

References

1. Althöfer, I.: On optimal realizations of finite metric spaces by graphs. Discret. Comput. Geom. **3**(2), 103–122 (1988). https://doi.org/10.1007/BF02187901
2. Anstee, R.P.: An algorithmic proof of Tutte's f-factor theorem. J. Algorithms **6**(1), 112–131 (1985)
3. Asano, T.: Graphical degree sequence problems with connectivity requirements. In: Ng, K.W., Raghavan, P., Balasubramanian, N.V., Chin, F.Y.L. (eds.) ISAAC 1993. LNCS, vol. 762, pp. 38–47. Springer, Heidelberg (1993). https://doi.org/10.1007/3-540-57568-5_233
4. Baldisserri, A.: Buneman's theorem for trees with exactly n vertices. CoRR (2014)
5. Bandelt, H.: Recognition of tree metrics. SIAM J. Discret. Math. **3**(1), 1–6 (1990)
6. Bar-Noy, A., Böhnlein, T., Peleg, D., Perry, M., Rawitz, D.: Relaxed and approximate graph realizations. In: 32nd IWOCA (2021)
7. Bar-Noy, A., Choudhary, K., Cohen, A., Peleg, D., Rawitz, D.: Minimum neighboring degree realization in graphs and trees. In: 28th ESA, vol. 173 of LIPIcs, pp. 10:1–10:15 (2020)
8. Bar-Noy, A., Choudhary, K., Peleg, D., Rawitz, D.: Realizability of graph specifications: characterizations and algorithms. In: Lotker, Z., Patt-Shamir, B. (eds.) SIROCCO 2018. LNCS, vol. 11085, pp. 3–13. Springer, Cham (2018). https://doi.org/10.1007/978-3-030-01325-7_1
9. Bar-Noy, A., Choudhary, K., Peleg, D., Rawitz, D.: Graph profile realizations and applications to social networks. In: Das, G.K., Mandal, P.S., Mukhopadhyaya, K., Nakano, S. (eds.) WALCOM 2019. LNCS, vol. 11355, pp. 3–14. Springer, Cham (2019). https://doi.org/10.1007/978-3-030-10564-8_1
10. Bar-Noy, A., Choudhary, K., Peleg, D., Rawitz, D.: Graph realizations: maximum degree in vertex neighborhoods. In: 17th SWAT, vol. 162 of LIPIcs, pp. 10:1–10:17 (2020)
11. Bar-Noy, A., Peleg, D., Perry, M., Rawitz, D., Schwartz, N.L.: Distance realization approximations. In: Preparation (2021)
12. Behzad, M., Simpson, J.E.: Eccentric sequences and eccentric sets in graphs. Discrete Math. **16**(3), 187–193 (1976)
13. Buneman, P.: A note on the metric properties of trees. J. Comb. Theory B **17**, 48–50 (1974)
14. Burstein, D., Rubin, J.: Sufficient conditions for graphicality of bidegree sequences. SIAM J. Discrete Math. **31**, 50–62 (2017)
15. Chung, F.R.K., Garrett, M.W., Graham, R.L., Shallcross, D.: Distance realization problems with applications to internet tomography. J. Comput. Syst. Sci. **63**(3), 432–448 (2001)
16. Culberson, J.C., Rudnicki, P.: A fast algorithm for constructing trees from distance matrices. Inf. Process. Lett. **30**(4), 215–220 (1989)
17. Dahlhaus, E.: Fast parallel recognition of ultrametrics and tree metrics. SIAM J. Discret. Math. **6**(4), 523–532 (1993)
18. Dress, A.W.M.: Trees, tight extensions of metric spaces, and the cohomological dimension of certain groups: a note on combinatorial properties of metric spaces. Adv. Math. **53**, 321–402 (1984)
19. Edmonds, J.: Existence of k-edge-connected ordinary graphs with prescribed degrees. J. Res. Natl. Bur. Stand. **68B**(2), 73–74 (1964)

20. Feder, T., Meyerson, A., Motwani, R., O'Callaghan, L., Panigrahy, R.: Representing graph metrics with fewest edges. In: Alt, H., Habib, M. (eds.) STACS 2003. LNCS, vol. 2607, pp. 355–366. Springer, Heidelberg (2003). https://doi.org/10.1007/3-540-36494-3_32

21. Frank, A.: Augmenting graphs to meet edge-connectivity requirements. SIAM J. Discrete Math. **5**, 25–43 (1992)

22. Frank, A.: Connectivity augmentation problems in network design. In: Mathematical Programming: State of the Art, pp. 34–63. University of Michigan (1994)

23. Frank, H., Chou, W.: Connectivity considerations in the design of survivable networks. IEEE Trans. Circuit Theory, CT **17**, 486–490 (1970)

24. Gale, D.: A theorem on flows in networks. Pac. J. Math. **7**, 1073–1082 (1957)

25. Gomory, R., Hu, T.: Multi-terminal network flows. J. Soc. Ind. Appl. Math. **9**, 551–570 (1961)

26. Gupta, G., Joshi, P., Tripathi, A.: Graphic sequences of trees and a problem of Frobenius. Czechoslovak Math. J. **57**, 49–52 (2007)

27. Hakimi, S.L., Yau, S.S.: Distance matrix of a graph and its realizability. Q. Appl. Math. **22**, 305–317 (1965)

28. Imrich, W., Simões-Pereira, J.M.S., Zamfirescu, C.: On optimal embeddings of metrics in graphs. J. Comb. Theory Ser. B **36**(1), 1–15 (1984)

29. Lesniak, L.: Eccentric sequences in graphs. Periodica Mathematica Hungarica **6**(4), 287–293 (1975)

30. Nieminen, J.: Realizing the distance matrix of a graph. J. Inf. Process. Cybern. **12**(1/2), 29–31 (1976)

31. Patrinos, A.N., Hakimi, S.L.: The distance matrix of a graph and its tree realizability. Q. Appl. Math. **30**, 255 (1972)

32. Rao, S.B.: A survey of the theory of potentially P-graphic and forcibly P-graphic degree sequences. In: Rao, S.B. (ed.) Combinatorics and Graph Theory. LNM, vol. 885, pp. 417–440. Springer, Heidelberg (1981). https://doi.org/10.1007/BFb0092288

33. Rubei, E.: Weighted graphs with distances in given ranges. J. Classif. **33**, 282–297 (2016)

34. Ryser, H.: Combinatorial properties of matrices of zeros and ones. Can. J. Math. **9**, 371–377 (1957)

35. Simões-Pereira, J.M.S.: A note on the tree realizability of a distance matrix. J. Comb. Theory B **6**, 303–310 (1969)

36. Simões-Pereira, J.M.S.: A note on distance matrices with unicyclic graph realizations. Discret. Math. **65**, 277–287 (1987)

37. Simões-Pereira, J.M.S.: An optimality criterion for graph embeddings of metrics. SIAM J. Discret. Math. **1**(2), 223–229 (1988)

38. Simões-Pereira, J.M.S.: An algorithm and its role in the study of optimal graph realizations of distance matrices. Discret. Math. **79**(3), 299–312 (1990)

39. Tamura, H., Sengoku, M., Shinoda, S., Abe, T.: Realization of a network from the upper and lower bounds of the distances (or capacities) between vertices. In: Proceedings of IEEE International Symposium on Circuits and Systems (ISCAS), pp. 2545–2548. IEEE (1993)

40. Varone, S.C.: A constructive algorithm for realizing a distance matrix. Eur. J. Oper. Res. **174**(1), 102–111 (2006)

41. Wang, D., Kleitman, D.: On the existence of n-connected graphs with prescribed degrees ($n >$). Networks **3**, 225–239 (1973)
42. Zaretskii, K.A.: Constructing a tree on the basis of a set of distances between the hanging vertices. Uspekhi Mat. Nauk **20**, 90–92 (1965)
43. Zverovich, I.E., Zverovich, V.E.: Contributions to the theory of graphic sequences. Discret. Math. **105**(1–3), 293–303 (1992)

All Subgraphs of a Wheel
Are 5-Coupled-Choosable

Sam Barr[(✉)] and Therese Biedl[(✉)]

David R. Cheriton School of Computer Science, University of Waterloo,
Waterloo, Canada
{s4barr,biedl}@uwaterloo.ca

Abstract. A wheel graph consists of a cycle along with a center vertex connected to every vertex in the cycle. In this paper we show that every subgraph of a wheel graph has list coupled chromatic number at most 5, and this coloring can be found in linear time. We further show that '5' is tight for every wheel graph with at least 5 vertices, and briefly discuss possible generalizations to planar graphs of treewidth 3.

1 Introduction

In this paper we study the problem of *coupled choosability*, the problem of finding a valid coloring given list assignments to every vertex and face of a planar graph. The problem is of great relevance to list coloring 1-planar graphs, as list coupled coloring a planar graph corresponds to list coloring an optimal 1-planar graph. (Detailed definitions will be given in Sect. 2.) Wang and Lih [20] show that every planar graph is 7-coupled-choosable, and hence every optimal 1-planar graph is 7-choosable. They further show that maximal planar graphs are 6-coupled-choosable, planar graphs of maximum degree 3 are 6-coupled-choosable, and outerplanar graphs (and more generally, all K_4-minor free graphs) are 5-coupled-choosable. Hetherington [11] also proves that K_4-minor free graphs are 5-coupled-choosable, and further shows that $K_{2,3}$-minor free graphs (and more generally, all $(\overline{K}_2 + (K_1 \cup K_2))$-minor free graphs) are also 5-coupled-choosable. We note here that wheel graphs are not included in any of these classes of graphs.

The problem of coupled coloring a planar graph has also been extensively studied, see [1,2,4,5,15].

The result by Wang and Lih and by Hetherington settles the coupled choosability for planar partial 2-trees (which are the same as K_4-minor free graphs). Initially wishing to investigate the coupled choosability of planar partial 3-trees, in this paper we investigate the coupled choosability of wheel graphs and their subgraphs. In Theorem 2, we show that any subgraph of a wheel is 5-coupled-choosable, and the coloring can be found in linear time. (Prior papers such as [20] and [11] did not address the run-time of finding their colorings; it can clearly be done in polynomial time by following the steps of their proofs but linear time

Research of TB supported by NSERC.

P. Flocchini and L. Moura (Eds.): IWOCA 2021, LNCS 12757, pp. 78–91, 2021.
https://doi.org/10.1007/978-3-030-79987-8_6

is not obvious.) In Theorem 3, we characterize the coupled choosability of wheel graphs by showing that 5 is tight for wheel graphs with at least 5 vertices. In the last section of the paper, we touch upon how these results could be relevant in finding the coupled choosability of planar partial 3-trees.

As for related results, the (non-coupled) choosability of wheel graphs was characterized in a different paper by Wang and Lih [19]: wheels of even order have list chromatic number 4, while wheels of odd order have list chromatic number 3. This stands in contrast to our result, as the parity of the number of vertices in the graph does not affect the coupled choosability of wheel graphs. Wang and Lih also show that Halin graphs that are not wheels have list chromatic number 3, while in Theorem 4 we prove the existence of a Halin graph that is not 5-coupled-choosable (in fact, it is not 5-coupled-colorable).

Our paper is structured as follows: In Sect. 2 we will go over the necessary definitions and terminology for graphs and graph coloring. In Sect. 3 we investigate the coupled choosability of wheel graphs. In Sect. 4 we examine how coupled choosability behaves under certain graph operations. In Sect. 5 we extend our analysis of wheel graphs to subgraphs of wheels, along with lower-bounding the coupled choosability of wheel graphs. In Sect. 6 we go over several possible extensions to our results, in particular some conjectures about the coupled-choosability of planar partial 3-trees.

2 Definitions

We assume basic familiarity with graph theory (see [7]). In this paper all graphs are finite and connected.

The *complete graph* K_4 consists of four vertices and all possible edges between them. A *subdivision* of a graph G is formed by repeatedly taking some edge $uv \in E(G)$, removing e from G, adding a new vertex x, and adding edges ux and xv. A graph is called K_4-*minor free* if none of its subgraphs is a subdivision of K_4.

We recall that a graph G is called *planar* if it can be drawn in the plane without edges crossing, and *plane* if a specific planar drawing Γ is given. The maximal regions of $\mathbb{R} \backslash \Gamma$ are called *faces*; the unbounded region is known as the *outer face* and all other faces are *inner faces*. An *outerplanar* graph is a graph that can be drawn in the plane such that every vertex is on the outer face; such a graph is K_4-minor free. A *bigon* is a face that is bounded by two duplicate edges between a pair of vertices. For a plane graph G, we use $V(G)$, $E(G)$, and $F(G)$ to denote the set of *vertices*, the set of *edges*, and the set of *faces* of G, respectively. The *dual graph* G^* of a plane graph is obtained by exchanging the roles of vertices and faces, i.e., G^* has a vertex for every face of G, and an edge (f_1, f_2) for every common edge of the two corresponding faces f_1, f_2 in G. However, we must make one exception to this definition of the dual: If G has a *bridge* (an edge e which if removed would disconnected the graph), then the unique face f incident to e does *not* receive a loop in the dual graph, because this would make the dual graph uncolorable. (This exception appears to have been made, without being stated explicitly, in previous papers as well.)

A *list assignment* is a map L that assigns a set of *colors* for each vertex or face in $V(G) \cup F(G)$. A *coupled coloring with respect to* L is a map c such that $c(x) \in L(x)$ for every $x \in V(G) \cup F(G)$, and $c(x) \neq c(y)$ for incident or adjacent elements $x, y \in V(G) \cup F(G)$. If such a map c exists, then we say that G is *L-coupled-choosable*. If G is L-coupled-choosable for every L such that $|L(x)| = k$ for every $x \in V(G) \cup F(G)$, then we say that G is *k-coupled-choosable*. The smallest integer k such that G is k-coupled-choosable is called the *list coupled chromatic number* of G and denoted $\chi_{vf}^{L}(G)$. Observe that a list coupled coloring of a graph G implies a list coupled coloring of the dual graph G^*, since the roles of the vertices and the faces is exchanged but incidences/adjacencies stay the same. Hence, we have $\chi_{vf}^{L}(G) = \chi_{vf}^{L}(G^*)$.

A natural way to express the list coupled chromatic number is to define a new graph $X(G)$ with vertices for all vertices and faces of G and edges whenever the vertices and faces G are adjacent/incident. We again assume that this graph has no loops. This graph $X(G)$ is *1-planar*, i.e., can be drawn in the plane with at most one crossing per edge. In fact, if G is 3-connected then $X(G)$ is an *optimal 1-planar graph*, i.e., it is simple and has the maximum-possible $4n - 8$ edges. (All optimal 1-planar graphs can be obtained in this fashion [16].) A coupled coloring of G corresponds to a *vertex coloring* of $X(G)$, i.e., a coloring of the vertices such that adjacent vertices have different colors. When restricting a vertex coloring to given lists L, then the respective terms are *L-choosable*, *k-choosable*, and the *list chromatic number* $\chi^{L}(X)$.

The *wheel graph* W_n is formed by starting with a cycle C_{n-1} on $n - 1$ vertices (the *outer cycle*), adding a *center vertex* inside the cycle and adding a *spoke-edge* from the center vertex to every vertex on the cycle. We will label the center vertex and the outer face of the wheel graph as x_0 and f_0, respectively. We further label the vertices in the outer cycle as $x_1, \ldots, x_{n-1}$, and label the inner faces as $f_1, \ldots, f_{n-1}$ such that x_i is incident to f_i and f_{i+1} for $1 \leq i < n - 1$, and x_{n-1} is adjacent to f_{n-1} and f_1 (see Fig. 1).

3   Coupled Choosability of Wheel Graphs

In order to prove the desired result for all subgraphs of the wheel graph, we first determine the coupled choosability of the wheel graph itself. It will be helpful to recall the following result relating the choosability of a graph to the maximum degree; it is an analogue to Brook's theorem and similarly upper-bounds the chromatic number of a graph by its maximum degree.

Lemma 1. *(Erdős, Rubin, and Taylor [9]). Let G be a connected graph that is neither an odd cycle nor a complete graph. Then G is $\Delta(G)$-choosable.*

Our main result in this section is:

Lemma 2. *Every wheel graph W_n, $n \geq 4$, is 5-coupled-choosable.*

Proof. For $n = 4$, W_4 is the complete graph K_4. Wang and Lih [20] proved that $\chi_{vf}^{L}(K_4) = 4$, so we assume $n \geq 5$. Let L be a color assignment for W_n such that

$|L(y)| = 5$ for every $y \in V(W_n) \cup F(W_n)$. Our goal is to find a coupled coloring with respect to L. Since x_0 and f_0 are both adjacent to all remaining vertices, we will color them first and then color the rest of $X(W_n)$. We will use X_n as a shortcut for $X(W_n)\backslash\{x_0, f_0\}$. Observe that $|V(X_n)| = 2n - 2$ and that X_n is 4-regular (see Fig. 1). Furthermore, it suffices to find a vertex-colouring of X_n with respect to L, plus two suitable colors in $L(x_0)$ and $L(f_0)$ for x_0 and f_0. We have two cases:

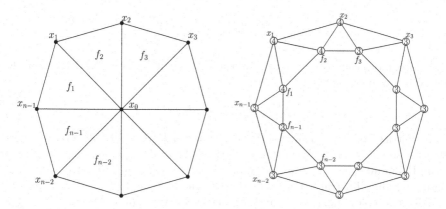

Fig. 1. The graph W_9 (left) and X_9 (right). Circled numbers indicate a lower bound on the list-length in L'.

Case 1: $L(x_0) \cap L(f_0) \neq \emptyset$. Let $a \in L(x_0) \cap L(f_0)$, and assign color a to x_0 and f_0. Observe that $|L(y)\backslash\{a\}| \geq 4$ for every $y \in V(X_n)$ and X_n has maximum degree 4. Moreover $|X_n| = 2n - 2$ is even, so X_n is not an odd cycle. Also x_1 and x_3 are not adjacent by $n \geq 5$, so X_n is not a complete graph. Therefore, by Lemma 1, we have a list coloring of the vertices of X_n that only uses colors in $L\backslash\{a\}$, which in turn implies an L-list-coloring of the vertices and faces of W_n.

Case 2: $L(x_0) \cap L(f_0) = \emptyset$. We find suitable colors for x_0 and f_0 by imitating the method used for K_4 in [20] (but adapted here to five colors). Define *color-pairs* $S := \{\{a, b\} : a \in L(x_0), b \in L(f_0)\}$. By case-assumption $|S| = 25$.

We claim that $|\{s \in S : s \subseteq L(y)\}| \leq 6$ for any $y \in V(X_n)$. To see this, let $y \in V(X_n)$, and consider the disjoint sub-lists $L_1 := L(y) \cap L(x_0)$ and $L_2 := L(y) \cap L(f_0)$. Since $|L_1| + |L_2| \leq |L(y)| = 5$, and $|L_1|$ and $|L_2|$ are integers, we have

$$|\{s \in S : s \subseteq L(y)\}| = |L_1 \times L_2| = |L_1| \cdot |L_2| \leq 6.$$

Therefore, color-pairs of S appear as subsets of lists in X_n at most

$$\sum_{y \in X_n} |\{s \in S : s \subseteq L(y)\}| \leq (2n - 2) \cdot 6 = 12n - 12$$

times. By $|S| = 25$, some element $\{a', b'\}$ of S appears at most

$$\frac{12n - 12}{25} < \frac{n - 1}{2}$$

times as a subset of a list in X_n. Color x_0 with a' and f_0 with b'. For $y \in V(X_n)$, define $L'(y) := L(y)\backslash\{a', b'\}$. For any $y \in V(X_n)$, we have $3 \leq |L'(y)| \leq 5$. We call y a *3-vertex* if $|L'(y)| = 3$ (this implies $\{a', b'\} \subset L(y)$), and a *4-vertex* otherwise. From our choice of colors a' and b', we have

$$\frac{|\{y \in V(X_n) : y \text{ is a 3-vertex}\}|}{|V(X_n)|} < \frac{(n-1)/2}{2n-2} = \frac{1}{4}$$

Therefore, more than three quarters of the vertices of X_n are 4-vertices. Consider the cyclic enumeration

$$\sigma := \langle f_1, x_1, f_2, x_2, \ldots, f_{n-1}, x_{n-1} \rangle$$

of the vertices of X_n. Since strictly more than $\frac{3}{4}|V(X_n)|$ of the vertices are 4-vertices, we have four consecutive 4-vertices in σ. Up to exchange of f_i and x_i and renumbering, we may assume that f_1, x_1, f_2, and x_2 are 4-vertices. Figure 1(right) illustrates the lower bounds on the size of L'.

We next color f_{n-1}, x_{n-1} and x_1 and have two sub-cases. If $L'(f_{n-1}) \cap L'(x_1) \neq \emptyset$, then color f_{n-1} and x_1 with the same color. Otherwise, since $|L'(f_{n-1}) \cup L'(x_1)| \geq 7 > |L(f_1)|$, there are colors p and q for f_{n-1} and x_1 respectively such that at least one of them is not in $L(f_1)$, i.e., $|L(f_1) \cap \{p, q\}| \leq 1$. Pick these colors for f_{n-1} and x_1. In either case, two vertices adjacent to x_{n-1} have been colored, and $|L'(x_{n-1})| \geq 3$, so x_{n-1} will have at least one valid color left, and we pick this color for x_{n-1}.

We now have colors p, q, and r for f_{n-1}, x_1, and x_{n-1} (respectively) such that $|L'(f_1) \cap \{p, q, r\}| \leq 2$. Removing these colors from the lists of their neighbors produces new lists L'' such that

$$|L''(f_1)| = |L'(f_1) \setminus \{p, q, r\}| \geq 4 - 2 = 2$$
$$|L''(f_2)| = |L'(f_2) \setminus \{q\}| \geq 4 - 1 = 3$$
$$|L''(x_2)| = |L'(x_2) \setminus \{q\}| \geq 4 - 1 = 3$$
$$|L''(x_{n-2})| = |L'(x_{n-2}) \setminus \{p, r\}| \geq 3 - 2 = 1$$
$$|L''(f_{n-2})| = |L'(f_{n-2}) \setminus \{p\}| \geq 3 - 1 = 2$$
$$|L''(x_i)| \geq 3 \quad (\text{for all } 3 \leq i \leq n - 3)$$
$$|L''(f_i)| \geq 3 \quad (\text{for all } 3 \leq i \leq n - 3)$$

The figure on the right illustrates these lower bounds on the list-lengths in L''.

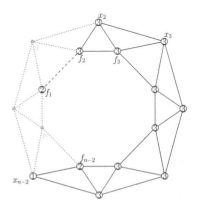

Let $X'_n := X_n \backslash \{f_{n-1}, x_{n-1}, f_1, x_1\}$ (X'_n is solid in the above figure) and color it with respect to list assignment L''. This is feasible since X'_n is outerplanar and outerplanar graphs are 3-choosable even if the colors of two consecutive vertices on the outer face are fixed [13] (here we fix the colors for x_{n-2} and f_{n-2}). This colors all vertices except for f_1, but $|L''(f_1)| \geq 2$ and f_1 has only one neighbor in X'_n, so we can give it a color not used by f_2. Therefore, we have a list vertex-coloring of X_n that is compatible with the colors for x_0, f_0 chosen earlier and so implies a list coupled coloring of W_n. □

Note that this coloring can easily be found in linear time. This is obvious in Case 1 since the coloring of Lemma 1 can be found in linear time [17]. Determining the colors a, b for Case 2 takes linear time since all list-lengths are constant, and then we mostly appeal to list-coloring an outer-planar graph, which can be done in linear time since outer-planar graphs are 2-degenerate.

4   Coupled Choosability Under Graph Operations

In Sect. 5 we seek to prove that all subgraphs of a wheel are 5-coupled-choosable. In pursuit of this, we examine how various graph operations affect the list-coupled-chromatic number. First, in contrast to list-vertex-coloring, there is no clear relationship between the list coupled chromatic number of a graph and the list coupled chromatic number of its subgraphs. Indeed, a subgraph may have larger list coupled chromatic number.

Observation 1. *There exists a plane graph G with subgraph $H \subseteq G$ such that* $\chi^L_{vf}(H) > \chi^L_{vf}(G)$

Proof. Let H be the graph obtained by deleting one edge of K_4; see Fig. 2. From Theorem 10 of [20], we know that the graph K_4 is 4-coupled-choosable, i.e., $\chi^L_{vf}(K_4) = 4$. But in graph H, the incidences and adjacencies between x_0, x_1, x_2, f_2, and f' form a K_5, and therefore $\chi^L_{vf}(H) \geq 5 > 4 = \chi^L_{vf}(K_4)$. □

Fig. 2. K_4 and subgraph H. Observe that $\chi^L_{vf}(H) = 5$ since it is outerplanar.

Other graph operations are better behaved in this respect. For instance, there is a clear relationship between the coupled choosability of some graph G and the coupled choosability of any subdivision of G.

Lemma 3. *For any plane graph G, any subdivision H of G is $\max\{5, \chi^L_{vf}(G)\}$-coupled-choosable.*

Proof. Let L be a list assignment for H such that $|L(x)| = \max\{5, \chi_{vf}^L(G)\}$ for every vertex and face of H. We prove the statement by induction on the number of subdivisions performed on G to obtain H. If H is the result of subdividing the edges of G zero times, then $H = G$ and so trivially any L-coupled-coloring of G is an L-coupled-coloring of H.

Otherwise, H was the result of performing $k + 1$ subdivisions on G for some $k \geq 0$. In particular, H is the result of subdividing a single edge of some graph H', where H' was the result of performing k subdivisions on G. Let $uv \in E(H')$ be the edge of H' that was subdivided, and let x be the vertex which was added. By the inductive hypothesis, H' is $\max\{5, \chi_{vf}^L(G)\}$-coupled-choosable. Color the faces of H and the vertices $V(H)\setminus\{x\}$ according to how they would be colored in H'. Then we only need to color the remaining vertex x. Note that x has degree two with neighbors u and v. Let f_1 and f_2 be the two faces adjacent to the edge uv in H'. Then u, v, f_1, and f_2 are the only vertices and faces that are adjacent (respectively incident) to x. Hence, after coloring the vertices and faces from H', x still has at least $|L(x)| - 4 \geq 5 - 4 = 1$ color left and can be colored.   □

This implies another result. For a planar graph G, subdividing an edge corresponds in the dual graph G^* to duplicating edges to form bigons. Since $\chi_{vf}^L(G) = \chi_{vf}^L(G^*)$ we therefore have:

Corollary 1. *Let G be a plane graph, and H the result of duplicating some edges of G to form bigons. Then H is $\max\{5, \chi_{vf}^L(G)\}$-coupled-choosable.*

A similar result can also be had for adding a vertex of degree one to a graph.

Lemma 4. *Let G be a planar graph, and let H be G plus a new vertex of degree one. Then H is $\max\{3, \chi_{vf}^L(G)\}$-coupled-choosable.*

Proof. Let x be the new vertex, and let L be a list assignment for H such that $|L(y)| = \max\{3, \chi_{vf}^L(G)\}$ for every vertex and face of H. Color the faces and vertices of $H - x$ according to how they would be colored in G. It remains to color x. Since x is adjacent to only one vertex and incident to only one face in H, after coloring the vertices and face of $H - x$, x still has at least $|L(x)| - 2 \geq 3 - 2 = 1$ color left and can be colored.   □

Note that for all three of the above lemmas, the coloring of H can be found in constant time, given a suitable coloring of G.

Wang and Lih [20] and Hetherington [11] proved that all K_4-minor free graphs are 5-coupled-choosable, but it is not clear whether their proofs lead to a linear-time algorithm to find the coloring. With the above two results, such an algorithm is immediate.

Theorem 1. *All K_4-minor free graphs are 5-coupled-choosable, and the coloring can be found in linear time.*

Proof. It is known (see [8]) that every K_4-minor free graph G can be obtained from some tree T via a series of duplicating edges, subdividing edges, and adding vertices of degree one. Then by Lemmas 3 and 4, Corollary 1, and the 3-coupled-choosability of trees, we have that G is 5-coupled-choosable.

To find the coloring efficiently, first split G into its 2-connected components $C_1, \ldots, C_d$ [12]. Then run on each component C_i the algorithm that recognizes so-called series-parallel graphs in linear time [18]. Since 2-connected K_4-minor free graphs are series-parallel graphs, this algorithm will succeed on each C_i, and following the trace of its execution one obtains how to construct C_i from a single edge via a series of duplicating edges and subdividing edges. Combining this with the tree of 2-connected components shows how to obtain G. Since trees are trivially 3-coupled-colorable (choose a color for the unique face, then find a 2-coloring of the vertices), and each of our expansion steps takes constant time, we can find the coloring of G in linear time.   □

5   Subgraphs of Wheels

We now turn to graphs that are subgraphs of wheels. As demonstrated in Observation 1, non-trivial work is required to demonstrate that any subgraph of a wheel graph is also 5-coupled-choosable. The result comes quickly from the results proved in the previous section.

Theorem 2. *Let G be a subgraph of a wheel graph W_n, $n \geq 4$. Then G is 5-coupled-choosable and the coloring can be found in linear time.*

Proof. We examine several possibilities of the structure of G.

Case 1: $G = W_n$. Then by Lemma 2 G is 5-coupled-choosable, and the coloring can be found in linear time.

Case 2: G is the result of deleting at least one edge or vertex of W_n that is on the outer face. Then G is outerplanar and therefore K_4-minor free, and so by Theorem 1, G is 5-coupled-choosable and the coloring can be found in linear time.

Case 3: G is the result of removing the center vertex of W_n. Then $G = C_{n-1}$ is outerplanar and (as in the previous case) 5-coupled choosable.

Case 4: None of the above. Then all vertices of W_n belong to G, but we deleted some edges which were not on the outer face. So G is the result of deleting some of the spoke-edges incident to the center vertex. If at most two spokes remain, then G has at most 3 faces and therefore is K_4-minor free, and hence is 5-coupled-choosable by Theorem 1. If at least three spokes remain, then G is a subdivision of some W_k for $k \geq 4$. By Lemmas 2 and 3 G is 5-coupled-choosable, and we can find the coloring in linear time since we can detect all subdivision-vertices by scanning for vertices in linear time.   □

Having established an upper bound on the list coupled chromatic number of wheel graphs in Lemma 2, one might wonder whether this bound is tight. In [20], it is shown that the graph $K_4 = W_4$ is 4-coupled-choosable. In fact, this is the only wheel graph that is 4-coupled-choosable. For all other wheel graphs, the bound of 5-coupled-choosability is tight.

Theorem 3. $\chi_{vf}^L(W_n) = 5$, for $n \geq 5$.

Proof. From Lemma 2, we know that all wheel graphs are 5-coupled-choosable. It remains to show that they are not 4-coupled-choosable for $n \geq 5$.

For $n = 5, 6$, we consider the list assignment L such that $L(y) = \{1, 2, 3, 4\}$ for every $y \in V(W_n) \cup F(W_n)$. (So these graphs are not even 4-coupled-colorable.) Assume for contradiction that we have an L-coupled-coloring c of W_n. If $c(x_0) \neq c(f_0)$, then this leaves two colors for coloring the triangle x_1, f_1, f_2 in X_n, impossible. Hence $c(x_0) = c(f_0)$, say they are both colored 4. Then we have an L'-coloring of X_n with lists $L'(y) := L(y) \backslash \{4\} = \{1, 2, 3\}$.

Observe that for X_5 and X_6, any putative L'-coloring would be unique up to renaming the colors, since once we have colored one triangle, every other vertex can be reached via a sequence of triangles. One verifies that for these graphs (and indeed every X_k where $k - 1$ is not divisible by 3), attempting such a 3-coloring leads to a contradiction (see Fig. 3). This proves Theorem 3 for $n = 5, 6$.

For $n \geq 7$, we construct a list assignment L such that W_n is not L-coupled-choosable. Set $L(x_0) = \{1, 2, 3, 4\}$ and $L(f_0) = \{5, 6, 7, 8\}$. We further define:

$$L(f_1) = L(x_1) = L(f_2) = \{1, 2, 5, 6\}$$
$$L(x_2) = L(f_3) = L(x_3) = \{1, 2, 7, 8\}$$
$$L(f_4) = L(x_4) = L(f_5) = \{3, 4, 5, 6\}$$
$$L(x_5) = L(f_6) = L(x_6) = \{3, 4, 7, 8\}$$

Observe that each of these triples forms a triangle in X_n, and for any $a \in \{1, 2, 3, 4\}$ and $b \in \{5, 6, 7, 8\}$, one of these triangles has colors $\{a, b, x, y\}$ for some colors x, y. Assume for contradiction that we have an L-coupled-coloring c of W_n. Up to symmetry, assume $c(x_0) = 1$ and $c(f_0) = 5$. But then f_1, x_1, and f_2 have two colors left, and therefore cannot be colored, a contradiction. $\qquad \square$

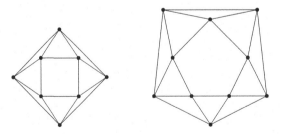

Fig. 3. The graphs X_5 (left) and X_6 (right).

With this, we can characterize the coupled choosability of wheel graphs.

Corollary 2. *For a wheel graph* W_n, *we have* $\chi_{vf}^L(W_n) = \min\{5, n\}$.

6   Towards Partial 3-Trees

Our investigation of wheel graphs was motivated by wanting to determine the coupled choosability number of planar partial 3-trees. To define these, we first define *Apollonian networks* recursively as follows. A triangle is an Apollonian network. If G is an Apollonian network, and f is a face of G (necessarily a triangle) that is not the outer-face, then the graph obtained by stellating face f is also an Apollonian network. Here *stellating* means the operation of inserting a new vertex v inside face f and making it adjacent to all vertices of f. A *planar partial 3-tree* is a graph that is a subgraph of an Apollonian network (see Fig. 4). (This definition is different, but equivalent, to the "standard" definition of partial 3-trees via treewidth or via chordal supergraphs with clique-size 4 [3].) We offer the following conjecture:

Conjecture 1. Every planar partial 3-tree is 6-coupled-choosable.

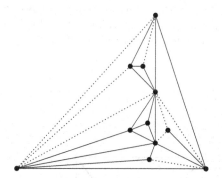

Fig. 4. A planar partial 3-tree. Dotted edges show the Apollonian network.

Note that the conjecture holds for Apollonian networks, since these are maximal planar graphs and these are known to be 6-coupled-choosable [20]. But this does not imply 6-coupled-choosability of subgraphs, and so the conjecture remains open.

Towards the conjecture, we studied several graph classes that are planar partial 3-trees (and generalize wheels). One such class of graphs is the class of *Halin graphs*, which are defined by starting with a tree T and adding a cycle between the leaves of T. See also the solid edges in Fig. 5. Wheel graphs are the special case of Halin graphs where T is a star graph. A second class of planar partial 3-trees are the *stellated outer-planar graphs*, obtained by starting with some outerplanar graph G, and stellating the outer-face. See also the dashed edges in Fig. 5. Wheel graphs are the special case of stellated outerplanar graphs where the outerplanar graph is a cycle.

One can easily see that Halin graphs are exactly the duals of stellated outerplanar graphs. Therefore, any list coupled coloring of a stellated outerplanar graph corresponds to a list coupled coloring of a Halin graph. Unfortunately, our upper bound for the coupled choosability of wheel graphs does not in general extend to Halin graphs.

Theorem 4. *There exists a stellated outerplanar graph (equivalently a Halin graph) that is not 5-coupled-colorable (in particular therefore it is not 5-coupled-choosable).*

Proof. The Halin-graph G is the triangular prism, see Fig. 5 where we also show the dual graph G^* and the 1-planar graph $X(G)$. The claim holds if we show that there is no 5-coloring of the vertices of $X(G)$.

Assume for contradiction that $X(G)$ had a 5-coloring; up to symmetry we may assume that the triangle formed by the three degree-4-faces of G is colored $1, 2, 3$. Let (t, t') be the edge that crosses the edge colored with 2 and 3. Vertices t, t' are colored with 1, 4 or 5; up to renaming of colors 4 and 5 hence one of them is colored 4.

Starting with this coloring, propagate restrictions on the possible colors to other vertices of $X(G)$ along the numerous copies of K_4 (note that all vertices other than t, t' are adjacent to the one colored 1). This leads to a triangle that has only two possible colors left, a contradiction.   □

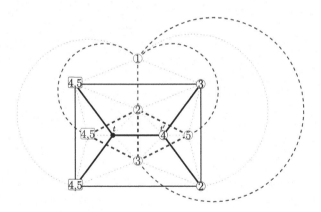

Fig. 5. A Halin-graph G (black solid; the tree is bold), and the dual graph G^* (blue dashed) which is a stellated outerplanar graph (the outerplanar graph is bold). Taking both, and adding the face-vertex incidences (red dotted) gives graph $X(G)$. We also show the only possible 5-coloring (up to symmetry) of $X(G)$, which leads to a contradiction since a triangle would have to be colored with 2 colors. (Color figure online)

In particular, this shows that we cannot replace '6' by '5' in Conjecture 1. We also remark that, in line with Observation 1, a supergraph of the triangular prism *is* 5-coupled-colorable. Namely, one can insert diagonals in the degree-4 faces and obtain the octahedron. An octahedron is 3-colorable because all faces are triangles and the vertex-degrees are even. The dual graph (which is the cube) is bipartite and hence 2-colorable. Therefore, using disjoint sets of colors for the primal and the dual graph, we get a 5-coupled-coloring of the octahedron.

Returning to wheel graphs, Theorem 4 shows that wheels are strictly better (as far as coupled choosability is concerned) than Halin-graphs. Now we study a second graph class that lies between the wheels and the planar partial 3-trees. These are the *IO-graphs*, which are the planar graphs that can be obtained by adding an independent set to the interior faces of an outerplanar graph (see Fig. 6). Certainly any subgraph of a wheel is an IO graph.

Conjecture 2. Every IO-graph is 5-coupled choosable.

We studied subgraphs of wheel graphs because they may be an important stepping stone towards Conjecture 2. In particular, consider some IO-graph G obtained from an outerplanar graph O and independent set I. Let G^+ be a maximal IO-graph containing G, i.e., add edges to G for as long as the result is simple and an IO-graph. Then G^+ is a tree of wheels, where each wheel consists of a vertex $x \in I$ with its neighbours, and the wheels have been glued together at edges. Correspondingly G is a tree of subgraphs of wheels. It may be possible to use Theorem 2 (enhanced with further restrictions on the coloring of some parts) to prove Conjecture 2 by building a coloring of G incrementally in this tree, but this remains future work.

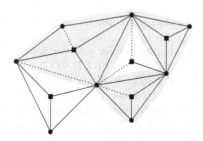

Fig. 6. An IO graph G consists of an outerplanar graph (circles) and an independent set (squares). Dotted edges are added to obtain G^+, and some of the wheels used to build G^+ are shaded.

We end with some other open questions surrounding list-colorability and list-coupled-colorability. Foremost, is every 1-planar graph 7-list-colorable? Borodin states this to be true [6], but quotes the paper by Wang and Lih [20] which only deals with 7-coupled-choosability. Hence all optimal 1-planar graphs are 7-list-colorable but to our knowledge the problem remains open for 1-planar graphs that are not subgraphs of optimal 1-planar graphs (e.g. any 1-planar graph that contains K_6 as a subgraph). Second, how easy is it to test whether a planar graph G is k-coupled choosable, or whether it is L-coupled-choosable?

The latter is easily shown to be NP-hard, and it can be solved in polynomial time if the treewidth t is bounded. (Specifically, the non-coupled version can be solved in $n^{O(t)}$ time in graphs of treewidth t [14].) One can also easily argue that for a planar graph G, the treewidth of $X(G)$ can be bounded by a constant times the treewidth of G—details are left to the reader. Hence, the same result also holds for coupled choosability.) But what is the dependency on the treewidth? It is known that L-choosability is $W[1]$-hard when parameterized by treewidth [10]. (This is in contrast to k-choosability, which surprisingly enough is fixed-parameter tractable in the treewidth [10].) But the reduction for $W[1]$-hardness does not construct planar graphs since it contains a large clique as a minor. Is L-coupled-choosability $W[1]$-hard with respect to the treewidth?

References

1. Archdeacon, D.: Coupled colorings of planar maps. In: Proceedings of the Fourteenth Southeastern Conference on Combinatorics, Graph Theory and Computing, Boca Raton, Florida, 1983, vol. 39, pp. 89–94 (1983)
2. Berman, K.A., Shank, H.: Full 4-colorings of 4-regular maps. J. Graph Theory **3**(3), 291–294 (1979). https://doi.org/10.1002/jgt.3190030312
3. Biedl, T., Velázquez, L.R.: Drawing planar 3-trees with given face areas. Comput. Geom.: Theory Appl. **46**(3), 276–285 (2013). https://doi.org/10.1016/j.comgeo.2012.09.004
4. Borodin, O.V.: Solution of the Ringel problem on vertex-face coloring of planar graphs and coloring of 1-planar graphs. Metody Diskret. Analiz. **41**(12–26), 108 (1984)
5. Borodin, O.V.: Coupled colorings of graphs on a plane. Metody Diskret. Analiz. **45**, 21–27 (1987)
6. Borodin, O.V.: Colorings of plane graphs: a survey. Discret. Math. **313**(4), 517–539 (2013). https://doi.org/10.1016/j.disc.2012.11.011
7. Diestel, R.: Graph Theory, Graduate Texts in Mathematics, vol. 173, 5th edn. Springer, Berlin (2018). https://doi.org/10.1007/978-3-662-53622-3
8. Doczkal, C., Pous, D.: Treewidth-two graphs as a free algebra. In: Potapov, I., Spirakis, P., Worrell, J. (eds.) 43rd International Symposium on Mathematical Foundations of Computer Science (MFCS 2018). Leibniz International Proceedings in Informatics (LIPIcs), vol. 117, pp. 60:1–60:15. Schloss Dagstuhl-Leibniz-Zentrum fuer Informatik, Dagstuhl, Germany (2018). https://doi.org/10.4230/LIPIcs.MFCS.2018.60, http://drops.dagstuhl.de/opus/volltexte/2018/9642
9. Erdős, P., Rubin, A.L., Taylor, H.: Choosability in graphs. In: Proceedings of the West Coast Conference on Combinatorics, Graph Theory and Computing (Humboldt State Univ., Arcata, Calif., 1979), pp. 125–157. Congress. Numer., XXVI, Utilitas Math., Winnipeg, Man. (1980)
10. Fellows, M.R., et al.: On the complexity of some colorful problems parameterized by treewidth. Inf. Comput. **209**(2), 143–153 (2011). https://doi.org/10.1016/j.ic.2010.11.026
11. Hetherington, T.J.: Coupled choosability of near-outerplane graphs. Ars Combin. **113**, 23–32 (2014)
12. Hopcroft, J.E., Tarjan, R.E.: Efficient algorithms for graph manipulation. Commun. ACM **16**(6), 372–378 (1973)

13. Hutchinson, J.P.: On list-coloring extendable outerplanar graphs. Ars Math. Contemp. **5**(1), 171–184 (2012). https://doi.org/10.26493/1855-3974.179.189

14. Jansen, K., Scheffler, P.: Generalized coloring for tree-like graphs. Discret. Appl. Math. **75**(2), 135–155 (1997). https://doi.org/10.1016/S0166-218X(96)00085-6

15. Ringel, G.: Ein Sechsfarbenproblem auf der Kugel. Abhandlungen aus dem Mathematischen Seminar der Universität Hamburg **29**, 107–117 (1965). https://doi.org/10.1007/BF02996313

16. Schumacher, H.: Zur Struktur 1-planarer Graphen. Mathematische Nachrichten **125**, 291–300 (1986)

17. Skulrattanakulchai, S.: Δ-list vertex coloring in linear time. Inf. Process. Lett. **98**(3), 101–106 (2006). https://doi.org/10.1016/j.ipl.2005.12.007

18. Valdes, J., Tarjan, R.E., Lawler, E.L.: The recognition of series parallel digraphs. SIAM J. Comput. **11**(2), 298–313 (1982). https://doi.org/10.1137/0211023

19. Wang, W., Lih, K.W.: List coloring Halin graphs. Ars Combin. **77**, 53–63 (2005)

20. Wang, W., Lih, K.W.: Coupled choosability of plane graphs. J. Graph Theory **58**(1), 27–44 (2008). https://doi.org/10.1002/jgt.20292

Conflict-Free Coloring: Graphs of Bounded Clique Width and Intersection Graphs

Sriram Bhyravarapu[1], Tim A. Hartmann[2],
Subrahmanyam Kalyanasundaram[1(✉)], and I. Vinod Reddy[3]

[1] Department of Computer Science and Engineering, IIT Hyderabad,
Sangareddy, India
{cs16resch11001,subruk}@iith.ac.in
[2] Department of Computer Science, RWTH Aachen, Aachen, Germany
hartmann@algo.rwth-aachen.de
[3] Department of Electrical Engineering and Computer Science,
IIT Bhilai, Bhilai, India
vinod@iitbhilai.ac.in

Abstract. Given an undirected graph, a conflict-free coloring (CFON*)
is an assignment of colors to a subset of the vertices of the graph such that
for every vertex there exists a color that is assigned to exactly one vertex
in its open neighborhood. The minimum number of colors required for
such a coloring is called the conflict-free chromatic number. The decision
version of the CFON* problem is NP-complete even on planar graphs.
In this paper, we show the following results.

- The CFON* problem is fixed-parameter tractable with respect to
  the combined parameters clique width and the solution size.
- We study the problem on block graphs and cographs, which have
  bounded clique width. For both graph classes, we give tight bounds
  of three and two respectively for the CFON* chromatic number.
- We study the problem on the following intersection graphs: interval graphs, unit square graphs and unit disk graphs. We give tight
  bounds of two and three for the CFON* chromatic number for
  proper interval graphs and interval graphs. Moreover, we give upper
  bounds or the CFON* chromatic number on unit square and unit
  disk graphs.
- We also study the problem on split graphs and Kneser graphs. For
  split graphs, we show that the problem is NP-complete. For Kneser
  graphs $K(n, k)$, when $n \geq k(k + 1)^2 + 1$, we show that the CFON*
  chromatic number is $k + 1$.

We also study the closed neighborhood variant of the problem denoted
by CFCN*, and obtain analogous results in some of the above cases.

1 Introduction

Given an undirected graph $G = (V, E)$, a *conflict-free coloring* is an assignment
of colors to a subset of the vertices of G such that every vertex in G has a uniquely

© Springer Nature Switzerland AG 2021
P. Flocchini and L. Moura (Eds.): IWOCA 2021, LNCS 12757, pp. 92–106, 2021.
https://doi.org/10.1007/978-3-030-79987-8_7

colored vertex in its neighborhood. The minimum number of colors required for such a coloring is called the *conflict-free chromatic number*. This problem was introduced in 2002 by Even, Lotker, Ron and Smorodinsky [8], motivated by the frequency assignment problem in cellular networks where base stations and clients communicate with one another. To avoid interference, we require that there exists a base station with a unique frequency in the neighborhood of each client. Since the number of frequencies are limited and expensive, it is ideal to minimize the number of frequencies used.

This problem has been well studied [1,5,11,16,18] for nearly 20 years. Several variants of the problem have been studied. We focus on the following variant of the problem with respect to both closed and open neighborhoods, which are defined as follows.

Definition 1 (Conflict-Free Coloring). *A CFON* coloring of a graph $G = (V, E)$ using k colors is an assignment $C : V(G) \rightarrow \{0\} \cup \{1, 2, \ldots, k\}$ such that for every $v \in V(G)$, there exists a color $i \in \{1, 2, \ldots, k\}$ such that $|N(v) \cap C^{-1}(i)| = 1$. The smallest number of colors required for a CFON* coloring of G is called the CFON* chromatic number of G, denoted by $\chi^*_{ON}(G)$.*

The closed neighborhood variant, CFCN* coloring, is obtained by replacing the open neighborhood $N(v)$ by the closed neighborhood $N[v]$ in the above. The corresponding chromatic number is denoted by $\chi^*_{CN}(G)$.

In the above definition, vertices assigned the color 0 are treated as "uncolored". Hence in a CFON* coloring (or CFCN* coloring), no vertex can have a vertex colored 0 as its uniquely colored neighbor. The *CFON* problem* (resp. *CFCN* problem*) is to compute the minimum number of colors required for a CFON* coloring (resp. CFCN* coloring) of a graph. Abel et al. in [1] showed that both the problems are NP-complete even for planar graphs. They also showed that eight colors are sufficient to CFON* color planar graphs, which was improved to four colors [12]. Further these problems have been studied on outerplanar graphs [4], and intersection graphs like string graphs, circle graphs [13], disk graphs, square graphs and interval graphs [9]. Continuing this line of work, we study these problems on various restricted graph classes such as block graphs, cographs, intervals graphs, unit square graphs, unit disk graphs, Kneser graphs and split graphs.

The parameterized complexity of conflict-free coloring, for both neighborhoods, has been of recent research interest. They are fixed-parameter tractable (FPT) when parameterized by tree width [2,5], distance to cluster (distance to disjoint union of cliques) [17] and neighborhood diversity [11]. Further, with respect to distance to threshold graphs there is an additive approximation algorithm in FPT-time [17].[1]

We study CFON* and CFCN* problems for the parameter clique width, which generalizes all the above parameters. Specifically, for every graph G,

---

[1] Some of the above FPT results are shown for the "full-coloring variant" of the problem (as defined in Definition 2). Our clique width result can also be adapted for the full-coloring variant.

$cw(G) \leq 3 \cdot 2^{tw(G)-1}$, where $tw(G)$ and $cw(G)$ denote the tree width of G and the clique width of G respectively [7]. Graphs with distance to cluster at most $k \in \mathbb{N}$, have clique width of at most $O(2^k)$ [19]. We show that the CFON* and CFCN* problems are FPT with respect to the combined parameters clique width and the number of colors used. Note that the previously mentioned FPT-results [2,5,11,17] do not additionally need the solution size as a parameter.

1.1 Results

- In Sect. 3, we show fixed-parameter tractable algorithms for both CFON* CFCN* problems with respect to the combined parameters clique width w and the solution size k, that runs in $2^{O(w3^k)}n^{O(1)}$ time where n is the number of vertices of G.
- In Sect. 4, we discuss the results on block graphs and cographs. Both the graph classes are solvable in polynomial time, which follows from the clique width result.
  - For block graphs G, we show that $\chi^*_{ON}(G) \leq 3$. We show a block graph G that requires three colors making the above bound tight.
  - For cographs, we show that two colors are sufficient for a CFON* coloring. We also characterize cographs for which one color suffices.
- In Sect. 5, we show that for interval graphs G, $\chi^*_{ON}(G) \leq 3$. We show an interval graph that requires three colors making the above bound tight. Moreover, two colors are sufficient to CFON* color proper interval graphs.
  We also show that the CFCN* problem is polynomial time solvable on interval graphs.
- In Sect. 6, we study the problem on geometric intersection graphs like unit square graphs and unit disk graphs.
  We show that $\chi^*_{ON}(G) \leq 27$ for unit square graphs G. For unit disk graphs G, we show that $\chi^*_{ON}(G) \leq 51$. No upper bound was previously known.
- In Sect. 7, we study both the problems on Kneser graphs and split graphs.
  - We show that $k+1$ colors are sufficient to CFON* color the Kneser graphs $K(n,k)$, when $n \geq 3k-1$. We also show that $\chi^*_{ON}(K(n,k)) \geq k+1$ when $n \geq k(k+1)^2 + 1$, thereby proving that $\chi^*_{ON}(K(n,k)) = k+1$ when $n \geq k(k+1)^2 + 1$.
    We also show that k colors are sufficient to CFCN* color a Kneser graph $K(n,k)$, when $n \geq 2k+1$.
  - On split graphs, we show that the CFON* problem is NP-complete and the CFCN* problem is polynomial time solvable.

2 Preliminaries

Throughout the paper, we assume that the graph G is connected. Otherwise, we apply the algorithm on each component independently. We also assume that G does not contain any isolated vertices as the CFON* problem is not defined for an isolated vertex. We use $[k]$ to denote the set $\{1, 2, \ldots, k\}$ and $C : V(G) \rightarrow \{0\} \cup [k]$ to denote the color assigned to a vertex. A *universal vertex* is a vertex that is adjacent to all other vertices of the graph. In some of our algorithms and proofs, it is convenient to distinguish between vertices that are intentionally left uncolored, and the vertices that are yet to be assigned any color. The assignment of color 0 is used to denote that a vertex is left "uncolored".

To avoid clutter and to simplify notation, we use the shorthand notation vw to denote the edge $\{v, w\}$. The open neighborhood of a vertex $v \in V(G)$ is the set of vertices $\{w : vw \in E(G)\}$ and is denoted by $N(v)$. Given a conflict-free coloring C, a vertex $w \in N(v)$ is called a *uniquely colored neighbor* of v if $C(w) \neq 0$ and $\forall x \in N(v) \setminus \{w\}, C(w) \neq C(x)$. The closed neighborhood of v is the set $N(v) \cup \{v\}$, denoted by $N[v]$. The notion of uniquely colored neighbor in the closed neighborhood variant is analogous to the open neighborhood variant, and is obtained by replacing $N(v)$ by $N[v]$. We sometimes use the mapping $h : V \to V$ to denote the uniquely colored neighbor of a vertex. We also extend C for vertex sets by defining $C(V') = \bigcup_{v \in V'} C(v)$ for $V' \subseteq V(G)$. To refer to the multi-set of colors used in V', we use $C_{\{\}\}}(V')$. The difference between $C_{\{\}\}}(V')$ and $C(V')$ is that we use multiset union in the former.

In many of the sections, we also refer to the full coloring variant of the conflict-free coloring problem, which is defined below.

Definition 2 (Conflict-Free Coloring – Full Coloring Variant). *A CFON coloring of a graph $G = (V, E)$ using k colors is an assignment $C : V(G) \to \{1, 2, \ldots, k\}$ such that for every $v \in V(G)$, there exists an $i \in \{1, 2, \ldots, k\}$ such that $|N(v) \cap C^{-1}(i)| = 1$. The smallest number of colors required for a CFON coloring of G is called the CFON chromatic number of G, denoted by $\chi_{ON}(G)$.*

The corresponding closed neighborhood variant is denoted CFCN coloring, and the chromatic number is denoted $\chi_{CN}(G)$.

A full conflict-free coloring, where all the vertices are colored with a non-zero color, is also a partial conflict-free coloring (as defined in Definition 1) while the converse is not true. It is clear that one extra color suffices to obtain a full coloring variant from a partial coloring variant. However, it is not always clear if the extra color is actually necessary.

For the theorems marked $(\star)$, the full proofs are omitted due to space constraints.

3   FPT with Clique Width and Number of Colors

In this section, we study the conflict-free coloring problem with respect to the combined parameters clique width $cw(G)$ and number of colors k. We present FPT algorithms for both the CFON* and CFCN* problems.

Definition 3 (Clique width [7]). *Let $w \in \mathbb{N}$. A w-expression Φ defines a graph G_Φ where each vertex receives a label from $[w]$, using the following four recursive operations with indices $i, j \in [w]$, $i \neq j$:*

1. *Introduce, $\Phi = v(i)$: G_Φ is a graph consisting a single vertex v with label i.*
2. *Disjoint union, $\Phi = \Phi' \oplus \Phi''$: G_Φ is a disjoint union of $G_{\Phi'}$ and $G_{\Phi''}$.*

3. *Relabel, $\Phi = \rho_{i \to j}(\Phi')$: G_Φ is the graph $G_{\Phi'}$ where each vertex labeled i in $G_{\Phi'}$ now has label j.*
4. *Join, $\Phi = \eta_{i,j}(\Phi')$: G_Φ is the graph $G_{\Phi'}$ with additional edges between each pair of vertices u of label i and v of label j.*

The clique width *of a graph G denoted by* cw(G) *is the minimum number w such that there is a w-expression Φ that defines G.*

In the following, we assume that a w-expression Ψ of G is given. There is an FPT-algorithm that, given a graph G and integer w, either reports that $\text{cw}(G) > w$ or outputs a $(2^{3w+2} - 1)$-expression of G [15].

A w-expression Ψ is an *irredundant w-expression of G*, if no edge is introduced twice in Ψ. Given a w-expression of G, it is possible to get an irredundant w-expression of G in polynomial time [7]. For a coloring of G, a vertex v is said to be *conflict-free dominated* by the color c, if exactly one vertex in $N(v)$ is assigned the color c. In general, a vertex v is said to be conflict-free dominated by a set of colors S, if each color in S conflict-free dominates v. Also, a vertex v is said to *miss the color c* if there exists no vertex in $N(v)$ that is assigned the color c. In general, a vertex v is said to miss a set of colors T, if every color in T is missed by v.

Now, we prove the main theorem of this section.

Theorem 4. *Given a graph G, a w-expression of G and an integer k, it is possible to decide if $\chi^*_{ON}(G) \leq k$ in $2^{O(w3^k)}n^{O(1)}$ time.*

Proof. We give a dynamic program that works bottom-up over a given irredundant w-expression Ψ of G. For each subexpression Φ of Ψ and a coloring $C : V(G_\Phi) \to \{0, 1, \dots, k\}$ of G_Φ, we have a boolean table entry $d[\Phi; N; M]$ with

$$N = n_{1,0}, \dots, n_{1,k}, \dots, n_{w,0}, \dots, n_{w,k}, \text{ and}$$

$$M = M_1, \dots, M_w \quad \text{where for every } a \in [w], \quad M_a = m_{a,S_1,T_1}, \dots, m_{a,S_{3^k},T_{3^k}}$$

where S_ℓ, T_ℓ are all the possible disjoint subsets of the set of colors $[k]$. Note that there are 3^k many disjoint subsets $S_\ell, T_\ell \in [k]$.

Given some vertex-coloring of G_Φ, values of M and N have the following meaning.

N: For each label $a \in [w]$ and color $q \in \{0\} \cup [k]$, the variable $n_{a,q} \in \{0, 1, 2\}$. Let $n^*_{a,q}$ be the number of vertices with label a that are colored q. Then $n_{a,q}$ is equal to $n^*_{a,q}$ when limited to a maximum of two, in other words $n_{a,q} = \min\{2, n^*_{a,q}\}$.

M: For each label $a \in [w]$, and disjoint sets $S, T \subseteq [k]$, the variable $m_{a,S,T} \in \{0, 1\}$. The variable $m_{a,S,T}$ is equal to 1 if there is at least one vertex v with label a which is conflict-free dominated by exactly colors S and the set of colors that misses v is exactly T. If there is no such vertex, then $m_{a,S,T}$ is equal to 0.

For each subexpression Φ of Ψ, the boolean entry $d[\Phi; N; M]$ is set to TRUE if and only if there exists a vertex-coloring $C : V(G_\Phi) \to \{0\} \cup [k]$ that satisfies the variables $n_{a,q}$ and $m_{a,S,T}$, for each label $a \in [w]$, color $q \in \{0\} \cup [k]$ and disjoint subsets $S, T \subseteq [k]$. To decide if k colors are sufficient to CFON* color G, we consider the expression Ψ with $G_\Psi = G$. We answer 'yes' if and only if there exists an entry $d[\Psi; N; M]$ set to TRUE where $m_{a,\{\},T} = 0$ for each $a \in [w]$ and for each $T \subseteq [k]$. This means there exists a coloring such that there is no label $a \in [w]$ with a vertex v that is not conflict-free dominated.

Now, we show how to compute $d[\Phi; N; M]$ at each operation.

1. $\Phi = v(i)$.

   The graph G_Φ represents a node with one vertex v that is labelled $i \in [w]$. For each color $q \in \{0\} \cup [k]$, we set the entry $d[\Phi; N; M] =$ TRUE if and only if $n_{i,q} = 1$, $m_{i,\{\},[k]} = 1$ and all other entries of N and M are 0.

2. $\Phi = \Phi' \oplus \Phi''$.

   The graph G_Φ results from the disjoint union of graphs $G_{\Phi'}$ and $G_{\Phi''}$.

   We set $d[\Phi; N; M] =$ TRUE if and only if there exist entries $d[\Phi'; N'; M']$ and $d[\Phi''; N''; M'']$ such that $d[\Phi'; N'; M'] =$ TRUE, $d[\Phi''; N''; M''] =$ TRUE and the following conditions are satisfied:

   (a) For each label $a \in [w]$ and color $q \in \{0\} \cup [k]$, $n_{a,q} = \min\{2, n'_{a,q} + n''_{a,q}\}$.

   (b) For each label $a \in [w]$ and disjoint $S, T \subseteq [k]$, $m_{a,S,T} = \min\{1, m'_{a,S,T} + m''_{a,S,T}\}$.

   We may determine each table entry of $d[\Phi; N, M]$ for every N, M as follows. We initially set $d[\Phi; N, M]$ to FALSE for all N, M. We iterate over all combinations of table entries $d[\Phi'; N'; M']$ and $d[\Phi''; N''; M'']$. For each combination of TRUE entries $d[\Phi'; N'; M']$ and $d[\Phi''; N''; M'']$, we update the corresponding entry $d[\Phi; N; M]$ to TRUE. The corresponding entry $d[\Phi; N; M]$ has variables $n_{a,q}$ which is the sum of $n'_{a,q}$ and $n''_{a,q}$ limited by two, and variables $m_{a,S,T}$ which is the sum of $m'_{a,S,T}$ and $m''_{a,S,T}$ limited by one. Thus, to compute every entry for $d[\Phi; ;]$ we visit at most $(3^{w(k+1)}2^{w3^k})^2$ combinations of table entries and for each of those compute $w(k+1) + w3^k$ values for M and N.

3. $\Phi = \rho_{i \to j}(\Phi')$.

   The graph G_Φ is obtained from the graph $G_{\Phi'}$ by relabelling the vertices of label i in $G_{\Phi'}$ with label j where $i, j \in [w]$. Hence, $n_{i,q} = 0$ for each $q \in \{0\} \cup [k]$ and $m_{i,S,T} = 0$ for each disjoint $S, T \subseteq [k]$.

   We set $d[\Phi; N; M] =$ TRUE if and only if there exists an entry $d[\Phi'; N'; M']$ such that $d[\Phi'; N'; M'] =$ TRUE in $G_{\Phi'}$ that satisfies the following conditions:

   (a) For each color $q \in \{0\} \cup [k]$, each label $a \in [w] \setminus \{i, j\}$ and disjoint $S, T \subseteq [k]$, $n_{a,q} = n'_{a,q}$ and $m_{a,S,T} = m'_{a,S,T}$.

   (b) For each color $q \in \{0\} \cup [k]$, $n_{j,q} = \min\{2, n'_{i,q} + n'_{j,q}\}$ and $n_{i,q} = 0$.

   (c) For each disjoint $S, T \subseteq [k]$, $m_{j,S,T} = \min\{1, m'_{i,S,T} + m'_{j,S,T}\}$ and $m_{i,S,T} = 0$.

   We may determine each table entry of $d[\Phi; N; M]$ for every N, M as follows. We initially set $d[\Phi; N; M]$ to FALSE for all N, M. We iterate over all the TRUE table entries $d[\Phi'; N'; M']$, and for each such entry we update the corresponding entry $d[\Phi; N; M]$ to TRUE, if applicable. To compute every entry for $d[\Phi; ;]$ we visit at most $3^{w(k+1)}2^{w3^k}$ table entries $d[\Phi'; ;]$ and for each of those compute $w(k+1) + w3^k$ values for M and N.

4. $\Phi = \eta_{i,j}(\Phi')$.

   The graph G_Φ is obtained from the graph $G_{\Phi'}$ by connecting each vertex with label i with each vertex with label j where $i, j \in [w]$. Consider a vertex v labelled i in $G_{\Phi'}$ and let v contribute to the variable $m'_{i,\widehat{S},\widehat{T}}$, which is v is conflict-free dominated by exactly $\widehat{S}$ and the set of colors that misses v is exactly $\widehat{T}$. After this operation, the vertex v may contribute to the variable $m_{i,S,T}$ in G_Φ where the choice of the set S in G_Φ depends on the colors assigned to the vertices labelled j in $G_{\Phi'}$.

   We set $d[\Phi; N; M] = \text{TRUE}$ if and only if there exists an entry $d[\Phi'; N'; M']$ such that $d[\Phi'; N'; M'] = \text{TRUE}$ in $G_{\Phi'}$ that satisfies the following conditions:

   (a) For each label $a \in [w]$ and color $q \in \{0\} \cup [k]$, $n_{a,q} = n'_{a,q}$.
   (b) For each label $a \in [w] \setminus \{i, j\}$ and disjoint $S, T \subseteq [k]$, $m_{a,S,T} = m'_{a,S,T}$.
   (c) For the label i and disjoint $S, T \subseteq [k]$, $m_{i,S,T} = 1$ if and only if there are disjoint subsets $\widehat{S}, \widehat{T} \subseteq [k]$ with $m'_{i,\widehat{S},\widehat{T}} = 1$ such that
       i. For each color $q \in S \cap \widehat{S}$, variable $n'_{j,q} = 0$.
       ii. For each color $q \in S \setminus \widehat{S}$, variable $n'_{j,q} = 1$.
       iii. For each color $q \in \widehat{S} \setminus S$, variable $n'_{j,q} \geq 1$.
       iv. $S \setminus \widehat{S} \subseteq \widehat{T}$ and $T \subseteq \widehat{T}$.
       v. For each color $q \in \widehat{T} \setminus (T \cup S)$, $n'_{j,q} = 2$.
   (d) For the label j, entry $m_{j,S,T}$ is computed in a symmetric fashion by swapping the labels i and j in (c).

   It can be observed that each TRUE table entry $d[\Phi'; N'; M']$ sets exactly one entry $d[\Phi; N; M]$ to TRUE. We can determine each table entry of $d[\Phi; N; M]$ as follows. We initially set $d[\Phi; N, M]$ to FALSE for all N, M. We iterate over all the TRUE table entries $d[\Phi'; N'; M']$, and for each such entry we update the corresponding entry $d[\Phi; N; M]$ to TRUE, if applicable. To compute every entry for $d[\Phi; ;]$ we visit at most $3^{w(k+1)} 2^{w3^k}$ table entries $d[\Phi'; ;]$ and for each of those compute $w(k+1) + w3^k$ values for M and N.

   We described the recursive formula at each operation, that computes the value of each entry $d[; ;]$. The correctness of the algorithm easily follows from the description of the algorithm. The DP table consists of $3^{w(k+1)} 2^{w3^k}$ entries at each node of the w-expression. The running time is dominated by the operations at the disjoint union node that requires $O(3^{2w(k+1)} 2^{2w3^k} w(k + 1 + 3^k) n^{O(1)})$ time. $\square$

We described similarly, we obtain the following result for the CFCN* problem:

Theorem 5 ($\star$). *Given a graph G, a w-expression and an integer k, it is possible to decide if $\chi^*_{CN}(G) \leq k$ in $2^{O(w3^k)} n^{O(1)}$ time.*

By modifying the above algorithm, it is possible to obtain FPT algorithms for the full coloring variants (CFON and CFCN) of the problem. We merely have to restrict the entries of the dynamic program to entries without color 0.

Theorem 6. *The CFON and the CFCN problems are FPT when parameterized by the combined parameters clique width and the solution size.*

4 Block Graphs and Cographs

In this section, we study the problems on block graphs and cographs. Note that block graphs have clique width at most 3, and cographs have clique width at most 2. Hence, CFON* and CFCN* problems are polynomial time solvable on block graphs and cographs by Theorems 4 and 5 respectively. However, we present direct proofs for these problems on these graph classes. In particular we show that $\chi^*_{ON}(G) \leq 3$ and $\chi^*_{CN}(G) \leq 2$, for block graphs G. We show a block graph G such that $\chi^*_{ON}(G) = 3$, making the above bound tight. Next, we show that $\chi^*_{ON}(G), \chi^*_{CN}(G) \leq 2$, for cographs G.

Definition 7 (Block Graph). *A block graph is a graph in which every 2-connected component is a clique.*

For the CFON* problem, we give a tight upper bound of 3, in the following sense: we present a graph (see Fig. 1) that is not CFON*-colorable with colors $\{0, 1, 2\}$, where 0 is the dummy-color. Complementing this result, we show that there is an algorithm that colors a given block-graph with colors $\{1, 2, 3\}$, thus without the need of a dummy-color 0.

Lemma 8 ($\star$). *If G is a block graph, $\chi_{ON}(G) \leq 3$, hence $\chi^*_{ON}(G) \leq 3$.*

Proof (Proof Sketch). We give a constructive algorithm that given a block graph G outputs a CFON-coloring C using at most three colors $1, 2, 3$. For convenience, let us also specify a mapping h that maps each vertex $v \in G$ to one of its uniquely colored neighbors $w \in N(v)$. We use the fact that block-graphs are exactly the diamond-free chordal graphs (a diamond is a K_4 with one edge removed) [3]. As usual, we assume that G is connected and contains at least one edge uv. Color $C(u) = 1$ and $C(v) = 2$. Color every vertex $w \in (N(u) \cup N(v)) \setminus \{u, v\}$ with $C(w) = 3$. Assign $h(w) = v$ for every $w \in N(v)$, and assign $h(w) = u$ for every $w \in N(u) \setminus N(v)$.

Let G_v contain every connected component of $G \setminus \{u, v\}$ that contains a vertex from $N(v)$. Similarly, let G_u contain every connected component of $G \setminus \{u, v\}$ that contains a vertex from $N(u) \setminus N(v)$.

Claim ($\star$). The sets $V(G_u)$ and $V(G_v)$ are disjoint.

We color every vertex $x \in V(G_v)$ in distance $2, 3, 4, 5, 6, 7, \ldots$ from v in graph G_v with colors $1, 2, 3, 1, 2, 3, \ldots$ periodically. We assign $h(x)$ for $x \in V(G_v)$ in distance $i \geq 2$ to v to an arbitrary neighbor $y \in N(x)$ that has distance $i - 1$ to v in graph G_v. Similarly we color every vertex $x \in V(G_u)$ in distance $2, 3, 4, 5, 6, 7, \ldots$ from u in G_u with colors $2, 1, 3, 2, 1, 3, \ldots$ periodically. Again, let $h(x)$ for $x \in V(G_u)$ in distance $i \geq 2$ to u map to an arbitrary neighbor $y \in N(x)$ in distance $i - 1$ to u in graph G_u. $\square$

Lemma 9. *There is block graph G with $\chi^*_{ON}(G) > 2$.*

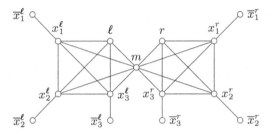

Fig. 1. A block graph G with $\chi^*_{ON}(G) > 2$.

Proof. Let G have vertex set $\{\ell, m, r\} \cup \bigcup_{i \in \{1,2,3\}} \{x^\ell_i, \overline{x}^\ell_i, x^r_i, \overline{x}^r_i\}$, see also Fig. 1. Let the edge set be defined by the set of maximal cliques $\{x^s_1, x^s_2, x^s_3, s, m\}$ and $\{x^i_s, \overline{x}^i_s\}$ for every $s \in \{\ell, r\}$ and $i \in \{1, 2, 3\}$. It is easy to see that G is a block graph. To prove that $\chi^*_{ON}(G) > 2$, assume, for the sake of contradiction, that there is χ^*_{ON}-coloring $C : V \rightarrow \{0, 1, 2\}$. Then there is a mapping h on V that assigns each vertex $v \in V(G)$ its uniquely colored neighbor $w \in N(v)$. Note that x^s_i, for $s \in \{\ell, r\}$ and $i \in \{1, 2, 3\}$, has to be colored 1 or 2, since it is the only neighbor of $\overline{x}^s_i$. Further, we may assume that $h(m) \in \{\ell, x^\ell_1\}$ and $C(h(m)) = 2$ because of symmetry.

First consider that $h(m) = \ell$ and $C(\ell) = 2$. Then $C(x^s_i) = 1$ for every $s \in \{\ell, r\}$ and $i \in \{1, 2, 3\}$. It follows that $h(\ell) = m$ and hence $C(m) = 2$. Then however $C_{\{\}}(N(x^\ell_1)) \supseteq \{\!\!\{1, 1, 2, 2\}\!\!\}$, a contradiction.

Thus it remains to consider that $h(m) = x^\ell_1$ and $C(x^\ell_1) = 2$. Then $C(x^s_i) = 1$ for every x^s_i with $(s, i) \in \{\ell, r\} \times [3] \setminus (\ell, 1)$. It follows that $h(r) = m$ and hence $C(m) = 2$. Then however $C_{\{\}}(N(\ell)) = \{\!\!\{1, 1, 2, 2\}\!\!\}$, also a contradiction.

Since both cases lead to a contradiction, it must be that $\chi^*_{ON}(G) > 2$.   □

Since a block graph G have clique width at most 3, and since $\chi^*_{ON}(G) \leq 3$, we may use Theorem 4 to decide the CFON* problem for block graphs in polynomial time.

Corollary 10. *For block graphs, CFON* is polynomial time solvable.*

By observing that the number of colors required is constant, we have the following analogous result on the CFCN* problem. However, we also present a direct proof using a characterization of block graphs G with $\chi^*_{CN}(G) = 1$.

Theorem 11 ($\star$). *If G is a block graph, then $\chi^*_{CN}(G) \leq 2$. The CFCN* problem is polynomial time solvable on block graphs.*

We now consider the problem on cographs, and obtain Theorem 13, the proof of which is omitted.

Definition 12 (Cograph [6]). *A graph G is a cograph if G consists of a single vertex, or if it can be constructed from a single vertex graph using the disjoint union and complement operations.*

Theorem 13 ($\star$). *The CFON* and the CFCN* problems are polynomial time solvable on cographs.*

Since constant bounds for the partial coloring variants imply constant bounds for the full coloring variants and since block graphs and cographs have clique width at most 3, we have the following.

Theorem 14. *The CFON and the CFCN problems are polynomial time solvable on block graphs and cographs.*

5   Interval Graphs

In this section, we show three colors are sufficient and sometimes necessary to CFON* color an interval graph. For *proper interval graphs*, we show that two colors are sufficient. We also show that the CFCN* problem is polynomial time solvable on interval graphs.

Definition 15 (Interval Graph). *A graph $G = (V, E)$ is an* interval graph *if there exists a set $\mathcal{I}$ of intervals on the real line such that there is a bijection $f : V \to \mathcal{I}$ satisfying the following: $\{v_1, v_2\} \in E$ if and only if $f(v_1) \cap f(v_2) \neq \emptyset$.*

For an interval graph G, we refer to the set of intervals $\mathcal{I}$ as the *interval representation* of G. An interval graph G is a *proper interval graph* if it has an interval representation $\mathcal{I}$ such that no interval in $\mathcal{I}$ is properly contained in any other interval of $\mathcal{I}$. An interval graph G is a *unit interval graph* if it has an interval representation $\mathcal{I}$ where all the intervals are of unit length. It is known that the class of proper interval graphs and unit interval graphs are the same [10].

Lemma 16 ($\star$). *If G is an interval graph, then $\chi_{ON}^*(G) \leq 3$.*

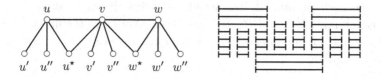

Fig. 2. On the left hand side, we have the graph G', and on the right hand side we have an interval graph representation of G, a graph where $\chi_{ON}(G) > 3$. The graph G is obtained by replacing each vertex $u, v, w, u^\star, v^\star$ of G' with a 3-clique and replacing $u', u''.v', v'', w', w''$ by a 4-clique.

The bound of $\chi_{ON}^*(G) \leq 3$ for interval graphs is tight. In particular, there is an interval graph G (see Fig. 2) that cannot be colored with three colors when excluding the dummy-color 0. That shows the stronger result $\chi_{ON}(G) > 3$, which implies that $\chi_{ON}^*(G) > 2$.

Lemma 17 ($\star$). *There is an interval graph G such that $\chi_{ON}(G) > 3$ (and thus $\chi_{ON}^*(G) \geq 3$).*

Lemma 18. *If G is a proper interval graph, then $\chi_{ON}^*(G) \leq 2$.*

Proof. Let $\mathcal{I}$ be a unit interval representation of G. We denote the left endpoint of an interval I by $L(I)$. We assign $C : \mathcal{I} \rightarrow \{1, 2, 0\}$ which will be a CFON* coloring.

At each iteration i, we pick two intervals $I_1^i, I_2^i \in \mathcal{I}$. The interval I_1^i is the interval whose $L(I_1^i)$ is the least among intervals for which C has not been assigned. The interval I_2^i is a neighbor of I_1^i, whose $L(I_2^i)$ is the greatest. It might be the case that C has been already assigned for all neighbors of I_1^i. This can happen only in the very last iteration of the algorithm. Depending on this, we have the following two cases.

- **Case 1: I_1^i has neighbors for which C is unassigned.**
  We assign $C(I_1^i) = 1$ and $C(I_2^i) = 2$. All other intervals adjacent to I_1^i and I_2^i are assigned the color 0.
  Now, we argue that C is a CFON* coloring. The intervals I_1^i and I_2^i act as the uniquely colored neighbors for each other. All intervals that are assigned 0 are adjacent to either I_1^i or I_2^i, and thus will have a uniquely colored neighbor. Notice that for every iteration i, the vertices I_1^i (or I_2^i) and I_1^{i+1} (or I_2^{i+1}) will have the same color. This is fine as there is no interval that intersects both I_1^i and I_1^{i+1}.
- **Case 2: C is already assigned for all the neighbors of I_1^i.**
  As mentioned before, this can happen only during the last iteration $i = j$. In this case, I_1^j is the only interval for which C is yet to be assigned. Choose an interval $I_m \in N(I_2^{j-1}) \cap N(I_1^j)$. Such an I_m exists, else $\mathcal{I}$ is disconnected. We reassign $C(I_1^{j-1}) = 0$, $C(I_2^{j-1}) = 1$, $C(I_m) = 2$ and assign $C(I_1^j) = 0$.
  The assignment of colors in iterations $1 \leq i \leq j - 2$ are unchanged. Though $C(I_1^{j-1})$ is changed to 0, this does not affect any interval, since there are no intervals which depend only on I_1^{j-1} for their uniquely colored neighbor. If there was such an interval, this would contradict the choice of I_1^{j-1}.
  For the intervals I_2^{j-1} and I_1^j, we have the interval I_m as the uniquely colored neighbor and for the interval I_m, we have the interval I_2^{j-1} as the uniquely colored neighbor. $\qquad\square$

It is known [9] that 2 colors suffice to CFCN* color an interval graph. We show that the CFCN* problem is polynomial time solvable on interval graphs using a characterization.

Theorem 19 ($\star$). *CFCN* problem is polynomial time solvable on interval graphs.*

6   Unit Square and Unit Disk Intersection Graphs

Unit square (respectively, unit disk) intersection graphs are intersection graphs of unit sized squares (resp., disks) in the Euclidean plane. It is shown in [9] that $\chi_{CN}^*(G) \leq 4$ for a unit square intersection graph G. They also showed that $\chi_{CN}^*(G) \leq 6$ for a unit disk intersection graph G. We study the CFON* problem on these graphs and get the following constant upper bounds. To the best of our knowledge, no upper bound was previously known on unit square and unit disk graphs for CFON* coloring. Due to space constraints, the proofs of the following theorems are omitted.

Theorem 20 (⋆). *If G is a unit square intersection graph, then $\chi_{ON}^*(G) \leq 27$.*

Theorem 21 (⋆). *If G is a unit disk intersection graph, then $\chi_{ON}^*(G) \leq 51$.*

7   Kneser Graphs and Split Graphs

In this section, we study the CFON* and the CFCN* colorings of Kneser graphs and split graphs. For Kneser graphs $K(n, k)$, we show that $\chi_{ON}^*(K(n,k)) = k+1$ when $n \geq k(k+1)^2+1$ and show bounds for $\chi_{CN}^*(K(n,k))$. For split graphs, we show that CFON* problem is NP-complete and CFCN* problem is polynomial time solvable.

Definition 22 (Kneser graph). *The Kneser graph $K(n,k)$ is the graph whose vertices are $\binom{[n]}{k}$, the k-sized subsets of $[n]$, and the vertices x and y are adjacent if and only if $x \cap y = \emptyset$ (when x and y are viewed as sets).*

Theorem 23 (⋆). $\chi_{ON}^*(K(n,k)) \leq k+1$, *for $n \geq 3k-1$. Further when $n \geq k(k+1)^2+1$, $\chi_{ON}^*(K(n,k)) = k+1$.*

It is easy to see that a proper coloring of a graph G is also a CFCN* coloring. Since $\chi(K(n,k)) \leq n-2k+2$ [14], we have that $\chi_{CN}^*(K(n,k)) \leq n-2k+2$. We show the following:

Theorem 24 (⋆). $\chi_{CN}^*(K(n,k)) \leq n-2k+1$, *for $2k+1 \leq n \leq 3k-1$. For the case when $n \geq 3k$, we have $\chi_{CN}^*(K(n,k)) \leq k$.*

Definition 25 (Split Graph). *A graph $G = (V,E)$ is a split graph if there exists a partition of $V = K \cup I$ such that the graph induced by K is a clique and the graph induced by I is an independent set.*

Theorem 26 (⋆). *The CFON* problem is NP-complete on split graphs.*

Theorem 27. *The CFCN* problem is polynomial time solvable on split graphs.*

The proof of Theorem 27 is through a characterization. We first show that for split graphs G, $\chi_{CN}^*(G) \leq 2$. Then we characterize split graphs G for which $\chi_{CN}^*(G) = 1$ thereby proving Theorem 27.

Lemma 28. *If $G = (V,E)$ is a split graph, then $\chi_{CN}^*(G) \leq 2$.*

Proof. Let $V = K \cup I$ be a partition of vertices into a clique K and an independent set I. We use $C : V \rightarrow \{1, 2, 0\}$ to assign colors to the vertices of V. Choose an arbitrary vertex $u \in K$ and assign $C(u) = 2$. The remaining vertices (if any) in $K \setminus \{u\}$ are assigned the color 0. For every vertex $v \in I$, we assign $C(v) = 1$. Each vertex in I will have itself as the uniquely colored neighbor and every vertex in K will have the vertex u as the uniquely colored neighbor.   □

We now characterize split graphs that are CFCN* colorable using one color.

Lemma 29. *Let $G = (V, E)$ be a split graph with $V = K \cup I$, where K and I are the clique and independent sets respectively. We have $\chi_{CN}^*(G) = 1$ if and only if at least one of the following is true: (i) G has a universal vertex, or (ii) $\forall v \in K, |N(v) \cap I| = 1$.*

Proof. We first prove the reverse direction. If there exists a universal vertex $u \in V$, then we assign the color 1 to u and assign the color 0 to all vertices in $V \setminus \{u\}$. This is a CFCN* coloring.

Suppose[2] $\forall v \in K, |N(v) \cap I| = 1$. We assign the color 1 to each vertex in I and color 0 to the vertices in K. Each vertex in I acts as the uniquely colored neighbor for itself and for its neighbor(s) in K.

For the forward direction, let $C : V \rightarrow \{1, 0\}$ be a CFCN* coloring of G. We further assume that $\exists y \in K, |N(y) \cap I| \neq 1$ and show that there exists a universal vertex. We assume that $|K| \geq 2$ and $|I| \geq 1$ (if either assumption is violated, G has a universal vertex). We first prove the following claim.

Claim. Exactly one vertex in K is assigned the color 1.

Proof. Suppose not. Let two vertices $v, v' \in K$ be such that $C(v) = C(v') = 1$. Then none of the vertices in K have a uniquely colored neighbor.

Suppose if all vertices in K are assigned the color 0. For vertices in I to have a uniquely colored neighbor, each vertex in I has to be assigned the color 1. By assumption, $\exists y \in K$ such that $|N(y) \cap I| \neq 1$. This means that y does not have a uniquely colored neighbor.   □

Now we show that there is a universal vertex in K.

By the above claim, there is a unique vertex $v \in K$ such that $C(v) = 1$. We will show that v is a universal vertex. Suppose not. Let $w' \notin N(v) \cap I$. For w' to have a uniquely colored neighbor, either w' or one of its neighbors in K has to be assigned the color 1. The latter is not possible because v is the lone vertex in K that is colored 1. If $C(w') = 1$, then its neighbor(s) in K does not have a uniquely colored neighbor because of the vertices w' and v. Hence, v is a universal vertex.   □

From Lemmas 28 and 29, we get Theorem 27.

---

[2] This case also captures the case when K is empty.

8 Conclusion

We gave an FPT algorithm for conflict-free coloring for the combined parameters clique width w and number of colors k. Since the problem is NP-hard for constant number of colors k, it is unlikely to be FPT with respect to k only. However an interesting open question is whether this result can be strengthened to an FPT algorithm for parameter clique width w only. To the best of our knowledge, it is open whether there is some bound of any conflict-free chromatic number by the clique width. If there exists such a bound, our algorithm would also be a fixed-parameter tractable algorithm for parameter w only.

Further we showed a constant upper bound of conflict-free chromatic numbers for several graph classes. For most of them we established matching or almost matching lower and upper bounds for their conflict-free chromatic numbers. For unit square and square disk graphs there still is a wide gap, and it would be interesting to improve those bounds.

Acknowledgment. We would like to thank Rogers Mathew for helpful discussions. We would also like to thank Alexander Hermans for his help on finding a lower bound example for interval graphs. The third author acknowledges DST-SERB (MTR/2020/000497) for supporting this research. The last author acknowledges DST-SERB (SRG/2020/001162) for funding to support this research.

References

1. Abel, Z., et al.: Conflict-free coloring of graphs. SIAM J. Discret. Math. **32**(4), 2675–2702 (2018)
2. Agrawal, A., Ashok, P., Reddy, M.M., Saurabh, S., Yadav, D.: FPT algorithms for conflict-free coloring of graphs and chromatic terrain guarding. CoRR, abs/1905.01822 (2019)
3. Bandelt, H.-J., Mulder, H.M.: Distance-hereditary graphs. J. Comb. Theory, Ser. B **41**(2), 182–208 (1986)
4. Bhyravarapu, S., Kalyanasundaram, S.: Combinatorial bounds for conflict-free coloring on open neighborhoods. In: Adler, I., Müller, H. (eds.) WG 2020. LNCS, vol. 12301, pp. 1–13. Springer, Cham (2020). https://doi.org/10.1007/978-3-030-60440-0_1
5. Bodlaender, H.L., Kolay, S., Pieterse, A.: Parameterized complexity of conflict-free graph coloring. In: Friggstad, Z., Sack, J.-R., Salavatipour, M.R. (eds.) WADS 2019. LNCS, vol. 11646, pp. 168–180. Springer, Cham (2019). https://doi.org/10.1007/978-3-030-24766-9_13
6. Corneil, D.G., Lerchs, H., Stewart Burlingham, L.: Complement reducible graphs. Discrete Appl. Math. **3**(3), 163–174 (1981)
7. Courcelle, B., Olariu, S.: Upper bounds to the clique width of graphs. Discrete Appl. Math. **101**(1–3), 77–114 (2000)
8. Even, G., Lotker, Z., Ron, D., Smorodinsky, S.: Conflict-free colorings of simple geometric regions with applications to frequency assignment in cellular networks. SIAM J. Comput. **33**(1), 94–136 (2004)
9. Fekete, S.P., Keldenich, P.: Conflict-free coloring of intersection graphs. In: 28th International Symposium on Algorithms and Computation, ISAAC 2017, 9–12 December 2017, vol. 92, pp. 31:1–31:12 (2017)

10. Gardi, F.: The roberts characterization of proper and unit interval graphs. Discrete Math. **307**(22), 2906–2908 (2007)
11. Gargano, L., Rescigno, A.A.: Complexity of conflict-free colorings of graphs. Theor. Comput. Sci. **566**(C), 39–49 (2015)
12. Huang, F., Guo, S., Yuan, J.: A short note on open-neighborhood conflict-free colorings of graphs. SIAM J. Discret. Math. **34**(3), 2009–2015 (2020)
13. Keller, C., Rok, A., Smorodinsky, S.: Conflict-free coloring of string graphs. Discrete Comput. Geom., 1–36 (2020)
14. Lovász, L.: Kneser's conjecture, chromatic number, and homotopy. J. Comb. Theor. Ser. A **25**(3), 319–324 (1978)
15. Oum, S.-I., Seymour, P.D.: Approximating clique-width and branch-width. J. Comb. Theor. Ser. B **96**(4), 514–528 (2006)
16. Pach, J., Tardos, G.: Conflict-free colourings of graphs and hypergraphs. Comb. Probab. Comput. **18**(5), 819–834 (2009)
17. Vinod Reddy, I.: Parameterized algorithms for conflict-free colorings of graphs. Theor. Comput. Sci. **745**, 53–62 (2018)
18. Smorodinsky, S.: Conflict-free coloring and its applications. In: Bárány, I., Böröczky, K.J., Tóth, G.F., Pach, J. (eds.) Geometry — Intuitive, Discrete, and Convex. BSMS, vol. 24, pp. 331–389. Springer, Heidelberg (2013). https://doi.org/10.1007/978-3-642-41498-5_12
19. Sorge, M., Weller, M.: The graph parameter hierarchy. https://manyu.pro/assets/parameter-hierarchy.pdf. Accessed 9 Mar 2021

Edge Exploration of Temporal Graphs

Benjamin Merlin Bumpus[(✉)] and Kitty Meeks

School of Computing Science, University of Glasgow, Glasgow, UK
b.bumpus.1@research.gla.ac.uk, kitty.meeks@glasgow.ac.uk

Abstract. We introduce a natural temporal analogue of Eulerian circuits and prove that, in contrast with the static case, it is NP-hard to determine whether a given temporal graph is temporally Eulerian even if strong restrictions are placed on the structure of the underlying graph and each edge is active at only three times. However, we do obtain an FPT-algorithm with respect to a new parameter called *interval-membership-width* which restricts the times assigned to different edges; we believe that this parameter will be of independent interest for other temporal graph problems. Our techniques also allow us to resolve two open questions of Akrida, Mertzios and Spirakis [CIAC 2019] concerning a related problem of exploring temporal stars.

Keywords: Temporal graphs · Temporal exploration · Temporal Eulerian circuit · Fixed parameter tractability

1 Introduction

Many real-world problems can be formulated and modeled in the language of graph theory. However, real-world networks are often not *static*. They change over time and their edges may appear or disappear (for instance friendships may change over time in a social network). Such networks are called *dynamic* or *evolving* or *temporal* and their structural and algorithmic properties have been the subject of active study in recent years [1,6,14,15,19]. Some of the most natural and most studied topics in the theory of temporal graphs are temporal walks (in which consecutive edges appear at increasing times), paths and corresponding notions of temporal reachability [2,4,5,7,16,17,21,22]. Related to these notions is the study of explorability of a temporal graph which asks whether it is possible to visit all vertices or edges of a temporal graph via some temporal walk.

Temporal vertex-exploration problems (such as temporal variants of the Travelling Salesman problem) have already been thoroughly studied [3,10,20]. In contrast, here we focus on temporal *edge*-exploration and specifically we study *temporally Eulerian graphs*. Informally, these are temporal graphs admitting a

---

B. M. Bumpus—Supported by an EPSRC doctoral training account.

K. Meeks—Supported by a Royal Society of Edinburgh Personal Research Fellowship, funded by the Scottish Government, and EPSRC grant EP/T004878/1.

P. Flocchini and L. Moura (Eds.): IWOCA 2021, LNCS 12757, pp. 107–121, 2021.
https://doi.org/10.1007/978-3-030-79987-8_8

temporal circuit that visits every edge at exactly one time (i.e. a temporal circuit that yields an Euler circuit in the underlying static graph).

Deciding whether a static graph is Eulerian is a prototypical example of a polynomial time solvable problem. In fact this follows from Euler's characterization of Eulerian graphs dating back to the 18th century [11]. In contrast, here we show that, unless $P = NP$, a characterization of this kind cannot exist for temporal graphs. In particular we show that deciding whether a temporal graph is *temporally Eulerian* is NP-complete even if strong restrictions are placed on the structure of the underlying graph and each edge is active at only three times.

The existence of problems that are tractable on static graphs, but NP-complete on temporal graphs is well-known [3,6,18,19]. In fact there are examples of problems whose temporal analogues remain hard even on trees [3,18]. Thus the need for parameters that take into account the temporal structure of the input is clear. Some measures of this kind (such as temporal variants of feedback vertex number and tree-width) have already been studied [7,12]. Unfortunately we shall see that these parameters will be of no use to us since the problems we consider here remain NP-complete even when these measures are bounded by constants on the underlying static graph. To overcome these difficulties, we introduce a new purely-temporal parameter called *interval-membership-width*. Parameterizing by this measure we find that the problem of determining whether a temporal graph is temporally Eulerian is in FPT.

Temporal graphs of low interval-membership-width are 'temporally sparse' in the sense that only few edges are allowed to appear both before and after any given time. We point out that this parameter does *not* depend on the structure of the underlying static graph, but it is instead influenced only by the temporal structure. We believe that interval-membership-width will be a parameter of independent interest for other temporal graph problems in the future.

It turns out that our study of temporally Eulerian graphs is closely related to a temporal variant of the Travelling Salesman Problem concerning the exploration of temporal stars via a temporal circuit which starts at the center of the star and which visits all leaves. This problem was introduced and proven to be NP-complete by Akrida, Mertzios and Spirakis on temporal stars in which every edge has at most k appearances times for all $k \geq 6$ [3]. Although they also showed that the problem is polynomial-time solvable whenever each edge of the input temporal star has at most 3 appearances, they left open the question of determining the hardness of the problem when each edge has at most 4 or 5 appearances. We resolve this open problem in the course of proving our results about temporally Eulerian graphs. Combined with Akrida, Mertzios and Spirakis' results, this gives a complete dichotomy: their temporal star-exploration problem is in P if each edge has at most 3 appearances and is NP-complete otherwise.

As a potential 'island of tractability', Akrida, Mertzios and Spirakis proposed to restrict the input to their temporal star-exploration problem by requiring consecutive appearances of the edges to be evenly spaced (by some globally defined spacing). Using our new notion of interval-membership-width we are able to show that this restriction does indeed yield tractability parameterized by the maximum number of times per edge (thus partially resolving their open problem). Furthermore, we show that a slightly weaker result also holds for the problem of determining whether a temporal graph is temporally Eulerian in the setting with evenly-spaced edge-times.

Outline. We fix notation and provide background definitions in Sect. 2. We prove our hardness results in Sect. 3. Section 4 contains the definition of interval-membership-width as well as our FPT algorithms parameterized by this measure. In Sect. 5 we show that Akrida, Mertzios and Spirakis' temporal star-exploration problem is in FPT parameterized by the maximum number of appearances of any edge in the input whenever the input temporal star has evenly-spaced times on all edges. We also show a similar result for our temporally Eulerian problem. Finally we provide concluding remarks and open problems in Sect. 6. Due to space constraints, only sketch proofs are given for most results (we link the arXiv version here for full details).

2   Background and Notation

For any graph-theoretic notation not defined here, we refer the reader to Diestel's textbook [9]; similarly, for any terminology in parameterized complexity, we refer the reader to the textbook by Cygan et al. [8].

The formalism for the notion of dynamic or time-evolving graphs originated from the work of Kempe, Kleinberg, and Kumar [16]. Formally, if $\tau : E(G) \rightarrow 2^{\mathbb{N}}$ is a function mapping edges of a graph $G = (V(G), E(G))$ to sets of integers, then we call the pair $\mathcal{G} := (G, \tau)$ a *temporal graph*. We shall assume all temporal graphs to be finite and simple in this paper.

For any edge e in G, we call the set $\tau(e)$ the *time-set* of e. For any time $t \in \tau(e)$ we say that e is *active* at time t and we call the pair (e, t) a *time-edge*. The set of all edges active at any given time t is denoted $E_t(G, \tau) := \{e \in E(G) : t \in \tau(e)\}$. The latest time Λ for which $E_\Lambda(G, \tau)$ is non-empty is called the *lifetime* of a temporal graph (G, τ) (or equivalently $\Lambda := \max_{e \in E(G)} \max \tau(e)$). Here we will only consider temporal graphs with finite lifetime.

In a temporal graph there are two natural notions of walk: one is the familiar notion of a walk in static graphs and the other is a truly temporal notion where we require consecutive edges in walks to appear at non-decreasing times. Formally, given vertices x and y in a temporal graph $\mathcal{G}$, a *temporal (x, y)-walk* is a sequence $W = (e_1, t_1), \ldots, (e_n, t_n)$ of time-edges such that $e_1, \ldots, e_n$ is a walk in G starting at x and ending at y and such that $t_1 \leq t_2 \leq \ldots \leq t_n$. If $n > 1$, we denote by $W - (e_n, t_n)$ the temporal walk $(e_1, t_1), \ldots, (e_{n-1}, t_{n-1})$. We call a temporal (x, y)-walk *closed* if $x = y$ and we call it a *strict temporal walk* if the times of the walk form a strictly increasing sequence. Hereafter we will assume all temporal walks to be strict.

Recall that an Euler circuit in a static graph G is a circuit $e_1 \ldots, e_m$ which traverses every edge of G exactly once. In this paper we are interested in the natural temporal analogue of this notion.

Definition 1. *A temporal Eulerian circuit in a temporal graph* (G, τ) *is a closed temporal walk* $(e_1, t_1), \ldots, (e_m, t_m)$ *such that* $e_1 \ldots, e_m$ *is an Euler circuit in the underlying static graph* G. *If there exists a temporal Eulerian circuit in* (G, τ), *then we call* (G, τ) *temporally Eulerian.*

Note that if (G, τ) is a temporal graph in which every edge appears at exactly one time, then we can determine whether (G, τ) is temporally Eulerian in time linear in $|E(G)|$. To see this, note that, since every edge is active at precisely one time, there is only one candidate ordering of the edges (which may or may not give rise to an Eulerian circuit). Thus it is clear that the number of times per edge is relevant to the complexity of the associated decision problem – which we state as follows.

TempEuler(k)
Input: A temporal graph (G, τ) where $|\tau(e)| \leq k$ for every edge e in the graph G.
Question: Is (G, τ) *temporally Eulerian?*

As we mentioned in Sect. 1, here we will show that TEMPEULER(k) is related to an analogue of the Travelling Salesman problem on temporal stars [3]. This problem (denoted as STAREXP(k)) was introduced by Akrida, Mertzios and Spirakis [3]. It asks whether a given temporal star (S_n, τ) (where S_n denotes the n-leaf star) with at most k times on each edge admits a closed temporal walk starting at the center of the star and which visits every leaf of S_n. We call such a walk an *exploration* of (S_n, τ). A temporal star that admits an exploration is called *explorable*. Formally we have the following decision problem.

StarExp(k)
Input: A temporal star (S_n, τ) where $|\tau(e)| \leq k$ for every edge e in the star S_n.
Question: Is (S_n, τ) explorable?

---

3   Hardness of Temporal Edge Exploration

In this section we will show that TEMPEULER(k) is NP-complete for all k at least 3 (Corollary 2) and that STAREXP(k) is NP-complete for all $k \geq 4$ (Corollary 1). This last result resolves an open problem of Akrida, Mertzios and Spirakis which asked to determine the complexity of STAREXP(4) and STAREXP(5) [3].

We begin by showing that STAREXP(4) is NP-hard. We will do so via a reduction from the 3-COLORING problem (see for instance Garey and Johnson [13] for a proof of NP-completeness) which asks whether an input graph G is 3-colorable.

3-Coloring
Input: A finite simple graph G.
Question: Does G admit a proper 3-coloring?

Throughout, for an edge e of a temporal star $(S_n \tau)$, we call any pair of times $(t_1, t_2) \in \tau(e)^2$ with $t_1 < t_2$ a *visit of* e. We say that e is *visited at* (t_1, t_2) in a temporal walk if the walk proceeds from the center of the star along e at time t_1 and then back to the center at time t_2. We say that two visits (x_1, x_2) and (y_1, y_2) of two edges e_x and e_y are *in conflict* with one another (or that 'there is a conflict between them') if there exists some time t with $x_1 \leq t \leq x_2$ and $y_1 \leq t \leq y_2$. Note that a complete set of visits (one visit for each edge of the star) which has no pairwise conflicts is in fact an exploration.

Theorem 1. STAREXP(4) *is NP-hard.*

Proof (Sketch). Take any 3-COLORING instance G with vertices $\{x_1, \ldots, x_n\}$. We will construct a STAREXP(4) instance (S_p, τ) (where $p = n + 3m$) from G (see Fig. 1).

Defining S_p. The star S_p is defined as follows: for each vertex x_i in G, we make one edge e_i in S_p while, for each edge $x_i x_j$ with $i < j$ in G, we make three edges e_{ij}^0, e_{ij}^1 and e_{ij}^2 in S_p.

Defining τ. For $i \in [n]$ and any non-negative integer $\psi \in \{0, 1, 2, \ldots\}$, let t_ξ^i be the integer

$$t_\xi^i := 2in^2 + 2\psi(n+1) \tag{1}$$

and take any edge $x_j x_k$ in G with $j < k$. Using the times defined in Eq. (1) and taking $\xi \in \{0, 1, 2\}$, we then define $\tau(e_i)$ and $\tau(e_{jk}^\xi)$ as

$$\tau(e_i) := \{t_0^i, t_1^i, t_2^i, t_3^i\} \text{ and} \tag{2}$$

$$\tau(e_{jk}^\xi) := \{t_\xi^j + 2k - 1, \quad t_\xi^j + 2k, \quad t_\xi^k + 2j - 1, \quad t_\xi^k + 2j\}. \tag{3}$$

Note that the elements of these sets are written in increasing order (see Fig. 1). Intuitively, the times associated to each edge $e_i \in E(S_p)$ corresponding to a vertex $x_i \in V(G)$ (Eq. (2)) encode the possible colorings of x_i via the three possible starting times of a visit of e_i. The three edges e_{ij}^0, e_{ij}^1 and e_{ij}^2 corresponding to some $x_i x_j \in E(G)$ are instead used to 'force the colorings to be proper' in G. That is to say that, for a color $\xi \in \{0, 1, 2\}$, the times associated with the edge e_{ij}^ξ (Eq. (3)) will prohibit us from entering e_i at its ξ-th appearance and also entering e_j at its ξ-th appearance (i.e. 'coloring x_i and x_j the same color'). $\qquad \square$

Observe that increasing the maximum number of times per edge cannot make the problem easier: we can easily extend the hardness result to any $k' > 4$ by simply adding a new edge with k' times all prior to the times that are already

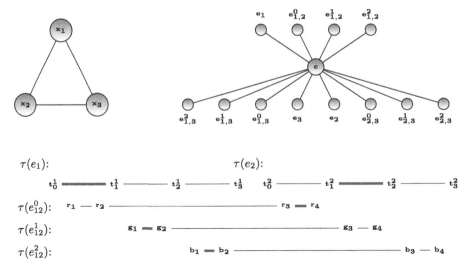

Fig. 1. Top left: K^3. Top right: star constructed from K^3. Bottom: times (and corresponding intervals) associated with the edges e_1, e_2 and $e^0_{1,2}$, $e^1_{1,2}$, $e^2_{1,2}$ (time progresses left-to-right and intervals are not drawn to scale). We write r_1, r_2, r_3, r_4 as shorthand for the entries of $\tau(e^0_{1,2})$ (similarly, for $i \in [4]$, we write g_i and b_i with respect to $\tau(e^1_{1,2})$ and $\tau(e^2_{1,2})$). The red and thick intervals correspond to visits defined by the coloring $x_i \mapsto i - 1$ of the K^3.

in the star. This, together with the fact that Akrida, Mertzios and Spirakis [3] showed that STAREXP(k) is in NP for all k, allows us to conclude the following corollary.

Corollary 1. *For all k at least* 4, *STAREXP(k) is NP-complete.*

Next we shall reduce STAREXP(k) to TEMPEULER($k-1$). We point out that, for our purposes within this section, only the first point of the statement of the following result is needed. However, later (in the proof of Corollary 3) we shall make use of the properties stated in the second point of Lemma 1 (this is also why we allow any k times per edge rather than just considering the case $k = 4$). Thus we include full details here.

Lemma 1. *For all $k \geq 2$ there is a polynomial-time-computable mapping taking every STAREXP(k) instance (S_n, τ) to a TEMPEULER($k - 1$) instance (D_n, σ) such that*

1. *(S_n, τ) is a yes instance for STAREXP(k) if and only if (D_n, σ) is a yes instance for TEMPEULER($k - 1$) and*
2. *D_n is a graph obtained by identifying n-copies $\{K^3_1, \dots, K^3_n\}$ of a cycle on three vertices along one center vertex (see Fig. 2) and such that*

$$\max_{t \in \mathbb{N}} |\{e \in E(D_n) : \min(\sigma(e)) \leq t \leq \max(\sigma(e))\}|$$

$$\leq 3 \max_{t \in \mathbb{N}} |\{e \in E(S_n) : \min(\tau(e)) \leq t \leq \max(\tau(e))\}|.$$

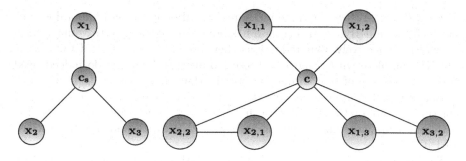

Fig. 2. The graph D_3 built from S_3 in Lemma 1

Since TEMPEULER(k) is clearly in NP (where the circuit acts as a certificate), our desired NP-completeness result follows immediately from Lemma 1 and Corollary 1.

Corollary 2. TEMPEULER(k) *is* NP-*complete for all k at least* 3.

As we noted earlier, TEMPEULER(1) is trivially solvable in time linear in the number of edges of the underlying static graph. Thus, towards obtaining a complexity dichotomy for TEMPEULER(k), the only case remaining open is when $k = 2$.

Observe that the reduction in Lemma 1 rules out FPT algorithms with respect to many standard parameters describing the structure of the underlying graph (for instance the path-width is 2 and feedback vertex number is 1). In fact we can strengthen these intractability results even further by showing that TEMPEULER(k) is hard even for instances whose underlying static graph has vertex-cover number 2. This motivates our search in Sect. 4 for parameters that describe the structure of the times assigned to edges rather than just the underlying static structure.

Notice that this time we will reduce from STAREXP(k) to TEMPEULER(k) (rather than from STAREXP($k+1$) as in Lemma 1), so, in contrast to our previous reduction (Lemma 1), the proof of the following result cannot be used to show hardness of TEMPEULER(3).

Theorem 2. *For all* $k \geq 4$, *the* TEMPEULER(k) *problem is* NP-*complete even on temporal graphs whose underlying static graph has vertex-cover number* 2.

4 Interval-Membership-Width

As we saw in the previous section, both TEMPEULER(k) and STAREXP($k + 1$) are NP-complete for all $k \geq 3$ even on instances whose underlying static graphs are very sparse (for instance even on graphs with vertex cover number 2). Clearly this means that any useful parameterization must take into account the *temporal structure* of the input. Other authors have already proposed measures of this kind such as the temporal feedback vertex number [7] or temporal analogues of tree-width [12]. However these measures are all bounded on temporal graphs for which the underlying static graph has bounded feedback vertex number and tree-width respectively. Our reductions therefore show that TEMPEULER(k) is para-NP-complete with respect to these parameters. Thus we do indeed need some new measure of temporal structure. To that end, here we introduce such a parameter called *interval-membership-width* which depends only on temporal structure and not on the structure of the underlying static graph. Parameterizing by this measure, we will show that both TEMPEULER(k) and STAREXP(k) lie in FPT.

To first convey the intuition behind our width measure, consider again the TEMPEULER(1) problem. As we noted earlier, this is trivially solvable in time linear in $|E(G)|$. The same is true for any TEMPEULER(k)-instance (G, τ) in which every edge is assigned a 'private' interval of times: that is to say that, for all distinct edges e and f in G, either $\max \tau(f) < \min \tau(e)$ or $\max \tau(e) < \min \tau(f)$. This holds because, on instances of this kind, there is only one possible relative ordering of edges available for an edge-exploration. It is thus natural to expect that, for graphs whose edges have intervals that are 'almost private' (defined formally below), we should be able to deduce similar tractability results.

Towards a formalization of this intuition, suppose that we are given a temporal graph (G, τ) which has precisely two edges e and f such that there is a time t with $\min \tau(e) \leq t \leq \max \tau(e)$ and $\min \tau(f) \leq t \leq \max \tau(f)$. It is easy to see that the TEMPEULER(k) problem is still tractable on graphs such as (G, τ) since there are only two possible relative edge-orderings for an edge exploration of (G, τ) (depending on whether we choose to explore e before f or f before e). These observations lead to the following definition of *interval-membership-width* of a temporal graph (see Fig. 3).

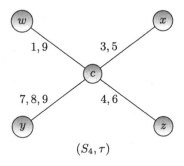

(S_4, τ)

Fig. 3. A temporal star (S_4, τ) with interval-membership-sequence: $F_1 = F_2 = \{cw\}$, $F_3 = \{cw, cx\}$, $F_4 = F_5 = \{cw, cx, cy\}$, $F_6 = \{cw, cy\}$ and $F_7 = F_8 = F_9 = \{cw, cz\}$.

Definition 2. *The interval membership sequence of a temporal graph (G, τ) is the sequence $(F_t)_{t \in [\Lambda]}$ of edge-subsets of G where $F_t := \{e \in E(G) : \min \tau(e) \leq t \leq \max \tau(e)\}$ and Λ is the lifetime of (G, τ). The interval-membership-width of (G, τ) is the integer $\boldsymbol{imw}(G, \tau) := \max_{t \in \mathbb{N}} |F_t|$.*

Note that a temporal graph has unit interval-membership-width if and only if every edge is active at times spanning a 'private interval'. Furthermore, we point out that the interval membership sequence of a temporal graph is not the same as the sequence $(E_t(G, \tau))_{t \in \mathbb{N}}$. In fact, although $\max_{t \in \mathbb{N}} |E_t(G, \tau)| \leq \mathbf{imw}(G, \tau)$, there exist classes $\mathcal{C}$ of temporal graphs with unbounded interval-membership-width but such that every temporal graph in $\mathcal{C}$ satisfies the property that at most one edge is active at any given time. To see this consider any graph H with edges $e_1, \ldots, e_m$ and let (H, ν) be the temporal graph defined by $\nu(e_i) := \{i, m + i\}$. Clearly $\max_{i \in \mathbb{N}} |E_i(H, \nu)| = 1$, but we have $\mathbf{imw}(H, \nu) = m$.

Note that the interval membership sequence of a temporal graph (G, τ) can be computed in time $\mathcal{O}(\mathbf{imw}(G, \tau)\Lambda)$ by iterating over the edges of G.

Armed with the notion of interval-membership-width, we will now show that both TEMPEULER(k) and STAREXP(k) are in FPT when parameterized by this measure. We will do so first for TEMPEULER(k) (Theorem 3) and then we will leverage the reduction of Lemma 1 to deduce the fixed-parameter-tractability of STAREXP(k) as well (Corollary 3).

Theorem 3. *There is an algorithm that decides whether any temporal graph (G, τ) with n vertices and lifetime Λ is a yes-instance of* TEMPEULER(k) *in time $\mathcal{O}(w^3 2^w \Lambda)$ where $w = \mathbf{imw}(G, \tau)$ is the interval-membership-width of (G, τ).*

Proof (Sketch). Let $(F_t)_{t \in [\Lambda]}$ be the interval membership sequence of (G, τ) and suppose without loss of generality that F_1 is not empty.

We will now describe an algorithm that proceeds by dynamic programming over the sequence $(F_i)_{i \in [\Lambda]}$ to determine whether (G, τ) is temporally Eulerian. For each set F_i we will compute a set $L_i \subseteq F_i^{\{0,1\}} \times V(G) \times V(G)$ consisting of triples of the form (f, s, x) where s and x are vertices in G and f is a function mapping each edge in F_i to an element of $\{0, 1\}$. Intuitively each entry (f, s, x) of L_i corresponds to the existence of a temporal walk starting at s and ending at x at time at most i and such that, for any edge $e \in F_i$, we will have $f(e) = 1$ if and only if e was traversed during this walk.

We will now define the entries L_i recursively starting from the dummy set $L_0 := \{(\mathbf{0}, x, x) : \exists e \in F_1 \text{ incident with } x\}$ where $\mathbf{0} : e \in F_1 \mapsto 0$ is the function mapping every element in F_1 to 0. Take any (f, s, y) in $F_i^{\{0,1\}} \times V(G) \times V(G)$. For (f, s, y) to be in L_i we will require there to be an entry (g, s, x) of L_{i-1} such that

$$g(e) = 1 \text{ for all } e \in F_{i-1} \setminus F_i \qquad (4)$$

and such that the one of the following cases holds: either

C1 $y = x$ and $f(e) = 1$ if and only if $e \in F_{i-1} \cap F_i$ and $g(e) = 1$,
  or
C2 there exists an edge xy in G such that:
C2.P1 $xy \in E_i(G, \tau) \setminus \{e \in F_i : g(e) = 1\}$ and
C2.P2 $f(e) = 1$ if and only if $g(e) = 1$ or $e = xy$.

The Cases **C1** and **C2** correspond to the the two available choices we have when extending a temporal (s, x)-walk at time i: either we stay put at x (Case

C1) or we find some new edge xy active at time i (Case **C2**) which has never been used before (Property **C2.P1**) and add it to the walk (Property **C2.P2**). Equation (4) ensures that we filter out partial solutions that we already know cannot be extended to a Eulerian circuit. To see this, note that, if an edge e will never appear again after time $i - 1$ and we have $g(e) = 0$, then there is no way of extending the temporal walk represented by the triple (g, s, x) to an Eulerian circuit in (G, τ) because one edge will always be left out (namely the edge e). □

As a corollary of Theorem 3, we can leverage the reduction of Lemma 1 to deduce that $\textsc{StarExp}(k)$ is in FPT parameterized by the interval-membership-width.

Corollary 3. *There is an algorithm that decides whether a $\textsc{StarExp}(k)$ instance (S_n, τ) is explorable in time $\mathcal{O}(w^3 2^{3w} \Lambda)$ where $w = \boldsymbol{imw}(S_n, \tau)$ and Λ is the lifetime of the input.*

Proof. By Lemma 1, we know that there is a polynomial-time reduction that maps any $\textsc{StarExp}(k)$ instance (S_n, τ) to a $\textsc{TempEuler}(k-1)$-instance (D_n, σ) such that

$$\max_t |\{e \in E(D_n) : \min(\sigma(e)) \leq t \leq \max(\sigma(e))\}|$$
$$\leq 3 \max_t |\{e \in E(S_n) : \min(\tau(e)) \leq t \leq \max(\tau(e))\}|.$$

In particular this implies that $\mathbf{imw}(D_n, \sigma) \leq 3w$. Thus we can decide whether (S_n, τ) is explorable in time $\mathcal{O}(w^3 2^{3w} \Lambda)$ by applying the algorithm of Theorem 3 to (D_n, σ). □

5   Win-Win Approach to Regularly Spaced Times

In this section we will find necessary conditions for edge-explorability of temporal graphs with respect to their interval-membership-width. This will allow us to conclude that either we are given a no-instance or that the interval-membership-width is small (in which case we can employ our algorithmic results from the previous section).

We will apply this bidimensional approach to a variants of $\textsc{TempEuler}(k)$ and $\textsc{StarExp}(k)$ in which we are given upper and lower bounds (u and ℓ respectively) on the difference between any two consecutive times at any edge. Specifically we will show that $\textsc{StarExp}(k)$ is in FPT parameterized by k, ℓ and u (Theorem 4) and that $\textsc{TempEuler}(k)$ is in FPT parameterized by k and u (Theorem 5). In other words, these results allow us to trade in the dependences on the interval-membership-width of Corollary 3 and Theorem 3 for a dependences on k, ℓ, u and k, u respectively.

We note that, for $\textsc{StarExp}$ instances, the closer ℓ and u get, the more restricted the structure becomes to the point that the dependence on ℓ and u in the running time of our algorithm vanishes when $\ell = u$. In particular this shows that the problem of determining the explorability of $\textsc{StarExp}(k)$-instances for which consecutive times at each edge are exactly λ time-steps apart (for some $\lambda \in \mathbb{N}$) is in FPT parameterized solely by k (Corollary 4). This partially resolves an open problem of Akrida, Mertzios and Spirakis [3] which asked to determine the complexity of exploring $\textsc{StarExp}(k)$-instances with evenly-spaced times.

Towards these results, we will first provide sufficient conditions for non-explorability of any $\textsc{StarExp}(k)$ instance (Lemma 2). These conditions will depend only on: (1) knowledge of the maximum and minimum differences between any two successive appearances of any edge, (2) on the interval-membership-width and (3) on k.

Lemma 2. *Let (S_n, τ) be a temporal star with at most k times at any edge and such that every two consecutive times at any edge differ at least by ℓ and at most by u. If (S_n, τ) is explorable, then $\textbf{imw}(S_n, \tau) \leq 2(ku+1)/(\ell+1)$.*

Proof. Let Λ be the lifetime of (S_n, τ), let $(F_t)_{t \in [\Lambda]}$ be the interval membership sequence of (S_n, τ) and choose any $n \in [\Lambda]$ such that $|F_n| = \textbf{imw}(S_n, \tau)$. Let m and M be respectively the earliest and latest times at which there are edges in F_n which are active and chose representatives e_m and e_M in F_n such that $m = \min \tau(e_m)$ and $M = \max \tau(e_M)$.

Recall that visiting any edge e in S_n requires us to us pick two appearances (which differ by at least $\ell + 1$ time-steps) of e (one appearance to go along e from the center of S_n to the leaf and another appearance to return to the center of the star). Thus, whenever we specify how to visit an edge e of F_n, we remove at least $\ell + 1$ time-steps from the available time-set $\{m, \ldots, M\}$ at which any other edge in F_n can be visited. Furthermore, since any exploration of (S_n, τ) must explore all of the edges in F_n, for (S_n, τ) to be explorable, we must have $|F_n|(\ell+1) \leq M - m + 1$. This concludes the proof since $\textbf{imw}(S_n, \tau) = |F_n|$ and $M - m \leq |\max \tau(e_M) - \min \tau(e_m)| + |\max \tau(e_m) - \min \tau(e_m)|$ (since, by the definition of F_n, n is in the intervals of any two elements of F_n) which is at most $2ku + 1$ (since consecutive times at any edge differ by at most u). $\qquad\square$

Notice that nearly-identical arguments yield the following slightly weaker result with respect to the $\textsc{TempEuler}(k)$ problem.

Lemma 3. *Let (G, τ) be a $\textsc{TempEuler}(k)$ instance such that every two consecutive times at any edge differ at most by u. If (G, τ) is temporally Eulerian, then $\textbf{imw}(G, \tau) \leq 2(ku+1)$.*

The reason that the we can only bound $\textbf{imw}(G, \tau)$ above by $2(ku+1)$ (rather than $2(ku+1)/(\ell+1)$ as in the $\textsc{StarExp}(k)$ case of Lemma 2) is that temporal Euler circuits only visit each edge once (so exploring each edge only removes exactly one available time).

Lemma 2 allows us to employ a 'win-win' approach for $\textsc{StarExp}(k)$ when we know the maximum difference between consecutive times at any edge: either the considered instance does not satisfy the conditions of Lemma 2 (in which case we have a no-instance) or the interval-membership-width is 'small' (in which case we apply the algorithm given by Corollary 3). These ideas allow us to conclude the following result.

Theorem 4. *Let (S_n, τ) be a temporal star with at most k times at any edge and such that every two consecutive times at any edge differ at least by ℓ and at most by u. There is an algorithm deciding whether (S_n, τ) is explorable in time $2^{\mathcal{O}(ku/\ell)}\Lambda$ where Λ is the lifetime of the input.*

Proof. The algorithm proceeds as follows. First determine $\mathbf{imw}(S_n, \tau)$ (this can be done in time $\mathcal{O}(\Lambda n)$ where Λ is the lifetime of the input). If $\mathbf{imw}(S_n, \tau) > 2(ku+1)/(\ell+1)$, then (S_n, τ) is not explorable by Lemma 2. Otherwise run the algorithm given in Corollary 3. In this case, since $w := \mathbf{imw}(S_n, \tau) \leq 2(ku+1)/(\ell+1)$, we know that the algorithm of Corollary 3 will run on (S_n, τ) in time $2^{\mathcal{O}(ku/\ell)}\Lambda$. $\qquad\square$

Once again arguing by bidimensionality (this time using Lemma 3 and Theorem 3) we can deduce the following fixed-parameter tractability result for TEMPEULER.

Theorem 5. *Let (G, τ) be a TEMPEULER(k) instance such that every two consecutive times at any edge differ at most by u. There is an algorithm deciding whether (G, τ) is temporally Eulerian in time $2^{\mathcal{O}(ku)}\Lambda$ where Λ is the lifetime of the input.*

As a special case of Theorem 4 (i.e. the case where $\ell = u$) we resolve an open problem of Akrida, Mertzios and Spirakis [3] which asked to determine the complexity of exploring STAREXP(k)-instances with evenly-spaced times. In particular we show that the problem of deciding the explorability of such evenly-spaced STAREXP(k)-instances is in FPT when parameterized by k.

Corollary 4. *There is an algorithm which, given any STAREXP(k) instance (S_n, τ) with lifetime Λ and in which every two pairs of consecutive times assigned to any edge differ by the same amount, decides whether (S_n, τ) is explorable in time $2^{\mathcal{O}(k)}\Lambda$.*

6 Discussion

We introduced a natural temporal analogue of Eulerian circuits and proved that, in contrast to the static case, TEMPEULER(k) is NP-complete for all $k \geq 3$. In fact we showed that the problem remains hard even when the underlying static graph has path-width 2, feedback vertex number 1 or vertex cover number 2. Along the way, we resolved an open problem of Akrida, Mertzios and Spirakis [3] by showing that STAREXP(k) is NP-complete for all $k \geq 4$. This result yields a complete complexity dichotomy with respect to k when combined with Akrida, Mertzios and Spirakis' results [3]; however, a similar dichotomy for TEMPEULER(k) still eludes us. In fact, although we know that TEMPEULER(1) is in P, our reduction cannot be extended to obtain a complete dichotomy result. Thus to determine the complexity of the only remaining open case (i.e. TEMPEULER(2)) new ideas are needed.

Our hardness results rule out the possibility of obtaining FPT algorithms for TEMPEULER(k) and STAREXP(k) with respect to many standard parameters describing the structure of the underlying graph (such as path-width, feedback vertex number and vertex-cover number). We thus introduced a new width measure which captures structural information that is purely temporal; we call this the *interval-membership-width*. In contrast to our hardness results, we showed that TEMPEULER(k) and STAREXP(k) can be solved in times $\mathcal{O}(w^3 2^w \Lambda)$ and $\mathcal{O}(w^3 2^{3w} \Lambda)$ respectively where w is our new parameter and Λ is the lifetime of the input.

Our fixed-parameter-tractability results parameterized by interval-membership-width can also be leveraged via a win-win approach to obtain tractability results for both TEMPEULER(k) and STAREXP(k) parameterized solely by k and the spacing of appearances of edges in the input. These results allow us to partially resolve another open problem of Akrida, Mertzios and Spirakis concerning the complexity of STAREXP(k): we showed that it can be solved in time $2^{\mathcal{O}(k)} \Lambda$ when the input has evenly spaces appearances of each edge and lifetime Λ. We note, however, that it remains an open problem to determine the complexity of the evenly-spaced STAREXP(k) problem when k is unbounded.

Finally we point out that all of our hardness reductions hold also for the case of non-strict temporal walks and, with slightly more work, this also holds for our tractability results.

Acknowledgements. The authors would like to thank Samuel Hand for spotting a slight inaccuracy in the preliminary version of this article and the anonymous reviewers for their helpful comments and suggestions.

References

1. Akrida, E.C., Gąsieniec, L., Mertzios, G.B., Spirakis, P.G.: The complexity of optimal design of temporally connected graphs. Theory Comput. Syst. **61**(3), 907–944 (2017). https://doi.org/10.1007/s00224-017-9757-x
2. Akrida, E.C., Mertzios, G.B., Nikoletseas, S., Raptopoulos, C., Spirakis, P.G., Zamaraev, V.: How fast can we reach a target vertex in stochastic temporal graphs? J. Comput. Syst. Sci. **114**, 65–83 (2020). https://doi.org/10.1016/j.jcss.2020.05.005
3. Akrida, E.C., Mertzios, G.B., Spirakis, P.G.: The temporal explorer who returns to the base. In: Heggernes, P. (ed.) CIAC 2019. LNCS, vol. 11485, pp. 13–24. Springer, Cham (2019). https://doi.org/10.1007/978-3-030-17402-6_2
4. Axiotis, K., Fotakis, D.: On the size and the approximability of minimum temporally connected subgraphs. In: Chatzigiannakis, I., Mitzenmacher, M., Rabani, Y., Sangiorgi, D. (eds.) 43rd International Colloquium on Automata, Languages, and Programming (ICALP 2016). Leibniz International Proceedings in Informatics (LIPIcs), vol. 55, pp. 149:1–149:14. Schloss Dagstuhl-Leibniz-Zentrum fuer Informatik, Dagstuhl (2016). https://doi.org/10.4230/LIPIcs.ICALP.2016.149

5. Bhadra, S., Ferreira, A.: Complexity of connected components in evolving graphs and the computation of multicast trees in dynamic networks. In: Pierre, S., Barbeau, M., Kranakis, E. (eds.) ADHOC-NOW 2003. LNCS, vol. 2865, pp. 259–270. Springer, Heidelberg (2003). https://doi.org/10.1007/978-3-540-39611-6_23

6. Casteigts, A., Flocchini, P., Quattrociocchi, W., Santoro, N.: Time-varying graphs and dynamic networks. In: Frey, H., Li, X., Ruehrup, S. (eds.) ADHOC-NOW 2011. LNCS, vol. 6811, pp. 346–359. Springer, Heidelberg (2011). https://doi.org/10.1007/978-3-642-22450-8_27

7. Casteigts, A., Himmel, A.S., Molter, H., Zschoche, P.: Finding temporal paths under waiting time constraints. In: Cao, Y., Cheng, S.W., Li, M. (eds.) 31st International Symposium on Algorithms and Computation (ISAAC 2020). Leibniz International Proceedings in Informatics (LIPIcs), vol. 181, pp. 30:1–30:18. Schloss Dagstuhl-Leibniz-Zentrum für Informatik, Dagstuhl (2020). https://doi.org/10.4230/LIPIcs.ISAAC.2020.30

8. Cygan, M., et al.: Parameterized Algorithms. Springer, Cham (2015). https://doi.org/10.1007/978-3-319-21275-3

9. Diestel, R.: Graph Theory. GTM, vol. 173. Springer, Heidelberg (2017). https://doi.org/10.1007/978-3-662-53622-3

10. Erlebach, T., Hoffmann, M., Kammer, F.: On temporal graph exploration. In: Halldórsson, M.M., Iwama, K., Kobayashi, N., Speckmann, B. (eds.) ICALP 2015. LNCS, vol. 9134, pp. 444–455. Springer, Heidelberg (2015). https://doi.org/10.1007/978-3-662-47672-7_36

11. Euler, L.: Solutio problematis ad geometriam situs pertinentis. Commentarii academiae scientiarum Petropolitanae, pp. 128–140 (1741)

12. Fluschnik, T., Molter, H., Niedermeier, R., Renken, M., Zschoche, P.: As time goes by: reflections on treewidth for temporal graphs. In: Fomin, F.V., Kratsch, S., van Leeuwen, E.J. (eds.) Treewidth, Kernels, and Algorithms. LNCS, vol. 12160, pp. 49–77. Springer, Cham (2020). https://doi.org/10.1007/978-3-030-42071-0_6

13. Garey, M.R., Johnson, D.S.: Computers and Intractability: A Guide to the Theory of NP-Completeness. W. H. Freeman, San Francisco (1979)

14. Himmel, A.-S., Molter, H., Niedermeier, R., Sorge, M.: Adapting the Bron–Kerbosch algorithm for enumerating maximal cliques in temporal graphs. Soc. Netw. Anal. Min. **7**(1), 1–16 (2017). https://doi.org/10.1007/s13278-017-0455-0

15. Holme, P., Saramäki, J.: Temporal networks. Phys. Rep. **519**(3), 97–125 (2012). https://doi.org/10.1016/j.physrep.2012.03.001. Temporal Networks

16. Kempe, D., Kleinberg, J., Kumar, A.: Connectivity and inference problems for temporal networks. J. Comput. Syst. Sci. **64**(4), 820–842 (2002). https://doi.org/10.1006/jcss.2002.1829

17. Mertzios, G.B., Michail, O., Spirakis, P.G.: Temporal network optimization subject to connectivity constraints. Algorithmica **81**(4), 1416–1449 (2018). https://doi.org/10.1007/s00453-018-0478-6

18. Mertzios, G.B., Molter, H., Niedermeier, R., Zamaraev, V., Zschoche, P.: Computing maximum matchings in temporal graphs. arXiv preprint arXiv:1905.05304 (2019)

19. Michail, O.: An introduction to temporal graphs: an algorithmic perspective. Internet Math. **12**(4), 239–280 (2016). https://doi.org/10.1080/15427951.2016.1177801

20. Michail, O., Spirakis, P.G.: Traveling salesman problems in temporal graphs. In: Csuhaj-Varjú, E., Dietzfelbinger, M., Ésik, Z. (eds.) MFCS 2014. LNCS, vol. 8635, pp. 553–564. Springer, Heidelberg (2014). https://doi.org/10.1007/978-3-662-44465-8_47

21. Wu, H., Cheng, J., Ke, Y., Huang, S., Huang, Y., Wu, H.: Efficient algorithms for temporal path computation. IEEE Trans. Knowl. Data Eng. **28**(11), 2927–2942 (2016). https://doi.org/10.1109/TKDE.2016.2594065
22. Xuan, B.B., Ferreira, A., Jarry, A.: Computing shortest, fastest, and foremost journeys in dynamic networks. Int. J. Found. Comput. Sci. **14**(02), 267–285 (2003). https://doi.org/10.1142/S0129054103001728

Optimal Monomial Quadratization
for ODE Systems

Andrey Bychkov[1] and Gleb Pogudin[2]($\boxtimes$)

[1] Higher School of Economics, Myasnitskaya str., 101978 Moscow, Russia
abychkov@edu.hse.ru
[2] LIX, CNRS, École Polytechnique, Institute Polytechnique de Paris,
Palaiseau, France
gleb.pogudin@polytechnique.edu

Abstract. Quadratization is a transform of a system of ODEs with polynomial right-hand side into a system of ODEs with at most quadratic right-hand side via the introduction of new variables. Quadratization problem is, given a system of ODEs with polynomial right-hand side, transform the system to a system with quadratic right-hand side by introducing new variables. Such transformations have been used, for example, as a preprocessing step by model order reduction methods and for transforming chemical reaction networks.

We present an algorithm that, given a system of polynomial ODEs, finds a transformation into a quadratic ODE system by introducing new variables which are monomials in the original variables. The algorithm is guaranteed to produce an optimal transformation of this form (that is, the number of new variables is as small as possible), and it is the first algorithm with such a guarantee we are aware of. Its performance compares favorably with the existing software, and it is capable to tackle problems that were out of reach before.

Keywords: Differential equations · Branch-and-bound · Quadratization

1 Introduction

The *quadratization* problem considered in this paper is, given a system of ordinary differential equations (ODEs) with polynomial right-hand side, transform

The article was prepared within the framework of the HSE University Basic Research Program. GP was partially supported by NSF grants DMS-1853482, DMS-1760448, DMS-1853650, CCF-1564132, and CCF-1563942 and by the Paris Ile-de-France region. The authors are grateful to Mathieu Hemery, François Fages, and Sylvain Soliman for helpful discussions. The work has started when G. Pogudin worked at the Higher School of Economics, Moscow. The authors would like to thank the referees for their comments, which helped us improve the manuscript.

© Springer Nature Switzerland AG 2021
P. Flocchini and L. Moura (Eds.): IWOCA 2021, LNCS 12757, pp. 122–136, 2021.
https://doi.org/10.1007/978-3-030-79987-8_9

it into a system with quadratic right-hand side (see Definition 1). We illustrate the problem on a simple example of a scalar ODE:

$$x' = x^5. \tag{1}$$

The right-hand side has degree larger than two but if we introduce a new variable $y := x^4$, then we can write:

$$x' = xy, \quad \text{and} \quad y' = 4x^3x' = 4x^4y = 4y^2. \tag{2}$$

The right-hand sides of (2) are of degree at most two, and every solution of (1) is the x-component of some solution of (2).

A problem of finding such a transformation (*quadratization*) for an ODE system has appeared recently in several contexts:

- One of the recent approaches to *model order reduction* [11] uses quadratization as follows. For the ODE systems with quadratic right-hand side, there are dedicated model order reduction methods which can produce a better reduction than the general ones. Therefore, it can be beneficial to perform a quadratization first and then use the dedicated methods. For further details and examples of applications, we refer to [11,15,16,20].
- Quadratization has been used as a preprocessing step for *solving differential equations numerically* [6,12,14].
- Applied to *chemical reaction networks*, quadratization allows one to transform a given chemical reaction network into a bimolecular one [13].

It is known (e.g. [11, Theorem 3]) that it is always possible to perform quadratization with new variables being monomials in the original variables (like x^4 in the example above). We will call such quadratization *monomial* (see Definition 2). An algorithm for finding some monomial quadratization has been described in [11, Section G.]. In [13], the authors have shown that the problem of finding an optimal (i.e. of the smallest possible dimension) monomial quadratization is NP-hard. They also designed and implemented an algorithm for finding a monomial quadratization which is practical and yields an optimal monomial quadratization in many cases (but not always, see Sect. 3).

In this paper, we present an algorithm that computes an optimal monomial quadratization for a given system of ODEs. To the best of our knowledge, this is the first practical algorithm with the optimality guarantee. In terms of efficiency, our implementation compares favorably to the existing software [13] (see Table 3). The implementation is publicly available at https://github.com/ AndreyBychkov/QBee/. Our algorithm follows the classical Branch-and-Bound approach [17] together with problem-specific search and branching strategies and pruning rules (with one using the extremal graph theory, see Sect. 5.2).

Note that, according to [2], one may be able to find a quadratization of lower dimension by allowing the new variables to be arbitrary polynomials, not just monomials. We restrict ourselves to the monomial case because it is already challenging (e.g., includes an APX-hard [2]-sumset cover problem, see Remark 6) and monomial transformations are relevant for some application areas [13].

The rest of the paper is organized as follows. In Sect. 2, we state the problem precisely. In Sect. 3, we review the prior approaches, most notably [13]. Sections 4 and 5 describe our algorithm. Its performance is demonstrated and compared to [13] in Sect. 6. Sections 7 and 8 contain remarks on the complexity and conclusions/open problems, respectively.

2   Problem Statement

Definition 1. *Consider a system of ODEs*

$$x_1' = f_1(\bar{x}), \quad \dots, \quad x_n' = f_n(\bar{x}), \tag{3}$$

where $\mathbf{x} = (x_1, \dots, x_n)$ *and* $f_1, \dots, f_n \in \mathbb{C}[\mathbf{x}]$. *Then a list of new variables*

$$y_1 = g_1(\bar{x}), \dots, y_m = g_m(\bar{x}), \tag{4}$$

is said to be a quadratization *of* (3) *if there exist polynomials* $h_1, \dots, h_{m+n} \in \mathbb{C}[\bar{x}, \bar{y}]$ *of degree at most two such that*

- $x_i' = h_i(\bar{x}, \bar{y})$ *for every* $1 \leqslant i \leqslant n$;
- $y_j' = h_{j+n}(\bar{x}, \bar{y})$ *for every* $1 \leqslant j \leqslant m$.

The number m *will be called* the order of quadratization. *A* quadratization of *the smallest possible order will be called* an optimal quadratization.

Definition 2. *If all the polynomials* $g_1, \dots, g_m$ *are monomials, the quadratization is called* a monomial quadratization. *If a monomial quadratization of a system has the smallest possible order among all the monomial quadratizations of the system, it is called* an optimal monomial quadratization.

Now we are ready to precisely state the main problem we tackle:

Input A system of ODEs of the form (3).
Output An optimal monomial quadratization of the system.

Example 1. Consider a single scalar ODE $x' = x^5$ from (1), that is $f_1(x) = x^5$. As has been show in (2), $y = x^4$ is a quadratization of the ODE with $g(x) = x^4$, $h_1(x, y) = xy$, and $h_2(x, y) = 4y^2$. Moreover, this is a monomial quadratization.
Since the original ODE is not quadratic, the quadratization is optimal, so it is also an optimal monomial quadratization.

Example 2. The Rabinovich-Fabrikant system [19, Eq. (2)] is defined as follows:

$$x' = y(z - 1 + x^2) + ax, \quad y' = x(3z + 1 - x^2) + ay, \quad z' = -2z(b + xy).$$

Our algorithm finds an optimal monomial quadratization of order three: $z_1 = x^2, z_2 = xy, z_3 = y^2$. The resulting quadratic system is:

$$x' = y(z_1 + z - 1) + ax, \quad z_1' = 2z_1(a + z_2) + 2z_2(z - 1),$$
$$y' = x(3z + 1 - z_1) + ay, \quad z_2' = 2az_2 + z_1(3z + 1 - z_1 + z_3) + z_3(z - 1)$$
$$z' = -2z(b + z_2), \quad z_3' = 2az_3 + 2z_2(3z + 1 - z_1).$$

3 Discussion of Prior Approaches

To the best of our knowledge, the existing algorithms for quadratization are [11, Algotirhm 2] and [13, Algorithm 2]. The former has not been implemented and is not aimed at producing an optimal quadratization: it simply adds new variables until the system is quadratized, and its termination is based on [11, Theorem 2].

It has been shown [13, Theorem 2] that finding an optimal quadratization is NP-hard. The authors designed and implemented an algorithm for finding a small (but not necessarily optimal) monomial quadratization which proceeds as follows. For an n-dimensional system $\bar{x}' = \bar{f}(\bar{x})$, define, for every $1 \leqslant i \leqslant n$,

$$D_i := \max_{1 \leqslant j \leqslant n} \deg_{x_i} f_j.$$

Then consider the set

$$M := \{x_1^{d_1} \ldots x_n^{d_n} \mid 0 \leqslant d_1 \leqslant D_1, \ldots, 0 \leqslant d_n \leqslant D_n\}. \tag{5}$$

[4, Proof of Theorem 1] implies that there always exists a monomial quadratization with the new variables from M. The idea behind [13, Algorithm 2] is to search for an optimal quadratization *inside* M. This is done by an elegant encoding into a MAX-SAT problem.

However, it turns out that the set M does not necessarily contain an optimal monomial quadratization. As our algorithm shows, this happens, for example, for some of the benchmark problems from [13] (Hard and Monom series, see Table 3). Below we show a simpler example illustrating this phenomenon.

Example 3. Consider a system

$$x_1' = x_2^4, \quad x_2' = x_1^2. \tag{6}$$

Our algorithm shows that it has a unique optimal monomial quadratization

$$z_1 = x_1 x_2^2, \quad z_2 = x_2^3, \quad z_3 = x_1^3 \tag{7}$$

yielding the following quadratic ODE system:

$$x_1' = x_2 z_2, \qquad z_1' = x_2^6 + 2x_1^3 x_2 = z_2^2 + 2x_2 z_3, \qquad z_3' = 3x_1^2 x_2^4 = 3z_1^2,$$
$$x_2' = x_1^2, \qquad z_2' = 3x_1^2 x_2^2 = 3x_1 z_1.$$

The degree of (7) with respect to x_1 is larger than the x_1-degree of the original system (6), so such a quadratization will not be found by the algorithm [13].

It would be interesting to find an analogue of the set M from (5) always containing an optimal monomial quadratization as this would allow using powerful SAT-solvers. For all the examples we have considered, the following set worked

$$\widetilde{M} := \{x_1^{d_1} \ldots x_n^{d_n} \mid 0 \leqslant d_1, \ldots, d_n \leqslant D\}, \quad \text{where } D := \max_{1 \leqslant i \leqslant n} D_i.$$

4   Outline of the Algorithm

Our algorithm follows the general *Branch-and-Bound (B&B)* paradigm [17]. We will describe our algorithm using the standard B&B terminology (see, e.g., [17, Section 2.1]).

Definition 3 (B&B formulation for the quadratization problem).

- The search space *is a set of all monomial quadratizations of the input system* $\bar{x}' = \bar{f}(\bar{x})$.
- The objective function *to be minimized is the number of new variables introduced by a quadratization.*
- *Each* subproblem *is defined by a set of new monomial variables* $z_1(\bar{x}), \ldots, z_\ell(\bar{x})$ *and the corresponding subset of the search space is the set of all quadratizations including the variables* $z_1(\bar{x}), \ldots, z_\ell(\bar{x})$.

Definition 4 (Properties of a subproblem). *To each subproblem (see Definition 3) defined by new variables* $z_1(\bar{x}), \ldots, z_\ell(\bar{x})$, *we assign:*

1. *the set of* generalized variables, *denoted by V, consisting of the polynomials* $1, x_1, \ldots, x_n, z_1(\bar{x}), \ldots, z_\ell(\bar{x})$;
2. *the set of* nonsquares, *denoted by* NS, *consisting of all the monomials in the derivatives of the generalized variables which do not belong to* $V^2 := \{v_1 v_2 \mid v_1, v_2 \in V\}$. *In particular, a subproblem is a quadratization iff* NS $= \varnothing$.

Example 4. We will illustrate the notation introduced in Definition 4 on a system $x' = x^4 + x^3$ and a new variable $z_1(x) = x^3$. We have $z_1' = 3x^2 x' = 3x^6 + 3x^5$. Therefore, for this subproblem, we have:

$$V = \{1, x, x^3\}, \quad V^2 = \{1, x, x^2, x^3, x^4, x^6\}, \quad \text{NS} = \{x^5\}.$$

In order to organize a B&B search in the search space defined above, we define several subroutines/strategies answering the following questions:

- *How to set the original bound?* [4, Theorem 1] implies that the set M from (5) gives a quadratization of the original system, so it can be used as the starting incumbent solution.
- *How to explore the search space?* There are two subquestions:
  - *What are the child subproblems of a given subproblem (branching strategy)?* This is described in Sect. 4.1.
  - *In what order we traverse the tree of the subproblems?* We use DFS (to make new incumbents appear earlier) guided by a heuristic as described in Algorithm 1.
- *How to prune the search tree (pruning strategy)?* We use two algorithms for computing a lower bound for the objective function in a given subtree, they are described and justified in Sect. 5.

4.1 Branching Strategy

Let $\bar{x}' = \bar{f}(\bar{x})$ be the input system. Consider a subproblem defined by new monomial variables $z_1(\bar{x}), \dots, z_\ell(\bar{x})$. The child subproblems will be constructed as follows:

1. among the nonsquares (NS, see Definition 4), choose any monomial $m = x_1^{d_1} \dots x_n^{d_n}$ with the value $\prod_{i=1}^{n}(d_i + 1)$ the smallest possible;
2. for every decomposition $m = m_1 m_2$ as a product of two monomials, define a new subproblem by adding the elements of $\{m_1, m_2\} \setminus V$ (see Definition 4) as new variables. Since $m \in$ NS, at least one new variable will be added.

The score function $\prod_{i=1}^{n}(d_i + 1)$ is twice the number of representations $m = m_1 m_2$, so this way we reduce the branching factor of the algorithm.

Lemma 1. *Any optimal subproblem $z_1(\bar{x}), \dots, z_\ell(\bar{x})$ is a solution of at least one of the children subproblems generated by the procedure above.*

Proof. Let $z_1(\bar{x}), \dots, z_n(\bar{x})$ be any solution of the subproblem. Since m must be either of the form $z_i z_j$ or z_j, it will be a solution of the child subproblem corresponding to the decomposition $m = z_i z_j$ or $m = 1 \cdot z_j$, respectively.

Example 5. Figure 1 below show the graph representation of system $x' = x^4 + x^3$ from Example 4. The starting vertex is $\varnothing$. The underlined vertices correspond to optimal quadratizations, so the algorithm will return one of them. On the first step, the algorithm chooses the monomial x^3 which has two decompositions $x^3 = x \cdot x^2$ and $x^3 = 1 \cdot x^3$ yielding the left and the right children of the root, respectively. The subproblem $\{x^3\}$ was described in more details in Example 4.

The score function $\prod_{i=1}^{n}(d_i + 1)$ for the decompositions $x^3 = x \cdot x^2$ and $x^3 = 1 \cdot x^3$ takes values 6 and 4, respectively. Hence the algorithm will first explore the branch on the right.

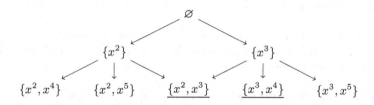

Fig. 1. Graph illustration for equation $x' = x^4 + x^3$

4.2   Recursive Step of the Algorithm

The recursive step of our algorithm can be described as follows.

Algorithm 1: Branch and Bound recursive step

Input
- polynomial ODE system $\bar{x}' = \bar{f}(\bar{x})$;
- set of new variables $z_1(\bar{x}), \ldots, z_\ell(\bar{x})$;
- an optimal quadratization found so far (incumbent) with N new variables.

Output the algorithm replaces the incumbent with a more optimal quadratization containing $z_1(\bar{x}), \ldots, z_\ell(\bar{x})$ if such quadratization exists.

(Step 1) if $z_1(\bar{x}), \ldots, z_\ell(\bar{x})$ is a quadratization
      (a) if $\ell < N$, replace the incumbent with $z_1(\bar{x}), \ldots, z_\ell(\bar{x})$;
      (b) **return**;
(Step 2) if any of the pruning rules (Algorithm 2 or 3) applied to $z_1(\bar{x}), \ldots, z_\ell(\bar{x})$ and N return True, **return**;
(Step 3) generate set C of child subproblems as described in Section 4.1
(Step 4) sort C in increasing order w.r.t. $S + n|V|$, where S is the sum of the degrees of the elements in V (V is different for different subproblems as defined in Definition 4);
(Step 5) for each element of C, call Algorithm 1 on it.

---

5   Pruning Rules

In this section, we present two pruning rules yielding a substantial speedup of the algorithm: based on a quadratic upper bound and based on squarefree graphs.

Property 1. Each pruning rule has the following input-output specification:

Input:

- the original ODE system $\bar{x}' = \bar{f}(\bar{x})$;
- already added new variables $z_1(\bar{x}), \ldots, z_\ell(\bar{x})$ which are monomials in $\bar{x}$;
- positive integer N.

Output: True if it is guaranteed that the set of new variables $z_1(\bar{x}), \ldots, z_s(\bar{x})$ cannot be extended to a monomial quadratization of $\bar{x}' = \bar{f}(\bar{x})$ of order less then N. False otherwise.

Note that, if False is returned, it does not imply that the set of new variables can be extended.

Remark 1. Both pruning rules presented here actually check a stronger condition: whether the set of new variables can be extended by at most $N - s$ variables so that all the monomials NS in the *current* subproblem can be written as a product of two generalized variables. It would be very interesting to strengthen these rules by taking into account the derivatives of the extra new variables.

5.1 Rule Based on Quadratic Upper Bound

Remark 2 (Intuition behind the rule). Consider a subproblem with the generalized variables V and set of nonsquares NS (see Definition 4). Assume that it can be quadratized by adding a set W of variables. This would imply that $\mathrm{NS} \subseteq (V \cup W)^2$. This yields a bound

$$|\mathrm{NS}| \leqslant \frac{(|V| + |W|)(|V| + |W| + 1)}{2}. \tag{8}$$

The general ideal of the rule is: since $|V|$ and $|\mathrm{NS}|$ are known, (8) can be used to find a lower bound for $|W|$. However, a straightforward application of (8) does not lead to noticeable performance improvements. We found that one can do much better by first estimating the number of elements of $\mathrm{NS} \cap (V \cdot W)$ and then applying an argument as in (8) to $\mathrm{NS} \setminus (V \cdot W)$ and W.

Algorithm 2: Pruning rule: based on a quadratic upper bound

(Step 1) Compute the following multiset of monomials in $\bar{x}$

$$D := \{m/v \mid m \in \mathrm{NS}, v \in V, v \mid m\}.$$

(Step 2) Let `mult` be the list of multiplicities of the elements of D sorted in the descending order.

(Step 3) Find the smallest integer k such that

$$|\mathrm{NS}| \leqslant \sum_{i=1}^{k} \mathrm{mult}[i] + \frac{k(k+1)}{2}. \tag{9}$$

(We use 1-based indexing and set $\mathrm{mult}[i] = 0$ for $i > |\mathrm{mult}|$)

(Step 4) If $k + \ell \geqslant N$, return `True`. Otherwise, return `False`.

Lemma 2. *Algorithm 2 satisfied the specification described in Property 1.*

Proof. Assume that Algorithm 2 has returned `True`. Consider any quadratization $z_1, \ldots, z_{\ell+r}$ of $\bar{x}' = \bar{f}(\bar{x})$ extending $z_1, \ldots, z_\ell$. We define $\widetilde{V}$, a superset of V, as $\{1, x_1, \ldots, x_n, z_1, \ldots, z_{\ell+r}\}$. By the definition of quadratization, $\mathrm{NS} \subseteq \widetilde{V}^2$. We split NS into two subsets $\mathrm{NS}_0 := \mathrm{NS} \cap (V \cdot \widetilde{V})$ and $\mathrm{NS}_1 := \mathrm{NS} \setminus \mathrm{NS}_0$. For every $1 \leqslant i \leqslant r$, the cardinality of $\mathrm{NS} \cap (z_{\ell+i} \cdot V)$ does not exceed the multiplicity of $z_{\ell+i}$ in the multiset D constructed at **(Step 1)**. Therefore, $|\mathrm{NS}_0| \leqslant \sum_{i=1}^{r} \mathrm{mult}[i]$. The number of products of the form $z_{\ell+i} z_{\ell+j}$ with $1 \leqslant i \leqslant j \leqslant r$ does not exceed $\frac{r(r+1)}{2}$. Therefore, we have

$$|\mathrm{NS}| = |\mathrm{NS}_0| + |\mathrm{NS}_1| \leqslant \sum_{i=1}^{r} \mathrm{mult}[i] + \frac{r(r+1)}{2},$$

so r satisfies (9). The minimality of k implies $r \geqslant k$. Thus, $r + \ell \geqslant N$, so $z_1, \ldots, z_\ell$ cannot be extended to a quadratization of order less than N.

5.2 Rule Based on Squarefree Graphs

Remark 3 (Intuition behind the rule). We will illustrate the idea behind the rule in a simple example. Assume that we have five monomials $m_1, \ldots, m_5$ such that none of them is a square. Assume also that there is a set V of monomial new variables such that $|V| = 4$ and $m_i \in V^2$ for every i. Since none of m_i's is a square, it can be written as $m_i = z_{i,1} z_{i,2}$ for distinct $z_{i,1}, z_{i,2} \in V$. We can therefore think about a graph with vertices being elements of V and edges given by $m_1, \ldots, m_5$. One can check that every graph with four vertices and five edges must contain a four-cycle. Let the cycle consist of edges m_1, m_2, m_3, m_4 in this order. Then, for some numbering of elements in V, we have:

$$m_1 = z_1 z_2, \; m_2 = z_2 z_3, \; m_3 = z_3 z_4, \; m_4 = z_4 z_1 \implies m_1 m_3 = m_2 m_4.$$

Thus, by checking that all pairwise product of $m_1, \ldots, m_5$ are distinct, we can verify that $m_1, \ldots, m_5 \in V^2$ implies that $|V| > 4$.

In order to take into account the monomials which are squares, we consider not just graphs but pseudographs. We also employ the separation strategy NS = $(\text{NS} \cap (V \cdot W)) \cup (\text{NS} \setminus (V \cdot W))$ as described in Remark 2.

Definition 5. *A pseudograph G (i.e., a graph with loops and multiple edges allowed) is called $C4^*$-free if there is no cycle of length four in G with every two adjacent edges being distinct (repetition of edges and/or vertices is allowed).*

Example 6. A $C4^*$-free pseudograph cannot contain:

- *A vertex with two loops.* If the loops are ℓ_1 and ℓ_2 then the cycle $\ell_1, \ell_2, \ell_1, \ell_2$ will violate $C4^*$-freeness.
- *Multiple edges.* If e_1 and e_2 are edges with the same endpoints, then e_1, e_2, e_1, e_2 will violate $C4^*$-freeness.
- *Two vertices with loops connected by an edge.* If the loops are ℓ_1 and ℓ_2 and the edge is e, then ℓ_1, e, ℓ_2, e will violate $C4^*$-freeness.

Definition 6. *By $C(n, m)$ we denote the largest possible number of edges in a $C4^*$-free pseudograph G with n vertices and at most m loops.*

Remark 4. Note that the example above implies that $C(n, n + k) = C(n, n)$ for every positive integer k because a $C4^*$-free pseudograph cannot contain more than n loops.

The number $C(n, 0)$ is the maximal number of edges in a $C4$-free graph and has been extensively studied (e.g. [1,5,7,9]). Values for $n \leqslant 31$ are available as a sequence A006855 in OEIS [18].

In Algorithm 3, we use the exact values for $C(n, m)$ found by an exhaustive search and collected in Table 1 for $n \leqslant 7$. The script for the search is available at https://github.com/AndreyBychkov/QBee/blob/0.5.0/qbee/no_C4_count.py. For $n > 7$, we use the following bound

$$C(n, m) \leqslant C(n, 0) + m \leqslant \frac{n}{2}(1 + \sqrt{4n - 3}) + m,$$

Table 1. Exact values for $C(n, m)$ (see Definition 6).

n	m							
	0	1	2	3	4	5	6	7
1	0	1						
2	1	2	2					
3	3	3	4	4				
4	4	5	5	6	6			
5	6	6	7	7	8	8		
6	7	8	9	9	9	10	10	
7	9	10	11	12	12	12	12	12

where the bound for $C(n, 0)$ is due to [10, Chapter 23, Theorem 1.3.3].

Algorithm 3: Pruning rule: based on squarefree graphs

(Step 1) Compute a subset $E = \{m_1, \ldots, m_e\} \subseteq$ NS such that all the products $m_i m_j$ for $1 \leqslant i \leqslant j \leqslant e$ are distinct.
(done by traversing NS in a descending order w.r.t. the total degree and appending each monomial if it does not violate the property)

(Step 2) Compute the following multiset of monomials in **x**

$$D := \{m/v \mid m \in E, v \in V, v \mid m\}.$$

(Step 3) Let `mult` be the list of multiplicities of the elements of D sorted in descending order.

(Step 4) Let c be the number of elements in E with all the degrees being even.

(Step 5) Find the smallest integer k such that

$$|E| \leqslant \sum_{i=1}^{k} \text{mult}[i] + C(k, c). \tag{10}$$

(We use 1-based indexing and set $\text{mult}[i] = 0$ for $i > |\text{mult}|$)

(Step 6) If $k + \ell \geqslant N$, return `True`. Otherwise, return `False`.

Lemma 3. *Algorithm 3 satisfied the specification described in Property 1.*

Proof. Assume that Algorithm 2 has returned `True`. Consider any quadratization $z_1, \ldots, z_{\ell+r}$ of $\bar{x}' = \bar{f}(\bar{x})$ extending $z_1, \ldots, z_\ell$. We define $\widetilde{V}$, a superset of V, as $\{1, x_1, \ldots, x_n, z_1, \ldots, z_{\ell+r}\}$. By the definition of quadratization, $E \subseteq$ NS $\subseteq \widetilde{V}^2$. Similarly to the proof of Lemma 2, we split E into two subsets

$$E_0 := E \cap (V \cdot \widetilde{V}) \quad \text{and} \quad E_1 := E \setminus E_0.$$

For every $1 \leqslant i \leqslant r$, the cardinality of $E \cap (z_{\ell+i} \cdot V)$ does not exceed the multiplicity of $z_{\ell+i}$ in the multiset D from **(Step 2)**. Therefore, $|E_0| \leqslant \sum_{i=1}^{r} \text{mult}[i]$.

Consider a pseudograph G with r vertices numbered from 1 to r corresponding to $z_{\ell+1}, \ldots, z_{\ell+r}$, respectively. For every element $m \in E_1$, we fix a representation $m = z_{\ell+i} z_{\ell+j}$, and add an edge connecting vertices i and j in G (this will be a loop of $i = j$). We claim that pseudograph G will be $C4^*$-free. Indeed, if there is a cycle formed by edges $m_1, m_2, m_3, m_4 \in E_0$, then we will have $m_1 \cdot m_3 = m_2 \cdot m_4$. Moreover, $\{m_1, m_3\} \cap \{m_2, m_4\} = \varnothing$, so such a relation contradicts the condition on E imposed by **(Step 1)**. Finally, a monomial $m \in E$ can correspond to a loop in G only if it is a square, that is, all the degrees in m are even. Hence E_1, the total number of edges in G, does not exceed $C(r, c)$

In total, we have

$$|E| = |E_0| + |E_1| \leqslant \sum_{i=1}^{r} \text{mult}[i] + C(r, c),$$

so r satisfies (10). The minimality of k implies that $r \geqslant k$. Thus, $r + \ell \geqslant N$, so $z_1, \ldots, z_\ell$ cannot be extended to a quadratization of order less than N.

Remark 5 (Cycles of even length). One can modify this rule to use graphs not containing cycles of even length. In this case, the set E from **(Step1)** of Algorithm 3 would satisfy the condition that there are no multi-subsets of equal cardinality and with equal product. However, this approach did not work that well in practice, in particular, due to the overhead for finding such E.

5.3   Performance of the Pruning Rules

Table 2 below shows the performance of our algorithm with a different combination of the pruning rules employed. It shows that the rules substantially speed up the computation and that Algorithm 3 is particularly successful in higher dimensions.

Table 2. Comparison of the pruning rules used by our algorithm. Values in the cells represent an average time with the standard deviation in seconds.

ODE system	Dimension	No pruning	Algorithm 2	Algorithm 3	Algorithm 2 & 3
Circular(8)	2	4293 ± 445	497 ± 5	526 ± 8	453 ± 7
Hill(20)	3	3.4 ± 0.1	3.0 ± 0.1	2.4 ± 0.1	2.4 ± 0.1
Hard(2)	3	106.3 ± 1.0	19.6 ± 1.1	20.1 ± 0.6	16.7 ± 0.6
Hard(4)	3	360.1 ± 5.6	107.5 ± 2.4	108.8 ± 2.1	96.6 ± 1.5
Monom(3)	3	552.9 ± 10.9	85.7 ± 4.2	124.7 ± 5.5	84.2 ± 3.3
Cubic Cycle(6)	6	187.3 ± 0.8	43.6 ± 0.6	20.0 ± 0.5	20.1 ± 0.3
Cubic Cycle(7)	7	2002 ± 6.4	360.7 ± 1.1	150.2 ± 1.3	160.9 ± 5.9
Cubic Bicycle(7)	7	1742 ± 89	73.2 ± 0.6	29.8 ± 0.3	30.5 ± 0.2
Cubic Bicycle(8)	8	$4440+$	175.4 ± 4.0	64.8 ± 0.5	68.9 ± 0.7

6   Performance and Results

We have implemented our algorithm in Python, and the implementation is available at https://github.com/AndreyBychkov/QBee/tree/0.5.0. We compare our algorithm with the one proposed in [13]. For the comparison, we use the set of benchmarks from [13] and add a couple of new ones (described in the Appendix).

The results of the comparison are collected in Table 3. All computation times are given either in milliseconds or in seconds and were obtained on a laptop with the following parameters: Intel(R) Core(TM) i7-8750H CPU @ 2.20 GHz, WSL Windows 10 Ubuntu 20.04, CPython 3.8.5. From the table, we see that the only cases when the algorithm from [13] runs faster are when it does not produce an optimal quadratization (while we do). Also, cases, when the algorithm from [13] is not able to terminate, are marked as "—" symbol.

Table 3. Comparison of our implementation with the algorithm [13] on a set benchmarks .

ODE system	Biocham time	Biocham order	Our time	Our order
Circular(3), **ms**	83.2 ± 0.1	3	5.1 ± 0.1	3
Circular(4), **ms**	106.7 ± 2.3	4	164.8 ± 32.3	4
Circular(5), **ms**	596.2 ± 10.9	4	20.0 ± 0.1	4
Circular(6), **s**	37.6 ± 0.4	5	4.2 ± 0.1	5
Circular(8), **s**	—	—	453.3 ± 6.9	6
Hard(3), **s**	1.09 ± 0.01	11	8.6 ± 0.2	9
Hard(4), **s**	20.2 ± 0.3	13	96.9 ± 1.5	10
Hill(5), **ms**	87.8 ± 0.9	2	4.6 ± 0.0	2
Hill(10), **ms**	409.8 ± 5.6	4	49.7 ± 1.3	4
Hill(15), **s**	64.1 ± 0.4	5	0.34 ± 0.1	5
Hill(20),**s**	—	—	2.4 ± 0.1	6
Monom(2), **ms**	96.4 ± 1.6	4	15 ± 0.1	3
Monom(3), **s**	0.44 ± 0	13	84.2 ± 3.3	10
Cubic Cycle(6), **s**	—	—	20.1 ± 0.3	12
Cubic Cycle(7), **s**	—	—	160.9 ± 5.9	14
Cubic Bicycle(7), **s**	—	—	30.5 ± 0.2	14
Cubic Bicycle(8), **s**	—	—	68.9 ± 0.7	16

7   Remarks on the Complexity

It has been conjectured in [13, Conjecture 1] that the size of an optimal monomial quadratization may be exponential in the number of monomials of the input system in the worst case. Interestingly, this is not the case if one allows monomials with negative powers (i.e., Laurent monomials): Proposition 1 shows that there exists a quadratization with the number of new variables being linear in the number of monomials in the system.

Proposition 1. *Let $\bar{x}' = \bar{f}(\bar{x})$, where $\bar{x} = (x_1, \ldots, x_n)$, be a system of ODEs with polynomial right hand sides. For every $1 \leqslant i \leqslant n$, we denote the monomials in the right-hand side of the i-th equation by $m_{i,1}, \ldots, m_{i,k_i}$. Then the following set of new variables (given by Laurent monomials) is a quadratization of the original system:*

$$z_{i,j} := \frac{m_{i,j}}{x_i} \text{ for every } 1 \leqslant i \leqslant n, \ 1 \leqslant j \leqslant k_i.$$

Proof. Since $m_{i,j} = z_{i,j}x_i$, the original equations can be written as quadratic in the new variables. Let the coefficient in the original system in front of $m_{i,j}$ be denoted by $c_{i,j}$. We consider any $1 \leqslant i \leqslant n, \ 1 \leqslant j \leqslant k_j$:

$$z'_{i,j} = \sum_{s=1}^{n} f_s(\mathbf{x}) \frac{\partial z_{i,j}}{\partial x_s} = \sum_{s=1}^{n} \sum_{r=1}^{k_s} c_{s,r} m_{s,r} \frac{\partial z_{i,j}}{\partial x_s}.$$

Since $\frac{\partial z_{i,j}}{\partial x_s}$ is proportional to $\frac{z_{i,j}}{x_s}$, the monomial $m_{s,r} \frac{\partial z_{i,j}}{\partial x_s}$ is proportional to a quadratic monomial $z_{s,r} z_{i,j}$, so we are done.

Remark 6 (Relation to the [2]-sumset cover problem). The [2]-sumset cover problem [3] is, given a finite set $S \subset \mathbb{Z}_{>0}$ of positive integers, find a smallest set $X \subset \mathbb{Z}_{>0}$ such that $S \subset X \cup \{x_i + x_j \mid x_i, x_j \in X\}$. It has been shown in [8, Proposition 1] that the [2]-sumset cover problem is APX-hard, moreover, the set S used in the proof contains 1. We will show how to encode this problem into the optimal monomial quadratization problem thus showing that the latter is also APX-hard (in the number of monomials, but not necessarily in the size of the input). For $S = \{s_1, \ldots, s_n\} \subset \mathbb{Z}_{>0}$ with $s_1 = 1$, we define a system

$$x'_1 = 0, \quad x'_2 = \sum_{i=1}^{n} x_1^{s_i}.$$

Then a set $X = \{1, a_1, \ldots, a_\ell\}$ is a minimal [2]-sumset cover of S iff $x_1^{a_1}, \ldots, x_1^{a_\ell}$ is an optimal monomial quadratization of the system.

8   Conclusions and Open Problems

In this paper, we have presented the first practical algorithm for finding an optimal monomial quadratization. Our implementation compares favorably with the existing software and allows us to find better quadratizations for already used benchmark problems. We were able to compute quadratization for ODE systems which could not be tackled before.

We would like to mention several interesting open problems:

1. Is it possible to describe a finite set of monomials which must contain an optimal quadratization? This would allow using SAT-solving techniques of [13] as described in Sect. 3.
2. As has been shown in [2], general polynomial quadratization may be of a smaller dimension than an optimal monomial quadratization. This poses a challenge: design an algorithm for finding optimal polynomial quadratization (or at least a smaller one than an optimal monomial).
3. How to search for optimal monomial quadratizations if negative powers are allowed (see Sect. 7)?
4. How to design a faster algorithm for approximate quadratization (that is, finding a quadratization which is close to the optimal) with guarantees on the quality of the approximation?

Appendix: Benchmark Systems

Most of the benchmark systems used in this paper (in Tables 2 and 3) are described in [13]. Here we show additional benchmarks we have introduced:

1. *Cubic Cycle(n).* For every integer $n > 1$, we define a system in variables $x_1, \ldots, x_n$ by
$$x_1' = x_2^3, \ x_2' = x_3^3, \ \ldots, \ x_n' = x_1^3.$$

2. *Cubic Bicycle(n).* For every integer $n > 1$, we define a system in variables $x_1, \ldots, x_n$ by

$$x_1' = x_n^3 + x_2^3, \ x_2' = x_1^3 + x_3^3, \ \ldots, \ x_n' = x_{n-1}^3 + x_1^3.$$

References

1. Abreu, M., Balbuena, C., Labbate, D.: Adjacency matrices of polarity graphs and of other C4-free graphs of large size. Des. Codes Crypt. **55**(2–3), 221–233 (2010). https://doi.org/10.1007/s10623-010-9364-1
2. Alauddin, F.: Quadratization of ODEs: monomial vs. non-monomial. SIAM Undergraduate Res. Online **14** (2021). https://doi.org/10.1137/20s1360578
3. Bulteau, L., Fertin, G., Rizzi, R., Vialette, S.: Some algorithmic results for [2]-sumset covers. Inf. Process. Lett. **115**(1), 1–5 (2015). https://doi.org/10.1016/j.ipl.2014.07.008
4. Carothers, D.C., Parker, G.E., Sochacki, J.S., Warne, P.G.: Some properties of solutions to polynomial systems of differential equations. Electron. J. Diff. Eqns. **2005**(40), 1–17 (2005). http://emis.impa.br/EMIS/journals/EJDE/Volumes/2005/40/carothers.pdf
5. Clapham, C.R.J., Flockhart, A., Sheehan, J.: Graphs without four-cycles. J. Graph Theor. **13**(1), 29–47 (1989). https://doi.org/10.1002/jgt.3190130107
6. Cochelin, B., Vergez, C.: A high order purely frequency-based harmonic balance formulation for continuation of periodic solutions. J. Sound Vib. **324**(1–2), 243–262 (2009). https://doi.org/10.1016/j.jsv.2009.01.054

7. Erdös, P., Rényi, A., Sós, V.: On a problem of graph theory. Studia Sci. Math. Hungar. **1**, 215–235 (1966)
8. Fagnot, I., Fertin, G., Vialette, S.: On finding small 2-generating sets. In: Ngo, H.Q. (ed.) COCOON 2009. LNCS, vol. 5609, pp. 378–387. Springer, Heidelberg (2009). https://doi.org/10.1007/978-3-642-02882-3_38
9. Füredi, Z.: On the number of edges of quadrilateral-free graphs. J. Comb. Theor. Ser. B **68**(1), 1–6 (1996). https://doi.org/10.1006/jctb.1996.0052
10. Graham, R., Grotschel, M., Lovász, L.: Handbook of Combinatorics, vol. 2. North Holland (1995)
11. Gu, C.: QLMOR: a projection-based nonlinear model order reduction approach using quadratic-linear representation of nonlinear systems. IEEE Trans. Comput. Aided Des. Integr. Circuits Syst. **30**(9), 1307–1320 (2011). https://doi.org/10.1109/TCAD.2011.2142184
12. Guillot, L., Cochelin, B., Vergez, C.: A Taylor series-based continuation method for solutions of dynamical systems. Nonlinear Dyn. **98**(4), 2827–2845 (2019). https://doi.org/10.1007/s11071-019-04989-5
13. Hemery, M., Fages, F., Soliman, S.: On the complexity of quadratization for polynomial differential equations. In: Abate, A., Petrov, T., Wolf, V. (eds.) CMSB 2020. LNCS, vol. 12314, pp. 120–140. Springer, Cham (2020). https://doi.org/10.1007/978-3-030-60327-4_7
14. Karkar, S., Cochelin, B., Vergez, C.: A high-order, purely frequency based harmonic balance formulation for continuation of periodic solutions: the case of non-polynomial nonlinearities. J. Sound Vib. **332**(4), 968–977 (2013). https://doi.org/10.1016/j.jsv.2012.09.033
15. Kramer, B., Willcox, K.E.: Balanced truncation model reduction for lifted nonlinear systems (2019). https://arxiv.org/abs/1907.12084
16. Kramer, B., Willcox, K.E.: Nonlinear model order reduction via lifting transformations and proper orthogonal decomposition. AIAA J. **57**(6), 2297–2307 (2019). https://doi.org/10.2514/1.J057791
17. Morrison, D.R., Jacobson, S.H., Sauppe, J.J., Sewell, E.C.: Branch-and-bound algorithms: a survey of recent advances in searching, branching, and pruning. Discret. Optim. **19**, 79–102 (2016). https://doi.org/10.1016/j.disopt.2016.01.005
18. OEIS Foundation Inc.: The on-line encyclopedia of integer sequences. http://oeis.org
19. Rabinovich, M.I., Fabrikant, A.L.: Stochastic self-modulation of waves in nonequilibrium media. J. Exp. Theor. Phys. **77**, 617–629 (1979)
20. Ritschel, T.K., Weiß, F., Baumann, M., Grundel, S.: Nonlinear model reduction of dynamical power grid models using quadratization and balanced truncation. at-Automatisierungstechnik **68**(12), 1022–1034 (2020). https://doi.org/10.1515/auto-2020-0070

A Hamilton Cycle in the k-Sided Pancake Network

B. Cameron[1], J. Sawada[1(✉)], and A. Williams[2]

[1] University of Guelph, Guelph, Canada
jsawada@uoguelph.ca
[2] Williams College, Williamstown, USA

Abstract. We present a Hamilton cycle in the k-sided pancake network and four combinatorial algorithms to traverse the cycle. The network's vertices are coloured permutations $\pi = p_1 p_2 \cdots p_n$, where each p_i has an associated colour in $\{0, 1, \ldots, k-1\}$. There is a directed edge (π_1, π_2) if π_2 can be obtained from π_1 by a "flip" of length j, which reverses the first j elements and increments their colour modulo k. Our particular cycle is created using a greedy min-flip strategy, and the average flip length of the edges we use is bounded by a constant. By reinterpreting the order recursively, we can generate successive coloured permutations in $O(1)$-amortized time, or each successive flip by a loop-free algorithm. We also show how to compute the successor of any coloured permutation in $O(n)$-time. Our greedy min-flip construction generalizes known Hamilton cycles for the pancake network (where $k = 1$) and the burnt pancake network (where $k = 2$). Interestingly, a greedy max-flip strategy works on the pancake and burnt pancake networks, but it does not work on the k-sided network when $k > 2$.

1 Introduction

Many readers will be familiar with the story of Harry Dweighter, the harried waiter who sorts stacks of pancakes for his customers. He does this by repeatedly grabbing some number of pancakes from the top of the stack, and flipping them over. For example, if the chef in the kitchen creates the stack 🥞, then Harry can sort it by flipping over all four pancakes 🥞, and then the top two 🥞.

This story came from the imagination of Jacob E. Goodman [9], who was inspired by sorting folded towels [24]. His original interest was an upper bound on the number of flips required to sort a stack of n pancakes. Despite its whimsical origins, the problem attached interest from many mathematicians and computer scientists, including a young Bill Gates [12]. Eventually, it also found serious applications, including genomics [11].

A variation of the original story involves burnt pancakes. In this case, each pancake has two distinct sides: burnt and unburnt. When Harry flips the pancakes, the pancakes involved in the flip also turn over, and Harry wants to sort the pancakes so that the unburnt sides are facing up. For example, Harry could sort the stack 🥞 by flipping all four 🥞, then the top two 🥞, and the top

© Springer Nature Switzerland AG 2021
P. Flocchini and L. Moura (Eds.): IWOCA 2021, LNCS 12757, pp. 137–151, 2021.
https://doi.org/10.1007/978-3-030-79987-8_10

one 🥞. Similar lines of research developed around this problem (e.g. [6,11]). The physical model breaks down beyond two sides, however, many of the same applications do generalize to "k-sided pancakes".

1.1 Pancake Networks

Interconnection networks connect single processors, or groups of processors, together. In this context, the underlying graph is known as the network, and classic graph measurements (e.g. diameter, girth, connectivity) translate to different performance metrics. Two networks related to pancake flipping are in Fig. 1.

The pancake network $\mathbb{G}(n)$ was introduced in the 1980s [1] and various measurements were established (e.g. [14]). Its vertex set is the set of permutations of $\{1, 2, \ldots, n\}$ in one-line notation, which is denoted $\mathbb{P}(n)$. For example, $\mathbb{P}(2) = \{12, 21\}$. There is an edge between permutations that differ by a *prefix-reversal of length* ℓ, which reverses the first ℓ symbols. For example, $(3421, 4321)$ is the $\ell = 2$ edge between 🥞 and 🥞. Goodman's original problem is finding the maximum shortest path length to the identity permutation. Since $\mathbb{G}(n)$ is vertex-transitive, this value is simply its diameter.

The burnt pancake network $\overline{\mathbb{G}}(n)$ was introduced in the 1990s [6]. Its vertex set is the set of signed permutations of $\{1, 2, \ldots, n\}$, which is denoted $\overline{\mathbb{P}}(n)$. For example, $\overline{\mathbb{P}}(2) = \{12, 1\overline{2}, \overline{1}2, \overline{1}\overline{2}, 21, 2\overline{1}, \overline{2}1, \overline{2}\overline{1}\}$ where overlines denote negative symbols. There is an edge between signed permutations that differ by a *sign-complementing prefix-reversal of length* ℓ, which reverses the order and sign of the first ℓ symbols. For example, $(\overline{2}134, 1234)$ is the $\ell = 2$ edge between 🥞 and 🥞.

The k-*sided pancake network* $\mathbb{G}_k(n)$ is a directed graph that was first studied in the 2000s [15]. Its vertex set is the set of k-coloured permutations of $\{1, 2, \ldots, n\}$ in one-line notation, which is denoted $\mathbb{P}_k(n)$. For example, $\mathbb{P}_3(2)$ is illustrated below, where colours the 0, 1, 2 are denoted using superscripts, or in **black**, red, **blue**.

$$\mathbb{P}_3(2) = \{\mathbf{12}, \mathbf{12}, \mathbf{12}, \mathbf{12}, \mathbf{12}, \mathbf{12}, \mathbf{12}, \mathbf{12}, \mathbf{12}, \mathbf{21}, \mathbf{21}, \mathbf{21}, \mathbf{21}, \mathbf{21}, \mathbf{21}, \mathbf{21}, \mathbf{21}\}$$
$$= \{1^0 2^0, 1^0 2^1, 1^0 2^2, 1^1 2^0, 1^1 2^1, 1^1 2^2, 1^2 2^0, 1^2 2^1, 1^2 2^2, 2^0 1^0, \ldots, 2^2 1^1, 2^2 1^2\}.$$

There is a directed edge from $\pi_1 \in \mathbb{P}_k(n)$ to $\pi_2 \in \mathbb{P}_k(n)$ if π_1 can be transformed into π_2 by a *colour-incrementing prefix-reversal of length* ℓ, which reverses the order and increments the colour modulo k of the first ℓ symbols. For example, $(2134, 1234) = (2^1 1^2 3^0 4^0, 1^0 2^2 3^0 4^0)$ is a directed $\ell = 2$ edge.

Notice that $\mathbb{G}(n)$ and $\mathbb{G}_1(n)$ are isomorphic, while $\overline{\mathbb{G}}(n)$ and $\mathbb{G}_2(n)$ are isomorphic, so long as we view each undirected edge as two opposing directed edges. It also bears mentioning that $\mathbb{G}_k(n)$ is a (connected) directed Cayley graph, and its underlying group is the wreath product of the cyclic group of order k and the symmetric group of order n.

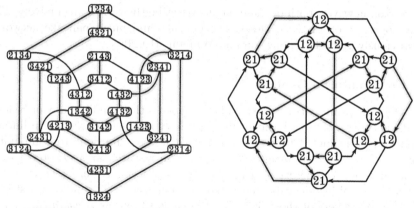

(a) The pancake network $\mathbb{G}(4)$.   (b) The 3-sided pancake network $\mathbb{G}_3(2)$.

Fig. 1. Hamilton cycles in a pancake network and a coloured pancake network. The highlighted cycles start at $12\cdots n$ (or $1^0 2^0 \cdots n^0$) and are constructed by the greedy min-flip strategy. The colours $0, 1, 2$ in (b) correspond to black, red, and blue. (Color figure online)

When the context is clear, or the distinction is not necessary, we use the term *flip* for prefix-reversal (when $k = 1$), sign-complementing prefix-reversal (when $k = 2$), and colour-incrementing prefix-reversal (when $k > 2$).

1.2   (Greedy) Hamilton Cycles

In this paper, we are not interested in shortest paths in pancake networks, but rather Hamilton cycles. There are myriad ways that researchers attempt to build Hamilton cycles in highly-symmetric graphs, and the greedy approach is perhaps the simplest (see Williams [26]). This approach initializes a path at a specific vertex, then repeatedly extends the path by a single edge. More specifically, it uses the highest priority edge (according to some criteria) that leads to a vertex that is not on the path. The path stops growing when the current vertex is only adjacent to vertices on the path. A Hamilton cycle has been found if every vertex is on the path, and there is an edge from the final vertex to the first vertex. Despite its simplicity, the approach is known to work on many well-known graphs [26].

We show that the greedy approach works for the coloured pancake network $\mathbb{G}_k(n)$ when we prioritize the edges by shortest flip length. More specifically, we start a path at $1^0 2^0 \cdots n^0 \in \mathbb{P}_k(n)$, then repeatedly extend it to a new vertex along the edge that corresponds to the shortest colour-incrementing prefix-reversal. We refer to this as the *greedy min-flip construction*, denoted $\mathsf{GreedyMin}_k(n)$, and it is illustrated in Fig. 1. When $k = 1$, the cycle that we create is identical to the one given by Zaks [27], and when $k = 2$, our cycle in the burnt pancake network was previously produced by Suzuki, N. Sawada, and

Kaneko [16]; however, both of these papers describe their cycles recursively. The greedy construction of the cycles in the pancake and burnt pancake networks was previously given by J. Sawada and Williams [10,20].

1.3   Combinatorial Generation

Ostensibly, the primary contribution of this paper is the Hamiltonicity of k-sided pancake networks. However, the authors' primary motivation was not in finding a Hamilton cycle, but rather in investigating its contributions to combinatorial generation. *Combinatorial generation* is the research area devoted to the efficient and clever generation of combinatorial objects. By *efficient* we mean that successive objects can be generated in amortized $O(1)$-time or worst-case $O(1)$-time, regardless of their size. The former is known as *constant amortized time (CAT)*, while the latter is known as *loop-free*. By *clever* we mean that non-lexicographic orders are often desirable. When describing these alternate orders, the authors make liberal use of the term *Gray code*—in reference to the eponymous binary reflected Gray code patented by Frank Gray [13])—and we refer to our Hamilton cycle as a *colour-incrementing prefix-reversal Gray code* for coloured permutations. Informally, it is a *flip Gray code*.

There are dozens of publications on the efficient generation of permutation Gray codes. In fact, comprehensive discussions on this topic date back to Sedgewick's survey in 1977 [22], with more modern coverage in Volume 4 of Knuth's *The Art of Computer Programming* [17]. However, to our knowledge, there are no published Gray codes for coloured permutations. This is surprising as the combinatorial [4,5,8,18,19] and algebraic [2,3,23] properties of coloured permutations have been of considerable interest. Work on the latter is due to the group theoretic interpretation of $\mathbb{P}_k(n)$ as the wreath product of the cyclic and symmetric group, $\mathbb{Z}_k \wr S_n$. We find our new Gray code of interest for two additional reasons.

1. Other greedy approaches for generating $\mathbb{P}(n)$ do not seem to generalize to $\mathbb{P}_k(n)$.
2. Flips are natural and efficient operations in certain contexts.

To expand on the first point, consider the Steinhaus-Johnson-Trotter (SJT) order of permutations, which dates back to the1600s [17]. In this order, successive permutations differ by an *adjacent-transition* (or *swap*) meaning that adjacent values in the permutations change place. In other words, the order for $\mathbb{P}(n)$ traces a Hamilton path in the permutohedron of order n. For example, SJT order for $n = 4$ appears below

$$1234, 1243, 1423, 4123, 4132, 1432, 1342, 1324, 3124, 3142, 3412, 4312,$$
$$4321, 3421, 3241, 3214, 2314, 2341, 2431, 4231, 4213, 2413, 2143, 2134.$$

The symbols that are swapped to create the next permutation are underlined, and the larger value is in bold. The latter demarcation shows the order's underlying greedy priorities: Swap the largest value. For example, consider the fourth permutation in the list, 4123. The largest value 4 cannot be swapped to the left (since it is in the leftmost position) or the right (since 1423 is already in the order), so the next option is to consider 3, and it can only be swapped to the left, which gives the fifth permutation 4132. If this description is perhaps too brief, then we refer the reader to [26].

Now consider greedy generalizations of SJT to signed permutations. The most natural generalization would involve the use of *sign-complementing adjacent-transpositions* which swap and complement the sign of two adjacent values. Unfortunately, any approach using these operations is doomed to fail. This is because the operation does not change the parity of positive and negative values. The authors experimented with other types of signed swaps—complementing the leftmost or rightmost value in the swap, or the larger or small value in the swap—without success.

More surprising is the fact that our greedy min-flip strategy works for coloured permutations, but the analogous max-flip strategy does not. For example, the max-flip strategy creates the following path in $\mathbb{G}_3(2)$ before getting stuck.

$$1^0 2^0,\ 2^1 1^1,\ 1^2 2^2,\ 2^0 1^0,\ 1^1 2^1,\ 2^2 1^2,\ 2^0 1^2,\ 1^0 2^1,\ 2^2 1^1,\ 1^2 2^0,\ 2^1 1^0,\ 1^1 2^2,\ \not{f}$$

The issue is that the neighbors of last coloured permutation in the path are already on the path. More specifically, a flip of length one transforms $1^1 2^2$ into $1^2 2^2$, and a flip of length two transforms $1^1 2^2$ into $2^0 1^2$, both of which appear earlier. The failure of the max-flip strategy on coloured permutations is surprising due to the fact that it works for both permutations and signed-permutations [10, 20].

To expand on the second point, note that the time required to flip a prefix is proportional to its length. In particular, if a permutation over $\{1, 2, \ldots, n\}$ is stored in an array or linked list of length n, then it takes $O(m)$-time to flip a prefix of length m[1]. Our min-flip strategy ensures that the shortest possible flips are used. In fact, the average flip length used in our Gray codes is bounded by $e = 2.71828 \cdots$ when $k = 1$, and the average is even smaller for $k > 1$.

We also note that flips can be the most efficient operation in certain situations. For example, consider a brute force approach to the undirected travelling salesman problem, wherein every Hamilton path of the n cities is represented by a permutation in $\mathbb{P}(n)$. If we iterate over the permutations using a prefix-reversal Gray code, then successive Hamilton paths differ in a single edge. For example, the edges in 12345678 and 43215678 are identical, except that the former includes $(4, 5)$ while the latter includes $(1, 5)$. Thus, the cost of each Hamilton cycle can be

---

[1] Some unusual data structures can support flips of any lengths in constant-time [25].

updated from permutation to permutation using one addition and subtraction. More generally, flip Gray codes are the most efficient choice when the cost (or value) of each permutation depends on its unordered pairs of adjacent symbols. Similarly, our generalization will be the most efficient choice when the cost (or value) of each coloured permutation depends on its unordered pairs of adjacent symbols *and* the minimum distance between their colours.

1.4   New Results

We present a flip Gray code for $\mathbb{P}_k(n)$ that corresponds to a Hamilton cycle in the k-sided pancake network. We present the following four different combinatorial algorithms for traversing the Hamilton cycle, each having unique and interesting properties:

1. A greedy algorithm that is easy to describe, but requires an exponential amount of memory.
2. A recursive algorithm, that reveals the structure of the listing and can be implemented in $O(1)$-amortized time.
3. A simple successor rule approach that allows the cycle to start from any vertex (coloured permutation) and takes on average $O(1)$-time amortized over the entire listing.
4. A loop-free algorithm to generate the flip-sequence iteratively.

Before we present these algorithms in Sect. 3, we first present some notation in Sect. 2. We conclude with a summary and related work in Sect. 4.

2   Notation

Let $\pi = p_1 p_2 \cdots p_n$ be a coloured permutation where each $p_i = v_i^{c_i}$ has value $v_i \in \{1, 2, \ldots, n\}$ and colour $c_i \in \{0, 1, \ldots, k-1\}$. Recall that $\mathbb{P}_k(n)$ denotes the set of k-coloured permutations of $\{1, 2, \ldots, n\}$. Observe that $\mathbb{P}_1(n)$ corresponds to regular permutations and $\mathbb{P}_2(n)$ corresponds to signed permutations. For the remainder of this paper, it is assumed that all permutations are coloured.

As mentioned earlier, a flip of a permutation π, denoted $\mathsf{flip}_i(\pi)$, applies a prefix-reversal of length i on π that also increments the colour of the flipped elements by 1 (modulo k). As an example for $k = 3$:

$$\mathsf{flip}_4(7^0 1^2 6^1 5^0 3^1 4^1 2^1) = \mathbf{5^1 6^2 1^0 7^1} 3^1 4^1 2^1.$$

A *pre-perm* is any prefix of a permutation in $\mathbb{P}_k(n)$, i.e. $\mathbf{p} = p_1 p_2 \cdots p_j$ is a pre-perm if there exist $p_{j+1}, \ldots p_n$ such that $p_1 p_2 \cdots p_n$ is a permutation. Note that if $j = n$, then the pre-perm is a permutation. Let $\mathbf{p} = p_1 p_2 \cdots p_j$ be an arbitrary pre-perm for given a k. For a given element $p_i = v_i^{c_i}$, let $p_i^{+s} = v_i^{(c_i+s) \ (\mathrm{mod} \ k)}$. For $0 \leq i < k$, let $\mathbf{p}^{+\mathbf{i}}$ denote $p_1^{+i} p_2^{+i} \cdots p_j^{+i}$, i.e. $\mathbf{p}$ with the colour of each element incremented by i modulo k. Note, $\mathbf{p}^{+0} = \mathbf{p}$. Furthermore, let $\rho(\mathbf{p}) = \mathbf{p}^{+(k-1)} \cdot \mathbf{p}^{+(k-2)} \cdots \mathbf{p}^{+0} = r_1 r_2 \cdots r_m$ be a circular string of length $m = kj$ where $\cdot$ denotes the concatenation of symbols. Let $\rho(\mathbf{p})_i$ denote the length $j - 1$ subword ending with r_{i-1}.

Example 1   Consider a pre-perm $\mathbf{p} = 1^0 2^0 3^2$ where $j = 3$ and $k = 4$. Then

$$\rho(\mathbf{p}) = 1^3 2^3 3^1 \cdot 1^2 2^2 3^0 \cdot 1^1 2^1 3^3 \cdot 1^0 2^0 3^2 \quad \text{and} \quad \rho(\mathbf{p})_2 = 3^2 1^3.$$

For any pre-perm $\mathbf{p} = p_1 p_2 \cdots p_j$, let $\overleftarrow{\mathbf{p}}$ denote the reverse of $\mathbf{p}$. i.e. $\overleftarrow{\mathbf{p}} = p_j p_{j-1} \cdots p_2 p_1$. Note that $\overleftarrow{\mathbf{p}}$ is not equivalent to applying a flip of length j to $\mathbf{p}$ when $k > 1$ as the colours of each symbol do not change in $\overleftarrow{\mathbf{p}}$. For the remainder of this paper we will use $\mathbf{p}$ to denote a pre-perm, and when it is clear we will use π to denote a permutation.

3   Constructions of a Cyclic Flip Gray Code for $\mathbb{P}_k(n)$

In this section we present four different combinatorial algorithms for generating the same cyclic flip Gray code for $\mathbb{P}_k(n)$. We begin by studying the listing of permutations generated by a greedy min-flip algorithm. We define the *flip-sequence* of a listing of permutations as the sequence of the flip lengths used to generate the listing beginning with the first permutation. By studying the underlying recursive structure of the greedy listing, we provide a recursive description and its corresponding flip-sequence and prove it is equivalent to the flip-sequence generated by the greedy algorithm. This proves that the greedy algorithm generates all permutations in $\mathbb{P}_k(n)$. We then present a successor-rule that determines the successor of a given permutation in the greedy min-flip listing in expected $O(1)$-time. We conclude by showing how the flip-sequence can be generated via a loop-free algorithm.

3.1   Greedy Algorithm

Recall that $\mathsf{GreedyMin}_k(n)$ denotes the greedy algorithm on $\mathbb{P}_k(n)$ that starts at permutation $1^0 2^0 \cdots n^0$ and prioritizes the neighbors of each permutation in the k-sided pancake network by increasing flip length.

Example 2   The following listing (left of the vertical bar) denotes the output
of GreedyMin$_3$(3) (read top to bottom, then left to right), where black, red and
blue correspond to the colours 0,1 and 2 respectively. This listing is exhaustive
and cyclic; the last permutation differs from the first permutation by a flip of
length $n = 3$. To the right of the vertical line is the flip length required to get
from the permutation in that position to its successor.

123	312	231	123	312	231	123	312	231	1 1 1 1 1 1 1 1 1
123	312	231	123	312	231	123	312	231	1 1 1 1 1 1 1 1 1
123	312	231	123	312	231	123	312	231	2 2 2 2 2 2 2 2 2
213	132	321	213	132	321	213	132	321	1 1 1 1 1 1 1 1 1
213	132	321	213	132	321	213	132	321	1 1 1 1 1 1 1 1 1
213	132	321	213	132	321	213	132	321	2 2 2 2 2 2 2 2 2
123	312	231	123	312	231	123	312	231	1 1 1 1 1 1 1 1 1
123	312	231	123	312	231	123	312	231	1 1 1 1 1 1 1 1 1
123	312	231	123	312	231	123	312	231	2 2 2 2 2 2 2 2 2
213	132	321	213	132	321	213	132	321	1 1 1 1 1 1 1 1 1
213	132	321	213	132	321	213	132	321	1 1 1 1 1 1 1 1 1
213	132	321	213	132	321	213	132	321	2 2 2 2 2 2 2 2 2
123	312	231	123	312	231	123	312	231	1 1 1 1 1 1 1 1 1
123	312	231	123	312	231	123	312	231	1 1 1 1 1 1 1 1 1
123	312	231	123	312	231	123	312	231	2 2 2 2 2 2 2 2 2
213	132	321	213	132	321	213	132	321	1 1 1 1 1 1 1 1 1
213	132	321	213	132	321	213	132	321	1 1 1 1 1 1 1 1 1
213	132	321	213	132	321	213	132	321	3 3 3 3 3 3 3 3 3

Observe that each column of permutations ends with the same element. Furthermore, the last permutation in each column is a subword of the cyclic word
321321321.

Unlike the max-flip approach, we will prove that GreedyMin$_k$(n) exhaustively
generates all permutations in $\mathbb{P}_k(n)$ for all $n, k \geq 1$. We also show that the last
permutation in the listing differs by a flip of length n from the first permutation,
so the listing is a cyclic flip Gray code. To prove this result, we study the underlying recursive structure of the resulting listings and examine the flip-sequences.

3.2   Recursive Construction

By applying the two observations made following the listing of GreedyMin$_3$(3)
in Example 2, we arrive at the following recursive definition for a listing of pre-
perms, given a pre-perm $\mathbf{p}$ of a permutation in $\mathbb{P}_k(n)$:

$$\mathbf{Rec}_k(\mathbf{p}) = \mathbf{Rec}_k(\rho(\mathbf{p})_\mathbf{m}) \cdot r_m, \ \mathbf{Rec}_k(\rho(\mathbf{p})_{\mathbf{m-1}}) \cdot r_{m-1}, \ldots, \ \mathbf{Rec}_k(\rho(\mathbf{p})_\mathbf{1}) \cdot r_1, \ (1)$$

where $\mathbf{Rec}_k(p_x) = p_x^{+0}, p_x^{+1}, p_x^{+2}, \ldots, p_x^{+(k-1)}$ and $\rho(p) = r_1 \cdots r_m$. Here, the
operation $\mathcal{L} \cdot r$ denotes the listing $\mathcal{L}$ with r appended to every element in the
listing. We prove that $\mathbf{Rec}_k(1^0 2^0 \cdots n^0)$ generates the same (exhaustive) listing
of permutations as GreedyMin$_k$(n).

Lemma 1. *Let* $\mathbf{p} = p_1 p_2 p_3 \cdots p_j$ *be a pre-perm of a permutation in* $\mathbb{P}_k(n)$ *for some* $n \geq j$. *Then the first and last pre-perms in the listing* $\mathbf{Rec}_k(\mathbf{p})$ *are* $\mathbf{p}$ *and* $\overleftarrow{\mathbf{p}}^{+(\mathbf{k-1})}$, *respectively.*

Proof. The proof proceeds by induction on j. When $j = 1$, we have $\mathbf{p} = \overleftarrow{\mathbf{p}} = p_1$, so $\mathbf{Rec}_k(\mathbf{p}) = \mathbf{p}, \mathbf{p}^{+1}, \mathbf{p}^{+2}, \ldots, \mathbf{p}^{+(k-1)}$. Since $\mathbf{p}^{+(k-1)} = \overleftarrow{\mathbf{p}}^{+(k-1)}$ the claim holds. Now for $1 \leq j < n$ and any pre-perm $\mathbf{p} = p_1 p_2 \cdots p_j$ of a permutation in $\mathbb{P}_k(n)$, suppose that the first and last pre-perms in $\mathbf{Rec}_k(\mathbf{p})$ are $\mathbf{p}$ and $\overleftarrow{\mathbf{p}}^{+(k-1)}$ respectively. Let $\mathbf{p} = p_1 p_2 \cdots p_j p_{j+1}$ be a pre-perm of a permutation in $\mathbb{P}_k(n)$. By definition, the first pre-perm of $\mathbf{Rec}_k(\mathbf{p})$ is the first pre-perm of $\mathbf{Rec}_k(\rho(\mathbf{p})_{\mathbf{m}}) \cdot r_m$ where $m = (j+1)k$. By definition of $\rho(\mathbf{p})$ and $\rho(\mathbf{p})_{\mathbf{m}}$, it is clear that $r_m = p_{j+1}$ and $\rho(\mathbf{p})_{\mathbf{m}} = p_1 p_2 \cdots p_{j-1} p_j$. Applying the inductive hypothesis, the first pre-perm of $\mathbf{Rec}_k(p_1 p_2 \cdots p_{j-1} p_j)$ is $p_1 p_2 \cdots p_{j-1} p_j$. Therefore, the first pre-perm of $\mathbf{Rec}_k(\mathbf{p})$ is $p_1 p_2 \cdots p_{j-1} p_j \cdot p_{j+1} = \mathbf{p}$. Similarly, the last pre-perm of $\mathbf{Rec}_k(\mathbf{p})$ is the last pre-perm of $\mathbf{Rec}_k(\rho(\mathbf{p})_1) \cdot r_1$. Now, $r_1 = p_1^{+(k-1)}$ and $\rho(\mathbf{p})_1 = p_2 p_3 \cdots p_j p_{j+1}$ and, by the inductive hypothesis, the last pre-perm in $\mathbf{Rec}_k(\rho(\mathbf{p})_1)$ is $p_{j+1}^{+(k-1)} p_j^{+(k-1)} \cdots p_2^{+(k-1)}$. Therefore, the last pre-perm of $\mathbf{Rec}_k(\mathbf{p})$ is $\overleftarrow{\mathbf{p}}^{+(k-1)}$. $\square$

Define the sequence $\sigma_{k,n}$ recursively as

$$\sigma_{k,n} = \begin{cases} 1^{k-1} & \text{if } n = 1 \\ (\sigma_{k,n-1}, n)^{kn-1}, \sigma_{k,n-1} & \text{if } n > 1, \end{cases} \qquad (2)$$

where given a sequence τ, τ^j denotes j copies of τ concatenated together. We will show that $\sigma_{k,n}$ is the flip-sequence for both $\mathbf{Rec}_k(\mathbf{p})$ and $\mathsf{GreedyMin}_k(n)$. This flip-sequence is a straightforward generalization of the recurrences for non-coloured permutations [27] and signed permutations [20]. Note that $\sigma_{3,3}$ is shown to the right of the vertical bar in Example 2.

Lemma 2. *For* $n \geq 1$, $k \geq 1$, *and* $\pi \in \mathbb{P}_k(n)$, *the flip-sequence for* $\mathbf{Rec}_k(\pi)$ *is* $\sigma_{k,n}$.

Proof. By induction on n. In the base case $\mathbf{Rec}_k(p_1) = p_1, p_1^{+1}, p_1^{+2}, \ldots, p_1^{+(k-1)}$ and the flip-sequence is $\sigma_{k,1} = 1^{k-1}$. For $n \geq 1$ assume that the sequence of flips used to create $\mathbf{Rec}_k(p_1 p_2 p_3 \cdots p_n)$ is given by $\sigma_{k,n}$. Consider $\mathbf{Rec}_k(\pi)$ where $\pi = p_1 p_2 p_3 \cdots p_{n+1} \in \mathbb{P}_k(n+1)$. By the inductive hypothesis, it suffices to show that the last permutation of $\mathbf{Rec}_k(\rho(\pi)_i) \cdot r_i$ and the first permutation of $\mathbf{Rec}_k(\rho(\pi)_{i-1}) \cdot r_{i-1}$ differ by a flip of length $n + 1$ for $i = 2, 3, \ldots, m$ ($= k(n+1)$). By definition, $\rho(\pi)_i = r_{i-n} r_{i-(n-1)} \cdots r_{i-2} r_{i-1}$ where the indices are taken modulo m. Therefore, by Lemma 1, the last permutation in $\mathbf{Rec}_k(\rho(\pi)_i)$ is $(r_{i-1} r_{i-2} \cdots r_{i-(n-1)} r_{i-n})^{+(k-1)}$. Applying a flip of length $n + 1$ to $\mathbf{Rec}_k(\rho(\pi)_i) \cdot r_i$ yields

$$r_i^{+1} r_{i-n} r_{i-(n-1)} \cdots r_{i-2} r_{i-1}. \qquad (3)$$

By Lemma 1, the first permutation of $\mathbf{Rec}_k(\rho(\pi)_{i-1})$ is $r_{i-(n+1)}r_{i-n}\cdots r_{i-3}r_{i-2}$. By the definition of $\rho(\pi)$, it follows that $r_{i-(n+1)} = r_i^{+1}$. Thus, from (3), it follows that $\mathbf{Rec}_k(\rho(\pi)_i)\cdot r_i$ and the first permutation of $\mathbf{Rec}_k(\rho(\pi)_{i-1})\cdot r_{i-1}$ differ by a flip of length $n+1$. By applying the inductive hypothesis, the flip-sequence for $\mathbf{Rec}_k(\pi)$ is $(\sigma_{k,n}, n+1)^{k(n+1)-1}, \sigma_{k,n}$ which is exactly $\sigma_{k,n+1}$. □

Theorem 1. *For $n \geq 1$, $k \geq 1$, and $\pi \in \mathbb{P}_k(n)$, $\mathbf{Rec}_k(\pi)$ is a cyclic flip Gray code for $\mathbb{P}_k(n)$, where the first and last permutations differ by a flip of length n.*

Proof. From Lemma 2, the flip-sequence for $\mathbf{Rec}_k(\pi)$ is given by $\sigma_{k,n}$. Inductively, it is easy to see that the length of the flip-sequence $\sigma_{k,n}$ is $k^n n! - 1$ and that each permutation of $\mathbf{Rec}_k(\pi)$ is unique. Thus, each of the $k^n n!$ permutations must be listed exactly once and, from Lemma 1, the first and last permutations of the listing differ by a flip of length n, making $\mathbf{Rec}_k(\pi)$ a cyclic flip Gray code for permutations. □

Lemma 3. *For $n \geq 1$ and $k \geq 1$, the flip-sequence for $\mathsf{GreedyMin}_k(n)$ is $\sigma_{k,n}$.*

Proof. By contradiction. Suppose the sequence of flips used by $\mathsf{GreedyMin}_k(n)$ differs from $\sigma_{k,n}$ and let j be the smallest value such that the j-th flip used to create $\mathsf{GreedyMin}_k(n)$ differs from the j-th value of $\sigma_{k,n}$. Let these flip lengths be s and t respectively. Since $\mathsf{GreedyMin}_k(n)$ follows a greedy minimum-flip strategy and because $\sigma_{k,n}$ produces a valid flip Gray code for permutations by Theorem 1 where no permutation is repeated, it must be that $s < t$. Let $\pi = p_1 p_2 p_3 \cdots p_n$ denote the j-th permutation in the listing $\mathsf{GreedyMin}_k(n)$, i.e. the permutation immediately prior to the j-th flip. Since $\sigma_{k,n}$ is the flip-sequence for $\mathbf{Rec}_k(1^0 2^0 \cdots n^0)$ by Lemma 2, from the recursive definition it follows inductively that all other permutations with suffix $p_t p_{t+1} \cdots p_n$ appear before π in $\mathbf{Rec}_k(1^0 2^0 \cdots n^0)$, since no permutations are repeated by Theorem 1. Since $\sigma_{k,n}$ and the sequence of flips used by $\mathsf{GreedyMin}_k(n)$ agree until the j-th value, all other permutations with suffix $p_t p_{t+1} \cdots p_n$ appear before π in $\mathsf{GreedyMin}_k(n)$. Therefore, flipping π by a flip of length $s < t$ results in a permutation already visited in $\mathsf{GreedyMin}_k(n)$ before index j contradicting the fact that $\mathsf{GreedyMin}_k(n)$ produces a list of permutations without repetition. □

By definition, $\mathsf{GreedyMin}_k(n)$ starts with the permutation $1^0 2^0 \cdots n^0$ and by Lemma 1, $\mathbf{Rec}_k(1^0 2^0 \cdots n^0)$ also starts with $1^0 2^0 \cdots n^0$. Since they are each created by the same flip-sequence by Lemma 2 and Lemma 3, we get the following corollary.

Corollary 1. *For $n \geq 1$ and $k \geq 1$, the listings $\mathsf{GreedyMin}_k(n)$ and $\mathbf{Rec}_k(1^0 2^0 \cdots n^0)$ are equivalent.*

3.3 Successor Rule

In this section, we will generalize the successor rules found for non-coloured permutations and signed permutations in [21] for $\mathsf{GreedyMin}_k(n)$ for $k > 2$. We say a permutation in $\mathbb{P}_k(n)$ is *increasing* if it corresponds to a length n subword of the circular string $\rho(1^0 2^0 \cdots n^0)$. For example if $n = 6$ and $k = 4$, then the following permutations are all increasing:

$$2^3 3^3 4^3 5^3 6^3 1^2 \qquad 5^1 6^1 1^0 2^0 3^0 4^0 \qquad 1^0 2^0 3^0 4^0 5^0 6^0 \qquad 5^0 6^0 1^3 2^3 3^3 4^3.$$

A permutation is *decreasing* if it is a reversal of an increasing permutation. A pre-perm is *increasing* (*decreasing*) if it corresponds to a subsequence of an increasing (decreasing) permutation (when the permutation is thought of as a sequence). For example, $5^1 6^1 2^0 4^0$ is an increasing pre-perm, but $5^1 2^0 4^0 6^0$ and $1^2 2^2 3^1 4^0$ are not. Given a permutation π_2, let $\mathsf{succ}(\pi_2)$ denote the successor of π_2 in $\mathbf{Rec}_k(\pi)$ when the listing is considered to be cyclic.

Lemma 4. *Let $\pi_2 = q_1 q_2 \cdots q_n$ be a permutation in the (cyclic) listing $\mathbf{Rec}_k(\pi)$, where $\pi = p_1 p_2 \cdots p_n$ is increasing. Let $q_1 q_2 \cdots q_j$ be the longest prefix of π_2 that is decreasing. Then $\mathsf{succ}(\pi_2) = \mathsf{flip}_j(\pi_2)$.*

Proof. By induction on n. When $n = 1$, the result follows trivially as only flips of length 1 can be applied. Now, for $n > 1$, we focus on the permutations whose successor is the result of a flip of length n and the result will follow inductively by the recursive definition of $\mathbf{Rec}_k(\pi)$. By Lemma 2, the successor of π_2 will be $\mathsf{flip}_n(\pi_2)$ if and only if it is the last permutation in one of the recursive listings of the form $\mathbf{Rec}_k(\rho(\pi)_\mathbf{i}) \cdot r_i$. Recall that r_i is the i-th element in $\rho(\pi)$ when indexed from $r_1 = p_1^{+(k-1)}$ to $r_m = p_n$. As it is clear that at most one permutation is decreasing in each recursive sublist, it suffices to show that the last permutation in each sublist is decreasing to prove the successor rule holds for flips of length n. By Lemma 1, the last permutation in $\mathbf{Rec}_k(\rho(\pi)_\mathbf{i}) \cdot r_i$ is $\overleftarrow{\mathbf{s}} \cdot r_i$ where $\mathbf{s} = \rho(\pi)_\mathbf{i}^{+(k-1)}$. Since π is increasing, it is clear that $\rho(\pi)_\mathbf{i}$ is increasing and therefore that $\mathbf{s}$ is increasing. Hence, $\overleftarrow{\mathbf{s}}$ is decreasing by definition. Furthermore, by the definition of the circular word $\rho(\pi)$, the element immediately before $r_{i-1-(n-1)}^{+(k-1)}$ in $\rho(\pi)$ is r_i (note the subscript $i - 1 - (n - 1)$ is considered modulo nk here). Therefore, $\overleftarrow{\mathbf{s}} \cdot r_i$ is decreasing. Therefore, the successor rule holds for flips of length n and thus for flips of all lengths by induction. $\square$

Example 3   With respect to the listing $\mathbf{Rec}_{10}(1^0 2^0 3^0 4^0 5^0 6^0)$,

$$\mathsf{succ}(3^8 2^8 5^9 4^9 1^7 6^3) = \mathsf{flip}_4(3^8 2^8 5^9 4^9 1^7 6^3) = 4^0 5^0 2^9 3^9 1^7 6^3$$

and

$$\mathsf{succ}(1^8 3^7 2^6 5^5 4^3 6^2) = \mathsf{flip}_1(1^8 3^7 2^6 5^5 4^3 6^2) = 1^9 3^7 2^6 5^5 4^3 6^2.$$

By applying the previous lemma, computing $\mathsf{succ}(\pi_2)$ for a permutation in the listing $\mathbf{Rec}_k(\pi)$ can easily be done in $O(n)$-time as described in the pseudocode given in Algorithm 1.

Algorithm 1. Computing the successor of π in the listing $\mathbf{Rec}_k(1^0 2^0 \cdots n^0)$

1: **function** SUCCESSOR(π)
2:    $incr \leftarrow 0$
3:    **for** $j \leftarrow 1$ **to** $n - 1$ **do**
4:        **if** $v_j < v_{j+1}$ **then** $incr \leftarrow incr + 1$
5:        **if** $incr = 2$ **or** $(incr = 1$ **and** $v_{j+1} < v_1)$ **then** **return** $\mathsf{flip}_j(\pi)$
6:        **if** $k > 1$ **and** $v_j < v_{j+1}$ **and** ($(c_{j+1} - c_j + k) \bmod k \neq 1$) **then** **return** $\mathsf{flip}_j(\pi)$
7:        **if** $k > 1$ **and** $v_j > v_{j+1}$ **and** ($c_j \neq c_{j+1}$) **then** **return** $\mathsf{flip}_j(\pi)$
8:    **return** $\mathsf{flip}_n(\pi)$

Theorem 2. SUCCESSOR(π) *returns the length of the flip required to obtain the successor of π in the listing* $\mathbf{Rec}_k(1^0 2^0 \cdots n^0)$ *in* $O(n)$-*time.*

Though the worst case performance of SUCCESSOR(π) is $O(n)$-time, on average it is much better. Let $\overline{\sigma}_{k,n}$ denote $(\sigma_{k,n}, n)$, i.e. the sequence of flips used to generate the listing $\mathbf{Rec}_k(\pi)$ with an extra flip of length n at the end to return to the starting permutation. Our goal is to determine the average flip length of $\overline{\sigma}_{k,n}$, denoted AVG(k, n). Note that our analysis generalize the results for AVG$(1, n)$ [27] and AVG$(2, n)$ [20].

Lemma 5. *For* $n \geq 1$ *and* $k \geq 1$,

$$\mathrm{AVG}(k, n) = \sum_{j=0}^{n-1} \frac{1}{k^j j!}.$$

Moreover, AVG$(k, n) < \sqrt[k]{e}$.

Proof. By definition of $\sigma_{k,n}$, it is not difficult to see that $\overline{\sigma}_{k,n+1}$ is equivalent to the concatenation of $(n + 1)k$ copies of $\overline{\sigma}_{k,n}$ with the last element in every copy of $\overline{\sigma}_{k,n}$ incremented by 1. Therefore, we have

$$\mathrm{AVG}(k, n + 1) = \frac{\left(1 + \sum_{f \in \overline{\sigma}_{k,n}} f\right)(n + 1)k}{(n + 1)! k^{n+1}}$$

$$= \frac{\sum_{f \in \overline{\sigma}_{k,n}} f}{n! k^n} + \frac{1}{n! k^n}$$

$$= \mathrm{AVG}(k, n) + \frac{1}{n! k^n}.$$

Hence, with the trivial base case that AVG$(k, 1) = 1$, we have

$$\mathrm{AVG}(k, n) = \sum_{j=0}^{n-1} \frac{1}{k^j j!}.$$

Therefore,

$$\mathrm{AVG}(k, n) < \sum_{j=0}^{\infty} \frac{1}{k^j j!} = \sqrt[k]{e}$$

by applying the well-known Maclaurin series expansion for e^x.   □

Observe that the SUCCESSOR function runs in expected $O(1)$-time when the permutation is passed by reference because the average flip length is bounded above by the constant $\sqrt[k]{e}$ as proved in Lemma 5. Thus, by repeatedly applying the successor rule, we obtain a CAT algorithm for generating $\mathbf{Rec}_k(1^0 2^0 \cdots n^0)$.

3.4   Loop-Free Generation of the Flip-Sequence $\sigma_{k,n}$

Based on the recursive definition of the flip-sequence $\sigma_{k,n}$ given in (2), Algorithm 2 will generate $\sigma_{k,n}$ in a loop-free manner. The algorithm generalizes a similar algorithm presented by Zaks for non-coloured permutations [27]. The next flip length x is computed using an array of counters $c_1, c_2, \ldots, c_{n+1}$ initialized to 0, and an array of flip lengths $f_1, f_2, \ldots, f_{n+1}$ with each f_i initialized to i. For a formal proof of correctness, we invite the readers to see the simple inductive proof for the non-coloured case in [27], and note the primary changes required to generalize to coloured permutations are in handling of the minimum allowable flip lengths (when $k = 1$, the smallest allowable flip length is 2) corresponding to lines 5–6 and adding a factor of k to line 8.

Algorithm 2. Loop-free generation of the flip-sequence $\sigma_{k,n}$

```
 1: procedure FLIPSEQ
 2:     c_1, c_2, ..., c_{n+1} ← 0, 0, ..., 0
 3:     f_1, f_2, ..., f_{n+1} ← 1, 2, ..., n + 1
 4:     repeat
 5:         if k = 1 then x ← f_2;   f_2 ← 2
 6:         else  x ← f_1;   f_1 ← 1
 7:         c_x ← c_x + 1
 8:         if c_x = kx−1 then
 9:             c_x ← 0
10:             f_x ← f_{x+1}
11:             f_{x+1} ← x + 1
12:         OUTPUT(x)
13:     until x > n
```

Theorem 3. *The algorithm* FLIPSEQ *is a loop-free algorithm to generate the flip-sequence $\sigma_{k,n}$ one element at a time.*

Since the average flip length in $\sigma_{k,n}$ is bounded by a constant, as determined in the previous subsection, Algorithm 2 can be modified to generate $\mathbf{Rec}_k(\pi)$ by passing the initial permutation π as a parameter, outputting π at the start of the **repeat** loop, and updating $\pi \leftarrow \mathrm{flip}_x(\pi)$ at the end of the loop instead of outputting the flip length.

Corollary 2. *The algorithm* FLIPSEQ *can be modified to generate successive permutations in the listing* $\mathbf{Rec}_k(\pi)$ *in* $O(1)$-*amortized time.*

4   Summary and Related Work

We presented four different combinatorial algorithms for traversing a specific Hamilton cycle in the k-sided pancake network. The Hamilton cycle corresponds to a flip Gray code listing of coloured permutations. Given such combinatorial listings, it is desirable to have associated ranking and unranking algorithms. Based on the recursive description of the listing in (1), such algorithms are relatively straightforward to derive and implement in $O(n^2)$-time. A complete C implementation of our algorithms is available on The Combinatorial Object Server [7].

References

1. Akers, S., Krishnamurthy, B.: A group-theoretic model for symmetric interconnection networks. IEEE Trans. Comput. **38**(4), 555–566 (1989)
2. Athanasiadis, C.A.: Binomial Eulerian polynomials for colored permutations. J. Comb. Theory Ser. A **173**, 105214 (2020)
3. Bagno, E., Garber, D., Mansour, T.: On the group of alternating colored permutations. Electron. J. Comb. **21**(2), 2.29 (2014)
4. Borodin, A.: Longest increasing subsequences of random colored permutations. Electron. J. Comb. **6**(13), 12 (1999)
5. Chen, W.Y.C., Gao, H.Y., He, J.: Labeled partitions with colored permutations. Discret. Math. **309**(21), 6235–6244 (2009)
6. Cohen, D.S., Blum, M.: On the problem of sorting burnt pancakes. Discret. Appl. Math. **61**(2), 105–120 (1995)
7. COS++: The Combinatorial Object Server. http://combos.org/cperm
8. Duane, A., Remmel, J.: Minimal overlapping patterns in colored permutations. Electron. J. Comb. **18**(2), 38 (2011). Paper 25
9. Dweighter, H.: Problem E2569. Am. Math. Mon. **82**, 1010 (1975)
10. Essed, H., Therese, W.: The harassed waitress problem. In: Ferro, A., Luccio, F., Widmayer, P. (eds.) Fun with Algorithms. FUN 2014. Lecture Notes in Computer Science, vol. 8496. Springer, Cham (2014). https://doi.org/10.1007/978-3-319-07890-8_28
11. Fertin, G., Labarre, A., Rusu, I., Vialette, S., Tannier, E.: Combinatorics of Genome Rearrangements. MIT Press, Cambridge (2009)
12. Gates, W.H., Papadimitriou, C.H.: Bounds for sorting by prefix reversal. Discret. Math. **27**(1), 47–57 (1979)
13. Gray, F.: Pulse code communication. U.S. Patent 2,632,058 (1947)
14. Heydari, M.H., Sudborough, I.H.: On the diameter of the pancake network. J. Algorithms **25**(1), 67–94 (1997)
15. Justan, M.P., Muga, F.P., Sudborough, I.H.: On the generalization of the pancake network. In: Proceedings International Symposium on Parallel Architectures, Algorithms and Networks. I-SPAN 2002, pp. 173–178 (2002)
16. Kaneko, K.: Hamiltonian cycles and Hamiltonian paths in faulty burnt pancake graphs. IEICE - Trans. Inf. Syst. **E90-D**(4), 716–721 (2007)

17. Knuth, D.E.: The Art of Computer Programming, volume 4: Combinatorial Algorithms, Part 1. Addison-Wesley (2010)
18. Mansour, T.: Pattern avoidance in coloured permutations. Sém. Lothar. Combin. **46**, B46g-12 (2001)
19. Mansour, T.: Coloured permutations containing and avoiding certain patterns. Ann. Comb. **7**(3), 349–355 (2003)
20. Sawada, J., Williams, A.: Greedy flipping of pancakes and burnt pancakes. Discret. Appl. Math. **210**, 61–74 (2016)
21. Sawada, J., Williams, A.: Successor rules for flipping pancakes and burnt pancakes. Theoret. Comput. Sci. **609**(part 1), 60–75 (2016)
22. Sedgewick, R.: Permutations generation methods. ACM Comput. Surv. **9**(2), 137–164 (1977)
23. Shin, H., Zeng, J.: Symmetric unimodal expansions of excedances in colored permutations. Eur. J. Comb. **52**(part A), 174–196 (2016)
24. Singh, S.: Flipping pancakes with mathematics. The Guardian (2013)
25. Williams, A.: O(1)-time unsorting by prefix-reversals in a boustrophedon linked list. In: Boldi, P., Gargano, L. (eds.) FUN 2010. LNCS, vol. 6099, pp. 368–379. Springer, Heidelberg (2010). https://doi.org/10.1007/978-3-642-13122-6_35
26. Williams, A.: The greedy gray code algorithm. In: Dehne, F., Solis-Oba, R., Sack, J.-R. (eds.) WADS 2013. LNCS, vol. 8037, pp. 525–536. Springer, Heidelberg (2013). https://doi.org/10.1007/978-3-642-40104-6_46
27. Zaks, S.: A new algorithm for generation of permutations. BIT **24**(2), 196–204 (1984)

Algorithms and Complexity of *s*-Club Cluster Vertex Deletion

Dibyayan Chakraborty[1,2], L. Sunil Chandran[1,2], Sajith Padinhatteeri[1,2],
and Raji R. Pillai[1,2(✉)]

[1] Indian Institute of Science, Bangalore, India
rajpillai@iisc.ac.in
[2] BITS-Pilani, Hyderabad, India

Abstract. An *s-club* is a graph which has diameter at most *s*. Let *G* be a graph. A set of vertices $D \subseteq V(G)$ is an *s-club deleting (s-CD) set* if each connected component of $G - D$ is an *s*-club. In the *s*-CLUB CLUSTER VERTEX DELETION (*s*-CVD) problem, the goal is to find an *s*-CD set with minimum cardinality. When $s = 1$, the *s*-CVD is equivalent to the well-studied CLUSTER VERTEX DELETION problem. On the negative side, we show that unless the *Unique Games Conjecture* is false, there is no $(2 - \varepsilon)$-algorithm for 2-CVD on split graphs, for any $\varepsilon > 0$. This contrast the polynomial-time solvability of CLUSTER VERTEX DELETION on split graphs. We show that for each $s \geq 2$, *s*-CVD is NP-hard on bounded degree planar bipartite graphs and APX-hard on bounded degree bipartite graphs. On the positive side, we give a polynomial-time algorithm to solve *s*-CVD on trapezoid graphs, for each $s \geq 1$.

Keywords: Vertex deletion problem · *s*-Club · Split graphs · (Planar) bipartite graphs · NP-hardness · APX-hardness · Trapezoid graphs · Polynomial-time algorithms

1 Introduction and Results

Vertex deletion problems form a core topic in algorithmic graph theory with many applications. Typically, the objective of a vertex deletion problem is to delete the minimum number of vertices so that the remaining graph satisfies some property. Many classic optimization problems like MAXIMUM CLIQUE, MAXIMUM INDEPENDENT SET, VERTEX COVER are examples of vertex deletion problems. In this paper, we study the *s*-CLUB CLUSTER VERTEX DELETION problem.

For an integer, an *s-club* is a graph with diameter at most *s*. Let *G* be a graph with vertex set $V(G)$ and edge set $E(G)$. A set of vertices *D* is an *s-club deleting set* (*s*-CD set) if each connected component of $G - D$ is an *s*-club. In *s*-CLUB CLUSTER VERTEX DELETION(*s*-CVD), the input is an undirected graph and integers s, k. The objective is to decide if there exists an *s*-CD set of cardinality at most *k*.

s-CLUB CLUSTER VERTEX DELETION was first introduced by Schäfer [24]. He proved NP-completeness of *s*-CVD on planar graphs and gave polynomial-time algorithm for trees. In this paper, we strengthen Schäfer's hardness result by proving the following result.

© Springer Nature Switzerland AG 2021
P. Flocchini and L. Moura (Eds.): IWOCA 2021, LNCS 12757, pp. 152–164, 2021.
https://doi.org/10.1007/978-3-030-79987-8_11

Theorem 1. *For each $s \geq 2$, s-CVD is NP-hard on planar bipartite graphs with maximum degree 7.*

We use the proof techniques for Theorem 1 to prove the following.

Theorem 2. *For each $s \geq 2$, s-CVD is APX-hard on bipartite graphs with maximum degree 7.*

Liu et al. [20] studied the special case of 2-CVD and gave a fixed parameter algorithms with respect to the solution size. In this paper, we show that 2-CVD is difficult to solve even on *split graphs* i.e. graphs whose vertex set can be partitioned into a clique and independent set.

Theorem 3. *Unless the* Unique Games Conjecture *is false, there is no $(2 - \varepsilon)$-approximation algorithm for 2-CVD on split graphs, for any $\varepsilon > 0$.*

On the other hand, 1-CVD on split graphs can be solved in polynomial-time. Note that, 1-CVD is equivalent to the well-studied CLUSTER VERTEX DELETION [14].

Given a collection $\mathscr{C} = \{C_1, C_2, \ldots, C_n\}$ of sets, the *intersection graph* of $\mathscr{C}$ is the graph with vertex set $[n]$ and two vertices i and j are adjacent if and only if the corresponding sets C_i and C_j have non-empty intersection. *Trapezoid graphs* are special class of intersection graphs. Let ℓ_t, ℓ_b be two horizontal lines. In the case of *trapezoid graphs*, the collection of sets $\mathscr{C}$ is restricted to contain only trapezoids with two corner points on ℓ_t(top line) and other two corner points on ℓ_b(bottom line). In this paper, we study the computational complexity of s-CVD on *trapezoid graphs*. Specifically, we prove the following theorem.

Theorem 4. *For each $s \geq 1$, there is an $O(n^8)$-time algorithm on s-CVD on trapezoid graphs on n vertices.*

Theorem 4 generalises a result of Cao et al. [6] where they gave a polynomial-time algorithm for CLUSTER VERTEX DELETION on *interval graphs*, intersection graphs of intervals on the real line. To the best of our knowledge, Theorem 4 provides the first polynomial-time algorithm for CLUSTER VERTEX DELETION on trapezoid graphs.

Related Works and Significance of Our Results: The notion of s-clubs is important in the contexts of biological networks [22], protein interaction networks [3], and social networks. The special case of 1-clubs i.e. cliques and 1-CVD i.e. CLUSTER VERTEX DELETION has been widely studied in literature since it's inception due to Hüffner et al. [14]. Being an hereditary property, the work of Lewis and Yannakakis [19] provide a dichotomy result for CLUSTER VERTEX DELETION. A result of Yannakakis [27] implies that that CLUSTER VERTEX DELETION is NP-Complete on planar graphs and bipartite graphs. To the best of our knowledge, the complexity of CLUSTER VERTEX DELETION on planar bipartite graphs is unknown. Theorem 1 proves that s-CVD is NP-complete even on planar bipartite graphs for each $s \geq 2$. Note that, for $s \geq 2$, the property "diameter at most s" is not hereditary and therefore meta theorems that work for CLUSTER VERTEX DELETION do not hold for s-CVD.

Observe that, solving CLUSTER VERTEX DELETION is equivalent to hitting all induced paths of length 3. Therefore, the general dynamic programming approach given by Sau and Souza [23] provides a polynomial-time algorithm for CLUSTER VERTEX DELETION on bounded *tree-width* graphs. However, popular graph classes like *chordal graphs* and split graphs have unbounded tree-width. Contrasting the polynomial-time solvability of CLUSTER VERTEX DELETION on split graphs, Theorem 3 implies that 2-CVD is even hard to approximate beyond a factor of 2 on split graphs.

Often intersection graphs have high tree-width and therefore separate attention is required to formulate polynomial-time algorithms. Indeed, the complexity of CLUSTER VERTEX DELETION on interval graphs remained open until Cao et al. [6] gave a quadratic time optimal algorithm in 2018. However, to the best of our knowledge complexity of CLUSTER VERTEX DELETION remained open on related graph classes like *permutation graphs* [18], *triangle graphs* [21], *simple triangle graphs* [26]. *Trapezoid graph* is an important class of graphs that contain aforementioned graph classes as subclasses. Computational complexities of many algorithmic problems have been studied on trapezoid graphs [4,5,13]. Theorem 4 settles the complexity of s-CVD on trapezoid graphs, for each $s \geq 1$.

Our algorithm for s-CVD relies on a connection between *maximum weight* of an *independent set* of *cocomparability graphs* and dual of the s-CD set.

Even though the problem of finding the 2-club of maximum cardinality has been extensively studied [2,3,7,12,25], the computational complexity of 2-CVD is much less understood. Liu et al. [20] studied the fixed parameter tractability of 2-CVD and gave an $O^*(3.31^k)$ algorithm where k is the solution size. Figiel et al. [10] proved that for 2-CVD there does not exists a polynomial kernel with respect to the solution size k, unless $NP \subseteq CoNP \setminus Poly$. Our results make significant advancements towards understanding of the computational complexity of 2-CVD on various important classes of graphs.

Notations: We define some notations that will be used throughout the paper. Let G be a graph and $v \in V(G)$ be a vertex. The sets $N[v]$ and $N(v)$ denote the *closed* and *open* neighbourhood of v. For a subset S of vertices, $N[S]$ is the union of the closed neighbourhoods of the vertices in S and $N(S)$ is the union of the open neighbourhoods of the vertices in S. The graph $\overline{G}$ denotes the complement of G. For two sets X, Y, the set of elements in X but not in Y is denoted by either $X - Y$ or $X \setminus Y$. For a graph G and a set S of vertices, $G - S$ shall denote the subgraph induced by $V(G) - S$. For a finite collection of finite sets $\mathscr{S}$, let $\min\{\mathscr{S}\}$ denote any set $Z \in \mathscr{S}$ with minimum cardinality.

Organisation of the Paper: In Sect. 2 we prove Theorems 1 and 2. In Sect. 3, we prove the hardness result Theorem 3. In Sect. 4 we give a polynomial-time algorithm for s-CVD on trapezoid graphs for each $s \geq 1$.

2   Hardness on (Planar) Bipartite Graphs

First we shall prove Theorem 1. For each $s \geq 2$, we shall reduce the MINIMUM VERTEX COVER on cubic planar graphs which is known to be NP-Complete [11] to s-CVD on planar bipartite graphs of maximum degree 7. A *vertex cover* of a graph G is a set S of vertices such that each edge of G is incident on at least one vertex in S. Given a graph G

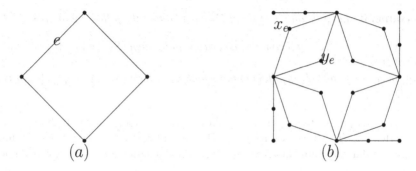

Fig. 1. Reduction for Theorem 1. (a) The graph G, (b) The graph H for $s = 4$

and an integer k, MINIMUM VERTEX COVER is to decide whether there exists a vertex cover of G with cardinality at most k.

Let G be a cubic planar graph and $s \geq 2$ be an integer. For each vertex $v \in V(G)$ introduce a path P_v having $s - 2$ edges and introduce an edge between v and one of the pendent vertices of P_v. Note that P_v contains $s - 1$ vertices. For each edge $e = uv \in E(G)$ introduce two new vertices x_e, y_e and add the edges $\{x_e u, x_e v, y_e u, y_e v\}$. Finally delete the edge e from G (See Fig. 1). Let H be the graph obtained after applying the above operations on G. Note that if G is a cubic planar graph, then H is a planar bipartite graph of maximum degree at most 7.

Lemma 1. *If G has a vertex cover of size k, then H has an s-CD set of size k.*

Proof. Let D be a vertex cover of G of size k. We shall show that D is an s-CD set of H. For a vertex $u \in G$, let $E_G(u)$ denote the edges incident on u in G. Let C be the vertices of a connected component of $H - D$. Suppose C contains a vertex u that also belongs to G. This means $u \notin D$ and therefore $N_G(u) \subseteq D$. Therefore $C = V(P_u) \cup \{u\} \cup \{x_e, y_e\}_{e \in E_G(u)}$. This implies that the diameter of C is at most s. Now suppose C does not contain any vertex u of G. Then either C is $V(P_v)$ for some vertex $v \in D$ or C is a singleton set $\{w\}$ where $w \in \{x_e, y_e\}_{e \in E_G(u)}$. In either case, the diameter of C is at most $s - 1$. □

Observation A. *Let S be an s-CD set of H. Let $e = uv$ be an edge of G such that S contains none of its end points. That is, $S \cap \{u, v\} = \emptyset$. Then at least one of the following is true:*

1. *The set S contains at least one vertex each from P_v and P_u, or*
2. *The set S contains both x_e and y_e.*

Proof. Suppose for contradiction, that for an edge $e = uv$ of G none of the conditions listed in the statement of this observation are satisfied. Then $H - S$ contains P_u, P_v, u and v. Moreover, $H - S$ also contains one of the vertex from the set $\{x_e, y_e\}$. Without loss of generality assume that x_e is in $H - S$. Then $H - S$ has a connected component C that contains a path $P = (P_u, u, x_e, v, P_v)$. Note that the distance between the endpoints of P in C is at least $s + 1$. Clearly the diameter of C is at least $s + 1$, and therefore S is not an s-CD set of H. □

Observation B. *Let S be an s-CD set of H. Then both of the following are true.*

1. *For a vertex $v \in V(G)$, if S contains a vertex w of P_v, then $(S - \{w\}) \cup \{v\}$ is an s-CD set of H.*
2. *For an edge $e = uv \in E(G)$, if S contains both x_e, y_e then $(S - \{x_e, y_e\}) \cup \{u, v\}$ is an s-CD set of H.*

Proof. The first part of the observation follows from the fact that any path between w and a vertex $z \in H - P_v$ goes through v. The second part of the observation follows from the fact that any path between $w \in \{x_e, y_e\}$ and a vertex $z \in V(H - \{x_e, y_e\})$ goes through u or v.   □

Lemma 2. *Let H has an s-CD of size at most k. Then G has a vertex cover of size k.*

Proof. Let S be an s-CD of H of size at most k. Observations A and B imply that there is an s-CD set S' of H such that for each edge $uv \in E(G)$ either u or v belongs to S' and the cardinality of S' is at most k. Therefore, S' is a vertex cover of G of size at most k.   □

Since G is a cubic planar graph, then H is a planar bipartite graph with maximum degree at most 7. Now Lemma 1 and 2 together imply Theorem 1.

Proof of Theorem 2: Since MINIMUM VERTEX COVER is APX-hard on cubic graphs [1], the above reduction when applied on cubic graphs we have the proof of Theorem 2.

3   APX-Hardness on Split Graphs

In this section, we shall prove Theorem 3. We shall reduce MINIMUM VERTEX COVER (MVC) on general graphs to 2-CVD on split graphs. Let $\langle G, k \rangle$ be an instance of MINIMUM VERTEX COVER such that maximum degree of G is at most $n - 3$. Let $\overline{G}$ denote the complement of G. Now construct a split graph G_{split} from G as follows. For each vertex of $v \in V(G)$, we introduce a new vertex x_v in G_{split} and for each edge $e \in E(\overline{G})$ we introduce a new vertex y_e in G_{split}. For each pair of edges $e_1, e_2 \in E(\overline{G})$ we introduce an edge between y_{e_1} and y_{e_2} in G_{split}. For each edge $e = uv \in E(\overline{G})$, we introduce the edges $x_u y_e$ and $x_v y_e$ in G_{split} (See Fig. 2). Observe that $C = \{y_e\}_{e \in E(\overline{G})}$ is the clique and $I = \{x_v\}_{v \in V(G)}$ is the independent set of G_{split}. We shall show that G has a vertex cover of size k if and only if G_{split} has a 2-CD set of size k.

Observation C. *For each vertex $v \in C$, $|N[v] \cap I| = 2$ and for each vertex $u \in I$, the degree of u is at least two.*

Lemma 3. *Let D be a subset of I and let $T = \{u \in V(G) : x_u \in D\}$. The set D is a 2-CD set of G_{split} if and only if T is a vertex cover of G.*

Proof. Let $D' = I - D$ and $T' = \{u \in V(G) : x_u \in D'\}$ (note that $T = V(G) - T'$). Note that the induced subgraph $G_{\mathsf{split}} - D$ forms a single connected component since there are no isolated vertices by observation C. Therefore, for any two vertices $x_u, x_v \in D'$ the distance between x_u, x_v is 2 if and only if there is an edge between u, v in $\overline{G}$. Therefore,

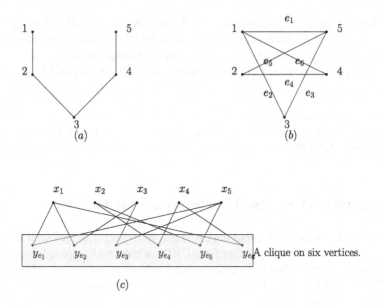

Fig. 2. Reduction for Theorem 3. (a) The graph G, (b) The graph $\overline{G}$, and (c) The graph G_{split}.

distance between any two pair of vertices in D' is 2 if and only if T' induces a clique in $\overline{G}$ and therefore an independent set in G. Since $T = V(G) - T'$, we have that distance between any two pair of vertices in D' is 2 if and only if T is a vertex cover of G. Since $D' = I - D$ we have that D is a 2-CD set of G_{split} if and only if T is a vertex cover of G.   □

Lemma 4. *There is a subset of I which is a minimum* 2-CD *set of* G_{split}.

Proof. Let S be a minimum 2-CD set of G_{split} such that $|S \cap I|$ is maximum. We claim that $S \subseteq I$. Suppose for contradiction this is not true. That is, $S \cap C \neq \emptyset$. Consider the collection $\mathscr{C}$ of connected components of $G_{\mathsf{split}} - S$. First, observe that there exists at most one connected component in $\mathscr{C}$ that intersects C (the clique of G_{split}). We shall call such a component as the *big component* and let X be the set of vertices of the big component. In fact I itself is a 2-CD set and observation C implies $|I| \leq |C|$. Therefore, without loss of generality we can assume that $C \not\subseteq S$ and indeed such a big component exists.

Let Y denote those vertices of $G_{\mathsf{split}} - S$ that belongs to $I - X$. Let $S_C = S \cap C$ and $S_I = S \cap I$. Recall that by assumption, $S_C \neq \emptyset$.

If there is a vertex $v \in S_C$ such that $|N[v] \cap Y| = 0$, then $S - \{v\}$ is a 2-CD set with $X \cup \{v\}$ as corresponding big component with diameter less than or equal to 2. This contradicts the minimality of S. Similarly, if there exists a vertex $v \in S_C$ such that $N[v] \cap Y = \{u\}$, a singleton set then $S' = S \cup \{u\} - \{v\}$ is a new 2-CD set with $X \cup \{v\}$ as corresponding new big component. This contradicts the assumption that S is a minimum 2-CD set with $|S \cap I|$ is maximum. Hence together with observation C we infer that $|N(v) \cap Y| = 2$, for each $v \in S_C$. Observation C also implies that for each vertex $u \in Y$,

$|N(u) \cap S_C| \geq 2$, since $Y \subseteq I$ for each $u \in Y$ we have $N(u) \subseteq S_C$. Therefore, $|Y| \leq |S_C|$ and $S' = (S - S_C) \cup Y$ is a minimum 2-CD set with $X \cup S_C$ as the corresponding new big component and $|S' \cap I| > |S \cap I|$. This contradicts the assumption for S.

Hence we conclude that S is indeed a minimum 2-CD set such that $S \subseteq I$.   □

Lemmas 3 and 4 imply that G has a vertex cover of size k if and only if G_{split} has a 2-CD set of size k. Now Theorem 3 follows from a result of Khot and Regev [15], where they showed that unless the Unique Games Conjecture is false, there is no $(2 - \varepsilon)$-approximation algorithm for MINIMUM VERTEX COVER on general graphs, for any $\varepsilon > 0$.

4   Polynomial-Time Algorithm for Trapezoid Graphs

Recall that we have discussed about trapezoid graphs in Sect. 1. In this section we shall propose a polynomial-time algorithm for s-CVD on trapezoid graphs, for each $s \geq 1$. We shall consider the following maximisation problem called WEAK MAXIMUM s-CLUB (WMs-CLUB). A *weak s-club* of G is a subgraph of G such that each connected component of it is an s-club. In WMs-CLUB the input is a graph and an integer k. The objective is to decide if G has a weak s-club having at least k vertices. We have the following observation.

Observation D. *For each $s \geq 1$, a graph G has an s-CD set of size k if and only if G has a weak s-club having $|V(G)| - k$ vertices.*

In the rest of the section, we present a polynomial-time algorithm for WMs-CLUB on trapezoid graphs.

Every trapezoid graph is a *cocomparability* graph. A graph G is a cocomparability graph if the edges of $\bar{G}$ admits a transitive orientation[1]. We shall use the following structural and algorithmic results on cocomparability graphs. For a graph G and an integer $k \geq 1$, the kth *power* of G, denoted by G^k, is the graph whose vertex set is same as G, and two vertices u and v are adjacent if and only if the distance between u and v in G is at most k.

Theorem 5 [8]. *Let G be a cocomparability graph. For each $k \geq 1$, G^k is a cocomparability graph.*

Theorem 6 [16]. *There is an $O(n + m)$-time algorithm to compute the maximum clique of a cocomparability graph with n vertices and m edges.*

Theorem 7 [17]. *There is an $O(n + m)$-time algorithm to compute the weighted maximum independent set of a cocomparability graph with n vertices and m edges.*

---

[1] Let H be a graph and $E(H)$ be the edge set of H. A transitive orientation of H is an orientation of the edges in $E(H)$ such that if $(a, b), (b, c) \in E(G)$ and are oriented from a to b and b to c respectively then $(a, c) \in E(H)$ and is oriented from a to c.

Now we formally define a *trapezoid representation* of a graph. Let G be a graph, ℓ_t, ℓ_b be two horizontal lines. For each $v \in G$, let $I_t(v) = [t^-(v), t^+(v)]$ and $I_b(v) = [b^-(v), b^+(v)]$ be two intervals on ℓ_t, ℓ_b, respectively. Let $I(v)$ denote the trapezoid defined by $I_t(v)$ and $I_b(v)$ so that $I_t(v)$ and $I_b(v)$ form a pair of parallel sides of $I(v)$. The collection of trapezoids $\mathscr{I} = \{I(v)\}_{v \in V(G)}$ is a *trapezoid representation* of G if two trapezoids $I(v)$ and $I(u)$ intersect if and only if $uv \in E(G)$. For an induced subgraph H, let

$$t^-(H) = \min\{t^-(v) : v \in V(H)\}$$

$$t^+(H) = \max\{t^+(v) : v \in V(H)\}$$

$$b^-(H) = \min\{b^-(v) : v \in V(H)\}$$

$$b^+(H) = \max\{b^+(v) : v \in V(H)\}$$

A vertex $v \in V(H)$ is a *corner vertex* of H if

$$\{t^-(v), t^+(v), b^-(v), b^+(v)\} \cap \{t^-(H), t^+(H), b^-(H), b^+(H)\} \neq \emptyset$$

Observe that any induced subgraph of G has at most four corner vertices. For the remainder of this section, s is a fixed positive integer, G denotes a fixed trapezoid graph with n vertices and m edges and the set $\mathscr{I} = \{I(v)\}_{v \in V(G)}$ is a fixed trapezoid representation of G where corners of the trapezoids are distinct[2]. Let H_1 and H_2 be two induced subgraphs of G. We say that H_1 is *left of* H_2 if $t^+(H_1) < t^-(H_2)$ and $b^+(H_1) < b^-(H_2)$. The subgraphs H_1 and H_2 of G are *neighbour disjoint* if (i) $V(H_1) \cap V(H_2) = \emptyset$ and (ii) there is no edge $uv \in E(G)$ such that $u \in V(H_1)$ and $v \in V(H_2)$.

The subgraph H is *valid* if (i) whenever we have a vertex $v \in V(G)$ such that $t(v) \in [t^-(H), t^+(H)]$ and $b(v) \in [b^-(H), b^+(H)]$ we have $v \in V(H)$ and (ii) distance between the corner vertices of H in G is at most s. Observe that there are at most $O(n^4)$ many valid induced subgraphs of G. We shall use the following observation about trapezoid graphs.

Observation E. *Let H_1 and H_2 be two valid induced subgraphs of G. Then H_1 and H_2 are neighbour disjoint if and only if either H_1 is left of H_2 or H_2 is left of H_1.*

Proof. First to see the easy direction, without loss of generality assume that H_1 is *left of* H_2. Clearly for any vetex $u \in H_1$ and $v \in H_2$ we have $t^+(u) \leq t^+(H_1) < t^-(H_2) \leq t^-(v)$ and $b^+(u) \leq b^+(H_1) < b^-(H_2) \leq b^-(v)$ and hence the vertices u and v are not adjacent in G. Therefore H_1 and H_2 are neighbour disjoint. For the other direction, assume that H_1 and H_2 are neighbour disjoint. For the sake of contradiction suppose H_1 is not *left of* H_2 and H_2 is not *left of* H_1. Then we have two cases:

1. $t^+(H_1) > t^+(H_2)$ and $b^+(H_1) > b^+(H_2)$. Let $u, u' \in V(H_1)$ be the corner vertices of H_1 such that $t^+(u) = t^+(H_1)$ and $b^+(u') = b^+(H_1)$, possibly $u = u'$. Clearly $t^-(u) \geq t^-(H_1)$ and $b^-(u') \geq b^-(H_1)$. If $t^-(H_1) > t^+(H_2)$ and $b^-(H_1) > b^+(H_2)$ then H_2 is *left of* H_1, contradicting the assumption (See Fig. 3 (a)). Hence assume

---

[2] Such representation of *trapezoid graphs* are possible, see [9].

that one of the above condition is false and without loss of generality assume that $t^-(H_1) < t^+(H_2)$. If $t^-(u) < t^+(H_2)$ then $I_t(u) \cap I_t(v) \neq \emptyset$ where $v \in V(H_2)$ is the corner vertex such that $t^+(v) = t^+(H_2)$, in which case the trapezoids corresponding to the vertices u and v intersects. So assume that $t^-(u) > t^+(H_2)$. Now consider the corner vertex $u'' \in V(H_1)$ such that $t^-(u'') = t^-(H_1)$. We infer that $t^+(u'') < t^-(v) \leq t^+(v)$. Otherwise, $I_t(u'') \cap I_t(v) \neq \emptyset$ and the trapezoid u'' intersects with the trapezoid v. Now we have two sub cases depending on the position of $b^-(H_1)$. If $b^-(H_1) > b^+(H_2)$, then $b^-(u'') > b^+(v)$. Since $t^-(v) > t^+(u'')$ the trapezoid corresponding to the vertex v intersects with the trapezoid corresponding to the vertex u'' and hence $(u'',v) \in E(G)$ (See Fig. 3 (b)). This contradicts the assumption that H_1 and H_2 are neighbour disjoint. Now consider the other sub case, $b^-(H_1) < b^+(H_2)$. Since $t^-(v) > t^+(u'')$ we must have $b^+(v) \geq b^-(v) > b^+(u'') \geq b^-(u'')$(See Fig. 3 (c)). Otherwise, $I_b(u'') \cap I_b(v) \neq \emptyset$ and the trapezoid u'' intersects with the trapezoid v. Thus $t(v) \in [t^-(H_1), t^+(H_1)]$ and $b(v) \in [b^-(H_1), b^+(H_1)]$. Since by assumption H_1 is valid, we have $v \in V(H_1)$. But $v \in V(H_2)$ and this implies H_1 and H_2 are neighbour disjoint, a contradiction to the assumption.

2. $t^+(H_1) > t^+(H_2)$ and $b^+(H_2) > b^+(H_1)$. Let $x \in V(H_2)$ be a corner vertex of H_2 such that $b^+(x) = b^+(H_2)$ and $y \in V(H_1)$ be a corner vertex of H_1 such that $t^+(y) = t^+(H_1)$. Observe that $t^+(x) \leq t^+(H_2)$ and $b^+(y) \leq b^+(H_1)$. Therefore, $t^+(H_2) < t^+(H_1)$ implies $t^+(x) < t^+(y)$ and $b^+(H_2) > b^+(H_1)$ implies $b^+(y) < b^+(x)$. This clearly shows that the trapezoid corresponding to the vertex x intersects with the trapezoid corresponding to the vertex y and hence $(x,y) \in E(G)$ (See Fig. 3 (d)). This contradicts the assumption that H_1 and H_2 are neighbour disjoint.   □

Now construct a graph $\mathcal{H}$ whose vertex set is all valid induced subgraphs of G and two vertices H_1, H_2 are adjacent if and only if H_1 and H_2 are not neighbour disjoint in G. Clearly, $\mathcal{H}$ has $O(n^4)$ vertices, $O(n^8)$ edges and can be constructed in polynomial-time. We prove the following lemma.

Lemma 5. *The graph $\mathcal{H}$ is a cocomparability graph.*

Proof. To prove that $\mathcal{H}$ is a cocomparability graph we show that the edges of $\overline{\mathcal{H}}$ can be transitively oriented. For an edge $(H_1, H_2) \in E(\overline{\mathcal{H}})$, clearly H_1 and H_2 are neighbor disjoint. From Observation E either H_1 is *left of* H_2 or H_2 is *left of* H_1. If H_1 is *left of* H_2, we orient the edge (H_1, H_2) in $\overline{\mathcal{H}}$ from H_1 to H_2. Otherwise, we orient the edge from H_2 to H_1. Now to verify that this orientation is indeed a transitive orientation of $\overline{\mathcal{H}}$, consider two directed edges $(H_1, H_2), (H_2, H_3) \in E(\overline{\mathcal{H}})$. This means that H_1 is *left of* H_2 and H_2 is *left of* H_3. From the definition of *left of* it is easily follows that H_1 is *left of* H_3. By invoking Observation E, we can see that H_1 and H_3 are neighbour disjoint and therefore H_1 and H_3 are adjacent in $\overline{\mathcal{H}}$. Therefore, by the rule of our orientation the edge (H_1, H_3) is oriented from H_1 to H_3, as required.   □

Let $\omega_s : V(\mathcal{H}) \to \mathbb{Z}^+$ be a positive integer valued weight function where $\omega_s(H)$ denote the cardinality of the maximum s-CLUB of H. From now on, we will consider $\mathcal{H}$ as a weighted graph where weights are defined by ω_s.

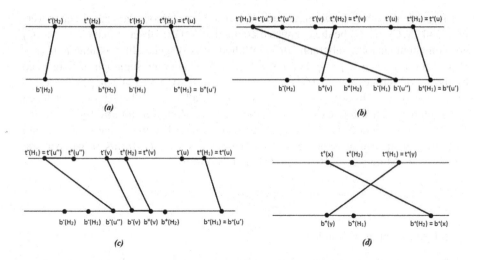

Fig. 3. Trapezoid representation of corner vertices of induced subgraphs H_1 and H_2. (a) $t^+(H_1) > t^-(H_1) > t^+(H_2)$ and $b^+(H_1) > b^-(H_1) > b^+(H_2)$, (b) $t^+(H_1) > t^+(H_2) > t^-(H_1)$ and $b^+(H_1) > b^-(H_1) > b^+(H_2)$, (c) $t^+(H_1) > t^+(H_2) > t^-(H_1)$ and $b^+(H_1) > b^+(H_2) > b^-(H_1)$ and (d) $t^+(H_1) > t^+(H_2)$ and $b^-(H_2) > b^+(H_1)$

Lemma 6. *The weight function ω_s can be determined in polynomial-time.*

Proof. Let H be a valid induced subgraph of G and H^s be the sth power of H. Observe that, an induced subgraph F of H is an s-club if and only if $V(F)$ induces a complete subgraph in H^s. Hence, the maximum s-club of H corresponds to the maximum clique in H^s. Since the class of trapezoid graphs is closed under taking induced subgraphs, H is a trapezoid graph. Therefore H is a cocomparability graph and Theorem 5 implies that H^s is a cocomparability graph. Theorem 6 implies that the maximum clique of H^s can be found in polynomial-time. Hence, the maximum s-club of H and $\omega_s(H)$ can be found in polynomial-time. □

In the following lemma, we shall relate a weak s-club of G with an independent set of $\mathscr{H}$.

Lemma 7. *The trapezoid graph G has a weak s-club with at least k vertices if and only if $\mathscr{H}$ has an independent set of weight at least k.*

Proof. Let $\mathscr{I} = \{H_1, H_2, \ldots, H_t\}$ be an independent set of $\mathscr{H}$ of weight k. For $1 \le i \le t$, let H_i' denote the maximum s-club of H_i. Then $|V(H_i')| = \omega_s(H_i)$ and therefore

$$\sum_{i=1}^{t} |V(H_i')| = \sum_{i=1}^{t} \omega_s(H_i) \ge k$$

Since $H_1, H_2, \ldots, H_t$ are pairwise neighbour disjoint, $H_1', H_2', \ldots, H_t'$ are also pairwise neighbour disjoint. Hence, $H_1', H_2', \ldots, H_t'$ forms a weak s-club of G with at least k vertices.

To prove the other direction, assume H is a weak s-club of G having at least k vertices and $C_1, C_2, \ldots, C_q$ be the connected components of H. Observe that $C_1, C_2, \ldots, C_q$ are neighbour disjoint subgraphs of G. Without loss of generality, assume $t^-(C_i) < t^-(C_j)$ for any $1 \leq i < j \leq q$. For $1 \leq i \leq q$, let C_i' denote the subgraph of G induced by the set of vertices $\{w \in V(G) : [t^-(w), t^+(w)] \subseteq [t^-(C_i), t^+(C_i)], [b^-(w), b^+(w)] \subseteq [b^-(C_i), b^+(C_i)]\}$. Observe that each C_i' is a supergraph of C_i and C_i' is a valid induced subgraph of G. Therefore each C_i' are vertices of $\mathscr{H}$. Moreover, for any $1 \leq i < j \leq q$, $t^+(C_i') < t^-(C_j')$ and $b^+(C_i') < b^-(C_j')$. By Observation E implies that C_i' and C_j' are valid neighbour disjoint subgraphs of G. Hence, $C_1', C_2', \ldots, C_q'$ forms an independent set $\mathscr{I}$ of $\mathscr{H}$. Moreover, $\omega_s(C_i') \geq |V(C_i)|$ and therefore

$$\sum_{i=1}^{q} \omega_s(C_i') \geq k$$

Therefore weight of $\mathscr{I}$ is at least k. □

Since a weighted maximum independent set of a cocomparability is possible in polynomial-time (Theorem 7), we have the following polynomial-time algorithm for solving WMs-CLUB on a trapezoid graph G. Given two integers $s \geq 1$, $k \geq 1$, a trapezoid graph G and a trapezoid representation $\mathscr{I}$ of G

– construct the graph $\mathscr{H}$ and the weight function ω_s, and
– decide if $\mathscr{H}$ has an independent set of weight k.

The above algorithm solves WMs-CLUB on a trapezoid graph G and due to Observation D we have a polynomial-time algorithm for s-CVD on trapezoid graphs.

5   Conclusion

In this paper we studied the computational complexity of s-CVD on three important graph classes: (planar) bipartite graphs, split graphs and trapezoid graphs. We gave a polynomial-time algorithm for s-CVD on trapezoid graphs for each $s \geq 1$. This raises the following question. Is there a polynomial-time algorithm for s-CVD on cocomparability or *asteroidal triple free* graphs? Note that, the computational complexity of CLUSTER VERTEX DELETION is unknown for both the aforementioned graph classes.

We proved that for each $s \geq 2$, s-CVD is NP-hard on planar bipartite graphs. On the other hand, the complexity of CLUSTER VERTEX DELETION on planar bipartite graphs remains unknown. Observe that, solving CLUSTER VERTEX DELETION on (planar) bipartite graphs is equivalent to finding the minimum number of vertices whose deletion cause the remaining graphs to have maximum degree 1. The aforementioned problem is known as MINIMUM DISSOCIATION NUMBER whose complexity on planar bipartite graphs is unknown.

We also showed that unless the *Unique Games Conjecture* is false, it is not possible to approximate 2-CVD on split graphs beyond a factor of 2. CLUSTER VERTEX DELETION is polynomial-time solvable on split graphs but it's computational complexity is unknown on chordal graphs. For each $s \geq 2$, is there a constant factor approximation algorithm for s-CVD on chordal graphs?

References

1. Alimonti, P., Kann, V.: Hardness of approximating problems on cubic graphs. In: Bongiovanni, G., Bovet, D.P., Di Battista, G. (eds.) CIAC 1997. LNCS, vol. 1203, pp. 288–298. Springer, Heidelberg (1997). https://doi.org/10.1007/3-540-62592-5_80
2. Asahiro, Y., Miyano, E., Samizo, K.: Approximating maximum diameter-bounded subgraphs. In: López-Ortiz, A. (ed.) LATIN 2010. LNCS, vol. 6034, pp. 615–626. Springer, Heidelberg (2010). https://doi.org/10.1007/978-3-642-12200-2_53
3. Balasundaram, B., Butenko, S., Trukhanov, S.: Novel approaches for analyzing biological networks. J. Comb. Optim. **10**(1), 23–39 (2005)
4. Bera, D., Pal, M., Pal, T.K.: An efficient algorithm for finding all hinge vertices on trapezoid graphs. Theor. Comput. Syst. **36**(1), 17–27 (2003)
5. Bodlaender, H.L., Kloks, T., Kratsch, D., Müller, H.: Treewidth and minimum fill-in on d-trapezoid graphs. In: Graph Algorithms And Applications I, pp. 139–161. World Scientific (2002)
6. Cao, Y., Ke, Y., Otachi, Y., You, J.: Vertex deletion problems on chordal graphs. Theor. Comput. Sci. **745**, 75–86 (2018)
7. Chang, M., Hung, L., Lin, C., Su, P.: Finding large *k*-clubs in undirected graphs. Computing **95**(9), 739–758 (2013)
8. Damaschke, P.: Distances in cocomparability graphs and their powers. Discrete Appl. Math. **35**(1), 67–72 (1992)
9. Felsner, S., Müller, R., Wernisch, L.: Trapezoid graphs and generalizations, geometry and algorithms. In: Schmidt, E.M., Skyum, S. (eds.) SWAT 1994. LNCS, vol. 824, pp. 143–154. Springer, Heidelberg (1994). https://doi.org/10.1007/3-540-58218-5_13
10. Figiel, A., Himmel, A., Nichterlein, A., Niedermeier, R.: On 2-clubs in graph-based data clustering: theory and algorithm engineering. arXiv preprint arXiv:2006.14972 (2020)
11. Garey, M., Johnson, D.: Computers and Intractability: A Guide to the Theory of NP-Completeness (1978)
12. Hartung, S., Komusiewicz, C., Nichterlein, A., Suchỳ, O.: On structural parameterizations for the 2-club problem. Discrete Appl. Math. **185**, 79–92 (2015)
13. Hota, M., Pal, M., Pal, T.K.: An efficient algorithm for finding a maximum weight k-independent set on trapezoid graphs. Comput. Optim. Appl. **18**(1), 49–62 (2001)
14. Hüffner, F., Komusiewicz, C., Moser, H., Niedermeier, R.: Fixed-parameter algorithms for cluster vertex deletion. Theor. Comput. Syst. **47**(1), 196–217 (2010)
15. Khot, S., Regev, O.: Vertex cover might be hard to approximate to within 2- ε. J. Comput. Syst. Sci. **74**(3), 335–349 (2008)
16. Köhler, E., Mouatadid, L.: Linear time LexDFS on cocomparability graphs. In: Ravi, R., Gørtz, I.L. (eds.) SWAT 2014. LNCS, vol. 8503, pp. 319–330. Springer, Cham (2014). https://doi.org/10.1007/978-3-319-08404-6_28
17. Köhler, E., Mouatadid, L.: A linear time algorithm to compute a maximum weighted independent set on cocomparability graphs. Inf. Process. Lett. **116**(6), 391–395 (2016)
18. Kratsch, D., McConnell, R.M., Mehlhorn, K., Spinrad, J.P.: Certifying algorithms for recognizing interval graphs and permutation graphs. SIAM J. Comput. **36**(2), 326–353 (2006)
19. Lewis, J.M., Yannakakis, M.: The node-deletion problem for hereditary properties is NP-complete. J. Comput. Syst. Sci. **20**(2), 219–230 (1980)
20. Liu, H., Zhang, P., Zhu, D.: On editing graphs into 2-club clusters. In: Snoeyink, J., Lu, P., Su, K., Wang, L. (eds.) AAIM/FAW -2012. LNCS, vol. 7285, pp. 235–246. Springer, Heidelberg (2012). https://doi.org/10.1007/978-3-642-29700-7_22
21. Mertzios, G.B.: The recognition of triangle graphs. Theor. Comput. Sci. **438**, 34–47 (2012)

22. Pasupuleti, S.: Detection of protein complexes in protein interaction networks using n-clubs. In: Marchiori, E., Moore, J.H. (eds.) EvoBIO 2008. LNCS, vol. 4973, pp. 153–164. Springer, Heidelberg (2008). https://doi.org/10.1007/978-3-540-78757-0_14
23. Sau, I., Souza, U.S.: Hitting forbidden induced subgraphs on bounded treewidth graphs. arXiv preprint arXiv:2004.08324 (2020)
24. Schäfer, A.: Exact algorithms for s-club finding and related problems. PhD thesis, Friedrich-Schiller-University Jena (2009)
25. Schäfer, A., Komusiewicz, C., Moser, H., Niedermeier, R.: Parameterized computational complexity of finding small-diameter subgraphs. Optim. Lett. **6**(5), 883–891 (2012)
26. Takaoka, A.: A recognition algorithm for simple-triangle graphs. Discrete Appl. Math. **282**, 196–207 (2020)
27. Yannakakis, M.: Node-and edge-deletion NP-complete problems. In: Proceedings of the Tenth Annual ACM Symposium on Theory of Computing, pp. 253–264 (1978)

Covering Convex Polygons by Two Congruent Disks

Jongmin Choi[1($\boxtimes$)], Dahye Jeong[1($\boxtimes$)], and Hee-Kap Ahn[2($\boxtimes$)] (iD)

[1] Department of Computer Science and Engineering, Pohang University of Science and Technology, Pohang, Korea
{icothos,dahyejeong}@postech.ac.kr
[2] Department of Computer Science and Engineering, Graduate School of Artificial Intelligence, Pohang University of Science and Technology, Pohang, Korea
heekap@postech.ac.kr

Abstract. We consider the planar two-center problem for a convex polygon: given a convex polygon in the plane, find two congruent disks of minimum radius whose union contains the polygon. We present an $O(n \log n)$-time algorithm for the two-center problem for a convex polygon, where n is the number of vertices of the polygon. This improves upon the previous best algorithm for the problem.

Keywords: Two-center problem · Covering · Convex polygon

1 Introduction

The problem of covering a region R by a predefined shape Q (such as a disk, a square, a rectangle, a convex polygon, etc.) in the plane is to find k homothets[1] of Q with the same homothety ratio such that their union contains R and the homothety ratio is minimized. The homothets in the covering are allowed to overlap, as long as their union contains the region. This is a fundamental optimization problem [2,4,19] arising in analyzing and recognizing shapes, and it has real-world applications, including computer vision and data mining.

The covering problem has been extensively studied in the context of the *k-center problem* and the *facility location* problem when the region to cover is a set of points and the predefined shape is a disk in the plane. In last decades, there have been a lot of works, including exact algorithms for $k = 2$ [3,11,13,14,33,35], exact

---

[1] For a shape Q in the plane, a (positive) homothet of Q is a set of the form $\lambda Q + v :=$ $\{\lambda q + v \mid q \in Q\}$, where $\lambda > 0$ is the homothety ratio, and $v \in \mathbb{R}^2$ is a translation vector.

This research was supported by the Institute of Information & Communications Technology Planning & Evaluation (IITP) grant funded by the Korea government (MSIT) (No. 2017-0-00905, Software Star Lab (Optimal Data Structure and Algorithmic Applications in Dynamic Geometric Environment)) and (No. 2019-0-01906, Artificial Intelligence Graduate School Program(POSTECH)).

P. Flocchini and L. Moura (Eds.): IWOCA 2021, LNCS 12757, pp. 165–178, 2021.
https://doi.org/10.1007/978-3-030-79987-8_12

and approximation algorithms for large k [2,19,21,24], algorithms in higher dimensional spaces [1,2,29], and approximation algorithms for streaming points [5,6,12, 22,25,37]. There are also some works on the k-center problem for small k when the region to cover is a set of disks in the plane, for $k = 1$ [20,27,28] and $k = 2$ [8].

In the context of the facility location, there have also been some works on the *geodesic k-center* problem for simple polygons [7,30] and polygonal domains [9], in which we find k points (centers) in order to minimize the maximum geodesic distance from any point in the domain to its closest center.

In this paper we consider the covering problem for a convex polygon in which we find two congruent disks of minimum radius whose union contains the convex polygon. Thus, our problem can be considered as the *(geodesic) two-center problem for a convex polygon*. See Fig. 1 for an illustration.

Previous Works. For a convex polygon with n vertices, Shin et al. [34] gave an $O(n^2 \log^3 n)$-time algorithm using parametric search for the two-center problem. They also gave an $O(n \log^3 n)$-time algorithm for the restricted case of the two-center problem in which the centers must lie at polygon vertices. Later, Kim and Shin [26] improved the results and gave an $O(n \log^3 n \log \log n)$-time algorithm for the two-center problem and an $O(n \log^2 n)$-time algorithm for the restricted case of the problem.

There has been a series of work dedicated to variations of the k-center problem for a convex polygon, most of which require certain constraints on the centers, including the centers restricted to lie on the polygon boundary [31] and on a given polygon edge(s) [17,31]. For large k, there are quite a few approximation algorithms. For $k \geq 3$, Das et al. [17] gave an $(1 + \epsilon)$-approximation algorithm with the centers restricted to lie on the same polygon edge, along with a heuristic algorithm without such restriction. Basappa et al. [10] gave a $(2 + \epsilon)$-approximation algorithm for $k \geq 7$, where the centers are restricted to lie on the polygon boundary. There is a 2-approximation algorithm for the two-center problem for a convex polygon that supports insertions and deletions of points in $O(\log n)$ time per operation [32].

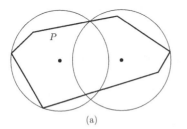 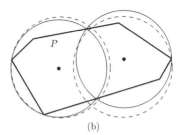

(a)   (b)

Fig. 1. (a) Two congruent disks whose union covers a convex polygon P. (b) P can be covered by two congruent disks of smaller radius.

Our Results. We present an $O(n \log n)$-time deterministic algorithm for the two-center problem for a convex polygon P with n vertices. That is, given a convex polygon with n vertices, we can find in $O(n \log n)$ time two congruent disks of minimum radius whose union covers the polygon. This improves upon the $O(n \log^3 n \log \log n)$ time bound of Kim and Shin [26].

Sketch of Our Algorithm. Our algorithm is twofold. First we solve the sequential decision problem in $O(n)$ time. That is, given a real value r, decide whether $r \geq r^*$, where r^* is the optimal radius value. Then we present a parallel algorithm for the decision problem which takes $O(\log n)$ time using $O(n)$ processors, after an $O(n \log n)$-time preprocessing. Using these decision algorithms and applying Cole's parametric search [16], we solve the optimization problem, the two centers for P, in $O(n \log n)$ deterministic time.

We observe that if P is covered by two congruent disks D_1 and D_2 of radius r, D_1 covers a connected subchain P_1 of the boundary of P and D_2 covers the remaining subchain P_2 of the boundary of P. Thus, in the sequential decision algorithm, we compute for any point x on the boundary of P, the longest subchain of the boundary of P from x in counterclockwise direction that is covered by a disk of radius r, and the longest subchain of the boundary P from x in clockwise direction that is covered by a disk of radius r. We show that the determinators of the disks that define the two longest subchains change $O(n)$ times while x moves along the boundary of P. We also show that the disks and the longest subchains can be represented by $O(n)$ algebraic functions. Our sequential decision algorithm computes the longest subchains in $O(n)$ time. Finally, the sequential decision algorithm determines whether there is a point x' in P such that the two longest subchains from x', one in counterclockwise direction and one in clockwise direction, cover the polygon boundary in $O(n)$ time.

Our parallel decision algorithm computes the longest subchains in parallel and determines whether there is a point x' in P such that the two longest subchains from x' covers the polygon boundary in $O(\log n)$ parallel steps using $O(n)$ processors after $O(n \log n)$-time preprocessing. For this purpose, the algorithm finds rough bounds of the longest subchains, by modifying the parallel decision algorithm for the planar two-center problem of points in convex position [14] and applying it for the vertices of P. Then the algorithm computes $O(n)$ algebraic functions of the longest subchains in $O(\log n)$ time using $O(n)$ processors. Finally, it determines in parallel computation whether there is a point x' in P such that the two longest subchains from x covers the polygon boundary.

We can compute the optimal radius value r^* using Cole's parametric search [16]. For a sequential decision algorithm of running time T_S and a parallel decision algorithm of parallel running time T_P using N processors, Cole's parametric search is a technique that computes an optimal value in $O(N T_P + T_S(T_P + \log N))$ time. In our case, $T_S = O(n)$, $T_P = O(\log n)$, and $N = O(n)$. Therefore, we get a deterministic $O(n \log n)$-time algorithm for the two-center problem for a convex polygon P.

Due to lack of space, some of the proofs and details are omitted. A full version of this paper is available in [15].

2   Preliminaries

For any two sets X and Y in the plane, we say X *covers* Y if $Y \subseteq X$. We say a set X is r-*coverable* if there is a disk D of radius r covering X. For a compact set A, we use ∂A to denote the boundary of A. We simply say x moves *along* ∂A when x moves in the counterclockwise direction along ∂A. Otherwise, we explicitly mention the direction.

Let P be a convex polygon with n vertices $v_1, v_2, \ldots, v_n$ in counterclockwise order along the boundary of P. Throughout the paper, we assume general circular position on the vertices of P, meaning no four vertices are cocircular. We denote the subchain of ∂P from a point x to a point y in ∂P in counterclockwise order as $P_{x,y} = \langle x, v_i, v_{i+1}, \ldots, v_j, y \rangle$, where $v_i, v_{i+1}, \ldots, v_j$ are the vertices of P that are contained in the subchain. We call $x, v_i, v_{i+1}, \ldots, v_j, y$ the *vertices* of $P_{x,y}$. By $|P_{x,y}|$, we denote the number of distinct vertices of $P_{x,y}$.

We can define an order on the points of ∂P, with respect to a point $p \in \partial P$. For two points x and y of ∂P, we use $x <_p y$ if y is farther from p than x in the counterclockwise direction along ∂P. We define $\leq_p, >_p, \geq_p$ accordingly.

For a subchain C of ∂P, we denote by $I_r(C)$ the intersection of the disks of radius r, each centered at a point in C. See Fig. 2(a). Observe that any disk of radius r centered at a point $p \in I_r(C)$ covers the entire chain C. Hence, $I_r(C) \neq \emptyset$ if and only if C is r-coverable. The *circular hull* of a set X, denoted by $\alpha_r(X)$, is the intersection of all disks of radius r covering X. See Fig. 2(b). Let S be the set of vertices of a subchain C of ∂P. If a disk covers C, it also covers S. If a disk covers S, it covers C since it covers every line segment induced by pairs of the points in S, due to the convexity of a disk. Therefore, $\alpha_r(C)$ and $\alpha_r(S)$ are the same and $I_r(C)$ and $I_r(S)$ are the same.

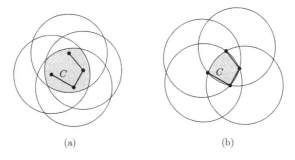

(a)   (b)

Fig. 2. C is a subchain of ∂P and S is the vertex set of C. (a) $I_r(S) = I_r(C)$ (b) $\alpha_r(S) = \alpha_r(C)$

Every vertex of $\alpha_r(C)$ is a vertex of C. The boundary of $\alpha_r(C)$ consists of arcs of radius r, each connecting two vertices of C. The circular hull $\alpha_r(C)$ is dual to the intersection $I_r(C)$, in the sense that every arc of $\alpha_r(C)$ is on the circle of radius r centered at a vertex of $I_r(C)$, and every arc of $I_r(C)$ is on the circle of radius r centered at a vertex of $\alpha_r(C)$. This implies that $\alpha_r(C) \neq \emptyset$ if and only if $I_r(C) \neq \emptyset$. Therefore, $\alpha_r(C)$ is nonempty if and only if C is r-coverable.

For a vertex v of $\alpha_r(C)$, we denote by $\mathrm{ccw}(v)$ its counterclockwise neighbor on $\partial\alpha_r(C)$, and by $\mathrm{cw}(v)$ its clockwise neighbor on $\partial\alpha_r(C)$. We denote by $\gamma(v)$ the arc of $\alpha_r(C)$ connecting v and $\mathrm{ccw}(v)$ of $\alpha_r(C)$. By $\delta(v)$, we denote the supporting disk of the arc $\gamma(v)$ of $\alpha_r(C)$, that is, the disk containing $\gamma(v)$ in its boundary. We may use $\alpha(C)$ and $I(C)$ to denote $\alpha_r(C)$ and $I_r(C)$, respectively, if it is understood from context. Since $\alpha(C)$ and $\alpha(S)$ are the same, we obtain the following observation on subchains from the observations on planar points [18,23].

Observation 1 [18,23]. For a subchain C of ∂P the followings hold.

1. For any subchain $C' \subseteq C$, $\alpha_r(C') \subseteq \alpha_r(C)$.
2. A vertex of C appears as a vertex in $\alpha_r(C)$ if and only if C is r-coverable by a disk containing the vertex on its boundary.
3. An arc of radius r connecting two vertices of C appears as an arc of $\alpha_r(C)$ if and only if C is r-coverable by the supporting disk of the arc.

For a point $x \in \partial P$, let $f_r(x)$ be the farthest point on ∂P from x in the counterclockwise direction along ∂P such that $P_{x,f_r(x)}$ is r-coverable. We denote by $D_1^r(x)$ the disk of radius r covering $P_{x,f_r(x)}$. Similarly, let $g_r(x)$ be the farthest point on ∂P from x in the clockwise direction such that $P_{g_r(x),x}$ is r-coverable, and denote by $D_2^r(x)$ the disk of radius r covering $P_{g_r(x),x}$. Note that x may not lie on the boundaries of D_1^r and D_2^r. We may use $f(x)$, $D_1(x)$, $g(x)$, and $D_2(x)$ by omitting the subscript and superscript r in the notations, if they are understood from context.

Since we can determine in $O(n)$ time whether P is r-coverable [29], we assume that P is not r-coverable in the remainder of the paper. For a fixed r, consider any two points t and t' in ∂P satisfying $t <_t t' <_t f(t)$. Then $P_{t',f(t)}$ is r-coverable, which implies $f(t) \leq_{t'} f(t')$. Thus, we have the following observation.

Observation 2. For a fixed r, as x moves along ∂P in the counterclockwise direction, both $f(x)$ and $g(x)$ move monotonically along ∂P in the counterclockwise direction.

3   Sequential Decision Algorithm

In this section, we consider the decision problem: given a real value r, decide whether $r \geq r^*$, that is, whether there are two congruent disks of radius r whose union covers P.

For a point x moving along ∂P, we consider two functions, $f(x)$ and $g(x)$. If there is a point $x \in \partial P$ such that $f(x) \geq_x g(x)$, the union of $P_{x,f(x)}$ and $P_{g(x),x}$ is ∂P. Thus there are two congruent disks of radius r whose union covers P, and the decision algorithm returns **yes**. Otherwise, we conclude that $r < r^*$, and the decision algorithm returns **no**. For a subchain $P_{x,y}$ of ∂P, we use $\alpha(x,y)$ to denote $\alpha(P_{x,y})$, and $I(x,y)$ to denote $I(P_{x,y})$.

3.1 Characterizations

For a fixed r, $I(x, f(x))$ is a point, and it is the center of $\alpha(x, f(x))$. Moreover, $\alpha(x, f(x))$ and $D_1(x)$ are the same. Observe that $D_1(x)$ is defined by two or three vertices of $P_{x,f(x)}$, which we call the *determinators* of $D_1(x)$. For our purpose, we define four *types* of $D_1(x)$ by its determinators: (T1) x, $f(x)$, and one vertex. (T2) x and $f(x)$. (T3) $f(x)$ and one vertex. (T4) $f(x)$ and two vertices. See Fig. 3 for an illustration of the four types.

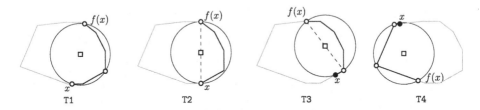

Fig. 3. Four types of $D_1(x)$ and its determinators (small circles).

We denote by $e(a)$ the edge of P containing a point $a \in \partial P$. If a is a vertex of P, $e(a)$ denotes the edge of P incident to a lying in the counterclockwise direction from a. For a point x moving along ∂P, the *combinatorial structure* of $f(x)$ is determined by $e(x)$, $e(f(x))$, and the determinators of $D_1(x)$. We call each point x in ∂P at which the combinatorial structure of $f(x)$ changes a *breakpoint* of $f(x)$. For $x \in \partial P$ lying in between two consecutive breakpoints, we can compute $f(x)$ using $e(x)$, $e(f(x))$, and $D_1(x)$.

Consider x moving along ∂P starting from x_0 on ∂P in counterclockwise direction. Let $x_1 = f(x_0)$, $x_2 = f(x_1)$ and $x_3 = f(x_2)$. We simply use the index i instead of x_i for $i = 0, \ldots, 3$ if it is understood from context. For instance, we use $P_{i,j}$ to denote P_{x_i, x_j}, and $\leq_i$ to denote $\leq_{x_i}$. For the rest of the section, we describe how to handle the case that x moves along $P_{0,1}$. The cases that x moves along $P_{1,2}$ and $P_{2,3}$ can be handle analogously. As x moves along $P_{0,1}$, $f(x)$ moves along $P_{1,2}$ in the same direction by Observation 2.

Lemma 1. *For any fixed $r \geq r^*$, the union of $P_{0,1}$, $P_{1,2}$, and $P_{2,3}$ is ∂P.*

The structure of a circular hull can be expressed by the circular sequence of arcs appearing on the boundary of the circular hull. There is a 1-to-1 correspondence between a breakpoint of $f(x)$ for x moving along $P_{0,1}$ and a structural change to $\alpha(x, f(x))$. This is because $D_1(x)$ and $\alpha(x, f(x))$ are the same. Thus, we maintain $D_1(x)$ for x moving along $P_{0,1}$ and capture every structural change to $\alpha(x, f(x))$. Observe that the boundary of $\alpha(x, f(x))$ consists of a connected boundary part of $\alpha(x, x_1)$, a connected boundary part of $\alpha(x_1, f(x))$, and two arcs of $D_1(x)$ connecting $\alpha(x, x_1)$ and $\alpha(x_1, f(x))$. See Fig. 4 for an illustration.

The following lemmas give some characterizations to the four types of $D_1(x)$. Recall that $\delta(v)$ is the supporting disk of the arc $\gamma(v)$ of an circular hull, that is, the disk containing $\gamma(v)$ on its boundary.

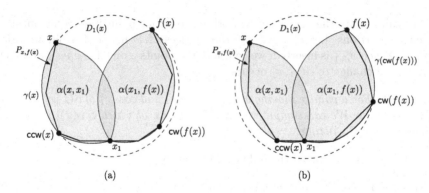

Fig. 4. Two cases of $D_1(x)$ of type **T1**. Two arcs (dashed) of $D_1(x)$ connecting $\alpha(x, x_1)$ and $\alpha(x_1, f(x))$. (a) If v is on the boundary of $\alpha(x, x_1)$, $D_1(x)$ is $\delta(x)$ of $\alpha(x, x_1)$. (b) If v is on the boundary of $\alpha(x_1, f(x))$, $D_1(x)$ is $\delta(\mathsf{cw}(f(x)))$.

Lemma 2. *The following characterizations hold for each type of* $D_1(x)$.

- *For* $D_1(x)$ *of type T1, it is* $\delta(x)$ *of* $\alpha(x, x_1)$ *or* $\delta(\mathsf{cw}(f(x)))$ *of* $\alpha(x_1, f(x))$.
- *For* $D_1(x)$ *of type T2, the Euclidean distance between* x *and* $f(x)$ *is* $2r$.
- *For* $D_1(x)$ *of type T3 or T4 containing* x *on its boundary, it is* $\delta(x)$ *of* $\alpha(x, x_1)$ *or* $\delta(\mathsf{cw}(f(x)))$ *of* $\alpha(x_1, f(x))$. *Moreover, for any point* y *in the interior of* $P_{x,v}$, $D_1(y)$ *has the same type as* $D_1(x)$, *where* v *is the determinator of* $D_1(x)$ *closest to* x *in counterclockwise order.*

If there is a change to $e(x)$, $e(f(x))$, $\mathsf{ccw}(x)$ of $\alpha(x, x_1)$ or $\mathsf{cw}(f(x))$ of $\alpha(x_1, f(x))$, the combinatorial structure of $f(x)$ changes. Therefore, we compute the changes to $e(x)$ and $\mathsf{ccw}(x)$ of $\alpha(x, x_1)$ for a point x moving along $P_{0,1}$, and compute the changes to $e(y)$ and $\mathsf{cw}(y)$ of $\alpha(x_1, y)$ for a point y moving along $P_{1,2}$. We call the points inducing these changes the *event points*. From this, we detect the combinatorial changes to $f(x)$.

3.2 Data Structures and Decision Algorithm

Wang [36] proposed a semi-dynamic (insertion-only) data structure for maintaining the circular hull for points in the plane that are inserted in increasing order of their x-coordinates. It is also mentioned that the algorithm can be modified to work for points that are inserted in the sorted order around a point. Since the vertices of P are already sorted around any point in the interior of P, we can use the algorithm for our purpose.

Lemma 3 (Theorem 5 in [36]). *We can maintain the circular hull of a set* Q *of points such that when a new point to the right of all points of* Q *is inserted, we can decide in* $O(1)$ *amortized time whether* $\alpha(Q)$ *is nonempty, and update* $\alpha(Q)$.

We can modify the algorithm to work not only for point insertions, but also for edge insertions. Let $v_1, \ldots, v_i$ be the vertices of P inserted so far in order from v_1. When v_{i+1} is inserted, we compute the points z on edge $v_i v_{i+1}$ at which a structural change to $\alpha(v_1, z)$ occurs.

Lemma 4. *For a point x moving along $P_{0,1}$, $e(x)$ and ccw(x) of $\alpha(x, x_1)$ change $O(|P_{0,1}|)$ times. We can compute the event points x at which ccw(x) of $\alpha(x, x_1)$ changes in $O(|P_{0,1}|)$ time.*

From Lemma 4, we obtain the following Corollary.

Corollary 1. *For a point y moving along $P_{1,2}$, $e(y)$ and cw(y) of $\alpha(x_1, y)$ change $O(|P_{1,2}|)$ times. We can compute the event points y at which cw(y) of $\alpha(x_1, y)$ changes in $O(|P_{1,2}|)$ time.*

The event points subdivide $P_{0,1}$ and $P_{1,2}$ into $O(|P_{0,1}|)$ and $O(|P_{1,2}|)$ pieces, respectively. Since the vertices of $P_{0,1}$ and $P_{1,2}$ are also event points (defined by the changes to $e(x)$ and $e(y)$), each piece is a segment contained in an edge. Moreover, any point x in a segment of $P_{0,1}$ has the same ccw(x) of $\alpha(x, x_1)$, and any point y in a segment of $P_{1,2}$ has the same cw(y) of $\alpha(x_1, y)$.

Let T be a maximal segment contained in an edge of $P_{0,1}$ such that $e(x)$, $e(f(x))$, ccw(x) of $\alpha(x, x_1)$, and cw$(f(x))$ of $\alpha(x_1, f(x))$ remain the same for any $x \in T$. We count the breakpoints of $f(x)$ in the interior of T. There are $O(n)$ such segments by Lemmas 1, 4 and Corollary 1. We count the breakpoints of $f(x)$ by computing point x where the type of $D_1(x)$ or the determinators of $D_1(x)$ changes. We show that there are at most $O(1)$ breakpoints in the interior of each maximal segment, and therefore there are $O(n)$ breakpoints in total. In order to compute $f(x)$, we first compute $f(x_0) = x_1$. Then starting from $x = x_0$ and $f(x) = x_1$, we compute $f(x)$ as x moves along ∂P by maintaining the two maximal segments such that $e(x)$ and ccw(x) of $\alpha(x, x_1)$ remain the same and $e(f(x))$ and cw$(f(x))$ of $\alpha(x_1, f(x))$ remain the same. By repeating this process over maximal segments, we get the following lemma. The details of the process can be found in [15].

Lemma 5. *For a fixed $r \geq r^*$, there are $O(n)$ breakpoints of $f(x)$ and $g(x)$, and they can be computed in $O(n)$ time.*

Recall that our algorithm returns yes if there exists a point $x \in \partial P$ such that $f(x) \geq_x g(x)$, otherwise it returns no. Hence, using Lemma 5, we have the following theorem.

Theorem 1. *Given a convex polygon P with n vertices in the plane and a radius r, we can decide whether there are two congruent disks of radius r covering P in $O(n)$ time.*

4 Parallel Decision Algorithm

Given a real value r, our parallel decision algorithm computes $f(x)$ and $g(x)$ that define the longest subchains of ∂P from x covered by disks of radius r, and determines whether there is a point $x \in \partial P$ such that $f(x) \geq_x g(x)$, in parallel. To do this efficiently, our algorithm first finds rough bounds of $f(x)$ and $g(x)$ by modifying the parallel decision algorithm for the two-center problem for points in convex position by Choi and Ahn [14] and applying it for the vertices of P. Then our algorithm computes $f(x)$ and $g(x)$ exactly.

The parallel decision algorithm by Choi and Ahn runs in two phases: the preprocessing phase and the decision phase. In the preprocessing phase, their algorithm runs sequentially without knowing r. In the decision phase, their algorithm runs in parallel for a given value r. It constructs a data structure that supports intersection queries of a subset of disks centered at input points in $O(\log n)$ parallel time using $O(n)$ processors after $O(n \log n)$-time preprocessing. In our problem, two congruent disks must cover the edges of P as well as the vertices of P, and thus we modify the preprocessing phase.

In the preprocessing phase, their algorithm partitions the vertices of P into two subsets $S_1 = \{v_1, \ldots, v_k\}$ and $S_2 = \{v_{k+1}, \ldots, v_n\}$, each consisting of consecutive vertices along ∂P such that there are $v_i \in S_1$ and $v_j \in S_2$ satisfying $\{v_i, v_{i+1}, \ldots, v_{j-1}\} \subset D_1$ and $\{v_j, v_{j+1}, \ldots, v_{i-1}\} \subset D_2$ for an optimal pair (D_1, D_2) of disks for the vertices of P. The indices of vertices are cyclic such that $n + k \equiv k$ for any integer k. Then in $O(n \log n)$ time, it finds $O(n/\log^6 n)$ pairs of subsets, each consisting of $O(\log^6 n)$ consecutive vertices such that there is one pair (U, W) of sets with $v_i \in U$ and $v_j \in W$, where v_i and v_j are the vertices that determine the optimal partition.

In the preprocessing phase, our algorithm partitions ∂P into two subchains. Then, we partition ∂P into $O(n/\log^6 n)$ subchains, each consisting of $O(\log^6 n)$ consecutive vertices, and compute $O(n/\log^6 n)$ pairs of the subchains such that at least one pair has x in one subchain and x' in the other subchain, and $P_{x,x'}$ and $P_{x',x}$ is r^*-coverable.

In the decision phase, their algorithm constructs a data structure in $O(\log n)$ parallel time with $O(n)$ processors, that for a query with r computes $I_r(u, w)$, where $u \in U', w \in W'$ for any pair (U', W') among the $O(n/\log^6 n)$ pairs. Then it computes $I(u, w)$ in $O(\log n)$ time and determines if $I(u, w) = \emptyset$ in $O(\log^3 \log n)$ time using the data structure.

In our case, our algorithm constructs a data structure that for a query with r computes $I_r(v_i, v_j)$ and $I_r(v_j, v_i)$ for $v_i \in P_1$, $v_j \in P_2$, where (P_1, P_2) is one of the $O(n/\log^6 n)$ pairs of subchains computed in our preprocessing phase. Our data structure also determines if $I(v_i, v_j) = \emptyset$.

Using the data structure, our algorithm gets rough bounds of $f(x)$ and $g(x)$. Then it computes $f(x)$ and $g(x)$ exactly. In doing so, it computes all breakpoints of $f(x)$ and $g(x)$, and their corresponding combinatorial structures, and determines whether there exists $x \in \partial P$ such that $f(x) \geq_x g(x)$.

4.1 Preprocessing Phase

We use $f^*(x)$ and $g^*(x)$ to denote $f_{r^*}(x)$ and $g_{r^*}(x)$, respectively. Our algorithm partitions ∂P into two subchains such that $P_{x,x'}$ and $P_{x',x}$ are r^*-coverable, for x and x' contained in each subchain. Then it computes $O(n/\log^6 n)$ pairs of subchains of ∂P, each consisting of $O(\log^6 n)$ consecutive vertices.

More precisely, for any two points $x, y \in \partial P$, let $\tau(x, y)$ be the smallest value such that $P_{x,y}$ is $\tau(x, y)$-coverable. For a point $p \in \partial P$, let $h(p)$ be the farthest point from p in counterclockwise direction along ∂P that satisfies $\tau(p, h(p)) \leq \tau(h(p), p)$. Then for any vertex v of P, $P_{v,h(v)}$ and $P_{h(v),v}$ form a partition of ∂P such that there are $x \in P_{v,h(v)}$ and $x' \in P_{h(v),v}$, and $P_{x,x'}$ and $P_{x',x}$ are r^*-coverable. The details on the partition of ∂P into two subchains can be found in [15].

We consider x moving along $P_{v_1,h(v_1)}$ and $f(x)$ moving along $P_{h(v_1),v_1}$. Also, from now on, we use $<$ instead of $<_{v_1}$. The same goes for $\leq_{v_1}, >_{v_1}, \geq_{v_1}$. To compute rough bounds of $f(x)$ and $g(x)$, our algorithm computes a step function $F(x)$ approximating $f^*(x)$ and a step function $G(x)$ approximating $g^*(x)$ on the same set of intervals of the same length. More precisely, at every $(\log^6 n)$-th vertex v from v_1 along ∂P, it evaluates step functions $F(v)$ and $G(v)$ on r^* such that $f^*(v) \leq F(v)$ and $g^*(v) \geq G(v)$. See Fig. 5(a). In each interval, the region bounded by $F(x)$ from above and by $G(x)$ from below is a rectangular cell. Thus, there is a sequence of $O(n/\log^6 n)$ rectangular cells of width at most $\log^6 n$. See Fig. 5(b). Observe that every intersection of $f^*(x)$ and $g^*(x)$ is contained in one of the rectangular cells. Thus, we focus on the sequence of rectangular cells bounded in between $F(x)$ and $G(x)$, which we call the *region of interest* (ROI shortly). In addition, we require $F(x)$ and $G(x)$ to approximate $f^*(x)$ and $g^*(x)$ tight enough such that each rectangular cell can be partitioned further by horizontal lines into disjoint rectangular cells of height at most $\log^6 n$, and in total there are $O(n/\log^6 n)$ disjoint rectangular cells of width and height at most $\log^6 n$ in ROI. See Fig. 5(c).

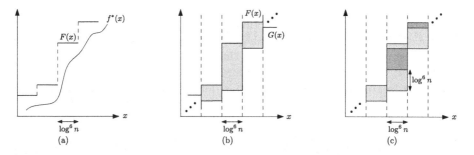

Fig. 5. (a) Step function $F(x)$ satisfying $f^*(x) \leq F(x)$, with intervals, each consisting of $\log^6 n$ consecutive vertices. (b) Sequence of rectangular cells bounded in between $F(x)$ and $G(x)$. (c) Disjoint rectangular cells of width and height at most $\log^6 n$.

We say that a vertex pair (v_i, v_j) is in ROI if and only if $G(v_i) \leq v_j \leq F(v_i)$. Also, we say an edge pair $(v_i v_{i+1}, v_j v_{j+1})$ is in ROI if and only if vertex pairs (v_i, v_j), (v_i, v_{j+1}), (v_{i+1}, v_j) and (v_{i+1}, v_{j+1}) are all in ROI. The details on computing ROI can be found in [15].

4.2 Decision Phase

Recall that our parallel decision algorithm finds, for a given r, the intersections of the graphs of $f(x)$ and $g(x)$ in ROI. We use the data structure of the parallel decision algorithm of the two-center problem for points in convex position [14]. To evaluate $f(x)$ for a given r, we first find $O(n)$ edge pairs $(e(x), e(f(x)))$ in ROI. Then we assign a processor to each edge pair to compute the event points. Then we assign a processor to each event point to compute the breakpoints and the corresponding combinatorial structures of $f(x)$. We also do this for $g(x)$. Lastly, for each combinatorial structure, we determine whether there exists $x \in \partial P$ such that $f(x) \geq g(x)$. This process can be done in $O(\log n)$ parallel steps using $O(n)$ processors, after $O(n \log n)$-time preprocessing.

Data Structures. We adopt the data structure for the two-center problem for points in convex position by Choi and Ahn [14]. To construct the data structure, they store the frequently used intersections of disks for all $r > 0$. Then, they find a range of radii $(r_1, r_2]$ containing the optimal radius r' for the two center problem for points in convex position. To do this they use binary search and the sequential decision algorithm for points in convex position. In our case, we compute a range of radii $(r_1, r_2]$ containing the optimal radius r^* using binary search and the sequential decision algorithm in Sect. 3 running in $O(n)$ time. For $r \in (r_1, r_2]$, we construct a data structure that supports the following.

Lemma 6 [14]. *After $O(n \log n)$-time preprocessing, we can construct a data structure in $O(\log n)$ parallel steps using $O(n)$ processors that supports the following queries with $r \in (r_1, r_2]$: (1) For any vertex v_i in $P_{v_1, h(v_1)}$, compute $I(v_i, h(v_1))$ represented in a binary search tree with height $O(\log n)$ in $O(\log n)$ time. (2) For any pair (v_i, v_j) of vertices in ROI, determine if $I(v_i, v_j) = \emptyset$ in $O(\log^3 \log n)$ time.*

Computing Edge Pairs. Using the data structure in Lemma 6, we get the following lemma.

Lemma 7. *Given $r \in (r_1, r_2]$, we can compute all edge pairs $(e(x), e(f(x)))$ in ROI in $O(\log n)$ parallel time using $O(n)$ processors, after $O(n \log n)$-time preprocessing.*

Computing the Combinatorial Structure. After computing the edge pairs using Lemma 7, we compute the breakpoints and the corresponding combinatorial structures of $f(x)$. To do this, we compute event points and find breakpoints from the event points for each edge pair. For $D_1(x)$ of type T3 or T4, its determinators never change for an edge pair $(e(x), e(f(x)))$ by Lemma 2. Thus, for each edge pair we find candidates of the determinators of $D_1(x)$ of type T3 or T4 in $O(\log n)$ time.

For $D_1(x)$ of type T1 or T2, we find the event points of $\mathsf{ccw}(x)$ of $\alpha(x, h(v_1))$, and the event points of $\mathsf{cw}(f(x))$ of $\alpha(h(v_1), f(x))$. Consider an edge pair $(u'u, vv')$ in ROI such that $f(x) \in vv'$ for some $x \in u'u$. The edge pair $(u'u, vv')$ may have $O(n)$ event points at which $\mathsf{ccw}(x)$ of $\alpha(x, h(v_1))$ or $\mathsf{cw}(f(x))$ of $\alpha(h(v_1), f(x))$ changes, while the total number of event points is $O(n)$. We find the event points of $\mathsf{ccw}(x)$ of $\alpha(x, h(v_1))$ represented in a binary search tree using $I(u, h(v_1))$ in $O(\log n)$ time. Thus, we can find the event points of $\mathsf{ccw}(x)$ of $\alpha(x, h(v_1))$ for all edge pairs in $O(\log n)$ parallel steps using $O(n)$ processors. For two consecutive event points of $\mathsf{ccw}(x)$ of $\alpha(x, h(v_1))$, we compute the corresponding event points of $\mathsf{cw}(f(x))$ of $\alpha(h(v_1), f(x))$ in $O(\log n)$ time. For a segment T such that $e(x)$, $\mathsf{ccw}(x)$ of $\alpha(x, h(v_1))$, $e(f(x))$, and $\mathsf{cw}(f(x))$ of $\alpha(h(v_1), f(x))$ remain the same for any $x \in T$, we compute $f(x)$.

Lemma 8. *Given $r \in (r_1, r_2]$, we can compute $f(x)$ for all $x \in \partial P$ such that $(e(x), e(f(x)))$ is an edge pair in ROI, represented as a binary search tree of height $O(\log n)$ consisting of $O(n)$ nodes, in $O(\log n)$ parallel steps using $O(n)$ processors, after $O(n \log n)$-time preprocessing.*

Now, we have $f(x)$ and $g(x)$ within ROI, each represented as a binary search tree of height $O(\log n)$ and size $O(n)$. For two consecutive breakpoints t and t' of $f(x)$, we find the corresponding combinatorial structures of $g(t)$ and $g(t')$. Then we determine whether there exists $x \in tt'$ such that $f(x) \geq g(x)$ for all combinatorial structures of $g(x)$. Since $f(x)$ and $g(x)$ have $O(n)$ breakpoints by Lemma 5, we can determine whether two disks of radius r cover P in $O(\log n)$ parallel steps using $O(n)$ processors, after $O(n \log n)$-time preprocessing. Therefore, using Lemma 8, we get the following theorem.

Theorem 2. *Given a real value r, we can determine whether $r \geq r^*$ in $O(\log n)$ parallel steps using $O(n)$ processors, after $O(n \log n)$-time preprocessing.*

We use Cole's parametric search technique [16] to compute the optimal radius r^*. For a sequential decision algorithm of running time T_S and a parallel decision algorithm of parallel running time T_P using N processors, we can apply Cole's parametric search to compute r^* in $O(NT_P + T_S(T_P + \log N))$ time. To apply Cole's parametric search, the parallel decision algorithm must satisfy a bounded fan-in/bounded fan-out requirement. Our parallel decision algorithm satisfies such requirement. In our case, $T_S = O(n)$, $T_P = O(\log n)$, and $N = O(n)$. Therefore, by applying Cole's technique, r^* can be computed in $O(n \log n)$ time.

Theorem 3. *Given a convex polygon with n vertices in the plane, we can find in $O(n \log n)$ time two congruent disks of minimum radius whose union covers the polygon.*

References

1. Agarwal, P.K., Ben Avraham, R., Sharir, M.: The 2-center problem in three dimensions. Computat. Geom. Theor. Appl. **46**(6), 734–746 (2013)
2. Agarwal, P.K., Procopiuc, C.M.: Exact and approximation algorithms for clustering. Algorithmica **33**(2), 201–226 (2002). https://doi.org/10.1007/s00453-001-0110-y
3. Agarwal, P.K., Sharir, M.: Planar geometric location problems. Algorithmica **11**(2), 185–195 (1994). https://doi.org/10.1007/BF01182774
4. Agarwal, P.K., Sharir, M.: Efficient algorithms for geometric optimization. ACM Comput. Surv. **30**(4), 412–458 (1998)
5. Agarwal, P., Sharathkumar, R.: Streaming algorithms for extent problems in high dimensions. Algorithmica **72**(1), 83–98 (2015)
6. Ahn, H.K., Kim, H.S., Kim, S.S., Son, W.: Computing k centers over streaming data for small k. Int. J. Comput. Geom. Appl. **24**(2), 107–123 (2014)
7. Ahn, H.K., Barba, L., Bose, P., Carufel, J.L.D., Korman, M., Oh, E.: A linear-time algorithm for the geodesic center of a simple polygon. Discrete Comput. Geom. **56**(4), 836–859 (2016). https://doi.org/10.1007/s00454-016-9796-0
8. Ahn, H.K., Kim, S.S., Knauer, C., Schlipf, L., Shin, C.S., Vigneron, A.: Covering and piercing disks with two centers. Comput. Geom. Theor. Appl. **46**(3), 253–262 (2013)
9. Bae, S.W.: L_1 geodesic farthest neighbors in a simple polygon and related problems. Discrete Comput. Geom. **62**(4), 743–774 (2019)
10. Basappa, M., Jallu, R.K., Das, G.K.: Constrained k-center problem on a convex polygon. Int. J. Found. Comput. Sci. **31**(02), 275–291 (2020)
11. Chan, T.M.: More planar two-center algorithms. Comput. Geom. Theor. Appl. **13**(3), 189–198 (1999)
12. Chan, T., Pathak, V.: Streaming and dynamic algorithms for minimum enclosing balls in high dimensions. Comput. Geom. Theor. Appl. **47**(2), 240–247 (2014)
13. Cho, K., Oh, E.: Optimal algorithm for the planar two-center problem. arXiv:2007.08784 (2020)
14. Choi, J., Ahn, H.K.: Efficient planar two-center algorithms. Comput. Geom. **97**, 101768 (2021)
15. Choi, J., Jeong, D., Ahn, H.K.: Covering convex polygons by two congruent disks. arXiv:2105.02483 (2021)
16. Cole, R.: Slowing down sorting networks to obtain faster sorting algorithms. J. ACM **34**(1), 200–208 (1987)
17. Das, G.K., Roy, S., Das, S., Nandy, S.C.: Variations of base-station placement problem on the boundary of a convex region. Int. J. Found. Comput. Sci. **19**(02), 405–427 (2008)
18. Edelsbrunner, H., Kirkpatrick, D., Seidel, R.: On the shape of a set of points in the plane. IEEE Trans. Inf. Theor. **29**(4), 551–559 (1983)
19. Feder, T., Greene, D.: Optimal algorithms for approximate clustering. In: Proceedings of the 20th Annual ACM Symposium on Theory of Computing (STOC 1988), pp. 434–444 (1988)
20. Fischer, K., Gärtner, B.: The smallest enclosing ball of balls: combinatorial structure and algorithms. Int. J. Comput. Geom. Appl. **4**(5), 341–378 (2004)
21. Gonzalez, T.F.: Clustering to minimize the maximum intercluster distance. Theor. Comput. Sci. **38**, 293–306 (1985)

22. Hershberger, J., Suri, S.: Adaptive sampling for geometric problems over data streams. Comput. Geom. Theor. Appl. **39**(3), 191–208 (2008)
23. Hershberger, J., Suri, S.: Finding tailored partitions. J. Algorithms **12**(3), 431–463 (1991)
24. Hwang, R., Lee, R.C.T., Chang, R.: The slab dividing approach to solve the Euclidean p-center problem. Algorithmica **9**(1), 1–22 (1993). https://doi.org/10.1007/BF01185335
25. Kim, S.S., Ahn, H.K.: An improved data stream algorithm for clustering. Comput. Geom. Theor. Appl. **48**(9), 635–645 (2015)
26. Kim, S.K., Shin, C.-S.: Efficient algorithms for two-center problems for a convex polygon. In: Du, D.-Z.-Z., Eades, P., Estivill-Castro, V., Lin, X., Sharma, A. (eds.) COCOON 2000. LNCS, vol. 1858, pp. 299–309. Springer, Heidelberg (2000). https://doi.org/10.1007/3-540-44968-X_30
27. Löffler, M., Van Kreveld, M.: Largest bounding box, smallest diameter, and related problems on imprecise points. Comput. Geom. Theor. Appl. **43**(4), 419–433 (2010)
28. Megiddo, N.: On the ball spanned by balls. Discrete Comput. Geom. **4**(6), 605–610 (1989). https://doi.org/10.1007/BF02187750
29. Megiddo, N.: Linear programming in linear time when the dimension is fixed. J. ACM **31**(1), 114–127 (1984)
30. Oh, E., De Carufel, J.L., Ahn, H.K.: The geodesic 2-center problem in a simple polygon. Comput. Geom. Theor. Appl. **74**, 21–37 (2018)
31. Roy, S., Bardhan, D., Das, S.: Base station placement on boundary of a convex polygon. J. Parallel Distrib. Comput. **68**(2), 265–273 (2008)
32. Sadhu, S., Roy, S., Nandi, S., Maheshwari, A., Nandy, S.C.: Two-center of the convex hull of a point set: dynamic model, and restricted streaming model. Fundamenta Informaticae **164**(1), 119–138 (2019)
33. Sharir, M.: A near-linear algorithm for the planar 2-center problem. Discrete Comput. Geom. **18**(2), 125–134 (1997)
34. Shin, C.-S., Kim, J.-H., Kim, S.K., Chwa, K.-Y.: Two-center problems for a convex polygon (extended abstract). In: Bilardi, G., Italiano, G.F., Pietracaprina, A., Pucci, G. (eds.) ESA 1998. LNCS, vol. 1461, pp. 199–210. Springer, Heidelberg (1998). https://doi.org/10.1007/3-540-68530-8_17
35. Wang, H.: On the planar two-center problem and circular hulls. In: Proceeding of the 36th International Symposium on Computational Geometry (SoCG 2020), vol. 164, pp. 68:1–68:14 (2020)
36. Wang, H.: On the planar two-center problem and circular hulls. arXiv preprint arXiv:2002.07945 (2020)
37. Zarrabi-Zadeh, H.: Core-preserving algorithms. In: Proceedings of the 20th Annual Canadian Conference on Computational Geometry (CCCG 2008), pp. 159–162 (2008)

The Tandem Duplication Distance Problem Is Hard over Bounded Alphabets

Ferdinando Cicalese$^{(\boxtimes)}$ and Nicolò Pilati

Department of Computer Science, University of Verona, Verona, Italy
`ferdinando.cicalese@univr.it, nicolo.pilati@studenti.univr.it`

Abstract. A tandem duplication is an operation that converts a string $S = AXB$ into a string $T = AXXB$. As they appear to be involved in genetic disorders, tandem duplications are widely studied in computational biology. Also, tandem duplication mechanisms have been recently studied in different contexts, from formal languages, to information theory, to error-correcting codes for DNA storage systems. The question of determining the complexity of computing the tandem duplication distance between two given strings was posed by (Leupold *et al.*, 2004) and, very recently, the problem was shown to be NP-hard for the case of unbounded alphabets (Lafond et al., STACS 2020).

In this paper, we significantly improve this result and show that the tandem duplication distance problem is NP-hard already for the case of strings over an alphabet of size 5.

For a restricted class of strings, we establish the tractability of the *existence problem*: given strings S and T over the same alphabet, decide whether there exists a sequence of duplications converting S into T. A polynomial time algorithm that solves this existence problem was only known for the case of the binary alphabet.

1 Introduction

Since the draft sequence of the human genome was published, it has been known that a very large part of it consists of repeated substrings [10]. One talks about a tandem repeat when a pattern of one or more nucleotides occurs twice and the two occurrences are adjacent. For instance, in the word CTACTAGTCA, the substring CTACTA is a tandem repeat. In this case, we say that CTACTAGTCA is generated from CTAGTCA by a tandem duplication of length three. As tandem repeats appear to be correlated to several genetic disorders [16,17], the study of tandem duplication mechanisms has attracted the interest of different communities also outside the specific area of computational biology [1,3,5,8,15].

Problem Definition. Formally, a *tandem duplication* (TD)—later simply referred to as a duplication—is an operation on a string S that replaces a substring X with its *square* XX. Given two strings S and T, we write $S \Rightarrow T$ if there exist strings A, B, X such that $S = AXB$ and $T = AXXB$.

© Springer Nature Switzerland AG 2021
P. Flocchini and L. Moura (Eds.): IWOCA 2021, LNCS 12757, pp. 179–193, 2021.
https://doi.org/10.1007/978-3-030-79987-8_13

The *tandem duplication distance from S to T*, denoted by $dist_{TD}(S,T)$, is the minimum value of k such that S can be turned into T by a sequence of k duplications. If no such k exists, then $dist(S,T) = \infty$.

For example, $dist_{TD}(0121, 0101211) = 2$, since: (i) we have the two duplications $0121 \Rightarrow 010121 \Rightarrow 0101211$; (ii) and it is easy to verify that no single duplication can turn 0121 into 0101211.

We consider the following two problems about the possibility of converting a string S into a string T by using tandem duplications:

TANDEM DUPLICATION EXISTENCE (TD-EXIST)
Input: Strings S and T over the same alphabet Σ.
Question: Is $dist_{TD}(S,T) < \infty$?

TANDEM DUPLICATION DISTANCE (TD-DIST)
Input: Strings S and T over the same alphabet Σ and an integer k.
Question: Is $dist_{TD}(S,T) \leq k$?

The problem of determining the complexity of computing the tandem duplication distance between two given strings (the TD-DIST problem) was posed in [15]. Only very recently, the problem was shown to be NP-hard in the case of unbounded alphabets [9]. Here, we significantly improve this result by showing that TD-DIST is NP-hard for the case of bounded alphabets of size 5.

For both, the result of [9] and ours, it is assumed that the strings S and T satisfy $dist_{TD}(S,T) < \infty$. In general, the complexity of deciding if a string S can be turned into a string T by a sequence of tandem duplications (the TD-EXIST problem above) is still an open problem for alphabets of size > 2. In the second part of the paper we also consider this *existence problem* (TD-EXIST), focussing on a special class of strings—which we call *purely alternating* (see Sect. 2 for the definition)—that generalize the special structure of binary strings to larger alphabets. We show that a linear time algorithm for the TD-EXIST problem exists for every alphabet of size ≤ 5 if the strings are *purely alternating*. In a final section we also discuss the limit of the approach used here for larger alphabets $|\Sigma| > 5$.

Related Work. To the best of our knowledge, the first papers explicitly dealing with tandem duplication mechanisms are in the area of formal languages [2,4,6,12–15]. In [14,15], the authors study decidability and hierarchy issues of a duplication language, defined as the set of words generated via tandem duplications. In the same line of research, Jain et al. [7] proved that k-bounded duplication languages are regular for $k \leq 3$. More recently, the authors of [1] investigated extremal and information theoretic questions regarding the number of tandem duplications required to generate a binary word starting from its unique root (the square free word from which it can be generated via duplications). In the same paper, the authors also considered approximate duplication operations. In [5] Farnoud et al. began the study of the average information content of a k-bounded duplication language, referred to as the capacity of a duplication system. Motivated by problems arising from DNA storage applications, Jain et al. [8] proposed the study of codes that correct tandem duplications to improve the reliability of data storage, and gave optimal constructions for the case where tandem duplication length is at most two. In [3], Chee et al. investigated the question of confusability under duplications, i.e., whether, given words x and y, there exists a word z such that x and y can be transformed into z via duplications. They show that even for small duplication lengths, the solutions to this question are nontrivial, and exact solutions are provided for the case of tandem duplications of size at most three.

2 Notation and Basic Properties

We follow the terminology from [1,9]. For any positive integer n we denote by $[n]$ the set of the first n positive integers $\{1, 2, \ldots, n\}$. We also use $[n]_0$ to denote $[n] \cup \{0\}$. Given a string S we denote by $\Sigma(S)$ the alphabet of the string S, i.e., the set of characters occurring in S. If $|\Sigma(S)| = q$ we say that the string is q-ary. We say that S' is a subsequence of S, if S' can be obtained by deleting zero or more characters from S, and we denote this by $S' \subseteq S$. T is a substring of S if T is a subsequence of S whose characters occur contiguously in S. If the substring T occurs at the beginning (resp. end) of S that it is also called a prefix (resp. suffix) of S.

A string XX consisting of the concatenation of two identical strings is called a square. A string is square-free if it does not contain any substring which is a square. We say that a string S is exemplar if no two characters of S are equal. Given two strings S and T, we say that there is a duplication turning S into T, denoted by $S \Rightarrow T$, if there exist strings A, B, X such that $S = AXB$ and $T = AXXB$. We use $\Rightarrow_k$ to denote the existence of a sequence of k TD's, i.e., $S \Rightarrow_k T$ if there exist $S_1, \ldots, S_{k-1}$ such that $S \Rightarrow S_1 \Rightarrow \cdots \Rightarrow S_{k-1} \Rightarrow T$. We write $S \Rightarrow_* T$ if there exists some k such that $S \Rightarrow_k T$. In this case, we also say that T is a descendant of S and S is an ancestor of T. If S is a square free ancestor of T, we also say that S is a root of B.

The reverse operation of a tandem duplication consists of taking a square XX in S and deleting one of the two occurences of X. This operation is referred to as a contraction. We write $T \rightarrowtail S$ if there exist strings A, B, X such that $T = AXXB$ and $S = AXB$. For a sequence of k contractions, we write $T \rightarrowtail_k S$ and for a sequence of an arbitrary number of contractions we write $T \rightarrowtail_* S$. In particular, we have that $T \rightarrowtail_k S$ if and only if $S \Rightarrow_k T$, and $T \rightarrowtail_* S$ if and only if $S \Rightarrow_* T$.

For any string $A = a_1 \ldots a_n$ and set of indices $I = \{i_1, \ldots, i_k\}$, such that $1 \leq i_1 < i_2 < \cdots < i_k \leq n$ we define $dup(A, I)$ as the string obtained by duplicating character a_{i_j} for each $j = 1, \ldots, k$, i.e.,

$$dup(A, I) = a^{(1)} a_{i_1} a_{i_1} a^{(2)} a_{i_2} a_{i_2} a^{(3)} \cdots a^{(k)} a_{i_k} a_{i_k} a^{(k+1)},$$

where $a^{(j)} = a_{i_{j-1}+1} a_{i_{j-1}+2} \cdots a_{i_j-1}$, and $i_0 = 0, i_{k+1} = n+1$.

For a character a and a positive integer l, let a^l denote the string consisting of l copies of a. A run in a string S is a maximal substring of S consisting of copies of the same character. Given a string S containing k runs, the *run-length encoding* of S, denoted $RLE(S)$, is the sequence $(s_1, l_1), \ldots, (s_k, l_k)$, such that $S = s_1^{l_1} s_2^{l_2} \ldots s_k^{l_k}$, where, in particular, for each $i = 1, \ldots, k$, we have that $s_i^{l_i}$ is the i-th run of S, consisting of the symbol s_i repeated l_i times. We write $|RLE(S)|$ for the number of runs contained in S. For example, given the string $S = 111001222 = 1^3 0^2 1^1 2^3$, we have $RLE(S) = (1, 3), (0, 2), (1, 1), (2, 3)$ and $|RLE(S)| = 4$.

A q-ary string $S = s_1^{l_1} \cdots s_k^{l_k}$ is called *purely alternating* if, there is an order on the symbols of the alphabet $\Sigma(S) = \{\sigma_0 < \sigma_1 < \ldots \sigma_{q-1}\}$ and a $j \in [k]$, such that for each $i = 1, \ldots, k$ $s_i = \sigma_{j+i \bmod q}$, i.e., each run of the symbol σ_j is followed by a run of the symbol $\sigma_{j+1 \bmod q}$. Note that a purely alternating string is uniquely determined by the order on the alphabet, the initial character and the lengths of its runs. In general, we will assume, w.l.o.g., that, for a q-ary purely alternating string the alphabet is the set $\{0, 1, \ldots, q-1\}$ with the natural order and the first run is a run of 0's.

For example, the string 0001220112 is purely alternating, but the string 01202 is not. Note that all binary strings (that up to relabelling we assume to start with 0) are purely alternating.

Given a q-ary purely alternating string S, a *group* of S is a substring X of S containing exactly q runs of S.

Given a string $S = s_1 s_2 \ldots s_n$, we denote by S^{dup} the string obtained by duplicating each single character in S, i.e., $S^{dup} = dup(S, [|S|]) = s_1 s_1 s_2 s_2 \ldots s_n s_n$.

A string S is *almost square-free* if there exists a square-free string S_{SF} such that $S_{SF} \subseteq S \subseteq S_{SF}^{dup}$.

For example, the string 01120022 is almost square-free, while the strings 01122201 and 0012212 are not.

The following lemma states a useful immediate consequence of the definition of an almost square free string.

Lemma 1. *Z is an almost square free string if and only if there exists a square free string Z_{SF} and a set $I \subseteq [|Z_{SF}|]$ such that $Z = dup(Z_{SF}, I)$. Moreover, the only contractions possible on Z are of size 1, i.e., those that remove one of two consecutive equal characters.*

Due to the space limitation, several proofs are deferred to the extended version of the paper.

3   The Hardness of TD-DIST Over Alphabets of Size 5

In this section we show that given two strings S and T, over a 5-ary alphabet Σ and such that $S \Rightarrow_* T$, finding the minimum number of duplications required to transform S into T is NP-complete.

To see that the problem is in NP, note that, if $S \Rightarrow_* T$, then $dist_{TD}(S, T) \leq |T|$ because each duplication from S to T adds at least one character. Thus a certificate for the problem consisting of the sequence of duplications turning S into T has polynomial size and can also be clearly verified in polynomial time.

The Block-Exemplar Tandem Duplication Distance Problem. For the hardness proof, we reduce from a variant of the TANDEM DUPLICATION DISTANCE problem over unbounded alphabets, which we was originally presented in [9] in a more specialized setting and we repropose here in a somehow streamlined form. In this variant, only instances made of strings of a special structure are allowed. We call the allowed instances *block exemplar pairs*. We provide an operative definition of how a block-exemplar pair of strings is built. Recall that a string is exemplar if its characters are all distinct.

Definition 1 (Block-Exemplar Pair). *Fix an integer $t > 1$ and $t + 2$ exemplar strings $X, B_0, B_1, \ldots, B_t$ over pairwise disjoint alphabets, such that $|X|, |B_0|, |B_1| > 1$ and $|B_i| = 1$ for each $i = 2, \ldots, t$. Fix a second positive integer $p < t$ and p distinct subsets $\emptyset \neq I_1, \ldots, I_p \subsetneq [|X|]$. Finally, fix a character L which does not appear in any of the strings $X, B_0, \ldots, B_t$.*

The strings $L, X, B_0, B_1, \ldots, B_t$ and the sets $I_1, \ldots, I_p$, determine a block-exemplar pair of strings S and T as follows:

For each $j = 1, \ldots, p$, define the distinct strings $X_1, \ldots, X_p$, where X_j is obtained from X by single character duplications, $X_j = dup(X, I_j)$, hence $X \subsetneq X_i \subsetneq X^{dup}$. Define the strings

$$\mathcal{B}_t^0 = B_t B_{t-1} \ldots B_2 B_1 B_0^{dup} \qquad \mathcal{B}_t^1 = B_t B_{t-1} \ldots B_2 B_1^{dup} B_0,$$

and for each $i = 1, \ldots, t$, define the strings

$$\mathcal{B}_i = B_i B_{i-1} \ldots B_2 B_1 B_0 \qquad \mathcal{B}_i^{01} = B_i B_{i-1} \ldots B_2 B_1^{dup} B_0^{dup}.$$

Then set

$$S = \mathcal{B}_t X L = B_t B_{t-1} \ldots B_2 B_1 B_0 X L, \tag{1}$$
$$T = \mathcal{B}_t^0 X^{dup} L \mathcal{B}_t^1 X L \ \mathcal{B}_1^{01} X_1 L \mathcal{B}_t^1 X L \ \mathcal{B}_2^{01} X_2 L \mathcal{B}_t^1 X L \ \ldots \mathcal{B}_p^{01} X_p L \mathcal{B}_t^1 X L. \tag{2}$$

More generally, we say that S, T is a block-exemplar pair, if there exist integers t, p, strings $L, X, B_0, \ldots, B_t$, and sets $I_1, \ldots, I_p$ such that S and T are given by the above construction.

We can now define the following variant of the TD-DIST problem.

BLOCK-EXEMPLAR TANDEM DUPLICATION DISTANCE PROBLEM (B-EX-TD)
Input: A Block-Exemplar pair of strings S and T, an integer bound k.
Question: Is $dist_{TD}(S, T) \leq k$?

In [9] the authors showed how to map instances (G, k) of the NP-complete problem CLIQUE to choices of strings $X, B_0, \ldots, B_t$, and sets $I_1, \ldots, I_p$ and a parameter k' such that G has clique of size k if and only if $S \Rightarrow_{k'} T$, where S, T is the block-exemplar pair given by $X, B_0, \ldots, B_t, I_1, \ldots, I_p$. Hence, the following result is implicit in [9].

Theorem 1. *The* B-Ex-TD *problem is NP-complete.*

From B-Ex-TD to TD-Dist on Alphabets of Size 5.
The basic idea is to map a block-exemplar pair S and T to a pair of 5-ary strings $\hat{S}, \hat{T}$ having a structure analogous to the one of S and T (as given in (1)-(2)). However, instead of using exemplar strings for the building blocks $X, B_0, \ldots, B_t$ we use substrings of a square free ternary string. The crucial point is to carefully choose such substrings so that we can control the squares appearing in the resulting $\hat{S}$ and $\hat{T}$. We want that the only duplications allowed in any process transforming $\hat{S}$ into $\hat{T}$ can be one-to-one mapped to some duplication sequence from S to T. For this, we need to overcome significant technical hurdles, that make the hard part of the reduction non-trivial (i.e., the equivalence in the direction from the 5-ary variant considered here to the block-exemplar variant of the problem considered in [9]).

The Mapping from the Block-Exemplar Pair (S, T) **to the 5-ary Pair** $(\hat{S}, \hat{T})$. We first show how to map the different substrings that define the structure of the pair S and T (see Def. 1) into 5-ary strings. Then we will show that the resulting strings $\hat{S}, \hat{T}$ preserve the properties of the original strings with respect to the possible duplication sequences $\hat{S} \Rightarrow_* \hat{T}$.

The strings $\hat{S}$ and $\hat{T}$ will be defined on the alphabet $\{0, 1, 2, L, \$\}$. For each $i = 0, 1, \ldots, t$ we define a string $\hat{B}_i \in \{0, 1, 2, \$\}^*$ that will be used to encode the string B_i. Also we define the encoding of strings $X, X_1, \ldots, X_p$ into strings $\hat{X}, \hat{X}_1, \ldots, \hat{X}_p$ over the alphabet $\{0, 1, 2\}$.

The encodings $\hat{B}_i$ and $\hat{X}$ are obtained by iteratively "slicing off" suffixes from a long square free string, denoted by $\mathcal{O}$. In addition each $\hat{B}_i$ is also extended with a single occurrence of the character $\$$.

More precisely, let us define $\mathcal{O}$ to be a square free string of length at least $|X| + t^2 + t(5 + \max\{|B_0|, |B_1|\})$ over the alphabet $\{0, 1, 2\}$. Note that creating $\mathcal{O}$ takes polynomial time in $|S| + |T|$ using a square-free morphism, for example Leech's morphism ([11]). Let us now proceed to define the building blocks $\tilde{B}_t, \ldots, \tilde{B}_0, \tilde{X}$. Refer to Fig. 1 for a pictorial description of this process. First, we define the string $\tilde{X}$ to be the suffix of length $|X|$ of $\mathcal{O}$. Let $\mathcal{O}^{(0)}$ be the string obtained from $\mathcal{O}$ after removing the suffix $\tilde{X}$. Let $\tilde{B}_0$ be the shortest suffix of $\mathcal{O}^{(0)}$ that does not start with 0 and has length at least $\max(|B_0|, |B_1|)$. We then define $\tilde{B}_1$ as the shortest suffix of the string obtained by removing the suffix $\tilde{B}_0$ from $\mathcal{O}^{(0)}$ such that $|\tilde{B}_1| > |\tilde{B}_0|$ and that doesn't start with 0.

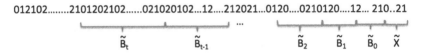

Fig. 1. The way the blocks $\tilde{B}_t, \tilde{B}_{t-1}, \ldots, \tilde{B}_1 \tilde{B}_0, \tilde{X}$ are consecutively extracted from the square free ternary string $\mathcal{O}$.

For $i = 2, \ldots t$ we define the string $\tilde{B}_i$ as the minimum suffix of the string $\mathcal{O}^{(i-1)}$ obtained from $\mathcal{O}$ after removing the suffix $\tilde{B}_{i-1}\tilde{B}_{i-2} \ldots \tilde{B}_1 \tilde{B}_0 \tilde{X}$ and such that: (i) $|\tilde{B}_i| > |\tilde{B}_{i-1}|$; (ii) if $i \neq t$ then $\tilde{B}_i$ does not start with 0; (iii) if $i = t$ then $\tilde{B}_i$ starts with 0. Thus, the lengths of the strings $\tilde{B}_0, \tilde{B}_1, \tilde{B}_2, \ldots, \tilde{B}_t$ are monotonically increasing, all the strings are ternary square-free and only for $j = t$ we have that $\tilde{B}_j$ starts with the character 0. Note that, in order to guarantee that the starting character of the strings $\tilde{B}_i$ is (resp. is not) 0, it is enough to consider a suffix of $\mathcal{O}^{(i-1)}$ of size at most $|\tilde{B}_{i-1}| + 5$. This follows from $\mathcal{O}^{(i-1)}$ being a ternary square free string, hence it cannot contain substrings of size 4 using only two characters, since every string of size 4 over only two characters necessarily contains a square. Therefore, the prefix of length 4 of the suffix of length $|\tilde{B}_{i-1}|+5$ of $\mathcal{O}^{(i-1)}$ contains an occurrence of any character in $\{0, 1, 2\}$ that can be chosen as the desired starting character of $\tilde{B}_i$. Therefore, the number of character we use for the strings $\tilde{X}, \tilde{B}_0, \ldots \tilde{B}_t$, is at most $|X| + \sum_{i=0}^{t} (\max\{|B_0|, |B_1|\} + i + 4) \leq |X| + t^2 + t(4 + \max\{|B_0|, |B_1|\}) = |\mathcal{O}|$.

We now use these strings and the two additional characters L, \$ as building blocks to create larger strings which will be our 5-ary "analog" of the strings $\mathcal{B}_i$.

Let Ω be the following set of strings:

$$\Omega = \{\mathrm{L}\} \cup \{B_2, B_3, \ldots, B_{2p}\} \cup \mathbb{B}_0 \cup \mathbb{B}_1 \cup \mathbb{X},$$

where $\mathbb{B}_0 = \{B_0' \mid B_0 \subseteq B_0' \subseteq B_0^{dup}\}$, $\mathbb{B}_1 = \{B_1' \mid B_1 \subseteq B_1' \subseteq B_1^{dup}\}$, $\mathbb{X} = \{X' \mid X \subseteq X' \subseteq X^{dup}\}$. Note that Ω contains all the almost square free strings that either appear among the building blocks of T and S (1)-(2) or can be obtained from such building blocks via single letter duplications. It turns out that (see Fact 2 below) any string encountered in a sequence of contractions $T \longmapsto_* S$ is a concatenation of elements of Ω. For this reason we refer to the elements of Ω as the macro-characters or *chunks* of S and T.

Note that, by Lemma 1, for any $C' \in \mathbb{B}_0 \cup \mathbb{B}_1 \cup \mathbb{X}$, there exists a unique $C \in \{B_0, B_1, X\}$ and a unique set $I \subseteq [|C|]$ such that $C' = dup(C, I)$.

We now define an encoding of the elements of Ω. Let $\mu : \Omega \mapsto \{0, 1, 2, \$, \mathrm{L}\}^*$ be defined as follows (for the sake of highlighting the factors involved in the formulas, $a \cdot b$ will denote concatenation of strings a and b):

- $\mu(\mathrm{L}) = \mathrm{L}$, and $\mu(X) = \tilde{X}$,
- for each $i = 2, \ldots, t$, we set $\mu(B_i) = \tilde{B}_i \cdot \$$
- for each $B' \in \mathbb{B}_0 \cup \mathbb{B}_1$, we let $\mu(B') = dup(\mu(B), I) \cdot \$$, where, by Lemma 1, $B \in \{B_0, B_1\}$ and $I \subseteq [|B|]$ are uniquely defined by $B' = dup(B, I)$;
- for each $X' \in \mathbb{X}$ we let $\mu(X') = dup(\mu(X), I)$, where, by Lemma 1, $I \subseteq [|X|]$ is uniquely defined by $X' = dup(X, I)$.

The mapping μ is naturally extended to concatenations of elements of Ω by setting $\mu(A_1 \cdot A_2 \cdots A_r) = \mu(A_1) \cdot \mu(A_2) \cdots \mu(A_r)$, for each $A_1, \ldots, A_r \in \Omega$.

For each $i = 1, \ldots, t$, let $\hat{B}_i = \mu(B_i)$ and for $i = 0, 1$, let $\hat{B}_i^* = \mu(B_i^{dup})$. Finally, let $\hat{X} = \mu(X)$ and $\hat{X}^* = \mu(X^{dup})$. The set of macro-characters $\hat{\Omega}$ that constitute the range of μ is then given by:

$$\hat{\Omega} = \{\mathrm{L}\} \cup \{\hat{B}_2, \hat{B}_3, \ldots, \hat{B}_t\} \cup \hat{\mathbb{B}}_0 \cup \hat{\mathbb{B}}_1 \cup \hat{\mathbb{X}},$$

where $\mathbb{B}_0 = \{\hat{B}_0' \mid \hat{B}_0 \subseteq \hat{B}_0' \subseteq \hat{B}_0^*\}$, $\mathbb{B}_1 = \cup\{\hat{B}_1' \mid \hat{B}_1 \subseteq \hat{B}_1' \subseteq \hat{B}_1^*\}$, and $\mathbb{X} = \{\hat{X}' \mid \hat{X} \subseteq \hat{X}' \subseteq \hat{X}^{dup}\}$. It is easy to see that μ is one-one from Ω to $\hat{\Omega}$.

We are ready to define the 5-ary strings $\hat{S}$ and $\hat{T}$ that we use in our reduction:

$$\hat{S} = \mu(S) = \hat{B}_t\hat{B}_{t-1}\ldots\hat{B}_2\hat{B}_1\hat{B}_0\hat{X}Ł = \tilde{B}_t\$\tilde{B}_{t-1}\$\ldots\$\tilde{B}_2\$\tilde{B}_1\$\tilde{B}_0\$\tilde{X}Ł \tag{3}$$

$$\hat{T} = \mu(T) = \hat{B}_t^0\hat{X}^*Ł\,\hat{B}_1^1\hat{X}Ł\,\hat{B}_1^{01}\hat{X}_1Ł\hat{B}_t^1\hat{X}Ł\,\hat{B}_2^{01}\hat{X}_2Ł\hat{B}_t^1\hat{X}Ł\,\ldots\,\hat{B}_p^{01}\hat{X}_pŁ\hat{B}_t^1\hat{X}Ł, \tag{4}$$

where for each $i = 1, \ldots, t$, we let $\hat{X}_i = \mu(X_i)$, and for $q \in [t]$, and $a \in \{0, 1, 01\}$ we let $\hat{B}_q = \mu(\mathcal{B}_q)$, and $\hat{B}_q^a = \mu(\mathcal{B}_q^a)$. Note that $\hat{S}$ is square-free. In analogy with S and T, we refer to the substrings from $\hat{\Omega}$ that constitute the building blocks of $\hat{S}$ and $\hat{T}$ as *chunks*.

The Final Steps of the Hardness Proof. Exploiting the properties of μ and the structure of $S, T, \hat{S}, \hat{T}$, we have the following result.

Theorem 2. *Let S, T be a block exemplar pair of strings and let $\hat{S}$ and $\hat{T}$ be the corresponding pair of 5-ary strings built as above. Then $S \Rightarrow_k T$ iff $\hat{S} \Rightarrow_k \hat{T}$.*

The hardness of the TD-DIST problem over a 5-ary alphabet, immediately follows from Theorem 1 and Theorem 2.   □

3.1   Sketch of the Proof of Theorem 2: The Main Analytic Tools

The first properties we use are summarized in the following fact. They come from the interdependence between μ and the operator *dup* and their effect on concatenations of a prefix of a chunk with a suffix of a chunk. They basically say that the equality of the prefixes (resp. suffixes) of pairs of chunks from Ω is preserved also by the images of such chunks via the map μ. Moreover, the concatenation of prefixes and suffixes of chunks from Ω result in a chunk from Ω iff the same holds true for their images via μ. For a string $C = c_1c_2\ldots c_n$, and indices $i, j \in [n]$ we denote by $C[i, j]$ the substring $c_i c_{i+1}\ldots c_j$, if $i \leq j$ and the empty string otherwise.

Fact 1. *Let $C \in \{B_0, B_1, X\}$ and $C \subseteq C' \subseteq C^{dup}$. Let I be the unique set of indices such that $C' = dup(C, I)$. Let $\hat{C} = \mu(C)$ and denote by $\hat{C}_h = \hat{C}[1, |C|]$ and $\hat{C}_t = \hat{C}[|C| + 1, |\hat{C}|]$. Let $c = |C|, c' = |C'|, \hat{c} = |\hat{C}|$. Then, it holds that*

1. $\mu(C') = dup(\hat{C}, I) = dup(\hat{C}_h, I) \cdot \hat{C}_t$;
2. *for any $I', I'' \subseteq [c]$ and $i \in [c + |I'|]_0$, $j \in [c + |I''|]_0$, we have*
   - $dup(C, I')[1, i] = dup(C, I'')[1, i] \iff dup(\hat{C}, I')[1, i] = dup(\hat{C}, I'')[1, i]$
   - $dup(C, I')[i + 1, c + |I'|] = dup(C, I'')[j + 1, c + |I''|]$
   $\iff dup(\hat{C}, I')[i + 1, \hat{c} + |I'|] = dup(\hat{C}, I'')[j + 1, \hat{c} + |I''|]$
3. *for any $I', I'', I''' \subseteq [|C|]$, and $i \in [\hat{c} + |I'|], j \in [\hat{c} + |I''|]$, we have*
   - $dup(\hat{C}, I')[1, i] \cdot dup(\hat{C}, I'')[j + 1, \hat{c} + |I''|] = dup(\hat{C}, I''')$
   $\Rightarrow dup(C, I')[1, \min\{i, c + |I'|\}] \cdot dup(C, I'')[j + 1, c + |I''|] = dup(C, I''')$

The first property directly follows from the definition of μ. The following properties are immediate consequences of Property 1 and the one-one correspondence between the position of the duplications in a chunk $C' = dup(C, I)$ and the position of the duplication in its image $\mu(C') = dup(\mu(C), I)$. This correspondence comes from the fact that both strings C and $\mu(C)$ are square free.

Part 1 - The Proof that $S \Rightarrow_k T$ Implies $\hat{S} \Rightarrow_k \hat{T}$. Assume that $S \Rightarrow_k T$ and let $T = T(0) \rightarrowtail T(1) \rightarrowtail \cdots \rightarrowtail T(k) = S$ be the corresponding series of contractions leading from T to S. We recall a result that directly follows from [9, Appendix, Claim 2].

Fact 2. *For each $\ell = 0, 1, \ldots, k - 1$, the string $T(\ell)$ has a factorization into elements of Ω, i.e., $T(\ell) = A_1 A_2 \ldots, A_r$ where for each $i = 1, \ldots, r$, we have $A_i \in \Omega$. Moreover, adjacent chunks are over disjoint alphabets, i.e., for each $i = 1, \ldots, r - 1$, $\Sigma(A_i) \neq \Sigma(A_{i+1})$.*

Exploiting the factorization of $T(\ell)$ into chunks, guaranteed by Fact 2, and using Property 2 in Fact 1 we have that the following claim holds.

Claim 1. *For $\ell = 0, 1, \ldots, k - 1$, the contraction $T(\ell) \rightarrowtail T(\ell + 1)$ implies the existence of a contraction $\mu(T(\ell)) \rightarrowtail \mu(T(\ell + 1))$.*

Applying this claim for each contraction from T to S, we have that a sequence of contractions exists from $\hat{T}$ to $\hat{S}$, i.e., $\hat{T} = \mu(T(0)) \rightarrowtail \mu(T(1)) \rightarrowtail \cdots \rightarrowtail \mu(T(k)) = \hat{S}$. Hence $\hat{S} \Rightarrow_k \hat{T}$, proving the "only if" part of Theorem 2.

The proof of Claim 1 is based on the fact that each contraction $T(\ell) \rightarrowtail T(\ell + 1)$ is either a single character within a chunk A_i of the factorization of $T(\ell)$ given by Fact 2 or it is a contraction that starts in a chunk A_i and spans at least $A_i A_{i+1} A_{i+2} A_{i+3}$. To see this let DD be the square removed by the contraction and let us denote by D^L and D^R the left and right copy of D, with the first character of D^L being in A_i.

(i) If the first character of D^R is also in A_i, then D^L is a substring of A_i. Since $\Sigma(A_i) \cap \Sigma(A_{i+1}) = \emptyset$, D^R, being equal to D^L cannot extend to A_{i+1}. Since A_i is almost square free, the only square are pairs of single characters, hence $D^L D^R$ must be a pair of adjacent single characters. In this case, by 1. in Fact 1, a single letter square is present in the chunk $\mu(A_i)$ and the contraction that removes it implies $\mu(T(\ell)) \rightarrowtail \mu(T(\ell + 1))$.

(ii) If the first character of D^R is not in A_i, the first character of D^L cannot be from A_{i+1} which does not have any character occurring in A_i. The analogous argument about the last characters of D^R and D^L implies that $D^L D^R$ spans at least four chunks. Then, $D^L = A_i[u+1, |C_1|]D'A_j[1, u']$, i.e., it starts with a suffix $A_i[u+1, |A_i|]$ of chunk A_i and ends with a (possibly empty) prefix $A_j[1, u']$ of a chunk A_j with $j \geq i+2$ and D^R starts with the suffix of $A_j[u'+1, |A_j|]$ of A_j and ends with the (possibly empty) prefix $A_{j'}[1, u']$ of a chunk $A_{j'}$. The remaining parts of D^L and D^R must be equal and coincide with some sequence of chunks D', i.e., $D^L = A_i[u+1, |A_i|] \cdot D' \cdot A_j[1, u'] = A_j[u'+1, |A_j|]D'A_{j'}[1, u'] = D^R$. Then, by using 2. of Fact 1 we show that there is an equivalent square $\hat{D}^L \hat{D}^R$ in

$\mu(T(\ell)$ corresponding to $D^L = \mu(C_1)[i+1, |\mu(C_1)|]\mu(D')\mu(C_2)[1,j] = \mu(C_2)[j+ 1, |\mu(C_2)|]\mu(D')\mu(C_3)[1,j] = \hat{D}^R$. Contracting it, implies $\mu(T(\ell)) \rightarrowtail \mu(T(\ell+1))$ also in this case. The complete proof is deferred to the extended version of the paper.

Part 2 - The Proof that $\hat{S} \Rightarrow_k \hat{T}$ Implies $S \Rightarrow_k T$. In this case, we show the possibiliy of mapping every sequence of contractions $\hat{T} \rightarrowtail \hat{T}(1) \rightarrowtail \hat{T}(2) \rightarrowtail \cdots \rightarrowtail \hat{S}$ to a sequence of contraction $T \rightarrowtail_k S$. Finding a characterization of the strings $\hat{T}(\ell)$ analogous to Fact 2 is significantly more involved. Although by construction $\hat{T}$ and $\hat{S}$ also have a factorization into chunks of $\hat{\Omega}$, most of these chunks (also adjacent ones) are from the same 4-ary alphabet. Therefore, showing that every intermediate string $\hat{T}(\ell)$ is also factorizable into chunks requires more care. Moreover, this fact, together with the invertibility of μ, shows only that we can find strings $T(0) = \mu^{-1}(\hat{T}), T(1) = \mu^{-1}(\hat{T}(1)), \ldots, T(k-1) = \mu^{-1}(\hat{T}(k-1))$, each of which is factorizable into chunks of Ω. Since a contraction $\hat{T}(\ell) \rightarrowtail \hat{T}(\ell+1)$ can involve suffixes and prefixes of chunks, we need a deeper analysis of such strings to characterize precisely the structure of the possible contractions $\hat{T}(\ell) \rightarrowtail \hat{T}(\ell+1)$ (Proposition 1 in appendix). Then, we show that 3. in Fact 1 guarantees the existence of a contraction $T(\ell) \rightarrowtail T(\ell + 1)$. Due to the space limitations, this part of the proof is deferred to the extended version of the paper.

4   TD-distance for Purely Alternating Strings

In this section, we investigate the existence of polynomial time algorithms to decide whether a purely alternating string S can be transformed into another purely alternating string T through a series of duplications, i.e., if $S \Rightarrow_* T$.

Definition 2. *Fix strings $S = s_1^{l_1} s_2^{l_2} \ldots s_n^{l_n}$ and $T = t_1^{l'_1} t_2^{l'_2} \ldots t_m^{l'_m}$. We say that the run $s_i^{l_i}$ matches the run $t_j^{l'_j}$ if $s_i = t_j$ and $l_i \leq l'_j$. We also say that S matches T (and write $S \preceq T$.) if $n = m$ and for $i = 1, \ldots, n$ we have $s_i^{l_i}$ matches $t_j^{l'_j}$.*

The existence of a string S' that matches T and that satisfies $S \Rightarrow_* S'$, implies that $S \Rightarrow T$: we can convert S' into T by duplications on single letters.

Definition 3. *Given two q-ary strings S and T, we say that the operation $S = AXB \Rightarrow AXXB = T$ is a normal duplication if one of the following conditions holds: (i) X is a q-ary string with exactly q runs such that the first and the last run are of length 1; (ii) X is a single character.*

We write $S \Rightarrow^N T$ if there exists a normal duplication converting S into T. We write $S \Rightarrow_k^N T$ if there exist $S_1, \ldots, S_{k-1}$ such that $S \Rightarrow^N S_1 \Rightarrow^N \cdots \Rightarrow^N S_{k-1} \Rightarrow^N T$. We write $S \Rightarrow_*^N T$ if there exists some k such that $S \Rightarrow_k^N T$.

In perfect analogy with the definition of contractions given in Sect. 2, we define *normal contractions*: $T \rightarrowtail_k^N S$ and $T \rightarrowtail_*^N S$ by $T \rightarrowtail_k^N S$ if and only if $S \Rightarrow_k^N T$ and $T \rightarrowtail_*^N S$ if and only if $S \Rightarrow_*^N T$.

Intuitively, normal duplications are effective in converting a string S into a string S' that matches a string T because they keep the resulting string purely alternating and create new runs that are as small as possible; these runs allow the string S' to match many strings.

We proceed to characterizing pairs of strings S and T such that $S \Rightarrow_*^N T$.

Lemma 2. *Fix $2 \le q \le 5$. Let $S = s_0^{l_1} s_1^{l_2} \ldots s_n^{l_n}$ and $T = t_0^{l'_1} t_1^{l'_2} \ldots t_m^{l'_m}$ be purely alternating strings over the same q-ary alphabet $\Sigma = \{0, 1, 2, \ldots, q - 1\}$. Then, $S \Rightarrow_*^N T$ if and only if there exists a function $f : \{1, \ldots, n-q+2\} \mapsto \{1, \ldots, m - q + 2\}$ such that: (1.) $f(1) = 1$ and $f(n - q + 2) = m - q + 2$; (2.) $f(i) = j \implies s_i = t_j$ and for each $u = 0, \ldots, q - 2$ we have that $l_{i+u} \le l'_{j+u}$; (3.) $f(i) = j$ and $f(i') = j'$ and $i < i' \implies j < j'$; (4.) if $q = 5$ and $f(i) = j$ and $f(i+1) = j' \ne j+1 \implies$ there exists a substring M in T starting in a position p such that $j \le p \le j'$ with the form $M = s_{i+3}^{l'_p}, s_{i+4}^{l'_{p+1}}, \ldots s_{i+q}^{l'_{p+q-3}}, s_{i+q+1}^{l'_{p+q-2}}$ such that for each $u = 0, 1, \ldots, q - 3$ it holds that $l_{i+3+u} \le l'_{p+u}$ and $l_{i+1} \le l'_{p+q-2}$.*

It turns out that for alphabets of size ≤ 5 the set of normal duplications is as "expressive" as the set of all possible duplications. We use the following claim (the proof is omitted here because of the space limitation).

Claim 2. *Let S and T be q-ary purely alternating strings such that $S \Rightarrow_* T$. Let $AXA' \Rightarrow AXXA'$ be one of the duplications of the sequence leading from S to T. Then, $|RLE(X)| \bmod q \le 1$.*

Lemma 3. *Let S and T be purely alternating strings over the same alphabet Σ of size ≤ 5. Then $S \Rightarrow_* T$ if and only if $S \Rightarrow_*^N T$.*

Proof. Using Claim 2, we show that any duplication can be simulated by normal duplications. We give the complete argument only for the case $|\Sigma| = 5$. The cases $|\Sigma| \in \{2, 3, 4\}$ can be showed analogously.

Claim 3. *Let S and T be a 5-ary purely alternating strings over the alphabet $\Sigma = \{0, 1, 2, 3, 4\}$. If there exists a duplication $S \Rightarrow T$, then we can create a series of normal duplications $S \Rightarrow \cdots \Rightarrow T'$ such that T' matches T.*

Proof. Let $S = AXB \Rightarrow AXXB = T$ be the original duplication. Like before, we create a string that matches XX starting from X through normal duplications, depending on how many runs are contained in X. By Claim 2 the only possible cases are $|RLE(X)| \bmod 5 \in \{0, 1\}$

Case 1. $|RLE(X)| = 1$. It means that the only effect of the original duplication is to extend one of the runs of S. For this reason S must already match T.

Case 2. $|RLE(X)| \bmod 5 = 0$, we suppose that the string X starts with a 0 (rotate the characters if it starts with any other symbol), so X has the form $X = 0^{l_1} 1^{l_2} 2^{l_3} 3^{l_4} 4^{l_5} 0^{l_6} \ldots 3^{l_{n-1}} 4^{l_n}$. If $|RLE(X)|$ is equal to 5 then it is trivially equivalent to some normal duplication plus possibly additional duplications of single characters.

If $|RLE(X)| = 10$, we consider the following two sequences of normal duplications to match XX (the duplicated part is underlined):

(i) $X = \underline{0^{l_1}1^{l_2}2^{l_3}3^{l_4}4^{l_5}}0^{l_6}1^{l_7}2^{l_8}3^{l_9}4^{l_{10}} \Rightarrow 0^{l_1}1^{l_2}\underline{2^{l_3}3^{l_4}4^{l_5}0^1}1^{l_2}2^{l_3}3^{l_4}4^{l_5}0^{l_6}1^{l_7}2^{l_8}3^{l_9}4^{l_{10}}$

$\Rightarrow 0^{l_1}1^{l_2}2^{l_3}3^{l_4}4^1 0^1 1^1 2^1 \boxed{3^{l_4}} 4^1 0^1 1^{l_2}2^{l_3}3^{l_4}4^{l_5}0^{l_6}1^{l_7}2^{l_8}3^{l_9}4^{l_{10}} = X'$

(ii) $X = 0^{l_1}1^{l_2}2^{l_3}3^{l_4}4^{l_5}\underline{0^{l_6}1^{l_7}2^{l_8}3^{l_9}4^{l_{10}}} \Rightarrow 0^{l_1}1^{l_2}2^{l_3}3^{l_4}4^{l_5}0^{l_6}1^{l_7}\underline{2^{l_8}3^{l_9}4^1 0^1}1^{l_7}2^{l_8}3^{l_9}4^{l_{10}}$

$\Rightarrow 0^{l_1}1^{l_2}2^{l_3}3^{l_4}4^{l_5}0^{l_6}1^{l_7}2^{l_8}3^{l_9}4^1 0^1 1^1 2^1 \boxed{3^{l_9}} 4^1 0^1 1^{l_7}2^{l_8}3^{l_9}4^{l_{10}} = X''$

$XX = 0^{l_1}1^{l_2}2^{l_3}3^{l_4}4^{l_5}0^{l_6}1^{l_7}2^{l_8} \boxed{3^{l_9}} 4^{l_{10}}0^{l_1}1^{l_2}2^{l_3}3^{l_4}4^{l_5}0^{l_6}1^{l_7}2^{l_8}3^{l_9}4^{l_{10}}$

It is easy to see that according to whether $l_4 \leq l_9$ or not, we have that either X' or X'' matches XX. Hence, in either case, we have the desired sequence of duplications proving the claim.

Finally, let us assume that X contains $5r$ runs for some $r > 2$. Then, in order to produce a string through normal duplications that matches XX, it suffices to execute the duplications explained before, then continue with $|RLE(X)|/5 - 2$ normal duplications containing the four adjacent runs of length 1 plus another adjacent run. This pushes to the right the original runs of X remaining, together with $\boxed{3^{l_4}}$ (in the first case) or $\boxed{3^{l_9}}$ (in the second case). We can see that $\boxed{3^{l_4}}$ in X' will be in the same position as $\boxed{3^{l_9}}$ in XX and vice versa.

The previous two lemmas imply the following

Theorem 3. *Let Σ be an alphabet of size ≤ 5. There exists a algorithm that for every pair of purely alternating strings S and T over Σ can decide in linear time whether $S \Rightarrow_* T$.*

Proof. The algorithm computes the run length encoding of S and T and then decides about the existence of the function f satisfying the properties of Lemma 2. By Lemma 3 we have that $S \Rightarrow_* T$ *if and only if* $S \Rightarrow_*^N T$. By Lemma 2 this latter condition holds *if and only if* there exists a function f satisfying the conditions 1–4 in Lemma 2. Therefore, to prove the claim it is enough to show that the existence of such a function f can be decided in linear time. This is easily attained by employing the following greedy approach: once the values of $f(1) = 1, \ldots, f(i - 1) = j$ have been fixed, sets the assignment $f(i) = j'$ to the smallest j' such that $l_{i+u} \leq l'_{j'+u}$ for each $u = 0, \ldots k - 1$ and if this condition does not hold for $j' = j + 1$, then j' is the smallest integer $> j$ that guarantees the existence of a $j < p < j'$ satisfying condition 4 in Lemma 2. The correctness of this approach can be easily shown by a standard exchange argument and it is deferred to the appendix for the sake of the space limitations. The resulting algorithm takes $O(|S| + |T|)$ time, since it only scans for a constant number of times each component of the run length encoding of T and S.

Final Remarks on 6-ary Strings and Some Open Problems. The technique we used to prove Lemmas 2, 3 is not generalizable to the case of larger alphabets: For purely alternating strings with $|\Sigma(S)| = 6$, we cannot always simulate a general duplication with normal duplications. Take, e.g., the duplication $AXB \Rightarrow AXXB$ where X is the string: $X = 0^2 1^1 2^2 3^1 4^2 5^1 0^1 1^2 2^1 3^2 4^1 5^2$. One can show that no pair of normal duplications can convert the string X into a string $T' \preceq XX$. We do not know whether this has an implication on the polynomial time solvability of the problem TD-EXIST already on 6-ary alphabets, and we leave it as a first step for future research. More generally, the main algorithmic problems that are left open by our results are the complexity of TS-DIST for binary alphabets (more generally, whether our hardness result can be extended to smaller alphabets) and the complexity of TD-EXIST for arbitrary ternary alphabets. Also, on the basis of the hardness result, approximation algorithms for the distance problem is another interesting direction for future research.

APPENDIX

The Basic Ingredients for the Proof of Part 2 ("if" part) of Theorem 2

Let us recall the definition of $\hat{T}$ and $\hat{S}$:

$$\hat{S} = \hat{\mathcal{B}}_{2p}\hat{X}\text{Ł} = \hat{\mathcal{B}}_{2p}\$\hat{\mathcal{B}}_{2p-1}\$\ldots\$\hat{\mathcal{B}}_2\$\hat{\mathcal{B}}_1\$\hat{\mathcal{B}}_0\$\hat{X}\text{Ł} \tag{5}$$
$$\hat{T} = \hat{\mathcal{B}}_{2p}^0\hat{X}^{dup}\text{Ł}\,\hat{\mathcal{B}}_{2p}^1\hat{X}\text{Ł}\,\hat{\mathcal{B}}_1^{01}\hat{X}_1\text{Ł}\hat{\mathcal{B}}_{2p}^1\hat{X}\text{Ł}\,\hat{\mathcal{B}}_2^{01}\hat{X}_2\text{Ł}\hat{\mathcal{B}}_{2p}^1\hat{X}\text{Ł}\,\ldots\,\hat{\mathcal{B}}_p^{01}\hat{X}_p\text{Ł}\hat{\mathcal{B}}_{2p}^1\hat{X}\text{Ł}. \tag{6}$$

Assume that there exists a sequence of k contractions

$$\hat{T} = \mu(T) = \hat{T}(0) \rightarrowtail \hat{T}(1) \rightarrowtail \hat{T}(2) \rightarrowtail \cdots \rightarrowtail \hat{T}(k) = \hat{S},$$

where $\hat{T}(\ell)$ denotes the string obtained from $\hat{T}$ after the first ℓ contractions have been performed. We tacitly assume that in each contraction $\hat{T}(\ell) = ADDA' \rightarrowtail ADA' = \hat{T}(\ell + 1)$, it is the right copy of D which is removed.

A *block* of $\hat{T}(\ell)$ is a *maximal* substring P of $\hat{T}(\ell)$ satisfying the following properties: the last character of P is Ł; and this is the only occurrence of Ł in P. Hence, the first character of P is either preceded by Ł or it is the first character of $\hat{T}(\ell)$.

We denote by $\hat{B}\hat{X}\text{Ł}$ the first, leftmost block $\hat{\mathcal{B}}_{2p}^0\hat{X}\text{Ł}$ of $\hat{T}$. For $i = 1, \ldots, p$, we also denote by E_i the block $\hat{\mathcal{B}}_i^{01}\hat{X}_i\text{Ł}$ in $\hat{T}$. We let $E_i(\ell)$ be the substring of $\hat{T}(\ell)$ formed by all the characters that belong to E_i. Note that $E_i(\ell)$ is any possibly empty subsequence of E_i.

For any $a \in \{0, 1, 01\}$ and $j \in [2p]$, a block $\hat{\mathcal{B}}'\hat{X}'\text{Ł}$ is called a $\hat{\mathcal{B}}_j^a\hat{X}\text{Ł}$-*block* if $\hat{\mathcal{B}}_j\hat{X}\text{Ł} \subseteq \hat{\mathcal{B}}'\hat{X}'\text{Ł} \subseteq \hat{\mathcal{B}}_j^a\hat{X}^{dup}\text{Ł}$. In other words, $\hat{\mathcal{B}}'\hat{X}'\text{Ł}$ has the same runs of $\hat{\mathcal{B}}_j^a\hat{X}^{dup}\text{Ł}$ in the same order, but some of the duplicated characters may have been contracted into a single character. A $\hat{\mathcal{B}}_{2p}^1\hat{X}\text{Ł}$-*cluster* is a string obtained by concatenating an arbitrary number of $\hat{\mathcal{B}}_{2p}^1\hat{X}\text{Ł}$-blocks. We write $(\hat{\mathcal{B}}_{2p}^1\hat{X}\text{Ł})^*$ to denote a possibly empty $\hat{\mathcal{B}}_j^1\hat{X}\text{Ł}$-cluster.

Proposition 1. *For each* $\ell = 0, \ldots, k$

1. $\hat{T}(\ell)$ *has the form:*

$$\hat{B}\hat{X}L\,(\hat{\mathcal{B}}^1_{2p}\hat{X}L)^* E_{i_1}(\ell)(\hat{\mathcal{B}}^1_{2p}\hat{X}L)^* E_{i_2}(\ell)(\hat{\mathcal{B}}^1_{2p}\hat{X}L)^* \ldots E_{i_h}(\ell)(\hat{\mathcal{B}}^1_{2p}\hat{X}L)^* \quad (7)$$

where:
- $\hat{B}\hat{X}L$ *is a* $\hat{\mathcal{B}}^0_{2p}\hat{X}L$*-block*
- $i_1 < i_2 < \cdots < i_h$
- *each* $(\hat{\mathcal{B}}^1_{2p}\hat{X}L)^*$ *is a* $\hat{\mathcal{B}}^1_{2p}\hat{X}L$*-cluster*
- *for each* $j \in \{i_1, \ldots, i_h\}$, $E_j(\ell)$ *is a* $\hat{\mathcal{B}}^{01}_j\hat{X}L$*-block.*

2. the contraction $\hat{T}(\ell) = ADDA' \rightarrowtail ADA' = \hat{T}(\ell + 1)$ *satisfies one of the following possibilities*

  (a) D *is a single character, hence, necessarily one of the characters in some substring* $\hat{X}' \in \hat{\mathbb{X}}$, $\hat{B}'_0 \in \hat{\mathbb{B}}_0$, $\hat{B}'_1 \in \hat{\mathbb{B}}_1$ —*these are the substrings corresponding to the elements of* $\hat{\Omega}$, *which appear in the above factorization of* $\hat{T}(\ell)$ *which are not square free;*

  (b) there exists a contraction $\hat{T}(\ell) = A\tilde{D}\tilde{D}A' \rightarrowtail A\tilde{D}A' = \hat{T}(\ell + 1)$ *such that* $\tilde{D}$ *is a sequence of whole chunks (note that this case includes the possibility* $\tilde{D} = D$, *i.e., already* D *is a sequence of whole chunks)*

  (c) there are chunks $\hat{C}_1 = \hat{C}'_1\hat{C}''_1$; $\hat{C}_2 = \hat{C}'_2\hat{C}''_2$; $\hat{C}_3 = \hat{C}'_3\hat{C}''_3$ *such that for some* $\hat{C} \in \{\hat{B}_0, \hat{B}_1, \hat{X}\}$ *for each* $i = 1, 2, 3$, *it holds that* $\hat{C} \subseteq \hat{C}_i \subseteq \hat{C}^*$ *and* $D = C''_1 D' C'_2 = C''_2 D' C'_3$, *and* C'_1 *is a suffix of* A *and* C''_3 *is a prefix of* A'.

The proof of this proposition is deferred to the extended version of the paper.

References

1. Alon, N., Bruck, J., Farnoud Hassanzadeh, F., Jain, S.: Duplication distance to the root for binary sequences. IEEE Trans. Inf. Theory **63**(12), 7793–7803 (2017)
2. Bovet, D.P., Varricchio, S.: On the regularity of languages on a binary alphabet generated by copying systems. Inf. Process. Lett. **44**(3), 119–123 (1992)
3. Chee, Y.M., Chrisnata, J., Kiah, H.M., Nguyen, T.T.: Deciding the confusability of words under tandem repeats in linear time. ACM Trans. Algorithms **15**(3) (2019)
4. Ehrenfeucht, A., Rozenberg, G.: On regularity of languages generated by copying systems. Discret. Appl. Math. **8**(3), 313–317 (1984)
5. Farnoud, F., Schwartz, M., Bruck, J.: The capacity of string-duplication systems. IEEE Trans. Inf. Theory **62**(2), 811–824 (2016)
6. Ito, Masami, Leupold, Peter, Shikishima-Tsuji, Kayoko: Closure of language classes under bounded duplication. In: Ibarra, Oscar H.., Dang, Zhe (eds.) DLT 2006. LNCS, vol. 4036, pp. 238–247. Springer, Heidelberg (2006). https://doi.org/10.1007/11779148_22
7. Jain, S., Farnoud Hassanzadeh, F., Bruck, J.: Capacity and expressiveness of genomic tandem duplication. IEEE Trans. Inf. Theory **63**(10), 6129–6138 (2017)
8. Jain, S., Farnoud Hassanzadeh, F., Schwartz, M., Bruck, J.: Duplication-correcting codes for data storage in the DNA of living organisms. IEEE Trans. Inf. Theory **63**(8), 4996–5010 (2017)

9. Lafond, M., Zhu, B., Zou, P.: The tandem duplication distance is np-hard. In: Paul, C., Bläser, M. (eds.) 37th International Symposium on Theoretical Aspects of Computer Science, STACS 2020, 10–13 March 2020, Montpellier, France. LIPIcs, vol. 154, pp. 15:1–15:15. Schloss Dagstuhl - Leibniz-Zentrum für Informatik (2020)
10. Lander, E., et al.: Initial sequencing and analysis of the human genome. Nature **409** (2001)
11. Leech, J.: A problem on strings of beads. Math. Gaz. **41**, 277–278 (1957)
12. Leupold, Peter: Duplication roots. In: Harju, Tero, Karhumäki, Juhani, Lepistö, Arto (eds.) DLT 2007. LNCS, vol. 4588, pp. 290–299. Springer, Heidelberg (2007). https://doi.org/10.1007/978-3-540-73208-2_28
13. Leupold, P.: Languages generated by iterated idempotency. Theor. Comput. Sci. **370**(1–3), 170–185 (2007)
14. Leupold, P., Martín-Vide, C., Mitrana, V.: Uniformly bounded duplication languages. Discret. Appl. Math. **146**(3), 301–310 (2005)
15. Leupold, Peter, Mitrana, Victor, Sempere, José M..: Formal languages arising from gene repeated duplication. In: Jonoska, Nataša, Păun, Gheorghe, Rozenberg, Grzegorz (eds.) Aspects of Molecular Computing. LNCS, vol. 2950, pp. 297–308. Springer, Heidelberg (2003). https://doi.org/10.1007/978-3-540-24635-0_22
16. Sutherland, G.R., Richards, R.I.: Simple tandem DNA repeats and human genetic disease. Proc. Natl. Acad. Sci. **92**(9), 3636–3641 (1995)
17. Usdin, K.: The biological effects of simple tandem repeats: Lessons from the repeat expansion diseases. Genome Res. **18**, 1011–1019 (2008)

On the Oriented Coloring of the Disjoint Union of Graphs

Erika Morais Martins Coelho[1], Hebert Coelho[1], Luerbio Faria[2],
Mateus de Paula Ferreira[1(✉)], Sylvain Gravier[4], and Sulamita Klein[3]

[1] Universidade Federal de Goiás, Goiânia, Brazil
{erikamorais,hebert,mateuspaula}@inf.ufg.br
[2] Universidade do Estado do Rio de Janeiro, Rio de Janeiro, Brazil
luerbio@ime.uerj.br
[3] Universidade Federal do Rio de Janeiro, Rio de Janeiro, Brazil
sula@cos.ufrj.br
[4] Université Grenoble Alpes, Grenoble, France
sylvain.gravier@ujf-grenoble.fr

Abstract. Let $\overrightarrow{G} = (V, A)$ be an oriented graph and G the underlying graph of $\overrightarrow{G}$. An *oriented k-coloring* of $\overrightarrow{G}$ is a partition of V into k subsets such that there are no two adjacent vertices belonging to the same subset, and all the arcs between a pair of subsets have the same orientation. The *oriented chromatic number* $\chi_o(\overrightarrow{G})$ of $\overrightarrow{G}$ is the smallest k, such that $\overrightarrow{G}$ admits an oriented k-coloring. The *oriented chromatic number* of G, denoted by $\chi_o(G)$, is the maximum of $\chi_o(\overrightarrow{G})$ for all orientations $\overrightarrow{G}$ of G. Oriented chromatic number of product of graphs were widely studied, but the disjoint union has not being considered. In this article we study oriented coloring for the disjoint union of graphs. We establish the exact values of the union: of two complete graphs, of one complete with a forest graph, and of one complete and one cycle. Given a positive integer k, we denote by $\mathcal{CN}_k$ the class of graphs G such that $\chi_o(G) \leq k$. We use those results to characterize the class of graphs $\mathcal{CN}_3$. We evaluate, as far as we know for the first time, the value of $\chi_o(W_n)$ and we yield with this value an upper bound for the union of one complete and one wheel graph W_n.

Keywords: Oriented graph · Oriented chromatic number · Disconnected graphs · Graph classes · Disjoint union of graphs

1 Introduction

Given a graph $G = (V, E)$, the *orientation* of an edge $e = \{u, v\} \in E$ is one of the two possible ordered pairs uv or vu called *arcs*. If $uv \in E$ we say that u *dominates* v. An *oriented graph* $\overrightarrow{G}$ is obtained from G by orienting each edge of E, $\overrightarrow{G}$ is called an *orientation* of G, and G is called the *underlying graph* of $\overrightarrow{G}$.

The authors are grateful to FAPEG, FAPERJ, CNPq and CAPES for their support of this research.

© Springer Nature Switzerland AG 2021
P. Flocchini and L. Moura (Eds.): IWOCA 2021, LNCS 12757, pp. 194–207, 2021.
https://doi.org/10.1007/978-3-030-79987-8_14

Note that an oriented graph is a digraph without opposite arcs or loops. Given an arc $uv \in E(\overrightarrow{G})$, v is called the *successor* of u and u is called the *predecessor* of v. A vertex without predecessors is called *source* and a vertex without successors is called *sink*. Let G and H be a pair of graphs. If H is a *subgraph* of G we say that G *contains* H as a subgraph, otherwise we say that G is H-*free*. Two graphs are *disjoint* if they have no vertex in common. If G and H are disjoint, their *disjoint union* graph denoted by $G \cup H$, has $V(G \cup H) = V(G) \cup V(H)$ and $E(G \cup H) = E(G) \cup E(H)$.

A *directed path* is the orientation of a path, a *directed cycle* is the orientation of a cycle. If for each pair u, v of consecutive vertices in a directed cycle we have the arc uv, then this orientation called *cyclic*, otherwise is called *acyclic*. A *tournament* $\overrightarrow{K}_n$ with n vertices is an orientation of a complete graph K_n. A tournament is called *transitive* if and only if whenever uv and vw are arcs, uw is also an arc. The complete bipartite graph $G = K_{1,n}$ is a *star*. A *wheel* graph W_n has $V(W_n) = \{v_1, v_2, \ldots, v_n, c\}$ and $E(W_n) = \{v_i v_{i+1} : i \in \{1, 2, \ldots, n - 1\}\} \cup \{v_n v_1\} \cup \{v_i c : i \in \{1, 2, \ldots, n\}\}$. We say that $\overrightarrow{G}$ is an *oriented star* (the same for a tree, forest, cycle, and wheel).

Let $\overrightarrow{G}$ be an oriented graph, $xy, zt \in E(\overrightarrow{G})$ and $C = \{1, 2, \ldots, k\}$ be a set of colors. An *oriented k-coloring* of $\overrightarrow{G}$ is a function $c : V(\overrightarrow{G}) \to C$, such that $c(x) \neq c(y)$, and if $c(x) = c(t)$, then $c(y) \neq c(z)$. The *oriented chromatic number* of $\overrightarrow{G}$ denoted by $\chi_o(\overrightarrow{G})$ is the smallest k such that $\overrightarrow{G}$ admits an oriented k-coloring. An *oriented absolute clique* or *o-clique* [7] is an oriented graph $\overrightarrow{G}$ for which $\chi_o(\overrightarrow{G}) = |V(\overrightarrow{G})|$.

Let $\overrightarrow{G}$ and $\overrightarrow{H}$ be oriented graphs, a *homomorphism* of $\overrightarrow{G}$ into $\overrightarrow{H}$ is a mapping $f : V(\overrightarrow{G}) \to V(\overrightarrow{H})$ such that $f(u)f(v) \in E(\overrightarrow{H})$ for all $uv \in E(\overrightarrow{G})$. When $\overrightarrow{H}$ is an oriented graph on k vertices, a homomorphism from $\overrightarrow{G}$ into $\overrightarrow{H}$ is an oriented k-coloring of $\overrightarrow{G}$.

We can extend the definition of oriented chromatic number to graphs. The oriented chromatic number of a graph G denoted by $\chi_o(G)$, is the maximum $\chi_o(\overrightarrow{G})$ for all orientations $\overrightarrow{G}$ of G. Given a positive integer k, we denote by $\mathcal{CN}_k$ the class of graphs G such that $\chi_o(G) \leq k$.

Oriented coloring has been studied by many authors. A survey on oriented coloring can be seen in [13]. Subsequently, many other papers have been published on oriented coloring. See for instance [3] and [7] on complexity aspects and approximation algorithms, and [8–10] for bounds on oriented coloring.

It is NP-complete [3,6,7] to decide whether a graph belongs to $\mathcal{CN}_k$ for all $k \geq 4$. In [2] it was shown that $\mathcal{CN}_k$ for all $k \geq 4$ is NP-complete even for acyclic oriented graph such that the underlying graph has maximum degree 3 and it is at the same time connected, planar and bipartite. Already, it can be decided in polynomial time [7] whether a graph belongs to $\mathcal{CN}_k$. So, in the Sect. 2 we characterize the class of connected and disconnected graphs that belong to the $\mathcal{CN}_3$ class.

The works of [1,4,5,12] presents various bounds for oriented chromatic number on the product of graphs. In spite of the vast amount of literature dedicated to the product of graphs, we don't have many results on the disjoint union.

Assume the 3-oriented coloring for $\overrightarrow{K}_3$ in Fig. 1 (a), where colors 1, 2 and 3 are assigned respectively, to vertices a, b and c. Notice that, by definition of oriented coloring, if the P_4 in Fig. 1 (b) is colored with the three colors 1, 2 and 3, then necessarily to vertices d, e, f are assigned, respectively, colors 1, 2 and 3. Hence, it is required a fourth color to assign to vertex g.

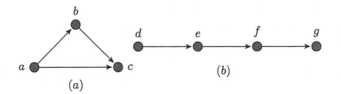

Fig. 1. Graph $\overrightarrow{K}_3 \cup \overrightarrow{P}_4$.

From the coloring given to the graph of Fig. 1 we can notice that, different from the usual coloring, the oriented coloring given to a connected component interferes in the coloring of another connected component in graphs formed by the disjoint union of two other graphs. Motivated by this fact, in Sect. 4 we determine the oriented chromatic number of the disjoint union between complete graphs and others graphs, such as stars, trees, forests, cycles and an upper bound for the union of one complete and one wheel. In Sect. 3 we show the oriented chromatic number of wheel graphs, for the first time as far as we know.

2   The Chromatic Number of the Class $\mathcal{CN}_3$

In this section, we characterize the class of graphs $\mathcal{CN}_3 = \{G; \chi_o(G) \leq 3\}$. First, we consider the case when the graph G is connected.

Lemma 1. *Let $G = (V, E)$ be a connected graph, $|V| \geq 4$. If G contains a K_3 as a subgraph, then $\chi_o(G) \geq 4$.*

Proof. Let $G = (V, E)$ be a connected graph with $|V| \geq 4$ and u, v, w be the vertices of a K_3 subgraph of G. As G is connect there is a vertex $t \notin \{u, v, w\}$ in V such that t is adjacent to a vertex in $\{u, v, w\}$. Assume $\{t, u\} \in E$. Consider an orientation $\overrightarrow{G} = (V, \overrightarrow{E})$ of G where $uv, vw, uw, tu \in \overrightarrow{E}$. We need 3 different colors to vertices u, v, w since u, v, w belong to K_3. As there is a path of size at most 2 from t to each vertex in $\{u, v, w\}$ by the oriented k-coloring definition, an additional fourth color is necessary to t. Hence, $\chi_o(G) \geq 4$. $\square$

From Lemma 1, we know that the connected not K_3-free graphs on 4 vertices or more do not belong to the class $\mathcal{CN}_3$. Sopena [11] proved that for oriented graphs with maximum degree 2, the oriented chromatic number is at most 5. He also proved that the cycle on 5 vertices has oriented chromatic number 5, this result is presented in Lemma 2. We use these results to propose Lemma 3.

Lemma 2 ([11]). *If C_5 is the cycle on 5 vertices, then $\chi_o(C_5) = 5$.*

Lemma 3. *If a connected graph G contains C_k as a subgraph, with $k \geq 4$, then $\chi_o(G) \geq 4$. In particular, if G contains C_5 as a subgraph, then $\chi_o(G) \geq 5$.*

Now we can describe the class of connected graphs that belongs to $\mathcal{CN}_3$.

Theorem 1. *The connected graph $G \in \mathcal{CN}_3$ if and only if, G is either a K_3 or a tree.*

Proof. Let $G \in \mathcal{CN}_3$ be a connected graph. If G is acyclic, then G is a tree. If G is not acyclic, then from Lemmas 1 and 3, it follows that $G = K_3$. We conclude that G is either a K_3 or a tree. Suppose that G is a K_3 or a tree. If G is a K_3 then $\chi_o(G) = 3$. If G is a tree, then $\chi_o(G) = 3$ by [3]. Therefore $G \in \mathcal{CN}_3$. □

Now will consider the case when G is a disconnected graph.

Lemma 4. *Let G be a graph with q connected components $X_1, X_2, \ldots, X_q$, $q \geq 2$, such that X_i contains K_3 as a subgraph, for some $i \in \{1, 2, \ldots, q\}$. If there is a component X_j, $i \neq j$, containing K_3 or P_4 as a subgraph, then $\chi_o(G) \geq 4$.*

Proof. Consider a graph G with q connected components $X_1, X_2, \ldots, X_q$, $q \geq 2$. Suppose there are two connected components X_i and X_j, $i \neq j$, such that both contains K_3 as a subgraph.

We can obtain an oriented graph $\overrightarrow{G}$ from G with $\chi_o(\overrightarrow{G}) \geq 4$, by defining the orientation of the subgraph K_3 of component X_i as a directed cycle and the subgraph K_3 of the component X_j as a transitive tournament. Let c be an oriented coloring for the subgraph K_3 of the component X_i, c has 3 colors, suppose $\{1, 2, 3\}$ and the property that no color dominates the two others. Let c_1 be an oriented coloring of K_3 of the component X_j. In c one color dominates the two others, thus one fourth color is required in the component X_j and therefore $\chi_o(G) \geq 4$.

Now suppose that the component X_j contains P_4 as a subgraph. In the oriented graph $\overrightarrow{G}$ obtained from G, we choose the transitive orientation $\overrightarrow{K}_3$ for the subgraph K_3 of the component X_i and the directed path $\overrightarrow{P}_4$ for the subgraph P_4 of the component X_j. We know that $\chi_o(\overrightarrow{K}_3) = 3$, we use colors 1, 2 and 3 in the oriented coloring of $\overrightarrow{K}_3$ of the component X_i. We choose the oriented coloring of $\overrightarrow{K}_3$ such as the vertex with color 1 is the source and the vertex with color 2 is the sink. We will show that, using the constraints obtained in the oriented coloring of the subgraph $\overrightarrow{K}_3$ in the component X_i, we cannot color the subgraph $\overrightarrow{P}_4$ of the component X_j only with colors 1, 2 and 3.

We consider three cases:

Case 1: (Assign color 1 to the source of $\overrightarrow{P}_4$) Since the vertex with color 1 is a predecessor of the vertex with color 2 in the oriented coloring of $\overrightarrow{K}_3$, we can assign color 2 to the successor of the source in $\overrightarrow{P}_4$. The vertex with color 2 in $\overrightarrow{K}_3$ is the sink, so we cannot assign any of the colors 1, 2 or 3 to the successor of the vertex with color 2 in $\overrightarrow{P}_4$. A fourth color is needed in component X_j.

Another sub-case is to assign color 3 to the successor of the source in $\overrightarrow{P}_4$, because the vertex with the color 1 also precedes a vertex with color 3 in an oriented coloring of $\overrightarrow{K}_3$. We can assign color 2 to the successor of the vertex with color 3 in $\overrightarrow{P}_4$, but again the color 2 is assigned to a vertex that is not sink in $\overrightarrow{P}_4$ and a fourth color is needed in component X_j.

Case 2: (Assign color 2 to the source of $\overrightarrow{P}_4$) The vertex with color 2 in the oriented coloring of $\overrightarrow{K}_3$ is a sink, so none of the colors 1, 2 or 3 can be assigned to the successor of the source in $\overrightarrow{P}_4$. A fourth color is required in component X_j.

Case 3: (Assign color 3 to the source of $\overrightarrow{P}_4$) Respecting the constraints on the coloring of $\overrightarrow{K}_3$, we can assign color 2 to the successor of the source in $\overrightarrow{P}_4$. Again, the successor of the vertex with color 2 in $\overrightarrow{P}_4$ cannot be colored with any color used in $\overrightarrow{K}_3$. A fourth color is required in component X_j.

We conclude that $\chi_o(G) \geq 4$.   □

It follows from Lemma 4 that the graph $G = K_3 \cup P_4 \notin \mathcal{CN}_3$. In Fig. 1 we have an orientation of graph G such that $\chi_o(G) = 4$. If we consider the graph G to be a forest, we have the following results.

Lemma 5. *Let F be a forest with a collection $\{T_1, T_2, \ldots, T_q\}$ of q disjoint trees, then $\chi_o(F) = \max\{\chi_o(T_i); i = 1, 2, \ldots, q\}$.*

From Lemma 5 we can show that every oriented forest has a homomorphism to a directed cycle, as we show on Corollary 1.

Corollary 1. *Every oriented forest $\overrightarrow{F}$ has a homomorphism into a directed cycle $\overrightarrow{C}_3$.*

Finally in Theorem 2 we can characterize the class $\mathcal{CN}_3$.

Theorem 2. *Let G be a graph. $G \in \mathcal{CN}_3$ if and only if, G is either a forest or a $K_3 \cup S$, where S is a forest of stars.*

Proof. Suppose that $G \in \mathcal{CN}_3$. If G has a cycle, then by Lemmas 1, 3 and 4 there is at most one connected component G_i of G which has a cycle as a subgraph, and in this case $G_i = K_3$. Still by Lemma 4 the remaining components have a diameter that is less than 3, and hence G is a disjoint union of K_3 and a forest of stars.

If G is acyclic, then G is a forest and by Lemma 5 and [3] we have $\chi_o(\overrightarrow{G}) \leq 3$. Conversely, first suppose that G is a forest. For every tree T_i of G we know that $\chi_o(G_i) \leq 3$, by [3]. By Lemma 5 we conclude that $\chi_o(G) \leq 3$.

Now suppose that $G = K_3 \cup S$. The connected component K_3 can be oriented in two different ways, with circular orientation or transitive orientation. If the component K_3 have a circular orientation $\overrightarrow{K}_3$, we know by Corollary 1 that there is a homomorphism from $\overrightarrow{S}$ into $\overrightarrow{K}_3$ and $\chi_o(G) \leq 3$. Now consider the component K_3 with a transitive orientation $\overrightarrow{K}'_3$. We choose the oriented coloring of $\overrightarrow{K}'_3$ with the colors 1, 2 and 3, so that the vertices with color 1 are predecessors of vertices with color 2 and the vertices with color 2 are predecessors of vertices with color 3.

We define a homomorphism from $\overrightarrow{S}$ into $\overrightarrow{K}'_3$ where all sources in $\overrightarrow{S}$ are mapped into the vertex with color 1 in $\overrightarrow{K}'_3$, and all sinks in $\overrightarrow{S}$ are mapped into the vertex with color 3 in $\overrightarrow{K}'_3$, if the vertex is neither a source nor a sink in $\overrightarrow{S}$, then it is mapped into a vertex with color 2 in $\overrightarrow{K}'_3$. This homomorphism is easily verified, since only one vertex that has more than one neighbor in $\overrightarrow{S}$ can be mapped into the vertex with color 2 in $\overrightarrow{K}'_3$. $\qquad\square$

3   The Oriented Chromatic Number of Wheel Graphs

In this section we establish that the family of wheel graphs W_q with $q \geq 8$ has its oriented chromatic number 8. We use this value, in Sect. 4, in order to establish an upper bound for the disjoint union of a wheel with a complete graph.

Theorem 3. *Let $q \geq 8$, be a positive integer. Then $\chi_o(W_q) = 8$.*

Proof. We consider $q \mod 3$, i.e., $q = 3k + 1, 3k + 2, k \geq 2$ and $q = 3k, k \geq 3$. We prove first that 8 colors are sufficient to color every orientation ω of W_q. Consider an orientation ω for W_q. We construct an 8–oriented color for this orientation. Let $V(W_q) = \{v_1, v_2, \ldots, v_q, c\}$ and $E(W_q) = \{v_i v_{i+1}, v_i c : i \in \{1, 2, 3, \ldots, q - 1\}\} \cup \{v_q v_1, v_q c\}$.

In order to yield an 8–oriented coloring for ω we consider a key property of an orientation ω that is when there is a 4-oriented coloring for the corresponding C_q, such that there is one color, say color 4, that occurs just in one vertex $v \in V(C_q)$.

From this 4–oriented coloring of C_q, we give the following recipe to color C_q in W_q, with at most 7 colors, and hence W_q with 8 colors. For each $x \in \{1, 2, 3\}$ of the 3 colors that can be repeated, consider the oriented bipartite graph B_x induced of W_q by the vertices with color x and vertex c. If there are sinks and sources in $B_x \setminus \{c\}$, then If $v \in V(B_x) \setminus \{c\}$ and v is a sink, set to $x + 4$ the color of v. If the orientation ω in C_q is acyclic, then there is a sink vertex v_i, hence we color the path $v_{i+1}, \ldots, v_n, v_1, \ldots, v_{i-1}$ with 6 colors in $\{1, 2, 3, 5, 6, 7\}$, color v_i with color 4, and c with color 8. Hence, when ω is acyclic, there is an 8–oriented coloring for W_q.

The remaining case is when the orientation ω is cyclic in C_q. Next we consider $q = 3k, k \geq 3$ and $q = 3k + 1, k \geq 2$, and prove that there is a 4–oriented coloring for the corresponding C_q where there is a color class with at most one vertex $v \in C_q$, say color 4.

1. If $q = 3k, k \geq 3$, in this case we color $v_1, v_2, \ldots, v_n$, respectively, with colors $1, 2, 3, \ldots, 1, 2, 3$.
2. If $q = 3k + 1, k \geq 2$, in this case we color $v_1, v_2, \ldots, v_{n-1}$, respectively, with colors $1, 2, 3, \ldots, 1, 2, 3$, and color v_n with color 4.

Hence, when ω is cyclic in C_q, $q = 3k, k \geq 3$ or $q = 3k + 1, k \geq 2$, there is an 8–oriented coloring for W_q.

We prove that if the orientation is cyclic, and $q = 3k + 2, k \geq 3$, then there is a 5–oriented coloring such that exactly 2 colors appear once. For that we color $v_1, v_2, \ldots, v_{n-2}$, respectively, with colors $1, 2, 3, \ldots, 1, 2, 3$, color v_{n-1} with color 4, and v_n with color 5. From this 5–oriented coloring of C_q, we give the following recipe to color C_q in W_q, with at most 7 colors, and hence W_q with 8 colors.

We consider 2 cases:

1. Vertex c is a sink or a source of W_q. In this case we can color W_q with 6 colors.
2. Vertex c is neither a sink nor a source of W_q. In this case we assume that $cv_n, v_1 c \in \omega$. We can assume that because the orientation of C_q is cyclic.

   First, for each $x \in \{1, 2, 3\}$ of the 3 colors that can be repeated in C_q, consider the oriented bipartite graph B_x induced by the vertices with color x and vertex c. If there are sinks and sources in $B_x \setminus \{c\}$, then If $v \in V(B_x) \setminus \{c\}$ and v is a sink, set to $x + 5$ the color of v. Hence, we have an 8-oriented coloring of C_q in W_q, which is an 9-oriented coloring of W_q, that we will reduce to a 8-oriented coloring of W_q.

   Hence, we set to 6 the color of vertex v_n. This can be done, since v_1 has color 1, and every other vertex in C_q with color 6, has a distance to v_n of at least 3. And thus, we have a coloring of C_q with colors $1, 2, 3, 4, 6, 7, 8$, and we can give the color 5 to vertex c.

Now we prove that 8 colors are necessary. For that we show an example of W_8 that requires 8 colors. For the convenience of the reader we exhibit this example in Fig. 2 and ask the reader to follow the Figure with the next items. Let ϕ be an 8-coloring of W_8. The set of vertices $\{v_1, v_2, v_4, v_5, v_6, v_8, c\}$ is an o-clique, thus the colors of this vertices are different, respectively $\{0, 1, 2, 3, 4, 5, 6\}$. Hence, we know from the orientation of W_8 that $\phi(v_3) \notin \{0, 1, 2, 3, 5, 6\}$ because all of the vertices with these colors are adjacent or have a path of size two to v_3. We can color v_3 with the color 4. Again from the orientation of W_8 we have that $\phi(v_7) \notin \{0, 2, 3, 4, 5, 6\}$ because all of the vertices with these colors are adjacent or have a path of size two to v_7. We also can not color v_7 with the color 1 because we have $v_3 v_2 \in E(\overrightarrow{W_8})$ and $\phi(v_3) = 4$, so we need an eighth color for v_7.   □

4   On the Oriented Chromatic Number of the Union of Graphs

The study of the class $\mathcal{CN}_3$ motivated us to study the oriented chromatic number of disconnected graphs. We show an example in Fig. 1, where the oriented

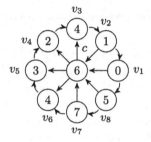

Fig. 2. An orientation of W_8 that has $\chi_o(\overrightarrow{W_8}) = 8$.

chromatic number of a graph $G = K_3 \cup P_4$ is greater than the oriented chromatic number of each of its connected components separately.

In Fig. 3, where $G = K_4 \cup P_5$, consider the orientation $\overrightarrow{G}$ of G in which $\overrightarrow{K}_4$ is the transitive tournament and $\overrightarrow{P}_5$ is the directed path.

So we have another example in which $\chi_o(G) > \max\{K_4; P_5\}$, where K_4 and P_5 are components of G. Since $\chi_o(\overrightarrow{K}_4) = 4$, we assign a 4-oriented coloring of $\overrightarrow{K}_4$. Using the constraints of 4-oriented coloring of $\overrightarrow{K}_4$ in the component $\overrightarrow{P}_5$, we prove that $\overrightarrow{P}_5$ cannot be colored only with four colors and one fifth color is required, so the graph $G = K_4 \cup P_5 \notin \mathcal{CN}_4$.

Fig. 3. Graph $\overrightarrow{K}_4 \cup \overrightarrow{P}_5$.

Now, we will obtain the oriented chromatic number of the disjoint union between the complete graph and others graphs, such as graphs that can be colored by the path $\overrightarrow{P}_3$ or the cycle $\overrightarrow{C}_3$, stars, trees, forests and cycles. First we analyse the case of graphs that have a homomorphism to the path $\overrightarrow{P}_3$ or the cycle $\overrightarrow{C}_3$

Theorem 4. *Let G be a graph with two connected components G_1 and G_2, where G_1 is a complete graph K_p, $p \geq 3$, and G_2 is a graph such that all oriented graphs $\overrightarrow{G}_2$ have a homomorphism f into a directed path $\overrightarrow{P}_3$, then $\chi_o(G) = p$.*

Proof. Since $\chi_o(\overrightarrow{P}_3) = 3$ (by definition of oriented coloring), considering an oriented coloring c of $\overrightarrow{P}_3$, in which we assign color 1 to the source, color 3 to the sink, and color 2 to the remaining vertex (successor of color 1 and predecessor of color 3). By hypothesis, all oriented graphs $\overrightarrow{G}_2$ have a homomorphism f

into a directed path $\overrightarrow{P}_3$. Thus, we can assign an oriented coloring for $\overrightarrow{G}_2$ using $c \circ f : V(\overrightarrow{G}_2) \rightarrow \{1, 2, 3\}$.

We will assign an oriented coloring with p colors to any oriented graph $\overrightarrow{G}_1$ from G_1 respecting the constraints used in $\overrightarrow{G}_2$. As $G_1 = K_p$, $p \geq 3$, $\chi_o(G_1) = p$ and all oriented graphs $\overrightarrow{G}_1$ from G_1 contains either a transitive or a circular $\overrightarrow{K}_3$. In both cases, there exists a directed path P_3 as a subgraph. This directed path can be colored with the same constraints used in G_1. There are no restrictions for the remaining $p - 3$ colors and therefore we can assign these colors to the other vertices not yet colored without conflict.   □

Theorem 5. *Let G be a graph with two connected components G_1 and G_2, where G_1 is a complete graph K_p, $p \geq 3$, and G_2 is a graph such that all oriented graphs $\overrightarrow{G}_2$ have a homomorphism into a directed cycle $\overrightarrow{C}_3$ and diameter greater than p. Then $\chi_o(G) = p + 1$.*

Proof. By hypothesis, $\overrightarrow{G}_2$ requires three colors $1, 2, 3$ to an oriented coloring, with the property that no color dominates the two others. We can obtain an oriented graph $\overrightarrow{G}$ from G with $\chi_o(\overrightarrow{G}) \geq p + 1$, in the following way: orient $\overrightarrow{G}_1$ as a transitive tournament. It follows that all subgraphs $\overrightarrow{K}_3$ of $\overrightarrow{G}_1$ are transitive. As in an oriented coloring of $\overrightarrow{G}_1$, for all $\overrightarrow{K}_3$ of $\overrightarrow{G}_1$ one color dominates the two others, at least one different color from $1, 2, 3$ is required in some component $\overrightarrow{K}_3$. Then $\chi_o(G) \geq p + 1$.

Conversely, we show that $\chi_o(G) \leq p + 1$. Let $\overrightarrow{K}_p$ be any orientation for G_1. As $\chi_o(\overrightarrow{K}_p) = p$, without loss of generality, we admit a coloring of $\overrightarrow{K}_p$ using the colors from 1 to p. We add a vertex v to the graph $\overrightarrow{K}_p$, and if there is source f or sink s in $\overrightarrow{K}_p$ we add the arcs vf and sv, we call the resulting graph of $\overrightarrow{K}'_{p+1}$, the remaining edges assume any orientation so that v is neither source nor sink in the new graph. We assign the color $p + 1$ to the vertex v. Note that $\overrightarrow{K}'_{p+1}$ has neither sources nor sinks. On the other hand, considers the directed cycle $\overrightarrow{C}_3$. We assign an oriented coloring of $\overrightarrow{C}_3$ respecting the constraint on the coloring of $\overrightarrow{K}'_{p+1}$.

We start by assigning a color $p + 1$ to any vertex v_1 of $\overrightarrow{C}_3$. By the construction of $\overrightarrow{K}'_{p+1}$ the vertex v with color $p + 1$ is neither source nor sink, so we divide the neighbors of v into two disjoint sets, a set of successors of v denoted by $Suc(v)$ and a set of predecessors of v denoted by $Pred(v)$. We will assign the same color as the successor v_2 of v_1 in $\overrightarrow{C}_3$ of a vertex $r \in Suc(v)$ who has a successor in $t \in Pred(v)$, the same color for predecessor v_3 of v_1 in $\overrightarrow{C}_3$ of the vertex of $t \in Pred(v)$.

By construction, there exists at least one vertex in $r \in Suc(v)$ such that rt is an arc in $\overrightarrow{K}'_{p+1}$, where $t \in Pred(v)$. So we can assign colors to $\overrightarrow{C}_3$ with the $p + 1$ colors of $\overrightarrow{K}'_{p+1}$ and as $\overrightarrow{K}_p$ is a subgraph of $\overrightarrow{K}'_{p+1}$ then $\chi_o(G) \leq \chi_o(\overrightarrow{K}'_{p+1} \cup \overrightarrow{C}_3) = p + 1$.   □

Corollary 2 follows directly from Theorem 5 and Corollary 1. We also show an upper bound for the disjoint union of complete graphs and stars on Corollary 3.

Corollary 2. *Given* $G = K_p \cup P_q$ *or* $G = K_p \cup T_q$ *or* $G = K_p \cup F_q$, *then* $\chi_o(G) = p + 1$. *Where* $p \geq 3$ *and* P_q, T_q, F_q *be respectively a path, a tree and a forest on* q *vertices and diameter greater than 2.*

Corollary 3. *Given* $G = K_p \cup S_q$, *then* $\chi_o(G) = p$, *where* $p \geq 3$ *and* S_q *is a star on* q *vertices.*

Now we define a special tournament on 5 vertices that we will use to describe the union of cycles and a few other graph classes. Let T_5^U be the tournament where $V(T_5^U) = \{v_1, v_2, v_3, v_4, v_5\}$, and $E(T_5^U) = \{v_1v_2, v_2v_3, v_2v_5, v_3v_1, v_3v_4, v_3v_5, v_4v_2, v_4v_5, v_1v_4, v_5v_1\}$. Also for this purpose we show that every tournament in 4 vertices has a sub-tournament which has a homomorphism to the acyclic tournament in 3 vertices.

Lemma 6. *Every tournament with 4 vertices has a homomorphism into* T_5^U.

Proof. We can verify by exhaustion that every 4-vertex tournament has a homomorphism into T_5^U.

Corollary 4. *Every tournament in 4 vertices has a sub-tournament which has a homomorphism to the acyclic tournament in 3 vertices.*

Now we define the chromatic number of the disjoint union of graphs that belongs to the class $\mathcal{CN}_4$ and cycles.

Theorem 6. *Let* $G \in \mathcal{CN}_4$ *be a graph and* C *be a cycle. Then* $\chi_o(G \cup C) = 5$.

Proof. Let $\overrightarrow{C_5^d}$ be a directed cycle with 5 vertices, then $\chi_o(C_5) = 5$, see Lemma 2. By Lemma 3 and because any other orientation of C_5 has a 4-oriented coloring, the class $C \backslash \overrightarrow{C_5^d} \in \mathcal{CN}_4$. By Lemma 6 every $G \in \mathcal{CN}_4$ has a homomorphism into T_5^U. The cycle $\overrightarrow{C_5^d}$ also has homomorphism in T_5^U, see that T_5^U has a directed cycle 1, 2, 3, 4, 5, 1. Therefore, $\chi_o(G \cup C) = 5$ with T_5^U as a color graph. □

Corollary 5. *Let* $G = C \cup C$ *or* $G = C \cup P$ *or* $G = C \cup T$ *or* $G = C \cup K4$, *then* $\chi_o(G) = 5$, *where* C, P, T, K_4 *be respectively a cycle, a path, a tree and the complete graph with 4 vertices.*

We also define the chromatic number of the disjoint union of complete graphs and cycles.

Theorem 7. *Let* p *and* q *be a pair of integers with* $p \geq 2$ *and* $q \geq 3$, *then*

$$\chi_o(K_p \cup C_q) = \begin{cases} 3, & \text{if } p = 2 \quad \text{and} \quad \chi_o(C_q) = 3 \\ 4, & \begin{cases} \text{if } p = 2 & \text{and} \quad \chi_o(C_q) = 4 \\ \text{if } p = 3 & \text{and} \quad (\chi_o(C_q) = 3 \text{ or } \chi_o(C_q) = 4) \end{cases} \\ 5, & \begin{cases} \text{if } p = 2 & \text{and} \quad \chi_o(C_q) = 5 \\ \text{if } p = 3 & \text{and} \quad \chi_o(C_q) = 5 \end{cases} \\ p + 1, & \text{if } p \geq 4 \end{cases}$$

Finally we will analyse the chromatic number of the disjoint union of two complete graphs.

Lemma 7. *Let c be an oriented coloring of $\overrightarrow{K}_p \cup \overrightarrow{K}_q$. Given $\overrightarrow{G}_1$ and $\overrightarrow{G}_2$ subgraphs induced of $\overrightarrow{K}_p$ and $\overrightarrow{K}_q$ respectively, such that $\exists\, u \in V(\overrightarrow{G}_1)$ if and only if $\exists\, a \in V(\overrightarrow{G}_2)$ with $c(u) = c(a)$. Then $\overrightarrow{G}_1$ and $\overrightarrow{G}_2$ are isomorphic.*

Proof. As $\overrightarrow{G}_1$ and $\overrightarrow{G}_2$ are induced subgraphs by vertices of tournaments, then $\overrightarrow{G}_1$ and $\overrightarrow{G}_2$ are also tournaments. Thus, in an oriented coloring c of $\overrightarrow{K}_p \cup \overrightarrow{K}_q$ there are no identical colors between the vertices of $\overrightarrow{G}_1$, as well as between the vertices of $\overrightarrow{G}_2$, then by hypothesis we know that $|V(\overrightarrow{G}_1)| = |V(\overrightarrow{G}_2)|$.

Case $|V(\overrightarrow{G}_1)| = |V(\overrightarrow{G}_2)| \leq 2$ then $\overrightarrow{G}_1$ and $\overrightarrow{G}_2$ are isomorphic.

Suppose that $|V(\overrightarrow{G}_1)| = |V(\overrightarrow{G}_2)| \geq 2$. Let $u, v \in V(\overrightarrow{G}_1)$ and $a, b \in V(\overrightarrow{G}_2)$ such that $c(u) = c(a)$ and $c(v) = c(b)$. We define $f : V(\overrightarrow{G}_1) \to V(\overrightarrow{G}_2)$ such that $f(u) \mapsto a$ and $f(v) \mapsto b$.

Let $f(u) = f(v)$. As $\overrightarrow{G}_2$ is a tournament then $c(f(u)) = c(f(v))$. By function f we have that $c(f(u)) = c(u)$ we get by replacing $c(u) = c(v)$. Like $\overrightarrow{G}_2$ also is a tournament, then $u = v$. We conclude that the function f is injective. As $|V(\overrightarrow{G}_1)| = |V(\overrightarrow{G}_2)|$ and $\overrightarrow{G}_1$, $\overrightarrow{G}_2$ are tournaments, then the function f is sobrejective. $\square$

Theorem 8. *Let K_p and K_q be complete graphs, and $\overrightarrow{K}$ be the collection of all tournaments. Consider the sets P and Q consisting of all orientations of K_p and K_q respectively. Define the set $L = \{\overrightarrow{K}^l \in K; |V(\overrightarrow{K}^l)| = \max\{|V(\overrightarrow{K}^j)|; \overrightarrow{K}^j \subseteq \overrightarrow{K}'_p, \overrightarrow{K}^j \subseteq \overrightarrow{K}'_q\}, \forall \overrightarrow{K}'_p \in P$ and $\overrightarrow{K}'_q \in Q\}$. Let $r = \min\{|V(\overrightarrow{K}^l)|; \forall \overrightarrow{K}^l \in L\}$. Then $\chi_o(K_p \cup K_q) = p + q - r$.*

Proof. Let $\overrightarrow{K}_r$ a tournament on r vertices, where $r = \min\{|V(\overrightarrow{K}^l)|; \forall \overrightarrow{K}^l \in L\}$. We denote by $\overrightarrow{K}_r^p$ a subgraph $\overrightarrow{K}_r$ of $\overrightarrow{K}_p$ and $\overrightarrow{K}_r^q$ a subgraph $\overrightarrow{K}_r$ of $\overrightarrow{K}_q$. Since $\overrightarrow{K}_r^p$ and $\overrightarrow{K}_r^q$ are isomorphic, we can assign identical r colors to the vertices of both graphs. As $r \leq q \leq p$ remain $p + q - r$ vertices to be colored. Then $\chi_o(K_p \cup K_q) \leq p + q - r$.

By Lemma 7, the maximum number of colors used in both $\overrightarrow{K}_p$ and $\overrightarrow{K}_q$ is r, otherwise we contradict the cardinality of $\overrightarrow{K}_r$. Hence $\chi_o(K_p \cup K_q) = p + q - r$. $\square$

We also analyse some specific disjoint unions of K_5 with another K_5 and with complete graphs.

Theorem 9. *Given the union $K_5 \cup K_5$, set $L = \{\overrightarrow{K}^l \in K; |V(\overrightarrow{K}^l)| = \max\{|V(\overrightarrow{K}^j)|; \overrightarrow{K}^j \subseteq \overrightarrow{K}'_5, \overrightarrow{K}^j \subseteq \overrightarrow{K}'_5\}, \forall \overrightarrow{K}'_p \in P$ and $\overrightarrow{K}'_q \in Q\}$, then $r = \min\{|V(\overrightarrow{K}^l)|; \forall \overrightarrow{K}^l \in L\} = 3$.*

Corollary 6. *Given the union* $K_p \cup K_5, p \geq 5$, *set* $L = \{\overrightarrow{K}^l \in K; |V(\overrightarrow{K}^l)| = \max\{|V(\overrightarrow{K}^j)|; \overrightarrow{K}^j \subseteq \overrightarrow{K}_5', \overrightarrow{K}^j \subseteq \overrightarrow{K}_5'\}, \forall \overrightarrow{K}_p' \in P$ *and* $\overrightarrow{K}_q' \in Q\}$, *then* $r = \min\{|V(\overrightarrow{K}^l)|; \forall \overrightarrow{K}^l \in L\} = 3$.

We have done some computational experiments, that drove us to Conjecture 1.

Conjecture 1. Let K_p, K_q be 2 complete graphs with $p, q \geq 4$. Then $\chi_o(K_p \cup K_q) = p + q - 3$.

Lastly we show an upper bound for the disjoint union of wheel graphs and complete graphs.

Theorem 10. *Let* p, q, $p \geq 4, q \geq 3$ *be positive integers. Then* $\chi_o(K_p + W_q) \leq p + 5$.

Proof. Let $\overrightarrow{K}_3$ be the transitive orientation of the tournament with 3 vertices. We consider 2 cases:

1. $\overrightarrow{K}_3$ is not a subgraph of W_q. In this case we can color W_q with 3 colors. Hence, 2 colors of the graph K_p can be used with color $p + 1$ to color W_q.
2. $\overrightarrow{K}_3$ is a subgraph of W_q. In this case according to Theorem 3 we can color W_q with 8 colors. From Corollary 4 we know that we can use 3 colors of the graph K_p plus additional 5 colors to color W_q.   □

5   Conclusions

In this paper, we prove that if $q \geq 8$ then $\chi_o(W_q) = 8$ and for every forest F, $\chi_o(F)$ is determined by the connected component of F with the largest oriented chromatic number of its connected components, what is an exception to the general case of disconnected graphs.

We characterized the class $\mathcal{CN}_3$ of the graphs with $\chi_o(G) \leq 3$. This characterization motivated us to study the oriented chromatic number of disconnected graphs. We have established $\chi_o(K_p \cup P_q)$, $\chi_o(K_p \cup F)$, $\chi_o(K_p \cup C_q)$, and an upper bound for $\chi_o(K_p \cup W_q)$.

We establish the oriented chromatic number of the union of two complete graphs K_p, K_q as $\chi_o(K_p \cup K_q) = p + q - r$, where r is the size of the maximum tournament contained in all orientations of K_p and K_q. We have conjectured that $r = 3$ for every pair $4 \leq p, q$.

Table 1 presents the results obtained in this paper regarding to the union of complete graphs with other graph classes. For future works we intend to expand our Table of results where most of the important classes be added in the firsts column and row of the Table, besides considering the cases when we have more than 2 components.

Table 1. Oriented chromatic number of the union $\chi_o(G \cup H)$.

G \ H	Forest diameter $d \leq 2$	Forest diameter $d \geq 3$	C_q, $q \geq 3$	K_q	W_q
K_p, $p = 2$	$d + 1$	3 (Corol. 1)	3, if $\chi_o(C_q) = 3$ 4, if $\chi_o(C_q) = 4$ 5, if $\chi_o(C_q) = 5$	$p + q - r$ (Corol. 8)	$q + 1$, if $3 \leq q \leq 6$ 8, if $q \geq 8$
K_p, $p = 3$	3 (Thm. 2)	4 (Corol. 8)	4, if $\chi_o(C_q) = 3$ 4, if $\chi_o(C_q) = 4$ 5, if $\chi_o(C_q) = 5$	$p + q - r$ (Thm. 8)	$q + 1$, if $3 \leq q \leq 6$ 8, if $q \geq 8$
K_p, $p \geq 4$	p (Corol. 3)	$p + 1$ (Corol. 2)	$p + 1$ (Thm. 7)	$p + q - r$ (Corol. 8)	$\leq p + 5$ (Thm. 10)

References

1. Aravind, N.R., Narayanan, N., Subramanian, C.R.: Oriented colouring of some graph products. Discuss. Math. Graph Theory **31**(4), 675–686 (2011). https://doi.org/10.7151/dmgt.1572

2. Coelho, H., Faria, L., Gravier, S., Klein, S.: Oriented coloring in planar, bipartite, bounded degree 3 acyclic oriented graphs. Discret. Appl. Math. **198**, 109–117 (2016). https://doi.org/10.1016/j.dam.2015.06.023

3. Culus, J.-F., Demange, M.: Oriented coloring: complexity and approximation. In: Wiedermann, J., Tel, G., Pokorný, J., Bieliková, M., Štuller, J. (eds.) SOFSEM 2006. LNCS, vol. 3831, pp. 226–236. Springer, Heidelberg (2006). https://doi.org/10.1007/11611257_20

4. Dybizbanski, J., Nenca, A.: Oriented chromatic number of cartesian products and strong products of paths. Discuss. Math. Graph Theory **39**(1), 211–223 (2019). https://doi.org/10.7151/dmgt.2074

5. Fertin, G., Raspaud, A., Roychowdhury, A.: On the oriented chromatic number of grids. Inf. Proc. Lett. **85**(5), 261–266 (2003). https://doi.org/10.1016/S0020-0190(02)00405-2

6. Ganian, R., Hliněný, P.: New results on the complexity of oriented colouring on restricted digraph classes. In: van Leeuwen, J., Muscholl, A., Peleg, D., Pokorný, J., Rumpe, B. (eds.) SOFSEM 2010. LNCS, vol. 5901, pp. 428–439. Springer, Heidelberg (2010). https://doi.org/10.1007/978-3-642-11266-9_36

7. Klostermeyer, W., MacGillivray, G.: Homomorphisms and oriented colorings of equivalence classes of oriented graphs. Discret. Math. **274**(1–3), 161–172 (2004). https://doi.org/10.1016/S0012-365X(03)00086-4

8. Marshall, T.H.: Homomorphism bounds for oriented planar graphs of given minimum girth. Graphs Comb. **29**(5), 1489–1499 (2013). https://doi.org/10.1007/s00373-012-1202-y

9. Ochem, P., Pinlou, A.: Oriented colorings of partial 2-trees. Inf. Proc. Lett. **108**(2), 82–86 (2008). https://doi.org/10.1016/j.ipl.2008.04.007

10. Raspaud, A., Sopena, E.: Good and semi-strong colorings of oriented planar graphs. Inf. Proc. Lett. **51**(4), 171–174 (1994). https://doi.org/10.1016/0020-0190(94)00088-3

11. Sopena, É.: The chromatic number of oriented graphs. J. Graph Theory **25**(3), 191–205 (1997). https://doi.org/10.1002/(SICI)1097-0118(199707)25:3⟨191::AID-JGT3⟩3.0.CO;2-G
12. Sopena, É.: Upper oriented chromatic number of undirected graphs and oriented colorings of product graphs. Discuss. Math. Graph Theory **32**(3), 517–533 (2012). https://doi.org/10.7151/dmgt.1624
13. Sopena, É.: Homomorphisms and colourings of oriented graphs: An updated survey. Discret. Math. **339**(7), 1993–2005 (2016). https://doi.org/10.1016/j.disc.2015.03.018

The Pony Express Communication Problem

Jared Coleman[1]([⊠]), Evangelos Kranakis[2], Danny Krizanc[3],
and Oscar Morales-Ponce[4]

[1] University of Southern California, California, USA
jaredcol@usc.edu
[2] Carleton University, Ottawa, ON, Canada
kranakis@scs.carleton.ca
[3] Wesleyan University, Middletown CT, USA
dkrizanc@wesleyan.edu
[4] California State University, Long Beach, USA
Oscar.MoralesPonce@csulb.edu

Abstract. We introduce a new problem which we call the Pony Express problem. n robots with differing speeds are situated over some domain. A message is placed at some commonly known point. Robots can acquire the message either by visiting its initial position, or by encountering another robot that has already acquired it. The robots must collaborate to deliver the message to a given destination (We restrict our attention to message transmission rather than package delivery, which differs from message transmission in that packages cannot be replicated.). The objective is to deliver the message in minimum time. In this paper we study the Pony Express problem on the line where n robots are arbitrarily deployed along a finite segment. We are interested in both offline centralized and online distributed algorithms. In the online case, we assume the robots have limited knowledge of the initial configuration. In particular, the robots do not know the initial positions and speeds of the other robots nor even their own position and speed. They do, however, know the direction on the line in which to find the message and have the ability to compare speeds when they meet.

First, we study the Pony Express problem where the message is initially placed at one endpoint (labeled 0) of a segment and must be delivered to the other endpoint (labeled 1). We provide an $O(n \log n)$ running time offline algorithm as well as an optimal (competitive ratio 1) online algorithm. Then we study the Half-Broadcast problem where the message is at the center (at 0) and must be delivered to either one of the endpoints of the segment $[-1, +1]$. We provide an offline algorithm running in $O(n^2 \log n)$ time and we provide an online algorithm that attains a competitive ratio of $\frac{3}{2}$ which we show is the best possible. Finally, we study the Broadcast problem where the message is at the center (at 0) and must be delivered to both endpoints of the segment $[-1, +1]$. Here we give an FPTAS in the offline case and an online algorithm that attains a competitive ratio of $\frac{9}{5}$, which we show is tight.

E. Kranakis—Research supported in part by NSERC Discovery grant.

P. Flocchini and L. Moura (Eds.): IWOCA 2021, LNCS 12757, pp. 208–222, 2021.
https://doi.org/10.1007/978-3-030-79987-8_15

Keywords: Delivery · Robots · Competitive ratio · Pony express

1   Introduction

The Pony Express refers to the well-known mail delivery service performed by continuous horse-and-rider relays between a source and a destination point. It was employed in the US for a short period (1860 to 1861) to deliver mail between Missouri and California.

The problem considered in this paper is motivated by the above. If one thinks of the horses as robots of differing speeds operating over a continuous domain then the Pony Express can serve as a suitable paradigm for message delivery from a source to a destination by robots passing messages from one robot to the next upon contact. In particular, consider the following problem: Initially a piece of information is placed at a certain location, referred to as the source. A group of robots are required to deliver the information from the source to another location referred to as the destination. The problem is one of designing a message delivery algorithm that delivers the message by selecting a sequence of robots and their movements that relay the message from a source to a destination in optimal time.

As will be seen, designing such algorithms can be a challenge given that the robots do not necessarily have the same speed and the overall delivery time may depend on what knowledge the agents possess concerning the location and speeds of the other robots. Further, the communication exchange model is face-to-face (F2F) in that two robots can exchange a message only when they are at the same location at the same time.

The problem itself can be studied over any domain. In this paper, we restrict our attention to a finite interval which already offers some interesting questions to resolve.

1.1   Model

We consider a set R of n robots initially scattered along a finite interval. Each robot r has a speed $v(r)$ and unique initial position $p(r)$. Note that robots with the same initial position can be handled through some tie-breaking mechanism, adding minor perturbations to the robots' positions, etc. The goal is to use the robots to deliver a message to one or both of the interval endpoints. Robots acquire the message through face-to-face contact either with the message at its initial location, or by encountering another robot with the message. We consider three variants of the Pony Express problem in which the message is initially placed at the point 0.

1. **Pony Express.** On the interval $[0, 1]$, the message must reach the endpoint 1.

2. **Half-Broadcast.** On the interval $[-1, 1]$, the message must reach one of the endpoints ± 1.
3. **Broadcast-Problem.** On the interval $[-1, 1]$, the message must reach both endpoints ± 1.

In each case, the goal is to solve the problem in the minimum amount of time. We consider both the offline and online settings. In the offline setting, all information regarding the robots (their initial positions and speeds) are available and a centralized algorithm provides a sequence of robot meetings that relay the message from the source to the destination(s) in optimal time.

In the online setting, we consider a model where robots do not know their own location nor their own speeds. Further, the agents do no have any information about other agents (initial positions, speeds) or even the number of robots in the system. The robots do however know the direction of the origin from their current position. When two robots meet, they can compare their speeds and decide which is faster.

To measure the performance of our online algorithms, we consider their competitive ratios. Let $t^*(I)$ be the optimal delivery time for an instance I of a given problem and $t_A(I)$ be the time needed by some online algorithm A for the same instance. Then the competitive ratio of A is

$$\max_I \frac{t_A(I)}{t^*(I)}.$$

Our goal is to find online algorithms that minimize this competitive ratio.

1.2   Related Work

There are many applications in a communication network where message passing (see [6]) is used by agents so as to solve such problems as search, exploration, broadcasting and converge-casting, connectivity, and area coverage. For example, the authors of [4] address the issue of how well a group of collaborating robots with limited communication range is able to monitor a given geographical space. In particular, they study broadcasting and coverage resilience, which refers to the minimum number of robots whose removal may disconnect the network and result in uncovered areas, respectively. Another application may be patrolling whereby many agents are required to keep perpetually moving along a specified domain so as to minimize the time a point in the domain is left unvisited by an agent, e.g., see [11] for a related survey.

A general energy-aware data delivery problem was posed by [1], whereby n identical, mobile agents equipped with power sources (batteries) are deployed in a weighted network. Agents can move along network edges as far as their batteries permit and use their batteries in linear proportion to the distance traveled. At the start the agents possess some initial information which they can exchange upon meeting at a node. The authors investigate the minimal amount of power, initially available to all agents, necessary so that convergecast may be achieved. They study the question in the centralized and the distributed setting.

Two related communication problems: data delivery and convergecast are presented for a centralized scheduler which has full knowledge of the input in [10]. The authors show that if the agents are allowed to exchange energy, both problems have linear-time solutions on trees but for general undirected and directed graphs they show that these problems are NP-complete.

A restricted version of the problem above concerns n mobile agents of limited energy that are placed on a straight line and which need to collectively deliver a single piece of data from a given source point s to a given target point t on the line. In [5] the authors show that the decision problem is NP-hard for a single source and also present a 2-approximation algorithm for the problem of finding the minimum energy that can be assigned to each agent so that the agents can deliver the data. In [7] it is shown that deciding whether the agents can deliver the data is (weakly) NP-complete, while for instances where all input values are integers, a quasi-, pseudo-polynomial time algorithm in the distance between s and t is presented.

Additional research under various conditions and topological assumptions can be found in [2] which studies the game-theoretic task of selecting mobile agents to deliver multiple items on a network and optimizing or approximating the total energy consumption over all selected agents, in [3] which studies data delivery and combines energy and time efficiency, and in [12,13] which is concerned with collaborative exploration in various topologies.

The focus of our current study is on finding offline and online algorithms for message delivery from a source to a destination on a line segment where the goal is to minimize the time needed. This differs from the work outlined above which focuses on energy transfer and consumption to perform either a delivery or broadcast. To the best of our knowledge, the problem and analysis considered in this paper has not been considered before.

1.3   Outline and Results of the Paper

In Sect. 2 we discuss the Pony Express variant of the problem and present optimal online and offline algorithms. In Sect. 3, we discuss the Half-Broadcast variant of the problem. We provide an optimal offline algorithm and an online algorithm with a $\frac{3}{2}$ competitive ratio and show this ratio is the best possible. In Sect. 4, we discuss the Broadcast variant of the problem and provide an online algorithm with a competitive ratio of $\frac{9}{5}$ which we show is the best possible. We also present an offline FPTAS for the Broadcast variant. (Note: the offline algorithm is not exact but depends upon performing binary search over a real interval). Due to space limitations some proofs are omitted. A complete version of the paper can be found on arXiv [9].

2   Pony Express

In this section, we discuss the solution for the Pony Express variant of the problem over the segment $[0, 1]$, wherein the message is placed initially at 0 and must be delivered to 1.

2.1   Online

First, we propose an online algorithm for the Pony Express variant. The robots start at the same time and move towards the origin. The first robot to reach 0 acquires the message. A slower robot with the message meeting a faster one, transfers the message to the faster which then moves towards 1. The algorithm is as follows.

Algorithm 1. Pony Express Online Algorithm

1: All robots start at the same time and move with their own speeds towards the endpoint 0;
2: **if** a robot r reaches 0 **then**
3:     robot r acquires the message and moves towards 1;
4: **if** a robot with the message r meets a robot r' such that $v(r) < v(r')$ **then**
5:     robot r transmits the message to robot r';
6:     robot r' changes direction and moves towards 1;
7: **else**
8:     continue moving;
9: Stop when destination 1 is reached;

Next we prove the optimality of Algorithm 1.

Theorem 2.1. *Algorithm 1 delivers the message in optimal time.*

Proof. Let m_i be the ith handover point between robots r_{i-1} and r_i at time t_i. Observe that since r_{i-1} participated in the $i-1^{\text{th}}$ handover, it must be slower than r_i, or $v(r_{i-1}) < v(r_i)$ (or else there would not be a handover). For simplicity and consistency of notation, let r_0 be an additional robot with initial position and velocity 0. In other words, robot r_0 simply holds the message at its initial position until robot r_1 arrives at 0 to perform the first handover. Note this does not change the problem at all, since robot r_0 will not carry the message any distance.

We show by induction and use the following inductive hypothesis: "Each participating robot r to the left of m_i has speed $v(r) < v(r_i)$ and t_i is the earliest time the message can be delivered to m_i." We say the message is *delivered* to a point m as soon as any robot that has acquired the message (excluding the additional robot r_0) reaches point m.

For the base case consider m_1. Observe robot r_1 is the first robot to reach the source, since any robot with speed greater than 0 would satisfy the condition to participate in a handover. It is clear then, that every robot to the left of $m_1 = 0$ has speed less than $v(r_1)$, since otherwise it would not be the first to arrive at 0. Also, t_1 is the first time the message is *delivered* to 0 since r_1 is the first robot to arrive at 0.

Assume the inductive hypothesis holds for m_{i-1}. Observe that since there is a handover at m_i, $v(r) < v(r_{i-1}) < v(r_i)$ for all robots r to the left of r_{i-1}. Therefore, all participating robots to the left of m_i are slower than r_{i-1}. Furthermore, the message reaches m_i at the earliest possible time, since otherwise a slower robot must not have handed the message over to a faster available robot. Finally, observe that it is the fastest robot delivers the message to the destination point.

$\square$

2.2 Offline

In this section, we present an offline algorithm for computing the optimal delivery time for the Pony Express variant (see Algorithm 2). In the previous section, we discussed the behavior of the robots in an optimal solution (i.e. they move toward 0 until encountering the message and then turn around and move toward the endpoint). The goal for an offline algorithm, then, is to compute all the meeting points where a handover occurs. We could consider all n^2 possible meeting points, but that would be inefficient. The key observation is that every robot must encounter one of its neighbors (either from its left or from its right) before encountering any other robot. When two robots meet, either both robots are moving toward 0 (and neither have the message) or one robot is traveling toward the endpoint with the message and the other is traveling toward 0 to acquire it. In either case, the meeting robots' neighbors and/or directions change, so new meeting points must be computed. This is the idea behind the algorithm. We keep track of potential $O(n)$ meeting points in a priority queue and examine them one-by-one to see how they affect the system.

Theorem 2.2. *Algorithm 2 finds an optimal solution to the Pony Express problem and runs in $O(n \log n)$ time.*

Proof. Observe that q is a Priority Queue whose operations add(r, p) for adding element r with a priority p, remove(r) for removing element r from the queue, remove_front() for removing and returning the element with the highest priority, and update(r, p) for updating an element's priority in the queue each have a time-complexity of $O(\log n)$.

In the first step of the algorithm, each robot is added to the priority queue, using its meeting time with its left-hand neighbor as a priority. Note this could be ∞ if the robot's left-hand neighbor either does not exist or moves at a faster speed away from it. This step has time-complexity $O(n \log n)$.

Next, notice on each iteration of the loop in line 15, the size of the queue is decremented by at least one and thus terminates after at most n iterations. Therefore this part of the algorithm has time complexity $O(n \log n)$.

Finally, observe that robots change direction if and when they meet a slower robot with the message and meeting times are updated appropriately when a change in direction occurs. This behavior is equivalent to that of Algorithm 1 and therefore is optimal by Theorem 2.1.

$\square$

Algorithm 2 returns only the final delivery time of the message to its destination. Observe though, that the algorithm could easily be made to return the entire sequence of handover meeting times (each r.meet in line 18).

Algorithm 2. Pony Express Offline Algorithm

Input: r, array of n robot structs sorted by initial position, $r[i].p$

1: q ← PriorityQueue()
2: left ← Robot(p=0, v=0, meet=∞)   ▷ Additional robot to represent source
3: **for** $i \leftarrow 1 \ldots n$ **do**
4:     **if** $r[i].v > $ left.v **then**
5:         $r[i].$meet ← $\frac{r[i].p - \text{left.p}}{r[i].v - \text{left.v}}$ ▷ Meeting time when moving the same direction
6:     **else**
7:         $r[i].$meet ← ∞
8:     **if** $i \leq n - 1$ **then**
9:         $r[i].$right ← $r[i+1]$
10:    $r[i].$left ← left
11:    left ← $r[i]$
12:    q.add(r, $-r[i].$meet)   ▷ Add robot to queue with meet-time-based priority
13: dst ← Robot(p=1, v=0, meet=∞)   ▷ Additional robot to represent destination
14: q.add(dst, $-\infty$)
15: **while** $q.size > 0$ **do**
16:     r ← q.remove_front()   ▷ Get robot with first meeting time
17:     **if** r.left.has_message **then**
18:         r.has_message ← True
19:         **if** r.left.v ≤ r.v **then**
20:             q.remove(r.left)
21:         **if** r.right **then**
22:             **if** r.left.v ≤ r.v **then**
23:                 r.right.left ← r
24:             **else**
25:                 r.left.right ← r.right
26:                 r.right.left ← r.left
27:             r.right.meet ← $\frac{\text{r.right.p} - \text{r.p} + 2 \cdot \text{r.meet} \cdot \text{r.v}}{\text{r.v} + \text{r.right.v}}$ ▷ Compute new meeting time
28:             q.update(r.right, -r.right.meet)
29:     **else**   ▷ robot r passes non-participating robot
30:         q.remove(r.left)
31:         r.left ← r.left.left
32:         **if** r.left **then**
33:             r.left.right ← r
34:         **if** r.left.has_message **then**
35:             r.meet ← $\frac{\text{r.p} - \text{r.left.p} + 2 \cdot \text{r.left.meet} \cdot \text{r.left.v}}{\text{r.left.v} + \text{r.v}}$
36:         **else if** r.v > r.left.v **then**
37:             r.meet ← $\frac{\text{r.p} - \text{r.left.p}}{\text{r.v} - \text{r.left.v}}$
38:         **else**
39:             r.meet ← ∞
40:         q.add(r, -r.meet)
41: **return** dst.meet

---

3 Half-Broadcast

In this section we consider the Half-Broadcast variant of the problem in which a message initially placed at 0 must be delivered to one of the endpoints of the interval $[-1, 1]$.

3.1 Online

First, we show a lower bound of $\frac{3}{2}$ on the competitive ratio for any algorithm to solve this problem.

Theorem 3.1. *The competitive ratio for the Half-Broadcast problem is at least* $\frac{3}{2}$.

Proof. Consider two robots r and r' with speeds $v(r) = \frac{1}{2}$ and $v(r') = 1$. Initially r is placed at $p(r) = 0$. The initial position of r', $p(r')$ will be determined below. Let A be any online algorithm for two robots. Observe the movement of r during the time period $[0, 1]$. Without loss of generality, assume that the final position of r in this time period is $x \in \left[0, \frac{1}{2}\right]$. In this case, we let $p(r') = -1$. (Note that if r ends up in $\left[0, -\frac{1}{2}\right]$, we let $p(r') = 1$ and a symmetric argument will follow.)

Observe that the trajectories taken by r and r' cannot overlap during the time period $[0, 1]$. Indeed, at time $0 \le t \le 1$, r is in the range $\left[x - \frac{1-t}{2}, x + \frac{1-t}{2}\right]$ (as it must reach x by time 1) and r' is in the range $[-1, -1 + t]$. These ranges do not overlap for $x \in \left[0, \frac{1}{2}\right]$ and $t \in [0, 1]$ except at $x = 0$ and $t = 1$, in which case both robots are at 0 at time 1. Thus it is not possible for r' to receive the message before time 1. At time 1, r is at $x \in \left[0, \frac{1}{2}\right]$ and can not make it to either endpoint (-1 or 1) sooner than time 2. Let the position of r' be $-y \in [-1, 0]$ at time 1. If r' receives no help from r when delivering the message then it cannot obtain the message before an additional y units of time to travel from $-y$ to the message source 0 and 1 unit of time to bring the message to either endpoint, i.e., $2 + y \ge 2$ ($y \ge 0$ units of time). If r' does receive help from r, it cannot receive the message before time $\frac{x-y}{\frac{3}{2}}$ and deliver the message before time $1 + 1 + \frac{2(x-y)}{\frac{3}{2}} \ge 2$ for $x \in \left[0, \frac{1}{2}\right]$ and $y \in [-1, 0]$.

Thus any online algorithm A must take time at least 2 units of time to solve this instance of the problem. But the optimal offline algorithm can complete the task in time $\frac{4}{3}$ by having the two robots meet at time $\frac{2}{3}$ at position $-\frac{1}{3}$ and then having r' deliver the message to -1. Therefore the competitive ratio for any algorithm is at least $\frac{3}{2}$. $\qquad \square$

Next we provide an online algorithm that achieves the competitive ratio $\frac{3}{2}$. We consider the very simple algorithm that essentially partitions the line segment (and robots) into two instances of the Pony Express Problem (over $[-1, 0]$ and $[0, 1]$) solves them independently. The delivery time is given by whichever instance delivers the message first.

Algorithm 3. Half-Broadcast Online Algorithm
___
1: All robots start at the same time and move within their own subinterval at their
   own speeds towards the endpoint 0;
2: **if** a robot r reaches 0 **then**
3:     robot r acquires the message and moves towards the endpoint closest to its
       original position;
4: **if** a robot with the message r meets a robot r' such that $v(r) < v(r')$ **then**
5:     robot r transmits the message to robot r';
6:     robot r' changes direction and moves towards the nearest endpoint;
7: **else**
8:     continue moving;
9: Stop when either endpoint is reached by robot;

First, we show that Algorithm 3 guarantees a competitive ratio of $\frac{3}{2}$ when only two robots participate. Then, we extend the result to systems of n robots. Note that our algorithm is clearly optimal in the case where there is only one robot.

Lemma 3.1. *Algorithm 3 solves the Half-Broadcast problem for the case $n = 2$ with competitive ratio at most $\frac{3}{2}$.*

Proof. Consider two robots r and r'. Without loss of generality, assume that $v(r) \leq v(r')$ and that in the optimal algorithm the message is delivered at 1. Considering an optimal algorithm, observe that either robot r' delivers the message without any help or both collaborate to deliver the message. In the second case, suppose that m is the optimal meeting point between robots r and r'. Since $v(r) \leq v(r')$, r' must deliver the message. Thus, the delivery time is at least

$$\min\left(\frac{|p(r)| + 1}{v(r)}, \frac{|p(r')| + 1}{v(r')}, \frac{m + |p(r)|}{v(r)} + \frac{1 - m}{v(r')}\right).$$

Observe that in cases where either robot delivers the message without collaboration, Algorithm 3 is optimal. If the optimal algorithm requires the two robots to collaborate, however, Algorithm 3 is not optimal. Observe that the algorithm terminates when either of the two robots arrives at an endpoint. Therefore, the delivery time is maximized when $\frac{|p(r)|+1}{v(r)} = \frac{|p(r')|+1}{v(r')}$. Thus, the competitive ratio is given by:

$$\frac{\frac{|p(r')|+1}{v(r')}}{\frac{m+|p(r)|}{v(r)} + \frac{1-m}{v(r')}} = \frac{|p(r')| + 1}{\frac{v(r')}{v(r)}(m + |p(r)|) + 1 - m}$$

$$= \frac{v(r)(|p(r')| + 1)}{v(r')(m + |p(r)|) + v(r)(1 - m)}$$

$$= \frac{v(r)(m + |p(r')|) + v(r)(1 - m)}{v(r')(m + |p(r)|) + v(r)(1 - m)}$$

Observe that $p(r) = 0$ and $p(r') = 1$ maximizes the ratio. Thus, the competitive ratio is at most $\frac{2v(r)}{mv(r')+v(r)(1-m)}$ and $v(r') = 2v(r)$. Then:

$$\frac{2v(r)}{mv(r')+v(r)(1-m)} = \frac{v(r')}{mv(r')+(1-m)\frac{v(r')}{2}}$$

$$= \frac{2v(r')}{v(r')(m+1)}$$

$$= \frac{2}{m+1}$$

Since $2v(r) = v(r')$, $m = \frac{1}{3}$ and the competitive ratio is bounded by $3/2$ for any chosen speed of $v(r')$. □

Now we are ready to present the main result of the section in the following theorem. We show that the competitive ratio of Algorithm 3 for the Half-Broadcast problem is at most $\frac{3}{2}$ when n robots are participating. By Theorem 3.1, this is best possible.

Theorem 3.2. *The competitive ratio of Algorithm 3 is at most $\frac{3}{2}$ for systems of n robots.*

Proof. Without loss of generality, assume that the message is delivered to 1 in both the online and optimal offline algorithm. (Otherwise a symmetric argument can be used.) Let $\mu_1, \mu_2, \ldots, \mu_k$ be the $k < n$ meeting points of the optimal centralized algorithm where robot $r_{\pi(i)}$ carries the message between μ_{i-1} and μ_i. Let $\mu_0 = 0$. Let $m_1, m_2, \ldots m_l$ be the $l < n$ meeting points of the Algorithm 3 where robot $r_{\sigma(i)}$ traverses between m_{i-1} and m_i. Let $m_0 = 0$. The competitive ratio of our algorithm is

$$\frac{\frac{p(r_{\sigma(1)})}{v(r_{\sigma(1)})} + \frac{1-m_l}{v(r_{\sigma(l)})} + \sum_{i=1}^{l-1} \frac{m_i - m_{i-1}}{v(r_{\sigma(i)})}}{\frac{p(r_{\pi(1)})}{v(r_{\pi(1)})} + \frac{1-\mu_k}{v(r_{\pi(k)})} + \sum_{i=1}^{k-1} \frac{\mu_i - \mu_{i-1}}{v(r_{\pi(i)})}}$$

Observe that $m_l \le \mu_k$ since Algorithm 3 does not attain optimal time. Therefore, $\frac{1-m_l}{v(r_{\sigma(l)})} \le \frac{\mu_k - m_l}{v(r_{\pi(k-1)})} + \frac{1-\mu_k}{v(r_{\pi(k)})}$ since $v(r_{\pi(k-1)}) < v(r_{\pi(k)})$. Observe then, that we can trim the interval at m_l and solve the problem with $n-1$ robots. The key observation is that the online algorithm is actually "faster" at each intermediate handover *except* for the first handover. In other words, the first handover is the only segment that hurts the online algorithm. We have shown that the competitive ratio of the new problem is less than or equal to the competitive ratio of the original problem and therefore by induction (with Lemma 3.1 as the base case) the result follows. □

3.2   Offline

Next, we show an offline algorithm for computing the optimal solution. To do this, we make use of Algorithm 2, the offline algorithm for the Pony Express problem.

First we need the following lemma:

Lemma 3.2. *Let r and r' be the fastest robots in the subintervals $[-1,0]$ and $(0,1]$, respectively. Then either r or r' will deliver the message in optimal time.*

In an optimal solution, robots that initially start in the interval $[-1,0]$ can participate in the message delivery at the point 1, or vice-versa. The following lemma shows that even if many robots that initially start in an opposite interval can participate in delivering the message, there is always an identical solution where only one robot in the opposite interval participates in delivering the message.

Lemma 3.3. *There is an optimal solution such that at most one robot from the interval that does not contain the delivered-to endpoint participates.*

Theorem 3.3. *There exists an offline algorithm for finding an optimal solution to the Half-Broadcast problem with time-complexity $O(n^2 \log n)$.*

4   Broadcast

In this section we study the Broadcast variant of the problem. Recall that in the Broadcast problem, a message initially placed at 0 must be delivered by robots to *both* endpoints of the interval $[-1,1]$ in minimum time. We begin with the following lemma:

Lemma 4.1. *There is an optimal solution such that at most one robot participates in the message's delivery to both endpoints.*

4.1   Online

First we show that the competitive ratio of any online algorithm is at least $\frac{9}{5}$.

Theorem 4.1. *The competitive ratio for any Algorithm that solves the Broadcast problem is at least $\frac{9}{5}$.*

Proof. Consider two robots, r with speed 1 and initial location 0 and r' with speed and initial location to be determined below. Let A be any online algorithm for two robots. Observe the movement of r during the time period $[0,1]$ under algorithm A. Without loss of generality, assume the final position of r is $x \in [-1,0]$ and let y be the furthest that r progressed into $[0,1]$ during this time period. Observe that $0 \leq y \leq \frac{1}{2}$ since r is in $[-1,0]$ at time 1. Let $a = \frac{1-y}{2}$. In

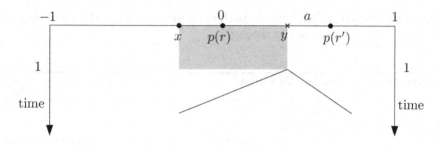

Fig. 1. Possible trajectory for the online algorithm. Robot r moves between x and y (the shaded region) during the time interval $[0,1]$.

this case, we set the r'''s speed $v(r') = a$ and its initial position $p(r') = y + a$ (Fig. 1). A symmetric argument can be used in the case that $x \in [0, 1]$.

Observe that the trajectories of r and r' do not overlap in the time period $[0, 1]$ for any $x \in [-1, 0]$ and $y \in \left[0, \frac{1}{2}\right]$ with the exception of the case where $y = x = 0$ and $t = 1$. Indeed, r' can only reach the position y at time 1. Prior to that time, its position must be to the right of y and therefore to the right of r. At time 1 it may reach y but by that time r is at $x \leq 0$. The only overlap occurs when $x = y = 0$.

At time 1, r' is at position $z \geq y \geq 0$. In order to deliver to the message to either end point, it must take time at least $\frac{1+z}{a} \geq \frac{2(1+y)}{1-y} \geq 2 + 2y$. Thus, if r' is the first to deliver the message to one of the end points, the algorithm must take at least time $3 + 2y$. On the other hand, if r is to deliver the message to both end points, it must take at least time y to reach position y, a further time y to return to 0, plus an additional time 3 to reach both end points. Therefore, the online algorithm A must take time at least $3 + 2y$ to solve this instance of the problem. (The case where r' delivers the message to both endpoints is clearly worse.)

Consider the following (offline) algorithm for the above instance: r and r' meet at position $\frac{y+a}{1+a}$ at time $\frac{y+a}{1+a}$ (they move toward each other until meeting). Then r delivers the message to -1 in a further $\frac{y+a}{1+a} + 1$ for a total of $1 + \frac{2(y+a)}{1+a}$ time. And r' delivers the message to 1 in a further $\frac{y+a}{1+a} + \frac{1-y-a}{a} = \frac{y+a}{1+a} + 1$ for a total of $1 + \frac{2(y+a)}{1+a}$ time.

Therefore, the competitive ratio of algorithm A on this instance is at least $\frac{3+2y}{1+\frac{2(y+a)}{1+a}} = \frac{(3+2y)(3-y)}{5+y} \geq \frac{9}{5}$ for $y \in \left[0, \frac{1}{2}\right]$. $\quad\square$

Now we show that there is an online algorithm that attains this competitive ratio. Algorithm 4 is very similar to the Half-Broadcast algorithm, in that we essentially partition the line segment (and robots) into two instances of the Pony Express Problem (over $[-1, 0]$ and $[0, 1]$). The difference is that every time a robot participates in a handover (at the source, endpoint, or with another robot), it turns around and moves in the opposite direction. This is necessary to ensure the message is delivered to both endpoints (consider the case where all robots start on one side of the message).

Algorithm 4. Broadcast Algorithm

1: All robots start at the same time and move within their own subinterval at their own speeds towards the endpoint 0;
2: **if** a robot r reaches 0 **then**
3:     robot r acquires the message and moves towards the endpoint closest to its original position;
4: **if** a robot with the message r meets a robot r' such that $v(r) < v(r')$ **then**
5:     robot r transmits the message to robot r', changes direction, and continues moving towards the opposite endpoint;
6:     robot r' changes direction and moves towards the nearest endpoint;
7: **else**
8:     continue moving;
9: **if** a robot with the message r reaches the endpoint the opposite endpoint **then**
10:     robot r changes direction and continues moving;
11: Stop when both endpoints have been reached by robot;

Lemma 4.2. *The competitive ratio of Algorithm 4 for the case $n = 2$ is at most $\frac{9}{5}$.*

Next, we show that Algorithm 4 attains optimal competitive ratio with $n \geq 3$ robots.

Theorem 4.2. *The competitive ratio of Algorithm 4 for systems of n robots is at most $\frac{9}{5}$.*

4.2   Offline

In this section, we provide an offline fully polynomial time approximation scheme (FPTAS).

Theorem 4.3. *For any $\epsilon > 0$, there exists an algorithm for finding a solution to within an additive factor of ϵ of optimal to the Broadcast problem with running time $O(n^2 \log n \log \frac{1}{\epsilon})$.*

Proof. According to Lemma 4.1, at most **one** robot must cross 0 and participate in the message's delivery to both endpoints. This robot, say r, may participate by delivering the message itself or handing it over to another robot. It's important to note that the receiving robot may not be the first encountered by r nor must it be faster than r. We must consider the scenarios where r delivers the message to each of the possible robots on the opposite subinterval. To facilitate the formulation of the solution, we assume there are robots with speed 0 at both endpoints -1 and 1 so that delivering to these robots is equivalent (in time and meaning) to delivering to the destination.

Suppose the optimal handover on the opposite side of the interval occurs at position m on the segment. Then, observe that since all robots must only participate in delivering the message to the nearest endpoint, there are essentially two instances of the regular PonyExpress problem to solve (one for each endpoint). One instance is on the interval $[-1, \min(m, 0)]$ and the other $[max(m, 0), 1]$. Also, the robots have shifted some distance toward 0, based on their speeds. This new instance can be constructed in linear time and solved in $O(n \log n)$ time by Algorithm 2.

All that remains, then, is to find m. Observe that two robots l and r with initial positions $p(l)$ and $p(r)$, and speeds $v(l)$ and $v(r)$, respectively can meet at any point on the interval

$$\left[0, \min \left(1, \frac{p(r) - p(l)}{v(l) - v(r)} \right) \right].$$

So the optimal solution can be described as:

$$\min_{l \in L, r \in \overline{L}} \quad \min_{m \in \left[0, \min\left(1, \frac{p(r)-p(l)}{v(l)-v(r)} \right) \right]} \quad \texttt{PonyExpress} \left(T \left(R, \frac{m - p(l)}{v(l)}, m \right) \right).$$

Notice the inner minimization is over a real domain where the function is bitonic and so it can be estimated using binary search. The runtime for this is therefore $O(n^2 \log n \log \frac{1}{\epsilon})$ where the computed meeting point is within ϵ of the optimal meeting point. $\square$

5 Conclusion

In this paper, we have introduced the Pony Express problem. We considered the case where the domain of interest is a line segment and the cases where a message must be delivered from one end to the other (Pony Express), from the center to one of the end points (Half-Broadcast) and from the center to both end points (Broadcast). For the first two problems we provide polynomial time offline algorithms and for the third an FPTAS. We provide online algorithms for each problem with best possible competitive ratio in each case.

A number of open problems are suggested by our study. First, it seems likely the runtime of our offline algorithms may be improved at least for the case of Half-Broadcast and Broadcast and that an exact algorithm exists for Broadcast. Second, it might be worth considering variations on the amount and type of information available to the agents in the online setting. Finally, another direction of study would be to consider domains other than a finite interval. Preliminary results for the plane can be found in [8].

References

1. Anaya, J., Chalopin, J., Czyzowicz, J., Labourel, A., Pelc, A., Vaxès, Y.: Collecting information by power-aware mobile agents. In: Aguilera, M.K. (ed.) DISC 2012. LNCS, vol. 7611, pp. 46–60. Springer, Heidelberg (2012). https://doi.org/10.1007/978-3-642-33651-5_4

2. Bärtschi, A., Graf, D., Penna, P.: Truthful mechanisms for delivery with mobile agents. arXiv preprint arXiv:1702.07665 (2017)

3. Bärtschi, A., Tschager, T.: Energy-efficient fast delivery by mobile agents. In: Klasing, R., Zeitoun, M. (eds.) FCT 2017. LNCS, vol. 10472, pp. 82–95. Springer, Heidelberg (2017). https://doi.org/10.1007/978-3-662-55751-8_8

4. Bereg, S., Brunner, A., Caraballo, L.E., Díaz-Báñez, J.M., Lopez, M.A.: On the robustness of a synchronized multi-robot system. J. Comb. Optim. 1–29 (2020)

5. Chalopin, J., Das, S., Mihal'ák, M., Penna, P., Widmayer, P.: Data delivery by energy-constrained mobile agents. In: Flocchini, P., Gao, J., Kranakis, E., Meyer auf der Heide, F. (eds.) ALGOSENSORS 2013. LNCS, vol. 8243, pp. 111–122. Springer, Heidelberg (2014). https://doi.org/10.1007/978-3-642-45346-5_9

6. Chalopin, J., Godard, E., Métivier, Y., Ossamy, R.: Mobile agent algorithms versus message passing algorithms. In: Shvartsman, M.M.A.A. (ed.) OPODIS 2006. LNCS, vol. 4305, pp. 187–201. Springer, Heidelberg (2006). https://doi.org/10.1007/11945529_14

7. Chalopin, J., Jacob, R., Mihalák, M., Widmayer, P.: Data delivery by energy-constrained mobile agents on a line. In: Esparza, J., Fraigniaud, P., Husfeldt, T., Koutsoupias, E. (eds.) ICALP 2014. LNCS, vol. 8573, pp. 423–434. Springer, Heidelberg (2014). https://doi.org/10.1007/978-3-662-43951-7_36

8. Coleman, J., Kranakis, E., Krizanc, D., Morales-Ponce, O.: The pony express communication problem on the plane. In: Preparation (2021)

9. Coleman, J., Kranakis, E., Krizanc, D., Ponce, O.M.: The pony express communication problem. arXiv preprint 2105.03545 (2021)

10. Czyzowicz, J., Diks, K., Moussi, J., Rytter, W.: Communication problems for mobile agents exchanging energy. In: Suomela, J. (ed.) SIROCCO 2016. LNCS, vol. 9988, pp. 275–288. Springer, Cham (2016). https://doi.org/10.1007/978-3-319-48314-6_18

11. Czyzowicz, J., Georgiou, K., Kranakis, E.: Patrolling. In: Distributed Computing by Mobile Entities, pp. 371–400. Springer (2019)

12. Das, S., Dereniowski, D., Karousatou, C.: Collaborative exploration by energy-constrained mobile robots. In: Scheideler, C. (ed.) SIROCCO 2014. LNCS, vol. 9439, pp. 357–369. Springer, Cham (2015). https://doi.org/10.1007/978-3-319-25258-2_25

13. Das, S., Dereniowski, D., Karousatou, C.: Collaborative exploration of trees by energy-constrained mobile robots. Theor. Comput. Syst. **62**(5), 1223–1240 (2018)

Skyline Groups Are Ideals. An Efficient Algorithm for Enumerating Skyline Groups

Simon Coumes[1], Tassadit Bouadi[2(✉)], Lhouari Nourine[3], and Alexandre Termier[2]

[1] ENS Rennes, 35170 Bruz, France
simon.coumes@ens-rennes.fr
[2] Univ Rennes, Inria, CNRS, IRISA, 35000 Rennes, France
{tassadit.bouadi,alexandre.termier}@irisa.fr
[3] Univ Clermont Auvergne, 63000 Clermont-Ferrand, France
lhouari.nourine@uca.fr

Abstract. Skyline queries are multicriteria queries that are of great interest for decision applications. Skyline Groups extend the idea of skyline to groups of objects. In the recent years, several algorithms have been proposed to extract, in an efficient way, the complete set of skyline groups. Due to the novelty of the skyline group concept, these algorithms use custom enumeration strategies. The first contribution of this paper is the observation that a skyline group corresponds to the notion of *ideal* of a partially ordered set. From this observation, our second contribution consists in proposing a novel and efficient algorithm for the enumeration of all ideals of a given size k (*i.e.* all skyline groups of size k) of a poset. This algorithm, called GENIDEALS, has a time delay complexity of $O(w^2)$, where w is the width of the poset, which improves the best known time output complexity for this problem: $O(n^3)$ where n is the number of elements in the poset. This work present new theoretical results and applications on skyline queries.

Keywords: Skyline queries · Ideal enumeration · Time delay complexity

1 Introduction

In decision making, one often wants to optimize simultaneously several characteristics. For example, consider a soccer coach who wants to recruit, into her team, a player who has both a low miss rate (corresponding to a high accuracy) and doesn't take long to cross the field. These characteristics (*i.e.* dimensions) are often multiple and conflicting: there is rarely a single solution optimizing all the characteristics at the same time. *Skyline queries* [3] solve this problem by considering the "best compromises" between the different dimensions. More formally, in a multidimensional space where the dimension domains are ordered

© Springer Nature Switzerland AG 2021
P. Flocchini and L. Moura (Eds.): IWOCA 2021, LNCS 12757, pp. 223–236, 2021.
https://doi.org/10.1007/978-3-030-79987-8_16

(totally or partially), *skyline queries* return the objects that are not dominated by any other object. An object dominates another object, *if it is as good or better in all dimensions and strictly better in at least one dimension.* This notion of dominance is also called *Pareto dominance.*

In the soccer example, results of skyline queries may be: (i) the best goal-scorer (whatever her running speed), (ii) the best runner (whatever her miss rate), or (iii) a player being average on both criteria, with no other player both missing less and taking less time to cross the field.

An interesting and challenging problem arises when, in some applications, users are interested in capturing skyline *groups* instead of points. This is for example the case when looking for the best soccer *team* and not the best soccer player. The Pareto dominance concept of one object over another is not directly applicable to the notion of groups of objects. For example, it is obvious that the best soccer team may not correspond to the group of the best individual players, nor necessarily to the group of the most average players. Recent works [6, 8,15,18] have considered the issue of *skyline group computation* by extending the dominance relation between points to groups of points: this is the *group dominance relation*, also called *g-dominance*, which allows comparison between groups of the same size. The *g-dominance* relation allows comparison between groups of the same size, and is defined by [8] as: "*Given two different groups G and G' with k points, we say that G g-dominates G', if we can find a permutation of the k points for G and G', such that either p_i dominates p'_i or $p_i=p'_i$ for all $1 \leq i \leq k$, and for at least one i, p_i dominates p'_i*".

Enumerating skyline groups defined by such group dominance relation is challenging given the huge size of the search space considered: in a set of n points, there are $\binom{n}{k}$ possible group skylines of size k. This prompted the need to design efficient enumeration algorithms dedicated to the discovery of group skylines. In the papers proposing g-dominance [8,9], the authors proposed to exploit a novel structure based on skyline layers (*i.e.* models the dominance links between the points), and two efficient search strategies to enumerate the skyline groups. Other works studied the g-dominance relation and proposed algorithmic improvements [7,15,17,19]. The state of the art approach is G-MDS [15], which is based on a structure called *minimum dominance graph (MDG)*, a directed acyclic graph representing the dominance relation among relevant points.

In this work, our objective is to get a finer understanding of the space of skyline groups as defined by the g-dominance relation, in order to propose an algorithm exploring only the space of solutions. By going back to enumeration theory concepts, we could show that the skyline group notion corresponds to the well known concept of *ideal*. This allows us to propose an elegant and efficient algorithm to compute skyline groups, that does not visit any unnecessary group. Furthermore, the algorithm that we propose to enumerate ideals of size k improves the state of the art for ideal enumeration, having a time delay complexity of $O(w^2)$ (with w the posets' width), where the best know time output complexity for this problem is $O(n^3)$ per ideal [16] (with n the number of elements in the poset).

We first recall that an ideal (or downset) of a poset is a subset that is closed by the order relation $\leq$, *i.e.* an ideal containing an element x, contains all elements inferior or equal to x. The notion of ideals of partially ordered sets has been found many times in several applications as (scheduling [13], verification of distributed systems [2]). Listing or enumerating all ideals of a poset simply means outputing them one after the other. Many algorithms have been proposed in the literature [1,4,10,12,13]. But all these algorithms cannot be used for our purposes, since they list all ideals. In addition their adaptation does not allow to have a good complexity.

In this work, we show that given the classical dominance relation for points of the data, group skylines of size k correspond exactly to ideals of size k for this relation (Proposition 1). We are then interested in the enumeration of ideals of a given size k. To the best of our knowledge the only works for enumerating ideals of a given size have been considered in [5] for particular cases of posets, and Wild's algorithm in [16] for the general case. We identify a directed graph whose vertices are ideals of size k and such that there is an edge between two ideals if there is a small transformation to obtain one from the others. We then propose an efficient algorithm to enumerate all ideals of a given size k using polynomial space.

The rest of the paper is organized as follows. Section 2 introduces the basic concepts related to skyline queries and the concept of skyline group. We present and detail, in Sect. 3, the established correspondence between key concepts from the field of skyline queries and the field of partial order theory. In Sect. 4, we develop the formal aspects, highlight new and useful properties, and present a way to organize the space of ideals (*i.e.* skyline groups) as a Depth First Search (DFS) tree. Our algorithm GENIDEALS for enumerating k-ideals based on this tree-shaped structuration of the space is then discussed in Sect. 5. In this same section, we present the detailed complexity analysis of GENIDEALS, showing its $O(w^2)$ complexity for k-ideals enumeration. Section 6 concludes the paper.

2   Preliminaries

In this section, we introduce the concept of skylines and extend it to the concept of group skylines. We then review a few useful results from the literature.

The various definitions are illustrated using the example of Table 1 which describes proposals for hotels according to the dimensions Price, Distance from the beach, and Distance from transportation. Without any loss of generality, we assume that it is always better if the values of the attributes are low.

Definition 1 (Pointset). *A pointset is a set of same size tuples of real numbers. We usually write the pointset using the letter D and assume said numbers to be positives. The elements of the pointset are called points, the elements of the points are its attributes.*

Example 1. In Table 1, $D = \{(10,1,4),(10,2,4),(10,4,1),(20,1,4),\ (40,1,1),\ (40,5,1),(50,4,1)\}$ is a pointset defined in a 3-dimensional space $F = $ (Price, Distance, Transportation). The domain of each attribute is totally ordered.

Table 1. A set of hotels

Hotel ID	Price	Distance	Transportation
a	10	1	4
b	10	4	4
c	10	1	5
d	20	2	1
e	40	1	2
f	40	3	2
g	50	2	3

Definition 2 (Point domination). *A point p dominates a different point p', denoted by $p \prec p'$, if p is lower or equal to p' on any attribute and p is strictly lower than p' on at least one attribute. We write $p \preceq p'$ to say "either p dominates p' or $p = p'$".*

Example 2. In Table 1, we have $(20, 2, 1) \preceq (50, 2, 3)$, since the price value 20 is lower than 50 and the other attribute values are equal. We also note that $(10, 1, 4) \not\preceq (20, 2, 1)$, since the transportation value 4 is higher than 1.

Definition 3 (Skyline). *The Skyline of a pointset D is the set of all points in D that are not dominated by any other point.*

2.1 Skyline Groups

In the hotel example (Table 1), one may consider the case of a travel agency, that wants to pre-book exactly k rooms (supposed in k different hotels for sake of simplicity). Each room has the same price and distances characteristics as before: among the many rooms available ($k << |D|$), the travel agency wants to identify the *best groups*, able to satisfy the needs of many potential customers.

The definition of Group domination and skyline group used in this paper was first introduced in [8]. It is the notion that, in our opinion, is the best adaptation of the idea of skylines to groups of points.

Definition 4 (Group domination). *A group (set) G containing k points dominates another group G' of size k, denoted by $G \prec_g G'$, if and only if there is a bijection f from G to G' such that: $\forall p \in G, p \preceq f(p) \land \exists p \in G$ s.t. $p \prec f(p)$.*

Example 3. For the pointset in Table 1, we have $\{a, e, g\} \prec_g \{b, f, g\}$, because $a \prec b$, $e \prec f$, and $g \preceq g$.

Definition 5 (Skyline group). *We say that a group G of size k is a skyline group if and only if it is dominated by no other group.*

Example 4. In Table 1, $\{a, c, e\}$ is a skyline group of size 3. Since the three points are skyline points, it is easy to verify that this group cannot be dominated by any other group of size 3.

Definition 6 (g-skyline). *The g-skyline of size k of a pointset D, denoted S_k, is the set of all groups of k elements of D that are not dominated by any other group, i.e. the set of all the skyline groups of size k of D.*

Example 5. In Table 1, $S_4 = \{\{a, b, c, d\}, \{a, b, c, e\}, \{a, b, d, e\}, \{a, c, d, e\}, \{a, c, e, g\}, \{a, c, e, f\}, \{c, e, f, g\}\}$

We also recall two useful notations for discussing partially ordered sets (*i.e.* posets) which we will need later. Those are the notions of *ideal* and that of the *width of a poset*.

Definition 7 (Ideal). *Given $(E, \leq)$ a partially ordered set, a subset I of E is an ideal if and only if: $\forall (x, y) \in E \times I$, if $x \leq y$, then $x \in I$.*
In other words, I is closed by the order relation $\leq$.

Definition 8 (Poset width). *Given $(E, \leq)$ a poset, the width of that poset is the size of the maximal subset I of E such that: $\forall (x, y) \in I^2$, $x \not< y$. I is called a maximal antichain of $(E, \leq)$.*

3 Skyline Groups Are Ideals

In this section, we present our first contribution: a new correspondence between concepts from the field of skyline queries and concepts from the field of partial order theory. Namely, we see that a skyline group is an ideal. This correspondence is in our eyes the most important in this paper. In [15], the authors introduced the concept of *Unit group* as follow:

Definition 9 (Unit group). *The unit group of point p is the set of all points that dominate it plus p. It is written $u(p)$. Formally : $u(p) = \{p' \in D \mid p' \preceq p\}$.*

The concept of *unit group* is equivalent to the concept of *principal ideal* in a partially ordered set (poset). For example, in Table 1, $u(b) = \{a, b\}$.

Proposition 1. *Consider a group G of size k. The following properties are equivalent: (1) G is a skyline group, (2) $\cup_{p \in G} u(p) = G$, (3) $\forall p \in G$, $u(p) \subseteq G$, (4) and G is an Ideal of $(D, \preceq)$.*

Because of the third property of Proposition 1, we know that a point with a unit group of size strictly superior to k cannot belong to a skyline group. Because domination is transitive, unit group size increases with domination (*i.e.*, $p \preceq p' \implies u(p) \subseteq u(p')$). Therefore, when searching for skyline groups, all points with an unit group of size strictly superior to k can be entirely removed without affecting in any way the information at hand on the other points.

The fourth property has important implications: it means that any algorithm suitable for the enumeration of ideals of size k of a poset is also suitable to be used for the enumeration of skyline groups of size k of a pointset, using $\preceq$ as an order relation.

In the rest of the paper, we draw on Proposition 1 and consider skyline groups on a given pointset D to be defined as *the ideals of the poset* $(D, \preceq)$. Thus, our problem becomes the enumeration of all ideals of a given size k of a poset.

We consider the particular case of skyline queries for our explanations and will make references to the canonical lexicographic order on points. However, all that follows is compatible with any poset with a topological order. Because calculating a topological order for a poset can be done in $O(n^2)$ with n the size of the poset, everything bellow this point and especially the algorithm we present can be adapted to list the ideals of size k of any given poset.

In the next section we will describe new properties to explore the graph of skyline groups. These properties allowed us to propose a simple and efficient k-ideal (*i.e.* skyline groups of size k) enumeration algorithm.

4   The Tree of Skyline Groups

Given a pointset D of dimension d and an integer k, we want to enumerate the elements of the set S_k of its skyline groups of size k, *i.e.* its g-skyline of parameter k. We identify a rooted covering tree $T = (S_k, \mathcal{P})$ which leads us to an efficient algorithm to search this tree in polynomial time and space. But first, we must introduce a few notions. We begin by some order relations on points and groups.

We define the usual lexicographical order relation $\leq$ on points of D as follows:

Definition 10. $p = (x_1, x_2, ..., x_d) \leq q = (y_1, y_2, ..., y_d)$ *if and only if $p = q$ or there is $i \in \{1, 2, ..., d\}$ such that for all $1 \leq j < i$, $x_j = y_j$ and $x_i < y_i$.*

It has the following interesting property.

Property 1. $\leq$ is a topological ordering for the dominance relation.

Because the notion of skyline group only depends on the dominance relation between points and not on their coordinates, we can rename points according to the lexicographic order and only remember which points dominate which ones.

Example 6. In Table 1, we have the following points order (according to $\leq$): $p_1 = (10, 1, 4)$, $p_2 = (10, 4, 4)$, $p_3 = (10, 1, 5)$, $p_4 = (20, 2, 1)$, $p_5 = (40, 1, 2)$, $p_6 = (40, 3, 2)$, $p_7 = (50, 2, 3)$

Fig. 1 shows the point domination relation between the points defined above. Such graphical representation helps to quickly identify skyline groups. Then, we can easily notice that (p_1, p_2, p_3, p_6) is not a skyline group because p_4 is not in it and $p_4 \preceq p_6$. Also, (p_1, p_2, p_3, p_4) is a skyline group since no point could be replaced by another that dominates it.

Since all elements of S_k have the same size, a group $G \in S_k$ is represented by a tuple $(p_1, p_2, ..., p_k)$ such that $p_1 \leq p_2 \leq ... \leq p_k$. In the same way as for D, we define a lexicographical ordering $\trianglelefteq$ on the set S_k.

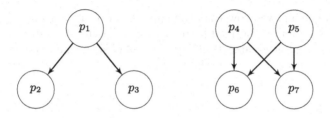

Fig. 1. Domination graph between points of Table 1

Example 7. In Table 1, we have the following skyline groups of size 4 (ordered according to $\leq$): $G_0 = (p_1, p_2, p_3, p_4)$, $G_1 = (p_1, p_2, p_3, p_5)$, $G_2 = (p_1, p_2, p_4, p_5)$, $G_3 = (p_1, p_3, p_4, p_5)$, $G_4 = (p_1, p_4, p_5, p_6)$, $G_5 = (p_1, p_4, p_5, p_7)$, $G_6 = (p_4, p_5, p_6, p_7)$

In the following examples, we fix the size of skyline groups to 4.

We denote by G_0 the smallest group according to the lexicographical ordering $\trianglelefteq$. We then define the parent relation.

Definition 11 (Parent). *Let G be a group in S_k with $G \neq G_0$. The parent of G, denoted by $Parent(G)$, is obtained from G by deleting the largest element b in G (w.r.t $\leq$) and adding the smallest element $a \in D \backslash G$.*

Proposition 2. *The parent of a skyline group is a skyline group of the same size.*

We say that G' is a *child* of G if $Parent(G') = G$, and denote by $Children(G)$ the set of all the children of G. Let $\mathcal{T} = (S_k, Parent)$ be the directed graph whose vertices are elements of S_k and edges correspond to the parent relation.

Proposition 3. *The directed graph $\mathcal{T} = (S_k, Parent)$ is a tree rooted at G_0*

Proof. Let $G \neq G_0$ be a group in S_k. We show that there is a unique path $G_0 \leq G_1 \leq G_2 \leq ... \leq G_m = G$ such that $Parent(G_i) = G_{i-1}$ for $0 < i \leq m$. Let $G' = Parent(G)$ with $G' = (G \backslash \{b\}) \cup \{a\}$. Since $G \neq G_0$, the smallest point not in G is smaller than the highest point in G. Hence, $a < b$ and $G' < G$. So for every group different from G_0, its parent is smaller for $\leq$ than it. By repetitive application of the relation parent we inevitably reach G_0 since S_k is finite. □

Our algorithm will simply run a tree exploration. However, we need to be able to enumerate the children of a given group.

We assume that the points in D are numbered according to the lexicographic order $\leq$. We have $p_1 \leq p_2 \leq ... \leq p_n$ such that $G_0 = \{p_1, p_2, ..., p_k\}$.

Definition 12 (Starting prefix). *Let G be a group in S_k. The starting prefix of G, denoted by $SPrefix(G)$ is the longest common prefix of G with G_0.*

Example 8. Starting prefixes are shown below in Fig. 2 in bold.

In the following we show that the children of a given group G are exactly the groups that can be obtained by removing one point of its starting prefix and adding another point greater than any point of G such that the new group is an ideal.

Proposition 4. *Let G_1 and G_2 be two skyline groups. G_2 is a child of G_1 iff G_2 is obtained by removing one element of the starting prefix of G_1 from G_1 and adding another element higher than any element of G_1.*

Proof. Let G_1 and G_2 be two groups in S_k.
Suppose that $G_2 = (G_1 \backslash \{a\}) \cup \{b\}$ where a is a point of the starting prefix of G_1 and $b \notin G_1$ is greater than every elements of G_1. We show that $G_1 = Parent(G_2)$. Clearly b is the largest element in G_2. Moreover by definition of the starting prefix, a is the smallest point (for $\leq$) that is not in $G_1 \backslash \{a\} = G_2 \backslash \{b\}$. Hence a is the smallest element that can be added to $G_2 \backslash \{b\}$, and thus $G_1 = Parent(G_2)$. Now, suppose that G_2 is child of G_1, i.e. $G_1 = Parent(G_2)$. By definition of the relation Parent, G_1 is obtained from G_2 by deleting the largest$_<$ element b in G_2 and adding the smallest$_<$ element a not in G_2. We show that a belongs to the starting prefix of G_1. For contradiction, suppose there is a' not in G_1, $a' < a$ and a' belongs to G_0. Then $a' \notin G_2 \backslash \{b\}$ and $b \nleq a'$. Thus a' can be added to $G_2 \backslash \{b\}$ which contradict that a is the smallest element that can be added. □

Moreover the depth of a group G in the tree $\mathcal{T} = (S_k, Parent)$ is equal to the size of $| G_0 \backslash G |$. So the depth of a leaf in the tree is bounded by k. Figure 2 shows the tree $\mathcal{T}$ for our example set of points.

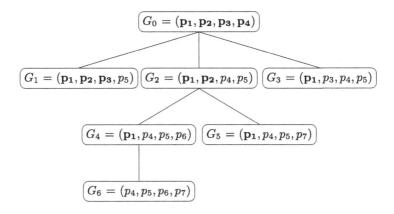

Fig. 2. Example tree of skyline groups

5   Algorithm

5.1   First Look at Our Algorithm

Our algorithm will be a DFS (Depth First Search) on the tree $\mathcal{T}$. First we show how to generate all the children of a given group G. Recall that a child is obtained by exchanging two elements (deleting one and adding another one).

Let $p_1, p_2, ..., p_n$ be a topological order of $(D, \leq)$ and G a group in S_k. We use $max(G)$ to denote the largest point in G w.r.t $\leq$. We denote by $Pred(p)$ the points that dominate a given point p and by $Succ(p)$ the points it dominates.

Let $SPrefix(G)$ be the set of elements in the starting prefix in G and $SPrefixToDel(G)$ the set of points in $SPrefix(G)$ that are maximal w.r.t. $\leq$ in G, i.e all points $a \in G$ such that $G \backslash \{a\}$ is still a skyline group (of size $k - 1$). No other elements of G can be removed to create a child of G.

We also consider the set of all potential candidates that could be added to G after deleting one element from $SPrefixToDel(G)$. Let $CandToAdd(G)$ be the set of points greater than $max(G)$ that are minimal in $D \backslash G$ w.r.t $\preceq$. The set of candidates when deleting an element $a \in SPrefixToDel(G)$ from G is obtained from $CandToAdd(G)$ by removing from it the successors of a in D, i.e. $CandToAdd(G, a) = CandToAdd(G) \backslash Succ(a)$.

Theorem 1. *Let G be a group in S_k. Then $Children(G) = \{(G \backslash \{a\}) \cup \{b\} \mid a \in SPrefixToDel(G), b \in CandToAdd(G, a)\}$.*

Proof. Let G be a group in S_k and let $G' \in \{(G \backslash \{a\}) \cup \{b\} \mid a \in SPrefix ToDel(G), b \in CandToAdd(G, a)\}$. We show that $Parent(G') = G$. We fix $G' = (G \backslash \{a\}) \cup \{b\}$ with $a \in SPrefixToDel(G)$ and $b \in CandToAdd(G, a)$. Since a belongs to the starting prefix of G then for all $a' \leq a$ we have a' in G. Thus a is the smallest point not in G'. Furthermore, by definition of $CandToAdd(G, a)$ we have $b \geq max(G)$ which implies that b is the largest point in G'. Hence G is the parent of G'. Conversely let G and G' be two groups $(G' \neq G_0)$ such that $G = Parent(G')$, we show that $G' \in \{(G \backslash \{a\}) \cup \{b\} \mid a \in SPrefixToDel(G), b \in CandToAdd(G, a)\}$. Again, we fix $G = (G' \backslash \{b\}) \cup \{a\}$ with b the largest point in G' and a the smallest one not in G'. We show that $a \in SPrefixToDel(G)$. a is the smallest point not in G' and therefore a belongs to the starting prefix of G. Since $G \cap G'$ is also an ideal, we conclude that a is maximal in G for $\preceq$ and thus $a \in SPrefixToDel(G)$. Since b is maximal in G' it is higher than any point in G. Because $b \notin G$ and because it can be added to G, we have $b \in min(D \backslash G)$. Hence $b \in CandToAdd(G)$. Moreover because b can be added to $G \backslash \{a\}$, we know $a \npreceq b$ (i.e. $b \notin Succ(a)$) and thus $b \in CandToAdd(G, a)$. □

Remark 1. It is worth noticing that if $G' \in Children(G)$ with $G' = (G \backslash \{a\}) \cup \{b\}$, then a cannot be added and b cannot be deleted from any group in the subtree rooted at G'. In other words, in any path of the execution tree, any point can be deleted or added at most once. This is because all deleted elements belong to the starting prefix (i.e. elements ranging from p_1 to p_k), and all added ones are greater than p_k in the topological ordering $\leq$.

We now describe an algorithm called GENIDEALS (Algorithm 2) that takes the starting prefix of a group G in S_k, the lists $SPrefixToDel(G)$ and $CandToAdd(G)$ and outputs all groups in the subtree of $\mathcal{T}$ rooted at G.
The first call is GENIDEALS($SPrefix(G_0), SPrefixToDel(G_0), CandToAdd(G_0), G_0$) which lists all groups in S_k.

5.2   Detailed Explanation of GENIDEALS: Data structures and algorithms

The GENIDEALS algorithm uses the following global data structures:

- $Succ[1..n]$ an array where $Succ[i]$ is a list of successors of p_i in $(D, \leq)$.
- $Pred[1..n]$ an array where $Pred[i]$ is a list of predecessors of p_i in $(D, \leq)$.
- $Tlook[1..n]$ an array where $Tlook[i] = 1$ if the point p_i is present and $Tlook[i] = 0$ otherwise.
- $PredCount[1..n]$ an array where $PredCount[i]$ is the number of predecessors of p_i not in the current group G. This allows us to check in $O(1)$ if a point p_i is ready to be added, i.e. minimal in $(D \backslash G)$.
- $SuccCount[1..n]$ an array where $SuccCount[i]$ is the number of successors of p_i in the current group. This counter let us check in $O(1)$ if a point p_i is maximal in the current group.

Remark 2. The sizes of $Succ[i], Pred[i], SPrefixToDel(G), CandToAdd(G)$ and $Children(G)$ are bounded in w the width of $(D, \leq)$.

The algorithm GENIDEALS_MAIN is split into two phases. We first call the *initialization* process to compute G_0 and then lists $SPrefixToDel(G_0)$ and $CandToAdd(G_0)$. Then we start the recursive process GENIDEALS that corresponds to the core of the algorithm.

We describe the function INITIALISATION (Algorithm 1) briefly here. The first **for** loop initializes the arrays $Tlook$ and $PredCount[1..n]$ and adds those points that are minimal in $(D, \leq)$ and greater than p_k to the list Min. This loop can be achieved in $O(n + m)$ time complexity where n is the number of points and m the size of the lists $Pred$.

The second **for** loop computes G_0 and updates the counter $PredCount$ and the list Min. At the end of this loop, Min contains the minimal points in $D \backslash G_0$ and $SuccCount[i]$ the number of successor points of P_i in G_0. The time spent by this loop is bounded by $O(kw)$.

The third **for** loop computes the maximal points in G_0 in $O(k)$. Thus the total complexity of the algorithm initialization is bounded by $O(n + m + kw)$ which is less than $O(n^2)$. Please mind that initialization isn't purely described by its output, it also initializes global variables in the form of the arrays $PredCount$ and $SuccCount$.

We now describe in details the recursive algorithm GENIDEALS (Algorithm 2). It takes a group G in S_k, the lists $SPrefixToDel(G)$ and $CandToAdd(G)$ and outputs all groups in the subtree of $\mathcal{T}$ rooted in G.

Algorithm 1. INITIALIZATION

```
 1: Min = ∅
 2: for i = 1 to n do
 3:     Tlook[i] = 0
 4:     PredCount[i] = sizeof(Pred[i])
 5:     SuccCount[i] = 0
 6:     if PredCount[i] = 0 and i > k then
 7:         Add i to Min
 8: { Now we initialize G₀}
 9: G₀ = ∅
10: for i = 1 to k do
11:     Add pᵢ to G₀
12:     for v ∈ Succ[i] do
13:         PredCount[v] = PredCount[v] − 1
14:         if v ≤ k then
15:             SuccCount[i] = SuccCount[i] + 1
16:         if PredCount[v] = 0 and v > k then
17:             Add v to Min
18: CandToAdd = Min
19: SPrefixToDel = ∅
20: for i = 1 to k do
21:     if SuccCount[i] = 0 then
22:         Add i to SPrefixToDel
23: return (SPrefixToDel, CandToAdd, G₀)
```

The outer loop **while** considers all points in the starting prefix of G that might be removed to create children of G. For each such $a \in SPrefixToDel(G)$, we delete a and compute the list $L = CandToAdd(G, a)$. This is done using the algorithm UPDATECANDTOADD1 (Algorithm 3). First we insert points in $Succ(a)$ into $Tlook$, and for each $v \in CandToAdd(G)$: we check if $Tlook[v] = 0$, then we add a to the output, else we delete it from $Tlook$ in order to keep the array $Tlook$ empty when entering and exiting the algorithm UPDATECAND-TOADD1. The time complexity of algorithm UPDATECANDTOADD1 is bounded by $O(w)$. Note that the list L may be empty for all $a \in SPrefixToDel(G)$. So the worst case is when G is a leaf and in this case the total cost is bounded by $O(w^2)$.

Then, the inner loop **while** takes any point b in the list L, calls the function *Print* and prepare the parameters for the new group $G' = (G\backslash\{a\})\cup\{b\}$. For the former step, we use two update functions UPDATECANDTOADD2 (Algorithm 4) and UPDATESPREFIXTODEL$(a, b, SPrefixtoDel)$ (Algorithm 5). The first function UPDATECANDTOADD2 adds the new candidates to be added when deleting the point b. The second one computes the starting prefix of G' and $SPrefixtoDel$, i.e. maximal elements in G' that can be deleted from G'.

The time complexity needed by the algorithms UPDATECANDTOADD2 and UPDATESPREFIXTODEL is bounded by $O(w)$.

Algorithm 2. GENIDEALS(SPrefixToDel, CandToAdd, G)

1: **while** $SPrefixToDel$ is not empty **do**
2:     $a =$ Delete and return an element in $SPrefixToDel$
3:     $L =$ UPDATECANDTOADD1$(a, CandToAdd)$
4:     **while** L is not empty **do**
5:         $b =$ Delete and return an element in L
6:         $G' = (G \setminus \{a\}) \cup \{b\}$
7:         Print (G')
8:         $SprefD =$ UPDATESPREFIXTODEL$(a, b, SPrefixetoDel)$
9:         $Cand =$ UPDATECANDTOADD2(b, L)
10:        GENIDEALS$(SprefD, Cand, G')$
11:        Cancel changes to $PredCount$ and $SuccCount$ done to add b
12:    Cancel changes to $PredCount$ and $SuccCount$ done to remove a

Algorithm 3. UPDATECANDTOADD1$(a, CandToAdd)$

1: $L = \emptyset$
2: **for** $v \in Succ[a]$ **do**
3:     $Tlook[v] = 1$
4: **for** $v \in CandToAdd$ **do**
5:     **if** $Tlook[v] = 0$ **then**
6:         Add v to L
7:     **else**
8:         $Tlook[v] = 0$
9: **return** L

Algorithm 4. UPDATECANDTOADD2$(b, CandToAdd)$

1: **for** $v \in Succ[b]$ **do**
2:     $PredCount[v] = PredCount[v] - 1$
3:     **if** $PredCount[v] = 0$ and $v > b$ **then**
4:         Add v to $CandToAdd$
5: **return** $CandToAdd$

Algorithm 5. UPDATESPREFIXTODEL $(a, b, SPrefixetoDel)$

1: **for** $v \in Pred[a]$ **do**
2:     $SuccCount[v] = SuccCount[v] - 1$
3:     **if** $SuccCount[v] = 0$ **then**
4:         Add v to $SPrefixetoDel$
5: **for** $v \in Pred[b]$ **do**
6:     **if** $v \leq k$ **then**
7:         $SuccCount[v] = SuccCount[v] + 1$
8:         Delete v from $SPrefixetoDel$
9: **return** $SPrefixetoDel$

Finally, the main algorithm GENIDEALS_MAIN is presented in Algorithm 6. It simply consists of a call to INITIALIZATION followed by the first call to the core function GENIDEALS.

Algorithm 6. GENIDEALS_MAIN

1: $(SPrefixToDel, CandToAdd, G_0)$ =INITIALIZATION()
2: GENIDEALS($SPrefixToDel, CandToAdd, G_0$)

Theorem 2. *The algorithm* GENIDEALS_MAIN *lists all groups of size k in $O(w^2)$ delay and polynomial space.*

Proof. The correctness of the algorithm GENIDEALS_MAIN comes from Theorem 1. The complexity of each call to GENIDEALS is dominated by the running time of the algorithm UPDATECANDTOADD1 which is bounded by $O(w^2)$ as discussed before. Thus the algorithm takes $O(w^2)$ time per ideal, but the delay between two outputs can be greater. This is the case when we output a group in depth k and the next outputted one is in depth 1, so the delay between them is $O(kw^2)$ since the depth is $O(k)$. To show $O(w^2)$ delay we use the idea in [11,14]. Indeed, the algorithm GENIDEALS_MAIN is internal output, that is it outputs an ideal at each node of the tree rather than outputing only for leaves. So we alternatively output depending on the parity of the depth of a node. As suggested in [14], we do the following two changes in algorithm GENIDEALS:

– We change the Line 7 by: *If the depth of the call is odd then output G'*
– We add after line 10 the instruction: *If the depth of the call is even then output G'.*

The space used by the algorithm is linear for each node of the searching tree. Moreover the depth of the tree is bounded by k.   □

6   Conclusion

This work is the first to show that skyline groups are objects set by partial order theory, called *ideals*. This allowed us to bring out interesting properties to ease the exploration of the graph of k-ideals (*i.e.* skyline groups of size k). Indeed, we presented a novel way to organize the space of skyline groups as a DFS tree. This helped us to propose a simple and efficient k-ideal enumeration algorithm: GENIDEALS_MAIN. The time delay complexity analysis that was performed highlights the relevance of our approach and shows that it outperforms the current state-of-the-art algorithm. Moreover, there are two directions to improve the result of this work. First one can improve the complexity of the algorithm UPDATECANDTOADD1 to $O(1)$ when its output is empty. Second, to improve the space complexity, one may find a lexicographic order on the groups to avoid the re-enumeration in the reverse search technique, but the time complexity may increase.

Acknowledgments. The third author is supported by the French government IDEX ISITE initiative 16-IDEX-0001 (CAP 20–25).

References

1. Abdo, M.: Efficient generation of the ideals of a poset in gray code order, part ii. Theor. Comput. Sci. **502**, 30–45 (2013), generation of Combinatorial Structures
2. Chang, Y., Garg, V.K.: Quicklex: a fast algorithm for consistent global states enumeration of distributed computations. In: Anceaume, E., Cachin, C., Potop-Butucaru, M.G. (eds.) 19th International Conference on Principles of Distributed Systems, OPODIS 2015, December 14–17, 2015, Rennes, France. LIPIcs, vol. 46, pp. 25:1–25:17. Schloss Dagstuhl - Leibniz-Zentrum für Informatik (2015)
3. Chomicki, J., Ciaccia, P., Meneghetti, N.: Skyline queries, front and back. ACM SIGMOD Rec. **42**(3), 6–18 (2013)
4. Habib, M., Medina, R., Nourine, L., Steiner, G.: Efficient algorithms on distributive lattices. Discrete Appl. Math. **110**(2), 169–187 (2001)
5. Habib, M., Nourine, L., Steiner, G.: Gray codes for the ideals of interval orders. J. Algorithms **25**, 52–66 (1997)
6. Im, H., Park, S.: Group skyline computation. Inf. Sci. **188**, 151–169 (2012)
7. Yang, Z., Xiao, G., Li, K., Li, K., et al.: Progressive approaches for pareto optimal groups computation. IEEE Trans. Knowl. Data Eng. **31**(3), 521–534 (2018)
8. Liu, J., Xiong, L., Pei, J., Luo, J., Zhang, H.: Finding pareto optimal groups: group-based skyline. Proc. VLDB Endowment **8**(13), 2086–2097 (2015)
9. Liu, J., Xiong, L., Pei, J., Luo, J., Zhang, H., Yu, W.: Group-based skyline for pareto optimal groups. IEEE Trans. Knowl. Data Eng. (2019)
10. Medina, R., Nourine, L.: Algorithme efficace de génération des ideaux d'un ensemble ordonné. C.R. Acad. Sci. Paris Sér. I Math. **319**, 1115–1120 (1994)
11. Nakano, S., Uno, T.: Constant time generation of trees with specified diameter. In: Hromkovič, J., Nagl, M., Westfechtel, B. (eds.) WG 2004. LNCS, vol. 3353, pp. 33–45. Springer, Heidelberg (2004). https://doi.org/10.1007/978-3-540-30559-0_3
12. Squire, M.: Enumerating the ideals of a poset. Preprint available electronically at (1995). http://citeseer.ist.psu.edu/465417.html
13. Steiner, G.: An algorithm for generating the ideals of a partial order. Oper. Res. Lett. **5**, 317–320 (1986)
14. Uno, T.: Two general methods to reduce delay and change of enumeration algorithms. NII Technical report (2003)
15. Wang, C., Wang, C., Guo, G., Ye, X., Philip, S.Y.: Efficient computation of g-skyline groups. IEEE Trans. Knowl. Data Eng. **30**(4), 674–688 (2017)
16. Wild, M.: Output-polynomial enumeration of all fixed-cardinality ideals of a poset, respectively all fixed-cardinality subtrees of a tree. Order **31**(1), 121–135 (2014)
17. Yang, Z., Zhou, X., Li, K., Xiao, G., Gao, Y., Li, K.: Efficient processing of top k group skyline queries. Knowl.-Based Syst. **182**, 104795 (2019)
18. Zhang, N., Li, C., Hassan, N., Rajasekaran, S., Das, G.: On skyline groups. IEEE Trans. Knowl. Data Eng. **26**(4), 942–956 (2013)
19. Zhou, X., Li, K., Yang, Z., Gao, Y., Li, K.: Efficient approaches to k representative g-skyline queries. ACM Trans. Knowl. Discov. Data (TKDD) **14**(5), 1–27 (2020)

Vertex Cover at Distance
on H-Free Graphs

Clément Dallard[1,2]($\boxtimes$) , Mirza Krbezlija[1] , and Martin Milanič[1,2]

[1] FAMNIT, University of Primorska, Koper, Slovenia
clement.dallard@famnit.upr.si, mirza.krbezlija.research@gmail.com,
martin.milanic@upr.si
[2] IAM, University of Primorska, Koper, Slovenia

Abstract. The question of characterizing graphs H such that the VER-
TEX COVER problem is solvable in polynomial time in the class of H-free
graphs is notoriously difficult and still widely open. We completely solve
the corresponding question for a distance-based generalization of vertex
cover called distance-k vertex cover, for any positive integer k. In this
problem the task is to determine, given a graph G and an integer ℓ,
whether G contains a set of at most ℓ vertices such that each edge of G
is at distance at most k from a vertex in the set. We show that for all
$k \geq 1$ and all graphs H, the distance-k vertex cover problem is solvable in
polynomial time in the class of H-free graphs if H is an induced subgraph
of $P_{2k+2} + sP_{\max\{k,2\}}$ for some $s \geq 0$, and NP-complete otherwise.

Keywords: Distance-k Vertex Cover · H-free graph ·
NP-completeness · Polynomial-time algorithm · Dichotomy

1 Introduction

Various theoretical and practical motivations have led to generalizations of many
classical graph optimization problems to their distance-based variants. Informally, this means that the adjacency property used to defined a feasible solution
to the problem is replaced with a relaxed property based on distances in the
graph.

For a concrete example, consider the VERTEX COVER problem. A *vertex cover*
in a graph G is a set of vertices intersecting all edges. For a non-negative integer
k, a *distance-k vertex cover* in a graph G is a set C of vertices such that every
edge has an endpoint which is at distance at most k from a vertex in C. Note that
a distance-0 vertex cover is the same thing as a vertex cover. The DISTANCE-k
VERTEX COVER problem is the problem of deciding, given a graph G and an
integer ℓ, whether G contains a distance-k vertex of size at most ℓ. Motivated by
an application in network monitoring, where links between hosts (edges in the

This work is supported in part by the Slovenian Research Agency (I0-0035, research
program P1-0285 and research projects J1-9110, N1-0102, N1-0160).

P. Flocchini and L. Moura (Eds.): IWOCA 2021, LNCS 12757, pp. 237–251, 2021.
https://doi.org/10.1007/978-3-030-79987-8_17

network graph) can be monitored from hosts (vertices) at distance at most k, the problem was first studied in 2008, under the name BEACON PLACEMENT PROBLEM, by Sasaki et al. [33]. They showed that the problem is NP-complete for all $k \geq 2$ and provided a greedy heuristic for the corresponding minimization problem. NP-completeness for $k = 1$ was established in 2003 by Horton and López-Ortiz [18]. Our terminology, DISTANCE-k VERTEX COVER, follows the work of Busch et al. [8] from 2010, where it was proved that for the class of dually chordal graphs, the problem is NP-complete for $k = 0$ but solvable in polynomial time for any fixed integer $k \geq 1$.

The case $k = 0$ corresponds to the VERTEX COVER problem, which is equivalent to the INDEPENDENT SET problem and has been the subject of great interest in algorithmic graph theory. One of the key questions in this area, investigated for decades, is whether for every positive integer p, the VERTEX COVER problem can be solved in polynomial time in the class of graphs not containing a p-vertex path as an induced subgraph. While the solution for $p = 4$ has been known at least since 1981 [12], the cases of $p = 5$ and $p = 6$ were settled only in 2014 and in 2019, respectively, by Lokshantov et al. [24] and Grzesik et al. [17]. The question is currently still open for all $p \geq 7$, but Gartland and Lokshtanov recently developed a quasi-polynomial time algorithm for every fixed p [16]. Going beyond forbidden induced paths, the following more general question is also open: Is the VERTEX COVER problem polynomial-time solvable in the class of H-free graphs whenever every component of H is either a path or a subdivision of the claw? While this restriction on H is necessary for the polynomial-time solvability (unless $\mathsf{P} = \mathsf{NP}$) [1], it is not known whether it is also sufficient. Also for the cases when H is not a path, only few partial results are known, see, e.g., [2,5].

Our Focus. We study the DISTANCE-k VERTEX COVER problem for $k \geq 1$. Contrary to the case $k = 0$, the distance-based generalizations have received only limited attention in the literature so far. We already mentioned the work of Busch et al. [8] and an application in network monitoring [33]. Canales et al. gave an extremal result regarding minimum distance 2-vertex covers of maximal outerplanar graphs [10]. This result was further generalized to all k by Alvarado et al. [3]. Nonetheless, the complexity of the problem for restricted inputs remains poorly understood. Our main goal is to fill this gap by providing a systematic study of the complexity of the problem. We do this by analyzing the DISTANCE-k VERTEX COVER problem for $k \geq 1$ in classes of H-free graphs, that is, in classes of graphs defined by a single forbidden induced subgraph H.

Our Results. For integers $k \geq 1$, $s \geq 0$, and $t \geq 1$, we denote by $P_k + sP_t$ the disjoint union of a k-vertex path and s copies of the t-vertex path. We develop the following computational complexity dichotomies for DISTANCE-k VERTEX COVER in classes of H-free graphs.

- A dichotomy for $k = 1$ and arbitrary graph H: the DISTANCE-1 VERTEX COVER problem is solvable in polynomial time in the class of H-free graphs if H is an induced subgraph of $P_4 + sP_2$ for some $s \geq 0$, and NP-complete otherwise.

- A dichotomy for any $k \geq 2$ and arbitrary graph H: the DISTANCE-k VERTEX COVER problem is solvable in polynomial time in the class of H-free graphs if H is an induced subgraph of $P_{2k+2} + sP_k$ for some $s \geq 0$, and NP-complete otherwise.

In polynomially solvable cases, the degrees of the polynomials expressing the running times of our algorithms depend on k and s. However, this is not the case when H is connected. We obtain the following result.

- A dichotomy for any $k \geq 1$ and connected graph H: the DISTANCE-k VERTEX COVER problem is solvable in time $\mathcal{O}((|V(G)| + |E(G)|)^2)$ in the class of H-free graphs if H is an induced subgraph of P_{2k+2}, and NP-complete otherwise.

To derive the NP-completeness results, we introduce a distance-based generalization of the notion of edge dominating set and establish the NP-completeness of the corresponding decision problems. As a corollary of our approach, we show that for all $k \geq 1$, polynomial-time solvability of the DISTANCE-k VERTEX COVER problem in the class of strongly chordal graphs established in [8] cannot be generalized to the class of chordal graphs, unless P = NP. Our polynomial-time algorithms are based on properties of minimum dominating sets in P_t-free graphs established by Camby and Schaudt [9].

Related Work. Besides vertex cover, distance-based generalizations of several other graph concepts were studied in the literature, including matchings [8,35], dominating sets [11,14,19,27], independent sets [4,11,14,15,21,26,30,31], and cliques [25,29].

To the best of our knowledge, the first systematic study aimed towards developing a complexity dichotomy for distance-based generalizations of some classical graph optimization problem was done in 2017 by Bacsó et al. [4]. They considered a natural distance generalization of the notion of independent set: given a positive integer k, a set S of vertices in a graph G is a *distance-k independent set* (also known as a *k-scattered set*) if any two distinct vertices in S are at distance at least k from each other. Bacsó et al. gave a complete characterization of graphs H such that, assuming ETH, the maximum size of a distance-k independent set on H-free graphs can be computed in subexponential time in the size of the input.

We remark that the notion of vertex cover has also been generalized with respect to the length of paths that need to be intersected. A *k-path vertex cover* is a set of vertices intersecting every path of order k [6,7,23,37]. An equivalent notion, known as *k-path transversal*, is defined as set of vertices whose removal leaves a graph that does not contain a path of order k [22].

Structure of the Paper. After summarizing the necessary preliminaries in Sect. 2, we develop NP-completeness results in Sect. 3 and polynomial-time algorithms in Sect. 4. The main results of the paper – the complexity dichotomies – are derived in Sect. 5. Due to lack of space, proofs of results marked by ♦ are omitted.

2   Preliminaries

Let G be a graph. The *order* of G is the number of vertices in it. For a vertex $v \in V(G)$, we denote by $N_G(v)$ (or simply $N(v)$ if the graph is clear from the context) the set of neighbors of v in G, by $N_G[v]$ (or simply $N[v]$) the set $N_G(v) \cup \{v\}$. The *degree* of a vertex v is the cardinality of $N(v)$. For a positive integer k, we denote by P_k the path graph of order k. The *length* of a path or a cycle is the number of edges in it. The *girth* of a graph G is the minimum length of a cycle in G (or ∞ is G is acyclic). The *distance* between two vertices u and v in G is defined as the length of a shortest path between u and v (or ∞ if there is no u, v-path in G). Given two sets $A, B \subseteq V(G)$, we denote by $\mathsf{dist}_G(A, B)$ the minimum over all distances in G between a vertex in A and a vertex in B. When clear from context, we may simply write $\mathsf{dist}(A, B)$. For simplicity, if A contains a unique element a, then we may simply write $\mathsf{dist}(a, B)$, and similarly for B.

A *distance-k vertex cover* in G is a set C of vertices such that for all edges $e \in E(G)$, it holds $\mathsf{dist}(e, C) \leq k$. We denote by $\tau_k(G)$ the size of a minimum distance-k vertex cover of G. For an integer $k \geq 0$, the DISTANCE-k VERTEX COVER problem is formally defined as follows.

DISTANCE-k VERTEX COVER

Instance: A graph G and an integer ℓ.

Question: Is there a distance-k vertex cover in G with size at most ℓ ?

In particular, VERTEX COVER is the same as DISTANCE-0 VERTEX COVER.

An *induced subgraph* of a graph G is any graph H such that $V(H) \subseteq V(G)$ and $E(H) = \{\{u, v\} \in E(G) : u, v \in V(H)\}$. Given a graph G and a set $S \subseteq V(G)$, we denote by $G[S]$ the *subgraph of G induced by S*, that is, the unique induced subgraph of G with vertex set S. Given two graphs G and H, we say that G is *H-free* if no induced subgraph of G is isomorphic to H. More generally, for graphs $H_1, \ldots, H_p$, we say that G is $\{H_1, \ldots, H_p\}$-free if G is H_i-free for all $i \in \{1, \ldots, p\}$. The *claw* is the graph with four vertices and three edges, all having an endpoint in common. Given two graphs G and H, we denote by $G + H$ their disjoint union. For a non-negative integer s, we denote by sG the disjoint union of s copies of G.

An *induced matching* in a graph G is a set M of edges of G such that no two of them share an endpoint (that is, M is a matching) and G contains no edge whose endpoints belong to different edges of M.

The *line graph* of a graph G is the graph, denoted by $L(G)$, with vertex set $E(G)$ in which two distinct vertices are adjacent if and only if the corresponding edges of G have an endpoint in common. It is well known, and easily observed, that line graphs are claw-free. A graph is *chordal* if it does not contain an induced cycle of length at least four, *bipartite* if its vertex set can be partitioned into two independent sets, *planar* if it can be drawn on the plane with no edges crossing, and *cubic* if every vertex has degree three. A *linear forest* is a disjoint union of paths. The operation of *subdividing* en edge $\{u, v\}$ in a graph G means deleting the edge and introducing a new vertex w adjacent precisely to u and v.

Given a positive integer k and a graph G, a set $S \subseteq V(G)$ is a *distance-k dominating set* in G if every vertex in G is at distance at most k from S. A *dominating set* in a graph G is a distance-1 dominating set, and a *connected dominating set* is a dominating set S such that the induced subgraph $G[S]$ is connected.

3 NP-Completeness Results

3.1 When H Is Not a Linear Forest

There are two reasons why the forbidden induced subgraph H may fail to be a linear forest: either because it is not a forest, that is, it contains a cycle, or because it contains an induced claw. We first consider the case when H contains a claw and later the case when H contains a cycle.

When H Contains a Claw. Note that DISTANCE-0 VERTEX COVER, that is, VERTEX COVER, can be solved in polynomial time for claw-free graphs [28, 34]. However, as we show next, this is not the case for DISTANCE-k VERTEX COVER when $k \geq 1$ (unless P = NP). The proof goes in two steps. First, we generalize the notion of *edge dominating set* to its distance-based variant, and consider the corresponding decision problem called DISTANCE-k EDGE DOMINATING SET. For $k = 0$ the problem coincides with the NP-complete EDGE DOMINATING SET problem. We establish NP-completeness also for all $k \geq 1$. Then we use this result to prove that, for every integer $k \geq 1$, DISTANCE-k VERTEX COVER is NP-complete on line graphs, which are claw-free.

For an edge e and a set of edges F, we denote by $\mathsf{dist}(e, F)$ the minimum over all distances between an endpoint of e and an endpoint of an edge in F. Given an integer $k \geq 0$ and a graph G, a set $F \subseteq E(G)$ is a *distance-k edge dominating set* if for every edge $e \in E(G)$, $\mathsf{dist}(e, F) \leq k$. The corresponding decision problem is defined as follows.

DISTANCE-k EDGE DOMINATING SET

Instance: A graph G and an integer ℓ.

Question: Does there exist a distance-k edge dominating set in G with
size at most ℓ?

The EDGE DOMINATING SET problem, which in our context is equivalent to the DISTANCE-0 EDGE DOMINATING SET problem, is known to be NP-complete.

Theorem 1 (Yannakakis and Gavril [36]). EDGE DOMINATING SET *is* NP-*complete, even for cubic bipartite graphs and cubic planar graphs.*

Construction 1. Given a graph G and an integer $k \geq 1$, we define a graph G' obtained from G as follows: for each edge $\{u, v\} \in E(G)$, create a path $P_{u,v}$ made of $2k$ new vertices and connect the endpoints of $P_{u,v}$ to u and v, respectively. The path $P_{u,v}$ together with the edge $\{u, v\}$ is called the u,v-*gadget*. Note that the u,v-gadget is an induced cycle in G' of length $2k+2$. In particular, there exists a unique edge $e' \in E(G')$ of the u,v-gadget such that $\mathrm{dist}_{G'}(e', u) = \mathrm{dist}_{G'}(e', v) = k$. We call the edge e' the *opposite edge* of the edge $\{u, v\}$. See Fig. 1 for an example.

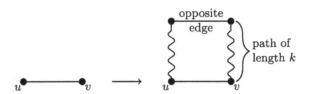

Fig. 1. An edge $\{u, v\}$ (left) and its corresponding u,v-gagdet (right).

♦ **Theorem 2.** *For every integer* $k \geq 1$, DISTANCE-k EDGE DOMINATING SET *is* NP-*complete, even for bipartite graphs with maximum degree 6 and for planar graphs with maximum degree 6.*

Proof sketch. Fix an integer $k \geq 1$. Since distances in graphs can be computed efficiently using breadth-first search, the DISTANCE-k EDGE DOMINATING SET problem is in NP. Let G' be the graph obtained from Construction 1 given G and k. Note that G' can be obtained in polynomial time. To complete the proof, it can be shown that G contains an edge dominating set of size at most ℓ if and only if G' contains a distance-k edge dominating set of size at most ℓ.   □

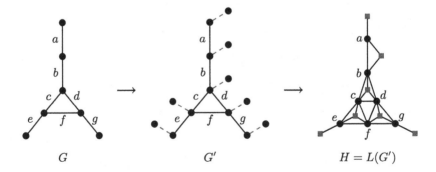

Fig. 2. An example of the graphs used in the proof of Theorem 3: the input graph G, the graph G' obtained from G, and H the line graph of G'. Dashed edges in G' correspond to square shaped vertices in H.

♦ **Theorem 3.** *For every fixed integer $k \geq 1$, DISTANCE-k VERTEX COVER is NP-complete for line graphs of bipartite graphs and line graphs of planar graphs, even if the maximum degree (of the line graphs) is at most 6 if $k = 1$ and at most 12 if $k \geq 2$.*

Proof sketch. Fix a positive integer k and let G be a graph. We construct a graph G' from G by adding for each vertex $u \in V(G)$ a new vertex u' and the edge $\{u, u'\}$ in G'. Let H be the line graph of G'. See Fig. 2 for an example. Note that H can be computed in polynomial time. The rest of the proof consists in showing that G has a distance-k edge dominating set with size at most ℓ if and only if H has a distance-k vertex cover with size at most ℓ.   □

Since every line graph is claw-free, Theorem 3 implies the following result.

Corollary 1. *Let H be a graph containing a claw as induced subgraph. Then for every fixed integer $k \geq 1$, DISTANCE-k VERTEX COVER is NP-complete on H-free graphs.*

When H Contains a Cycle. We start by generalizing a well-known fact, observed first in 1974 by Poljak [32], that a double subdivision of an edge increases the minimum size of a vertex cover by exactly one.

♦ **Lemma 1.** *Let G be a graph, let $e \in E(G)$, and let G' be the graph obtained from G by subdividing edge e exactly $2k + 2$ times, for some integer $k \geq 0$. Then $\tau_k(G') = \tau_k(G) + 1$.*

An iterative application of Lemma 1 leads to the following result.

♦ **Corollary 2.** *Let G be a graph, let $e \in E(G)$, and let G' be the graph obtained from G by subdividing edge e exactly $p(2k+2)$ times, for some two integers $k \geq 0$ and $p \geq 0$. Then $\tau_k(G') = \tau_k(G) + p$.*

Theorem 4. *Let H be a graph containing a cycle. Then for every fixed integer $k \geq 0$, DISTANCE-k VERTEX COVER is NP-complete on H-free graphs.*

Proof. Let G be any graph and k a non-negative integer. Denote by g the girth of H and let G' be the graph obtained from G by subdividing every edge of G exactly $g(2k + 2)$ times. Note that G' is obtained in polynomial time and by Corollary 2, $\tau_k(G') = \tau_k(G) + g|E(G)|$. Moreover, notice that G' has no cycle of length g, and thus is H-free. By Theorem 3, DISTANCE-k VERTEX COVER is NP-complete, and hence the problem remains NP-complete on H-free graphs when H contains a cycle.   □

Corollary 1 and Theorem 4 imply the following result.

Corollary 3. *Let H be a graph that is not a linear forest. Then for any fixed integer $k \geq 1$, DISTANCE-k VERTEX COVER is NP-complete on H-free graphs.*

3.2   When H Is a Linear Forest

We now show that DISTANCE-k VERTEX COVER remains NP-complete in the class of H-free graphs even for certain linear forests. In particular, this is the case when $H = 2P_{k+1}$ for $k \geq 2$ and when H is either P_5 or $2P_3$ for $k = 1$, even if the input graph is chordal.

Construction 2. Given a graph G containing at least one edge and an integer $k \geq 1$, we construct a graph G' as follows. First, we take a complete graph on a set Q of $|V(G)|$ new vertices such that for every vertex $u \in V(G)$ there exists a unique vertex u' in Q. Then, for each edge $\{u, v\} \in E(G)$, we create a u,v-*ladder* as follows. We create a path $P_{u,v}$ of order k and connect one of its endpoints to both u' and v'; then for each such vertex w of $P_{u,v}$ we add a new vertex w' and make it adjacent exactly to the vertices in $N[w]$ (in particular, this means that $N[w'] = N[w]$ in the resulting graph). We call the unique edge e of the u,v-ladder such that $\mathsf{dist}_{G'}(e, \{u, v\}) = k$ the *opposite edge* of the edge $\{u', v'\}$. See Fig. 3 for an example.

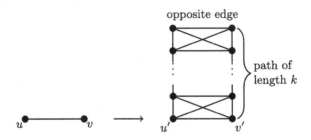

Fig. 3. An edge $\{u, v\}$ (left) and its corresponding u,v-ladder (right).

Theorem 5. DISTANCE-1 VERTEX COVER *is* NP-*complete on* $\{P_5, 2P_3\}$-*free chordal graphs and, for all* $k \geq 2$, DISTANCE-k VERTEX COVER *is* NP-*complete on* $2P_{k+1}$-*free chordal graphs.*

Proof. Let G be a graph containing at least one edge and k a positive integer. Let G' be the graph obtained from Construction 2 given G and k. Note that G' can be obtained in polynomial time. Besides, it is easily observed that G' is chordal. Notice that any induced subgraph of G' isomorphic to $P_{\max\{k+1,3\}}$ contains at least one vertex in the clique Q, which implies that G' cannot contain $2P_{\max\{k+1,3\}}$ as an induced subgraph, that is, if $k = 1$, then G' is $2P_3$-free, and if $k \geq 2$, then G' is $2P_{k+1}$-free. Furthermore, if $k = 1$, then G' is also P_5-free. To see this, consider an induced path P in G' of order 4. Then P has both its endpoints in $V(G) \setminus Q$ and its two internal vertices in Q, as otherwise P would not be induced. This readily implies that P is a maximal path, and thus that G' is P_5-free.

To complete the proof, we show that G has a vertex cover with size at most ℓ if and only if G' has a distance-k vertex cover with size at most ℓ.

Let C be a vertex cover in G with size at most ℓ and $C' = \{u' : u \in C\}$. Note that $C' \subseteq Q \subseteq V(G')$; we claim that C' is a distance-k vertex cover in G'. Suppose that this is not the case. Then there exists an edge $f \in E(G')$ at distance more than k from C'. Observe that $C' \subseteq Q$, and thus, every edge of G' having one endpoint in Q is at distance at most 1 from C'. Therefore, as f is at distance more than k from C', it must have both endpoints in $V(G') \setminus Q$, and hence belongs to an u,v-ladder for some edge $\{u, v\} \in E(G)$. Since C is a vertex cover in G, at least one of u or v belongs to C. We assume without loss of generality that u belongs to C. Since f belongs to the u,v-ladder and $u' \in C'$, we have $\text{dist}_{G'}(f, C') \leq \text{dist}_{G'}(f, u') \leq k$, a contradiction. Thus, C' is a distance-k vertex cover in G' with size $|C'| = |C| \leq \ell$.

Let C' be a distance-k vertex cover in G' with size at most ℓ. Observe that if $w \in V(G') \setminus Q$, then w is a vertex in a u,v-ladder for some $\{u, v\} \in E(G)$. Observe that, by construction of G', every edge f of G' with $\text{dist}_{G'}(f, w) \leq k$ is such that $\text{dist}_{G'}(f, u') \leq k$. Hence, if $w \in C'$, then the set $(C' \setminus \{w\}) \cup \{u'\}$ is also a distance-k vertex cover in G' with size at most ℓ. Hence, we may assume that $C' \subseteq Q$. Let $C = \{u \in V(G) : u' \in C'\}$. Suppose that C is not a vertex cover in G. Then there is an edge $\{u, v\} \in E(G)$ such that $u, v \notin C$. Therefore, $u', v' \notin C'$ but then the opposite edge e of the u,v-ladder is such that $\text{dist}_{G'}(e, C') > \text{dist}_{G'}(e, \{u', v'\}) = k$, a contradiction. Hence, C is a vertex cover in G with size $|C| = |C'| \leq \ell$.

Since VERTEX COVER is NP-complete [20], we obtain that DISTANCE-1 VERTEX COVER is NP-complete on $\{P_5, 2P_3\}$-free chordal graphs and that, for all $k \geq 2$, DISTANCE-k VERTEX COVER is NP-complete on $2P_{k+1}$-free chordal graphs. □

4   Polynomial Algorithms

In this section we identify, for each integer $k \geq 1$, an infinite family of graph classes in which DISTANCE-k VERTEX COVER can be solved in polynomial time. Our first result will be based on the following structural property of P_t-free graphs.

Theorem 6 (Camby and Schaudt [9]). *Let $t \geq 4$ be an integer, let G be a connected P_t-free graph, and let S be any minimum connected dominating set in G. Then the subgraph induced by S in G is either P_{t-2}-free or isomorphic to P_{t-2}.*

Theorem 6 has the following consequence for distance-k vertex covers in P_{2k+2}-free graphs.

Lemma 2. *For every integer $k \geq 1$, every connected P_{2k+2}-free graph G has a distance-k vertex cover that induces a path of order at most two.*

Proof. Fix a positive integer k. To prove the statement of the lemma, we will in fact establish the following stronger statement: every connected P_{2k+2}-free graph G has a distance-k dominating set that induces a path of order at most two. It follows immediately from the definitions that every distance-k dominating set is also a distance-k vertex cover, hence this will indeed suffice.

The proof is by induction on k. Suppose first that $k = 1$. In this case, the statement says that every connected P_4-free graph G has a dominating set that induces a path of order at most two. This follows from the well-known fact that for every connected P_4-free graph G with at least two vertices, the complement of G is disconnected (see, *e.g.*, [12]). Indeed, denoting by C any component of the complement of G and taking $u \in V(C)$ and $v \in V(G) \setminus V(C)$, the set $\{u, v\}$ is a dominating set in G that induces a path of order two. Suppose now that $k > 1$ and consider a connected P_{2k+2}-free graph G. Let S be a minimum connected dominating set in G and let G' be the subgraph of G induced by S. Following Theorem 6, we obtain that G' is either P_{2k}-free or isomorphic to P_{2k}. If G' is P_{2k}-free, then the induction hypothesis implies that G' has a distance-$(k-1)$ dominating set that induces a path of order at most two. If G' is isomorphic to P_{2k}, with vertices $v_1, \ldots, v_{2k}$ in order, then the edge $\{v_k, v_{k+1}\}$ is a distance-$(k-1)$ dominating set in G'. In either case, G' has a distance-$(k-1)$ dominating set S' that induces a path of order at most two. Since every vertex in G is either in S or has a neighbor in S, we infer that S' is a distance-k dominating set in G that induces a path of order at most two. $\square$

In the following theorem, the running time of the algorithm is independent of k, that is, the $\mathcal{O}$ notation does not hide any constants depending on k.

♦ Theorem 7. *For every integer $k \geq 1$, there is an algorithm with running time $\mathcal{O}((|V(G)| + |E(G)|)^2)$ that takes as input a P_{2k+2}-free graph G and computes a minimum distance-k vertex cover of G.*

Proof sketch. Fix a positive integer k and let G be a P_{2k+2}-free graph. To compute a minimum distance-k vertex cover of G, we first compute the connected components $G_1, \ldots, G_s$ of G, solve the problem in each connected component G_i, and combine the obtained solutions. By Lemma 2, each connected component G_i of G has a distance-k vertex cover that induces a path of order at most two. Thus, we immediately obtain a polynomial-time algorithm for computing a minimum distance-k vertex cover of a nontrivial component G_i. We first check if there exists a vertex $u \in V(G_i)$ such that $\{u\}$ is a distance-k vertex cover of G_i. If this is the case, then we have an optimal solution; otherwise we check for each edge $\{u, v\} \in E(G_i)$ if $\{u, v\}$ is a distance-k vertex cover of G_i. Once we find one, we return it. We can verify, in each G_i, if a vertex or an edge is a distance-k vertex cover using a breadth-first search, and the running time follows. $\square$

Remark 1. For $k = 1$, an improved running time of $\mathcal{O}(|V(G)| + |E(G)|)$ can be obtained using a different approach: the fact that P_4-free graphs have clique-width at most two, the fact that the defining property of distance-1 vertex covers can be expressed in MSO_1 logic, and a metatheorem for MSO_1 problems on graphs of bounded clique-width of Courcelle et al. [13].

We now consider the more general case of $(P_{2k+2} + sP_{\max\{k,2\}})$-free graphs for two integers $k \geq 1$ and $s \geq 0$. We first consider the case $k = 1$ and then the case $k \geq 2$.

♦ Lemma 3. *For every integer $s \geq 0$, every connected $(P_4 + sP_2)$-free graph G has a distance-1 vertex cover that induces a linear forest of order at most $2s + 2$.*

Lemma 4. *For every two integers $k \geq 2$ and $s \geq 0$, every connected $(P_{2k+2} + sP_k)$-free graph G has a distance-k vertex cover that induces a linear forest of order at most $f_k(s)$ where*

$$f_k(s) = \begin{cases} 2 & \text{if } s = 0, \\ (s+1)k + 2 & \text{if } s \geq 1. \end{cases}$$

Proof. Fix an integer $k \geq 2$. We use induction on s. For $s = 0$, the statement follows from Lemma 2.

Suppose now that $s \geq 1$ and that every connected $(P_{2k+2} + (s-1)P_k)$-free graph has a distance-k vertex cover that induces a linear forest of order at most $f_k(s-1)$. Let G be a connected $(P_{2k+2} + sP_k)$-free graph. If G is $(P_{2k+2} + (s-1)P_k)$-free, then G has a distance-k vertex cover that induces a linear forest of order at most $f_k(s-1) \leq f_k(s)$. On the other hand, if G is not $(P_{2k+2} + (s-1)P_k)$-free, then there exists a set $S \subseteq V(G)$ inducing a $P_{2k+2} + (s-1)P_k$. Note that S induces a linear forest of order $(s+1)k+2 = f_k(s)$. It thus suffices to show that S is a distance-k vertex cover in G. Let $X = N(S)$ be the set of vertices not in S and with a neighbor in S and $Y = V(G) \setminus (S \cup X)$ be the set of vertices not in S and without a neighbor in S. Let e be an edge of G. If e has an endpoint in $S \cup X$, then $\text{dist}_G(e, S) \leq 1 \leq k$. So let e be entirely contained in Y. Since G is connected, there exists a shortest path P between an endpoint of e and a vertex in S. Since G is $(P_{2k+2} + sP_k)$-free, the part of P entirely contained in Y has at most $k - 1$ vertices. Other than that, P has exactly one vertex in X and exactly one in S. Thus, the length of P is at most k, which implies $\text{dist}_G(e, S) \leq k$. This shows that S is a distance-k vertex cover in G and completes the proof. □

Lemmas 3 and 4 imply that for all integers $k \geq 0$ and $s \geq 0$ the minimum size of a distance-k vertex cover in a $(P_{2k+2} + sP_{\max\{k,2\}})$-free graph is bounded by a function depending only on k and s but independent of G. Thus, we can do a complete enumeration of small subsets of vertices to find a minimum distance-k vertex cover in such a graph, and essentially the same approach as the one used to prove Theorem 7 using Lemma 2 can be used to prove the following theorem using Lemmas 3 and 4.

Theorem 8. *For every two integers $k \geq 1$ and $s \geq 0$, there is a polynomial-time algorithm that takes as input a $(P_{2k+2} + sP_{\max\{k,2\}})$-free graph G and computes a minimum distance-k vertex cover of G.*

5   Complexity Dichotomies

Our results from Sects. 3 and 4 allow us to obtain a complexity dichotomy of DISTANCE-k VERTEX COVER on H-free graphs, for all $k \geq 1$.

Theorem 9. *For every graph H, the following holds:*

- DISTANCE-1 VERTEX COVER *is solvable in polynomial time in the class of H-free graphs if H is an induced subgraph of $P_4 + sP_2$, for some $s \geq 0$, and* NP-*complete otherwise.*
- *For every integer $k \geq 2$,* DISTANCE-k VERTEX COVER *is solvable in polynomial time in the class of H-free graphs if H is an induced subgraph of $P_{2k+2} + sP_k$ for some $s \geq 0$, and* NP-*complete otherwise.*

Proof. Fix a graph H and let $\mathcal{G}$ be the class of H-free graphs. If H is not a linear forest, then for all $k \geq 1$, Corollary 3 implies that DISTANCE-k VERTEX COVER is NP-complete on $\mathcal{G}$. Suppose that H is a linear forest.

Consider first the case when $k = 1$. If H contains P_5 or $2P_3$ as an induced subgraph, then $\mathcal{G}$ contains the class of $\{P_5, 2P_3\}$-free chordal graphs, and hence by Theorem 5 DISTANCE-1 VERTEX COVER is NP-complete on $\mathcal{G}$. Otherwise, we obtain that H is $\{P_5, 2P_3\}$-free. Recall that H is a linear forest. Let us denote by t be the maximum order of a component of H and let P be a component of H of order t. If $t \leq 2$, then every component of H has order at most two. If $t \geq 3$, then, since H is $2P_3$-free, every component of H other than P has order at most two. In either case, every component of H other than P has order at most two, which implies that H is an induced subgraph of $P_t + sP_2$ for some $s \geq 0$. Since H is P_5-free, we have $t \leq 4$, and hence every H-free graph is $(P_4 + sP_2)$-free. Thus, by Theorem 8 the problem can be solved in polynomial time for graphs in $\mathcal{G}$.

Suppose now that $k \geq 2$. If H contains $2P_{k+1}$ as an induced subgraph, then $\mathcal{G}$ contains the class of $2P_{k+1}$-free chordal graphs, and hence by Theorem 5 DISTANCE-k VERTEX COVER is NP-complete on $\mathcal{G}$. Again, let t denote the maximum order of a component of H and let P be a component of H of order t. If $t \leq k$, then every component of H has order at most k. If $t \geq k + 1$, then, since H is $2P_{k+1}$-free, every component of H other than P has order at most k. Thus, in either case, every component of H other than P has order at most k, and H is an induced subgraph of $P_t + sP_k$ for some $s \geq 0$. Since H is $2P_{k+1}$-free, it is also P_{2k+3}-free, which implies that $t \leq 2k + 2$, and thus H is an induced subgraph of $P_{2k+2} + sP_k$ for some $s \geq 0$. It follows that every H-free graph is $(P_{2k+2} + sP_k)$-free. Thus, by Theorem 8 the problem can be solved in polynomial time for graphs in $\mathcal{G}$. $\qquad\square$

For the case when the forbidden induced subgraph is connected, Theorems 7 and 9 imply the following dichotomy.

Corollary 4. *For every connected graph H and integer $k \geq 1$, the* DISTANCE-k VERTEX COVER *problem is solvable in time $\mathcal{O}((|V(G)|+|E(G)|)^2)$ in the class of H-free graphs if H is an induced subgraph of P_{2k+2}, and* NP-*complete otherwise.*

Furthermore, as explained in Remark 1, the running time can be improved to linear for the case $k = 1$.

Acknowledgments. The authors wish to thank Peter Muršič for valuable discussions.

References

1. Alekseev, V.E.: The effect of local constraints on the complexity of determination of the graph independence number. In: Combinatorial-Algebraic Methods in Applied Mathematics, pp. 3–13. Gor'kov. Gos. Univ., Gorki (1982)
2. Alekseev, V.E.: Polynomial algorithm for finding the largest independent sets in graphs without forks. Discrete Appl. Math. **135**(1–3), 3–16 (2004). https://doi.org/10.1016/S0166-218X(02)00290-1
3. Alvarado, J.D., Dantas, S., Rautenbach, D.: Distance k-domination, distance k-guarding, and distance k-vertex cover of maximal outerplanar graphs. Discrete Appl. Math. **194**, 154–159 (2015). https://doi.org/10.1016/j.dam.2015.05.010
4. Bacsó, G., Marx, D., Tuza, Z.: H-free graphs, independent sets, and subexponential-time algorithms. In: 11th International Symposium on Parameterized and Exact Computation, LIPIcs. Leibniz International Proceedings o Inform., vol. 63, pp. Art. No. 3, 12, Schloss Dagstuhl. Leibniz-Zent. Inform., Wadern (2017)
5. Brandstädt, A., Mosca, R.: Maximum weight independent set for ℓclaw-free graphs in polynomial time. Discrete Appl. Math. 237, 57–64 (2018). https://doi.org/10.1016/j.dam.2017.11.029
6. Brešar, B., Jakovac, M., Katrenič, J., Semanišin, G., Taranenko, A.: On the vertex k-path cover. Discrete Appl. Math. **161**(13–14), 1943–1949 (2013). https://doi.org/10.1016/j.dam.2013.02.024
7. Brešar, B., Kardoš, F., Katrenič, J., Semanišin, G.: Minimum k-path vertex cover. Discrete Appl. Math. **159**(12), 1189–1195 (2011). https://doi.org/10.1016/j.dam.2011.04.008
8. Busch, A.H., Dragan, F.F., Sritharan, R.: New min-max theorems for weakly chordal and dually chordal graphs. In: Wu, W., Daescu, O. (eds.) COCOA 2010. LNCS, vol. 6509, pp. 207–218. Springer, Heidelberg (2010). https://doi.org/10.1007/978-3-642-17461-2_17
9. Camby, E., Schaudt, O.: A new characterization of P_k-free graphs. Algorithmica **75**(1), 205–217 (2015). https://doi.org/10.1007/s00453-015-9989-6
10. Canales, S., Hernández, G., Martins, M., Matos, I.: Distance domination, guarding and covering of maximal outerplanar graphs. Discrete Appl. Math. **181**, 41–49 (2015). https://doi.org/10.1016/j.dam.2014.08.040
11. Chang, G.J., Nemhauser, G.L.: The k-domination and k-stability problems on sun-free chordal graphs. SIAM J. Algebraic Discrete Methods **5**(3), 332–345 (1984)
12. Corneil, D.G., Lerchs, H., Burlingham, L.S.: Complement reducible graphs. Discrete Appl. Math. **3**(3), 163–174 (1981). https://doi.org/10.1016/0166-218X(81)90013-5
13. Courcelle, B., Makowsky, J.A., Rotics, U.: Linear time solvable optimization problems on graphs of bounded clique-width. Theory Comput. Syst. **33**(2), 125–150 (2000). https://doi.org/10.1007/s002249910009
14. Dvořák, Z.: On distance r-dominating and $2r$-independent sets in sparse graphs. J. Graph Theory **91**(2), 162–173 (2019). https://doi.org/10.1002/jgt.22426

15. Eto, H., Guo, F., Miyano, E.: Distance-d independent set problems for bipartite and chordal graphs. J. Combin. Optimizat. **27**(1), 88–99 (2013). https://doi.org/10.1007/s10878-012-9594-4

16. Gartland, P., Lokshtanov, D.: Independent set on P_k-free graphs in quasi-polynomial time. In: 61st IEEE Annual Symposium on Foundations of Computer Science, FOCS 2020, Durham, NC, USA, November 16–19, 2020, pp. 613–624, IEEE (2020). https://doi.org/10.1109/FOCS46700.2020.00063

17. Grzesik, A., Klimošová, T., Pilipczuk, M., Pilipczuk, M.: Polynomial-time algorithm for maximum weight independent set on P_6-free graphs. In: Proceedings of the Thirtieth Annual ACM-SIAM Symposium on Discrete Algorithms, pp. 1257–1271, SIAM, Philadelphia, PA (2019). https://doi.org/10.1137/1.9781611975482.77

18. Horton, J.D., López-Ortiz, A.: On the number of distributed measurement points for network tomography. In: Proceedings of the 3rd ACM SIGCOMM Internet Measurement Conference, IMC 2003, Miami Beach, FL, USA, October 27–29, 2003, pp. 204–209, ACM (2003)

19. Jaffke, L., Kwon, O.J., Strømme, T.J.F., Telle, J.A.: Mim-width III. Graph powers and generalized distance domination problems. Theoret. Comput. Sci. **796**, 216–236 (2019). https://doi.org/10.1016/j.tcs.2019.09.012

20. Karp, R.M.: Reducibility among combinatorial problems. In: Complexity of computer computations, Proceedings of the Symposium IBM Thomas Journal Watson Research Center, Yorktown Heights, N.Y., 1972, pp. 85–103 (1972)

21. Katsikarelis, I., Lampis, M., Paschos, V.T.: Structurally parameterized d-scattered set. In: Brandstädt, A., Köhler, E., Meer, K. (eds.) Graph-Theoretic Concepts in Computer Science, LNCS, vol. 11159, pp. 292–305, Springer, Cham (2018). https://doi.org/10.1007/978-3-030-00256-5_24

22. Lee, E.: Partitioning a graph into small pieces with applications to path transversal. Math. Program. (1), 1–19 (2018). https://doi.org/10.1007/s10107-018-1255-7

23. Lemańska, M.: On the minimum vertex k-path cover of trees. Util. Math. **100**, 299–307 (2016)

24. Lokshantov, D., Vatshelle, M., Villanger, Y.: Independent set in P_5-free graphs in polynomial time. In: Proceedings of the Twenty-Fifth Annual ACM-SIAM Symposium on Discrete Algorithms, pp. 570–581. ACM, New York (2014). https://doi.org/10.1137/1.9781611973402.43

25. Luce, R.D.: Connectivity and generalized cliques in sociometric group structure. Psychometrika **15**, 169–190 (1950). https://doi.org/10.1007/BF02289199

26. Marx, D., Pilipczuk, M.: Optimal parameterized algorithms for planar facility location problems using Voronoi diagrams. In: Bansal, N., Finocchi, I. (eds.) ESA 2015. LNCS, vol. 9294, pp. 865–877. Springer, Heidelberg (2015). https://doi.org/10.1007/978-3-662-48350-3_72

27. Meir, A., Moon, J.W.: Relations between packing and covering numbers of a tree. Pacific J. Math. **61**(1), 225–233 (1975)

28. Minty, G.J.: On maximal independent sets of vertices in claw-free graphs. J. Combin. Theory Ser. B **28**(3), 284–304 (1980). https://doi.org/10.1016/0095-8956(80)90074-X

29. Mokken, R.J.: Cliques, clubs and clans. Qual. Quant. **13**(2), 161–173 (1979)

30. Montealegre, P., Todinca, I.: On distance-d independent set and other problems in graphs with "few" minimal separators. In: Heggernes, P. (ed.) WG 2016. LNCS, vol. 9941, pp. 183–194. Springer, Heidelberg (2016). https://doi.org/10.1007/978-3-662-53536-3_16

31. Pilipczuk, M., Siebertz, S.: Kernelization and approximation of distance-r independent sets on nowhere dense graphs. Eur. J. Combin. **94**, 103309, 19 (2021). https://doi.org/10.1016/j.ejc.2021.103309

32. Poljak, S.: A note on stable sets and colorings of graphs. Comment. Math. Univ. Carolinae **15**, 307–309 (1974)

33. Sasaki, M., Zhao, L., Nagamochi, H.: Security-aware beacon based network monitoring. In: 2008 11th IEEE Singapore International Conference on Communication Systems, pp. 527–531. IEEE (2008)

34. Sbihi, N.: Algorithme de recherche d'un stable de cardinalité maximum dans un graphe sans étoile. Discrete Math. **29**(1), 53–76 (1980). https://doi.org/10.1016/0012-365X(90)90287-R

35. Stockmeyer, L.J., Vazirani, V.V.: NP-completeness of some generalizations of the maximum matching problem. Inform. Process. Lett. **15**(1), 14–19 (1982). https://doi.org/10.1016/0020-0190(82)90077-1

36. Yannakakis, M., Gavril, F.: Edge dominating sets in graphs. SIAM J. Appl. Math. **38**(3), 364–372 (1980). https://doi.org/10.1137/0138030

37. Zhang, Z., Li, X., Shi, Y., Nie, H., Zhu, Y.: PTAS for minimum k-path vertex cover in ball graph. Inform. Process. Lett. **119**, 9–13 (2017). https://doi.org/10.1016/j.ipl.2016.11.003

On an Ordering Problem in Weighted Hypergraphs

Peter Damaschke[1,2]($\boxtimes$)

[1] Department of Computer Science and Engineering, Chalmers University,
41296 Göteborg, Sweden
ptr@chalmers.se
[2] Fraunhofer-Chalmers Research Centre for Industrial Mathematics,
Göteborg, Sweden

Abstract. We consider the problem of mapping the n vertices of an edge-weighted hypergraph to the points $1, \ldots, n$ on the real line, so as to minimize the weighted sum of the coordinates of right ends of all edges. This problem naturally appears in warehouse logistics: n shelves are arranged in one row, every shelf can host one type of items, the edges are sets of items requested together, their weights are the request frequencies, and items must be picked from the shelves and brought to a collection point at the left end of the row. The problem is to place all items so as to minimize the average length of the collection tours. It is NP-complete even for graphs, but it can be solved in $O^*(2^n)$ time by dynamic programming on subsets. In the present work we focus on hypergraphs with small connected components, which also has a practical motivation: Typical requests comprise related items from only one of many small disjoint groups. As a first result we solve, in polynomial time, an auxiliary problem with prescribed ordering in every component. For the unrestricted problem we conclude some worst-case time bounds that beat $O^*(2^n)$ for components of sizes up to 6. Some simple preprocessing can further reduce the time in many instances. Furthermore, the case of star graphs can be solved via bipartite matchings. Finally, there remain various interesting open problems.

Keywords: Hypergraph linear arrangement · Dynamic programming on subsets · Convex hull · Bipartite matching · Warehouse logistics

1 Introduction

Let $G = (V, E)$ be a given hypergraph, consisting of a set V of vertices and a set E of edges, which are non-empty subsets of V. (As this work mainly deals with hypergraphs, for brevity we prefer the term "edge" to "hyperedge".) Every edge $e \in E$ has a positive weight $w(e)$. For edges with only one vertex, such as $e = \{v\}$, we may simply write $w(v)$ instead of $w(e)$ or $w(\{v\})$, and we say that the vertex v has this weight, without risk of confusion. We will study the following arrangement problem. We first introduce it technically, because it is then easier to explain its motivation.

© Springer Nature Switzerland AG 2021
P. Flocchini and L. Moura (Eds.): IWOCA 2021, LNCS 12757, pp. 252–264, 2021.
https://doi.org/10.1007/978-3-030-79987-8_18

MinSumEnds

Given: a hypergraph $G = (V, E)$ with n vertices, and positive weights $w(e)$ of all edges $e \in E$.

Find: a bijective mapping π of V onto the set of integers $\{1, \ldots, n\}$ that minimizes the sum $\sum_{e \in E} w(e) \cdot \mu(e)$, where $\mu(e) := \max\{\pi(v) : v \in E\}$ for every edge $e \in E$.

Informally, our problem is to place the n vertices of V on n distinct points $1, \ldots, n$ on the real line, so as to minimize the weighted sum of the right ends of all edges. We can also view π as an ordering of V from left to right.

The objective function $\sum_{e \in E} w(e) \cdot \mu(e)$ can be rewritten as $\sum_{i=1}^{n} i \cdot L(i)$, where $L(i)$ denotes the total weight of all edges that end in point i, that is, $L(i) = \sum_{e: \mu(e)=i} w(e)$. For any ordering π we also define a function L_π on V by $L_\pi(v) := L(i)$ when $\pi(v) = i$. Hence the value $L_\pi(v)$ is the total weight of all edges having v as their rightmost vertex.

A hypergraph is called connected if, for any two vertices u and v, there exists a sequence $u = u_0, e_0, u_1, e_1, u_2, \ldots, u_{k-1}, e_{k-1}, u_k = v$ with $u_i \in e_i$ for all i with $0 \leq i \leq k - 1$, and $u_i \in e_{i-1}$ for all i with $1 \leq i \leq k$. For brevity, connected components of a hypergraph are just called *components* in this paper.

The main motivation of MinSumEnds comes from efficient warehouse operations. In one scenario, n types of items shall be located in n equidistant shelves along a wall. At a collection point to the left of all shelves, a collector (robot or human worker) is waiting for requests. When a subset of items is requested, the collector must retrieve the requested items from the shelves and bring them to the collection point. Hence the length of the walk is twice the distance to the farthest requested item, whereas the positions of the other items are irrelevant, and so are the amounts of requested items of each type. The weighted hypergraph models the typical requests and their frequencies. More precisely, every edge is a set of (types of) items in a request, and its weight is proportional to the frequency of exactly this request. Thus, placing the items in the shelves so as to minimize the average walking distance leads to the MinSumEnds problem.

One may doubt that the problem arises in exactly this form in practice, as a single line of shelves is a special case, the problem may be intertwined with other types of constraints, data on frequencies may only be rough estimates, etc. However, combinations of workflow and layout planning in factory halls and warehouses are definitely a subject of industrial projects, where users want to optimize layouts for work sequences and vice versa, possibly in several iterations. Extracting basic combinatorial optimization problems and trying to understand their complexity is a meaningful activity accompanying the practical developments.

MinSumEnds is NP-complete even for graphs, i.e., the case when all edges have at most two vertices. On the positive side, it can be solved in $O^*(2^n)$ time by dynamic programming on subsets [4]. Naturally, we are interested in relevant special cases that can be solved faster. One such practical case appears when the items are partitioned into small disjoint groups of related items, and only these items are typically requested together. In such cases, the sizes of the components of our hypergraphs are small integers, however, the frequencies of requests are still arbitrary numbers. Therefore we will study the complexity of MinSumEnds for weighted hypergraphs whose components have some limited fixed size k.

A problem being closely related to MinSumEnds has been studied much more extensively: The Minimum Linear Arrangement (MLA) problem asks to order the vertices of a graph or hypergraph so as to minimize the sum, over all edges, of the distances between the leftmost and rightmost vertex in the edge. Key results can be found, e.g., in [1–3,5,8–12]. (Here we do not summarize them all.) Apparently, the use of dynamic programming on subsets for that problem was first discovered in [3].

The vast majority of graph and hypergraph problems, including MLA, can be solved independently on the components. A remarkable issue it that this is no longer the case for MinSumEnds. Loosely speaking, since all vertices compete for the best positions, there is heavy interaction between the components, which makes the problem tricky even in hypergraphs with components of fixed size, which we mainly consider in this work.

Our contributions can be outlined as follows. Using some exchange arguments we solve, in polynomial time, a restricted version of MinSumEnds that we call MinSumEnds<. In that auxiliary problem, the vertices within every component must appear in some prescribed ordering (whereas all permutations are allowed in MinSumEnds). The MinSumEnds< problem also has some nice geometric interpretation in terms of convex functions. With the help of MinSumEnds< we obtain time bounds that beat the standard $O^*(2^n)$ bound in hypergraphs with components of at most 6 vertices. For concrete instances, the actual running times can be further improved by excluding several candidate orderings within the components, as candidates for optimal solutions have to pass some simple tests with linear inequalities in the given weights. Besides small components it is also worth considering structural restrictions. The case of star graphs can be solved via bipartite matchings, and this idea can be generalized. We conclude with various open problems.

2   An Exchange Property

The following lemma is simple, but it will be central to our approach.

Lemma 1. *Let π be an ordering, and let P and Q be the sets of vertices at points $i+1, \ldots, i+p$ and $i+p+1, \ldots, i+p+q$, respectively. Suppose that no $u \in P$, $v \in Q$, and $e \in E$ exist with $u, v \in e$. Then we have: If the inequality*

$$\sum_{u \in P} L_\pi(u)/p \geq \sum_{v \in Q} L_\pi(v)/q$$

is violated, then placing the vertices of Q at points $i+1, \ldots, i+q$ and the vertices of P at points $i+q+1, \ldots, i+q+p$ while preserving the orderings within P and Q, will make the objective smaller.

Proof. Let us swap P and Q. Since no edge contains vertices from both P and Q, every vertex $u \in P$ and $v \in Q$ is the rightmost vertex of the same edges as it was before the swap. Hence all $L_\pi(u)$ and $L_\pi(v)$ are preserved. Thus, the objective function changes by adding the amount $q \sum_{u \in P} L_\pi(u) - p \sum_{v \in Q} L_\pi(v)$. From this, the assertion obviously follows.   □

In words, Lemma 1 says that neighbored and consecutive sets P and Q of sizes $p = |P|$ and $q = |Q|$, such that no edge contains vertices from both P and Q, can be swapped if the mentioned inequality is violated. Hence, this inequality must hold in any optimal ordering π. If the inequality is an equation, then P and Q can also be swapped without destroying optimality.

3   An Auxiliary Problem with Ordered Components

Now we consider an auxiliary problem named MinSumEnds<. It is defined exactly as MinSumEnds, but with the extra condition that the vertices within every component C of the hypergraph must appear in π in some prescribed ordering. That is, $\pi(v_1) < \ldots < \pi(v_k)$ is required, where $\{v_1, \ldots, v_k\}$ is the ordered vertex set of C. (We do not repeat the entire formal definition, as the objective is the same as in MinSumEnds.) Since the orderings of vertices in the components are given, the components are already part of the input. Of course, they must be the true components of the input hypergraph G, but consistency can be checked in linear time.

Within any component C of the input hypergraph G, we call a subset M of vertices of C a *module* if the vertices of M appear consecutively in the prescribed ordering of C, and they remain consecutive also in every optimal ordering π of our hypergraph G.

Note that, due to the last condition, the modules are not "obvious" from the input; below we will show how to compute them.

We define the *density* of any subset M (module or not) of the vertex set of C as $D(M) := \sum_{v \in M} L_\pi(v)/|M|$. Since the ordering of C is fixed, actually the values $L_\pi(v)$, $v \in M \subseteq C$, do not depend on π, therefore $D(M)$ is well defined and easy to compute from the input. For single vertices v we write $D(v)$ instead of $D(\{v\}) = L_\pi(v)$.

Lemma 2. *Let M and N be disjoint modules in C with the following properties: $\pi(u) < \pi(v)$ for all $u \in M$ and all $v \in N$, no other vertex of C is between M and N in the ordering, and $D(M) < D(N)$. Then $M \cup N$ is a module, too.*

Proof. Assume for contradiction that π is some optimal ordering where the set I of vertices between M and N is non-empty. By assumption, all vertices in I belong to other components than C. We can uniquely partition I into subsets called bags, where every bag is a maximal subset of vertices of I that are consecutive in π and belong to the same component. M and N are considered bags as well. Let J and J' denote any two neighbored bags in this ordering (J to the left of J'). By Lemma 1, if $D(J) < D(J')$ then we can swap them to make the objective smaller, which contradicts the optimality of π. (Since J and J' are from different components, they have no common edges, hence they satisfy the assumptions of Lemma 1, and swapping also yields a valid solution, since the orderings in the components are not changed.) Hence the bags in the sequence from M to N have non-increasing densities. But this contradicts $D(M) < D(N)$. It follows $I = \emptyset$ in every optimal ordering π. By the definition of modules this implies that $M \cup N$ is a module.   □

Whenever two modules M and N satisfy the assumptions of Lemma 2, we can *merge* them to the module $M \cup N$ according to the conclusion of Lemma 2. Consider the following process based on this observation:

We start from the sequence of the single vertices of C (which are, trivially, modules) in the prescribed ordering, and we merge two arbitrarily chosen neighbored modules that satisfy the assumptions of Lemma 2. This step is repeated as long as possible. The final result is a sequence of modules that we call *blocks*.

We claim that the sequence of blocks does not depend on the arbitrary choices, that is, the resulting blocks are uniquely determined by the densities $D(v)$ of all single vertices v. Below we give a proof that also yields a geometric characterization of blocks.

The idea is to represent any partitioning of the ordered vertex set $\{v_1, \ldots, v_k\}$ of C into modules as a function f on the interval $[0, k)$, defined as follows: For every module $M = \{v_i, \ldots, v_j\}$ in this partitioning, let $f(x) := D(M)$ for all $x \in [i-1, j)$. Furthermore, let $g(x) := \int_0^x f(t)\, dt$.

Note that g is a monotone increasing and piecewise linear continuous function, where the slopes equal the densities. The initial function f denoted f_0 has values $f_0(x) = D(v) = L_\pi(v_i)$ for all $x \in [i-1, i)$, and for all i. Let $g_0(x) := \int_0^x f_0(t)\, dt$.

Yet another technical definition makes the proof convenient: On the graph of g we *mark* all endpoints of the modules. Then, the effect of merging two modules M and N to the graph of g can be figuratively described as follows. Let p and q be the left and right endpoint, respectively, of the straight-line segment corresponding to M and N, and let r denote the point separating these segments. Remember that the slope increases in the point r. We unmark r and move the two segments upwards, thereby transforming them continuously into the straight-line segment connecting p and q, Note that only unmarked points move upwards to the new segment. With these notations and observations we can state:

Lemma 3. *The endpoints of the blocks are exactly those points $(x, g_0(x))$ with integer x that are on the upper convex hull h of the graph of the function g_0.*

Proof. In the following, two functions are said to be in $\leq$ relation if their function values are in $\leq$ relation for every argument.

Initially, the marked points on g_0 are all points $(x, g_0(x))$ with integer values x, because the modules we start from are all single vertices. Trivially we have $g_0 \leq h$. If $g \leq h$, and we replace two incident straight-line segments in the graph of g with increasing slopes by one straight-line segment connecting the same endpoints, then this modified function g still satisfies $g \leq h$, since h is an upper convex hull (and hence a concave function).

As long as $g = h$ does not yet hold, another merge operation can be done, and the total number of merge operations is finite (actually, at most k). Thus we always eventually obtain $g = h$, regardless of the choice of merge operations in every step.

All marked points that were strictly below the graph of h got unmarked, since otherwise they could not have moved upwards to the graph of h. Furthermore, marked points on the graph of h never got unmarked. Thus, exactly those marked points that were already initially on the graph of h are still marked. Together this yields the assertion. □

Note that the blocks have non-increasing densities, in the prescribed ordering in the component C. We arrive at the complexity result for our auxiliary problem:

Theorem 1. MINSUMENDS< *can be solved in* $O(e + n \log n)$ *time, for arbitrary hypergraphs with n vertices, edges of total size e, and ordered components. Moreover, every sequence of the blocks from all components, sorted by non- increasing densities (where ties are broken arbitrarily), is an optimal ordering.*

Proof. First we compute all densities $D(v)$, $v \in V$, in $O(e)$ time. Then we compute the blocks in every component, by pairwise merging of modules, starting from the single vertices. The blocks are uniquely determined due to Lemma 3, and with some care this part can be implemented to run in $O(n)$ time in all components.

Recall that the blocks are modules. Hence, in an optimal ordering, every block is a consecutive set, and due to Lemma 1, also blocks from different components appear in non-increasing order of densities, where the order of blocks with equal densities is arbitrary. Thus, it only remains to sort the blocks by their densities and concatenate them. In the worst case of many small blocks this incurs a logarithmic factor. □

Due to the non-increasing densities we refer to the algorithm in Theorem 1 as the *sedimentation algorithm*.

The arbitrary tie breaking in Theorem 1 suggests to slightly re-define the notion of blocks as follows:

Whenever blocks from the same component have equal densities, we can place them consecutively and merge them to one block, without missing an optimal solution.

The advantage is that now the blocks from the same component appear with strictly decreasing densities. From now on we use the concept of *blocks* in this stricter sense, without risk of confusion.

4   Domination Relation

Next we apply the sedimentation algorithm for MINSUMENDS$<$ to the solution of the original MINSUMENDS problem. A very naive way would be to exhaustively try all combinations of orderings of the vertices in the components of the input hypergraph G, solve every case in polynomial time, and finally take the solution with the best objective value. In the following we try to reduce the large number of orderings to examine.

In an instance of MINSUMENDS$<$, let $v_1, \ldots, v_k$ again denote the vertices of some component C in their prescribed ordering, that is, $\pi(v_1) < \ldots < \pi(v_k)$ is required. Recall that the densities $D(v_i)$ uniquely determine the blocks of C, and the $D(v_i)$ in turn depend only on this internal ordering of C. We abbreviate the densities by $d_i := D(v_i)$.

Now let $(d_1, \ldots, d_k)$ and $(d'_1, \ldots, d'_k)$ be two different sequences of densities (resulting from different possible orderings of C in an instance of MIN-SUMENDS). We say that $(d_1, \ldots, d_k)$ *dominates* $(d'_1, \ldots, d'_k)$ if their prefix sums satisfy $\sum_{i=1}^{j} d_i \geq \sum_{i=1}^{j} d'_i$ for all $j = 1, \ldots, k$, and the inequality is strict for some index j. An equivalent characterization is that $(d_1, \ldots, d_k)$ is obtained from $(d'_1, \ldots, d'_k)$ by "moving some amounts" from some positions to other positions to the left, i.e., with smaller indices, while preserving the sum. We also say that the ordering of C with densities $(d_1, \ldots, d_k)$ dominates the ordering of C with densities $(d'_1, \ldots, d'_k)$. Finally, we call two orderings *equivalent* if they yield the same sequence of densities; note that they do not dominate each other.

From the objective function of MINSUMENDS, the following is obvious:

Lemma 4. *In an optimal ordering π solving an instance of* MINSUMENDS, *the ordering of the vertices of any component C is not dominated by any other ordering of the vertices of C.*

In simpler words, only non-dominated orderings of components may appear in an optimal solution π. Among equivalent orderings we need to consider only one, since the objective value solely depends on the densities. This suggest the following definition and observation:

A set of orderings of a component C is said to be a *candidate set* if every optimal solution to MINSUMENDS uses one of these orderings of C, or an equivalent ordering. We refer to the orderings in a fixed candidate set as *candidate orderings*.

Clearly, some candidate set for C can be obtained as follows: Take the set of all orderings not dominated by others, but for any equivalent orderings, keep only one arbitrarily selected representative.

An obvious idea is now to compute candidate sets in advance, which reduces the total number of orderings to consider. We elaborate on this idea in the next section.

5   Hypergraphs with Small Components

One use of the concept of blocks and of the above results would be to simplify some proofs from [6] (and rewrite them from a more general perspective). In that work we considered the case of unweighted graphs. However, in the following we derive new results for another class of hypergraphs, namely such with components whose size is bounded by some constant, but where the weights are arbitrary positive numbers.

Let $\{u, v\}$ be the vertex set of some component where $w(u) \geq w(v)$. Then, obviously, the ordering (u, v) dominates (v, u), unless $w(u) = w(v)$, in which case the orderings are equivalent. Thus, in either case we have to consider only one candidate ordering, and since $e = O(n)$, Theorem 1 yields immediately:

Theorem 2. MINSUMENDS *in weighted hypergraphs where all components have at most two vertices can be solved in* $O(n \log n)$ *time.*

Already components with three vertices turn out to be more tricky. However, we will obtain some upper bounds on the number of candidate orderings. We use the following convenient notation. Let $v_1, \ldots, v_k$ be the vertices of the considered component C. We abbreviate the weights by $w_i := w(v_i)$ and $w_{ij} := w(\{v_i, v_j\})$, and similarly, we use more subscripts for larger edges. Furthermore, we can assume $w_1 \leq \ldots \leq w_k$ by re-indexing.

Lemma 5. *Let c_k denote a number with the property that every component with k vertices admits a candidate set with at most c_k orderings. Then we have:*

$$c_2 = 1, \ c_3 \leq 2, \ c_4 \leq 5, \ c_5 \leq 16, \ c_6 \leq 62.$$

Moreover, for any fixed k and given weights, these candidate sets can be identified in constant time.

Proof. Before we stated Theorem 2 we have already shown $c_2 = 1$. Generalizing the *exchange argument* used there we see: Any ordering of C beginning with $(v_i, v_j, \ldots)$, where $i < j$, is dominated by or equivalent to the ordering $(v_j, v_i, \ldots)$ obtained by swapping the first two vertices.

Assume that some prefix of the ordering of C is already fixed. Let P denote the vertex set of this prefix. We consider the residual hypergraph defined as follows. Every vertex $v \in P$ becomes an edge whose weight is the sum of all original weights of edges with v as the rightmost vertex. Every subset $e \subseteq C \setminus P$ gets the weight $\sum_{Q \subseteq P} w(Q \cup e)$, where $w(.)$ denotes the original weights, and with the understanding that a non-existing edge is the same as an edge with zero weight. The rest of the hypergraph outside C is not affected. We remark that $C \setminus P$ in the residual hypergraph is not necessarily connected, but we will not make use of connectivity.

Since, in MINSUMENDS with the restriction that the ordering of C begins with the assumed prefix P, the costs of all vertices in $C \setminus P$ depend only on the residual hypergraph, the exchange argument also applies to $C \setminus P$ and the new weights. This allows us to bound the numbers c_k recursively as follows. The claimed constant time bound is trivial.

For $k = 3$, a candidate ordering starting with v_2 must continue with v_1, and any other candidate ordering starts with v_3, which yields $c_3 \leq 1 + c_2 \leq 1 + 1 = 2$. For $k = 4$, a candidate ordering starting with v_2 must continue with v_1, and the other candidate orderings start with either v_3 or v_4, which yields $c_4 \leq c_2 + 2c_3 \leq 1 + 4 = 5$. For $k = 5$, a candidate ordering must start with one of (v_2, v_1), (v_3, v_1), (v_3, v_2), (v_4), (v_5), which yields $c_5 \leq 3c_3 + 2c_4 \leq 6 + 10 = 16$. For $k = 6$, a candidate ordering must start with one of (v_2, v_1), (v_3, v_1), (v_3, v_2), (v_4, v_1), (v_4, v_2), (v_4, v_3), (v_5), (v_6), which yields $c_6 \leq 6c_4 + 2c_5 \leq 30 + 32 = 62$. $\quad\square$

Now we can beat the standard $O^*(2^n)$ time bound that holds for arbitrary hypergraphs (via dynamic programming on subsets), in the case when all components are small. In order to focus on the interesting exponential part only, we use the O^* notation that suppresses polynomial factors.

Theorem 3. MINSUMENDS *in weighted hypergraphs with n vertices, consisting only of components with at most k vertices, can be solved in $O^*(b_k^n)$ time, where $b_k = c_k^{1/k}$, in particular:*

$$b_3 \leq 1.26, \ b_4 \leq 1.5, \ b_5 \leq 1.7412, \ b_6 \leq 1.9895.$$

Proof. The numbers come from Lemma 5. In the special case when all components have exactly k vertices, we can apply the sedimentation algorithm to all $c_k^{n/k}$ combinations of candidate orderings of the components. The time bounds for the sedimentation algorithm remain valid also if all components have at most k vertices, by the monotonicity of the b_k and straightforward algebra. $\quad\square$

For $k > 6$, the time bounds obtained in this way would exceed $O^*(2^n)$. In order to avoid bases larger than 2 despite some large components we can, however, combine the benefits of small components with dynamic programming on subsets. The statement of the following result looks somewhat technical, but our aim was to make it as general as possible.

Theorem 4. *Let $C_1, \ldots, C_h$ be some components of a given hypergraph G, where C_i has k_i vertices. Suppose that we can, in polynomial time, compute for each component C_i a candidate set with $a_i < 2^{k_i}$ orderings. Then some optimal ordering of the given hypergraph can be computed in time $O^*(\prod_{i=1}^{h} a_i \cdot 2^{n - \sum_{i=1}^{h} k_i})$.*

Proof. First we compute the mentioned candidate sets and take all $\prod_{i=1}^{h} a_i$ combinations of the candidate orderings therein. For every such combination we apply Theorem 1 to compute an optimal ordering σ of $C_1 \cup \ldots \cup C_h$, using the candidate orderings as prescribed orderings within the components C_i.

We observe that the vertices of $C_1 \cup \ldots \cup C_h$ have the same ordering σ also within an optimal ordering of the entire hypergraph G. This "context-free" property holds since, by Theorem 1, an optimal ordering is characterized by the blocks having non-increasing densities, and neither blocks nor their densities depend on the remainder R of G outside $C_1 \cup \ldots \cup C_h$.

To R we apply dynamic programming on subsets: We generate all possible ordered subsets R incrementally from left to right, insert the resulting blocks into σ, discard solutions that violate the property that also blocks in R have non-increasing densities, and most importantly, whenever two ordered subsets of R are identical as sets and they end at the same position in σ, we keep only one ordering with minimal cost until that position. Correctness and time bound are straightforward. $\qquad\square$

Note that the time bound in Theorem 4 is never higher than $O^*(2^n)$, but it can be significantly smaller. Theorem 4 can be used, for instance, with $C_1, \ldots, C_h$ being the components with at most 6 vertices (due to Lemma 5), and Theorem 3 is the special case when larger components do not exist. But also large components might have much fewer than 2^{k_i} candidate orderings due to their specific edge weights.

6 Linear Inequalities Can Rule Out Candidate Orderings

Lemma 5 provides general upper bounds on the number of candidate orderings in components of sizes from 3 to 6. However, the given edge weights may rule out further candidate orderings, and thus some quick and simple preprocessing can make the main algorithms for MinSumEnds faster, not in the worst case but for many specific instances. In the following we illustrate these possibilities for size 3 only, but similar measures can be taken in larger components, too.

Remember from Sect. 5 that, in a component with three vertices with weights $w_1 \leq w_2 \leq w_3$, at most two candidate orderings exist: (v_2, v_1, v_3), and either (v_3, v_1, v_2) or (v_3, v_2, v_1). Assume that the candidate set is, in fact, of size 2, that is, the two sequences are neither equivalent nor is any of them dominated by another ordering. By the characterization of dominating sequences in Sect. 4, the resulting two sequences of densities of vertices must have the form $(a, b+c+d, e)$ and $(a + b, c, d + e)$, for some non-negative numbers a, b, c, d, e. (Actually, b and d are positive.) From these sequences we get the following necessary conditions for a candidate set of size 2:

$a = w_2$ and $a + b = w_3$, hence $b = w_3 - w_2$.
$e = w_3 + w_{13} + w_{23} + w_{123}$.
$b + c + d = w_1 + w_{12}$, hence $c + d = w_1 + w_2 + w_{12} - w_3$.
$w_3 + w_{23} \leq w_1 + w_{12}$, since otherwise (v_3, v_2, v_1) would dominate (v_2, v_1, v_3).

Case 1. The second candidate ordering is (v_3, v_1, v_2), since $w_2 + w_{23} \leq w_1 + w_{13}$. Then $c = w_1 + w_{13}$, hence $d = w_2 + w_{12} - w_3 - w_{13}$, thus also $w_3 + w_{13} < w_2 + w_{12}$.

Case 2. The second candidate ordering is (v_3, v_2, v_1), since $w_1 + w_{13} \leq w_2 + w_{23}$. Then $c = w_2 + w_{23}$, hence $d = w_1 + w_{12} - w_3 - w_{23}$, thus also $w_3 + w_{23} < w_1 + w_{12}$.

We conclude that the component has only one candidate ordering if some of the derived inequalities are violated.

Getting back to the case when two candidate orderings exist: Define x to be the distance (in π) between the first two vertices of the component, and define

y similarly for the last two vertices. Then (v_2, v_1, v_3) is strictly cheaper than the other candidate ordering (v_3, v_i, v_j), $\{i, j\} = \{1, 2\}$, if and only if $b \cdot x < d \cdot y$. In particular, the ordering in the component depends only on the ratios of these distances. Furthermore, in any optimal ordering using (v_2, v_1, v_3), since $w_2 \leq w_3 + w_{23} \leq w_1 + w_{12}$, the vertices v_2 and v_1 are in the same block, thus $x = 1$ and therefore $y > b/d$.

By checking these inequalities we can, in concrete instances, efficiently rule out certain partial solutions before or during dynamic programming when applying the algorithm from Theorem 4, and thus speed up its execution.

7   The Star Graph

So far we have focused attention on hypergraphs with small components. Another meaningful direction is to consider hypergraphs with structural restrictions. Here we provide such an example.

A *star graph* is a graph with n vertices where one vertex called the *center* is joined by $n-1$ edges to all other vertices called *leaves*. As a practical motivation of MINSUMENDS in the warehouse context, suppose that the center represents some main product, and the leaves represent optional accessoires only one of which may be chosen and used together with the main product. The main product as well as each accessoire may also be ordered separately. Hence all possible requests are the vertices and edges of a star graph.

We denote the center by c and the edge weights by $y(v) = w(\{c, v\})$. Recall that c and the leaves v also have vertex weights $w(c)$ and $w(v)$, respectively, as earlier.

Furthermore, let $B(n, m)$ denote a time bound of a minimum-weight perfect matching algorithm in bipartite graphs with n vertices on either side, and with m edges. For instance, we have $B(n, m) = O(n^2 \log n + nm)$ from [13].

Theorem 5. MINSUMENDS *on star graphs with n vertices can be solved witin* $O(n \cdot B(n, n^2))$ *time.*

Proof. For all n possible positions $\pi(c)$ of the center c, we compute an optimal solution with this fixed $\pi(c)$, and finally we take the best of these n solutions. For any fixed $\pi(c)$ we proceed as follows.

We construct a complete bipartite graph $(L, M; F)$ with vertex sets L and M of size $n-1$, and edge set $F = L \times M$. The vertices in L represent the leaves of the star, and the vertices in M represent the positions $i \in \{1, \ldots, n\} \setminus \{\pi(c)\}$. The weight $z(v, i)$ of any edge $\{v, i\} \in F$ is the cost of placing the leaf v at position $\pi(v) = i$. It is specified by $z(v, i) := i \cdot w(v) + \pi(c) \cdot y(v)$ if $i < \pi(c)$, and $z(v, i) := i \cdot (w(v) + y(v))$ if $i > \pi(c)$. The total cost of a solution is obviously $\pi(c) \cdot w(c) + \sum_{v \in L} z(v, \pi(v))$. Since $\pi(c)$ is fixed, it only remains to compute a minimum-weight perfect matching in $(L, M; F)$. $\square$

We conjecture that MINSUMENDS on star graphs can be solved faster than by doing n unrelated computations of bipartite matchings, encouraged by the observation that the n bipartite graphs for the different positions $\pi(c)$ are only slight variations of each other, and that their edge weights are also far from being arbitrary, e.g., they have obvious monotonicty properties. It should be possible to take advantage of that. In fact, the problem looks quite similar to the one in [7] (which has applications in scheduling), but the structure of edge weights is somewhat more complicated in our case, as M has vertices of two different types, namely those before and after $\pi(c)$. We must leave the question open.

On another front, Theorem 5 can be generalized straightforwardly: In graphs with a vertex cover of small fixed size γ we can decide on the γ positions of its vertices and then compute bipartite matchings for the other vertices, as they form an independent set, and thus the costs of each vertex depends only on its own position. This yields a time bound $O(n^\gamma \cdot B(n, n^2))$.

However, we do not see a way to solve MINSUMENDS on graphs with many such *simple* graphs as components, as we did with *small* components. (This would be interesting for the storage of several products with accessoires.)

8   More Open Problems

Driven by practical questions we have presented algorithms being faster than standard dynamic programming on subsets, but it remains the intriguing problem whether MINSUMENDS is NP-complete on weighted hypergraphs with components of at most three vertices (or any other constant size limit). There might exist a polynomial-time reduction from a "number problem" like PARTITIONING, but we did not manage to establish one.

A potentially interesting combinatorial question related to small components is whether the bounds in Lemma 5 are already tight. One may either refine the recurive argument or construct examples that enforce the obtained numbers of candidate orderings.

Besides graphs with small vertex covers, it would be worthwhile to find other natural cases (e.g., limited integer edge costs, or components with other restrictions such as tree structures) that can be solved in polynomial time or by FPT algorithms, using the concepts developed here.

Acknowledgment. This work extends initial ideas from Pedram Shirmohammad's master's thesis [14] supervised by the author, and it is also inspired by collaboration with Raad Salman and Fredrik Ekstedt at the Fraunhofer-Chalmers Research Centre for Industrial Mathematics.

References

1. Ambühl, C., Mastrolilli, M., Svensson, O.: Inapproximability results for maximum edge biclique, minimum linear arrangement, and sparsest cut. SIAM J. Comput. **40**, 567–596 (2011)
2. Arora, S., Frieze, A., Kaplan, H.: A new rounding procedure for the assignment problem with applications to dense graphs arrangements. Math. Progr. **92**, 1–36 (2002)

3. Bhasker, J., Sahni, S.: Optimal linear arrangement of circuit components. In: HICSS 1987, vol. 2, pp. 99–111 (1987)
4. Boysen, N., Stephan, K.: The deterministic product location problem under a pick-by-order policy. Discr. Appl. Math. **161**, 2862–2875 (2013)
5. Cohen, J., Fomin, F., Heggernes, P., Kratsch, D., Kucherov, G.: Optimal linear arrangement of interval graphs. In: Královič, R., Urzyczyn, P. (eds.) MFCS 2006. LNCS, vol. 4162, pp. 267–279. Springer, Heidelberg (2006). https://doi.org/10.1007/11821069_24
6. Damaschke, P.: Ordering a sparse graph to minimize the sum of right ends of edges. In: Gąsieniec, L., Klasing, R., Radzik, T. (eds.) IWOCA 2020. LNCS, vol. 12126, pp. 224–236. Springer, Cham (2020). https://doi.org/10.1007/978-3-030-48966-3_17
7. Domanic, N.O., Lam, D.K., Plaxton, C.G.: Bipartite Matching with Linear Edge Weights. In: Hong, S.K. (Ed.) ISAAC 2016. LIPIcs, vol. 64, pp. 28:1–28:13. Dagstuhl (2016)
8. Eikel, M., Scheideler, C., Setzer, A.: Minimum linear arrangement of series-parallel graphs. In: Bampis, E., Svensson, O. (eds.) WAOA 2014. LNCS, vol. 8952, pp. 168–180. Springer, Cham (2015). https://doi.org/10.1007/978-3-319-18263-6_15
9. Esteban, J.L., Ferrer-i-Cancho, R.: A correction on Shiloach's algorithm for minimum linear arrangement of trees. SIAM J. Comput. **46**, 1146–1151 (2017)
10. Feige, U., Lee, J.R.: An improved approximation ratio for the minimum linear arrangement problem. Info. Proc. Letters **101**, 26–29 (2007)
11. Fellows, M.R., Hermelin, D., Rosamond, F.A., Shachnai, H.: Tractable parameterizations for the minimum linear arrangement problem. ACM Trans. Comput. Theory **8**, 6:1-6:12 (2016)
12. Fernau, H.: Parameterized algorithmics for linear arrangement problems. Discr. Appl. Math. **156**, 3166–3177 (2008)
13. Fredman, M.L., Tarjan, R.E.: Fibonacci heaps and their uses in improved network optimization algorithms. J. ACM **34**, 596–615 (1987)
14. Shirmohammad, P.: Linear Arrangements with Closeness Constraints, Master's thesis, Chalmers and Univ. of Gothenburg (2020)

An Efficient Noisy Binary Search
in Graphs via Median Approximation

Dariusz Dereniowski[1] , Aleksander Łukasiewicz[2](✉) ,
and Przemysław Uznański[2]

[1] Faculty of Electronics, Telecommunications and Informatics,
Gdańsk University of Technology, Gdańsk, Poland
[2] Institute of Computer Science, University of Wrocław, Wrocław, Poland
aleksander.lukasiewicz@cs.uni.wroc.pl

Abstract. Consider a generalization of the classical binary search problem in linearly sorted data to the graph-theoretic setting. The goal is to design an adaptive query algorithm, called a *strategy*, that identifies an initially unknown *target* vertex in a graph by asking queries. Each query is conducted as follows: the strategy selects a vertex q and receives a reply v: if q is the target, then $v = q$, and if q is not the target, then v is a neighbor of q that lies on a shortest path to the target. Furthermore, there is a noise parameter $0 \leq p < \frac{1}{2}$ which means that each reply can be incorrect with probability p. The optimization criterion to be minimized is the overall number of queries asked by the strategy, called the *query complexity*. The query complexity is well understood to be $\mathcal{O}(\varepsilon^{-2} \log n)$ for general graphs, where n is the order of the graph and $\varepsilon = \frac{1}{2} - p$. However, implementing such a strategy is computationally expensive, with each query requiring possibly $\mathcal{O}(n^2)$ operations.

In this work we propose two efficient strategies that keep the optimal query complexity. The first strategy achieves the overall complexity of $\mathcal{O}(\varepsilon^{-1} n \log n)$ per a single query. The second strategy is dedicated to graphs of small diameter D and maximum degree Δ and has the average complexity of $\mathcal{O}(n + \varepsilon^{-2} D \Delta \log n)$ per query. We point out that we develop an algorithmic tool of graph median approximation that is of independent interest: the median can be efficiently approximated by finding a vertex minimizing the sum of distances to a randomly sampled vertex subset of size $\mathcal{O}(\varepsilon^{-2} \log n)$.

Keywords: Graph median · Graph searching · Noisy search

1 Introduction

Our research problems originate in the classical "twenty questions game" proposed by Rényi [36] and Ulam [42]. The classical problem of binary search with

Partially supported by National Science Centre (Poland) grant number 2018/31/B/ ST6/00820.

P. Flocchini and L. Moura (Eds.): IWOCA 2021, LNCS 12757, pp. 265–281, 2021.
https://doi.org/10.1007/978-3-030-79987-8_19

erroneous comparisons received a considerable attention and optimal query complexity algorithms are known, see e.g. [8,11,21,24,35] for asymptotically best results. The binary search in linearly ordered data can be re-casted as a search on a path, where each query selects a vertex q and reply gives whether the target element is q, or is to the left or to the right of q. This leads to the graph search problem introduced first for trees by Onak and Parys in [33] and then recently for general graphs by Emamjomeh-Zadeh et al. in [23]. We recall a following formal statement.

Problem formulation. Consider an arbitrary simple graph G whose one vertex v^* is marked as the *target*. The target is unknown to the query algorithm. Each query points to a vertex q, and a *correct* reply does the following: if $v^* = q$, then the reply returns q, and if $v^* \neq q$, then the reply returns a neighbor of q that belongs to a shortest path from q to v^*, breaking ties arbitrarily (and independently for each query). We further assume that some replies can be incorrect: each query receives an erroneous[1] reply (independently) with some fixed probability $0 \leq p < \frac{1}{2}$ (the value of the *noise parameter* p is known to the algorithm). We assume the strongest model in which an erroneous reply is adversarial, versus a randomly selected q. The goal is to design an algorithm, also called a *strategy* performing as few queries as possible.

Typically in the applications of the adaptive query problems the main concern is the number of queries to be performed, i.e., their *query complexity*. This is due to the fact that the queries usually model a time consuming and complex event like making a software check to verify whether it contains a malfunctioning piece of code, c.f. Ben-Asher et al. [7], or asking users for some sort of feedback c.f. Emamjomeh-Zadeh and Kempe [22]. However, as a second measure the computational complexity of each query comes into play and it is of practical interest to resolve the question of having an adaptive query algorithm that keeps an optimal query complexity and optimizes the computational cost as a second criterion. This may be especially useful in cases when queries are fast, like communication events over a noisy channel.

The asymptotics of the query complexity is quite well understood to be roughly $\frac{\log n}{1-H(p)} = \mathcal{O}(\varepsilon^{-2} \log n)$ (c.f. [20,23]), where n is the order of the graph, $\varepsilon = \frac{1}{2} - p$, and $H(p) = -p \log_2 p - (1-p) \log_2(1-p)$ is the entropy.[2] However, it is of theoretical and practical interest to know what is the optimal complexity of computing each particular query. This leads us to a general statement of the type of solution we seek.

Research question. By what amount can the computational complexity of an adaptive graph query algorithm be improved without worsening the query complexity?

---

[1] We note that such an erroneous reply may occur both when $v^* = q$ and $v^* \neq q$.

[2] By information-theoretic arguments, any algorithm that locates the target correctly with probability $1 - \delta$ has to make $\frac{\log_2 n}{1-H(p)} + \Omega(\frac{\log \delta^{-1}}{1-H(p)})$ queries. In the regime of *high probability algorithms* this becomes simply $\Omega(\frac{\log n}{\varepsilon^2})$ queries.

In this work we make the following assumption: a *distance oracle* is available to the algorithm and it gives the graph distance between any pair of vertices. This is dictated by the observation that the computation of multiple-pair shortest paths throughout the search would dominate the computational complexity. On the other hand, we note that this is only used to resolve (multiple times) the following for a query: given a vertex q, its neighbor v and an arbitrary vertex u, does v lie on a shortest path from q to u? Thus, some weaker oracles can be assumed instead. We further comment on this assumption in the next section.

1.1 Motivation

To sketch potential practical scenarios of using graph queries we mention one of examples given in [22]. These examples are anchored in the field of machine learning, and since they have the same flavor with respect as how graphs are used, we refer to one of them. Consider a situation in which a *system* wants to learn a clustering by asking queries. Each query presents a potential clustering to a user and if this is not the target clustering, then as a response the user either points two clusters that should be merged or points one cluster that should be split (but does not say how to split it). Thus, the goal is to construct a query algorithm to be used by the system. It turns out that learning the clustering can be done by asking queries on a graph: each vertex v corresponds to a clustering and a reply of the user for v will be aligned with one of the edges incident to v. In other words, the reply can be associated with an edge outgoing from v that lies on a shortest path to the desired target clustering. We emphasize some properties of this approach. First, the fact that the reply indeed reveals the shortest path to the target is an important property of the underlying graph used by the algorithm and thus the graph needs to be carefully defined to satisfy it. Second, the user is not aware of the fact that such a graph-theoretic approach is used, as only a series of proposed clustering is presented. Third, this approach is resilient to errors on the user side: the graph query algorithms easily handle the facts that some replies can be incorrect (the user may make a mistake, or may not be willing to reveal the truth).[3] It has been shown [22] that in a similar way one can approach the problems of learning a classifier or learning a ranking. We note that a well defined graph is used to prove that a particular query model is valid for a particular application, although in these applications it is not necessary to explicitly construct a graph to navigate the algorithm. Instead, it may be enough to have an abstract encoding that allow us to compute the next vertex to query and to conclude the reply vertex; we refer the reader for details to [22]. Moreover, it is desirable to approach a solution in this way due to a large search space.

From the standpoint of complexity we can approach such scenarios in two ways. First, one can derive an algorithm that specifically targets a particular application, in which it may not be necessary to construct the entire graph but instead reconstruct only what is necessary to perform each query. The second way

---

[3] Actually, our method is slightly more general: our proofs reveal that the algorithm is resilient to the placement of errors, and it will succeed if at most a p-fraction of replies is erroneous, which makes it usable against users that distribute lies adversarially.

is a general approach taken in this work: to consider the underlying graph as an abstract data structure out of the context of particular applications. We emphasize that examples like the ones mentioned above reveal that some applications may be burdened by the fact that the underlying graph is large, in which case the computational complexity, or local search procedures may be more crucial.

We finally comment on our assumption that a shortest path oracle is provided to the algorithm. In the machine learning applications [22], the graphs may be constructed in such a way that knowing which objects represent two vertices is sufficient to conclude the distance between them, i.e., a low-complexity distance oracle can be indeed implemented. This can be seen as a special case of a general approach to achieve distance oracles in practice through the so called distance-labeling schemes (c.f. Gavoille et al. [26] and for practical approaches, c.f. Abraham et al. and Kosowski and Viennot [3,30]). We finally note that having the exact distances between vertices is crucial for this problem: if the distance oracle is allowed to provide even just a 1-additive approximation of the exact distance, then each query algorithm needs to perform $\Omega(n)$ queries for some graphs c.f. Deligkas et al. [17]. We note that the distance oracle access can be replaced with a multi-source distance computation (e.g. using BFS), at the cost of replacing some of the $\mathcal{O}(n)$ factors in the cost functions with $\mathcal{O}(m)$ (the number of edges). Alternatively, a popular assumption borrowed from computational geometry is that we operate on a metric space with a metric (distance) function given.

1.2   Our Results and Techniques

For a query on a vertex q with a reply v, we say that a vertex u is *consistent* with the reply if $q = v = u$, or $q \neq v$ but v lies on a shortest path between u and q; the set of all such consistent vertices u is denoted by $N(q, v)$ (see Fig. 1).

Our method is based on a multiplicative weight update (MWU): the algorithm keeps the weights $\omega(v)$ for all vertices v, starting with a uniform assignment. The weight is representing the likelihood that a vertex is the target, although we point out that formally this is not a probability distribution. In MWU, the weight of each vertex that is not consistent with a reply is divided by an appropriately fixed constant Γ that depends on $\varepsilon = \frac{1}{2} - p$.

To keep the query complexity low, it is required that the queried vertex q fulfills a measure of 'centrality' in a graph in the sense that a query to such a central vertex results in an adequate decrease in the total weight. This is a graph-theoretic analogue of the 'central' element comparison in the classical binary search. Two functions that have been used [17,20,22] to formalize this are

$$\Phi(v) = \sum_{u \in V} d(u, v) \cdot \omega(u), \qquad \text{and} \qquad \Lambda(v) = \max_{u \in N(v)} \omega(N(v, u)),$$

where $N(v)$ is the set of neighbors of v in the graph, and $d(u, v)$ is the distance between u and v. For brevity, $\omega(S) = \sum_{u \in S} \omega(u)$ for any $S \subseteq V$, and $\omega = \omega(V)$.

Definition 1. *A vertex* $q = \arg\min_{v \in V} \Phi(v)$ *is called a* median.

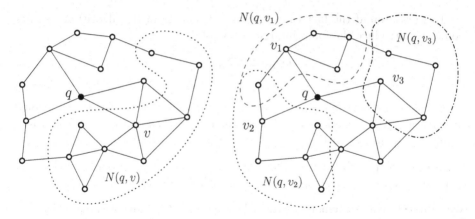

Fig. 1. Set $N(v, \cdot)$ for single neighbor u (on the left) and sets $N(v, \cdot)$ for the remaining neighbors of vertex v (on the right). Observe that in particular these sets might not be disjoint—e.g. $N(v, u_1)$ and $N(v, u_2)$ share one vertex.

We note a fundamental bisection property of a median:

Lemma 1 (c.f. [23] section 2). *If q is a median, then $\Lambda(q) \leq \omega(V)/2$.*

Such property is key for building efficient binary-search algorithms in graphs e.g., for the noiseless case, repeatedly querying a median of X, where $X \subseteq V$ is the subset of vertices that still can be a target, results in a strategy guaranteeing at most $\log_2 n$ queries. See also analysis of Algorithm 2 in [23] and the proof of Lemma 4 in [20].

A disadvantage of using median is that it is computationally costly to find, i.e., assuming the distance oracle it takes (deterministically) $\mathcal{O}(n^2)$ time. Moreover, using its multiplicative approximation, that is, through a function Φ' such that $\Phi'(q) = (1 \pm \varepsilon')\Phi(q)$ for any constant $\varepsilon' > 0$, blows up the strategy length exponentially [17] and thus this approach is not suitable. On the other hand, approximating a Λ-minimizer is feasible, as noted also by [17].

Hence, we work towards a method of efficient median approximation through Λ minimization. We believe that this algorithmic approach is of independent interest and can be used in different graph-theoretic problems. It turns out that we do not even need a multiplicative approximation of a Λ-minimizer but we only need that $\Lambda(q)$ is at most roughly half of the total weight. This is potentially usable in algorithms using generally understood graph bisection. (For an example of using such balanced separators for somewhat related search with persistent errors see e.g. Boczkowski et al. [10].) Formally, motivated by Lemma 1, we relax the notion of the median to the following.

Definition 2. *A vertex q^* is δ-close to a median, where $\delta > 0$, if $\Lambda(q^*) \leq \left(\frac{1}{2} + \delta\right) \cdot \omega$.*

To work-around the fact that Φ is not efficient from the algorithmic standpoint, we introduce the following relaxation of Φ:

$$\Phi^*(q) = \sum_{v \in S} d(q, v),$$

where S is a random sample of vertices with probability distribution proportional to ω (so $\Phi^*(q)$ is a random variable dependent on a choice of S). We can now formulate our main contribution in terms of new algorithmic tools:

Median approximation. The relaxation of Φ to Φ^* provides, with high probability, a sufficient approximation of the median vertex in a graph.

We formalize this statement in the following way. Consider a sample size $s = \frac{8 \ln n}{\delta^2}$, where n is the number of vertices of the graph and the parameter $\delta = \Theta(\varepsilon)$ is specified in Sect. 2. This allows us to say how to approximate the median efficiently through a local condition:

Theorem 1. *Let q be a vertex such that for each $v \in N(q)$ it holds $\Phi^*(q) \le \Phi^*(v) + \delta s$. Then, with high probability at least $1 - n^{-3}$, the vertex q is δ-close to a median.*

Corollary 1. *Let $q^* = \arg\min_{v \in V} \Phi^*(v)$. Then, the vertex q^* is δ-close to a median with high probability at least $1 - n^{-3}$.*

Returning to our search problem, these observations are enough to both find the right query vertex in each step, keep the strategy length low, and have a centrality measure that is efficient in terms of computational complexity. This leads us to the following theorem that is based on MWU with some appropriately fixed scaling factor Γ.

Theorem 2. *Let $p = \frac{1}{2} - \varepsilon$ be the noise parameter for some $0 < \varepsilon \le \frac{1}{2}$. There exists an adaptive query algorithm that after asking $\tau = \mathcal{O}(\frac{\log n}{\varepsilon^2})$ queries returns the target correctly with high probability. The computational complexity of the algorithm is $\mathcal{O}(\frac{n \log n}{\varepsilon})$ per query with high probability.*

The algorithm behind the theorem iterates over the entire vertex set to find a Φ^*-minimizer. We can refine this algorithm for graphs of low maximum degree Δ and diameter D. For that we use a local search whose direct application requires 'visiting' $D\Delta$ vertices to get to a Φ^*-minimizer. However, we introduce two ideas to speed it up. First, we add another approximation layer on top of Φ^*: it is not necessary to find the exact Φ^*-minimizer but its approximation, which we do as follows. Whenever the local search moves from one vertex u to its neighbor v and the improvement from $\Phi^*(u)$ to $\Phi^*(v)$ is sufficiently small, then v will do for the next query. The second one is to start the local search from the vertex queried in the previous step. These two ideas combined lead to the second main result.

Theorem 3. *Let* $p = \frac{1}{2} - \varepsilon$ *for some* $0 < \varepsilon \leq \frac{1}{2}$. *There exists an adaptive query algorithm that after asking* $\tau = \mathcal{O}(\frac{\log n}{\varepsilon^2})$ *queries returns the target correctly with high probability. The average computational complexity per query is* $\mathcal{O}(n + D\Delta\frac{\log n}{\varepsilon^2})$ *for graphs with diameter D and maximum degree Δ.*

Recall that our computational complexities are measured assuming oracle access to distances in graph, i.e., the distance between any pair of vertices is available in constant time. Additionally, if one is to measure bit-complexity of operations, the time grows by a factor of roughly $\mathcal{O}(\varepsilon^{-1})$ coming from the bound on bit-lengths of weights used in MWU.

1.3   Related Work

Median computation is one of the fundamental ways of finding central vertices of the graph, with huge impact on practical research [5,6,25,27,37,41]. A significant amount of research has been devoted to efficient algorithms for finding medians of networks [34,39,40] or approximating the notion [13,14]. We note the seminal work of Indyk [28] which includes $1 + \varepsilon$ (randomized) approximation to 1-median in time $\mathcal{O}(n/\varepsilon^5)$ in metric spaces – we note that the form of approximation there differs from ours, although the very-high level technique of using random sampling is common. Chechik et al. in [15] use (non-uniform) random sampling to answer queries on sum of distances to the queried vertices in graphs.

We also refer the reader to some recent work on the (deterministic) median computation in median graphs, see Beneteau et al. [9] and references therein. More related centrality measures of a graph are discussed in [1,2,12] in the context of fine-grained complexity, showing e.g. that efficient computation of a median vertex (in edge-weighted graphs) is equivalent under subcubic reductions to computation of All-Pairs Shortest Paths.

Substantial amount of research has been done on searching in sorted data (i.e., paths), which included investigations for fixed number of errors [4,35], optimal strategies for arbitrary number of errors and various error models, including linearly bounded [21], prefix-bounded [11] and noisy/probabilistic [8,29]. Also, a lot of research has been done on how different types of queries influence the search process—see [16] for a recent work and references therein. The mostly studied comparison queries for paths have been extended to graphs in two ways. The first one is a generalization to partial orders [7,31] (with some deterministic algorithms), although this does not further generalize well for arbitrary graphs [18]. It is worth noting that much work has been devoted to the computational complexity of finding error-less strategies [19,31,32]. The second extension is by using the vertex queries studied in this work, for which much less is known in terms of the complexity. This problem becomes equivalent to the vertex ranking problem for trees [38], but not for general graphs (see also [33]).

Similarly as in the case of the classical binary search, the graph structure guarantees that there always exists a vertex that adequately partitions the search space in the absence of errors [23]. The problem becomes much more challenging as this is no longer the case when errors are present. A centrality measure that works well for finding the right vertex to be queried is a median used in [20, 23]. However, as shown in [17], the median is sensitive to approximations in the following way. When the algorithm decides to query a $(1 + \varepsilon')$-approximation v of the median (minimizer of Φ' which is $1 + \varepsilon'$ approximation of Φ), then some graphs require $\mathcal{O}(\sqrt{n})$ queries, where the approximation is understood as $\Phi(v) \leq (1 + \varepsilon') \min_{u \in V} \Phi(u)$. This results holds for the error-less case. Furthermore, the authors introduce in [17] the potential Λ (denoted by Γ therein) and prove, also for the error-less case, that it guarantees $\frac{\log_2 n}{1 - \log_2(1 + \varepsilon)} \approx (1 + \varepsilon) \log_2 n$ queries, when in each step a $(1 + \varepsilon)$-approximation of the Λ-minimizer is queried. However, this issue has been considered from a theoretical perspective and no optimization considerations have been made. In particular, it was left open as to how to reduce the query complexity at an expense of working with such approximations. This, and the consideration of the noise are two our main improvements with respect to [17]. We also point out that our definition of δ-closeness to a median differs from $(1 + \varepsilon)$-approximations in the sense that our definition is much less strict: a vertex q^* that is δ-close to a median may have the property that $\Lambda(q^*)$ significantly deviates from $\min_{u \in V} \Lambda(u)$.

Some complexity considerations have been touched in [22], from the perspective of targeting specific machine learning applications, where already the above-mentioned Λ-minimizer has been used. To make the statements form that work comparable to our results, we have two distinguish two input size measures that apply. In [22], for a particular application an input consists of a specific machine learning instance, and denote its size by $\tilde{n}$. In order to find a solution for this instance, a graph G of size n is constructed and an adaptive query algorithm is being run on this graph. It is assumed that $\log_2 n$ is polynomial in $\tilde{n}$. The diameter D and maximum degree Δ of G are both assumed in [22] to be polylogarithmic in $\tilde{n}$. A local search is used to find a vertex that approximates the Λ-minimizer. For that, in each step a sampling is used for the approximation purposes: for each vertex v along the local search, all its neighbors u are tested for finding an approximation Λ, giving the complexity of $\mathcal{O}(D\Delta)$. It is concluded that the overall complexity of performing a single query is $\mathcal{O}(D\Delta \mathrm{poly}(\log n, \frac{1}{\varepsilon}))$.

1.4   Outline

We proceed in the paper as follows. Section 2 provides a 'template' strategy in which we simply query a vertex that is δ-close to a median. The strategy length is there fixed carefully to meet the tail bounds on the error probability. Then, in Sect. 3, we prove that our sample size is enough to ensure high success probability. Section 4 observes that the overall complexity of the algorithm can be reduced by avoiding recasting the entire sample in each step: it is enough to replace only a small fraction of the current sample when going from one step of the strategy to the next. We then combine these tools to prove our main theorems in Sect. 5, where for Theorem 3 we additionally make several observations on speeding-up the classical local search in a graph.

2 The Generic Strategy

Results in this section are an adaptation of ones presented in [20] to the setting of querying *approximate* median. We provide streamlined proofs for completeness.

As an intermediate convenient step, we recall the following adversarial error model: given a constant $r < 1/2$, if the strategy length is τ, then it is guaranteed that at most $r \cdot \tau$ errors occurred throughout the search (their distribution may be arbitrary). We set our parameters as follows: let $\eta = \varepsilon/2$, $r = \frac{1}{2} - \eta$, and assume without loss of generality that $\eta < 1/8$. (We can do this, since this increases the complexity of the algorithm by a constant factor at most.) Let $\delta = \eta/4$. With these parameters, we provide Algorithm LB-SEARCH that runs the multiplicative weight update with $\Gamma = \frac{1}{1-4\eta}$ for $\tau = \frac{10 \log_2 n}{\eta^2}$ steps. Then we prove (cf. Lemma 2) that this strategy length is sufficient for correct target detection in this error model.

We remark that the vertex q is selected in line 3 in two different ways to meet the complexities of Theorems 2 and 3. To obtain Theorem 2, we use a Φ^*-minimizer as stated in Corollary 1. For Theorem 3 we also maintain the function Φ^* but a local search is used to approximate a Φ^*-minimizer. In both cases, line 3 requires a resampling that is covered in Sect. 4. We write ω_t to denote the vertex weight in a step t. (So, ω_0 is the initial uniform weight assignment.)

Algorithm LB-Search: Always query a δ-close vertex to a median.

---

1   $\omega(v) \leftarrow \frac{1}{n}$ and $\ell_v \leftarrow 0$ for each $v \in V$

2   **for** $\tau = 10 \frac{\log_2 n}{\eta^2}$ steps **do**

3   Let q be any vertex that is δ-close to a median

4   Query the vertex q

5   **for** each vertex u not consistent with the answer **do**

6         $\omega(u) \leftarrow \omega(u)/\Gamma$, where $\Gamma = \frac{1}{1-4\eta}$

7         $\ell_u \leftarrow \ell_u + 1$

8   **return** the vertex v with the smallest ℓ_v

Lemma 2. *If during the execution of Algorithm* LB-SEARCH *over total* τ *queries there were at most* $r \cdot \tau$ *errors, then the algorithm outputs the target.*

Proof. If a vertex v at step t satisfies $\omega_t(v) > (\frac{1}{2} + \delta)\omega_t$, then we say that v is *heavy* at step t. We aim at proving that the overall weight decreases multiplicatively either by at least $(1 - \eta)^2$ or $\frac{\Gamma+1}{2\Gamma}$ per step. In the absence of a heavy vertex we get the first bound, and it is an immediate consequence of the Eq. (1) below. If we get a heavy vertex at some point, none of these bounds may be true in this particular step (this phenomenon is inherent to the graph query model itself) but we show below that the second one holds in an amortized way (cf. Lemma 4).

If at step t there is no heavy vertex, then assuming vertex q was queried, with u being an answer, we have $\omega_t(N(q, u)) \le (\frac{1}{2} + \delta)\omega_t$, since either $u \in N(q)$

and Lemma 1 applies bounding $\omega_t(N(q,u)) \leq \frac{1}{2}\omega_t$, or $u = q$ and we use the fact that $N(q,u) = \{q\}$. Then,

$$
\begin{aligned}
\omega_{t+1} &= \omega_t(N(q,u)) + \frac{\omega_t(V \setminus N(q,u))}{\Gamma} \\
&\leq \left(\frac{1}{2} + \delta + \frac{\frac{1}{2} - \delta}{\Gamma} \right) \omega_t \\
&= (1 - 2\eta + 4\eta\delta)\,\omega_t = (1 - \eta)^2 \omega_t.
\end{aligned}
\tag{1}
$$

Assume otherwise that there is vertex v that is heavy at step t.

Lemma 3. *If at any step t there is a heavy vertex v, then v is the only δ-close to a median vertex at this step.*

Proof. For any $u \neq v$, we have that $\Lambda(u) \geq \omega_t(v) > (\frac{1}{2} + \delta)\omega_t$, i.e., u is not δ-close to a median. On the other hand, $\Lambda(v) \leq \omega_t(V \setminus \{v\}) < (\frac{1}{2} - \delta)\omega_t$, i.e., v is δ-close to a median. □

The above lemma implies that if some v is heavy, then it will be queried in this particular step. The next lemma calculates (we omit the proof) the overall potential drop in a series of steps in which some vertex is heavy.

Lemma 4. *Consider the maximal consecutive segment of steps $\mathcal{I}$ where some q is heavy. That is, we pick t_1, t_2 such that q is heavy in all steps $t \in \mathcal{I} = \{t_1, \ldots, t_2 - 1\}$ and is not heavy in steps $t_1 - 1$ and t_2. Then, $\omega_{t_2} \leq \left(\frac{\Gamma + 1}{2\Gamma} \right)^{t_2 - t_1} \omega_{t_1}$.*

Let q be the target, and u be the output of Algorithm LB-SEARCH. Assume w.l.o.g. that the algorithm run for $\tau' \geq \tau$ steps. Since

$$
\tau' \geq 10 \frac{\log_2 n}{\eta^2} \geq \frac{\log_2 n}{r \log_2(1 - 4\eta) - 2\log_2(1 - \eta)},
$$

where the inequality follows from $(1/2 - \eta) \cdot \log_2(1 - 4\eta) - 2\log_2(1 - \eta) \geq \frac{1}{10}\eta^2$ when $0 \leq \eta \leq \frac{1}{8}$, we obtain a bound

$$
(1 - 4\eta)^{r\tau'} \geq (1 - \eta)^{2\tau'} \cdot n.
\tag{2}
$$

We assume that the algorithm outputs an incorrect vertex u, and show that it leads to a contradiction. We consider the state of the weights after τ' steps. We consider two cases.

In the first case suppose that there is no heavy vertex after τ' steps. We observe that the starting weight satisfies $\omega_0 = 1$, and by the bound on the number of errors accumulated on target vertex v^* (it cannot be more than $r\tau'$), we have $\omega_{\tau'} \geq \omega_{\tau'}(v^*) + \omega_{\tau'}(u) > \frac{1}{n} \left(\frac{1}{\Gamma} \right)^{r\tau'}$. By Eq. (1) and Lemma 4, we know that every step contributed at least a factor $(1 - \eta)^2$ or $(\Gamma + 1)/(2\Gamma) = (1 - 2\eta)$ multiplicatively to the total weight. Thus, by (2), $\omega_{\tau'} \leq (1 - \eta)^{2\tau'} \omega_0 \leq \frac{1}{n}(1 - 4\eta)^{r\tau'} = \frac{1}{n} \left(\frac{1}{\Gamma} \right)^{r\tau'}$, which leads to a contradiction.

Now consider the second case. The returned vertex u is heavy after τ' steps. In this case we cannot use Lemma 4 explicitly. We proceed as follow: we append at the end of the strategy a virtual sequence of k identical query-answers: algorithm queries u, and receives an no-answer pointing towards q. Here, k is chosen to be minimal such that after $\tau' + k$ steps u is no longer heavy (it exists, since each such query increases ℓ_u by 1, and leaves ℓ_q unchanged). However, at the end of $\tau' + k$ round ℓ_u is minimal (possibly not necessarily uniquely minimal). We note that appending those k steps did not increase the total number of errors from the answerer, and all of the queries were asked to a heavy vertex u. This reduces this case to the previous one, with increased value of τ'. □

We now transit from the adversarial search to the noisy setting. This is done by using Algorithm LB-SEARCH as a black box with η being fixed appropriately. Recall that $p = \frac{1}{2} - \varepsilon$, and we will use the following dependence of η on ε (note that by taking η smaller than ε we accommodate the necessary tail bound in the lemma below, i.e., we ensure that the event of having more than $r\tau$ errors is sufficiently unlikely).

Lemma 5. *Run Algorithm* LB-SEARCH *with* $r = \frac{1}{2} - \eta$, *where* $\eta = \varepsilon/2$. *If an answer to each query was erroneous with probability at most* p, *independently, then the algorithm outputs the target vertex with a high probability of at least* $1 - n^{-3}$.

Proof. Recall $\tau = 10\frac{\log_2 n}{\eta^2}$ in Algorithm LB-SEARCH. Denote by L the overall number of errors that have occurred during the execution of the algorithm. The expected number of errors is $p \cdot \tau$. By the Hoeffding inequality,

$$\Pr[L \geq r \cdot \tau] \leq \exp\left(-2\tau(r - p)^2\right) = \exp(-20\log_2 n) \leq n^{-3}.$$

Thus with high probability number of errors is bounded so that we can apply Lemma 2 (which in itself gives a deterministic guarantee). □

3   Sampling Guarantees

To take the 'random sampling' counterparts of Φ and Λ, consider a $S = \{m_1, \ldots, m_s\}$ to be a multiset of s vertices sampled from V with repetitions, with sampling probabilities $p(v) \sim \omega(v)$. That is, for each m_i, we have $\Pr(m_i = v) = \frac{\omega(v)}{\omega}$ and choices made for m_i are fully independent. To such an S, we refer as a *random sample*. We then define the following potentials

$$\Phi^*(v) = \sum_{u \in S} d(u, v) \qquad \text{and} \qquad \Lambda^*(v) = \max_{u \in N(v)} |S \cap N(v, u)|,$$

where the intersection of a multiset S with some set $X \subset V$ is defined as a multiset $S \cap X = \{m_i : i \in \{1, \ldots, s\} \wedge m_i \in X\}$ (see also Fig. 2 for an illustration).

We note a specific detail regarding these functions – we will prove and use the fact that in order to find a vertex that is δ-close to a median (a vertex we need to query), it is enough to pick an approximation of the Φ^*-minimizer. Note that

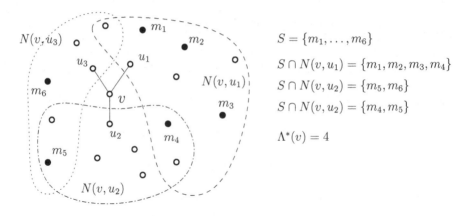

$$S = \{m_1, \ldots, m_6\}$$
$$S \cap N(v, u_1) = \{m_1, m_2, m_3, m_4\}$$
$$S \cap N(v, u_2) = \{m_5, m_6\}$$
$$S \cap N(v, u_2) = \{m_4, m_5\}$$
$$\Lambda^*(v) = 4$$

Fig. 2. Illustration for the definition of Λ^*.

δ-closeness is defined in terms of Λ which has a similar meaning to Λ^*. However, the subtlety here is due to a complexity issue—it is easier to recompute the Φ^* upon updating the sample S.

We denote $s = |S|$ and assume in the rest of the paper that $s = \frac{8 \ln n}{\delta^2}$. In this section we prove that this choice of s is sufficient, and then Sect. 4 deals with the complexity issues of the sampling method. The proofs of the following lemmas are omitted due to space limitations:

Lemma 6. *For each v, $\frac{\Lambda(v)}{\omega} \leq \frac{\Lambda^*(v)}{s} + \delta/2$ with a high probability at least $1 - n^{-3}$.*

Lemma 7. *Let q be such that $\forall_{v \in N(q)} \Phi^*(q) \leq \Phi^*(v) + \delta s$. Then, $\Lambda^*(q) \leq \frac{s(1+\delta)}{2}$.*

Hence, we can prove Theorem 1: Combining Lemma 7 and Lemma 6,

$$\Lambda(q) \leq \left(\frac{\Lambda^*(q)}{s} + \delta/2 \right) \cdot \omega \leq \left(\frac{1}{2} + \delta/2 + \delta/2 \right) \cdot \omega$$

with probability at least $1 - n^{-3}$.

4   Maintaining the Sample

We now discuss the complexity of maintaining the sample S upon the vertex weight updates. Given a sample set S_t at step t, the next sample S_{t+1} is computed by a call to Algorithm RESAMPLING below. The correctness of Algorithm RESAMPLING is given by Lemma 8 whose proof is omitted due to space limitations. Its proof follows the cases in the pseudo-code to show that both the vertices that remain in the sample and the new ones meet the probability requirements for a random sample.

Algorithm Resampling: Update of the sample after step t.

---

1 **foreach** $x_t \in S_t$ **do**
2   **if** x_t is consistent with the reply in step t **then**
3          $x_{t+1} \leftarrow x_t$
4   **else**
5   with probability $1/\Gamma$ do: $x_{t+1} \leftarrow x_t$
6   otherwise x_{t+1} is drawn randomly from V with
              distribution proportional to the weights ω_{t+1}
7   Insert x_{t+1} to S_{t+1}

Lemma 8. *Suppose that in Algorithm* LB-SEARCH, *after each weight update the current random sample S is recalculated by a call to Algorithm* RESAMPLING. *Then, with high probability at least $1 - n^{-3}$, at most $2\varepsilon|S|$ resampling operations occur at each step.*

5   Main Theorems

We start by proving Theorem 2.

Proof (Proof of Theorem 2). First, assume without loss of generality that $\frac{\log n}{\varepsilon^2} < n^2$, as otherwise the claimed one-step complexity is $\varepsilon^{-1} n \log n = \Omega(n^2)$. This can be met by an algorithm that at each step queries a median, see [23].

Run Algorithm LB-SEARCH that performs $\tau = \mathcal{O}(\frac{\log n}{\varepsilon^2})$ queries by Lemma 5. The algorithm maintains a sample S_t at each step t by using Algorithm RESAMPLING. By Corollary 1, the probability that each step of the algorithm indeed uses a vertex that is δ-close to a median is $1 - n^{-3}$. After each query, the algorithm updates the weights in time $\mathcal{O}(n)$, and $\mathcal{O}(\varepsilon s)$ vertices are re-sampled by Lemma 8, for the cost of $\mathcal{O}(n + \varepsilon s \log n)$ which is subsumed by other terms. Thus the cost of maintaining the values of Φ^* is $\mathcal{O}(\varepsilon s)$ per vertex, or $\mathcal{O}(n\varepsilon s)$ in total, which is the dominant cost for the algorithm, with the update being performed as:

$$\Phi^*(v) \leftarrow \Phi^*(v) - \sum_{u \in S_{t+1} \setminus S_t} d(u, v) + \sum_{u \in S_t \setminus S_{t+1}} d(u, v).$$

Taking a union bound over all steps, the success probability is $1 - \mathcal{O}(n^{-1})$.   □

Now we turn out attention to the proof of Theorem 3. Using local search is natural and gives an improvement for low-degree low-diameter graphs. The two 'twists' that we add are early termination (see the pseudo-code shown as Algorithm LOCAL-SEARCH) and resuming from the vertex that is the output of the previous local search (which is used in the proof of Theorem 3). The former allows us to directly bound the number of iterations; cf. Lemma 9.

Lemma 9. *If Algorithm* LOCAL-SEARCH *run with an input vertex v returns a vertex v', then the number of iterations is upper-bounded by $1 + \frac{\Phi^*(v) - \Phi^*(v')}{\delta s}$.*

Algorithm Local-Search: Find a local median starting
from an input vertex v

---

1 **while** true **do**

2      $q = \arg\min_{u \in N(v)} \Phi^*(u)$

3   **if** $\Phi^*(q) > \Phi^*(v) - \delta s$ **then**

4   **return** v

5   **else**

6          $v = q$

Now we prove Theorem 3. First, w.l.o.g. assume that $\log n/\varepsilon^2 < n$, by
the same reasoning as in the proof of Theorem 2. By Lemma 5, Algorithm
LB-SEARCH that performs $\tau = \mathcal{O}(\frac{\log n}{\varepsilon^2})$ queries. We consider the following modification to Algorithm LB-SEARCH. As before, the algorithm updates
weights in time $\mathcal{O}(n)$ and maintains a sample S_t at each step t (by using Algorithm RESAMPLING) in time $\mathcal{O}(n + \varepsilon s \log n)$ which is subsumed by other terms.
However, instead of choosing a vertex that is δ-close to a median in line 3, the
updated algorithm runs Algorithm LOCAL-SEARCH with the previously queried
vertex as an input, and sets the output vertex to be the vertex q to be queried.
In other words, at each step t, it uses Algorithm LOCAL-SEARCH with input v_{t-1}
which returns v_t, and queries $q = v_t$. The algorithm initializes v_0 arbitrarily.

By Theorem 1, v_t is δ-close to a median. By Observation 9, we bound the
total number of iterations K done by Algorithm LOCAL-SEARCH by

$$K \leq \sum_{t=0}^{\tau-1} \left(1 + \frac{\Phi^*_{t+1}(v_t) - \Phi^*_{t+1}(v_{t+1})}{\delta s}\right)$$

$$= \tau + \frac{\Phi^*_1(v_0) + \sum_{t=1}^{\tau-1}(\Phi^*_{t+1}(v_t) - \Phi^*_t(v_t)) - \Phi^*_\tau(v_\tau)}{\delta s}$$

$$\leq \tau + \frac{sD + 2\tau s \varepsilon D}{\delta s} = \mathcal{O}(D\tau),$$

where we used that $\Phi^*_{t+1}(v_t) - \Phi^*_t(v_t) \leq 2s\varepsilon D$ holds with high probability by
Lemma 8. Each iteration in Algorithm LOCAL-SEARCH has complexity $\mathcal{O}(\Delta s)$
making the total complexity of the algorithm to be $\mathcal{O}(\tau(n + D\Delta s))$. $\square$

6 Open Problems

Having an algorithm that keeps an optimal query complexity and obtains a low
computational complexity, one can ask what are the possible tradeoffs between
the two? Another question is - can we further decrease the computational complexity? Also, are there any possible lower bounds that can reveal the limits of
what is not achievable in the context of these problems? Regarding the centrality
measures we consider, we propose an efficient median approximation. Motivated
by this, another question is what are other possible vertex-functions that may
allow for further improvements, e.g. in the complexity?

References

1. Abboud, A., Grandoni, F., Williams, V.V.: Subcubic equivalences between graph centrality problems, APSP and diameter. In: SODA 2015, pp. 1681–1697 (2015). https://doi.org/10.1137/1.9781611973730.112
2. Abboud, A., Williams, V.V., Wang, J.R.: Approximation and fixed parameter subquadratic algorithms for radius and diameter in sparse graphs. In: SODA 2016, pp. 377–391 (2016). https://doi.org/10.1137/1.9781611974331.ch28
3. Abraham, I., Delling, D., Fiat, A., Goldberg, A.V., Werneck, R.F.: Highway dimension and provably efficient shortest path algorithms. J. ACM 63(5), 41:1–41:26 (2016). https://doi.org/10.1145/2985473
4. Aigner, M.: Searching with lies. J. Comb. Theory Ser. A 74(1), 43–56 (1996). https://doi.org/10.1006/jcta.1996.0036
5. Bavelas, A.: Communication patterns in task-oriented groups. J. Acoust. Soc. Am. 22(6), 725–730 (1950). https://doi.org/10.1121/1.1906679
6. Beauchamp, M.A.: An improved index of centrality. Behav. Sci. 10(2), 161–163 (1965). https://doi.org/10.1002/bs.3830100205
7. Ben-Asher, Y., Farchi, E., Newman, I.: Optimal search in trees. SIAM J. Comput. 28(6), 2090–2102 (1999). https://doi.org/10.1137/S009753979731858X
8. Ben-Or, M., Hassidim, A.: The Bayesian learner is optimal for noisy binary search (and pretty good for quantum as well). In: FOCS 2008, pp. 221–230 (2008). https://doi.org/10.1109/FOCS.2008.58
9. Bénéteau, L., Chalopin, J., Chepoi, V., Vaxès, Y.: Medians in median graphs in linear time. CoRR abs/1907.10398 (2019)
10. Boczkowski, L., Korman, A., Rodeh, Y.: Searching a tree with permanently noisy advice. In: ESA 2018, pp. 54:1–54:13 (2018). https://doi.org/10.4230/LIPIcs.ESA.2018.54
11. Borgstrom, R.S., Kosaraju, S.R.: Comparison-based search in the presence of errors. In: STOC 1993, pp. 130–136 (1993). https://doi.org/10.1145/167088.167129
12. Cabello, S.: Subquadratic algorithms for the diameter and the sum of pairwise distances in planar graphs. In: SODA 2017, pp. 2143–2152 (2017). https://doi.org/10.1137/1.9781611974782.139
13. Cantone, D., Cincotti, G., Ferro, A., Pulvirenti, A.: An efficient approximate algorithm for the 1-median problem in metric spaces. SIAM J. Optim. 16(2), 434–451 (2005). https://doi.org/10.1137/S1052623403424740
14. Chang, C.: Some results on approximate 1-median selection in metric spaces. Theor. Comput. Sci. 426, 1–12 (2012). https://doi.org/10.1016/j.tcs.2011.12.003
15. Chechik, S., Cohen, E., Kaplan, H.: Average distance queries through weighted samples in graphs and metric spaces: high scalability with tight statistical guarantees. In: APPROX-RANDOM 2015, pp. 659–679 (2015). https://doi.org/10.4230/LIPIcs.APPROX-RANDOM.2015.659
16. Dagan, Y., Filmus, Y., Gabizon, A., Moran, S.: Twenty (simple) questions. In: STOC 2017, pp. 9–21 (2017). https://doi.org/10.1145/3055399.3055422
17. Deligkas, A., Mertzios, G.B., Spirakis, P.G.: Binary search in graphs revisited. Algorithmica 81(5), 1757–1780 (2018). https://doi.org/10.1007/s00453-018-0501-y
18. Dereniowski, D.: Edge ranking and searching in partial orders. Discret. Appl. Math. 156(13), 2493–2500 (2008). https://doi.org/10.1016/j.dam.2008.03.007
19. Dereniowski, D., Kosowski, A., Uznański, P., Zou, M.: Approximation strategies for generalized binary search in weighted trees. In: ICALP 2017, pp. 84:1–84:14 (2017). https://doi.org/10.4230/LIPIcs.ICALP.2017.84

20. Dereniowski, D., Tiegel, S., Uznański, P., Wolleb-Graf, D.: A framework for searching in graphs in the presence of errors. In: SOSA@SODA 2019, pp. 4:1–4:17 (2019). https://doi.org/10.4230/OASIcs.SOSA.2019.4

21. Dhagat, A., Gács, P., Winkler, P.: On playing "twenty questions" with a liar. In: SODA 1992, pp. 16–22 (1992)

22. Emamjomeh-Zadeh, E., Kempe, D.: A general framework for robust interactive learning. In: NIPS 2017, pp. 7085–7094 (2007)

23. Emamjomeh-Zadeh, E., Kempe, D., Singhal, V.: Deterministic and probabilistic binary search in graphs. In: STOC 2016, pp. 519–532 (2016). https://doi.org/10.1145/2897518.2897656

24. Feige, U., Raghavan, P., Peleg, D., Upfal, E.: Computing with noisy information. SIAM J. Comput. **23**(5), 1001–1018 (1994). https://doi.org/10.1137/S0097539791195877

25. Freeman, L.C.: Centrality in social networks conceptual clarification. Soc. Netw. **1**(3), 215–239 (1978)

26. Gavoille, C., Peleg, D., Pérennes, S., Raz, R.: Distance labeling in graphs. J. Algorithms **53**(1), 85–112 (2004). https://doi.org/10.1016/j.jalgor.2004.05.002

27. Hakimi, S.L.: Optimum locations of switching centers and the absolute centers and medians of a graph. Oper. Res. **12**(3), 450–459 (1964)

28. Indyk, P.: Sublinear time algorithms for metric space problems. In: STOC 1999, pp. 428–434 (1999). https://doi.org/10.1145/301250.301366

29. Karp, R.M., Kleinberg, R.: Noisy binary search and its applications. In: SODA 2007, pp. 881–890 (2007)

30. Kosowski, A., Viennot, L.: Beyond highway dimension: Small distance labels using tree skeletons. In: SODA 2017, pp. 1462–1478 (2017). https://doi.org/10.1137/1.9781611974782.95

31. Lam, T.W., Yue, F.L.: Optimal edge ranking of trees in linear time. Algorithmica **30**(1), 12–33 (2001). https://doi.org/10.1007/s004530010076

32. Mozes, S., Onak, K., Weimann, O.: Finding an optimal tree searching strategy in linear time. In: SODA 2008, pp. 1096–1105 (2008)

33. Onak, K., Parys, P.: Generalization of binary search: searching in trees and forest-like partial orders. In: FOCS 2006, pp. 379–388 (2006). https://doi.org/10.1109/FOCS.2006.32

34. Ostresh, L.M.: On the convergence of a class of iterative methods for solving the weber location problem. Oper. Res. **26**(4), 597–609 (1978). https://doi.org/10.1287/opre.26.4.597

35. Rivest, R.L., Meyer, A.R., Kleitman, D.J., Winklmann, K., Spencer, J.: Coping with errors in binary search procedures. J. Comput. Syst. Sci. **20**(3), 396–404 (1980). https://doi.org/10.1016/0022-0000(80)90014-8

36. Rényi, A.: On a problem of information theory. MTA Mat. Kut. Int. Kozl. **6B**, 505–516 (1961)

37. Sabidussi, G.: The centrality index of a graph. Psychometrika **31**(4), 581–603 (1966). https://doi.org/10.1007/bf02289527

38. Schäffer, A.A.: Optimal node ranking of trees in linear time. Inf. Process. Lett. **33**(2), 91–96 (1989). https://doi.org/10.1016/0020-0190(89)90161-0

39. Tabata, K., Nakamura, A., Kudo, M.: Fast approximation algorithm for the 1-median problem. In: Ganascia, J.-G., Lenca, P., Petit, J.-M. (eds.) DS 2012. LNCS (LNAI), vol. 7569, pp. 169–183. Springer, Heidelberg (2012). https://doi.org/10.1007/978-3-642-33492-4_15

40. Tabata, K., Nakamura, A., Kudo, M.: An efficient approximate algorithm for the 1-median problem on a graph. IEICE Trans. Inf. Syst. **100–D**(5), 994–1002 (2017). https://doi.org/10.1587/transinf.2016EDP7398
41. Tansel, B.C., Francis, R.L., Lowe, T.J.: State of the art-location on networks: a survey. Part I: the p-center and p-median problems. Manag. Sci. **29**(4), 482–497 (1983)
42. Ulam, S.M.: Adventures of a Mathematician. Scribner, New York (1976)

A Study on the Existence of Null Labelling for 3-Hypergraphs

Niccolò Di Marco[1(✉)], Andrea Frosini[1(✉)], and William Lawrence Kocay[2]

[1] Dipartimento di Matematica e Informatica, Università di Firenze, Firenze, Italy
{niccolo.dimarco,andrea.frosini}@unifi.it
[2] Department of Computer Science and St. Pauls College, University of Manitoba, Winnipeg, MB, Canada

Abstract. A 3-uniform hypergraph H consists of a set V of vertices, and $E \subseteq \binom{V}{3}$ triples. Let a *null labelling* be an assignment of ± 1 to the triples such that each vertex has signed degree equal to zero. Assumed as necessary condition the degree of every vertex of H to be even, the Null Labelling Problem consists in determining whether H has a null labelling. Although the problem is NP-complete, the subclasses where the problem turns out to be polynomially solvable are of interest. In this study we define the notion of 2-intersection graph related to a 3-uniform hypergraph, and we prove that the existence of a Hamiltonian cycle there, is sufficient to obtain a null labelling in the related hypergraph. The proof we propose provides an efficient way of computing the null labelling.

Keywords: Discrete tomography · 3-hypergraph · Null labelling · Hamiltonian cycle

AMS Classification: 05C65 · 05C22 · 05C99

1 Introduction

The characterization of simple graphs from their degree sequences has been a challenging problem whose solution dates back to the well known result of Erdös and Gallai [7] in 1960: an integer sequence $d = (d_1, \ldots, d_n)$ is graphic if and only if $\sum_{i=1}^{n} d_i$ is even and

$$\sum_{i=1}^{k} d_i \leq k(k-1) + \sum_{i=k+1}^{n} \min\{k, d_i\}, \quad 1 \leq k \leq n.$$

This same problem related to hypergraphs remained open until 2018 when Deza et al. proved its NP-completeness [6] even in the simplest case of 3-uniform hypergraphs.

© Springer Nature Switzerland AG 2021
P. Flocchini and L. Moura (Eds.): IWOCA 2021, LNCS 12757, pp. 282–294, 2021.
https://doi.org/10.1007/978-3-030-79987-8_20

Before this result many necessary and a few sufficient conditions are present in the literature, and they mainly rely on a result by Dewdney [5]. As an example, Behrens et al. [1] proposed a sufficient and polynomially testable condition for a degree sequence to be k-graphic; their result still does not provide any information about the associated k-hypergraphs. Soon after, in [3,9,10], a series of polynomial time algorithms was proposed to reconstruct one of the k-hypergraphs associated to each degree sequence of some classes including that studied in [1]. These results use tools borrowed from discrete tomography (DT), a young discipline concerned with inverse problems. In particular, it deals with the determination of geometrical properties of unknown objects, usually modelled by binary matrices, from their projections, i.e., a quantitative measurements of the number of primary constituents along prescribed directions (see [12,13] for the main results and the open problems). So, the reconstruction of a k-hypergraph from its degree sequence can be translated in the DT framework as the reconstruction of a binary matrix, i.e., its incidence matrix, from horizontal and vertical projections, i.e., the constant vector of entries k and the degree sequence, respectively.

However, some relevant related questions remain open, in particular the study of the uniqueness (up to isomorphism) of k-hypergraphs sharing the same degree sequence. In our study, we focus on this problem by considering two hypergraphs H_1 and H_2 with the same degree sequence. Their symmetric difference $H_1 \ominus H_2$ produces a null hypergraph, when assigning a $+1$ label to each hyperedge of H_1 and H_2. Vice versa, given a null hypergraph H, call H_1 the hypergraph with the same vertex as H but only the positive hyperedges of H and H_2 the same but with only the negative hyperedges of H. It's easy to see that H_1 and H_2 have the same degree sequence.

In [14], the notion of null hypergraph has been used to study the changes of hyperedges that allow one to move through all the 3-hypergraphs with the same degree sequence. The present research links the null label of a k-hypergraph with its 2-intersection graph, showing that the existence of a Hamiltonian cycle in the 2-intersection graph is sufficient to define a null label of the related k-hypergraph.

In Sect. 2, we recall the basic definition of graphs and hypergraphs, then we introduce some useful notation used throughout the paper, together with the notion of 2-intersection graph. Some results about null hypergraphs are also provided. In Sect. 3, we prove some properties of the 2-intersection graph of a k-hypergraph. Finally, we define a null labelling of a 3-hypergraph from a given Hamiltonian cycle of its 2-intersection graph.

2   Definitions and Known Results

A graph G is defined as a pair $G = (V, E)$ such that $V = \{v_1, \ldots, v_n\}$ is the set of *vertices* and E is a collection of pairs of vertices called *edges*.

The notion of graph can be generalized to that of hypergraph by removing the constraint on the cardinality of the edges: $H = (V, E)$ is a hypergraph with $V = \{v_1, \ldots, v_n\}$ the set of vertices and E a collection of subsets of vertices called *hyperedges*. We choose to abbreviate the vertex notation v_i with its only index i, when no misunderstandings may arise.

In the sequel, we will consider only graphs and hypergraphs that are *simple*, i.e., they do not allow singleton (hyper)edges or (hyper)edges that are contained in or equal to other edges. The *degree* of a vertex $v \in V$ is the number of (hyper)edges containing v. The degree sequence $(d_1, d_2, \ldots, d_n)$ of a (hyper)graph is the list of its vertex degrees usually arranged in non-increasing order. A hypergraph whose hyperedges have fixed cardinality is called k-*uniform*, or simply a k-hypergraph, and its degree sequence is called a k-*graphic* sequence. Note that the case $k = 2$ corresponds to graphs, and a 2-graphic sequence is called graphic. Our study will focus on 3-hypergraphs.

Given a (hyper)graph, we can assign a labelling l with labels $+1$ or -1 to each (hyper)edge, resulting in positive and negative (hyper)edges. The positive degree of a vertex v is $d_l^+(v)$, the number of positive (hyper)edges containing v. The negative degree is $d_l^-(v)$, the number of negative (hyper)edges containing v. The signed degree of each vertex v is $d_l(v) = d_l^+(v) - d_l^-(v)$. The unsigned degree is $\deg(v) = d_l^+(v) + d_l^-(v)$. The subscript indicating the labelling will be omitted when no misunderstandings may arise.

An assignment of ± 1 to the (hyper)edges of a (hyper)graph is a *null labelling* if $d(v) = 0$, for all vertices v. A (hyper)graph with a null labelling is said to be a *null (hyper)graph*. An obvious necessary condition for a (hyper)graph to have a null labelling is that each vertex must have even degree, i.e., it is an *even (hyper)graph*.

The following lemma characterizes graphs with a null labelling.

Lemma 1. *A graph G has a null labelling if and only if every connected component is an Eulerian graph with an even number of edges.*

Proof. If a graph has a null labelling, then it must has an even number of edges. Therefore, the necessity follows from the comments preceding the lemma. To prove sufficiency, consider a connected component which is an Eulerian graph with an even number of edges. By following an Euler tour, assigning alternately ± 1 to alternate edges, a null labelling is obtained.   □

This lemma also characterizes the graphs with even degrees and an even number of edges that do not have a null labelling: they must be disconnected graphs such that at least two connected components have an odd number of edges. The smallest graph with a null labelling is a cycle on four vertices, which we denote by C_4.

Moving to hypergraphs the situation becomes more complex. Let H_1 and H_2 be two hypergraphs with the same vertex set, and the same degree sequence $(d_1, d_2, \ldots, d_n)$. Assign $+1$ to the triples of H_1 and -1 to the triples of H_2, and construct $H_1 \ominus H_2$. It is a hypergraph with a null labelling. This raises the question of whether there is a characterization of null hypergraphs. In [8] it is shown that the problem of finding a null labelling even for the simplest case of 3-hypergraphs is NP-complete.

Hypergraph Null Labelling Problem: Let H be a connected, even 3-hypergraph. When can ± 1 be assigned to the hyperedges of H to produce a null-labelled 3-hypergraph?

The *intersection graph* of a 3-hypergraph H is denoted $I(H)$. Its vertices are the hyperedges of H. Two hyperedges are adjacent if their intersection is non-empty. This is an extension of the idea of a line graph to 3-hypergraphs. In [8] the following result was proved.

Theorem 1. *Let H be a connected, even 3-hypergraph, in which every vertex has degree two. Then H has a null labelling if and only if $I(H)$ is bipartite.*

However, the inspection of the intersection graph does not provide evidence, in general, of the existence of a null labelling in the related 3-hypergraph as shown in the following example.

Example 1. Consider the following 3-hypergraphs H_1 and H_2 on six vertices and whose hyperedges, arranged in matrix form, are:

$$H_1 = \begin{bmatrix} 1 & 2 & 3 \\ 1 & 4 & 5 \\ 2 & 4 & 6 \\ 3 & 5 & 6 \end{bmatrix} \qquad H_2 = \begin{bmatrix} 1 & 2 & 5 \\ 2 & 3 & 5 \\ 2 & 3 & 4 \\ 1 & 2 & 4 \end{bmatrix}$$

It is easy to check that the vector of labels $l = (1, -1, 1, -1)$, where $l(i)$ is the label of the i-th hyperedge of the 3-hypergraph or, equivalently, of the i-th row in the matrix arrangement of its hyperedges, is a null label for H_2, while H_1 has no null labelling. However, H_1 and H_2 have the same intersection graph K_4, i.e. the complete graph on four vertices.

Relying on this fact, we modify the notion of intersection graph as follows: the *2-intersection graph* of a 3-hypergraph H is denoted $I_2(H) = (V_{2H}, E_{2H})$. Its vertices $V_{2H} = \{v_{e_1}, \ldots, v_{e_m}\}$ represents the hyperedges $E = \{e_1, \ldots, e_m\}$ of H. Two hyperedges are adjacent, i.e., they belong to the same edge $\{v_{e_i}, v_{e_j}\} \in E_{2H}$, if e_i and e_j share a pair of vertices of H (see Example 2 and the related Fig. 2). In the sequel, we label the edge $\{v_{e_i}, v_{e_j}\} \in E_{2H}$ with the pair of vertices that are shared by e_i and e_j, if needed.

3   Hypergraph and 2-Intersection Graph

Let $H = (V, E)$ be a 3-uniform hypergraph (3-hypergraph). We study some properties of the 2-intersection graph that are relevant to obtain information about the existence of a null labelling in the related 3-hypergraph. In particular, we will consider Hamiltonian cycles, as they appear to be relevant to null labellings.

Question: Consider the 2-intersection graph G of a connected, even 3-hypergraph H. Is it possible to construct a null-label from a Hamiltonian cycle in G? We want to prove that if G is Hamiltonian then H has a null-label.

We know that a connected, even graph G is Eulerian. Its line-graph $L(G)$ is the intersection graph of its edges. An Euler tour in G corresponds to a

Hamiltonian cycle in $L(G)$, and conversely. And it also corresponds to a null labelling of G. Thus, Hamiltonian cycles in $L(G)$ can be used to determine null labellings of G.

The 2-intersection graph of a 3-hypergraph H is similar to the line graph of a graph. It is well known (see Harary [11]) that a line graph has a unique decomposition into maximal cliques, with at most a vertex in common. Thus we consider cliques in 2-intersection graphs $I_2(H)$ of 3-hypergraphs.

Property 1. Let $H = (V, E)$ be a 3-hypergraph. The hyperedges $e_1, \ldots, e_n \in E$ sharing the same pair of elements form a clique in the 2-incidence graph $I_2(H)$.

This property is a direct consequence of the definition of 2-incidence graph. There is another kind of clique in a 2-intersection graph. Define a *triangle* in a 3-hypergraph H to be three hyperedges of the form $\{1, 2, 3\}, \{2, 3, 4\}, \{3, 4, 1\}$. Any two of them intersect in a pair, so they also form a triangle in $I_2(H)$.

Property 2. Let $H = (V, E)$ be a 3-hypergraph. There are two kinds of cliques in $I_2(H)$, namely those hyperedges all sharing a common pair, and those hyperedges deriving from a triangle in H.

Example 2. Consider the following hypergraph $H = (V, E)$ in which $V = \{1, \ldots, 6\}$ and $E = \{\{3, 4, 5\}, \{3, 5, 6\}, \{1, 3, 5\}, \{3, 4, 6\}, \{2, 4, 6\}, \{4, 5, 6\}, \{1, 2, 6\}, \{2, 3, 5\}, \{1, 2, 5\}, \{1, 2, 4\}, \{1, 2, 3\}, \{1, 4, 6\}\}$.

The 2-intersection graph of H is shown in Fig. 1. According to Property 1, one can check that the edges with the same label form a clique.

Our purpose here aims at detecting a strategy that starting from a Hamiltonian cycle C of $I_2(H)$ allows one to define a labelling of its vertices that is a null labelling of H.

A first naive strategy consists in alternately labelling ± 1 the vertices of C to obtain a null labelling of H. This is a strategy that works for line graphs of even graphs. Let v_{e_i} and v_{e_j} be two consecutive elements of C, with $e_i = \{u, x, y\}$ and $e_j = \{v, x, y\}$ hyperedges of H, we note that such a labelling maintains the signed degree of x and y. On the other hand, it increases/decreases by one the signed degree of u and v. So, we see that the alternating labelling of C does not always provide a null labelling of H, as witnessed by the following example.

Example 3. Consider the 3-hypergraph $H = (V, E)$ on six vertices and $E = \{e_1, \ldots, e_8\}$, where $e_1 = \{1, 2, 3\}, e_2 = \{1, 2, 4\}, e_3 = \{1, 2, 5\}, e_4 = \{1, 2, 6\}, e_5 = \{1, 3, 4\}, e_6 = \{1, 3, 5\}, e_7 = \{2, 3, 5\}, e_8 = \{2, 5, 6\}$.

The related 2-intersection graph $I_2(H)$ in Fig. 2 has the Hamiltonian cycle

$$C_1 = (v_{e_1}, v_{e_3}, v_{e_2}, v_{e_4}, v_{e_8}, v_{e_7}, v_{e_6}, v_{e_5}, v_{e_1})$$

It is easy to check that alternately labelling ± 1 the vertices of C_1, starting with $+1$, we obtain the null labelling $l_1 = (1, 1, -1, -1, -1, 1, -1, 1)$ on the eight hyperedges of H such that $l_1(i)$ is the label of e_i, with $1 \le i \le 8$.

Unfortunately, not every Hamiltonian cycle provides a null labelling. A second Hamiltonian cycle $C_2 = (v_{e_1}, v_{e_2}, v_{e_3}, v_{e_4}, v_{e_8}, v_{e_7}, v_{e_6}, v_{e_5}, v_{e_1})$ exists such that the alternating labelling $l_2 = (1, -1, 1, -1, -1, 1, -1, 1)$ is not null on H, as $d(v_4) = -2$ and $d(v_5) = +2$.

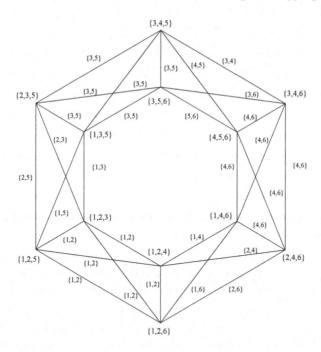

Fig. 1. The 2-intersection graph of the 3-hypergraph H in Example 2. The edges are labelled according to the pairs shared by their vertices.

Let us introduce some notation: suppose that v_{e_i} and v_{e_j} are two consecutive vertices of the Hamiltonian cycle C of $I_2(H)$, with $e_i = \{u, x, y\}$ and $e_j = \{v, x, y\}$. We say, by extension, that v_{e_j} contains the node $v \in H$. We see that $v \notin e_i$. There may be several consecutive vertices of C that contain v. Denote by $p_v = (v_{e_{j_1}}, \ldots, v_{e_{j_k}})$ the longest sub-path of C starting in v_{e_j} such that every vertex of p_v contains v. Let $l(p_v)$ denote the labels of the vertices of p_v, and let $\sigma(l(p_v))$ denote the sum of the elements of $l(p_v)$, and let $|p_v|$ denote the length $k - 1$ of p_v, i.e., its number of edges. In the case when p_v contains just a single vertex, $|p_v| = 0$.

In this example, $e_i = \{u, x, y\}$ is clearly the last vertex in a path p_u. Therefore we define $next(p_u) = p_v$, for this p_u and p_v, i.e., given a path p_u, $next(p_u)$ is the path beginning at the first vertex following the last vertex of p_u. In general, C may contain several different sub-paths of the form p_v, for each vertex v; we indicate them by $p_v^1, \ldots, p_v^n$. An example is shown in Fig. 3.

Property 3. Given an alternating labelling ± 1 on the vertices of a Hamiltonian cycle C of $I_2(H)$. For each sub-path $p_v = (v_{e_{j_1}}, \ldots, v_{e_{j_k}})$, the following holds:

– if p_v has odd length, then $\sigma(l(p_v)) = 0$, so that the labels of the hyperedges $e_{j_1}, \ldots, e_{j_k}$ containing v sum to zero in H. In this case the first and the last vertex of p_v have different labels;

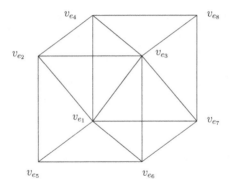

Fig. 2. The 2-intersection graph of the 3-hypergraph considered in Example 3.

- if p_v has even length, then $\sigma(l(p_v)) \neq 0$ and the sum of the labels of the hyperedges $e_{j_1}, \ldots, e_{j_k}$ containing v contribute $+1$ or -1 to the signed degree of v. In this case, the extremal vertices of p_v have the same label.

The proof of this property is straightforward. Figure 3 shows the 2-intersection graph of a 3-hypergraph on six vertices and eight hyperedges. One of its Hamiltonian cycles and the sub-paths related to the vertices of H are highlighted. Note that the paths in the diagram circle around the right edge of the diagram back to the left edge.

Let us continue analyzing the properties of the alternating labelling $l(C)$ of a Hamiltonian cycle C of $I_2(H)$. Property 3 assures that, if the labelling $l(C)$ produces a signed degree $d_l(v) = d \neq 0$ for vertex v, then there exists at least $p_v^1, \ldots, p_v^{|d|}$ subpaths with the same sum of labels. An example is seen in Fig. 3.

We define the *distance* between two paths p_u and p_v as the distance along C between the last point of p_u and the first point p_v. We observe that any two of the previous $|d|$ subpaths have even distance. The above observations lead to the following lemmas

Lemma 2. *Let H be an even 3-hypergraph and $I_2(H)$ its 2-intersection graph. If $I_2(H)$ has a Hamiltonian cycle C, an alternating ± 1 labelling $l(C)$ defines a null label of H if and only if, for each $v \in V$:*

i) each subpath p_v has odd length; OR
ii) the number of subpaths of v having even length is even and the sum of their labels is zero.

We emphasize that *ii)* expresses the condition that, for each vertex v of H, there are the same number of subpaths of p_v having label $+1$ as -1.

We also have the following lemma.

Lemma 3. *Let H be a 3-hypergraph and C a Hamiltonian cycle of $I_2(H)$, and let $v_e = \{u, x, y\} \in V_{2H}$ be a vertex of $I_2(H)$. There are exactly three subpaths containing v_e, namely $p_u, p_x,$ and p_y. One of them begins at v_e and one of them ends at v_e (possibly the same path begins and ends at v_e).*

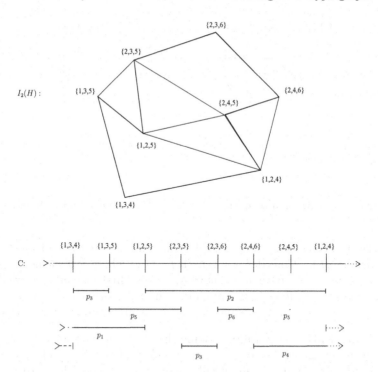

Fig. 3. A 2-intersection graph $I_2(H)$ and one of its Hamiltonian cycles C are shown. The sub-paths p_v related to the vertices of H are highlighted. Note that in each vertex of $I_2(H)$ starts and ends two sub-paths related to two (non necessarily distinct) vertices of H.

Proof. The next vertex of C intersects v_e in two vertices, say x and y. Then $e' = \{v, x, y\}$ is the next vertex of C. The previous vertex also intersects v_e in two vertices, wlog, either x, y or u, x. Let e'' be the previous vertex of C. Then either $e'' = \{w, x, y\}$ or $e'' = \{u, x, w\}$, for some w. In the first case, we have p_u begins and ends at v_e. In the second case, we have p_u ends at v_e and p_y begins. □

In the sequel, we describe an algorithm that modifies an alternating ± 1 labelling of a Hamiltonian cycle C not satisfying the conditions of Lemma 2 in order to obtain a null labelling of H. This algorithm relies on the $Switch()$ operator defined as follows: given two even sub-paths $p_u = (v_{e_{i_1}}, \ldots, v_{e_{i_k}})$ and $p_v = (v_{e_{j_1}}, \ldots, v_{e_{j_{k'}}})$, where $p_v = \text{next}(p_u)$, and $e_{i_k} \neq e_{j_1}$, the operator $Switch(p_u, p_v)$ produces a new labelling $l'(C)$ by changing the signs of e_{i_k} and e_{j_1}: $l'(e_{i_k}) = -l(e_{i_k})$ and $l'(e_{j_1}) = -l(e_{j_1})$; and keeping the remaining labels of $l(C)$ unchanged. Figure 4 shows an example of the action of $Switch(p_2, p_5)$.

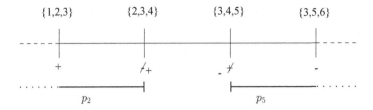

Fig. 4. Example of $Switch(p_2, p_5)$ between the two consecutive paths p_2 and p_5, i.e., such that $p_5 = next(p_2)$.

We will start with an alternating labelling $l(C)$, and gradually change it using $Switch()$.

Property 4. Let H be a 3-hypergraph, C a Hamiltonian cycle of $I_2(H)$, and l a ± 1 labelling of C. Consider a sub-path p_u of C whose last element v_{e_i} with label $+1$, and the sub-path $p_v = next(p_u)$ whose first element v_{e_j} with label -1. The operator $Switch(p_u, p_v)$ modifies l into l' so that $d_{l'}(u) = d_l(u) - 2$, $d_{l'}(v) = d_l(v) + 2$ and all the remaining signed degrees are left unchanged.

Proof. Without loss of generality, assume that $e_i = \{u, x, y\}$ and $e_j = \{v, x, y\}$. It is immediate that the change of the opposite labels of e_i and e_j keeps the signed degrees of x and y, while it subtracts 2 from u and adds 2 to v. As the starting labels of e_i and e_j are opposite, a symmetric result holds.   □

The algorithm $Balance()$ defined below modifies a labelling $l(C)$ of a Hamiltonian cycle C of $I_2(H)$ in order to change, after a sequence of successive applications of the $Switch()$ operator, the signed degree of two input vertices u and v of H, if possible, otherwise it gives failure.

First, we prove that the the algorithm $Balance()$ computes a null labelling starting from the alternating labelling $l(C)$ in the easiest case of having only two signed degrees u and v different from zero, in particular $+2$ and -2, respectively.

Lemma 4. *Let H be a 3-hypergraph, C a Hamiltonian cycle of $I_2(H)$, and l an alternating labelling of C. If u and v are the only nodes of H with signed degree different from zero, in particular $d_l(u) = +2$ and $d_l(v) = -2$, then $Balance(u, v, l(C))$ returns a null labelling $l'(C)$ of H.*

Proof. Since $d_l(u) = 2$, there exists at least one subpath p_u such that $|p_u|$ is even and $\sigma(l(p_u)) = +1$, i.e. it starts and ends with two elements labelled with $+1$.

The *While* cycle starts by performing the switch between p_u and $p_j = next(p_u)$ and l is updated to l'. Since l is an alternating labelling, the first element of p_j has label -1, so after $Switch(p_u, p_j)$, we have $d_{l'}(u) = 0$, $d_{l'}(j) = d_l(j) + 2$, and by Property 4, the other signed degrees do not change. Now, if $|p_j|$ is even and $j = v$, then $d_{l'}(p_v) = 0$ and l' is the desired null labelling.

On the other hand, if $|p_j|$ is odd, then it ends with a $+1$ label and its successor $next(p_j)$ starts with a -1. So, after updating $p_i = p_j$ and $p_j = next(p_i)$,

---

Algorithm 1. Balance($u, v, l(C)$)

Input: the label $l(C)$ of a Hamiltonian cycle C, and two vertices u and v of H with signed degrees $d_l(u) > 0$ and $d_l(v) < 0$.
Output: The label $l'(C)$ such that $d_{l'}(u) = d_l(u) - 2$ and $d_{l'}(v) = d_l(v) + 2$.

$p_i = p_u$ such that $|p_u|$ is even and $\sigma(l(p_u)) = +1$
while *true* **do**
    $p_j = next(p_i)$
    $Switch(p_i, p_j)$
    if $|p_j|$ is odd **then**
        $p_i = p_j$;
    else if a non already considered p'_j exists such that $|p'_j|$ is even and p'_j starts with $+1$ label **then**
        $p_i = p'_j$
    else
        FAILURE;
    end if
end while
return the final $l'(C)$ as OUTPUT.

$Switch(p_i, p_j)$ is again performed and the new l' changes back the signed degree of i to zero, while $+2$ is added to the signed degree of the new j. Note that in this case, even if $j = v$, we decide to continue with the algorithm until reaching a $|p_v|$ of even length and such that $\sigma(l'(p_v)) = -1$.

If the last case $|p_j|$ is even and $j \neq v$ occurs, then we move to another new even p'_j such that $\sigma(l(p'_j)) = +1$. Such a p'_j always exists since $d_l(j) = 0$, and consequently the number of even sub-paths containing j whose labels sum up to $+1$ equals those whose labels sum up to -1. So, FAILURE never occurs starting from an alternating labelling l.

Finally, the result is obtained by observing that $Balance(u, v, l(C))$ does not loop since the number of subpath of C is finite and each of them is involved in the *while* cycle at most once since the procedure always switches a sub-path that ends with a $+1$ with a sub-path that ends with -1. □

Lemma 5. *Let $H = (V, E)$ be a 3-hypergraph, C a Hamiltonian cycle of $I_2(H)$ and l an alternating labelling of C. If v_1 and v_2 are the only nodes of H with signed degree different from zero with respect to l, say $d_l(u) = +2k$ and $d_l(v) = -2k$, where $k \geq 1$, then H admits a null labelling.*

Proof. This is obtained by k successive runs of $Balance(u, v, l^i(C))$, with $0 \leq i < k$, where $l^{i+1}(C)$ is the labelling obtained as output of $Balance(u, v, l^i(C))$. We set $l^0 = l$; the output l^k of $Balance(u, v, l^{k-1}(C))$ provides a null labelling of H.

We emphasize that, from the second run of $Balance()$ until the last one, i.e., the k-th run, the choice of a new starting sub-path p_u is always possible. In fact, from $d_l(u) = +2k$ it follows that in $l(C)$ the number of sub-paths of u whose labels sum up to $+1$ exceeds exactly by k those whose labels sum up to -1. A last remark is required: since two different runs of $Balance()$ start from different sub-paths p_u, their computations do not involve the same sub-path twice. So, each call $Switch()$ in the k runs of $Balance()$ always modifies two elements of C whose labels are alternate, as set by $l(C)$.   □

This same reasoning can be generalized when more than two vertices of H have non null signed degree leading to our main result

Theorem 2. *Let H be a 3-hypergraph. If the 2-intersection graph $I_2(H)$ is Hamiltonian, then H admits a null labelling.*

The proof of this theorem and the related computation of the null label of H directly follow from the proofs of Lemmas 4 and 5, after observing that we can iterate the calls of $Balance(u, v, l(C))$ varying u among all the vertices with signed degree greater than zero until reaching the first vertex v among those having signed degree less than zero. The following example will clarify the situation.

Example 4. Consider the following 3−hypergraph $H = (V, E)$ with $V = \{1, \ldots, 8\}$ and $E = \{\{2, 3, 5\}, \{2, 5, 8\}, \{2, 4, 8\}, \{1, 4, 8\}, \{1, 4, 7\}, \{1, 6, 7\}, \{1, 4, 6\}, \{1, 5, 6\}, \{5, 6, 7\}, \{1, 5, 7\}, \{1, 2, 7\}, \{1, 2, 3\}, \{2, 3, 6\}, \{3, 6, 8\}, \{3, 7, 8\}, \{3, 5, 8\}\}$

Figure 5 shows a Hamiltonian cycle C of $I_2(H)$, and one of its alternating labellings $l(C)$.

The chosen labelling is not a null labelling of H. The vector of the signed degrees of the vertices of H is

$$d = (-2, 2, 0, 2, -2, 0, 2, -2).$$

Let us perform a sequence of runs of $Balance()$ to compute a null labelling of H starting from $l(C)$.

Let us start, as an example, the run $Balance(2, v, l(C))$ in the p_2 sub-path having $\{2, 3, 5\}$ as first element. It calls $Switch(p_2, p_1)$, with $p_1 = next(p_2)$ and $|p_1|$ even. Since $d_l(1) = -2$, we perform the choice $v = 1$, and the switchings of $\{2, 4, 8\}$ and $\{1, 4, 8\}$ leading to the labelling $l^1(C)$ such that $d_{l^1}(1) = d_{l^1}(2) = 0$, leaving the remaining labels unchanged.

Let us now arbitrarily choose the vertex 7 such that $d_{l^1}(7) = +2$ and run $Balance(7, v, l^1(C))$ with the starting p_7 sub-path whose first element is $\{5, 6, 7\}$. The sub-path $p_3 = next(p_7)$ has odd length so the labels of $\{1, 2, 7\}$ and $\{1, 2, 3\}$ are switched and we obtain $d(7) = 0$ and $d(3) = +2$. Now $p_8 = next(p_3)$ and the labels of $\{2, 3, 5\}$ and $\{2, 5, 8\}$ are switched obtaining $d(3) = d(8) = 0$. Since $|p_8|$ is even, the run $Balance(7, v, l^1(C))$ ends setting $v = 8$. A new labelling $l^2(C)$ is returned as output.

Two more vertices with signed degree different from zero are left, i.e., the vertices 4 and 5. A last run of $Balance(4, 5, l^2(C))$ is performed. Taking the p_4 subpath containing only $\{1, 4, 6\}$, we have $p_5 = next(p_4)$ with $|p_5|$ even.

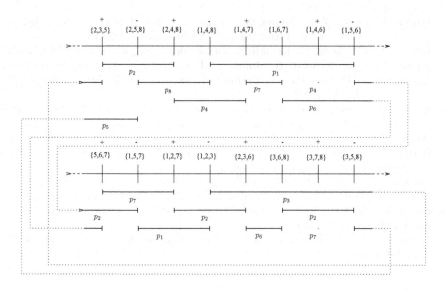

Fig. 5. A Hamiltonian cycle of $I_2(H)$ and its labelling.

Therefore, switching the sign of $\{1,4,6\}$ and $\{1,5,6\}$ we obtain a new labelling l^3 such that $d_{l^3}(4) = d_{l^3}(5) = 0$ and $Balance(4,5,l^2(C))$ ends. Therefore, the labelling

$$l^3 = (-1, 1, -1, 1, 1, -1, -1, 1, 1, -1, -1, 1, 1, -1, 1, -1)$$

is a null labelling of H. Note that the order of the calls of $Balance()$ is not relevant in order to obtain a null labelling of H.

References

1. Behrens, S., et al.: New results on degree sequences of uniform hypergraphs. Electron. J. Comb. **20**(4), 14 (2013)
2. Berge, C.: Hypergraphs, North-Holland, Amsterdam (1989)
3. Brlek, S., Frosini, A.: A tomographical interpretation of a sufficient condition on h-graphical sequences. In: Normand, N., Guédon, J., Autrusseau, F. (eds.) DGCI 2016. LNCS, vol. 9647, pp. 95–104. Springer, Cham (2016). https://doi.org/10.1007/978-3-319-32360-2_7
4. Colbourne, C.J., Kocay, W.L., Stinson, D.R.: Some NP-complete problems for hypergraph degree sequences. Discrete Appl. Math. **14**, 239–254 (1986)
5. Dewdney, A.K., Degree sequences in complexes and hypergraphs. Proc. Amer. Math. Soc. **53**(2), 535–540 (1975)
6. Deza, A., Levin, A., Meesum, S.M., Onn, S.: Optimization over degree sequences. SIAM J. Disc. Math. **32**(3), 2067–2079 (2018)
7. Erdös, P., Gallai, T.: Graphs with prescribed degrees of vertices (Hungarian). Math. Lapok. **11**, 264–274 (1960)

8. Frosini, A., Kocay, W.L., Palma, G., Tarsissi, L.: On null 3-hypergraphs. Discrete Appl. Math. Press (2021). https://doi.org/10.1016/j.dam.2020.10.020

9. Frosini, A., Picouleau, C., Rinaldi, S.: On the degree sequences of uniform hypergraphs. In: Gonzalez-Diaz, R., Jimenez, M.-J., Medrano, B. (eds.) DGCI 2013. LNCS, vol. 7749, pp. 300–310. Springer, Heidelberg (2013). https://doi.org/10.1007/978-3-642-37067-0_26

10. Frosini, A., Picouleau, C., Rinaldi, S.: New sufficient conditions on the degree sequences of uniform hypergraphs. Theoret. Comput. Sci. **868**, 97–111 (2021)

11. Harary, F.: Graph Theory. Addison Wesley Publishing Company, Boston (1972)

12. Herman, G.T., Kuba, A. (eds.): Discrete Tomography: Foundations Algorithms and Applications. Birkhauser, Boston (1999)

13. Herman, G.T., Kuba, A. (eds.): Advances in Discrete Tomography and Its Applications (Applied and Numerical Harmonic Analysis). Birkhauser, Boston (2007)

14. Kocay, W., Li, P.C.: On 3-hypergraphs with equal degree sequences. Ars Combinatoria **82**, 145–157 (2006)

15. Kocay, W.: A Note on Non-reconstructible 3-Hypergraphs. Graphs Comb. **32**(5), 1945–1963 (2016)

Piercing All Translates of a Set of Axis-Parallel Rectangles

Adrian Dumitrescu[1] and Josef Tkadlec[2]([⊠])

[1] Algoresearch L.L.C., Milwaukee, USA
[2] Department of Mathematics, Harvard University, Cambridge, MA 02138, USA

Abstract. For a given shape S in the plane, one can ask what is the lowest possible density of a point set P that pierces ("intersects", "hits") all translates of S. This is equivalent to determining the covering density of S and as such is well studied. Here we study the analogous question for families of shapes where the connection to covering no longer exists. That is, we require that a single point set P simultaneously pierces each translate of each shape from some family $\mathcal{F}$. We denote the lowest possible density of such an $\mathcal{F}$-piercing point set by $\pi_T(\mathcal{F})$. Specifically, we focus on families $\mathcal{F}$ consisting of axis-parallel rectangles. When $|\mathcal{F}| = 2$ we exactly solve the case when one rectangle is more squarish than 2×1, and give bounds (within 10% of each other) for the remaining case when one rectangle is wide and the other one is tall. When $|\mathcal{F}| \geq 2$ we present a linear-time constant-factor approximation algorithm for computing $\pi_T(\mathcal{F})$ (with ratio 1.895).

Keywords: Axis-parallel rectangle · Piercing · Approximation algorithm

1 Introduction

In a game of Battleship, the opponent secretly places ships of a fixed shape on an $n \times n$ board and your goal is to sink them by identifying all the cells the ships occupy (the ships are stationary). Consider now the following puzzle: If the opponent placed a single 2×3 ship, how many attempts do you need to surely hit the ship at least once? The answer depends on an extra assumption. If you know that the ship is placed, e.g., vertically, it is fairly easy to see that the answer is roughly $n^2/6$: When n is a multiple of 6, then one hit is needed per each of the $n^2/6$ interior-disjoint translates of the 2×3 rectangle that tile the board and, on the other hand, a lattice with basis $[2,0], [0,3]$ achieves the objective. The starting point of this paper was to answer the question when it is *not* known whether the ship is placed vertically or horizontally. It turns out that the answer is $n^2/5 + \mathcal{O}(n)$ hits (the main term comes from Theorem 1 (ii) whereas the $\mathcal{O}(n)$ correction term is due to the boundary effect).

Motivated by the above puzzle, we study the following problem: Given a family $\mathcal{F}$ of compact shapes in the plane, what is its translative piercing density

© Springer Nature Switzerland AG 2021
P. Flocchini and L. Moura (Eds.): IWOCA 2021, LNCS 12757, pp. 295–309, 2021.
https://doi.org/10.1007/978-3-030-79987-8_21

$\pi_T(\mathcal{F})$, that is, the lowest density of a point set that pierces ("intersects", "hits") every translate of each member of the family? Here the density of an infinite point set P (over the plane) is defined in the standard fashion as a limit of its density over a disk D_r of radius r, as r tends to infinity. The piercing density $\pi_T(\mathcal{F})$ of the family is then defined as the infimum over all point sets that pierce every translate of each member of the family [5, Ch. 1], [15]. (See Sect. 1.1 for precise definitions.) Note that unlike in the puzzle, we allow translations of each shape in the family by any, not necessarily integer, vector.

First, we cover the case when the family $\mathcal{F} = \{S\}$ consists of a single shape. The problem is then equivalent to the classical problem of determining the translative covering density $\vartheta_T(S)$ of the shape S: Indeed, determining the translative covering density $\vartheta_T(S)$ amounts to finding a (sparsest possible) point set P such that the translates $\{p + S \mid p \in P\}$ cover the plane, that is,

$$(\forall x \in \mathbb{R}^2)(\exists p \in P) \text{ such that } x \in p + S.$$

(Here "+" is the Minkowski sum.) This is the same as requiring that

$$(\forall x \in \mathbb{R}^2)(\exists p \in P) \text{ such that } p \in x + (-S),$$

that is, the point set P pierces all translates of the shape $-S$. Hence $\vartheta_T(S) = \pi_T(\{-S\}) = \pi_T(\{S\})$. Specifically, when S tiles the plane, then the answer is simply $\pi_T(\{S\}) = 1/\operatorname{Area}(S)$, where $\operatorname{Area}(S)$ is the area of S. We note that apart from the cases when S tiles the plane, the translative covering density $\vartheta_T(S)$ is known only for a few special shapes S such as a disk or a regular n-gon [5, Ch. 1].

For the rest of this work (apart from the Conclusions) we limit ourselves to the case when $\mathcal{F}$ consists of $n \geq 2$ axis-parallel rectangles. First we consider the special case $n = 2$ (Theorem 1 in Sect. 2), then we consider the case of arbitrary $n \geq 2$ (Theorem 2 in Sect. 3).

Related Work. There is a rich literature on related (but fundamentally different) fronts dealing with piercing *finite* collections. One broad direction is devoted to establishing combinatorial bounds on the piercing number as a function of other parameters of the collection, most notably the matching number [2,10,11,13,17,20,23,24,26–29] or in relation to Helly's theorem [12,21,23]; see also the survey articles [14,24]. Another broad direction deals with the problem of piercing a given set of shapes in the plane (for instance axis-parallel rectangles) by the minimum number of points and concentrates on devising algorithmic solutions, ideally exact but frequently approximate; see for instance [7–9]. Indeed, the problem of computing the piercing number corresponds to the hitting set problem in a combinatorial setting [19] and is known to be NP-hard even for the special case of axis-aligned unit squares [18]. The theory of ε-nets for planar point sets and axis-parallel rectangular ranges is yet another domain at the interface between algorithms and combinatorics in this area [3,32].

A third direction that appears to be most closely related to this paper is around the problem of estimating the area of the largest empty axis-parallel rectangle amidst n points in the unit square, namely, the quantity $A(n)$ defined below. Given a set S of n points in the unit square $U = [0,1]^2$, a rectangle $R \subset U$ is *empty* if it contains no points of S in its interior. Let $A(S)$ be the maximum volume of an empty box contained in U (also known as the *dispersion* of S), and let $A(n)$ be the minimum value of $A(S)$ over all sets S of n points in U. It is known that $1.504 \leq \lim_{n \to \infty} nA(n) \leq 1.895$; see also [1,30,34,36]. The lower bound is a recent result of Bukh and Chao [6] and the upper bound is another recent result of Kritzinger and Wiart [31]. It is worth noting that the upper bound $\varphi^4/(\varphi^2 + 1) = 1.8945\ldots$ can be expressed in terms of the golden ratio $\varphi = \frac{1}{2}(1 + \sqrt{5})$. The connection will be evident in Sect. 3.

1.1 Preliminaries

Throughout this paper, a *shape* is a Lebesque-measurable compact subset of the plane. Given a shape S, let $\mathrm{Area}(S)$ denote its area. We identify points in the plane with the corresponding vectors from the origin. Given two shapes $A, B \subset \mathbb{R}^2$, we denote by $A + B = \{a + b \mid a \in A, b \in B\}$ their *Minkowski sum*. A *translate* of a shape S by a point (vector) p is the shape $p + S = \{p + s \mid s \in S\}$.

In the next three definitions we introduce the (translative) piercing density $\pi_T(\mathcal{F})$ of a family $\mathcal{F}$ of shapes in the plane. Then we define a certain shorthand notation for the special case when $\mathcal{F}$ consists of two axis-parallel rectangles.

Definition 1 ($\mathcal{F}$-piercing sets). *Given a family $\mathcal{F}$ of shapes in the plane, we say that a point set P is $\mathcal{F}$-piercing if it intersects all translates of all the shapes in $\mathcal{F}$, that is, if*

$$(\forall S \in \mathcal{F})(\forall x \in \mathbb{R}^2)(\exists p \in P) \text{ such that } p \in x + S.$$

Definition 2 (Density of a point set). *Given a point set P and a bounded domain D with area $\mathrm{Area}(D)$, we define the density of P over D by $\delta(P, D) = \frac{|P \cap D|}{\mathrm{Area}(D)}$.*

Given a (possibly infinite and unbounded) point set P, we define its asymptotic upper and lower densities by

$$\overline{\delta}(P) = \limsup_{r \to \infty} \delta(P, D_r) \quad \text{and} \quad \underline{\delta}(P) = \liminf_{r \to \infty} \delta(P, D_r),$$

where D_r is the disk with radius r centered at the origin.

Definition 3 (Translative piercing density $\pi_T(\mathcal{F})$). *Fix a family $\mathcal{F}$ of shapes in the plane. Then we define the (translative) piercing density by*

$$\pi_T(\mathcal{F}) = \inf_{P \text{ is } \mathcal{F}\text{-piercing}} \{\underline{\delta}(P)\}$$

and the (translative) lattice piercing density $\pi_L(\mathcal{F})$ by

$$\pi_L(\mathcal{F}) = \inf_{P \text{ is an } \mathcal{F}\text{-piercing lattice}} \{\underline{\delta}(P)\}.$$

Pairs of Axis-Parallel Rectangles. Let $R_{w \times h}$ denote a rectangle with width w and height h. Here we introduce a shorthand notation for the case when $\mathcal{F} = \{R_{a \times b}, R_{c \times d}\}$ consists of two axis-parallel rectangles. If $a \leq c$ and $b \leq d$ then clearly $\pi_T(\mathcal{F}) = \pi_L(\mathcal{F}) = 1/(ac)$ as the lattice with basis $\{[a, 0], [0, c]\}$ that pierces all translates of the smaller rectangle also pierces all translates of the larger rectangle. Otherwise we can suppose $a \geq c$ and $b \leq d$. Stretching horizontally by a factor of c and then vertically by a factor of b, we have

$$\pi_T(\mathcal{F}) = c \cdot \pi_T(\{R_{\frac{a}{c} \times b}, R_{1 \times d}\}) = cd \cdot \pi_T(\{R_{\frac{a}{c} \times 1}, R_{1 \times \frac{d}{b}}\}).$$

and likewise for $\pi_L(\mathcal{F})$. Thus it suffices to determine

$$\pi_T(w, h) := \pi_T(\{R_{w \times 1}, R_{1 \times h}\}) \quad \text{and} \quad \pi_L(w, h) := \pi_L(\{R_{w \times 1}, R_{1 \times h}\})$$

for $w, h \geq 1$. We say that a point set (resp. a lattice) P is (w, h)-piercing if it is $\{R_{w \times 1}, R_{1 \times h}\}$-piercing. It is sometimes convenient to work with the reciprocals $A_T(w, h) = 1/\pi_T(w, h)$ (resp. $A_L(w, h) = 1/\pi_L(w, h)$) which correspond to the *largest possible per-point area* of a (w, h)-piercing point set (resp. lattice). Note that $A_L(w, h) \leq A_T(w, h)$, since the sparsest (w, h)-piercing point set perhaps does not have to be a lattice. Also, $A_T(w, h) \leq \min(w, h)$ as translates of the smaller rectangle tile the plane and each translate has to be pierced.

1.2   Results

The following theorem and its corollary summarize our results for piercing all translates of two axis-parallel rectangles in $\mathbb{R}^2$.

Theorem 1. *Fix $w, h \geq 1$.*

(i) When $\lfloor w \rfloor \neq \lfloor h \rfloor$ then $A_T(w, h) = A_L(w, h) = \min\{w, h\}$.
(ii) When $\lfloor w \rfloor = \lfloor h \rfloor = k \geq 1$, set $w = k + x$, $h = k + y$ for $x, y \in [0, 1)$. Then

$$\max\left\{k, k + xy - \frac{k-1}{k}(1-x)(1-y)\right\} \leq A_L(w, h) \leq A_T(w, h) \leq k + xy.$$

Note that the inequalities in (ii) become equalities in two different cases: When $k = 1$ then $A_L(w, h) = A_T(w, h) = k + xy$ and when $\min\{x, y\} = 0$ (that is, when w or h is an integer) then $A_L(w, h) = A_T(w, h) = \min\{w, h\} = k$.

Corollary 1. *Given a family $\mathcal{F} = \{R_1, R_2\}$ consisting of two axis-parallel rectangles, a 1.086-approximation of $\pi_T(\mathcal{F})$ can be computed in $\mathcal{O}(1)$ time. The output piercing set is a lattice with density at most $(\frac{5}{2} - \sqrt{2}) \cdot \pi_T(\mathcal{F})$.*

We then address the general case of piercing all translates of any finite collection of axis-parallel rectangles.

Theorem 2. *Given a family $\mathcal{F} = \{R_1, \ldots, R_n\}$ consisting of n axis-parallel rectangles, a 1.895-approximation of $\pi_T(\mathcal{F})$ can be computed in $\mathcal{O}(n)$ time. The output piercing set is a lattice with density at most $(1 + \frac{2}{5}\sqrt{5}) \cdot \pi_T(\mathcal{F})$.*

2   Piercing Two Rectangles

Proof (of Theorem 1). (i) Note that $A_T(w, h) \leq \min\{w, h\}$: Indeed, any (w, h)-piercing point set has to pierce all the translates of the rectangle with smaller area and certain copies of that smaller rectangle tile the plane. To complete the proof, it suffices to exhibit a suitable (w, h)-piercing lattice. Without loss of generality suppose that $\lfloor h \rfloor < \lfloor w \rfloor$. We will show that the lattice Λ_1 with basis $u_1 = [1, h - 1]$, $v_1 = [1, -1]$ (see Fig. 1(a)) is (w, h)-piercing. Note that the area of the fundamental parallelogram of the lattice is $(h - 1) + 1 = h$, as required.

We first show that Λ_1 pierces all $1 \times h$ rectangles. Observe that the $1 \times h$ rectangles centered at points in Λ_1 tile the plane. Denote this tiling by $\mathcal{T}$. Let now R be any $1 \times h$ rectangle. Its center is contained in one of the rectangles in $\mathcal{T}$, say σ. Then the center of σ pierces R, as required.

We next show that Λ_1 pierces all $w \times 1$ rectangles. It suffices to show that Λ_1 pierces all $w_0 \times 1$ rectangles, where $w_0 = \lfloor h \rfloor + 1$. Let R be any $w_0 \times 1$ rectangle. Assume that R is not pierced by Λ_1. Translate R downwards until it hits a point in Λ_1, say q, and then leftwards until it hits another point in Λ_1, say p. Let R' denote the resulting rectangle. Then p is the top left corner of R'. Observe that the top and the right side of R' are not incident to any other point in Λ_1. Consider the lattice point $s := p + u_1 + (w_0 - 1)v_1$; note that $x(s) - x(p) = w_0$ and $y(s) - y(p) = h - 1 - \lfloor h \rfloor \in [-1, 0)$. As such, s is contained in the right side of R', a contradiction. It follows that R is pierced by Λ_1, as required.

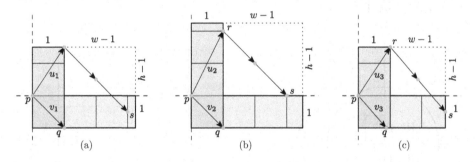

Fig. 1. (a) A lattice Λ_1 for the case $\lfloor w \rfloor \neq \lfloor h \rfloor$. Here $w = 3 + \frac{1}{4}$, $h = 2 + \frac{1}{2}$. (b) A lattice Λ_2 with basis $u_2 = [1, k - 1]$, $v_2 = [1, -1]$ attesting that $A_T(k + x, k + y) \geq k$. Here $w = 3 + \frac{1}{2}$, $h = 3 + \frac{1}{4}$. (c) A lattice Λ_3 with basis $u_3 = [1, h - 1]$, $v_3 = [(w - 1)/k, -1]$ attesting that $A_T(k + x, k + y) \geq k + xy - \frac{k-1}{k}(1 - x)(1 - y)$. Here $w = 2 + \frac{3}{4}$, $h = 2 + \frac{1}{2}$.

(ii) In order to prove the lower bound it suffices to exhibit suitable lattices. We will show that the following lattices do the job: The lattice Λ_2 with basis $u_2 = [1, k - 1]$, $v_2 = [1, -1]$ (see Fig. 1(b)) attests that $A_T(k + x, k + y) \geq k$. Note that the area of the fundamental parallelogram of Λ_2 is $(k - 1) + 1 = k$, as required. The lattice Λ_3 with basis $u_3 = [1, h - 1]$, $v_3 = [(w - 1)/k, -1]$ (see Fig. 1(c)) attests that $A_T(k + x, k + y) \geq k + xy - \frac{k-1}{k}(1 - x)(1 - y)$. Note that the area of the fundamental parallelogram of Λ_3 is $(w-1)(h-1)/k+1 = (k+xy) - \frac{k-1}{k}(1-x)(1-y)$, as required.

For both lattices, the proof proceeds by contradiction as in part (i). Assume that there exists an unpierced rectangle of dimensions either $w \times 1$ or $1 \times h$. Translate the rectangle downwards until it hits a point in the lattice, say q, and then leftwards until it hits another point in the lattice, say p. For Λ_2, note that $r := p + u_2$ lies on the right edge of the $1 \times h$ rectangle and that $s := p + u_2 + (k-1)v_2$ lies on the top edge of the $w \times 1$ rectangle. Similarly, for Λ_3 note that $r := p + u_3$ is the top right corner of the $1 \times h$ rectangle and that $s := p + u_3 + kv_3$ lies on the right edge of the $w \times 1$ rectangle. Either way, we get a contradiction.

Finally, we show the upper bound, that is, $A_T(w,h) \leq k+xy$. Recall that $A_T(w,h) \leq \min\{w,h\} = k + \min\{x,y\}$; we will obtain an improved bound $A_T(w,h) \leq k + xy$ by an integral calculus argument (which originates from a probabilistic argument). Let P be a (w,h)-piercing point set, where $w = k+x$, $h = k+y$ with $k \in \mathbb{N}$ and $x,y \in [0,1)$. The desired upper bound on $A_T(w,h)$ will follow from a lower bound on the density $\delta(P, D_r) = \frac{|P \cap D_r|}{\text{Area}(D_r)}$, where D_r is the disk with radius r centered at the origin. Fix a radius r and write $P_r = P \cap D_r$.

Given a point $a = (a_x, a_y) \in \mathbb{R}^2$, we denote by $R_a = [a_x - w, a_x] \times [a_y - h, a_y]$ the $w \times h$ rectangle whose top right corner is a. For brevity, we denote $R = R_{(w,h)} = [0,w] \times [0,h]$. We consider two sets of $w \times h$ rectangles: Those that intersect D_r and those that are contained within D_r. We denote the sets of their top right corners by $X = \{a \in \mathbb{R}^2 \mid R_a \cap D_r \neq \emptyset\}$ and $W = \{a \in \mathbb{R}^2 \mid R_a \subset D_r\}$, respectively. See Fig. 2(a).

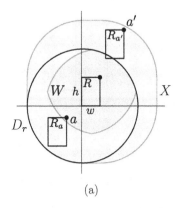
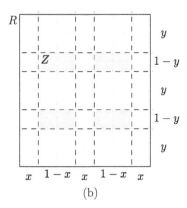

(a)   (b)

Fig. 2. (a) The top right corners of rectangles intersecting D_r form a region $X = D_r + R$. The top right corners of rectangles contained within D_r form a region $W = D_r \cap ([w,0] + D_r) \cap ([0,h] + D_r) \cap ([w,h] + D_r)$. Both regions are convex and their boundaries consist of circular arcs and line segments. (b) A $w \times h$ rectangle $R = R_{(w,h)}$ (here $k = 2$, $x = 1/3$, and $y = 2/3$, hence $w = k + x = 7/3$ and $h = k + y = 8/3$). Its zone Z is shaded.

Given a rectangle R_a, we define its *zone* Z_a to be a union of k^2 closed rectangles with sizes $(1-x) \times (1-y)$ each, arranged as in Fig. 2(b). Note that $\text{Area}(Z_a) = k^2(1-x)(1-y)$. Further, let $I_a := |P_r \cap Z_a|$ and $J_a := |P_r \cap (R_a \setminus Z_a)|$ be the number of points of P_r contained in R_a inside its zone and outside of it, respectively. We make two claims about I_a and J_a.

Claim 1. If $a \in W$ then $(k+1)I_a + kJ_a \ge k(k+1)$.

Proof. Fix $a \in W$. Since $R_a \subset D_r$, we have $P \cap R_a = P_r \cap R_a$. The key observation is that for any point $p \in R_a \setminus Z_a$, the set $R_a \setminus \{p\}$ contains k pairwise disjoint rectangles of dimensions either all $w \times 1$ or all $1 \times h$. We thus must have $|P_r \cap R_a| \ge k+1$, except when $P_r \cap R_a \subseteq Z_a$, in which case we must have $|P_r \cap R_a| \ge k$.

Denote $I = I_a$ and $J = J_a$. There are two simple cases:

1. $J \ge 1$: Then $I + J \ge k+1$, thus $(k+1)I + kJ \ge kI + kJ \ge k(k+1)$.
2. $J = 0$: Then $I \ge k$, thus $(k+1)I + kJ \ge k(k+1)$. $\qquad\qquad\qquad\square$

Claim 2. We have

$$\int_X \frac{I_a}{\text{Area}(Z)}\, \mathrm{d}a = \int_X \frac{J_a}{\text{Area}(R \setminus Z)}\, \mathrm{d}a = |P_r|.$$

Proof. Fix $p \in P_r$. Note that the set $X_p = \{a \in \mathbb{R}^2 \mid p \in Z_a\}$ of top right corners of $w \times h$ rectangles whose zone contains p is a subset of X congruent to Z. Thus $\text{Area}(X_p) = \text{Area}(Z)$. Summing over $p \in P_r$ we obtain

$$\int_X \frac{I_a}{\text{Area}(Z)}\, \mathrm{d}a = \frac{\sum_{p \in P_r} \text{Area}(X_p)}{\text{Area}(Z)} = |P_r|.$$

For J_a we proceed completely analogously. $\qquad\qquad\qquad\qquad\qquad\qquad\square$

Now we put the two claims together to get a lower bound on $|P_r|$.

Claim 3. We have $|P_r| \ge \frac{\text{Area}(W)}{k+xy}$.

Proof. First, applying Claim 1 to all $w \times h$ rectangles R_a with $a \in W$ and then invoking $W \subset X$, we obtain

$$\text{Area}(W) \cdot k(k+1) = \int_W k(k+1)\, \mathrm{d}a \le (k+1) \int_W I_a\, \mathrm{d}a + k \int_W J_a\, \mathrm{d}a$$

$$\le (k+1) \int_X I_a\, \mathrm{d}a + k \int_X J_a\, \mathrm{d}a.$$

By Claim 2 and straightforward algebra we further rewrite this as

$$(k+1) \int_X I_a\, \mathrm{d}a + k \int_X J_a\, \mathrm{d}a = |P_r| \cdot \big((k+1)\,\text{Area}(Z) + k\,\text{Area}(R \setminus Z)\big)$$

$$= |P_r| \cdot (k\,\text{Area}(R) + \text{Area}(Z)) = |P_r| \cdot k(k+1)(k+xy),$$

where the last equality follows from

$$k \operatorname{Area}(R) + \operatorname{Area}(Z) = k(k+x)(k+y) + k^2(1-x)(1-y) = k(k+1)(k+xy).$$

The bound $|P_r| \geq \frac{\operatorname{Area}(W)}{k+xy}$ follows by rearranging.   □

Consequently, by Claim 3 we have

$$\delta(P, D_r) = \frac{|P_r|}{\operatorname{Area}(D_r)} \geq \frac{\operatorname{Area}(W)}{\operatorname{Area}(D_r)} \cdot \frac{1}{k+xy} \rightarrow_{r \to \infty} \frac{1}{k+xy},$$

where we used that $\operatorname{Area}(W)/\operatorname{Area}(D_r) \to 1$ as $r \to \infty$. This in turn gives

$$\pi_T(w, h) = \inf_P \left\{ \liminf_{r \to \infty} \delta(P, D_r) \right\} \geq \frac{1}{k+xy} \quad \text{and} \quad A_T(w, h) = \frac{1}{\pi_T(w, h)} \leq k + xy$$

and completes the proof of Theorem 1.   □

For a visual illustration of our results, see Fig. 3.

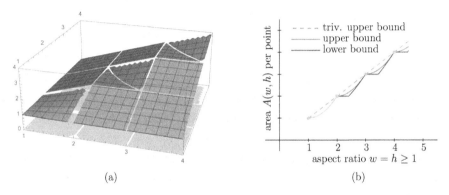

(a)   (b)

Fig. 3. (a) We plot $A_T(w, h)$ when $\lfloor w \rfloor \neq \lfloor h \rfloor$ and/or when $\lfloor w \rfloor = \lfloor h \rfloor = 1$ (orange). When $\lfloor w \rfloor = \lfloor h \rfloor \geq 2$ we plot the two lower bounds from Theorem 1, Item ii (blue). As $k \to \infty$, the two lower bounds coincide for $x + y = 1$. (b) A section corresponding to $w = h$. We plot the lower bounds (blue) and the upper bound (red) on $A_T(w, w)$ from Theorem 1, Item ii and the trivial upper bound $A_T(w, w) \leq w$ (red, dashed). (Color figure online)

Proof (of Corollary 1). It suffices to show that

$$\sup_{\substack{k \geq 2,\, k \in \mathbb{N} \\ x, y \in [0, 1)}} \frac{k + xy}{\max \left\{ k, k + xy - \frac{k-1}{k}(1-x)(1-y) \right\}} = \frac{5 - 2\sqrt{2}}{2} < 1.086.$$

A computer algebra system (such as Mathematica) shows that the supremum is attained when $k = 2$ and when x, y are both equal to a value that makes the two expressions inside the $\max\{\}$ operator equal. This happens for $x = y = \sqrt{2} - 1$ and the corresponding value is $(5 - 2\sqrt{2})/2 < 1.086$ as claimed.   □

3   Piercing n Rectangles

In this section we prove Theorem 2. Let $\varphi = \frac{1}{2}(1 + \sqrt{5})$ be the golden ratio. In Lemma 2 we show that a lattice Λ_φ with basis $u = [1, \varphi], v = [\varphi, -1]$ pierces all rectangles with area φ^4 or larger, irrespective of their aspect ratio. See Fig. 4 (a). Theorem 2 then follows easily by rescaling Λ_φ to match the smallest-area rectangle from the family.

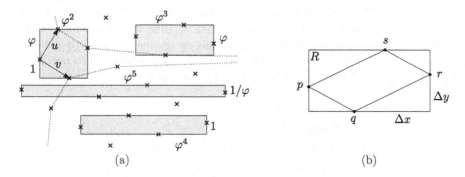

(a)   (b)

Fig. 4. (a) Empty rectangles amidst Λ_φ. (b) A generic empty rectangle R.

Recall the well-known sequence of Fibonacci numbers defined by the following recurrence:

$$F_i = F_{i-1} + F_{i-2}, \text{ with } F_1 = F_2 = 1. \tag{1}$$

The first few terms in the sequence are listed in Table 1 for easy reference; here it is convenient to extend this sequence by $F_{-1} = 1$ and $F_0 = 0$.

Table 1. The first few Fibonacci numbers.

m	-1	0	1	2	3	4	5	6	7	8	9	10
F_m	1	0	1	1	2	3	5	8	13	21	34	55

We first list several properties of Fibonacci numbers.

Lemma 1. *The following identities hold for every integer $m \geq 1$:*

1. *$F_m \varphi + F_{m-1} = \varphi^m$,*
2. *$F_m \varphi - F_{m+1} = (-1)^{m+1} \varphi^{-m}$,*
3. *$F_{2m+1} F_{2m-1} - (F_{2m})^2 = 1$.*

Proof. This is straightforward to verify, for instance using the well-known formula $F_m = \frac{1}{\sqrt{5}}(\varphi^m - \psi^m)$, where $\psi = -1/\varphi$. $\square$

Next we prove a lemma that establishes a key property of the lattice Λ_φ.

Lemma 2. *The area of every empty rectangle amidst the points in Λ_φ is at most φ^4.*

Proof. Let R be a maximal axis-parallel empty rectangle, bounded by four lattice points p, q, r, s from the left, below, right and top, respectively. Refer to Fig. 4 (b). Then $pqrs$ is a fundamental parallelogram of the lattice; as such, its area is $\varphi^2 + 1$. Clearly, we may assume $p = (0,0)$. Further, we may assume $q = cu + dv$, $s = au + bv$, where a, b, c, d are nonnegative integers: Indeed, since Λ_φ is invariant under rotation by $90°$, we can assume that the width of R is at least as large as its height. Points q, s thus lie on the "funnel" (depicted in Fig. 4 (a) dotted) within the angle formed by the vectors u, v. The coordinates of points s, q, r and the area of the parallelogram $pqrs$ are:

$$s = (a + b\varphi, a\varphi - b),$$
$$q = (c + d\varphi, c\varphi - d),$$
$$r = ((a + c) + (b + d)\varphi, (a + c)\varphi, -(b + d)),$$
$$\text{Area}(pqrs) = |(a + b\varphi)(c\varphi - d) - (a\varphi - b)(c + d\varphi)| = |(ad - bc)|(\varphi^2 + 1).$$

Since s lies above the horizontal line through p and since q lies below it, we have $a\varphi - b > 0$ and $c\varphi - d < 0$. This implies $a, b, c, d > 0$ and rewrites as $\frac{b}{a} < \varphi < \frac{d}{c}$, so in particular $ad > bc$. Together with the expression for $\text{Area}(pqrs) = \varphi^2 + 1$ this yields $|ad - bc| = ad - bc = 1$. To summarize, we have

$$\frac{b}{a} < \varphi < \frac{d}{c} \quad \text{and} \quad ad - bc = 1. \tag{2}$$

The relation $ad - bc = 1$ implies that $\gcd(a, b) = \gcd(c, d) = 1$. By a result from the theory of continued fractions [33], [22, Ch. 10], relation (2) implies that the fractions b/a and d/c are consecutive convergents of φ. Moreover, it is well known that the convergents of φ are ratios of consecutive Fibonacci numbers:

$$\frac{F_0}{F_{-1}} < \frac{F_2}{F_1} < \frac{F_4}{F_3} < \frac{F_6}{F_5} < \frac{F_8}{F_7} < \cdots < \varphi < \cdots < \frac{F_7}{F_6} < \frac{F_5}{F_4} < \frac{F_3}{F_2} < \frac{F_1}{F_0} = \infty,$$

$$\frac{0}{1} < \frac{1}{1} < \frac{3}{2} < \frac{8}{5} < \frac{21}{13} < \cdots < \varphi < \cdots < \frac{13}{8} < \frac{5}{3} < \frac{2}{1} < \frac{1}{0} = \infty.$$

Thus we can assume that $\frac{b}{a} = \frac{F_{2k}}{F_{2k-1}}$. There are two cases:

Case 1: $\frac{d}{c} = \frac{F_{2k-1}}{F_{2k-2}}$, and thus $\frac{F_{2k}}{F_{2k-1}} < \varphi < \frac{F_{2k-1}}{F_{2k-2}}$.

Case 2: $\frac{d}{c} = \frac{F_{2k+1}}{F_{2k}}$, and thus $\frac{F_{2k}}{F_{2k-1}} < \varphi < \frac{F_{2k+1}}{F_{2k}}$.

(Note that in both cases we indeed have $ad - bc = 1$ by Lemma 1, Item 3.) We compute $\text{Area}(R)$ using Items 1 to 2 of Lemma 1. Let Δx and Δy denote the side-lengths of R.

Case 1: We have

$$\Delta x = (a + c) + (b + d)\varphi = (F_{2k-1} + F_{2k-2}) + (F_{2k} + F_{2k-1})\varphi$$
$$= F_{2k} + F_{2k+1}\varphi = \varphi^{2k+1},$$
$$\Delta y = (a - c)\varphi - (b - d) = (F_{2k-1} - F_{2k-2})\varphi - (F_{2k} - F_{2k-1})$$
$$= F_{2k-3}\varphi - F_{2k-2} = \varphi^{-(2k-3)},$$

thus $\mathrm{Area}(R) = \Delta x \cdot \Delta y = \varphi^{2k+1}\varphi^{-(2k-3)} = \varphi^4$, as required.

Case 2: Similarly, we have

$$\Delta x = (a + c) + (b + d)\varphi = (F_{2k-1} + F_{2k}) + (F_{2k} + F_{2k+1})\varphi$$
$$= F_{2k+1} + F_{2k+2}\varphi = \varphi^{2k+2},$$
$$\Delta y = (a - c)\varphi - (b - d) = (F_{2k-1} - F_{2k})\varphi - (F_{2k} - F_{2k+1})$$
$$= -F_{2k-2}\varphi + F_{2k-1} = \varphi^{-(2k-2)},$$

thus $\mathrm{Area}(R) = \Delta x \cdot \Delta y = \varphi^{2k+2}\varphi^{-(2k-2)} = \varphi^4$, as required.   □

With Lemma 2 at hand, the proof of Theorem 2 is straightforward.

Proof (of Theorem 2). Let $R \in \mathcal{F}$ be the smallest-area rectangle among those in $\mathcal{F}$. By Lemma 2, the lattice Λ_φ pierces all axis-parallel rectangles with area at least φ^4. Thus the rescaled lattice $\Lambda'_\varphi = \sqrt{\mathrm{Area}(R)/\varphi^4} \cdot \Lambda_\varphi$ pierces all axis-parallel rectangles with area at least $\mathrm{Area}(R)$. In particular, it pierces all rectangles in $\mathcal{F}$. Since the fundamental parallelogram of Λ_φ has area $\varphi^2 + 1$, the fundamental parallelogram of Λ'_φ has area $(\varphi^2 + 1)/\varphi^4 \cdot \mathrm{Area}(R)$ and gives an approximation factor $\varphi^4/(\varphi^2 + 1) = 1 + \frac{2}{5}\sqrt{5} < 1.895$ as claimed. Note that computing the smallest-area rectangle and the rescaling only take $\mathcal{O}(n)$ time. □

Remarks. We have learned from the recent article of Kritzinger and Wiart [31] that a rescaled version of the lattice Λ_φ was considered several years ago by Thomas Lachmann (unpublished result) as a candidate for an upper bound on the minimum dispersion $A(n)$ of an n-point set in a unit square. Yet another lattice resembling Λ_φ was studied in the same context by Ismăilescu [25].

It is easy to check that the lattice Λ_φ yields the upper bound $\liminf_{n\to\infty} nA(n) \leq \varphi^4/(\varphi^2 + 1)$, i.e., matching exactly the dispersion bound obtained by Kritzinger and Wiart using a suitable modification of the so-called Fibonacci lattice [16]. It is worth noting that: (i) the lattice Λ_φ yields the above dispersion result with a cleaner and shorter proof; (ii) the Fibonacci lattice as well as its modification lead to this bound only by a limiting process; and perhaps more importantly, (iii) the upper bound in Lemma 2 on the maximum rectangle area amidst points in this lattice holds universally across the entire plane and not only inside a bounding box with n points (i.e., one does not need to worry about rectangles with a side supported by the bounding box boundary).

4   Conclusion

We list several open questions.

1. (Computational complexity.) Given a family $\mathcal{F}$ of n axis-parallel rectangles, what is the computational complexity of determining the optimal density $\pi_T(\mathcal{F})$ of a piercing set for all translates of all members in $\mathcal{F}$? (For the decision problem: given a threshold τ, is there a piercing set whose density is at most τ?) Is the problem algorithmically solvable? Is there a polynomial-time algorithm? And how about the complexity of determining the optimal density $\pi_L(\mathcal{F})$ of a piercing lattice for $\mathcal{F}$?

2. (Exact answer for two rectangles.) For two rectangles, what is the actual value of $\pi_T(w, h)$ (or its reciprocal $A_T(w, h)$) when $\lfloor w \rfloor = \lfloor h \rfloor \geq 2$? Is it the same as $\pi_L(w, h)$?

3. (Other shapes.) How about pairs of different shapes such as triangles?

4. (Congruent copies.) How about requiring that the point set pierces all congruent copies of a set of shapes, not only translates? See for instance [4] where it is shown that when $\mathcal{F}$ consists of a single square (or rectangle), a suitable triangular grid gives an upper bound on the density.

5. (Discrete version.) Consider a discrete version where: (i) each shape in the family $\mathcal{F}$ consists of cells of an infinite square grid; (ii) we consider translates by integer vectors only; and (iii) instead of piercing with a point set we pierce with a set of grid cells. What can be said about this (lowest possible) discrete hitting density $\pi_T^{\mathrm{disc}}(\mathcal{F})$ or its reciprocal $A_T^{\mathrm{disc}}(\mathcal{F})$? As one example, we note that Theorems 1 and 2 can be adapted to the discrete setting in a straightforward way: When $\mathcal{F}$ consists of two rectangles $R_{a \times b}$, $R_{b \times a}$, where $b = k \cdot a + r$ (with $k \geq 1$ and $r < a$), the adapted Theorem 1 yields an upper bound $A_T^{\mathrm{disc}}(\{R_{a \times b}, R_{b \times a}\}) \leq k \cdot a^2 + r^2$ that solves the puzzle we mentioned in the introduction. Furthermore, when $\mathcal{F}$ consists of all rectangles that have area at least K, then the adapted Theorem 2 gives an $\mathcal{F}$-piercing set of cells with density $(1 + \frac{2}{5}\sqrt{5})/K$. In the language of Fiat and Shamir [16], there is a probing strategy that locates a battleship of K squares in a rectangular sea of M squares (where $M \to \infty$) in at most $1.895\, M/K$ probes. This is a substantial improvement over the $3.065\, M/K$ bound from [16].

   As another example, when $\mathcal{F}$ consists of two L-triominoes that are centrally symmetric to each other, one can show that $A_T^{\mathrm{disc}}(\mathcal{F}) = 3$. In contrast, when $\mathcal{F}$ consists of two (or more) L-triominoes, one of which is obtained from another one by rotation by $90°$, one can show that $A_T^{\mathrm{disc}}(\mathcal{F}) = 2$.

6. (Higher dimensions.) Consider the problem in higher dimensions. When $d = 3$ and $|\mathcal{F}| = 2$, stretching along the coordinate axis yields a non-trivial case $\{(a, b, 1), (1, 1, c)\}$ where $a, b, c \geq 1$. The trivial upper bound $A_T(a, b, c) \leq \min\{ab, c\}$ can be matched in two "easy" cases:

   (i) When $c \geq \lceil a \rceil \cdot \lceil b \rceil$ then "piercing $1 \times 1 \times c$ is free": Briefly, in the plane $z = 0$ we use a lattice with basis $[a, 0], [0, b]$ and in the planes $z \in \mathbb{Z}$ we consider $\lceil a \rceil \cdot \lceil b \rceil$ "integer horizontal offsets" $[u, v]$, $u = 0, \ldots, \lceil a \rceil - 1$, $v = 0, \ldots, \lceil b \rceil - 1$ and use them periodically (in any order).

(ii) When $c \leq \lfloor a \rfloor \cdot \lfloor b \rfloor$ then "piercing $a \times b \times 1$ is free": Briefly, along the line $x = y = 0$ we put points at $k \cdot \lfloor a \rfloor \cdot \lfloor b \rfloor$ for $k \in \mathbb{Z}$. For other vertical lines through integer points we use $\lfloor a \rfloor \cdot \lfloor b \rfloor$ "integer vertical offsets" such that every $\lfloor a \rfloor \cdot \lfloor b \rfloor$ horizontal grid-rectangle contains all offsets.

Together, the easy cases (i) and (ii) cover the case when $a \in \mathbb{Z}$, $b \in \mathbb{Z}$, $c \in \mathbb{R}$ and the case when ab and c "differ by a lot". Another easy case is $a = 1$, when the planar bounds apply (for two rectangles with sizes $1 \times b$ and $c \times 1$). Finally, from the algorithmic standpoint, given a finite collection of axis-parallel boxes in $\mathbb{R}^d$, what approximations for the piercing density can be obtained?

7. (Disconnected shapes.) What can be said about disconnected shapes? The easiest variant seems to be when each shape is a set of integer points on the line (or in $\mathbb{Z}^d$) and we consider translates by integer vectors only. Note that the piercing density and the lattice piercing densities may differ in this case; for instance, when $S = \{0, 2\} \subset \mathbb{Z}$, these densities are $1/2$ and 1, respectively. In view of the connection to covering mentioned in Sect. 1, it is worth mentioning that the problem of tiling the infinite integer grid with finite clusters is only partially solved [35]; however, covering is generally easier than tiling.

Acknowledgments. We would like to thank Wolfgang Mulzer and Jakub Svoboda for helpful comments on an earlier version of this work.

References

1. Aistleitner, C., Hinrichs, A., Rudolf, D.: On the size of the largest empty box amidst a point set. Discret. Appl. Math. **230**, 146–150 (2017)
2. Alon, N., Kleitman, D.J.: Piercing convex sets and the Hadwiger-Debrunner (p, q)-problem. Adv. Math. **96**(1), 103–112 (1992)
3. Aronov, B., Ezra, E., Sharir, M.: Small-Size ε-nets for axis-parallel rectangles and boxes. SIAM J. Comput. **39**(7), 3248–3282 (2010)
4. Basic, B., Slivková, A.: On optimal piercing of a square. Discret. Appl. Math. **247**, 242–251 (2018)
5. Braß, P., Moser, W., Pach, J.: Research Problems in Discrete Geometry. Springer, New York (2005). https://doi.org/10.1007/0-387-29929-7
6. Bukh, B., Chao, T.-W.: Empty axis-parallel boxes, manuscript (2021). arXiv:2009.05820
7. Carmi, P., Katz, M.J., Lev-Tov, N.: Polynomial-time approximation schemes for piercing and covering with applications in wireless networks. Comput. Geom. Theory Appl. **39**, 209–218 (2008)
8. Chan, T.M.: Polynomial-time approximation schemes for packing and piercing fat objects. J. Algorithms **46**(2), 178–189 (2003)
9. Chan, T.M., Mahmood, A.-A.: Approximating the piercing number for unit-height rectangles. In: Proceedings of 17th Canadian Conference on Computational Geometry (CCCG 2005), University of Windsor, Ontario, Canada, pp. 15–18 (2005)
10. Chudnovsky, M., Spirkl, S., Zerbib, S.: Piercing axis-parallel boxes. Electron. J. Comb. **25**(1), #P1.70 (2018)

11. Correa, J.R., Feuilloley, L., Pérez-Lantero, P., Soto, J.A.: Independent and hitting sets of rectangles intersecting a diagonal line: algorithms and complexity. Discrete Comput. Geom. **53**(2), 344–365 (2015)
12. Danzer, L., Grünbaum, B., Klee, V.: Helly's theorem and its relatives. In: Proceedings of Symposia in Pure Mathematics, vol. 7, pp. 101–181. American Mathematical Society (1963)
13. Dumitrescu, A., Jiang, M.: Piercing translates and homothets of a convex body. Algorithmica **61**(1), 94–115 (2011)
14. Eckhoff, J.: A survey of the Hadwiger-Debrunner (p, q)-problem. In: Aronov, B., Basu, S., Pach, J., Sharir, M. (eds.) Discrete & Computational Geometry. Algorithms and Combinatorics, vol. 25, pp. 347–377. Springer, Heidelberg (2003). https://doi.org/10.1007/978-3-642-55566-4_16
15. Fejes Tóth, G.: Packing and covering. In: Goodman, J.E., O'Rourke, J., Tóth, C.D. (eds.) Handbook of Discrete and Computational Geometry (Chapter 2), 3rd edn, pp. 27–66. CRC Press, Boca Raton (2017)
16. Fiat, A., Shamir, A.: How to find a battleship. Networks **19**, 361–371 (1989)
17. Fon-Der-Flaass, D., Kostochka, A.V.: Covering boxes by points. Discret. Math. **120**(1–3), 269–275 (1993)
18. Fowler, R.J., Paterson, M.S., Tanimoto, S.L.: Optimal packing and covering in the plane are NP-complete. Inf. Process. Lett. **12**, 133–137 (1981)
19. Garey, M.R., Johnson, D.S.: Computers and Intractability: A Guide to the Theory of NP-Completeness. W. H. Freeman and Company, New York (1979)
20. Govindarajan, S., Nivasch, G.: A variant of the Hadwiger-Debrunner (p, q)-problem in the plane. Discrete Comput. Geom. **54**(3), 637–646 (2015)
21. Hadwiger, H., Debrunner, H.: Combinatorial Geometry in the Plane (English translation by Victor Klee). Holt, Rinehart and Winston, New York (1964)
22. Hardy, G.H., Wright, E.M.: An Introduction to the Theory of Numbers, 5th edn. Oxford University Press, Oxford (1979)
23. Helly, E.: Über Mengen konvexer Körper mit gemeinschaftlichen Punkten. Jahresber. Deutsch. Math.-Verein. **32**, 175–176 (1923)
24. Holmsen, A., Wenger, R.: Helly-type theorems and geometric transversals. In: Goodman, J.E., O'Rourke, J., Tóth, C.D. (eds.) Handbook of Discrete and Computational Geometry (Chapter 4), 3rd edn, pp. 91–123. CRC Press, Boca Raton (2017)
25. Ismăilescu, D.: Personal communication (2019)
26. Karasev, R.N.: Transversals for families of translates of a two-dimensional convex compact set. Discret. Comput. Geom. **24**, 345–353 (2000)
27. Károlyi, G.: On point covers of parallel rectangles. Period. Math. Hung. **23**(2), 105–107 (1991)
28. Károlyi, G., Tardos, G.: On point covers of multiple intervals and axis-parallel rectangles. Combinatorica **16**(2), 213–222 (1996)
29. Kim, S.-J., Nakprasit, K., Pelsmajer, M.J., Skokan, J.: Transversal numbers of translates of a convex body. Discret. Math. **306**, 2166–2173 (2006)
30. Krieg, D.: On the dispersion of sparse grids. J. Complex. **45**, 115–119 (2018)
31. Kritzinger, R., Wiart, J.: Improved dispersion bounds for modified Fibonacci lattices. J. Complex. **63**, 101522 (2021). https://doi.org/10.1016/j.jco.2020.101522
32. Mustafa, N., Varadarajan, K.: Epsilon-nets and epsilon-approximations. In: Goodman, J.E., O'Rourke, J., Tóth, C.D. (eds.) Handbook of Discrete and Computational Geometry (Chapter 47), 3rd edin. CRC Press, Boca Raton (2017)
33. Olds, C.D.: Continued Fractions. Random House, New York (1963)

34. Sosnovec, J.: A note on minimal dispersion of point sets in the unit cube. Eur. J. Comb. **69**, 255–259 (2018)
35. Szegedy, M.: Algorithms to tile the infinite grid with finite clusters. In: Proceedings of 39th Annual Symposium on Foundations of Computer Science (FOCS 1998), Palo Alto, California, USA, pp. 137–147. IEEE Computer Society (1998)
36. Ullrich, M., Vybíral, J.: An upper bound on the minimal dispersion. J. Complex. **45**, 120–126 (2018)

A Triangle Process on Regular Graphs

Colin Cooper[1]([✉]), Martin Dyer[2], and Catherine Greenhill[3]

[1] Department of Informatics, King's College London, London WC2R 2LS, UK
colin.cooper@kcl.ac.uk
[2] School of Computing, University of Leeds, Leeds LS2 9JT, UK
m.e.dyer@leeds.ac.uk
[3] School of Mathematics and Statistics, UNSW Sydney,
Kensington, NSW 2052, Australia
c.greenhill@unsw.edu.au

Abstract. Switches are operations which make local changes to the edges of a graph, usually with the aim of preserving the vertex degrees. We study a restricted set of switches, called triangle switches. Each triangle switch creates or deletes at least one triangle. Triangle switches can be used to define Markov chains which generate graphs with a given degree sequence and with many more triangles (3-cycles) than is typical in a uniformly random graph with the same degrees. We show that the set of triangle switches connects the set of all d-regular graphs on n vertices, for all $d \geq 3$. Hence, any Markov chain which assigns positive probability to all triangle switches is irreducible on these graphs. We also investigate this question for 2-regular graphs.

Keywords: Regular graphs · Triangles · Markov chains · Irreducibility

1 Introduction

Generating graphs at random from given classes and distributions has been the subject of considerable research. See, for example, [1–3,5,8,10,11,15–18,21]. Generation using Markov chains has been a topic of specific interest in this context, in particular Markov chains based on switches of various types, for example [2,5,8,10,15–18,21]. Switches delete a pair of edges from the graph and insert a different pair on the same four vertices. They have the important property that they preserve the degree sequence of the graph. Thus they are useful for generating regular graphs, or other graphs with a given degree sequence. Markov chains also give a dynamic reconfigurability property, which is useful in applications, for example [5,10,17]. For any such Markov chain, two questions arise. First, can it generate any graph in the chosen class? (Formally: is the Markov chain irreducible?) Secondly, we might wish to estimate its rate of convergence to the chosen distribution. (Formally: what is the mixing time of the chain?)

Dyer supported by EPSRC grant EP/S016562/1,"Sampling in hereditary classes", Greenhill by Australian Research Council grant DP190100977.

P. Flocchini and L. Moura (Eds.): IWOCA 2021, LNCS 12757, pp. 310–323, 2021.
https://doi.org/10.1007/978-3-030-79987-8_22

In the applied field of social networks, the existence of triangles (3-cycles) is seen as an indicator of mutual friendships [12,14]. Such networks can alter in a dynamic fashion as pairs of vertices friend or unfriend each other based on a mutual acquaintance. However, many random graph models, or processes for producing random graphs, tend to produce graphs with few triangles. This is true for any process which generates sparse graphs with a given degree sequence (approximately) uniformly at random. For example, the expected number of triangles is constant for d-regular graphs, when d is constant [4]. In this paper, we study a restricted set of switches, called *triangle switches*, and consider any reversible Markov chain whose transitions are exactly the triangle switches. Triangle switches were introduced in [6] in the context of cubic graphs, and examples were given of Markov chains using triangle switches with transition probabilities assigned to encourage the formation of triangles. It was proved in [6, Section 4] that it is possible to generate cubic graphs using this approach which have $\Omega(n)$ triangles in $O(n)$ steps of the Markov chain.

In this paper we address the first question posed above ("is the Markov chain irreducible?") for such chains on the state space of d-regular graphs, for any d. Note that the answer to this question is independent of the probabilities assigned to each triangle switch by the Markov chain, as it is a property of the undirected graph underlying the Markov chain. We leave the mixing question for future research, noting only that tight bounds on mixing time seem hard to come by in this setting. The recent paper [21] is a notable exception.

The proofs in [6] do not easily generalise to regular graphs of arbitrary d, though the main approach in our proof of irreducibility comes from [6]. If a component of a d-regular graph is a clique (that is, a complete subgraph) then it must be isomorphic to K_{d+1}. We call such a component a *clique component*. Our approach is to show that starting from an arbitrary d-regular graph, triangle switches can be used to increase the number of clique components. Furthermore, we show how to alter the set of vertices in a given clique component using triangle switches. After creating as many clique components as possible, there is at most one additional component C, which must satisfy $d + 1 < |C| < 2(d+1)$. We call such a component a *fragment*. We prove that triangle switches connect the set of all fragments on a given vertex set. In the cubic case, this last step is simpler as the only possible fragments are $K_{3,3}$ and $\overline{C}_6$.

Our result can be viewed as solving a particular *reconfiguration* problem for regular graphs. Reconfiguration is a topic of growing interest in discrete mathematics. For an introduction to the topic, and a survey of results, see [19]. We note that reconfiguration problems can be as hard as PSPACE-complete, in general. Our results show that there is a polynomial time algorithm to construct a path of triangle switches between any two d-regular graphs on n vertices.

The plan of the paper is as follows. In Sect. 1.1, we define and review switches and restricted switches, in particular triangle switches, and state our main result, Theorem 1. For most of the paper we assume that $d \geq 3$. In Sect. 2 we show that the set of all fragments with a given vertex set is connected under triangle switches. In Sect. 3, we show that triangle switches can be used to create a clique component, starting from any d-regular graph with at least $2(d+1)$ vertices. In

Sect. 4 we show how to relabel the vertices in clique components using triangle switches, and hence complete our proof of irreducibility. Finally, in Sect. 5, we consider the irreducibility question for d-regular graphs with $d \leq 2$. Many proofs are sketched or omitted here. For full details see [7].

Definitions, Notation and Terminology. The notation $[k]$ will denote the set $\{1, 2, \ldots, k\}$, for any integer k. Given a set V of vertices, let $V^{\{2\}}$ be the set of unordered pairs of distinct elements from V. A graph $G = (V, E)$ on vertex set $V(G)$ has edge set $E(G) \subseteq V^{\{2\}}$. We usually denote $|V|$ by n. We use the notation xy as a shorthand for the unordered pair $\{x, y\}$, whether or not this pair is an edge. If $E' \subseteq E$ and $V' = \{v \in V : v \in e \in E'\}$, then $G' = (V', E')$ is a *subgraph* of G. Given any vertex subset $U \subseteq V$, the subgraph $G[U]$ *induced* by U has vertex set U and edge set $E' = U^{(2)} \cap E$. If $|U| = k$ and $G[U]$ is a k-cycle, then we say that $G[U]$ is an *induced* C_k.

We will write $G \cong H$ to indicate that graphs G and H are isomorphic. Given a graph $G = (V, E)$, the *complement* of G is the graph $\overline{G} = (V, \overline{E})$ with $\overline{E} = V^{\{2\}} \setminus E$. An edge of $\overline{G}$ will be called a *non-edge* of G.

The *distance* $\operatorname{dist}(u, v)$ between two vertices u and v is the number of edges in a shortest path from u to v in G, with $\operatorname{dist}(u, v) := \infty$ if no such path exists. The maximum distance between two vertices in G is the *diameter* of G, and G is connected if it has finite diameter. The *component* C of G containing v is the largest connected induced subgraph of G which contains v.

Given a graph $G = (V, E)$ and vertex $v \in V$, let $\operatorname{N}_G(v) = \{u : uv \in E\}$ denote the neighbourhood of v, and let $\deg_G(v) = |\operatorname{N}_G(v)|$ denote the degree of v in G. The *closed neighbourhood* of v is $\operatorname{N}_G[v] := \operatorname{N}_G(v) \cup \{v\}$. We sometimes drop the subscript and write $\operatorname{N}(v)$ or $\operatorname{N}[v]$.

Say that G is *regular* if every vertex has the same degree, and if $\deg_G(v) = d$ for all $v \in V$ then we say that G is d-regular. As already stated, $\mathcal{G}_{n,d}$ will be the set of all d-regular graphs with vertex set $V = [n]$. Note that $\mathcal{G}_{n,d}$ is non-empty if and only if either d or n is even. This result seems to be folklore, but is easy to prove. Necessity is implied by edge counting, and sufficiency by a direct construction. An indirect proof can be found in [22, Prop. 1]. As usual, $K_{d+1} \in \mathcal{G}_{d+1,d}$ denotes the complete graph on $d + 1$ vertices, and $K_{d,d} \in \mathcal{G}_{2d,d}$ denotes the complete bipartite graph on $d + d$ vertices. A graph in $\mathcal{G}_{n,d}$ with $d + 1 < n < 2(d + 1)$ will be called a *fragment*. Note that K_{d+1} is not a fragment.

We often regard a graph $G \in \mathcal{G}_{n,d}$ as *layered*, in the following way. Let v be a given (fixed) vertex of a d-regular graph $G = (V, E)$, where $n = |V| \geq 2(d + 1)$, and let $C \subseteq V$ determine the component $G[C]$ of G such that $v \in C$. We regard C as partitioned by distance from v, with $V_i = \{u \in C : \operatorname{dist}(v, u) = i\}$. Thus $V_0 = \{v\}$ and $V_1 = \operatorname{N}(v)$, so $|V_0| = 1$ and $|V_1| = d$. Since V_2 appears frequently in the proof, we will denote $|V_2|$ by ℓ. By definition, C is a disjoint union $C = \bigcup_{i \geq 0} V_i$, and $|C| = \sum_{i \geq 0} |V_i|$. Let $G_i = G[V_i] = (V_i, E_i)$, and note that $G_0 = (\{v\}, \emptyset)$. Let $\operatorname{N}'(u)$ be the neighbourhood of $u \in V_i$ in G_i, i.e. $\operatorname{N}'(u) = \operatorname{N}(u) \cap V_i$, and let $d'(u) = |\operatorname{N}'(u)|$ be the degree of u in G_i. We omit explicit reference to i in this notation, since it is implicit from $u \in V_i$. Given $u \in V_i$, we denote the set of non-neighbours of u in G_i by $\overline{\operatorname{N}}'_i(u)$.

We will regard the edges of $G[C]$ from V_i to V_{i+1} as being directed away from the designated vertex v. Under this convention, for $u \in V_i$, $\text{In}(u)$ is the neighbour set of u in V_{i-1}, and $\text{Out}(u)$ is the neighbour set of u in V_{i+1}. Thus, if $u \in V_i$, $\text{In}(u) = \text{N}(u) \cap V_{i-1}$ and $\text{Out}(u) = \text{N}(u) \cap V_{i+1}$. Then let $\text{id}(u) = |\text{In}(u)|$ be the in-degree of $u \in V_i$ and $\text{od}(u) = |\text{Out}(u)|$ the out-degree of u. Thus $\text{N}(u) = \text{N}'(u) + \text{In}(u) + \text{Out}(u)$, and $\text{d}'(u) + \text{id}(u) + \text{od}(u) = d$. In particular, $\text{d}'(v) = \text{id}(v) = 0$, and $\text{od}(v) = d$. If $u \in V_1 = \text{N}(v)$ then $\text{id}(u) = 1$ and so $\text{d}'(u) + 1 + \text{od}(u) = d$, and thus $\text{od}(u) = d - 1 - \text{d}'(u) = |\overline{\text{N}}'_1(u)|$.

A pair of distinct vertices $x, y \in V_i$ will be *below* a pair of distinct vertices $a, b \in V_{i+1}$ if $a \in \text{Out}(x)$ and $b \in \text{Out}(y)$. In this case we also say that a, b are *above* x, y. Note that, if $a, b \in V_{i+1}$ is not above some pair $x, y \in V_i$, there must be a unique $z \in V_i$ with $a, b \in \text{Out}(z)$. We will be most interested in the case where $i = 1$ and $ab \notin E_2$.

For other graph-theoretic definitions and concepts not given here, see [24], for example.

1.1 Switches

As described above, an established approach to the generation of graphs with given degrees is to use local edge transformations known as switches. The process is *irreducible* if any graph in the class can be obtained from any other by a sequence of these local transformations. Here we will consider three possibilities for this local transformation.

In a switch, a pair of edges xy, wz of graph $G = (V, E)$ are chosen at random in some fashion, and replaced with the pair xw yz, provided these are currently non-edges and that the vertices x, y, w, z are distinct. See Fig. 1. We make no other assumptions about $G[\{w, x, y, z\}]$. Clearly switches preserve vertex degrees, since each vertex in the switch has one edge deleted and one added, and all other vertices are unaffected.

Taylor [20] proved that the set of graphs with given degrees is connected under switches. (See also [23], [24, Thm. 1.3.33], where switches are called "2-switches", and [16] for a more constructive proof.) Cooper, Dyer and Greenhill [5] showed rapid mixing of the switch Markov chain for regular graphs, and a generalisation to some (relatively sparse) irregular degree sequences was given in [13]. Switches can easily be restricted to preserve bipartiteness, by requiring that $\{w, y\}$ (or equivalently $\{x, z\}$) belong to the same side of the bipartition. In fact, the first use of switches as the transitions of a Markov chain was for bipartite graphs [15].

If we wish to generate only *connected* graphs, we may use the *flip*. This is defined in the same way as the switch, except that we specify that wy must also be an edge. See Fig. 2. Note that a flip is a restricted form of switch which cannot disconnect the graph. Tsuki [23] proved that the flip chain is irreducible for 3-regular connected graphs, while the corresponding result for d-regular graphs was proved by Mahlmann and Schindelhauer [17], for any $d \geq 3$. Cooper, Dyer, Greenhill and Handley [8] subsequently showed rapid mixing for regular graphs

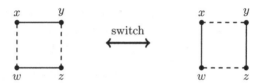

Fig. 1. A switch

of even degree. Note that flips are not well-defined on bipartite graphs, since $\{w, y\}$ clearly cannot be on the same side of a bipartition. Mahlmann and Schindelhauer also considered other restricted forms of switches, where there must be a k-edge path between w and y. The flip chain corresponds to $k = 1$, while the "2-flipper" with $k = 2$ preserves connected bipartite graphs.

Irreducibility of the 2-flipper was proved in [17], but the idea does not seem to have been considered subsequently.

Fig. 2. A flip

In [6], a different restriction of switches was introduced, designed to ensure that every switch changes the set of triangles in the graph. The definition is as for switches, except that x and w must have a common neighbour, which we denote by v. This is a *triangle switch*, which we abbreviate as Δ-switch. Every Δ-switch makes (creates) or breaks (destroys) at least one triangle. Again, we make no further assumption about $G[\{v, w, x, y, z\}]$. Clearly, Δ-switches do not preserve bipartiteness, since bipartite graphs have no triangles.

Specifically, if the 4-edge path $yxvwz$ is present in the graph and the edges xw, yz are absent, a *make* triangle switch at v, denoted Δ^+-switch, deletes the edges xy, wz and replaces them with edges xw, yz, forming a triangle on v, x, w. The Δ^+-switch is illustrated in Fig. 3, reading from left to right. Conversely, if the edge yz and the triangle on v, x, w are present in the graph, such that the edges xy, wz are both absent, then a *break* triangle switch at v, denoted Δ^--switch, deletes the edges xw, yz and replaces them with the edges xy, wz. This destroys the triangle on v, x, w. The Δ^--switch is illustrated in Fig. 3, reading from right to left. Note that a Δ^--switch reverses a Δ^+-switch and vice versa.

A Δ-switch which involves v and two incident edges, as in Fig. 3, will be called a Δ-switch at v. Note that this is equivalent to a switch in the graph $H = G[V_1 \cup V_2]$, if the graph is layered from v, and we will use this equivalence in our arguments below.

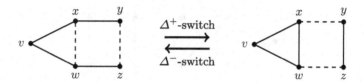

Fig. 3. The Δ^+-switch and Δ^--switch triangle switches at v

Now let $\mathcal{M}_{n,d}$ be the graph with $V(\mathcal{M}_{n,d}) = \mathcal{G}_{n,d}$ and $\{G, G'\} \in E(\mathcal{M}_{n,d})$ if and only if G' can be obtained from G by a single Δ-switch. Then a time-homogeneous Markov chain with state space $\mathcal{M}_{n,d}$ will be called a Δ-switch chain if its transition matrix P satisfies $P(G, G') > 0$ if and only if $\{G, G'\} \in E(\mathcal{M}_{n,d})$. That is, $\mathcal{M}_{n,d}$ is the transition graph underlying any Δ-switch chain. Then our main result is the following.

Theorem 1. *Suppose that $d \geq 3$. Then the graph $\mathcal{M}_{n,d}$ is connected. Equivalently, any Δ-switch chain is irreducible on $\mathcal{G}_{n,d}$.*

2   Small Regular Graphs

Here we show that Δ-switches connect the set of all fragments on a given vertex set. First we give some properties of fragments which are required in our proof.

Lemma 1. *Let G be a d-regular fragment, with $d \geq 3$. Then G is a connected graph with diameter 2.* □

Remark 1. We claim only connectedness, but fragments have higher connectivity. It is not difficult to prove 2-connectedness. For each d, we have examples with connectivity only $\lfloor d/2 \rfloor + 1$, and we believe this represents the lowest connectivity. However, we make no use of this, so we do not pursue it further here.

For an even integer $d \geq 2$, we construct the graph $T_{d,d,1}$ as follows. Take a copy of $K_{d,d}$ with vertex bipartition (A_d, B_d), where $A_d = \{a_i : i \in [d]\}$ and $B_d = \{b_i : i \in [d]\}$. Let M be the matching $\{a_i b_i : i \in [d/2]\}$ of size $d/2$ between $A_{d/2}$ and $B_{d/2}$. Form $T_{d,d,1}$ from the copy of $K_{d,d}$ by deleting the edges of M, adding a new vertex v and an edge from v to each a_i and b_i with $i \in [d/2]$. Then $T_{d,d,1}$ is a d-regular tripartite graph with $2d + 1$ vertices and vertex tripartition $\{v\} \cup A_d \cup B_d$.

For example, $T_{2,2,1}$ is a 5-cycle and $T_{4,4,1}$ is shown in Fig. 4.

Lemma 2. *Suppose that $G \in \mathcal{G}_{n,d}$ where $d \geq 3$ and $d + 1 < n < 2(d + 1)$. Let ab be an edge of G.*

(i) If $n < 2d$ then G has a triangle which contains the edge ab.
(ii) If $n = 2d$ then ab is contained in a triangle or an induced C_4 in G. Furthermore, if G is triangle-free then $G \cong K_{d,d}$.

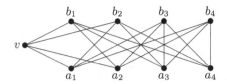

Fig. 4. The graph $T_{4,4,1}$

(iii) *If* $n = 2d + 1$ *then* ab *is contained in a triangle or an induced* C_4 *in* G. *Furthermore, if* G *is triangle-free then* $G \cong T_{d,d,1}$.   □

We use these properties to prove the following.

Lemma 3. *If* $d \geq 3$ *and* $d + 1 < n < 2(d+1)$ *then* $\mathcal{M}_{n,d}$ *is connected. Equivalently, the* Δ-*switch chain is irreducible on the set of fragments in* $\mathcal{G}_{n,d}$.

Proof (Sketch). The *flip* chain [8,17] is irreducible on all d-regular connected graphs. Furthermore, Lemma 1 proves that all fragments are connected graphs. We show that if $d + 1 < n < 2(d+1)$ then a flip can be performed using at most three Δ-switches. That is, if G is a fragment and G' is obtained from G by a flip, then there is a sequence of at most three Δ-switches which takes G to G'. The lemma then follows immediately.

There are two types of flip, as shown in Fig. 5, which we must consider separately. Solid lines are the edges and dashed lines the non-edges involved in the flip which deletes v_1v_2, v_3v_4 and inserts v_1v_3, v_2v_4, where v_1v_4 is an edge.   □

Fig. 5. The two types of flip

3   Creating a Clique Component Containing a Given Vertex

The next result, which we prove in this section, is the core of our proof of Theorem 1.

Theorem 2. *Suppose that* $d \geq 3$ *and* $n \geq 2(d + 1)$. *Given any* $G \in \mathcal{G}_{n,d}$ *and any vertex* v *of* G, *let* $S = \mathrm{N}_G[v]$ *be the closed neighbourhood of* v *in* G. *Then there is a sequence of* Δ-*switches which ends in a graph* G' *which has a clique component on the vertex set* S.

Note that $N_{G'}[v] = N_G[v]$, that is, the closed neighbourhood of v is preserved by this process. This property is used in Sect. 4.

3.1 Proof Strategy

Let C be the component of G containing v. We consider $G(C)$ as being layered from v, in a BFS manner, with layer V_{i+1} above layer V_i, and with v in the bottom layer (see Sect. 1). We prove that, provided $|C| \geq 2(d+1)$, there is a sequence of Δ-switches such that V_1 remains unchanged, but $|E_1|$ increases monotonically. We repeat the following steps to add edges to E_1 until $G[V_0 \cup V_1] \cong K_{d+1}$.

1. If, before any step below, the component C containing v is a fragment (that is, if $d + 1 < |C| < 2(d + 1)$), use a Δ^--switch to increase the size of C to at least $2(d + 1)$ without removing any edge in E_1. This can be done in such a way that G_2 now contains at least one non-edge, as we will prove in Lemma 4.
2. While there is a vertex $u \in V_1$ which is not adjacent to any vertex in V_1 (that is, with $d'(u) = 0$) make a Δ^+-switch to introduce an edge incident with u in G_1. That this is always possible will be proved in Lemma 5.

   After repeating this as many times as necessary, every vertex in V_1 will have an incident edge in G_1. Thus, every vertex $u \in V_1$ with a neighbour in V_2 will have $1 \leq d'(u) \leq d - 2$.
3. If $V_2 = \emptyset$ then $G[V_0 \cup V_1] \cong K_{d+1}$, return the current graph as G'. Otherwise, while there is a non-edge ab in G_2, insert edges into E_1 as follows:
   (a) Suppose that there is a unique $x \in V_1$ such that $a, b \in V_2$ are in $\mathrm{Out}(x)$ only. Thus $\mathrm{id}(a) = \mathrm{id}(b) = 1$. Use Lemma 6 to make a Δ^+-switch which replaces edge xb with yb for some $y \in V_1$, $y \neq x$, thus giving a pair x, y below non-edge ab.
   (b) Suppose that some pair x, y below ab is a non-edge of G_1. Use a Δ^+-switch at v to switch xa, yb to xy, ab; thus increasing the number of edges in G_1, as in Lemma 7.
   (c) Now suppose that every pair x, y below ab is an edge xy of G_1. Choose one such pair and use the Δ-switch at v of Lemma 8 to make xy a non-edge. Then use a Δ-switch at v to switch xa, yb to xy, ab.
4. If $\ell = |V_2| \geq d + 1$ then there are necessarily non-edges in G_2. If $\ell = d$ and $|C| = 2(d + 1)$, then $V_3 = \{u\}$, for some u, and $V_2 = N(u)$. Again there are necessarily non-edges in G_2. In either case go back to Step 3 above.
5. If we reach here, then $\ell \leq d$ and V_2 is a complete graph K_ℓ. If $V_3 = \emptyset$ then $|C| = 1 + d + \ell \leq 2d + 1$, a contradiction. Thus $V_3 \neq \emptyset$. The case $|C| = 2(d + 1)$ was covered in Step 4, so we assume that $|C| > 2(d + 1)$. Carry out the steps in Lemmas 9–11 to insert a non-edge into G_2, and go to Step 3 above.

3.2 Increasing the Size of C

Lemma 4. *Suppose that $d \geq 3$ and $n > 2(d + 1)$. If vertex v is in a fragment C then there is a Δ^--switch to increase the size of C to at least $2d + 3$ without changing the edges of G_1. After this switch, G_2 will contain a non-edge.* □

This procedure does not increase $|E_1|$ but, as we show next, the non-edge in G_2 allows us to increase $|E_1|$ with at most two further Δ-switches. Thus the process outlined in Sect. 3.1 must terminate in a finite number of steps.

3.3 G_2 Has a Non-edge

If G_2 has a non-edge, we use the following lemma to increase $|E_1|$.

Lemma 5. *Suppose that $d \geq 3$. Let C be the component of G which contains v, and suppose that $|C| \geq 2(d+1)$. If $u \in V_1$ has $d'(u) = 0$ then we can use at most two Δ-switches to insert an edge in V_1 at u, without altering other edges in E_1.* □

Let ab be a non-edge of V_2 above $x, y \in V_1$. We will show that we can rearrange the edges of G_1 as necessary to enable a Δ-switch $axvyb$, replacing xa, yb with xy, ab, inserting an edge xy into E_1. Lemma 6 deals with the case where ab lies uniquely within $\mathrm{Out}(u)$ for some $u \in V_1$. Lemmas 7 and 8 interchange edges and non-edges in G_1 if necessary. First, we show that we can assume that every non-adjacent pair $a, b \in V_2$ is above some pair in V_1.

Lemma 6. *Let $d \geq 3$ and $d'(u) \geq 1$ for all $u \in V_1$. Let $a, b \in V_2$ be a pair of distinct non-adjacent vertices such that $\mathrm{In}(a) = \mathrm{In}(b) = \{x\}$ for some $x \in V_1$. Then there is a Δ-switch at v to move b to $\mathrm{Out}(y)$ for some $y \in V_1$, $y \neq x$, without altering E_2, so that a, b is above x, y.*

Proof. In Fig. 6, xb is an edge and so $d'(x) \leq d - 2$. Hence there is a non-edge xw for some $w \in V_1$. As $d'(w) \geq 1$ there is some $y \in V_1$ such that wy is an edge. Clearly $y \neq x$. Note that yb is a non-edge because $\mathrm{id}(b) = 1$. (Pairs not shown as an edge or non-edge can be either.) Now switch xb, wy to xw, by. □

Fig. 6. The switch in Lemma 6, which changes $\mathrm{N}(b) \cap V_1$

Lemma 7. *Let ab be a non-edge of G_2, above a non-edge xy in G_1. Then there is a Δ^+-switch to put $xy \in E_1$ without altering any other edges of E_1 (Fig. 7).*

Proof. Clearly $axvyb$ is the required Δ^+-switch. □

Lemma 8. *Let $d \geq 3$, and let $xy \in E_1$ be such that $\mathrm{od}(y) \geq 1$. Then there is a Δ-switch which makes xy a non-edge, without changing E_2 or decreasing $|E_1|$.*

Fig. 7. The switch in Lemma 7 which inserts xy into E_1.

Proof. Consider the graph $H = G[V_1 \cup V_2]$. Note that all $u \in V_1$ have $\deg_H(u) = d-1$. Since $d'(y) = d-1-\mathrm{od}(y) \leq d-2$, but $V_1 \setminus \{y\} = d-1$, there exists $w \in V_1$ such that $yw \notin E_1$, as in Fig. 8. First suppose that $xw \notin E_1$. Let $W = \mathrm{N}(w) \setminus \{v\}$ and $X = \mathrm{N}(x) \setminus \{v, y\}$, so $|W| = d-1$, $|X| = d-2$. Thus there exists $z \in W \setminus X$. So $wz \in E(H)$ and $xz \notin E(H)$, and there is a switch replacing xy, wz by yw, xz. Now $|E_1|$ is unchanged but xy is a non-edge. If $z \in V_1$ then two edges in G_1 are added and two removed. If $z \in V_2$ then one edge of G_1 is added and one is removed. No edges of G_2 are changed, since only z can be in V_2. Finally, if $xw \in E_1$, let $W = \mathrm{N}(w) \setminus \{v, x\}$ and $X = \mathrm{N}(x) \setminus \{v, w, y\}$. Then $|W| = d-2$ and $|X| = d-3$. The argument now proceeds as above. □

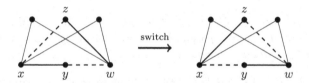

Fig. 8. The switch in Lemma 8, making xy a non-edge.

If ab was a non-edge above the edge xy, then both $\mathrm{od}(x), \mathrm{od}(y) \geq 1$, so Lemma 8 applies. After performing the Δ-switch from Lemma 8, ab will be above the non-edge xy. Then we use Lemma 7 to re-insert xy and increase $|E_1|$.

3.4 G_2 Has No Non-edges

If G_2 has no non-edges, we explain how to introduce one without changing E_1. After this we can apply the results of Sect. 3.3.

As usual, C denotes the component of G containing v. We assume that $|C| \geq 2(d+1)$ and that G_2 is complete. If (i) $V_4 \neq \emptyset$ and $\ell \geq 2$ or (ii) all vertices of G_2 have $d - \ell$ edges to V_3, then we can use Δ-switches to create a non-edge in G_2. This is proved in Lemmas 10 and 11 respectively. If these conditions are not met, then Lemma 9 describes a procedure which can be repeated until all vertices of V_2 have in-degree one. As a consequence $\ell \geq 2$, and all $u \in V_2$ have $\mathrm{od}(u) = d - (\ell - 1) - 1 = d - \ell$, thus satisfying Lemma 11.

If $\ell \geq d + 1$ then G_2 necessarily has a non-edge, so assume $\ell \leq d$. Then id$(u) \geq 1$, for any $u \in V_2$, and d$'(u) = \ell - 1$ as G_2 is complete. Thus od$(u) \leq d - \ell$, so there are at most $\ell(d - \ell)$ edges from V_2 to V_3. If $\ell = d$ then $V_3 = \emptyset$ and so $|C| = 2d + 1$, a contradiction. Similarly, if $|V_3| = 1$ then $\ell \geq d$ as $V_4 = \emptyset$, and again we have a non-edge in G_2 or a contradiction. If $|C| = 2(d+1)$ then $V_2 \cup V_3$ is a d-regular subgraph on $d + 1$ vertices, which must be isomorphic to K_{d+1}. But this contradicts the fact that all vertices $u \in V_2$ have id$(u) \geq 1$. Hence we may assume that $1 \leq \ell \leq d - 1$, $|C| > 2(d + 1)$, and $|V_3| \geq 2$.

Lemma 9. *Suppose that $d \geq 3$ and $|C| > 2(d+1)$. Further suppose that $1 \leq \ell \leq d - 1$ and $|V_3| \geq 2$, with G_2 complete and $V_4 = \emptyset$. If some $u \in V_2$ has id$(u) \geq 2$ then there is a Δ^+-switch which reduces id(u) by one and moves a vertex of V_3 to V_2, without altering E_1.* □

The above process can be repeated until there is a non-edge in V_2, in which case we proceed as in Sect. 3.3, or all vertices on V_2 have in-degree one. In this case $\ell \geq 2$, because there must be at least two vertices in V_1 with out-degree at least one, or else $G[V_1] = K_d$ and we are done.

Thus we may now assume that G_2 is complete with $2 \leq \ell \leq d - 1$, all vertices $u \in V_2$ have id$(u) = 1$, and $V_3 \neq \emptyset$ (or else C is a fragment). Hence all $u \in V_2$ have od$(u) = d - 1 - (\ell - 1) = d - \ell$, and there are exactly $\ell(d - \ell)$ edges between V_2 and V_3. Use the appropriate lemma below, and proceed as in Sect. 3.3.

Lemma 10. *Suppose that G_2 is complete and $2 \leq \ell \leq d - 1$, all vertices $u \in V_2$ have id$(u) = 1$, and $V_3 \neq \emptyset$. If $V_4 \neq \emptyset$ then we can apply a Δ^+-switch to create a non-edge in G_2 without altering E_1.* □

Lemma 11. *Let $d \geq 3$ and $|C| \geq 2(d+1)$. Suppose that G_2 is complete, $2 \leq \ell \leq d - 1$, and all $u \in V_2$ have id$(u) = 1$. Further suppose that $V_3 \neq \emptyset$ and $V_4 = \emptyset$. Then there is a Δ^--switch in $G[V_2 \cup V_3]$ which removes an edge of V_2 without altering E_1.* □

Remark 2. We can bound the number of Δ-switches required to create a clique component. The steps in Lemmas 5–8 require $\Theta(1)$ Δ-switches for each edge inserted in G_1, so $\Theta(d^2)$ in total. The steps in Lemmas 9–11 also require $O(1)$ Δ-switches, with the exception of Lemma 9, which could possibly be executed $\Theta(d)$ times between edge insertions in V_1. Thus the total number of Δ-switches required is $\Omega(d^2)$ and $O(d^3)$. Note that this is independent of n, since at most five layers of G are involved in the process.

4   Relabelling the Vertices of Clique Components

To complete the proof of Theorem 1, we need to show that any graph $X = (V, E_X) \in \mathcal{G}_{n,d}$ can be transformed to any other graph $Y = (V, E_Y) \in \mathcal{G}_{n,d}$ with a sequence of Δ-switches. We will do this by induction on n. It is trivially true for $n = d + 1$, since $\mathcal{G}_{n,d}$ contains only one labelled graph, K_{d+1}. We know from

Lemma 3 that $\mathcal{M}_{n,d}$ is connected for $d + 1 < n < 2(d + 1)$. For $n \geq 2(d+1)$, we will assume inductively that $\mathcal{M}_{n',d}$ is connected for all $n' < n$.

Choose any $v \in V$. First, suppose that $N_X(v) = N_Y(v)$. We know from Sect. 3 that we can perform a sequence of Δ-switches to transform X into a graph which is a disjoint union of a clique component on the vertex set $N_X[v]$ and a d-regular graph X' with $n - d - 1$ vertices. Similarly, we can perform a sequence of Δ-switches to transform Y into a disjoint union of a clique component on the vertex set $N_Y[v]$ and a d-regular graph Y' with $n-d-1$ vertices. Since $N_X(v) = N_Y(v)$, it follows that X' and Y' have the same vertex set. Hence, by induction, there is a sequence of Δ-switches that transforms X' into Y', as required.

Now suppose that $N_X(v) \neq N_Y(v)$. Using the above procedure, we can assume that $G[N_X[v]] \cong K_{d+1}$, and similarly for Y. We now show how to perform a sequence of switches, starting from X, to ensure that the neighbourhood of v matches $N_Y(v)$.

Let $x \in N_X(v) \setminus N_Y(v)$ and $y \in N_X(v) \setminus N_Y(v)$, where both x and y exist because $|N_X(v)| = |N_Y(v)|$. Since $y \notin N[v]$, it must be the case that y is a vertex of X'. Therefore, let yz be any edge of X' incident on y, and let w be any vertex of $N(v) \setminus \{x\}$, which exists since $d \geq 3$. Note that X has a triangle on the vertices v, w, x, since $G[N_X[v]] \cong K_{d+1}$. Perform the Δ^--switch shown in Fig. 9.

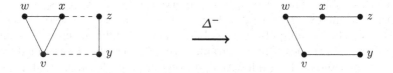

Fig. 9. Swapping x, y in $N_X(v)$

This creates a graph, which we rename as X, such that $N_X(v) \leftarrow N_X(v) \setminus \{x\} \cup \{y\}$. We then perform a sequence of Δ-switches to create a clique component with vertex set $N_X[v]$, using the method of Sect. 3. After this iteration, again renaming the new graph as X, we find that $N_X[v]$ spans a clique component and $|N_X[v] \cap N_Y[v]|$ has been increased by 1. After at most d repetitions of this process, we have reached a new graph X such that $N_X(v) = N_Y(v)$. We now follow the argument given above for that case, completing the proof.

Remark 3. It might be more efficient to incorporate this step into the procedure described in Sect. 3. In particular, we could show that x, y can be interchanged as soon as v and x have a common neighbour w. However, as we only need to show that $\mathcal{M}_{n,d}$ is connected, and not that we can find shortest paths in $\mathcal{M}_{n,d}$, we prefer to separate these two steps, for clarity.

Remark 4. We require only one Δ-switch to interchange x, y, but we may have to repeat this d times. Since $O(d^3)$ steps are needed to create a clique component (see Remark 2), this gives $O(d^4)$ steps in total for each inductive step. This must be repeated in graphs of order $n - i(d + 1)$ for $0 \leq i < \lfloor n/(d + 1) \rfloor$, that is, $O(n/d)$ times. Thus in total we may need $O(nd^3)$ Δ-switches to connect X with Y.

5   Regular Graphs of Degree at Most Two

Theorem 1 excludes the cases $d = 0, 1, 2$. The switch chain is irreducible in all these cases. We will briefly examine the question of connectedness of $\mathcal{M}_{n,d}$ (equivalently, irreducibility of Δ-switch chains on $\mathcal{G}_{n,d}$) when $d = 0, 1, 2$.

If $d = 0$ then the unique graph in $\mathcal{G}_{n,0}$ is a labelled independent set of order n. Hence $\mathcal{M}_{n,0}$ is trivially connected and any Δ-switch chain is trivially irreducible. If $d = 1$ then $G \in \mathcal{G}_{n,1}$ is a matching and n must be even. Now $|\mathcal{G}_{n,1}| > 1$ when $n \geq 4$ is even, but clearly no Δ-switch is possible as no element of $\mathcal{G}_{n,1}$ contains a triangle or a path of four edges. Thus $\mathcal{M}_{n,1}$ is not connected when $n \geq 4$ is even (indeed, $\mathcal{M}_{n,1}$ has no edges in this case).

For $d = 2$, it is not so obvious whether or not $\mathcal{M}_{n,2}$ is connected. We will now examine this case.

If $n < 3$ then $\mathcal{G}_{n,2}$ is empty. For $n \geq 3$, let c_i $(i = 1, 2)$ be the number of cycles in $G \in \mathcal{G}_{n,2}$ such that their length modulo 3 is i. Then $n \equiv c_1 + 2c_2$ (mod 3). We will say G has *class* (c_1, c_2).

If $n \geq 3$ then there is at least one class in $\mathcal{G}_{n,2}$. If $n \equiv i$ (mod 3) $(i \in \{0, 1, 2\})$, then the class $(i, 0)$ exists. Any class is preserved under Δ-switches, since a cycle can either be increased by length 3 by a Δ^+-switch or decreased by length 3 by a Δ^--switch. Nothing else is possible. Thus two different classes cannot be connected by Δ-switches. But any graph in the class can be transformed by Δ^--switches to a "canonical" graph with c_1 4-cycles, c_2 5-cycles and $(n - 4c_1 - 5c_2)/3$ triangles. Note that there are $(k - 1)!/2$ distinct labellings of the vertices of a k-cycle.

Lemma 12. *The graph $\mathcal{M}_{n,2}$ is connected if and only if $n \in \{3, 6, 7\}$. Hence any given Δ-switch chain is irreducible on $\mathcal{G}_{n,2}$ if and only if $n \in \{3, 6, 7\}$.*   □

We remark that for $n > 10$ there can be more than two classes in $\mathcal{G}_{n,2}$. For example, $\mathcal{G}_{20,2}$ contains the classes $(5, 0)$ and $(0, 4)$, as well as the two classes $(0, 1)$ and $(2, 0)$ inherited from $\mathcal{G}_{8,2}$.

References

1. Allen-Zhu, Z., Bhaskara, A., Lattanzi, S., Mirrokni, V., Orecchia, L.: Expanders via local edge flips. In: Proceedings of 27th Annual ACM-SIAM Symposium on Discrete Algorithms (SODA 2016), pp. 259–269 (2018)
2. Amanatidis, G., Kleer, P.: Rapid mixing of the switch Markov chain for strongly stable degree sequences. Random Struct. Algorithms **57**, 637–657 (2020)
3. Bayati, M., Kim, J.H., Saberi, A.: A sequential algorithm for generating random graphs. Algorithmica **58**, 860–910 (2010)
4. Bollobás, B.: Random Graphs, 2nd edn. Cambridge University Press, Cambridge (2001)

5. Cooper, C., Dyer, M., Greenhill, C.: Sampling regular graphs and a peer-to-peer network. Comb. Probab. Comput. **16**, 557–593 (2007)
6. Cooper, C., Dyer, M., Greenhill, C.: Triangle-creation processes on cubic graphs, arXiv:1905.04490 (2019)
7. Cooper, C., Dyer, M., Greenhill, C.: A triangle process on regular graphs, arXiv: 2012.12972 (2020)
8. Cooper, C., Dyer, M., Greenhill, C., Handley, A.: The flip Markov chain for connected regular graphs. Discret. Appl. Math. **254**, 56–79 (2019)
9. Erdős, P., Gallai, T.: Graphs with prescribed degree of vertices. Matematikai Lapok **11**, 264–274 (1960)
10. Feder, T., Guetz, A., Mihail, M., Saberi, A.: A local switch Markov chain on given degree graphs with application in connectivity of peer-to-peer networks. In: Proceedings of 47th Annual IEEE Symposium on Foundations of Computer Science (FOCS 2006), pp. 69–76 (2006)
11. Gao, P., Wormald, N.: Uniform generation of random regular graphs. SIAM J. Comput. **46**, 1395–1427 (2017)
12. Goodreau, S.M., Kitts, J.A., Morris, M.: Birds of a feather, or friend of a friend? Using exponential random graph models to investigate adolescent social networks. Demography **46**, 103–125 (2009)
13. Greenhill, C., Sfragara, M.: The switch Markov chain for sampling irregular graphs and digraphs. Theoret. Comput. Sci. **719**, 1–20 (2018)
14. Jin, E.M., Girvan, M., Newman, M.E.J.: Structure of growing social networks. Phys. Rev. E **64**, 046132 (2001)
15. Kannan, R., Tetali, P., Vempala, S.: Simple Markov chain algorithms for generating random bipartite graphs and tournaments. Random Struct. Algorithms **14**, 293–308 (1999)
16. Lowcay, C., Marsland, S., McCartin, C.: Constrained switching in graphs: a constructive proof. In: 2013 International Conference on Signal-Image Technology and Internet-Based Systems, pp. 599–604 (2013)
17. Mahlmann, P., Schindelhauer, C.: Peer-to-peer networks based on random transformations of connected regular undirected graphs. In: Proceedings of 17th Annual ACM Symposium on Parallelism in Algorithms and Architectures (SPAA 2005), pp. 155–164 (2005)
18. Miklos, I., Erdős, P., Soukup, L.: Towards random uniform sampling of bipartite graphs with given degree sequence. Electron. J. Comb. **20**, #P16 (2013)
19. Nishimura, N.: Introduction to reconfiguration. Algorithms **11**, 52 (2018)
20. Taylor, R.: Contrained switchings in graphs. In: McAvaney, K.L. (ed.) Combinatorial Mathematics VIII. LNM, vol. 884, pp. 314–336. Springer, Heidelberg (1981). https://doi.org/10.1007/BFb0091828
21. Tikhomirov, K., Youssef, P.: Sharp Poincaré and log-Sobolev inequalities for the switch chain on regular bipartite graphs, arXiv:2007.02729 (2020)
22. Tripathi, A., Tyagi, H.: A simple criterion on degree sequences of graphs. Discret. Appl. Math. **156**, 3513–3517 (2008)
23. Tsuki, T.: Transformations of cubic graphs. J. Franklin Inst. **33(B).4**, 565–575 (1996)
24. West, D.: Introduction to Graph Theory, 2nd edn. Prentice Hall, Hoboken (2000)
25. Wormald, N.: Models of random regular graphs. In: Surveys in Combinatorics 1999. London Mathematical Society Lecture Notes Series, vol. 267, pp. 239–298 (1999)

Complexity and Algorithms
for MUL-Tree Pruning

Mathieu Gascon[1], Riccardo Dondi[2], and Nadia El-Mabrouk[3(✉)]

[1] Département d'informatique et de recherche opérationnelle (DIRO),
Université de Montréal, CP 6128 succ Centre-Ville, Montreal, QC H3C 3J7, Canada
`mathieu.gascon.1@umontreal.ca`
[2] Dipartimento di Lettere, Filosofia, Comunicazione,
Università degli Studi di Bergamo, Bergamo, Italy
`riccardo.dondi@unibg.it`
[3] DIRO, Université de Montréal, Montreal, Canada
`mabrouk@iro.umontreal.ca`

Abstract. A multiply-labeled tree (or MUL-tree) is a rooted tree in which every leaf is labeled by an element from some set $\mathcal{X}$, but in which more than one leaf may be labeled by the same element of $\mathcal{X}$. MUL-trees have applications in many fields. In phylogenetics, they can represent the evolution of gene families, where genes are represented by the species they belong to, the non-uniqueness of leaf-labels coming from the fact that a given genome may contain many paralogous genes. In this paper, we consider two problems related to the leaf-pruning (leaf removal) of MUL-trees leading to single-labeled trees. First, given a set of MUL-trees, the *MUL-tree Set Pruning for Consistency* (MULSETPC) Problem asks for a pruning of each tree leading to a set of consistent trees, i.e. a collection of label-isomorphic single-labeled trees. Second, processing each gene tree at a time, the *MUL-tree Pruning for Reconciliation* (MULPR) Problem asks for a pruning minimizing a reconciliation cost with a given species tree. We show that MULTSETPC is NP-hard and that MULPR is W[2]-hard when parameterized by the duplication cost. We then develop a polynomial-time heuristic for MULPR and show its accuracy by comparing it to a brute-force exact method on a set of gene trees from the Ensembl Genome Browser.

Keywords: Multilabeled tree · Phylogeny · Gene tree · Duplication · Reconciliation

1 Introduction

In phylogenetics, leaf-labeled rooted trees are used to represent the vertical evolution of a collection $\mathcal{X}$ of species, genes or other units of heredity. In this context, the leaves of a tree T are labeled with elements of $\mathcal{X}$. A tree T is a multiply-labeled tree, or MUL-tree if each element of $\mathcal{X}$ may label many leaves of T. A single-labeled tree is a special case of a MUL-tree where no element of $\mathcal{X}$ labels

© Springer Nature Switzerland AG 2021
P. Flocchini and L. Moura (Eds.): IWOCA 2021, LNCS 12757, pp. 324–339, 2021.
https://doi.org/10.1007/978-3-030-79987-8_23

more than one leaf of T. Single-labeled trees are usually used to represent the evolutionary relationship between a set of species. On the other hand, a gene tree representing the evolutionary history of a gene family, i.e. a set of *homologous genes* deriving from the same ancestral gene, and belonging to a set $\mathcal{X}$ of species, may be represented by a tree leaf-labeled with the elements of $\mathcal{X}$, a label x indicating a gene belonging to species x [15]. Because, in addition to *orthologs* (genes diverging through speciation) a gene family may also contain *gene paralogs* (deriving from duplication) in the same genome, such a representation of a gene tree usually leads to a MUL-tree. Such trees with repeated leaf labels are also considered within approaches to construct phylogenetic networks [17], and have applications in other research fields such as biogeography [13], phylogenomics [10], the study of host-parasite cospeciation [21], data-mining [4] or string-matching [5].

Despite this variety of application contexts, MUL-trees remain relatively little studied compared with single-labeled trees, mainly due to the fact that many problems that are tractable for single-labeled trees become NP-hard when extended to MUL-trees. For example, most generalizations of the well-known greedy consensus tree methods (strict, majority rule and singular majority rule consensus) [3] are NP-hard in the case of MUL-trees [6, 16, 19].

In this paper, we focus on pruning (removing) leaves of MUL-trees in a way leading to single-labeled trees satisfying some properties. First, given a set of MUL-trees, we ask for a leaf-pruning of each tree leading to a set of consistent trees, i.e. a collection of label-isomorphic single-labeled trees. We call this problem the *MUL-tree Set Pruning for Consistency* (MULSETPC) Problem. Second, we process each tree at a time (gene tree), and we ask for a pruning minimizing a reconciliation cost with a given species tree. We call this problem the *MUL-tree Pruning for Reconciliation* (MULPR) Problem.

A straightforward application is to extract, from a set of gene trees, a coherent topological information that can then be used to produce a supertree. In a phylogenetic inferrence perspective this may lead to the prediction of the underlying species tree [2, 20, 23, 25]. Species tree reconstruction from a set of MUL-trees is the purpose in [24] where a problem related to MULSETPC, but relying on pruning subtrees rather than individual leaves, has been shown NP-hard [24]. A second application is the Super-Reconciliation [8, 9] problem aiming at reconstructing the evolutionary history of a set of syntenic regions, i.e. homologous regions in a set of genomes or chromosomes that have evolved from a common ancestor through rearrangements, but also segmental duplication and loss, and possibly Horizontal Gene Transfer (HGT) i.e. the exchange of genetic material between species. The model considered in [8, 9] does not allow for tandem duplications, while in real datasets, syntenies usually contain gene duplicates. Coping with this limitation is done, for the considered biological application on the opioid receptor genes [9], by manually pruning the gene trees seeking for a solution to the MULSETPC Problem. Considering the MULPR Problem on each gene tree independently, i.e. minimizing the reconciliation cost of a pruned gene tree with a given species tree, would be an alternative way of handling the problem of tandem duplications.

The paper is organized as follows. We begin by introducing the problems in Sect. 2. We then present our complexity results by showing in Sect. 3 that the MULTSETPC Problem is NP-complete and in Sect. 4 that the MULPR Problem is W[2]-hard when parameterized by the duplication cost. We then, in Sect. 5, develop a polynomial-time greedy heuristic for MULPR, and show its accuracy in Sect. 6 by comparing it to a brute-force exact method on a set of gene trees from the Ensembl Genome Browser. For space reason, most of the proofs are omitted.

2   Notations

All trees are considered binary and rooted, i.e. with a special node considered as the root. We denote by $r(T)$ the root, by $V(T)$ the node set, by $L(T) \subseteq V(T)$ the leaf set and by $E(T)$ the edge set of a tree T. The *size* of a tree T denotes the number $n = |L(T)|$ of its leaves. An *internal node* is a node of $V(T) \setminus L(T)$. An edge of $E(T)$ is written as a pair (x, y) of two adjacent nodes, where x, the closest to the root, is called *the parent* of y and y is called a *child* of x. The root is the only node of $V(T)$ with no parent.

In a binary tree, each internal node x has two children that we denote x_l and x_r, for "left" and "right" child of x, respectively. Notice that trees are unordered, and thus left and right is just a convenient notation for an arbitrary ordering of the children of x. For example, in Fig. 1, left and right will refer to the way the tree is represented (in the tree T, node 1 has left child $1_l = 2$ and right child $1_r = 4$). The node x_l (respec. x_r) is called *the sibling* of x_r (respec. x_l). A node x is an *ancestor* of a node y if x is on the path from y to the root; two nodes x and y are *separated* if no one is an ancestor of the other. Given a node x of T, we write $T[x]$ the subtree of T rooted at x.

The *lowest common ancestor* (LCA) in T of a subset L' of $L(T)$, denoted $lca_T(L')$, is the ancestor common to all nodes in L' that is the most distant from the root.

Pruning a leaf y of T consists in removing the leaf y together with its parent x. If x is the root, then the sibling v of y becomes the new root; otherwise, v is connected to the parent u of x, in other words the edges (u, x) and (x, y) are removed and an edge (u, v) is created (see trees T_1 and $T_{1,2}$ in Fig. 1 for an illustration of a leaf pruning). Conversely, *grafting* a leaf y consists in subdividing an edge (u, v) of T, thereby creating a new node x between u and v, then adding a leaf y with parent x. Given a binary tree T, an *extension of T* is a tree R obtained, beginning from T, by performing a series of leaf grafting.

In this paper, we consider leaf-labeled trees, i.e. trees with each leaf labeled with an element of the leaf-label-set $\tilde{L}(T) = \mathcal{X}$. We consider *multilabeled trees* or *MUL-trees*, i.e. trees where each element of $\mathcal{X}$ may label more than one leaf of a tree; a *single-labeled* tree is a special case of a MUL-tree where leaf labels are pairwise different.

We say that a tree T *displays* a tree T' if there is a subset L' of $L(T)$ such that the tree $T|_{L'}$, obtained by pruning the leaves of L' is label-isomorphic to

T'. A *perfect pruning of* T is a *single-labeled tree* T^* displayed by T such that $\tilde{L}(T^*) = \tilde{L}(T)$. Notice that, as T^* can be obtained from T by a set of leaf pruning, each node in T^* has a corresponding node in T. In other words, we can consider that $V(T^*) \subseteq V(T)$. A node $x \in V(T) \setminus V(T^*)$ is said to be *removed from* T. For example in Fig. 1, $T_{1,2}$ displays both T_1 and T_2 after pruning; the node u of $T_{1,2}$ corresponds to the nodes u in both T_1 and T_2; the node x is removed from T_1.

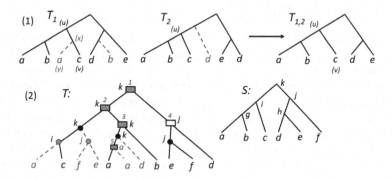

Fig. 1. (1) The MULSETPC Problem illustrated for two trees T_1 and T_2 on $\mathcal{X} = \{a, b, c, d, e\}$. The dotted and gray letters represent the leaf pruning in T_1 and T_2 leading to two consistent trees, represented by the tree $T_{1,2}$. (2) The MULPR Problem illustrated for the gene tree T and the species tree S. Letters at internal nodes of T represent the genome-labeling corresponding to the lca-mapping, while rectangles and circles represent the event-labeling of T corresponding to the lca-labeling, a rectangle representing a duplication and a circle a speciation event. We have $D(T, S) = 5$. The dotted and gray letters represent the leaf pruning leading to a perfect pruning T^* of minimum duplication cost. The only duplication remaining in T^* is the one filled in white.

A set $\mathcal{M}$ of single-labeled trees with possibly overlapping leaf-label-sets is *consistent* if we can find a tree on the union of leaf-label-sets displaying them all. For example, T_1 and T_2 after pruning in Fig. 1 are consistent as they are displayed by $T_{1,2}$. The consistency problem of rooted trees has been largely studied. For trees to be consistent, each triplet of data should exhibit the same topology in all trees. The BUILD algorithm [1] can be used to test, in polynomial-time, whether a collection of rooted trees is consistent.

2.1   Gene Tree, Species Tree and Reconciliation

A tree is a species tree if its set of leaves represent species, and is a gene tree if its set of leaves represent genes. We will make no difference between a leaf and the unit (genome or gene) it refers to. Thus, a tree S is a *species tree* if $L(S) = \Sigma$ is a set of species, and a tree T is a *gene tree* if $L(T) = \Gamma$ is a *gene family*, i.e. a set of genes where each gene x of Γ belongs to a species $s(x) \in \Sigma$, called the *genome-labeling* of x. Notice that any marker or unit can be considered instead of genes, as the theoretical results and algorithmic developments of this paper do not consider any specific feature about genes. Moreover, in this paper, the genes are simply identified by the genome they belong to. More precisely, two genes x and y of Γ such that $s(x) = s(y) = s$ are identified by two leaves of T labeled s. In other words, $\tilde{L}(T) = \mathcal{X}$ where $\mathcal{X} \subseteq \Sigma$ is the set of genomes with a gene in Γ. A gene tree is a MUL-tree and a species tree is a single-labeled tree. For example, in Fig. 1, the three leaves of the gene tree T labeled a correspond to three genes belonging to genome a, represented as one leaf of the species tree S.

Each internal node of S refers to a speciation event at the origin of the bifurcation, while each internal node of T may refer to a speciation (*Spec*), a duplication (*Dup*), a HGT event or other events, though in this paper we restrict the evolutionary model to speciations and duplications. When the type of event is known for each internal node, the gene tree T is said to be *event-labeled*. Inferring the event labeling of a gene tree T and the scenario of gene gain and loss explaining the difference between T and S is the purpose of the gene-tree-species-tree-reconciliation approach [12,14]. A *Reconciliation* of T with respect to S is usually defined as an event-labeled extension of T, where added branches represent lost (or missing) genes. In particular, a reconciliation minimizing the number $D(T, S)$ of Duplications (the D-distance) or the number $DL(T, S)$ of Duplications and Losses (the DL-distance) can be computed from the lca-labeling of T, defined as follows (see an illustration on the trees T and S of Fig. 1).

Definition 1 (DL-Reconciliation). *The lca-labeling of a gene tree T with respect to a species tree S (or simply lca-labeling of T if no ambiguity) is the triplet $\langle T, s, e_{lca} \rangle$ where s is an extension of the genome labeling to the internal nodes of T defined as $s(x) = lca_S(\{s(x') : x' \in L(T[x])\})$ (s is called the lca-mapping), and e_{lca} is the function from $V(T) \setminus L(T)$ to $\{Spe, Dup\}$ such that, for any $x \in V(T) \setminus L(T)$:*

- *If $s(x_l)$ and $s(x_r)$ are separated in S then $e_{lca}(x) = Spe$;*
- *otherwise $e_{lca}(x) = Dup$, representing a duplication in $s(x)$.*

A well known result of the reconciliation literature is that the D-distance is the number of nodes of T labeled as duplication from the lca-mapping, more precisely:

$$D(T, S) = |\{x \in V(T) \setminus L(T) : e_{lca}(x) = Dup\}|$$

Moreover, both the D-distance and the DL-distance can be computed in linear time [26].

2.2   Problem Statements

We consider two problems related to MUL-tree pruning. First, let $\mathcal{M} = \{T_{1 \leq i \leq n}\}$ be a set of trees where each T_i is *a tree on* $\mathcal{X}_i$, i.e. such that $\tilde{L}(T_i) = \mathcal{X}_i$. Then $\mathcal{M}$ is said to be a set of trees on $\mathcal{X} = \cup_{1 \leq i \leq n} \mathcal{X}_i$. The MULSETPC Problem for a set of MUL-trees, is defined as follows (Fig. 1(1)).

MUL-TREE SET PRUNING FOR CONSISTENCY (MULSETPC) PROBLEM:
Instance: A set $\mathcal{M}$ of MUL-trees on $\mathcal{X}$.
Question: $\exists$? a perfect pruning of each tree of $\mathcal{M}$ resulting in a set of consistent trees.

Second, focusing on a single tree, we seek for a perfect pruning minimizing the D-distance with respect to a given species tree S. The MULPR Problem (in its decision version) is formally defined as follows (Fig. 1(2)).

MUL-TREE PRUNING FOR RECONCILIATION (MULPR) PROBLEM:
Instance: A MUL-tree T on $\mathcal{X}$, a species tree S on $\mathcal{X}$, and a positive integer k.
Question: $\exists$? a perfect pruning of T having D-distance at most k with respect to S.

We begin by analysing, in the next two sections, the complexity of MULSETPC and MULPR. We then present a heuristic for the optimization version of MULPR.

3   Complexity of MUL-Tree Set Pruning for Consistency

In this section, we show, by reducing from the 3-SAT Problem, that MULSETPC is NP-complete. First, observe that the MULSETPC Problem is in NP. In fact, given a choice of leaves to prune in each tree to obtain perfect prunings, the pruned trees can be obtained in polynomial time, and the BUILD algorithm [1] can then be used, also in polynomial time, to check whether the pruned trees are consistent.

We recall here the definition of the 3-SAT Problem.

Problem 1. 3-SAT

Instance: A set of clauses $\mathcal{C} = (C_1 \wedge C_2 \wedge \cdots \wedge C_z)$ on a finite set L $= \{l_1, l_2, \ldots, l_m\}$ of variables where each C_i, $1 \leq i \leq z$, is a clause of the form $(x \vee y \vee w)$ with $\{x, y, w\} \subseteq \{l_1, l_2, \ldots, l_m, \overline{l_1}, \overline{l_2}, \ldots, \overline{l_m}\}$.

Question: $\exists$? a truth assignment satisfying $\mathcal{C}$.

We recall that $x \in \{l_1, l_2, \ldots, l_m, \overline{l_1}, \overline{l_2}, \ldots, \overline{l_m}\}$ is a literal. Given an instance $\mathcal{I} = (\mathcal{C}, L)$ of the 3-SAT problem, we compute in polynomial time a corresponding instance $\mathcal{I}' = (\mathcal{M}, \mathcal{X})$ of MULSETPC. First, the leaf-label-set $\mathcal{X}$ is computed as follows:

- For each literal $x \in \{l_1, l_2, \ldots, l_m, \overline{l_1}, \overline{l_2}, \ldots, \overline{l_m}\}$, $\mathcal{X}$ contains a label x;
- For each clause $C_i \in \mathcal{C}$, $1 \leq i \leq z$, $\mathcal{X}$ contains a label V_i, a label F_i and a label h_i.

Now, the set $\mathcal{M}$ of MUL-trees is computed as follows:

- We add to $\mathcal{M}$, for each variable $l_j \in L$, $1 \leq j \leq m$, the following MUL-trees $T_{l,j,1}$, $T_{l,j,2}$, $T_{l,j,3}$:

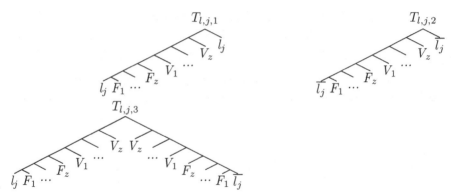

We also add, for each clause $C_i = (x \vee y \vee w) \in \mathcal{C}$, $1 \leq i \leq z$, the following MUL-trees $T_{C,i,1}$, $T_{C,i,2}$:

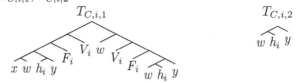

We next show that $\mathcal{I}$ is a satisfiable instance of the 3-SAT problem if and only if its corresponding instance $\mathcal{I}'$ of MULSETPC admits a perfect pruning in a set of consistent trees. The idea is that a pruning of the trees $T_{l,j,1}, T_{l,j,2}, T_{l,j,3}$, with $1 \leq j \leq m$, encodes a truth assignment to the variable l_j. In particular, we can show that a perfect pruning of $T_{l,j,3}$ removes either all the leaves with labels $F_1, \ldots, F_z, V_1, \ldots, V_z$ of the left subtree (corresponding to variable l_j set to true) or all the leaves with labels $F_1, \ldots, F_z, V_1, \ldots, V_z$ of the right subtree (corresponding to the variable l_j set to false). Then we show that, in order to obtain a perfect pruning of $T_{C,i,1}$, $1 \leq i \leq z$, consistent with the perfect pruning of other subtrees in $\mathcal{I}'$, there must exist a perfect pruning of a tree $T_{l,j,3}$, with $1 \leq j \leq m$, associated with a literal $x \in \{l_j, \overline{l_j}\}$ in C_i that encodes an assignment that satisfies x.

Lemma 1. *Let $\mathcal{I}$ be a satisfiable instance of the 3-SAT problem. Then its corresponding instance $\mathcal{I}'$ of* MULSETPC *admits a perfect pruning leading to a set of consistent trees.*

Lemma 2. *Let $\mathcal{I}$ be an unsatisfiable instance of the 3-SAT problem. Then its corresponding instance $\mathcal{I}'$ of* MULSETPC *does not admit a perfect pruning leading to a set of consistent trees.*

Note that, by construction, the instances of MULSETPC in the reduction only contain trees with at most three leaves labeled by the same label. From this remark, and since 3-SAT is NP-complete [18], these two lemmas lead to the following result.

Theorem 1. *The* MULSETPC *Problem is NP-complete even if each label is present at most 3 times in each tree.*

4   Complexity of MUL-Tree Pruning for Reconciliation

In this section, we study the parameterized complexity of MULPR. We denote by k the D-distance between a perfect pruning of T and S and we prove that MULPR is W[2]-hard when parameterized by k, by giving a parameterized reduction from the SET COVER Problem. We refer the reader to [7,11] for details on parameterized reductions. We recall here the definition of SET COVER.

Problem 2. SET COVER

Instance: A set $U = \{u_1, \ldots, u_z\}$ of z elements and a collection $\mathcal{I} = \{I_1, \ldots, I_m\}$ of sets, where each I_i, $1 \le i \le m$, is a subset of U, a positive integer h.

Question: $\exists$? a collection $\mathcal{I}' \subseteq \mathcal{I}$ consisting of at most h subsets such that for each element $u_i \in U$, with $1 \le i \le z$, there exists a set in $\mathcal{I}'$ containing u_i.

Next, given an instance of SET COVER, we construct a corresponding instance of MULPR. We first define the leaf-label-set $\mathcal{X}$ as follows:

$$\mathcal{X} = \{a, b, c\} \cup \{l_i : 1 \le i \le z\}$$

Now, the species tree S is a rooted caterpillar tree on $\mathcal{X}$, with the leaf-order $a, l_1, l_2, \ldots, l_z, b, c$:

The MUL-tree T is built from a set of subtrees that are then connected. First, for each set I_i, with $1 \le i \le m$, we define a subtree $T(I_i)$, which is rooted caterpillar on leaf-set $L_i = \{l_j : u_j \in I_i\}$ such that $T(I_i) = S|L_i$. Assuming $L_i = \{l_{i,1}, \ldots, l_{i,t}\}$, it holds that

$$T(I_i)$$

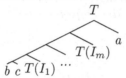

Finally, the tree T is defined by connecting subtrees $T(I_i)$, $1 \le i \le m$, together with leaves labeled by a, b, c, to a path from the root to the leaf labeled by b; this path is called the *spine* of T. The tree T is defined as follows:

$$T$$

First, we prove a property of tree T and of a pruning of T.

Lemma 3. *Given an instance $(U, \mathcal{I})$ of* SET COVER, *consider the corresponding instance (T, S) of* MULPR. *Then (1) in a perfect pruning T^* of T, each node on the spine of T^* between the ancestor of b, c (not included) and the root of T^* is a duplication and (2) each internal node of a subtree $T^*(I_i)$, $1 \leq i \leq m$, in T^* is a speciation.*

We are now able to prove the main result of the reduction. The idea is that the pruning removes all the subtrees $T(I_i)$, $1 \leq i \leq m$, that do not encode sets in the set cover.

Lemma 4. *Given an instance $(U, \mathcal{I})$ of* SET COVER, *consider the corresponding instance (T, S) of* MULPR. *There exists a cover $\mathcal{I}'$ of U consisting of at most h sets if and only if there exists a perfect pruning T^* of T that induces at most $h + 1$ duplications.*

Lemma 4 shows that we have designed a parameterized reduction from SET COVER, which is W[2]-hard when parameterized by h [22]. The described reduction is also a polynomial time many-one reduction. Thus it allows us to prove that MULPR is NP-hard, since SET COVER is known to be NP-hard [18], and W[2]-hard when parameterized by k.

Theorem 2. MULPR *is NP-hard and W[2]-hard when parameterized by k.*

5   A Greedy Heuristic for the MULPR Problem

In this section, we present a polynomial time heuristic for the MULPR Problem. It is based on a greedy approach selecting an appropriate duplication node x in the event-labeled gene tree T to be removed or turned into a speciation node (by removing leaves of $T[x]$), and iterating until no duplication can be removed or turned into speciation anymore.

Algorithm 1: $MULPR(T, S)$

---
1 $CurrentMinDupTree \leftarrow T; CurrentMinDup \leftarrow D(T, S);$
2 **Do**
3 $\quad$ $MinDup \leftarrow CurrentMinDup$
4 $\quad$ $MinDupTree \leftarrow CurrentMinDupTree$
5 $\quad$ **For** *all duplication nodes x of the lca-labeling $\langle MinDupTree, s, e_{lca} \rangle$*
   $\quad$ **Do**
6 $\quad\quad$ **If** $DupRem(T, S, x)$ **Then**
7 $\quad\quad\quad$ $T' \leftarrow DupRemTree(T, S, x)$
8 $\quad\quad\quad$ **If** $D(T', S) < CurrentMinDup$ **Then**
9 $\quad\quad\quad\quad$ $CurrentMinDup \leftarrow D(T', S);$
   $\quad\quad\quad\quad$ $CurrentMinDupTree \leftarrow T'$
10
11
12
13 **While** $CurrentMinDup \neq MinDup;$
14 Return $CurrentMinDupTree$

More precisely, the algorithm is based on a resolution of the following problem.

MUL-TREE DUPLICATION REMOVAL (DupRem) PROBLEM:

Instance: An lca-labeling $\langle T, s, e_{lca} \rangle$ of T with respect to S and a duplication node x, i.e. a node of T such that $e_{lca}(x) = Dup$.

Question: $\exists$? a perfect pruning T^* of T obtained by *fixing* x, i.e. such that x is either removed (node removal) or converted to a speciation node (label fixing) in the lca-labeling of T^*.

At each step, Algorithm 1 tests all the duplication nodes x of T and fixes the one leading, after fixing x, to a tree T' with the smallest $D(T', S)$ value. The algorithm stops when no duplication can be fixed. It uses the DupRem Algorithm returning the boolean value corresponding to the answer to the DupRem Problem for a given duplication node x of T, and the DupRemTree Algorithm returning the tree after fixing x, defined below.

5.1 Solving the DupRem Problem

The heart of the MULTPR Algorithm is the resolution of the DupRem decision problem for a given duplication node x of T. Denote $s = s(x)$ and α, β the children of s. Deciding whether x can be fixed requires considering all the following possibilities for leaves removal (the DupRem Algorithm):

- Removal of a full subtree $T[x_i]$, where $i \in \{l, r\}$, inducing the removal of the node x. This is possible if $\tilde{L}(T[x_i]) \subseteq \tilde{L}(T')$ where T' is the tree T with $T[x_i]$ removed. For example, in Fig. 1, the duplication node 3 of T can be removed by removing the whole left subtree of 3, as $\tilde{L}(T[3_l]) = \{a, d\}$, and both a and d are present elsewhere in the tree.
- Pruning leaves in $T[x_l]$ and $T[x_r]$ leading to two subtrees $T'[x_l]$ and $T'[x_r]$ containing only leaf labels in $S[\alpha]$ for one subtree, and only leaf labels in $S[\beta]$ for the other subtree, which makes x a speciation node. For example, in Fig. 1, the duplication node 1 of T can be turned to a speciation node by pruning the leaves labeled d, e and f in the subtree $T[1_l]$.
- Pruning leaves of $T[x_l]$ and $T[x_r]$ so that x is mapped lower in the species tree (either to α or β), in which case the algorithm is called recursively to test whether x can be converted to a speciation mapped lower than s. For example, in Fig. 1, the duplication node 2 of T mapped to k can neither be removed nor turned to a speciation node, but can be mapped to i by removing the leaves of $T[2]$ labeled by d, e and f. The node is then turned to a speciation node in a next iteration of the algorithm.

Lemma 5. *Given an lca-labeling $\langle T, s, e_{lca} \rangle$ of T with respect to S and a duplication node x of T, DupRem Algorithm returns True if and only if the answer to the DupRem Problem with input $(\langle T, s, e_{lca} \rangle, x)$ is True.*

Let n be the size of T (its number of leaves). To show that the DupRem Algorithm has linear running time, we show that it is possible to preprocess the trees T and S in $O(n)$ time so that we can then verify in $O(1)$ time if pruning a set of leaves of $T[x]$, for a given node x, leads to a tree on $\tilde{L}(T) = \mathcal{X}$. First, notice that $L(T)$, $L(T[x_l])$ and $L(T[x_r])$ can be computed in $O(n)$ with a postorder traversal of T. Now, consider a triplet $< B_1, B_2, B_3 >$ of boolean values assigned to each node v of $S[s(x)]$, defined as follows:

- B_1 is set to True if the tree T' obtained by pruning all the leaves of $T[x_l]$ with labels in $\tilde{L}(S[v])$ is leaf-labeled by $\mathcal{X}$ and false otherwise;
- B_2 is set to True if the tree T' obtained by pruning all leaves of $T[x_r]$ with labels in $\tilde{L}(S[v])$ is leaf-labeled by $\mathcal{X}$ and false otherwise;
- B_3 is set to True if the tree T' obtained by pruning all the leaves of $T[x_l]$ and $T[x_r]$ with labels in $\tilde{L}(S[v])$ is leaf-labeled by $\mathcal{X}$ and false otherwise.

Lemma 6. *Given an lca-labeling $\langle T, s, e_{lca} \rangle$ of T with respect to S and an internal node x of T, the triplet $< B_1, B_2, B_3 >$ associated to x can be computed in $O(n)$ time.*

Rewriting the conditions of the DupRem Algorithm using the values B_1, B_2, B_3 associated to the considered node x, we are now able to prove its linear-time computation.

Lemma 7. *The DupRem Algorithm computes a solution of DupRem in $O(n)$ time.*

5.2   The DupRemTree Algorithm

For a given duplication node x of T for which the DupRem Algorithm returns True, many pruning of T may lead to fixing x. For example, in Fig. 1, the duplication node 1 of T can be fixed, either by removing its right subtree, or by pruning the leaves labeled d, e and f in its left subtree. Notice that different pruning may lead to trees of different duplication cost, as fixing a node x may fix other duplications. For example, fixing the node 3 of T in Fig. 1 by pruning its left subtree leads to fixing the node 5 (5 is removed). The DupRemTree Algorithm tests all the possibilities (i.e. all the conditions in the DupRem Algorithm leading to the answer True), and chooses the pruning of minimum duplication cost, i.e. the one leading to fixing the most duplication nodes.

We are now ready to prove the main result of this section.

Theorem 3. *Algorithm 1 computes a pruning T^* of T in $O(nd^2)$ time, where n is the size of T and d is the number of duplication nodes of the lca-labeling $\langle T, s, e_{lca} \rangle$.*

Proof. At each step, computing the number of duplication nodes can be done in $O(n)$ [26]. For each of the d duplication nodes, Algorithm 1 applies the DupRemTree Algorithm for fixing (if it is possible) the largest number of duplication nodes. Finding if a certain duplication node can be fixed (the DupRem

Algorithm) can be done in $O(n)$ time from Lemma 7 and, if it is the case, prune the tree to fix it (the DupRemTree Algorithm) can also be done in time $O(n)$. Therefore, each iteration of Algorithm 1 can be computed in time $O(nd)$. In the worst case, Algorithm 1 executes d iterations and fixes one duplication node for each iteration, thus the overall time complexity of Algorithm 1 is $O(nd^2)$.   □

5.3   Gene Tree Preprocessing

To optimize Algorithm 1, we start by preprocessing T by removing duplication nodes for which the answer to the DupRem problem is obviously True, and such that there exists an optimal prefect pruning T^* of T with those duplications removed. It is the case, for example, of the lowest duplications inside a terminal branch of a species tree, such as the duplication 5 in the tree T of Fig. 1 leading to two gene copies in genome a. Obviously, removing one a or the other is equivalently good. We use Algorithm 2 to preprocess the tree T by calling $PreprocessT(T, S, r(T))$.

It is straightfoward to see that such leaf pruning is the optimal way to prune these subtrees and that it can be done in time $O(nd)$ where n is the size of T and d is the number of duplication nodes of the lca-labeling $\langle T, s, e_{lca} \rangle$.

Algorithm 2: $PreprocessT(T, S, x)$

---

1   **If** *x is not a leaf* **Then**
2 | $PreprocessT(T, S, x_l)$
3 | $PreprocessT(T, S, x_r)$
4 | **If** *x is a duplication node of the lca-labeling* $\langle T, s, e_{lca} \rangle$ **Then**
5 | | **If** $D(T[x_l], S) == 0$ *and* $D(T[x_r], S) == 0$ **Then**
6 | | | **If** $T[x_l]$ *displays* $T[x_r]$ **Then**
7 | | | | Prune all leaves of $T[x_r]$ in T
8 | | | |
9 | | | **If** $T[x_r]$ *displays* $T[x_l]$ **Then**
10 | | | | Prune all leaves of $T[x_l]$ in T
11 | | | |
12 | | |
13 | |
14 |

---

6   Results

We run the MULPR heuristic on trees from the Ensembl Genome Browser, the goal being to test its accuracy against the optimal solution computed by simply trying all possible perfect prunings T' of T after preprocessing T using Algorithm 2.

We tested our heuristic on a set of silverside fish gene trees from the Ensembl Genome Browser. To keep the execution time reasonable, we restricted the trees to a set of 12 species (*Nothobranchius furzeri, Kryptolebias marmoratus, Fundulus heteroclitus, Cyprinodon variegatus, Xiphophorus couchianus, Xiphophorus maculatus, Gambusia affinis, Poecilia mexicana, Poecilia formosa, Poecilia latipinna, Poecilia reticulata* and *Oryzias melastigma*). Out of the Ensembl Genome Browser gene trees (release 102), 16369 trees contain at least two of our species. We kept all those gene trees, pruned all the leaves that do not belong to our selected species, preprocessed them and then computed the solution obtained by MULPR (Algorithm 1) for each tree. The total number of duplication nodes for all the trees after the preprocessing was 49756. This number decreased to 22532 with the MULPR algorithm.

Due to the exponential running time of the exact algorithm (described above) we were able to run it in a reasonable time only on trees with at most 32 leaves after the preprocessing. This reduces the set of trees from 16369 to 15089 trees. Among those, the preprocessing returned a tree with at least one duplication node for 11902 trees, for which MULPR returned an optimal solution 90.2% of the time. The solution returned by the preprocessing was optimal for 10519 trees. Out of the 4570 trees for which the preprocessing did not returned an optimal solution, MULPR returned an optimal solution 3407 times so for 74.6% of those trees. It returned a solution with one more duplication node than the optimal solution for 21.3% of those trees, a solution with two more duplication nodes for 3.7% of those trees, and it never returned a solution with more than four more duplication nodes.

Figure 2 illustrates the accuracy of the heuristic on the considered trees, depending on the number of duplication nodes of the input trees after preprocessing. More precisely, for a given number of duplications d (x-axis), let

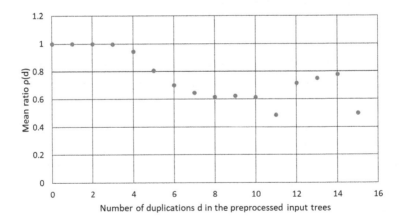

Fig. 2. Results of the exact and MULPR algorithms depending on the number of duplications (x-axis) in the input trees, after preprocessing (see the text for details). The y-axis gives the mean ratio $\rho(d)$ of the number of duplication nodes in the optimal pruning to the number of duplication nodes in the pruning returned by MULPR, for trees with d duplications after preprocessing (see text for details).

$\mathcal{T}_d = \{T_1, \cdots, T_{k_d}\}$ be the set of trees with d duplications remaining after the proprocessing and let $E(T_i)$ and $MULPR(T_i)$ be the Duplication-cost of the perfect pruning of T_i obtained, respectively, with the exact and MULPR algorithms. Let $\tau(T_i) = \frac{E(T_i)}{MULPR(T_i)}$ if $MULPR(T_i) \neq 0$ and $\tau(T_i) = 1$ otherwise (in that case we consider the ratio to be 1 because the solution is optimal). Then $\rho(d) = \frac{(\sum_{1 \leq i \leq k_d} \tau(T_i))}{k_d}$ (y-axis).

We see that the accuracy of the heuristic does decrease when the number of duplication nodes increases, but it still managed to return good solutions having most of the times less than 40% more duplication nodes than the optimal solution.

7 Conclusion

We have considered two problems related to the leaf-pruning of MUL-trees leading to single-labeled trees. MULSETPC asks for a pruning of each tree in a set of MUL-trees leading to a set of consistent trees, while MULPR asks for a pruning of a MUL-tree that minimizes the duplication cost with a given species tree. We have shown that MULTSETPC is NP-complete and that MULPR is W[2]-hard when parameterized by the duplication cost. We have developed a polynomial-time heuristic for MULPR and have shown that the method is accurate, presenting experiments on a set of gene trees from the Ensembl Genome Browser.

There are some interesting open problems related to the complexity of MULSETPC and MULPR. First, what is the computational complexity of MULSETPC when the set $\mathcal{M}$ consists of a constant number of trees? For MULPR, our parameterized reduction can be modified to prove that the problem is not approximable within factor $(1 - \varepsilon) \ln |\mathcal{X}|$, for any $\varepsilon > 0$, but it is not clear whether there are approximation algorithms reaching a $\ln |\mathcal{X}|$ approximation factor. Finally, since our heuristic for MULPR returns in many cases optimal or near-optimal solutions, it would be nice to understand if there are restrictions of the problem for which the heuristic is optimal.

From a biological point of view, it is important to further investigate whether the genes remaining after pruning are those best representing the evolution and functional conservation of gene families. While, for the MULSETPC Problem, this would allow applications for the reconstruction of species trees or synteny trees from gene families with duplicates, for the MULPR Problem could be used for the development of eukaryote-wide Hidden Markov Models (HMM), to allow most reliable protein identification.

Availability. Algorithm implementation available at http://www.iro.umontreal.ca/~mabrouk/.

References

1. Aho, A., Yehoshua, S., Szymanski, T., Ullman, J.: Inferring a tree from lowest common ancestors with an application to the optimization of relational expressions. SIAM J. Comput. **10**(3), 405–421 (1981)
2. Bininda-Emonds, O. (ed.): Phylogenetic Supertrees Combining Information to Reveal the Tree of Life. Computational Biology, Kluwer Academic, Dordrecht (2004)
3. Bryant, D.: A classification of consensus methods for phylogenetics. DIMACS Ser. Discrete Math. Theor. Comput. Sci. **61**, 163–184 (2003)
4. Chou, S., Hsu, C.L.: MMDT: a multi-valued and multi-labeled decision tree classifier for data mining. Expert Syst. Appl. **28**(4), 799–812 (2005)
5. Crochemore, M., Vérin, R.: Direct construction of compact directed acyclic word graphs. In: Apostolico, A., Hein, J. (eds.) CPM 1997. LNCS, vol. 1264, pp. 116–129. Springer, Heidelberg (1997). https://doi.org/10.1007/3-540-63220-4_55
6. Cui, Y., Jansson, J., Sung, W.: Polynomial-time algorithms for building a consensus MUL-tree. J. Comput. Biol. **19**(9), 10731088 (2012)
7. Cygan, M., et al.: Parameterized Algorithms. Springer, Cham (2015). https://doi.org/10.1007/978-3-319-21275-3
8. Delabre, M., et al.: Reconstructing the history of syntenies through super-reconciliation. In: Blanchette, M., Ouangraoua, A. (eds.) RECOMB-CG 2018. LNCS, vol. 11183, pp. 179–195. Springer, Cham (2018). https://doi.org/10.1007/978-3-030-00834-5_10
9. Delabre, M., et al.: Evolution through segmental duplications and losses: a super-reconciliation approach. Algorithms Mol. Biol. 499506 (2020)
10. Dondi, R., El-Mabrouk, N., Swenson, K.M.: Gene tree correction for reconciliation and species tree inference: Complexity and algorithms. J. Discrete Algorithms **25**, 51–65 (2014). https://doi.org/10.1016/j.jda.2013.06.001
11. Downey, R.G., Fellows, M.R.: Fundamentals of Parameterized Complexity. TCS. Springer, London (2013). https://doi.org/10.1007/978-1-4471-5559-1
12. El-Mabrouk, N., Noutahi, E.: Gene family evolution—an algorithmic framework. In: Warnow, T. (ed.) Bioinformatics and Phylogenetics. CB, vol. 29, pp. 87–119. Springer, Cham (2019). https://doi.org/10.1007/978-3-030-10837-3_5
13. Ganapathy, G., Goodson, B., Jansen, R., Le, H.S., Ramachandran, V., Warnow, T.: Pattern identification in biogeography. IEEE/ACM Trans. Comput. Biol. Bioinform. **3**(4), 334–346 (2006)
14. Goodman, M., Czelusniak, J., Moore, G., Romero-Herrera, A., Matsuda, G.: Fitting the gene lineage into its species lineage, a parsimony strategy illustrated by cladograms constructed from globin sequences. Syst. Zool. **28**, 132–163 (1979)
15. Gregg, W., Ather, S., Hahn, M.: Gene-tree reconciliation with MUL-trees to resolve polyploidy events. Syst. Biol. **66**(6), 10071018 (2017)
16. Huber, K., Moulton, V., Spillner, A.: Computing a consensus of multilabeled trees. In: 14th Workshop on Algorithm Engineering and Experiments (ALENEX 2012), pp. 84–92 (2012)
17. Huber, K., Moulton, V., Steel, M., Wu, T.: Folding and unfolding phylogenetic trees and networks. J. Math. Biol. **73**(6–7), 1761–1780 (2016)

18. Karp, R.M.: Reducibility among combinatorial problems. In: Miller, R.E., Thatcher, J.W. (eds.) Proceedings of a symposium on the Complexity of Computer Computations, held March 20–22, 1972, at the IBM Thomas J. Watson Research Center, Yorktown Heights, New York, USA, pp. 85–103. The IBM Research Symposia Series, Plenum Press, New York (1972). https://doi.org/10.1007/978-1-4684-2001-2_9

19. Lott, M., Spillner, A., Huber, K.: Inferring polyploid phylogenies from multiply-labeled gene trees. BMC Evol. Biol. **9**, 216 (2009)

20. Nguyen, N., Mirarab, S., Warnow, T.: MRL and SuperFine+MRL: new supertree methods. J. Algo. for Mol. Biol. **7**(3) (2012)

21. Page, R.: Parasites, phylogeny and cospeciation. Int. J. Parasitol. **23**, 499–506 (1993)

22. Paz, A., Moran, S.: Non deterministic polynomial optimization problems and their approximations. Theoret. Comput. Sci. **15**, 251–277 (1981). https://doi.org/10.1016/0304-3975(81)90081-5

23. Ranwez, V., et al.: PhySIC: a veto supertree method with desirable properties. Syst. Biol. **56**(5), 798–817 (2007)

24. Scornavacca, C., Berry, V., Ranwez, V.: From gene trees to species trees through a supertree approach. In: Dediu, A.H., Ionescu, A.M., Martín-Vide, C. (eds.) LATA 2009. LNCS, vol. 5457, pp. 702–714. Springer, Heidelberg (2009). https://doi.org/10.1007/978-3-642-00982-2_60

25. Semple, C., Steel, M.: A supertree method for rooted trees. Discret. Appl. Math. **105**(1), 147–158 (2000)

26. Zmasek, C.M., Eddy, S.R.: A simple algorithm to infer gene duplication and speciation events on a gene tree. Bioinformatics **17**, 821–828 (2001)

Makespan Trade-Offs for Visiting Triangle Edges
(Extended Abstract)

Konstantinos Georgiou[(⊠)], Somnath Kundu, and Paweł Prałat

Department of Mathematics, Ryerson University, Toronto, ON M5B 2K3, Canada
{konstantinos,somnath.kundu,pralat}@ryerson.ca

Abstract. We study a primitive vehicle routing-type problem in which a fleet of n unit speed robots start from a point within a non-obtuse triangle Δ, where $n \in \{1, 2, 3\}$. The goal is to design robots' trajectories so as to visit all edges of the triangle with the smallest visitation time makespan. We begin our study by introducing a framework for subdividing Δ into regions with respect to the type of optimal trajectory that each point P admits, pertaining to the order that edges are visited and to how the cost of the minimum makespan $R_n(P)$ is determined, for $n \in \{1, 2, 3\}$. These subdivisions are the starting points for our main result, which is to study makespan trade-offs with respect to the size of the fleet. In particular, we define $\mathscr{R}_{n,m}(\Delta) = \max_{P \in \Delta} R_n(P)/R_m(P)$, and we prove that, over all non-obtuse triangles Δ: (i) $\mathscr{R}_{1,3}(\Delta)$ ranges from $\sqrt{10}$ to 4, (ii) $\mathscr{R}_{2,3}(\Delta)$ ranges from $\sqrt{2}$ to 2, and (iii) $\mathscr{R}_{1,2}(\Delta)$ ranges from $5/2$ to 3. In every case, we pinpoint the starting points within every triangle Δ that maximize $\mathscr{R}_{n,m}(\Delta)$, as well as we identify the triangles that determine all $\inf_\Delta \mathscr{R}_{n,m}(\Delta)$ and $\sup_\Delta \mathscr{R}_{n,m}(\Delta)$ over the set of non-obtuse triangles.

Keywords: 2-dimensional search and navigation · Vehicle routing · Triangle · Make-span · Trade-offs

1 Introduction

Vehicle routing problems form a decades old paradigm of combinatorial optimization questions. In the simplest form, the input is a fleet of robots (vehicles) with some starting locations, together with stationary targets that need to be visited (served). Feasible solutions are robots' trajectories that eventually visit every target, while the objective is to minimize either the total length of traversed trajectories or the time that the last target is visited.

Vehicle routing problems are typically NP-hard in the number of targets. The case of 1 robot in a discrete topology corresponds to the celebrated Traveling Salesman Problem whose variations are treated in numerous papers and books. Similarly, numerous vehicle routing-type problems have been proposed and studied, varying with respect to the number of robots, the domain's topology and the solutions' specs, among others.

Research supported in part by NSERC.

A full version of the paper is available on arXiv [10].

P. Flocchini and L. Moura (Eds.): IWOCA 2021, LNCS 12757, pp. 340–355, 2021.
https://doi.org/10.1007/978-3-030-79987-8_24

We deviate from all previous approaches and we focus on efficiency trade-offs, with respect to the fleet size, of a seemingly simple geometric variation of a vehicle routing-type problem in which targets are the edges of a non-obtuse triangle. The optimization problem of visiting all these three targets (edges), with either 1, 2 or 3 robots, is computationally degenerate. Indeed, even in the most interesting case of 1 robot, an optimal solution for a given starting point can be found by comparing a small number of candidate optimal trajectories (that can be efficiently constructed geometrically). From a combinatorial geometric perspective, however, the question of characterizing the points of an arbitrary non-obtuse triangle with respect to optimal trajectories they admit when served by 1 or 2 robots, e.g. the order that targets are visited, is far from trivial (and in fact it is still eluding us in its generality).

In the same direction, we ask a more general question: Given an arbitrary non-obtuse triangle, what is the worst-case trade-off ratio of the cost of serving its edges with different number of robots, over all starting points? Moreover, what is the smallest and what is the largest such value as we range over all non-obtuse triangles? Our main contributions pertain to the development of a technical geometric framework that allows us to pinpoint exactly the best-case and worst-case non-obtuse triangles, along with the worst-case starting points that are responsible for the extreme values of these trade-off ratios. To the best of our knowledge, the study of efficiency trade-offs with respect to fleet sizes is novel, at least for vehicle routing type problems or even in the realm of combinatorial geometry.

1.1 Motivation and Related Work

Our problem is related to a number of topics including vehicle routing problems, the (geometric) traveling salesman problem, and search and exploration games. Indeed, the main motivation for our problem comes from the so-called shoreline search problem, first introduced in [4]. In this problem, a unit robot is searching for a hidden line on the plane (unlike our problem in which the triangle edges are visible). The objective is to visit the line as fast as possible, relative to the distance of the line to the initial placement of the robot. The best algorithm known for this problem has performance of roughly 13.81, and only very weak (unconditional) lower bounds are known [3]. Only recently, the problem of searching with multiple robots was revisited, and new lower bounds were proven in [1, 8].

As it is common in online problems, a typical argument for a lower bound for the shoreline problem lets an arbitrary algorithm perform for a certain time until the hidden item is placed at a location that cannot have been visited before by the robot. The lower bound then is obtained by adding the elapsed time with the distance of the robot to the hidden item (the line), since at this point one may only assume that the (online) algorithm has full knowledge of the input. Applying this strategy to the shoreline problem, one is left with the problem of identifying a number of lines, as close as possible to the starting point of the robot, and then computing the shortest trajectory of the robot that could visit them all, exactly as in our problem. In the simplest configuration that could result strong bounds, one would identify three lines, forming a non-obtuse triangle. The latter is also the motivating reason we restricted our attention to non-obtuse triangles (a second reason has to do with the optimal visitation cost of 3 robots, which for non-obtuse triangles is defined as the maximum distance over all triangle edges, treated as lines).

Our problem could also be classified as a vehicle routing-type problem, the first of which was introduced in [6] more than 60 years ago. The objective in vehicle routing problems is typically to minimize the visitation time (makespan) or the total distance traveled for serving a number of targets given a fleet of (usually capacitated) robots, see surveys [12, 15] for early results. Even though the underlying domain is usually discrete, geometric vehicle type problems have been studied extensively too, e.g. in [7]. Over time, the number of proposed vehicle routing variations is so vast that surveys for the problem are commonly subject-focused; see surveys [11, 13, 14] for three relatively recent examples.

Famously, vehicle routing problems generalize the celebrated traveling salesman problem (TSP) where a number of targets need to be toured efficiently by one vehicle. Similarly to vehicle routing, TSP has seen numerous variations, including geometric [2, 9], where in the latter work targets are lines. The natural extension of the problem to multiple vehicles is known as the multiple traveling salesman problem [5], a relaxation to vehicle routing problems where vehicles are un-capacitated. The latter problem has also seen variations where the initial deployment of the vehicles is either from a single location (single depot), as in our problem, or from multiple locations.

1.2   A Note on Our Contributions and Paper Organization

We introduce and study a novel concept of efficiency/fleet size trade-offs in a special geometric vehicle routing-type problem that we believe is interesting in its own right. Deviating from the standard combinatorial perspective of the problem, we focus on the seemingly simple case of visiting the three edges of a non-obtuse triangle with $n \in \{1, 2, 3\}$ robots. Interestingly, the problem of characterizing the starting points within arbitrary non-obtuse triangles with respect to structural properties of the optimal trajectories they admit is a challenging question. More specifically, one would expect that the latter characterization is a prerequisite in order to analyze efficiency trade-offs when serving with different number of robots, over all triangles. Contrary to this intuition, and without fully characterizing the starting points of arbitrary triangles, we develop a framework that allows us (a) to pinpoint the starting points of any triangle at which these (worst-case) trade-offs attain their maximum values, and (b) to identify the extreme cases of non-obtuse triangles that set the boundaries of the inf and sup values of these worst-case trade-offs.

This is an extended abstract of our work. Due to space limitations we only present the backbone of our arguments, along with the critical intermediate lemmata that are invoked toward proving our main results. The full version of the paper is available online [10]. The paper organization of the extended abstract is as follows. In Sect. 2.1 we give a formal definition of the problem we study, and we quantify our main contributions. Then, in Sect. 2.2 we establish some of the necessary terminology and we present some preliminary and important observations. The technical analysis starts in Sect. 3. First, in Sect. 3.1 we study the simpler problem of visiting only two triangle edges with one robot. It is followed by Sect. 3.2, were we find optimal trajectories for visiting all three triangle edges by one robot in a predetermined order. That brings us to Sect. 4 where we characterize triangle regions with respect to the optimal visitation strategies they admit, for 3 (Sect. 4.1), 2 (Sect. 4.2) and 1 robots (Sect. 4.1). Equipped with that machinery, we outline in Sect. 5 how our main contributions are proved. More specifically, Sect. 5.1, Sect. 5.2 and Sect. 5.3 discuss trade-offs between 1 and 3, 2 and 3, and 1 and 2 robots, respectively. Finally in Sect. 6 we conclude with some open questions.

2   Our Results and Basic Terminology and Observations

2.1   Problem Definition and Main Contributions

We consider the family of non-obtuse triangles $\mathscr{D}$, equipped with the Euclidean distance. For any $n \in \{1, 2, 3\}$, any given triangle $\Delta \in \mathscr{D}$, and any point P in the triangle, denoted by $P \in \Delta$, we consider a fleet of n unit speed robots starting at point P. A feasible solution to the triangle Δ visitation problem with n robots starting from P is given by robots' trajectories that eventually visit every edge of Δ, that is, each edge needs to be touched by at least one robot in any of its points including the endpoints. The visitation cost of a feasible solution is defined as the makespan of robots' trajectory lengths, or equivalently as the first time by which every edge is touched by some robot. By $R_n(\Delta, P)$ we denote the optimal visitation cost of n robots, starting from some point $P \in \Delta$. When the triangle Δ is clear from the context, we abbreviate $R_n(\Delta, P)$ simply by $R_n(P)$.

In this work we are interested in determining visitation cost trade-offs with respect to different fleet sizes. In particular, for some triangle $\Delta \in \mathscr{D}$ (which is a compact set as a subest of $\mathbb{R}^2$), and for $1 \leq n < m \leq 3$, we define

$$\mathscr{R}_{n,m}(\Delta) := \max_{P \in \Delta} \frac{R_n(\Delta, P)}{R_m(\Delta, P)}.$$

Our main technical results pertain to the study of $\mathscr{R}_{n,m}(\Delta)$ as Δ ranges over all non-obtuse triangles $\mathscr{D}$. In particular, we determine $\inf_{\Delta \in \mathscr{D}} \mathscr{R}_{n,m}(\Delta)$ and $\sup_{\Delta \in \mathscr{D}} \mathscr{R}_{n,m}(\Delta)$ for all pairs $(n, m) \in \{(1, 3), (2, 3), (1, 2)\}$. Our contributions are summarized in Table 1.[1]

Table 1. Our main contributions.

	$\mathscr{R}_{1,3}(\Delta)$	$\mathscr{R}_{2,3}(\Delta)$	$\mathscr{R}_{1,2}(\Delta)$
$\inf_{\Delta \in \mathscr{D}}$	$\sqrt{10}$	$\sqrt{2}$	2.5
$\sup_{\Delta \in \mathscr{D}}$	4	2	3

---

[1] Note that the entries in column 1 are not obtained by multiplying the entries of columns 2,3. This is because the triangles that realize the inf and sup values are not the same in each column.

For establishing the claims above, we observe that $\inf_{\Delta \in \mathscr{D}} \max_{P \in \Delta} \frac{R_n(\Delta, P)}{R_m(\Delta, P)} = \alpha$ is equivalent to that

$$\forall \Delta \in \mathscr{D}, \exists P \in \Delta, \frac{R_n(\Delta, P)}{R_m(\Delta, P)} \geq \alpha \quad \text{and} \quad \forall \epsilon > 0, \exists \Delta \in \mathscr{D}, \forall P \in \Delta, \frac{R_n(\Delta, P)}{R_m(\Delta, P)} \leq \alpha + \epsilon.$$

Similarly, $\sup_{\Delta \in \mathscr{D}} \max_{P \in \Delta} \frac{R_n(\Delta, P)}{R_m(\Delta P)} = \beta$ is equivalent to that

$$\forall \Delta \in \mathscr{D}, \forall P \in \Delta, \frac{R_n(\Delta, P)}{R_m(\Delta, P)} \leq \beta \quad \text{and} \quad \forall \epsilon > 0, \exists \Delta \in \mathscr{D}, \exists P \in \Delta, \frac{R_n(\Delta, P)}{R_m(\Delta, P)} \geq \beta - \epsilon.$$

Therefore, as a byproduct of our analysis, we also determine the best and the worst triangle cases of ratios $\mathscr{R}_{n,m}(\Delta)$, as well as the starting points that determine these ratios. In particular we show that (i) the extreme values of $\mathscr{R}_{1,3}(\Delta)$ are attained as Δ ranges between "thin" isosceles and equilateral triangles, and the worst starting point is the incenter, (ii) the extreme values of $\mathscr{R}_{2,3}(\Delta)$ are attained as Δ ranges between right isosceles and equilateral triangles, and the worst starting point is again the incenter, and (iii) the extreme values of $\mathscr{R}_{1,2}(\Delta)$ are attained as Δ ranges between equilateral and right isosceles triangles, and the worst starting point is the middle of the shortest altitude.

2.2   Basic Terminology and Some Useful Observations

The length of segment AB is denoted by $\|AB\|$. An arbitrary non-obtuse triangle will be usually denoted by $\triangle ABC$, which we assume is of bounded size. More specifically, without loss of generality, we often consider $\triangle ABC$ represented in the Cartesian plane in *standard analytic form*, with $A = (p, q)$, $B = (0, 0)$ and $C = (1, 0)$.

The cost of optimally visiting a collection of line segments $\mathscr{C}$ (triangle edges) with 1 robot starting from point P is denoted by $d(P, \mathscr{C})$. For example, when $\mathscr{C} = \{AB, BC\}$ we write $d(P, \{AB, BC\})$. When, for example, $\mathscr{C} = \{AB\}$ is a singleton set, we slightly abuse the notation and for simplicity write $d(P, AB)$ instead of $d(P, \{AB\})$. Note that if the projection P' of P onto the line defined by points A, B lies in segment AB, then $d(P, AB) = \|PP'\|$, and otherwise $d(P, AB) = \min\{\|PA\|, \|PB\|\}$. The following observation follows immediately from the definitions, and the fact that we restrict our study to non-obtuse triangles.

Observation 1. *For any non-obtuse triangle $\Delta = \triangle ABC$, and $P \in \Delta$, we have*

(i) $R_3(\Delta, P) = \max\{d(P, AB), d(P, BC), d(P, CA)\}.$

(ii) $R_2(\Delta, P) = \min \left\{ \begin{array}{l} \max\{d(P, AB), d(P, \{BC, CA\})\} \\ \max\{d(P, BC), d(P, \{AB, CA\})\} \\ \max\{d(P, CA), d(P, \{BC, AB\})\} \end{array} \right\}$

(iii) $R_1(\Delta, P) = d(P, \{AB, BC, CA\}).$

Motivated by our last observation, we also introduce notation for the cost of *ordered visitations*. Starting from point P, we may need to visit an *ordered list* of (2 or 3) line segments in a specific order. For example, we write $d(P, [AB, BC, AC])$ for the optimal cost of visiting the list of segments $[AB, BC, AC]$, in this order, with 1 robot. As we will be mainly concerned with $\triangle ABC$ edge visitations, and due to the already introduced standard analytic form, we refer to the trajectory realizing $d(P, [AB, BC, AC])$ as the (optimal) *LDR strategy* (L for "Left" edge AB, D for "Down" edge BC, and R for "Right" edge AC). We introduce analogous terminology for the remaining 5 permutations of the edges, i.e. LRD, RLD, RDL, DRL, DLR. Note that it may happen that in an optimal ordered visitation, robot visits a vertex of the triangle edges. In such a case we interpret the visitation order of the incident edges arbitrarily. For ordered visitation of 2 edges, we introduce similar terminology pertaining to (optimal) LD, LR, RL, RD, DR and DL strategies.

In order to obtain the results reported in Table 1, it is necessary to subdivide any triangle $\triangle$ into sets of points that admit the same optimal ordered visitations (e.g. all points P in which an optimal $R_1(\triangle, P)$ strategy is LRD). For $n \in \{2, 3\}$ robots, the subdivision is also with respect to the cost $R_n(\triangle, P)$. Specifically for $n = 2$, the subdivision is also with respect to whether the cost $R_2(\triangle, P)$ is determined by the robot that is visiting one or two edges (see Observation 1). We will refer to these subdivisions as the R_1, R_2, R_3 *regions*. For each $n \in \{1, 2, 3\}$, the R_n regions will be determined by collection (loci) of points between neighbouring regions that admit more than one optimal ordered visitations.

Angles are read counter-clockwise, so that for example for $\triangle ABC$ in standard analytic form, we have $\angle A = \angle BAC$. For aesthetic reasons, we may abuse notation and drop symbol $\angle$ from angles when we write trigonometric functions. Visitation trajectories will be denoted by a list of points $\langle A_1, \ldots, A_n \rangle$ $(n \geq 2)$, indicating a movement along line segments between consecutive points. Hence, the cost of such trajectory would be $\sum_{i=2}^{n} \|A_i A_{i-1}\|$.

3   Preliminary Results

3.1   Optimal Visitations of Two Triangle Edges

As a preparatory step, first consider the simpler problem of visiting two distinguished edges of a triangle $\triangle = ABC$, starting from a point within the triangle.

When $\angle A \geq \pi/3$, we define the concept of its *optimal bouncing subcone*, which is defined as a cone of angle $3\angle A - \pi$ and tip A, so that $\angle A$ and the subcone have the same angle bisector. When $\angle A = \pi/3$, then the optimal bouncing subcone is a ray with tip A that coincides with the angle bisector of $\angle A$. Whenever $\angle A < \pi/3$ we define its optimal bouncing subcone as the degenerate empty cone. The following two observations are used repeatedly in our results.

Observation 2. *If P is in the optimal bouncing subcone of $\angle A$, then $d(P, \{AB, AC\}) = \|PA\|$.*

For a point $P \in \triangle ABC$ outside the optimal bouncing subcone of $\angle A$, we define the (two) *optimal bouncing points* M, N *of the ordered* $[AB, AC]$ *visitation* as follows. Let C' be the reflections of C around AB. Let also P' be the projection of P onto AC'.

Then, M is the intersection of PP' with AB and N is the projection of M onto AC. Note that equivalently, M, N are determined uniquely by requiring that (i) $\angle BMP = \angle NMA$, and (ii) $\angle ANM = \pi/2$ (see also Fig. 1).

Observation 3. *If P is outside the optimal bouncing subcone of $\angle A$, then $d(P, \{AB, AC\}) = \|PM\| + \|MN\|$, where M, N are the optimal bouncing points of ordered $[AB, AC]$ visitation.*

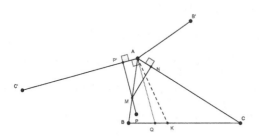
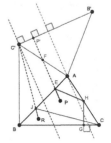

(a) Optimal trajectory for visiting $\{AB, AC\}$ for $\angle A \geq \pi/3$, starting outside the optimal bouncing subcone.

(b) Arbitrary non-obtuse $\triangle ABC$ shown with its LRD bounce indicator line and its LRD subopt indicator line.

Fig. 1. Figures 1a and 1b.

3.2 Optimal (Ordered) Visitation of Three Triangle Edges

In this section we discuss optimal LRD visitations of non-obtuse $\triangle ABC$, together with optimality conditions when serving with one robot. Optimality conditions for the remaining 5 ordered visitations are obtained similarly. In order to determine the optimal LRD visititation, we obtain reflection C' of C across AB, and reflection B' of B across $C'A$, see also Fig. 1b.

From C' and A, we draw a lines ϵ, ζ, both perpendicular to $C'B'$ which may (or may not) intersect $\triangle ABC$. We refer to line ϵ as the *LRD bounce indicator line*. We also refer to line ζ as the *LRD subopt indicator line*. Each of the lines identify a halfspace on the plane. The halfspace associated with ϵ on the side of vertex A will be called the *positive halfspace of the LRD bounce indicator line* or in short the *positive LRD bounce halfspace*, and its complement will be called the *negative LRD bounce halfspace*. The halfspace associated with ζ on the side of vertex B will be called the *positive halfspace of the LRD subopt indicator line*, or in short the *positive LRD subopt halfspace*, and its complement will be called the *negative LRD subopt halfspace*.

For a point P in the positive LRD bounce and subopt halfspaces, let P' be its projection onto $C'B'$. Let E, F be the intersections of PP' with AB, AC', respectively. Let also H be the reflection of F across AB. Points E, H, G will be called the *optimal LRD bouncing points* for point P. The points are also uniquely determined by requiring that $\angle BEP = \angle HEA$ and that HG is perpendicular to BC. For a point R in the negative

LRD bounce halfspace and in the positive subopt halfspace, let J be the intersection of RC' with AB. Point J will be called the *degenerate optimal LRD bouncing point*, which is also uniquely determined by the similar bouncing rule $\angle BJR = \angle CJA$. Finally, let A', A'' be the projection of A onto $B'C', BC$, respectively.

The next lemma refers to such points P, R together with the construction of Fig. 1b. Its proof uses the observations of Sect. 3.1 and follows easily by noticing that the optimal LRD visitation is in 1-1 correspondence with the optimal visitation of segment $B'C'$ using a trajectory that passes from segment AB.

Lemma 1. *The optimal LRD visitation trajectory, with starting points P, R, T, is:*

- *trajectory $\langle P, E, H, G \rangle$, provided that P is in the positive LRD bounce and subopt halfspaces,*
- *trajectory $\langle R, J, C \rangle$, provided that R is in the negative LRD bounce halfspace and in the positive subopt halfspace,*
- *trajectory $\langle T, A, A'' \rangle$, provided that T is in the negative LRD subopt halfspace.*

4 Computing the R_n Regions, $n = 1, 2, 3$

By Observations 2, 3 and Lemma 1, we see that optimal visitations of 2 or 3 edges have cost equal to (i) the distance of the starting point to a line (reflection of some triangle edge), or (ii) the distance of the starting point to some point (triangle vertex) or (iii) the distance of the starting point to some triangle vertex plus the length of some triangle altitude. In this section we describe the R_n regions of certain triangles, $n \in \{1, 2, 3\}$. For this, we compare optimal ordered strategies, and the subdivisions of the regions are determined by loci of points that induce ordered trajectories of the same cost. As these costs are of type (i), (ii), or (iii) above (and considering all their combinations) the loci of points in which two ordered strategies have the same cost will be either some line (line bisector or angle bisector), or some conic section (parabola or hyperbola).

4.1 Triangle Visitation with 3 Robots - The R_3 Regions

Consider $\Delta \in \mathscr{D}$ with vertices A, B, C. For every $P \in \Delta$, any trajectories require time at least the maximum distance of P from all edges, in order to visit all of them. This bound is achieved by having all robots moving along the projection of P onto the 3 edges, and so we have $R_3(P) = \max\{d(P, AB), d(P, BC), d(P, CA)\}$, as also in Observation 1. Next we show how to subdivide the region of Δ with respect to which of the 3 projections is responsible for the optimal visitation cost. For this, we let I denote the incenter (the intersection of angle bisectors) of Δ. Let also K, L, M be the intersections of the bisectors with edges BC, CA and AB, respectively, see also Fig. 2.

Lemma 2. *For every starting point $P \in \Delta$, we have that*

$$R_3(\Delta, P) = \begin{cases} d(P, AB) \text{, provided that} P \in CLIK \\ d(P, BC) \text{, provided that} P \in AMIL \\ d(P, CA) \text{, provided that} P \in BKIM \end{cases}.$$

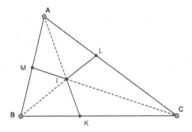

Fig. 2. The R_3 regions of an arbitrary non-obtuse $\triangle ABC$. AK, BL, CM are the angle bisectors of $\angle A, \angle B, \angle C$, respectively. Recall that the incenter I is equidistant from all triangle edges.

4.2   Triangle Visitation with 2 Robots - The R_2 Regions

In this section we show how to subdivide the region of any non-obtuse triangle $\Delta \in \mathscr{D}$ into subregions with respect to the optimal trajectories and their costs, for a fleet consisting of 2 robots. The following technical lemma describes a geometric construction.

Lemma 3. *Consider non-obtuse $\triangle ABC$ along with its incenter I. Let K, M be the intersections of angle bisectors of A, C with segments BC, AB respectively. From K, M we consider cones of angles A, C respectively, having direction toward the interior of the triangle, and placed so that their bisectors are perpendicular to BC, AB, respectively. Then, the extreme rays of the cones intersect at some point F in line segment BI.*

Motivated by Lemma 3, we will be referring to the subject point F in the line segment BI as the *separator of the angle B bisector*. Similarly, we obtain separators J, H of angles C, A bisectors, respectively, see also Fig. 3a. In what follows, we will be referring to the (possibly non-convex) hexagon $MFKJLH$ as the R_2 *(hexagon) separator of $\triangle ABC$*.

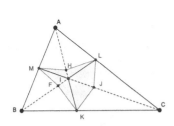

(a) The R_2 hexagon separator of $\triangle ABC$.

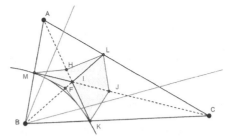

(b) The refined R_2 mixed-hexagon separator of $\triangle ABC$, where $\angle B > \pi/3$.

Fig. 3. Figures 3a and 3b.

The remaining of the section refers to non-obtuse triangle $\Delta = \triangle ABC$ as in Fig. 3a, where in particular $MFKJLH$ is the R_2 separator of Δ. Assume that $\angle B \geq \pi/3$. It can be shown that for every point P either in MF or FK which are outside the optimal bouncing subcone of angle B, we have that $d(P, AC) = d(P, \{BC, AB\})$. For points within the subcone, the optimal trajectory to visit $\{BC, AB\}$ would be to go directly to B. So for points P within the optimal bouncing subcone, condition $d(P, AC) = d(P, \{BC, AB\})$ translates into that P is equidistant from AC and B. Hence, P lies in a parabola with AC being the directrix and B being the focus. Next, we refer to that parabola as the *separating parabola of B*.

Motivated by the previous observation, we introduce the notion of the *refined R_2 mixed-hexagon separator of triangle* Δ as follows. For every angle of Δ which is more than $\pi/3$, we replace the portion of the R_2 hexagon separator within the optimal bouncing subcone of the same angle by the corresponding separating parabola. In Fig. 3b we display an example where only one angle is more than $\pi/3$. Combined with Observation 1 (ii), we can formalize our findings as follows.

Lemma 4. *For every starting point on the boundary of the refined R_2 mixed-hexagon separator of a triangle Δ, the cost of visiting only the opposite edge equals the cost of visiting the other two edges. For every starting point P outside the R_2 separator, $R_2(P)$ equals the distance of P to the opposite edge. Moreover, for every starting point P in the interior of the refined R_2 separator, $R_2(P)$ is determined by the cost of visiting two of the edges of Δ.*

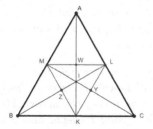

(a) The R_2 regions of the equilateral triangle, see also Corollary 1 and Lemma 5 for detailed description.

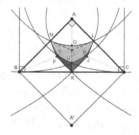

(b) The R_2 regions of the right isosceles, see also Corollary 2 and Lemma 6 for detailed description. The coloured region identifies the refined R_2 mixed-hexagon separator.

Fig. 4. Figures 4a and 4b. (Color figure online)

Lemma 4 implies the following corollaries pertaining to specific triangles $\triangle ABC$. In both statements, and the associated figures, I is the incenter of the triangles, and points K, L, M are defined as in Fig. 2.

Corollary 1 (Hexagon separator of equilateral triangle). *Consider equilateral $\Delta = \triangle ABC$, see Fig. 4a. Let W, Z, Y be the intersections of AK, BL, CM respectively (also the separators of angle A bisector, angle B bisector, and angle C bisector, respectively). Then, the R_2 hexagon separator of Δ is $MZKYLW$, which is also triangle MKL. More specifically, for all $P \in \triangle AML$, we have that $R_2(\Delta, P) = d(P, BC)$.*

Corollary 2 (Mixed-hexagon separator of right isosceles). *Consider right isosceles* $\Delta = \triangle ABC$, *see Fig. 4b. The separator of angle A bisector is incenter I. Let also* F, J *be the separators of angle B bisector and angle C bisector, respectively. Then, the* R_2 *hexagon separator of Δ is $IMFKJL$. The parabola with directix BC and focus* A, *intersecting AK at Q and passing through M, L is the separating parabola of A. Hence, for every point $P \in \Delta$ above the parabola, we have $R_2(\Delta, P) = d(P, BC)$, as well as for every point X in tetragon $MBKF$, we have $R_2(\Delta, X) = d(X, AC)$.*

Describing the subdivisions within the refined R_2 mixed-hexagon separator for arbitrary triangles is a challenging task. On the other hand, by Observation 1 (ii) and Lemma 4 the cost within the separator is determined by the cost of visiting just two edges. Also, by Observations 2, 3 the cost of such visitation can be described either as a distance to a line or to a point. We conclude that, within the R_2 separator, the subdivisions are determined by separators that are either parts of lines or parabolas (loci of points for which the cost of visiting some two edges are equal). Hence, for any fixed triangle, an extensive case analysis pertaining to pairwise comparisons of visitations costs can determine all R_2 subdivisions (and the challenging ones are within the refined separator). In what follows we summarize formally the subdivisions only of two triangle types, focusing on the visitation cost of all starting points within the (refined) hexagon separators.

Lemma 5 (R_2 regions of an equilateral triangle). *Consider equilateral* $\Delta = \triangle ABC$, *as in Corollary 1, see Fig. 4a. Then for every starting point $P \in \triangle MWI$, we have that $R_2(\Delta, P) = d(P, [AB, AC])$. The remaining cases of starting points within the hexagon separator $MZKYLW$ follow by symmetry.*

Lemma 6 (R_2 regions of a right isosceles triangle). *Consider right isosceles* $\Delta = \triangle ABC$, *as in Corollary 2, see Fig. 4b. Consider parabola with directrix the line passing through B that is perpendicular to BC (also the reflection of BC across AB) and focus A, passing through M, K and intersecting BL at point T (define also S as the symmetric point of T across AK). That parabola is the locus of points P for which* $\|PA\| = d(P, [AB, BC])$. *Let also A' be the reflection of A across BC. Consider parabola with directrix BA' and focus A, passing through T and intersecting AK at point U. That parabola is the locus of points P for which $\|PA\| = d(P, [BC, AB])$. Therefore, if P is a starting visitation point, we have that:*

– $R_2(\Delta, P) = \|PA\|$, *for all P in mixed closed shape $MTUSLQ$ (grey shape in Fig. 4b),*
– $R_2(\Delta, P) = d(P, [AB, BC])$, *for all P in mixed closed shape MFT (blue shape in Fig. 4b),*
– $R_2(\Delta, P) = d(P, [BC, AB])$, *for all P in mixed closed shape $FKUT$ (red shape in Fig. 4b).*

The visitation costs with starting points in the remaining subdivisions of the refined R_2 mixed-hexagon separator, green and purple regions in Fig. 4b, follow by symmetry.

4.3   Triangle Visitation with 1 Robot - The R_1 Regions

In this section we show how to partition the region of an arbitrary non-obtuse $\triangle ABC$ into sets of points P with respect to the optimal strategy of $R_1(P)$. There are 6 possible visitation strategies for $d(P, \{AB, AC, AB\})$, one for each permutation of the edges indicating the order they are visited (ordered visitations). Clearly, it is enough to describe, for each two ordered visitations, the borderline (separator) of points in which the two visitations have the same cost. By Lemma 1, any such ordered visitation cost is the distance of the starting point either to a point, or to a line, or a distance to a line plus the length of some altitude. Since the R_1 regions are determined by separators, i.e. loci of points in which different ordered visitations induce the same costs, it follows that these separators are either lines, or conic sections. Therefore, by exhaustively pairwise-comparing all ordered visitations along with their separators, we can determine the R_1 regions of any triangle. Next, we explicitly describe the R_1 regions only for three types of triangles that we will need for our main results. For the sake of avoiding redundancies, we omit any descriptions that are implied by symmetries.

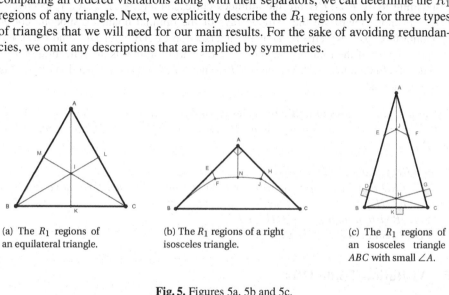

(a) The R_1 regions of an equilateral triangle.

(b) The R_1 regions of a right isosceles triangle.

(c) The R_1 regions of an isosceles triangle ABC with small $\angle A$.

Fig. 5. Figures 5a, 5b and 5c.

The next lemma describes the R_1 regions of an equilateral triangle, as in Fig. 5a.

Lemma 7 (R_1 regions of an equilateral triangle). *Consider equilateral triangle $\triangle ABC$ with angle bisectors AK, BL, CM and incenter I. Then, the angle bisectors are the loci of points in which optimal ordered visitations have the same cost. Moreover, for every starting point $P \in \triangle AMI$, the optimal strategy of $R_1(\triangle, P)$ is LRD visitation.*

The next lemma describes the R_1 regions of a right isosceles, as in Fig. 5b. Curve FJ is part of the parabola with directrix the relfection of BC across A and focus the reflection of A across BC. Curve BF is part of the parabola with directrix a line parallel to AB which is $\|AB\|$ away from AB, and focus the reflection of A across BC. Curve CJ is part of the parabola with directrix a line parallel to AC which is $\|AC\|$ away from AC, and focus the reflection of A across BC. CE (not shown in the figure) is the bisector of $\angle C$, and segment EF is part of the reflection of that bisector across AB.

BH is the bisector of $\angle C$, and segment HJ is part of the reflection of that bisector across AC. Segment AN is part of the altitude corresponding to A.

Lemma 8 (R_1 **regions of a right isosceles triangle**). *Consider right isosceles* $\Delta = \triangle ABC$, *and starting point* P. *Then, the optimal visitation strategy for* $R_1(\Delta, P)$ *is:*

– *an LRD visitation, if* $P \in AEFN$,
– *an LDR visitation if* $P \in BFE$, *and*
– *both an* DRL, DLR *visitation if* $P \in BCJF$ *(trajectory visits* $\{AB, AC\}$ *at point A).*

Next we consider a "thin" isosceles $\Delta = \triangle ABC$ with $\angle A \leq \pi/3$, as in Fig. 5c. (Eventually we will invoke the next lemma for $\angle A \to 0$.) AK is the altitude corresponding to A. CD, BG are the altitudes corresponding to AB, AC, respectively. CE, BF (not shown) are the extreme rays of the optimal bouncing subcone corresponding to C, B, respectively. H is the intersection of AK with BG (and CD), i.e. the orthocenter of the triangle. Segment EJ (as part of a line) is the reflection of EC (as part of a line) across AB. Segment FJ (as part of a line) is the reflection of BF (as part of a line) across AC.

Lemma 9 (R_1 **regions of a thin isosceles triangle**). *Consider isosceles* $\Delta = \triangle ABC$, *with* $\angle A \leq \pi/3$ *and starting point* P. *Then, the optimal visitation strategy for* $R_1(\Delta, P)$ *is:*

– *an LRD visitation if* $P \in AEJ$,
– *both LRD and LDR (optimal strategy is to visit first* AB *and then move to* C*), if* $P \in EDHJ$,
– *an LDR visitation, if* $P \in DBH$, *and*
– *a DLR visitation if* $P \in BKH$.

5   Visitation Trade-Offs

In this section we outline how we obtain our main results, as reported in Table 1. For this we invoke the lemmata we already established, along with the following claims (requiring lengthy and technical proofs) pertaining to optimal visitation costs of some special starting points. For the remaining of the section, we denote by I the incenter of $\triangle ABC$. All three following lemmata refer to non-obtuse $\triangle ABC$.

Lemma 10. *If* $\angle C$ *is the largest angle, then* $R_2(I) = \|IC\|$.

Lemma 11. *If* $\angle A$ *is the largest angle, then* $R_1(I) = \|IA'\|$, *where* A' *is the reflection of* A *across* BC.

Lemma 12. *Let* $\angle A \geq \angle B \geq \angle C$, *and* T *be the middle point of the altitude corresponding to the largest edge* BC. *Then the optimal* $R_1(T)$ *strategy is of* LRD *type, and has cost*

$$\frac{1}{2}(2 - \cos(2A)) \sin(B) \sin(C) \csc(B + C).$$

5.1   Searching with 1 vs 3 Robots

First we sketch the proof of $\sup_{\Delta \in \mathscr{D}} \mathscr{R}_{1,3}(\Delta) = 4$. The lower bound for $\sup_{\Delta \in \mathscr{D}} \mathscr{R}_{1,3}(\Delta)$ is given by the following lemma that utilizes Lemma 2 and Lemma 11.

Lemma 13. *Let Δ be an equilateral triangle. Then, $R_1(I)/R_3(I) = 4$.*

The remaining of the section is devoted in proving a tight upper bound for $\sup_{\Delta \in \mathscr{D}} \mathscr{R}_{1,3}(\Delta)$. Without loss of generality, we also assume that the starting point P lies within the tetragon (4-gon) $AMIL$, see also Fig. 2.

In order to provide the promised upper bound, we propose a heuristic upper bound for $R_1(P)$, as follows. Consider the projections P_1, P_2, P_3 of P onto AB, BC and CA respectively. Then, three (possibly) suboptimal visitation trajectories for one robot are $T_C(P) := \langle P, P_1, P, C, \rangle$, $T_A(P) := \langle P, P_2, P, A \rangle$, $T_B(P) := \langle P, P_3, P, B \rangle$, that is $R_1(P) \leq \min\{T_A(P), T_B(P), T_C(P)\}$. The upper bound proof follows by following lemma.

Lemma 14. *If $\angle A \leq \pi/3$, then $\min\{T_B(P), T_C(P)\}/R_3(P) \leq 4$. If $\angle A \geq \pi/3$, then $T_A(P)/R_3(P) \leq 4$.*

Next we outline how we obtain that $\inf_{\Delta \in \mathscr{D}} \mathscr{R}_{1,3}(\Delta) = \sqrt{10}$. First, using Lemma 2 and Lemma 9 we show that $\inf_{\Delta \in \mathscr{D}} \mathscr{R}_{1,3}(\Delta) \leq \sqrt{10}$.

Lemma 15. *For isosceles ABC with base BC, we have $\lim_{\angle A \to 0} \max_{P \in ABC} \frac{R_1(P)}{R_3(P)} = \sqrt{10}$.*

Next, we invoke Lemma 2 and Lemma 11 in order to show that $\inf_{\Delta \in \mathscr{D}} \mathscr{R}_{1,3}(\Delta) \geq \sqrt{10}$.

Lemma 16. *For any triangle $\Delta \in \mathscr{D}$ we have $R_1(I)/R_3(I) \geq \sqrt{10}$.*

5.2   Searching with 2 vs 3 Robots

First we outline the proof of that $\sup_{\Delta \in \mathscr{D}} \mathscr{R}_{2,3}(\Delta) = 2$. For this, and using Lemma 2 and Lemma 10, we establish that $\sup_{\Delta \in \mathscr{D}} \mathscr{R}_{2,3}(\Delta) \geq 2$.

Lemma 17. *For the equilateral triangle we have $R_2(I)/R_3(I) = 2$.*

The remaining of the section is devoted in proving that $\sup_{\Delta \in \mathscr{D}} \mathscr{R}_{2,3}(\Delta) \leq 2$. In that direction, we consider a triangle $\Delta = ABC$ along with its incenter I, see also Fig. 2.

In order to provide the promised upper bound, we propose a heuristic upper bound for $R_2(P)$. The two robots visit all edges as follows; one robot goes to the vertex corresponding to the largest angle (visiting the two incident edges), and the second robot visits the remaining edge moving along the projection of P along that edge. The heuristic is used the to show the following.

Lemma 18. *For any $\Delta \in \mathscr{D}$ and starting point P, we have $R_2(P)/R_3(P) \leq 2$.*

Next we outline how we prove that $\inf_{\Delta \in \mathscr{D}} \mathscr{R}_{2,3}(\Delta) = \sqrt{2}$. Using Lemma 2 and Lemma 6 we can show that $\inf_{\Delta \in \mathscr{D}} \mathscr{R}_{2,3}(\Delta) \le \sqrt{2}$.

Lemma 19. *Let* $\triangle ABC$ *be a right isosceles. Then, we have* $\max_{P \in ABC} \frac{R_2(P)}{R_3(P)} = \sqrt{2}$.

Then, by invoking Lemma 10 we show that $\inf_{\Delta \in \mathscr{D}} \mathscr{R}_{2,3}(\Delta) \ge \sqrt{2}$.

Lemma 20. *For any* $\Delta \in \mathscr{D}$, *we have* $R_2(I)/R_3(I) \ge \sqrt{2}$.

5.3 Searching with 1 vs 2 Robots

Finally, we outline how we prove that $\sup_{\Delta \in \mathscr{D}} \mathscr{R}_{1,2}(\Delta) = 3$. Using a simple heuristic upper bound for R_1, we can show the following.

Lemma 21. *For any* $\Delta \in \mathscr{D}$ *and any starting point* $P \in \Delta$, *we have* $R_1(P)/R_2(P) \le 3$.

The lower bound for $\sup_{\Delta \in \mathscr{D}} \mathscr{R}_{1,2}(\Delta)$ is attained for the right isosceles triangle (and for certain starting point). Indeed, using Lemma 12 we show that $\sup_{\Delta \in \mathscr{D}} \mathscr{R}_{1,2}(\Delta) \ge 3$.

Lemma 22. *Let* ABC *be a right isosceles triangle with right angle* A. *Let also* P *be the middle point of the altitude corresponding to angle* A. *Then,* $R_1(P)/R_2(P) = 3$.

It remains to sketch the proof of $\inf_{\Delta \in \mathscr{D}} \mathscr{R}_{1,2}(\Delta) = 5/2$. For this, using Lemma 5 and Lemma 7 we prove that $\inf_{\Delta \in \mathscr{D}} \mathscr{R}_{1,2}(\Delta) \le 5/2$.

Lemma 23. *For the equilateral triangle* Δ, *we have* $\max_{P \in \Delta} R_1(P)/R_2(P) = 5/2$.

Then, using Lemma 12, we prove that $\inf_{\Delta \in \mathscr{D}} \mathscr{R}_{1,2}(\Delta) \ge 5/2$.

Lemma 24. *For any* $\triangle ABC \in \mathscr{D}$, *let* T *be the middle point of the altitude corresponding to the largest edge. Then, we have* $R_1(T)/R_2(T) \ge 5/2$.

6 Conclusions

We considered a new vehicle routing-type problem in which (fleets of) robots visit all edges of a triangle. We proved tight bounds regarding visitation trade-offs with respect to the size of the available fleet. In order to avoid degenerate cases of visiting the edges with 3 robots, we only focused our study on non-obtuse triangles. The case of arbitrary triangles, as well as of other topologies, e.g. graphs, remains open. We believe the definition of our problem is of independent interest, and that the study of efficiency trade-offs in combinatorial problems with respect to the number of available processors (that may not be constant as in our case), e.g. vehicle routing type problems, will lead to new, deep and interesting questions.

References

1. Acharjee, S., Georgiou, K., Kundu, S., Srinivasan, A.: Lower bounds for shoreline searching with 2 or more robots. In: Felber, P., Friedman, R., Gilbert, S., Miller, A. (eds.) 23rd International Conference on Principles of Distributed Systems (OPODIS 2019). LIPIcs, vol. 153, pp. 26:1–26:11. Schloss Dagstuhl - Leibniz-Zentrum fur Informatik (2019)
2. Arora, S.: Polynomial time approximation schemes for Euclidean traveling salesman and other geometric problems. J. ACM **45**(5), 753–782 (1998)
3. Baeza-Yates, R., Schott, R.: Parallel searching in the plane. Comput. Geom. **5**(3), 143–154 (1995)
4. Baeza-Yates, R.A., Culberson, J.C., Rawlins, G.J.E.: Searching with uncertainty extended abstract. In: Karlsson, R., Lingas, A. (eds.) SWAT 1988. LNCS, vol. 318, pp. 176–189. Springer, Heidelberg (1988). https://doi.org/10.1007/3-540-19487-8_20
5. Bektas, T.: The multiple traveling salesman problem: an overview of formulations and solution procedures. Omega **34**(3), 209–219 (2006)
6. Dantzig, G.B., Ramser, J.H.: The truck dispatching problem. Manag. Sci. **6**(1), 80–91 (1959)
7. Das, A., Mathieu, C.: A quasipolynomial time approximation scheme for Euclidean capacitated vehicle routing. Algorithmica **73**(1), 115–142 (2015)
8. Richa, A.W., Scheideler, C. (eds.): SIROCCO 2020. LNCS, vol. 12156. Springer, Cham (2020). https://doi.org/10.1007/978-3-030-54921-3
9. Dumitrescu, A., Tóth, C.D.: The traveling salesman problem for lines, balls, and planes. ACM Trans. Algorithms **12**(3), 43:1–43:29 (2016)
10. Georgiou, K., Kundu, S., Pralat, P.: Makespan trade-offs for visiting triangle edges. CoRR, abs/2105.01191 (2021)
11. Kumar, S.N., Panneerselvam, R.: A survey on the vehicle routing problem and its variants. Intell. Inf. Manag. **4**(3), 66–74 (2012)
12. Laporte, G.: The vehicle routing problem: an overview of exact and approximate algorithms. Eur. J. Oper. Res. **59**, 345–358 (1992)
13. Mor, A., Speranza, M.G.: Vehicle routing problems over time: a survey. 4OR, pp. 1–21 (2020)
14. Ritzinger, U., Puchinger, J., Hartl, R.F.: A survey on dynamic and stochastic vehicle routing problems. Int. J. Prod. Res. **54**(1), 215–231 (2016)
15. Tóth, P., Vigo, D.: The vehicle routing problem. monographs. Discrete Mathematics and Applications, Society for Industrial and Applied Mathematics, Philadelphia (2002)

Augmenting a Tree to a k-Arbor-Connected Graph with Pagenumber k

Toru Hasunuma[✉][iD]

Department of Mathematical Sciences, Tokushima University, 2–1 Minamijosanjima,
Tokushima 770–8506, Japan
hasunuma@tokushima-u.ac.jp

Abstract. A tree is one of the most fundamental structures of networks and has good properties on layouts, while it is weak from a fault-tolerant point of view. Motivated by these points of view, we consider an augmentation problem for a tree to increase fault-tolerance while preserving its good properties on book-embeddings. A k-arbor-connected graph is defined to be a graph which has k spanning trees such that for any two vertices, the k paths between them in the spanning trees are pairwise edge-disjoint and internally vertex-disjoint. A minimally k-arbor-connected graph is a k-arbor-connected graph G such that deleting any edge from G does not preserve k-arbor-connectedness. A k-arbor-connected graph has the abilities to execute fault-tolerant broadcastings and protection routings as a communication network. The pagenumber of a graph is the minimum number of pages required for a book-embedding of the graph. We show that for any tree T of order n and for any k at most the radius of T, by adding new edges to T, a minimally k-arbor-connected graph T^* with pagenumber k can be obtained in $O(kn)$ time. Since any k-arbor-connected graph cannot be embedded in $k - 1$ pages, T^* is optimal with respect to not only the number of edges added to T but also the number of pages required for a book-embedding. We also show that the restriction on the upper bound on k can be removed if T is a caterpillar. Besides, we show that for $k \leq 3$ and for any tree T of order at least $2k$, a minimally k-arbor-connected graph with pagenumber k which contains T as a subgraph can be obtained in linear time. We moreover extend our result for a tree to a cactus for k greater than half of the maximum length of a cycle in the cactus, and to a unicyclic graph for any k at most the radius of the graph.

Keywords: Arbor-connectedness · Augmentation · Cactus · Pagenumber · Trees

1 Introduction

Throughout the paper, a graph means a simple undirected graph. Let $G = (V, E)$ be a graph. The order of G is $|V(G)|$. The *complement* $\overline{G}$ of G is the graph with vertex set $V(G)$ such that $uv \in E(\overline{G})$ if and only if $uv \notin E(G)$. For a subset E' of $E(\overline{G})$, we say that G is *augmented* to the graph $G' = (V, E \cup E')$ by adding every edge in E'. Given a condition on graphs, an *augmentation problem* for a graph G is to find a minimum set $E'' \subseteq E(\overline{G})$ such that the augmented graph $G'' = (V, E \cup E'')$ satisfies the condition.

© Springer Nature Switzerland AG 2021
P. Flocchini and L. Moura (Eds.): IWOCA 2021, LNCS 12757, pp. 356–369, 2021.
https://doi.org/10.1007/978-3-030-79987-8_25

For $S \subsetneq V(G)$ (respectively, $F \subseteq E(G)$), $G - S$ (respectively, $G - F$) denotes the graph obtained from G by deleting every element of S (respectively, F), where $G - \{a\}$ is abbreviated to $G - a$. A graph G is k-*connected* (respectively, k-*edge-connected*) if for any $S \subsetneq V(G)$ (respectively, $F \subseteq E(G)$) with $|S| < k < |V(G)|$ (respectively, $|F| < k$), $G - S$ (respectively, $G - F$) is connected. Since the notion of connectedness is fundamental and naturally has applications in fault-tolerance of networks, there are many augmentation results for the subject until now. In particular, augmenting a graph to be k-connected (respectively, k-edge-connected) was solved in [12] (respectively, [21]) for every fixed k. For any k, augmenting a k-connected graph to be $(k + 1)$-connected was shown to be polynomially solvable in [20]. Apart from connectedness, geometric properties are also fundamental in the study of graphs, and in fact there are many results on connectivity augmentation of graphs with geometric constraints (see [11]). In particular, Kant and Bodlaender [14] have shown that the problem of augmenting a connected planar graph to be 2-connected while preserving the planarity is NP-hard. Besides, Kant [13] has shown such a problem can be solved in linear time if we restrict ourselves to outerplanar graphs. Rutter and Wolff [19] have also proved that the problem of augmenting a connected planar graph to be 2-edge-connected while preserving the planarity is NP-hard. For a connected plane geometric graph G, Abellanas et al. [1] have studied bounds on the number of edges required to be added to G to obtain 2-connected or 2-edge-connected plane geometric graphs.

Concerning fault-tolerance of networks, the notion of completely independent spanning trees is known. Let $T_1, T_2, \ldots, T_k$ be spanning trees in a connected graph G. If for any two vertices of G, the paths between them in $T_1, T_2, \ldots, T_k$ are pairwise edge-disjoint and internally vertex-disjoint, then $T_1, T_2, \ldots, T_k$ are *completely independent spanning trees* in G. Completely independent spanning trees can be applied to fault-tolerant broadcastings [8] and protection routings [17]. A graph G is k-*arbor-connected* if there exist k completely independent spanning trees in G. A *minimally k-arbor-connected graph* is a k-arbor-connected graph G such that for any $e \in E(G)$, $G - e$ is not k-arbor-connected. Figure 1 illustrates an example of a 2-arbor-connected graph. Completely independent spanning trees $T_1, T_2, \ldots, T_k$ in G can be characterized as edge-disjoint spanning trees such that for any $v \in V(G)$, v has degree at least two in at most one spanning tree T_i. Thus, for any given spanning trees, we do not have to check the paths between every pair of vertices in order to confirm that the trees are completely independent. Arbor-connectedness of graphs has been studied for graph classes related to interconnection networks (e.g., see [5, 7, 16]). It has also been shown that every max-

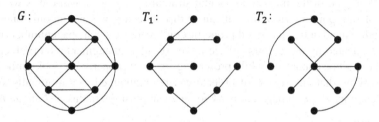

Fig. 1. A 2-arbor-connected graph G and completely independent spanning trees T_1 and T_2 in G.

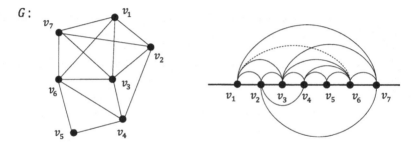

Fig. 2. A 3-page book-embedding of a graph G, where a line above the spine, a line below the spine, and the dotted line denote an edge assigned to the first page, the second page, and the third page, respectively. Since G is not planar, the pagenumber of G is determined to be 3.

imal 4-connected planar graph is 2-arbor-connected [9] and G is $\lfloor \frac{n}{k} \rfloor$-arbor-connected if the minimum degree of G is at least $n - k$, where $3 \leq k \leq \frac{n}{2}$ [10]. Although any k-arbor-connected graph is k-connected, it has been proved in [18] that for any $k \geq 2$, there exists a k-connected graph which is not 2-arbor-connected. This fact is in contrast to the theorem by Nash-Williams that any $2k$-edge-connected graph contains k edge-disjoint spanning trees. From an algorithmic point of view, it has been shown that the problem of deciding whether a given graph is 2-arbor-connected is NP-complete [9].

A *book* is a structure consisting of a line called the *spine* and half planes called *pages* sharing the spine as a common boundary. A *k-page book-embedding* of a graph G is defined by a placement of the vertices of G on the spine, i.e., a vertex-ordering σ of $V(G)$, and an assignment of the edges of G to pages so that no two edges assigned to the same page cross, where two edges uv and xy cross under σ if $\sigma(u) < \sigma(x) < \sigma(v) < \sigma(y)$. The *pagenumber* pn$(G)$ of G is the minimum number of pages for a book-embedding of G. Figure 2 illustrates an example of a 3-page book-embedding of a graph. Book-embeddings have applications in fault-tolerant VLSI layouts [6], graph layout, and other areas and there are many results on the subject until now. In particular, it has been shown in [4] that a graph G can be embedded in one page (respectively, two pages) if and only if G is outerplanar (respectively, a subgraph of a planar Hamiltonian graph). Besides, Yannakakis [23,24] proved that every planar graph can be embedded in four pages and there are planar graphs that require four pages in any book embedding[1]. The problem of deciding whether the pagenumber of a given planar graph is two is NP-complete [6,22].

A tree is one of the most fundamental structures of graphs, while it is weak from a fault-tolerant point of view since it can be disconnected by deleting only one vertex. On the other hand, a tree has good properties on layouts, e.g., the pagenumber of a tree is one. Motivated by these points of view, we consider an augmentation problem for a tree to be k-arbor-connected while preserving its good property on book-embeddings. We then show that any tree T of order n can be augmented to a minimally k-arbor-connected graph T^* with pagenumber k for any k at most the radius of T in $O(kn)$ time.

---

[1] After more than 30 years, the full details of the constructions of planar graphs that need four pages were presented in [25]. The lower bound of four was also proved independently in [3].

From the fact that any graph with pagenumber one is outerplanar, it follows that every graph of order n with m edges needs at least $\lceil \frac{m-n}{n-3} \rceil$ pages for its book-embedding. This means that any k-arbor-connected graph cannot be embedded in $k-1$ pages, i.e., the pagenumber of any k-arbor-connected graph is at least k. Thus, the augmented graph T^* is optimal with respect to not only the number of edges added to T but also the number of pages required for a book-embedding. The graph T^* also has a property that T^* is decomposed into completely independent spanning trees $T_1, T_2, \ldots, T_k$ such that each T_i can be embedded in one page under the same vertex-ordering. We also show that the restriction on the upper bound on k can be removed if T is a caterpillar. Besides, we show that for $k \leq 3$, any tree T of order at least $2k$ can be augmented to a minimally k-arbor-connected graph with pagenumber k in linear time. We moreover extend our result for a tree to a cactus for k greater than half of the maximum length of a cycle in the cactus, and to a unicyclic graph for any k at most the radius of the graph.

This paper is organized as follows. Section 2 presents terminology and fundamental results used in the paper. Our main augmentation result for trees is given in Sect. 3. Section 4 presents augmentation results for connected graphs for large k, caterpillars, and trees for $k \leq 3$. Sections 5 and 6 extend the result in Sect. 3 to cacti and unicyclic graphs, respectively. Section 7 concludes the paper with several remarks.

2 Preliminaries

For two sets A and B, $A \setminus B$ denotes the set difference $\{x \mid x \in A, x \notin B\}$. For a subset $S \subseteq V(G)$, the subgraph of G induced by S is denoted by $\langle S \rangle_G$, i.e., $\langle S \rangle_G = G - (V(G) \setminus S)$. Given a set F of edges, the graph induced by F is denoted by $\langle F \rangle$, i.e., $V(\langle F \rangle) = \{u \mid uv \in F\}$ and $E(\langle F \rangle) = F$. For a graph G and $F' \subseteq E(\overline{G})$, $G + F'$ denotes the graph $(V(G), E(G) \cup F')$. A *leaf* of G is a vertex with degree one, while an *internal vertex* of G is a vertex with degree greater than one. The set of leaves (respectively, internal vertices) in G is denoted by $V_L(G)$ (respectively, $V_I(G)$). Let $n'(G) = |V_I(G)|$.

In order to augment a tree to a minimally k-arbor-connected graph, we use the following characterization of completely independent spanning trees.

Theorem 1. [8] *Spanning trees $T_1, T_2, \ldots, T_k$ in G are completely independent if and only if for any $1 \leq i < j \leq k$, $E(T_i) \cap E(T_j) = \emptyset$ and $V_I(T_i) \cap V_I(T_j) = \emptyset$.*

The *distance* $d_G(u, v)$ of vertices u and v in a connected graph G is the length of a shortest path between u and v in G. The *eccentricity* $e_G(w)$ of a vertex w in G is defined to be $\max_{v \in V(G)} d_G(w, v)$. The *diameter* $\mathrm{diam}(G)$ of G is $\max_{w \in V(G)} e_G(w)$, while the *radius* $\mathrm{rad}(G)$ of G is $\min_{w \in V(G)} e_G(w)$. A *central vertex* of G is a vertex v with $e_G(v) = \mathrm{rad}(G)$. The *center* of G is the set of central vertices of G. The center of a tree T consists of one vertex (respectively, two vertices) if $\mathrm{diam}(T)$ is even (respectively, odd). A *star* (respectively, *double-star*) is a tree T with $n'(T) \leq 1$ (respectively, $n'(T) = 2$). A *caterpillar* is a tree T such that $T - V_L(T)$ is a path.

The complete graph of order n is denoted by K_n. Let $K_{p,q}$ denote the complete bipartite graph with partite sets of cardinalities p and q. For convenience, one of p and q may be zero. In such a case, the corresponding partite set is considered as an empty set and $K_{p,q}$ is regarded as an empty graph; namely, $K_{p,0} \cong \overline{K_p}$ and $K_{0,q} \cong \overline{K_q}$. A *cut-vertex* of G is a vertex whose deletion increases the number of components of G. A *block* of G is a maximal connected subgraph of G without a cut-vertex. A *cactus* is a graph whose every block is either a cycle or K_2. A cactus in which every cycle is a triangle is called a *triangular cactus*. A *unicyclic graph* is a graph with exactly one cycle. A *quasi-cycle* is a unicyclic graph G such that $G - V_L(G)$ is a cycle.

Let T be a tree rooted at a vertex r. The *ℓ-ancestor* $p_\ell(v)$ of a vertex v in T is a vertex w which is on the path from r to v such that $d_T(v, w) = \ell$. If w is the ℓ-ancestor of v, then v is an *ℓ-descendant* of w. In particular, the 1-ancestor of v is the parent of v and a 1-descendant of w is a child of w. The set of ℓ-descendants of w is denoted by $D_\ell(w)$. The *lowest common ancestor* $\text{lca}_T(u, v)$ of u and v in T is a common ancestor w of u and v in T such that there is no descendant of w which is a common ancestor of u and v. The *height* $h(T)$ of T is $\max_{v \in V(T)} d_T(r, v)$.

Let σ be a vertex-ordering of G, i.e., a bijection from $V(G)$ to $\{1, 2, \ldots, |V(G)|\}$. When $\sigma(u) < \sigma(v)$, we may write $u <_\sigma v$. For $u, v \in S \subseteq V(G)$, if $u <_\sigma v$ such that there is no vertex $w \in S$ with $u <_\sigma w <_\sigma v$, then u and v are *consecutive* in S under σ and we write $u <_{\sigma,S} v$. When $S = V(G)$, we may write $u <_\sigma v$. For any $u, v \in S$, if there is no vertex $w' \in V(G) \setminus S$ with $u <_\sigma w' <_\sigma v$, then the vertices in S are consecutive under σ.

3   Augmenting a Tree

Theorem 2. *Any tree T of order n can be augmented to a minimally k-arbor-connected graph T^* with pagenumber k for any $2 \leq k \leq \text{rad}(T)$ in $O(kn)$ time.*

Proof. If $\text{diam}(T)$ is odd, then let x and y be the central vertices of T. Note that $xy \in E(T)$. Otherwise, let x be the central vertex of T and let y a vertex adjacent to x such that y is on a path between x and a vertex v with $d_T(x, v) = \text{rad}(T)$. Let T^+ be the tree obtained from T by adding a new vertex z, joining it to x and y, and deleting the edge xy. In what follows, ancestors and descendants of a vertex are defined based on T^+ rooted at z unless otherwise stated. For any vertex u in T, T_u denotes the subtree rooted at u in T^+. By the definitions of x and y, it holds that $h(T_x) = \text{rad}(T) \geq h(T_y) \geq \text{rad}(T) - 1$.

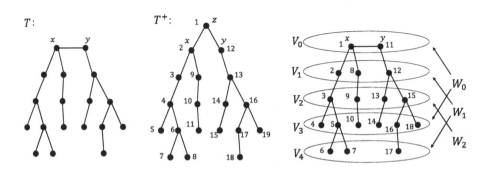

Fig. 3. Constructions of $T^+, \sigma^+, \sigma, V_i$, and W_i for $k = 3$.

Let $\sigma^+ : V(T^+) \mapsto \{1, 2, \ldots, n + 1\}$ be a depth-first search ordering of T^+ from z. Then, let $\sigma : V(T) \mapsto \{1, 2, \ldots, n\}$ be the vertex-ordering of T defined to be $\sigma(v) = \sigma^+(v) - 1$. Now we divide $V(T)$ into $\mathrm{rad}(T) + 1$ subsets based on the distance of a vertex and the root z. Namely, let $V_i = D_{i+1}(z)$ for $0 \leq i \leq \mathrm{rad}(T)$. Note that $|V_i \cap V(T_x)| \geq 1$ and $|V_i \cap V(T_y)| \geq 1$ for any $0 \leq i < \mathrm{rad}(T)$. Let $W_t = \bigcup_{i \bmod k = t} V_i$ for each $0 \leq t < k$. Figure 3 illustrates an example of T, T^+, σ^+, σ, V_i, and W_i for $k = 3$.

We next divide $E(T) \setminus \{xy\}$ into k subsets $E_1, E_2, \ldots, E_k$ defined as follows: for each $1 \leq i \leq k$,

– $E_i = \{vw \mid v \in W_{i-1},\ w \in D_1(v)\}$.

The set of added edges in our augmentation is divided into three types defined as follows: for each $1 \leq i \leq k$,

– $A_i = \{vw \mid v \in W_{i-1},\ w \in D_j(v),\ 2 \leq j \leq k\}$,
– $B_i = \{uw \mid u, v \in V_{i-1},\ u <_{\sigma, V_{i-1}} v,\ \sigma^{-1}(\max_{u' \in V(T_u)} \sigma(u')) <_\sigma w \leq_\sigma v\}$,
– $B'_i = \{uw \mid u = \sigma^{-1}(\max_{u' \in V(T_u)} \sigma(u')),\ w <_\sigma \sigma^{-1}(\min_{u' \in V_{i-1}} \sigma(u'))$ or $\sigma^{-1}(\max_{u' \in V(T_u)} \sigma(u')) <_\sigma w\}$.

Note that $B_1 = \{xy\}$. Thus, $T = \langle E_1 \cup E_2 \cup \cdots E_k \cup B_1 \rangle$. Based on these sets, we define T_i as $\langle E_i \cup A_i \cup B_i \cup B'_i \rangle$ for $1 \leq i \leq k$. We then show that $T_1, T_2, \ldots, T_k$ are completely independent spanning trees in $T^* = \langle E(T_1) \cup E(T_2) \cup \cdots \cup E(T_k) \rangle$ such that each T_i can be embedded in one page under σ, which implies that the augmented graph T^* is a minimally k-arbor-connected graph with pagenumber k which contains T.

The graph $\langle E_i \rangle$ is a disjoint union of trees with height 1 whose central vertices are in W_{i-1}. The augmented graph $\langle E_i \cup A_i \rangle$ is a disjoint union of $|V_{i-1}|$ trees, each of which is obtained from the trees with height 1 in $\langle E_i \rangle$ by joining each vertex in W_{i-1} and all its ℓ-descendants for $2 \leq \ell \leq k$. Thus, $V(\langle E_i \cup A_i \rangle) = V(T) \setminus \bigcup_{0 \leq j < i-1} V_j$. The $|V_{i-1}|$ trees are connected by the edges uv for $u <_{\sigma, V_{i-1}} v$ in B_i, and moreover all the vertices in $\bigcup_{0 \leq j < i-1} V_j$ are joined to a vertex in V_{i-1} by other edges in $B_i \cup B'_i$. Therefore, $\langle E_i \cup A_i \cup B_i \cup B'_i \rangle$ is a tree with vertex set $V(T)$. Note that any edge in $B_i \cup B'_i$ joins a vertex w in $\bigcup_{0 \leq j < i-1} V_j$ and a vertex in V_{i-1} which is not a descendant of w. In each T_i, every vertex in $V(T) \setminus W_{i-1}$ is directly joined to a vertex in W_{i-1} which means that every vertex in $V(T) \setminus W_{i-1}$ is a leaf

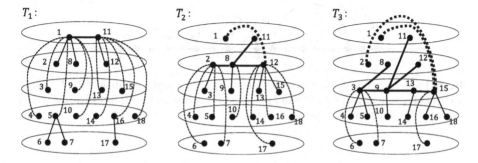

Fig. 4. Constructions of T_1, T_2, and T_3, where normal lines, dotted lines, bold lines, and bold dotted lines denote the edges in E_i, A_i, B_i, and B'_i, respectively.

of T_i and $V_l(T_i) \subseteq W_{i-1}$. Since $W_i \cap W_j = \emptyset$ for any $0 \le i < j < k$, $V_l(T_i) \cap V_l(T_j) = \emptyset$ for any $1 \le i < j \le k$. Now assume that $e \in E(T_i) \cap E(T_j)$ for some $i < j$. Then, e is incident to a vertex u in W_{i-1} and a vertex v in W_{j-1}. If $uv \in B_i \cup B_i'$, then $u \in V_{i-1}$ and v must be in V_ℓ where $0 \le \ell < i$ which is a contradiction. Thus, $uv \in A_i$ such that $u \in V_{kt+i-1}$, $v \in V_{kt+j-1}$ for some $t \ge 0$. This means that u is an ancestor of v. However, no ancestor of v is joined to v as a leaf of T_j. Therefore, $E(T_i) \cap E(T_j) = \emptyset$ for any $1 \le i < j \le k$. Consequently, $T_1, T_2, \ldots, T_k$ are completely independent spanning trees. Figure 4 illustrates T_1, T_2, and T_3 for the example in Fig. 3.

The graph $\langle E_i \cup A_i \rangle$ is a disjoint union of $|V_{i-1}|$ trees $S_1, S_2, \ldots, S_{|V_{i-1}|}$ such that the vertex set of each S_i corresponds to the vertex set of a subtree rooted at a vertex in V_{i-1}. From a property of a depth-first search, for any subtree S_i, the vertices in $V(S_i)$ are consecutive under σ. Thus, it can be inductively shown (on the height) that S_i can be embedded in one page under σ. No vertex in $\cup_{0 \le j < i-1} V_j$ is placed between two vertices of S_t for each t. Thus, any edge in $B_i \cup B_i'$ and any edge in $E_i \cup A_i$ do not cross. Besides, $\langle B_i \cup B_i' \rangle$ is a union of stars in which any internal vertex is in V_{i-1}. From the definitions of B_i and B_i', it follows that $\langle B_i \cup B_i' \rangle$ can be embedded in one page under σ. Therefore, each tree T_i can be embedded in one page under the same vertex-ordering σ. Figure 5 illustrates the 3-page book-embedding of $T^* = \langle E(T_1) \cup E(T_2) \cup E(T_3) \rangle$ for the trees T_1, T_2, T_3 in Fig. 4.

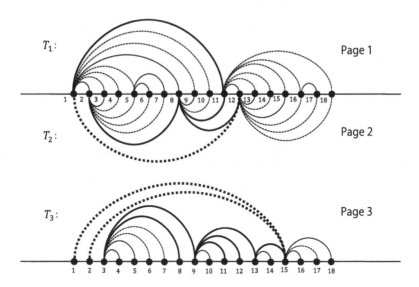

Fig. 5. The 3-page book-embedding of the 3-arbor-connected graph consisting of T_1, T_2, and T_3 shown in Fig. 4, where all the edges in T_1, T_2, and T_3 are assigned to the first page, the second page, and the third page, respectively.

The vertices x and y are on a longest path in T. Thus, they can be computed in linear time by applying a breadth-first search twice. The vertex ordering σ follows from σ^+ which is obtained by applying a depth-first search to T^+ from z. In the depth-first search, $p_1(v)$ and $\sigma^{-1}(\max_{u' \in V(T_v)} \sigma(u'))$ can also be found for each vertex v. Applying a breadth-first search from z, we can partition $V(T)$ into $V_0, V_1, \ldots, V_{\mathrm{rad}(T)}$. Thus, V_{i-1} and W_{i-1} for $1 \le i \le k$ can be found in $O(n)$ time. Based on $V_0, V_1, \ldots, V_{k-1}$, and σ, the ordering relation $<_{\sigma, V_{i-1}}$, $\sigma^{-1}(\min_{u' \in V_{i-1}} \sigma(u'))$, and $\sigma^{-1}(\max_{u' \in V_{i-1}} \sigma(u'))$ for $1 \le i \le k$ can be computed in $O(n)$ time. Thus, B_i and B'_i for $1 \le i \le k$ can be computed in $O(kn)$ time. Here, E_i and A_i can be rewritten as follows:

- $E_i = \{p_1(v)v \mid v \notin \{x, y\}, v \in W_{i \bmod k}\}$,
- $A_i = \{p_j(v)v \mid v \in V(T), \ p_j(v) \in W_{i-1}, \ 2 \le j \le k\}$.

For each vertex v, $p_j(v)$ for $1 \le i \le k$ can be found in $O(k)$ time. Therefore, E_i and A_i for $1 \le i \le k$ can be computed in $O(kn)$ time. Consequently, $T_1, T_2, \ldots, T_k$ can be found in $O(kn)$ time.   □

For a path of order n, the upper bound on k in Theorem 2 is $\lfloor \frac{n}{2} \rfloor$. Note that the complete graph of order n is $\lfloor \frac{n}{2} \rfloor$-arbor-connected; namely, there is no $(\lfloor \frac{n}{2} \rfloor + 1)$-arbor-connected graph of order n.

We here remark that in the proof of Theorem 2, other constructions can be employed if we do not insist on the upper bound on k. Select a path P with $|V(P)| \ge 3$ and consider the $|V(P)|$ subtrees each of which is rooted at a vertex in P (instead of two subtrees T_x and T_y). Then, we can construct a minimally k-arbor-connected graph where k is at most the maximum j such that there exist two vertices in P both of which have a $(j-1)$-descendant. In fact, we employ such a construction to prove Lemma 3 in Sect. 4 and Theorem 6 in Sect. 6.

4   Augmenting a Connected Graph for Large k

In this section, we first introduce a special k-arbor-connected graph. Based on the graph, we present a general augmentation result for large k.

Definition 1. Let k and ℓ be integers such that $1 \le k \le \ell \le \frac{n}{2}$. Let $G_{k,\ell,n}$ be the graph of order n obtained from two complete bipartite graphs $K_{k,\ell-k}$ with partite sets $V_k = \{v_0, v_1, \ldots, v_{k-1}\}$, $W_{\ell-k} = \{v_k, v_{k+1}, \ldots, v_{\ell-1}\}$ and $K_{k,n-\ell-k}$ with partite sets $V'_k = \{v_\ell, v_{\ell+1}, \ldots, v_{\ell+k-1}\}$, $W'_{n-\ell-k} = \{v_{\ell+k}, v_{\ell+k+1}, \ldots, v_{n-1}\}$ by connecting every pair of vertices in $V_k \cup V'_k$.

Lemma 1. The graph $G_{k,\ell,n}$ is a minimally k-arbor-connected graph with pagenumber k for any $2 \le k \le \frac{n}{2}$.

Proof. Let T be a tree of order n with vertex set $\{v_0, v_1, \ldots, v_{n-1}\}$ and edge set $\{v_0 v_i \mid 1 \le i \le \ell\} \cup \{v_\ell v_j \mid \ell + 1 \le j < n\}$. Namely, T is a double-star. Define the vertex ordering σ as $v_0 <_\sigma v_1 <_\sigma \cdots <_\sigma v_{n-1}$. Clearly, T can be embedded in one page under σ. Define $T_p = \langle \{v_{(i+p) \bmod n} v_{(j+p) \bmod n} \mid v_i v_j \in E(T)\} \rangle$ for $0 \le p < k$. Then, we can see that $\langle E(T_0) \cup E(T_1) \cup \cdots \cup E(T_{k-1}) \rangle = G_{k,\ell,n}$. (To see this, it is convenient to draw the spine as a circle and the edges inside the circle.) It can be checked that $T_0, T_1, \ldots, T_{k-1}$ are edge-disjoint and each T_i can be embedded in one page under σ. Since $V_I(T_i) = \{v_i, v_{\ell+i}\}$ for $0 \leq i < k$, $T_0, T_1, \ldots, T_{k-1}$ are completely independent. Therefore, $G_{k,\ell,n}$ is a minimally k-arbor-connected graph with pagenumber k. $\square$

Lemma 2. *Let G be a connected graph of order n. If $n'(G) \leq 2k \leq n$, then a positive integer q such that $G_{k,q,n}$ contains G' isomorphic to G can be found in linear time.*

Proof. If $n'(G) = 0$, then $n = 2$ and G is isomorphic to $G_{1,1,2}$. Suppose that $n'(G) \geq 1$. Partition $V_I(G)$ into two subsets X and X' so that $|X| \leq |X'| \leq |X| + 1$. If $n'(G)$ is odd, then we select a leaf u of G and add it to X. Let $r = \lceil \frac{n'(G)}{2} \rceil = |X| = |X'|$. Let Y and Y' be the sets of leaves (except for u) adjacent to a vertex in X and X', respectively. Let $s = |Y|$ and $s' = |Y'|$. Without loss of generality, we may assume that $s \leq s'$. By letting X, Y, X', and Y' correspond to V_r, W_s, V'_r, and $W'_{s'}$ of $G_{r,r+s,n}$, respectively, we can see that $G_{r,r+s,n}$ contains G' isomorphic to G. By definition, it holds that $G_{r,r+s,n} \subset G_{r+1,r+s,n} \subset \cdots \subset G_{r+s,r+s,n} \subset G_{r+s+1,r+s+1,n} \subset \cdots \subset G_{\lfloor \frac{n}{2} \rfloor, \lfloor \frac{n}{2} \rfloor, n}$. Since $n'(G) \leq 2k$, $r \leq k$. Thus, $G_{k,q,n}$ contains G' isomorphic to G for $q = r + s$ (respectively, $q = k$) if $k \leq r + s$ (respectively, $k > r + s$). Since r and s can be computed in linear time, we have q in linear time. $\square$

From Lemmas 1 and 2, we have the following theorem.

Theorem 3. *Any connected graph G of order n can be augmented to a minimally k-arbor-connected graph with pagenumber k for any $\frac{n'(G)}{2} \leq k \leq \frac{n}{2}$ in $O(kn)$ time.*

Note that the restriction on k in Theorem 3 is the lower bound, while that in Theorem 2 is the upper bound. The gap in the ranges of k in Theorems 2 and 3 disappears when $\frac{n'(T)}{2} \leq \operatorname{rad}(T) + 1$. For any caterpillar T, it holds that $\operatorname{rad}(T) = \lfloor \frac{n'(T)}{2} \rfloor + 1$. Thus, the following corollary is obtained.

Corollary 1. *Any caterpillar of order n can be augmented to a minimally k-arbor-connected graph with pagenumber k for any $2 \leq k \leq \frac{n}{2}$ in $O(kn)$ time.*

We next show that except for paths, the upper bound on k in Theorem 2 can be slightly improved for trees with even diameter.

Lemma 3. *Any tree T with even diameter except for a path can be augmented to a minimally $(\operatorname{rad}(T) + 1)$-arbor-connected graph with pagenumber $(\operatorname{rad}(T) + 1)$ in $O(kn)$ time.*

Proof. Let x be the central vertex of T and $k = \operatorname{rad}(T)$. Let ℓ be the minimum distance from x to a vertex with degree at least 3. Let x' be a vertex with degree at least 3 such that $d_T(x, x') = \ell$. Note that $x' = x$ if $\ell = 0$. Regarding x as the root of T, let $D_1(x') = \{w_1, \ldots, w_t\}$ (respectively, $\{w_1, \ldots, w_{t+1}\}$) if $\ell > 0$ (respectively, $\ell = 0$). Without loss of generality, we may assume that $D_{k-1}(w_{t+1}) \neq \emptyset$ if $\ell = 0$ and $\operatorname{rad}(T) \geq 2$. Let σ be a depth-first search ordering for T from x such that $\sigma(x') = \ell+1$, $w_i <_{\sigma,D_1(x')} w_{i+1}$ for $1 \leq i < t$, and $w_t <_{\sigma,D_1(x')} w_{t+1}$ if $\ell = 0$. Based on σ and T (instead of T^+), we construct $T_1, T_2, \ldots, T_k$ similarly to the proof of Theorem 2. Note that even if $|V_0| \geq 3$, the construction in Theorem 2 works well such that $\cup_{1 \leq i \leq k} E_i = E(T) \setminus \{xv \mid v \in D_1(x)\}$

and $V(T_i) = V(T) \setminus \{x\}$. Let $T_0 = \langle \{xv \mid v \leq_\sigma w_t\} \cup \{w_t v \mid w_t <_\sigma v\} \rangle$ and $T'_i = \langle E(T_i) \cup \{\sigma^{-1}(\max_{v \in V_{i-1}} \sigma(v))x\} \rangle$ for $1 \leq i \leq k$. Each of these trees can be embedded in one page under σ. Let $h = \ell \mod k + 1$. Since $V_I(T_0) = \{x, w_t\}$, $V_I(T_0) \cap V_I(T'_i) = \emptyset$ for any $i \neq h$. Any edge xv where $v \leq_\sigma w_t$ is not used in T'_i for any i, since $w_t <_\sigma \sigma^{-1}(\max_{v \in V_{i-1}} \sigma(v))$. Any edge $w_t v$ where $w_t <_\sigma v$ is also not used in T'_i for $i \neq h$. Thus, $E(T_0) \cap E(T'_i) = \emptyset$ for any $i \neq h$. Hence, $T_0, T'_1, \ldots, T'_k$ except for T'_h are k completely independent spanning trees in their union $T^\star$. Define T''_h by modifying T'_h as follows:

$$T''_h = \begin{cases} T'_h - \{w_t u \mid w_t u \in E_h \cup A_h \cup B_h\} + \{w_{t-1} u \mid w_t u \in E_h \cup A_h \cup B_h\} & \text{if } \ell < k, \\ T'_h - \{w_t u \mid w_t u \in E_h \cup A_h\} \cup \{p_k(w_t)w_t\} & \\ \quad + \{w_{t-1} u \mid w_t u \in E_h \cup A_h\} \cup \{w_{t-1} w_t\} & \text{if } \ell \geq k. \end{cases}$$

Then, $V_I(T''_h) \cap V_I(T'_i) = \emptyset$ for $i \neq h$ and $V_I(T''_h) \cap V_I(T_0) = \emptyset$. Note that any edge in $\{w_{t-1} u \mid w_t u \in E_h \cup A_h \cup B_h\}$ (respectively, $\{w_{t-1} u \mid w_t u \in E_h \cup A_h\} \cup \{w_{t-1} w_t\}$) is not used in T_0 and any T'_i for $i \neq h$ if $\ell < k$ (respectively, $\ell \geq k$). Since T''_h can be embedded in one page under σ and $E(T) \subset E(T''_h) \cup E(T^\star)$, $\langle E(T''_h) \cup E(T^\star) \rangle$ is a minimally $(k+1)$-arbor-connected graph with pagenumber $k+1$ which contains T.

Applying a breadth-first search from x, ℓ and x' can be found and w_t can be defined. By Theorem 2, $T_1, T_2, \ldots, T_k$ are constructed in $O(kn)$ time. According to the definition, T_0 can be constructed in $O(n)$ time. Each tree T'_i is obtained from T_i by adding only one edge, while the tree T''_h is obtained from T'_h by modifying $O(n)$ edges. Thus, we have a desired graph in $O(kn)$ time.   □

A tree T with radius 1 is a star, i.e., $n'(T) \leq 1$, while a tree T with radius 2 and odd diameter is a double-star, i.e., $n'(T) = 2$. For a tree with radius 2 and even diameter except for a path, we can apply Lemma 3. Thus, from Theorems 2 and 3 and Lemma 3, we have the following.

Theorem 4. *For $k \leq 3$, any tree of order $n \geq 2k$ can be augmented to a minimally k-arbor-connected graph with pagenumber k in linear time.*

In particular, the case $k = 2$ in Theorem 4 implies a planarity preserving augmentation of a tree.

Corollary 2. *Any tree of order $n \geq 4$ can be augmented to a planar minimally 2-arbor-connected graph in linear time.*

5   Augmenting a Cactus

Theorem 5. *Any cactus G of order n can be augmented to a minimally k-arbor-connected graph with pagenumber k for any $\frac{\ell_G}{2} < k \leq \text{rad}(G)$ in $O(kn)$ time, where ℓ_G is the maximum length of a cycle in G.*

Proof. Let G be a cactus of order n. Note that for any distinct cycles C and C' in G, $E(C) \cap E(C') = \emptyset$. Let x be a central vertex of G and y a vertex adjacent to x such that y is on a path between x and a vertex v with $d_G(x, v) = \text{rad}(G)$. Let G^+ be the graph obtained from $G - xy$ by adding a new vertex z with edges xz and yz. Besides, let T^+

be a breadth-first search tree of G^+ from z. For each cycle C in G, there is exactly one edge in $E(C) \setminus (E(T^+) \cup \{xy\})$ and we denote by $f(C)$ the edge. Consider a depth-first search ordering σ^+ for the vertices of T^+ from z such that for any $f(C) = a_C b_C$ with $d_{T^+}(z, a) \geq d_{T^+}(z, b)$, $a_C <_{\sigma^+} b_C$ and every vertex v with $a_C <_{\sigma^+} v <_{\sigma^+} b_C$ is either a descendant of a_C or an ancestor of b_C. Then, we define σ as $\sigma(v) = \sigma^+(v) - 1$ for any $v \in V(G)$. Similarly to the proof of Theorem 2, we also define E_i, A_i, B_i, B_i', and $T_i = \langle E_i \cup A_i \cup B_i \cup B_i' \rangle$ for $1 \leq i \leq k$. Consider $f(C) = a_C b_C$ with $a_C <_{\sigma} b_C$. If $a_C \in V_{i-1}$ where $1 \leq i \leq k$, then the edge $f(C)$ is used in T_i since $f(C) \in B_i$. Note that if xy is on a cycle C', then $f(C') \in \cup_{1 \leq i \leq k} B_i$. Suppose that $a_C \in V_{kt+i-1}$ where $t \geq 1$ and $1 \leq i \leq k$. Let r_C be the k-ancestor of a_C. Since $k \geq \lfloor \frac{\ell_C}{2} \rfloor + 1$, the subtree rooted at r_C contains the vertex $\text{lca}_{T^+}(a_C, b_C)$ where $r_C \neq \text{lca}_{T^+}(a_C, b_C)$. Now let $M(C) = \{w \mid \sigma^{-1}(\max_{v \in V(T_{a_C})} \sigma(v)) <_{\sigma} w \leq_{\sigma} b_C\}$. Note that $\{r_C w \mid w \in M(C)\} \subseteq A_i$. Replace T_i with $T_i - \{r_C w \mid w \in M(C)\} + \{a_C w \mid w \in M(C)\}$. Let $T_1', T_2', \ldots, T_k'$ be the spanning trees finally obtained by doing such a modification for each cycle C. Since any edge in $\{a_C w \mid w \in M(C)\}$ is not used in $E(T_1) \cup E(T_2) \cup \cdots \cup E(T_k)$, the resultant spanning trees $T_1', T_2', \ldots, T_k'$ are completely independent spanning trees such that their union contains G. Besides, each T_i' can be embedded in one page under σ.

It has been shown in [15] that the center of a cactus can be found in linear time. Thus, x (and also y) can be found in linear time. By applying a breadth-first search to G^+ from z, we can find $f(C)$ for each cycle C and label the end-vertices so that $d_{T^+}(z, a_C) \geq d_{T^+}(z, b_C)$. Besides, for each cycle C, we can find $\text{lca}_{T^+}(a_C, b_C)$ and recognize all the edges of C in $O(k)$ time. Let a_C' (respectively, b_C') be the vertex adjacent to $\text{lca}_{T^+}(a_C, b_C)$ on the path from $\text{lca}_{T^+}(a_C, b_C)$ to a_C (respectively, b_C). We then apply a depth-first search in which for each cycle C, each edge $p_1(v)v$ on the path from a_C' to a_C is traversed as the last edge in $\{p_1(v)w \mid w \in D_1(p_1(v))\}$ for the search of $T_{a_C'}$ and just after the search of $T_{a_C'}$, the traversal proceeds through b_C' and then the path from b_C' to b_C. Such a depth-first search generates σ^+ satisfying the above two properties in $O(n)$ time. For each edge $f(C)$, the corresponding modification can be done in $O(k)$ time. Since the number of cycles in G is at most $\lfloor \frac{n-1}{2} \rfloor$, we can obtain a desired graph in $O(kn)$ time. $\square$

As a corollary, we have an augmentation result for triangular cacti.

Corollary 3. *Any triangular cactus G of order n can be augmented to a minimally k-arbor-connected graph with pagenumber k for any $2 \leq k \leq \text{rad}(G)$.*

6 Augmenting a Unicyclic Graph

Theorem 6. *Any unicyclic graph G of order n can be augmented to a minimally k-arbor-connected graph with pagenumber k for any $2 \leq k \leq \text{rad}(G)$ in $O(kn)$ time.*

Proof. Let G be a unicyclic graph of order n. Let $C = (v_1, v_2, \ldots, v_m, v_1)$ be the cycle of G. Since a unicyclic graph is a cactus, it is sufficient to show the case that $k \leq \frac{m}{2}$. Let $T = G - v_1 v_m$. Suppose that m is even. Let $q = \frac{m-2}{2} \mod k$.

Case 1: $q \neq 0$. Let $x = v_{\frac{m}{2}}$ and $y = v_{\frac{m}{2}+1}$. In the proof of Theorem 2, by employing a depth-first search ordering σ such that $v_1 <_{\sigma} v_m$ and each edge $p_1(v)v$ on paths from x to v_1 and from y to v_m is traversed as the last edge in $\{p_1(v)w \mid w \in D_1(p_1(v))\}$, we construct a minimally k-arbor-connected graph T^*. For the spanning tree $T_{q+1} \subset T^*$, by replacing each edge $v_{\frac{m}{2}-q}w$ in B_{q+1} (respectively, $v_{\frac{m}{2}+q+1}w$ in B'_{q+1}) with the edge $v_1 w$ (respectively, $v_m w$) and then replacing the edge $v_1 w$ for $v_{\frac{m}{2}-q} <_{\sigma,V_q} w$ with $v_1 v_m$, we obtain a desired graph.

Case 2: $q = 0$. Define T_4^+ as the tree obtained from T by deleting the three edges in the path $(v_{\frac{m}{2}-1}, v_{\frac{m}{2}}, v_{\frac{m}{2}+1}, v_{\frac{m}{2}+2})$, adding the new vertex z, and joining z to each vertex in the path. Based on T_4^+ instead of T^+ in the proof of Theorem 2, we construct T^* under the condition $v_{\frac{m}{2}-1} <_{\sigma} v_{\frac{m}{2}} <_{\sigma} v_{\frac{m}{2}+1} <_{\sigma} v_{\frac{m}{2}+2}$ and each edge $p_1(v)v$ on paths from $v_{\frac{m}{2}-1}$ to v_1 and from $v_{\frac{m}{2}+2}$ to v_m is traversed as the last edge in $\{p_1(v)w \mid w \in D_1(p_1(v))\}$. Note that in this construction, $V_0 = \{v_{\frac{m}{2}-1}, v_{\frac{m}{2}}, v_{\frac{m}{2}+1}, v_{\frac{m}{2}+2}\}$, $B_1 = \{v_{\frac{m}{2}-1}v_{\frac{m}{2}}, v_{\frac{m}{2}}v_{\frac{m}{2}+1}, v_{\frac{m}{2}+1}v_{\frac{m}{2}+2}\}$ and $\{v_1, v_m\} \subseteq W_{k-1}$. By modifying T_k in a similar fashion for T_{q+1} in Case 1, we have a desired graph.

Suppose that m is odd. Let $r = \frac{m-3}{2} \bmod k$. Define T_3^+ (respectively, T_5^+) as the tree obtained from T by deleting the two edges in the path $(v_{\frac{m+1}{2}-1}, v_{\frac{m+1}{2}}, v_{\frac{m+1}{2}+1})$ (respectively, four edges in the path $(v_{\frac{m+1}{2}-2}, v_{\frac{m+1}{2}-1}, v_{\frac{m+1}{2}}, v_{\frac{m+1}{2}+1}, v_{\frac{m+1}{2}+2})$), adding the new vertex z, and joining z to each vertex in the path. Similarly to Case 1 (respectively, Case 2), we have the desired result by considering T_3^+ (respectively, T_5^+) and modifying T_{r+1} (respectively, T_k) if $r \neq 0$ (respectively, $r = 0$).

In any case, the corresponding modification can be done in $O(n)$ time. Thus, we have a desired graph in $O(kn)$ time. $\qquad\square$

For any quasi-cycle G, it holds that $\lfloor \frac{n'(G)}{2} \rfloor \leq \mathrm{rad}(G) \leq \lfloor \frac{n'(G)}{2} \rfloor + 1$. Thus from Theorems 3 and 6, we have the following corollary.

Corollary 4. *Any quasi-cycle of order n can be augmented to a minimally k-arbor-connected graph with pagenumber k for any $2 \leq k \leq \frac{n}{2}$.*

7 Concluding Remarks

For $S_1, S_2 \subset V(G)$ such that $S_1 \cap S_2 = \emptyset$, $\langle S_1, S_2 \rangle_G$ denotes the bipartite subgraph of G induced by partite sets S_1 and S_2, i.e., $\langle S_1, S_2 \rangle_G = \langle S_1 \cup S_2 \rangle_G - E(\langle S_1 \rangle_G) \cup E(\langle S_1 \rangle_G)$. A minimally k-arbor-connected graph can be characterized as follows.

Proposition 1. *A graph G is a minimally k-arbor-connected graph if and only if $V(G)$ is partitioned into k subsets $V_1, V_2, \ldots, V_k$ such that $\langle V_i \rangle_G$ is a tree for any $1 \leq i \leq k$ and $\langle V_i, V_j \rangle_G$ is a disjoint union of unicyclic graphs for any $1 \leq i < j \leq k$.*

Proof. Let G be a minimally k-arbor-connected graph. Then, G can be decomposed into completely independent spanning trees $T_1, T_2, \ldots, T_k$. Let $V_i = V_I(T_i)$ for $1 \leq i < k$ and $V_k = V_I(T_k) \cup (\cap_{1 \leq i \leq k} V_L(T_i))$. Since $V_I(T_i) \cap V_I(T_j) = \emptyset$ for any $i < j$, $V(G)$ is partitioned into $V_1, V_2, \ldots, V_k$. Clearly, $\langle V_i \rangle_G$ is a tree for each i. For any $v \in V_i$ and any $j \neq i$, v is a leaf of T_j and then we denote by $\rho_j(v)$ the neighbor of v in T_j. In $\langle V_i, V_j \rangle_G$, a walk $(v, \rho_j(v), \rho_i(\rho_j(v)), \rho_j(\rho_i(\rho_j(v))), \ldots)$ starting from $v \in V_i$ eventually reaches a cycle. This implies that every component of $\langle V_i, V_j \rangle_G$ is a unicyclic graph. Thus, $\langle V_i, V_j \rangle_G$ is a disjoint union of unicyclic graphs.

Suppose that $V(G)$ is partitioned into k subsets $V_1, V_2, \ldots, V_k$ such that $\langle V_i \rangle_G$ is a tree for any i and $\langle V_i, V_j \rangle_G$ is a disjoint union of unicyclic graphs for any $i < j$. For each component H in $\langle V_i, V_j \rangle_G$, we can orient every edge so that the outdegree of every vertex is one. Based on such an orientation, we can connect each vertex in V_i (respectively, V_j) to $\langle V_j \rangle_G$ (respectively, $\langle V_i \rangle_G$) as a leaf. In this way, we have k completely independent spanning trees in G. Since we use all the edges of G for such spanning trees, G is a minimally k-arbor-connected graph. □

As a corollary of Proposition 1, a characterization of a k-arbor-connected graph shown in [2] is obtained; namely, a graph G is k-arbor-connected if and only if $V(G)$ is partitioned into k subsets $V_1, V_2, \ldots, V_k$ such that $\langle V_i \rangle_G$ is connected for any $1 \leq i \leq k$ and $\langle V_i, V_j \rangle_G$ has no tree component for any $1 \leq i < j \leq k$.

Using the characterization in Proposition 1, we can obtain the following proposition.

Proposition 2. *Any tree of order n can be augmented to a minimally k-arbor-connected graph for any $2 \leq k \leq \frac{n}{2}$.*

Given a graph G with a vertex-ordering σ of a t-page book-embedding of G, let P be the path with $V(P) = V(G)$ and $E(P) = \{\sigma^{-1}(i)\sigma^{-1}(i + 1) \mid 1 \leq i < n, i \neq \lceil \frac{n}{2} \rceil \} \cup \{\sigma^{-1}(1)\sigma^{-1}(\lceil \frac{n}{2} \rceil + 1)\}$. Let P^* be the augmented graph according to the proof of Theorem 2. Then, $\langle E(G) \cup E(P^*) \rangle$ is a k-arbor-connected graph with pagenumber at most $t + k$. From this observation, we have the following results.

Proposition 3. *Any graph G of order n can be augmented to a k-arbor-connected graph with pagenumber at most $\mathrm{pn}(G) + k$ for any $2 \leq k \leq \frac{n}{2}$.*

Corollary 5. *Any tree of order n can be augmented to a k-arbor-connected graph with pagenumber at most $k + 1$ for any $2 \leq k \leq \frac{n}{2}$.*

Based on the results in this paper, we may pose the following conjecture which strengthens Proposition 2 and Corollary 5. Conjecture 1 holds for special trees such as caterpillars and trees with at most 8 internal vertices, and for the case that $k \leq 3$.

Conjecture 1. Any tree of order n can be augmented to a minimally k-arbor-connected graph with pagenumber k for any $2 \leq k \leq \frac{n}{2}$.

It would be interesting to consider augmentation problems for a tree to a k-arbor-connected graph while preserving other good properties on layouts.

Acknowledgments. The author is grateful to the reviewers for their helpful comments. This work was supported by JSPS KAKENHI Grant Number JP19K11829.

References

1. Abellanas, M., García, A., Hurtado, F., Tejel, J., Urrutia, J.: Augmenting the connectivity of geometric graphs. Comput. Geom. **40**, 220–230 (2008)
2. Araki, T.: Dirac's condition for completely independent spanning trees. J. Graph Theor. **77**, 171–179 (2014)
3. Bekos, M.A., Kaufmann, M., Klute, F., Pupyrev, S., Raftopoulou, C., Ueckerdt, T.: Four pages are indeed necessary for planar graphs. J. Comput. Geom. **11**, 332–353 (2020)

4. Bernhart, F., Kainen, P.C.: The book thickness of a graph. J. Combin. Theor. Ser. B **27**, 320–331 (1979)
5. Cheng, B., Wang, D., Fan, J.: Constructing completely independent spanning trees in crossed cubes. Discrete Appl. Math. **219**, 100–109 (2017)
6. Chung, F.R.K., Leighton, F.T., Rosenberg, A.L.: Embedding graphs in books: a layout problem with application to VLSI design. SIAM J. Algebraic Discrete Method **8**, 33–58 (1987)
7. Darties, B., Gastineau, N., Togni, O.: Completely independent spanning trees in some regular graphs. Discrete Appl. Math. **217**, 163–174 (2017)
8. Hasunuma, T.: Completely independent spanning trees in the underlying graph of a line digraph. Discrete Math. **234**, 149–157 (2001)
9. Hasunuma, T.: Completely independent spanning trees in maximal planar graphs. In: Goos, G., Hartmanis, J., van Leeuwen, J., Kučera, Luděk (eds.) WG 2002. LNCS, vol. 2573, pp. 235–245. Springer, Heidelberg (2002). https://doi.org/10.1007/3-540-36379-3_21
10. Hasunuma, T.: Minimum degree conditions and optimal graphs for completely independent spanning trees. In: Lipták, Z., Smyth, W.F. (eds.) IWOCA 2015. LNCS, vol. 9538, pp. 260–273. Springer, Cham (2016). https://doi.org/10.1007/978-3-319-29516-9_22
11. Hurtado, F., Tóth, C.: Plane geometric graph augmentation: a generic perspective. In: Pach, J. (ed.) Thirty Essays on Geometric Graph Theory, pp. 327–354. Springer, Heidelberg (2013). https://doi.org/10.1007/978-1-4614-0110-0_17
12. Jackson, B., Jordán, T.: Independence free graphs and vertex connectivity augmentation. J. Combin. Theor. Ser. B **94**, 31–77 (2005)
13. Kant, G.: Augmenting outplanar graphs. J. Algorithms **21**, 1–25 (1996)
14. Kant, G., Bodlaender, H.L.: Planar graph augmentation problems. In: Dehne, F., Sack, J.-R., Santoro, N. (eds.) WADS 1991. LNCS, vol. 519, pp. 286–298. Springer, Heidelberg (1991). https://doi.org/10.1007/BFb0028270
15. Lan, Y.-F., Wang, Y.-L., Suzuki, H.: A linear-time algorithm for solving the center problem on weighted cactus graphs. Inform. Process. Lett. **71**, 205–212 (1999)
16. Pai, K.-J., Chang, J.-M.: Constructing two completely independent spanning trees in hypercube-variant networks. Theor. Comput. Sci. **652**, 28–37 (2016)
17. Pai, K.-J., Chang, R.-S., Chang, J.-M.: A protection routing with secure mechanism in Möbius cubes. J. Parallel Distrib. Comput. **140**, 1–12 (2020)
18. Péterfalvi, F.: Two counterexamples on completely independent spanning trees. Discrete Math. **312**, 808–810 (2012)
19. Rutter, I., Wolff, A.: Augmenting the connectivity of planar and geometric graphs. J. Graph Algorithms Appl. **16**, 599–628 (2012)
20. Végh, L.A.: Augmenting undirected node-connectivity by one. SIAM J. Discrete Math. **25**, 695–718 (2011)
21. Watanabe, T., Nakamura, A.: Edge-connectivity augmentation problems. J. Comput. Syst. Sci. **35**, 96–144 (1987)
22. Wigderson, A.: The complexity of the Hamiltonian circuit problem for maximal planar graphs, Technical Report TR 298. Princeton University (1982)
23. Yannakakis, M.: Four pages are necessary and sufficient for planar graphs (extended abstract). In: Proceedings 18th Annual ACM Symposium on Theory of Computing, pp. 104–108 (1986)
24. Yannakakis, M.: Embedding planar graphs in four pages. J. Comput. Syst. Sci. **38**, 36–67 (1989)
25. Yannakakis, M.: Planar graphs that need four pages. J. Combin. Theor. Ser. B **145**, 241–263 (2020)

Approximation Algorithms for Hitting Subgraphs

Noah Brüstle, Tal Elbaz, Hamed Hatami$^{(\boxtimes)}$, Onur Kocer, and Bingchan Ma

School of Computer Science, McGill University, Montreal, Canada

Abstract. Let H be a fixed undirected graph on k vertices. The H-hitting set problem asks for deleting a minimum number of vertices from a given graph G in such a way that the resulting graph has no copies of H as a subgraph. This problem is a special case of the hypergraph vertex cover problem on k-uniform hypergraphs, and thus admits an efficient k-factor approximation algorithm. The purpose of this article is to investigate the question that for which graphs H this trivial approximation factor k can be improved.

Keywords: Hitting sets · Subgraph elimination · Vertex cover

1 Introduction

All graphs considered in this article are finite simple undirected graphs. Given a fixed graph H, a subset of the vertices of a graph G is called an H-*hitting set* if it intersects every (not necessarily induced) copy of H in G. In other words, removing these vertices from G results in an H-*free* graph. The H-*hitting set problem* asks for the size of the smallest H-hitting set in a given graph G. When H is a single edge, this is the infamous *vertex cover problem*, which is one of the most studied problems in the area of algorithmic graph theory. Another closely related problem is the *feedback vertex set problem*, in which the goal is to remove a smallest set of vertices from G so that the resulting graph contains no cycles. Note that here, instead of a single graph H, we wish to eliminate a family of graphs, namely all cycles. The vertex cover problem and the feedback vertex set problem are both NP-complete, however they both admit efficient 2-factor approximation algorithms.

The H-hitting set problem, as well as its analogue for the induced subgraph setting, have been studied for other specific graphs H such as paths [2, 3, 9, 11, 12], stars [7], and cliques [7]. It is not difficult to see that for any nonempty graph H, the H-hitting set problem is NP-complete (See Theorem 6 below). On the other hand, this problem is a special case of the hypergraph vertex cover problem for k-uniform hypergraphs where $k = |V(H)|$, and thus admits an efficient k-factor approximation. That is, while there is a copy of H in G, delete all the

This research was carried as an undergraduate research project under the supervision of the third author. The third authors' research is supported by an NSERC grant.

© Springer Nature Switzerland AG 2021
P. Flocchini and L. Moura (Eds.): IWOCA 2021, LNCS 12757, pp. 370–384, 2021.
https://doi.org/10.1007/978-3-030-79987-8_26

vertices of this copy, and repeat until the remaining graph becomes H-free. Since these detected copies of H are all vertex-disjoint, any H-hitting set needs to remove at least one vertex from each copy. Hence the number of vertices that are removed by the algorithm is at most k times the optimal solution. For the case of the vertex cover problem, it is widely believed that this simple algorithm is essentially optimal, in the sense that for no fixed constant $\epsilon > 0$, an efficient $(2 - \epsilon)$-factor approximation algorithm exists. In fact it is shown by Khot and Regev [10] that if the so called *unique games conjecture* (UGC for short) is true, then the existence of an efficient $(2 - \epsilon)$-factor approximation algorithm would imply P = NP. In fact their result overrules the existence of an efficient $(k - \epsilon)$-factor approximation algorithms for the k-uniform hypergraph vertex cover problem. This raises the following natural question.

Question 1. For which graphs H on k vertices, there is a constant $\epsilon > 0$ such that the H-hitting set problem admits an efficient $(k - \epsilon)$-factor approximation algorithm?

We shall refer to such graphs as *approximate-easy*. It is shown in [8] that there is an efficient $\frac{23}{11}$-factor approximation algorithm for the P_3-hitting set problem, where here and throughout the paper, P_k denotes *the path on k-vertices*. This was improved in [12] to a 2-factor algorithm by showing that the 2-factor primal-dual approximation algorithm of [5] for the feedback vertex set problem can be adapted to the P_3-hitting set problem. Subsequently, it is shown in [4] that the same ideas can be extended to give a 3-factor approximation algorithm for the P_4-hitting set problem. Lee [11] showed that for every k, there is an efficient $O(\log(k))$-approximation algorithm for the P_k-hitting set, and in particular for sufficiently large k, the path P_k is approximate-easy. Similarly, it is shown in [7] that the star S_k, consisting of a vertex that is connected to k other vertices, admits an $O(\log(k))$-approximation algorithm, and thus is approximate-easy provided that k is sufficiently large.

Let us now turn to negative results. The hardness of approximation for the H-hitting set problem has been studied extensively by Guruswami and Lee [7]. They prove that if H is a 2-vertex connected graph, then the H-hitting problem does not admit a $(k - 1 - \epsilon)$-approximation algorithms unless BPP $\neq$ NP. Since Guruswami and Lee's goal was not to classify the approximate-easy graphs, they preferred to focus on achieving the slightly weaker bound of $k - 1 - \epsilon$ and not rely upon the correctness of the UGC. However as they remark in their article, assuming the UGC conjecture, their approach can lead to the stronger $(k - \epsilon)$ lower-bound that is relevant to our investigation.

Theorem 1. *[7] Assuming the unique games conjecture and NP $\nsubseteq$ BPP, no 2-vertex connected graph is approximate-easy.*

The reason that Theorem 1 relies on the assumption NP $\nsubseteq$ BPP rather than P $\neq$ NP is that the reduction in the proof is obtained by a randomized algorithm. Since Theorem 1 was claimed in [7] without a proof, we present its proof in Sect. 3.2. Prior to this work, the only graphs that were known to be approximate-easy were paths and stars. In the following theorem, we show that in fact all trees are approximate-easy.

Theorem 2. *Let T be a tree on k nodes. The T-hitting set problem admits an efficient $(k - \frac{1}{2})$-factor approximation algorithm.*

Since our focus is only on the classification of approximate-easy graphs, we have not tried to optimize the approximation factor in Theorem 2. The proof of Theorem 2 can be applied to a wider class of graphs. These are the graphs that contain a vertex-cut that has certain properties (See Theorem 5). Inspired by these results and Theorem 1 we conjecture the following.

Conjecture 1. H is approximate-easy if and only if it is not 2-vertex connected.

The smallest example of an H for which we do not yet have a definite answer is the graph consisting of a triangle and a cycle of length 4 that share a single vertex.

Hitting Set Problem for Induced Subgraphs: The hitting set problem can be defined analogously for induced subgraphs. In this case, the goal is to remove the minimum number of vertices from G, so that the remaining graph does not have any *induced* copies of H. As in the case of the non-induced hitting set, there is a trivial $|V(H)|$-factor approximation algorithm, and thus one can analogously define the notion of *induced-approximate-easy*. In Sect. 3.3 we show that the proof of Theorem 1 can be modified to imply a similar result for the induced case.

Theorem 3. *Assuming the unique games conjecture and* NP $\not\subseteq$ BPP, *if H or its complement is a 2-vertex connected graph, then H is not induced-approximate-easy.*

In particular P_5 as well as many other trees are *not* induced-approximate-easy since their complements are 2-vertex connected, and thus the sets of approximate-easy and induced approximate-easy graphs are distinct.

2   The Algorithms

To develop our approximation algorithms, we need to consider the more general setting of the problem where G is a vertex-weighted graph. More precisely, we are given a graph G where every vertex has a non-negative weight, and the goal is to find the smallest possible total weight among H-hitting sets in G.

Phase I: Initial Simplification Using Good Subgraphs: Suppose that we are trying to develop a t-factor approximation algorithm for the H-hitting set problem. First note that we can remove the vertices with weight 0 at no cost. The next important idea is the concept of t-*good graphs* that is formally introduced in [6] but it is also implicit in some of the earlier algorithms.

Definition 1. *A graph K is called t-good for the H-hitting set problem if it is possible to assign non-negative weights to the vertices of K such that every H-hitting set in K has weight at least $\frac{1}{t}$ of the total weight.*

In other words, there is a choice of weights $w_K : V(K) \to \mathbb{R}^{\geq 0}$ for which, even picking all the vertices of K is a t-factor approximation of the H-hitting set problem[1]. The key idea behind this notion is that if a weighted graph (G, w_G) contains a copy of K on the vertices with strictly positive weights, then we can make progress on G in the following manner. With an abuse of notation let w_K also denote the extension of w_K to all the vertices of G by assigning weight 0 to the vertices that are not in that copy of K. Let $\lambda = \min_{v:w_K(v) \neq 0} \frac{w_G(v)}{w_K(v)}$, and let $w_1 = \lambda w_K$ and $w_2 = w_G - w_1$. Note that both w_1 and w_2 are non-negative functions, and furthermore w_2 assigns a weight of zero to at least one vertex in the copy of K. Let S be a t-factor H-hitting set for (G, w_2). Note that by the goodness property of K, S is also automatically a t-factor H-hitting set for (G, w_1). Since $w_G = w_1 + w_2$, we conclude that S is also a t-factor H-hitting set for (G, w_G). This suggests the following approach. Let $\mathcal{K} = K_1, \ldots, K_\ell$ be a set of t-good graphs for the H-hitting set problem.

Algorithm 1: Simplification of the problem using good subgraphs.

Data: On input (G, w)

while *there is a copy of some $K \in \mathcal{K}$ in G with strictly positive weights* **do**

    Set $\lambda = \min_{v:w_K(v) \neq 0} \frac{w(v)}{w_K(v)}$;

    Replace w with $w - \lambda w_K$;

end

Let w_{fin} denote the final weights.

Note that at every iteration of the algorithm, the weight of at least one more vertex of G decreases to 0, so the above algorithm terminates. The above discussion shows that in order to find a t-factor H-Hitting set for (G, w), it suffices to find a t-factor H-hitting set for (G, w_{fin}).

Let X be the set of the vertices that are assigned weight 0 by w_{fin}. We can include the vertices of X in a hitting set at no cost. Moreover $G - X$ is K-free for all $K \in \mathcal{K}$. Depending on $\mathcal{K}$, this could potentially restrict the structure of $G - X$ significantly, and allow us to find a t-factor H-hitting set Y for $(G - X, w_{\text{fin}})$ efficiently. Then we can output $X \cup Y$ as a t-factor H-hitting set for (G, w).

Phase II: Improved Factor Based On Colouring Hypergraphs: After the initial simplification in Phase I, we will end up with a weighted graph G that is K-free for all graphs $K \in \mathcal{K}$, where $\mathcal{K}$ is our set of good graphs. In the second phase, we will use the $\mathcal{K}$-freeness of G to find a desired colouring of the vertices of G that enables us to solve the hitting set problem efficiently with an approximation factor that is strictly less than $|V(H)|$. This is based on some known results for approximating the hypergraph vertex cover problem as described below.

---

[1] Our notion of goodness is slightly stronger than that of [6] as we do not consider minimality.

Consider the following setting. Let $\mathcal{H} = (V, E, w)$ be a vertex-weighted hypergraph where every edge is of size at most k, and $w : V \to \mathbb{R}^+$. The minimum vertex cover problem in this setting is the solution to the following integer linear program:

$$\min \sum_{v \in V} w(v) x_v$$
$$\text{s.t.} \sum_{v:v \in e} x_v \geq 1 \qquad \forall e \in E$$
$$x_v \in \{0, 1\} \qquad \forall v \in V$$

Let us denote the solution to this problem as $\tau(\mathcal{H})$. We can relax this to a linear program

$$\min \sum_{v \in V} w(v) x_v$$
$$\text{s.t.} \sum_{v:v \in e} x_v \geq 1 \qquad \forall e \in E$$
$$x_v \geq 0 \qquad \forall v \in V$$

Let $\tau^*(\mathcal{H})$ denote the cost of the optimal solution to this linear program. This is known as the fractional cover number of $\mathcal{H}$. Finally note that by the linear program duality, this is equal to the solution to the following linear program, which solves the maximal fractional matching problem.

$$\max \sum_{e \in E} y_e$$
$$\text{s.t.} \sum_{e:v \in e} y_e \leq w(v) \qquad \forall v \in V$$
$$y_e \geq 0 \qquad \forall e \in E$$

Definition 2. *A hypergraph $\mathcal{H}$ is called t-colourable if there exists a t-colouring of the vertices of $\mathcal{H}$ such that every edge of size at least 2 contains at least 2 different colours.*

The following theorem is adapted from Aharoni et al. [1], who demonstrated a bound on the ratio of τ and τ^* for t-colourable hypergraphs. We modify their arguments to present an explicit efficient approximation algorithm for τ, and furthermore generalize it to the case of weighted vertices.

Theorem 4. *Let H be a graph on k vertices, and $t \geq k$ be an integer. There is an efficient $k(1 - \frac{1}{t})$-factor approximation algorithm that solves the H-hitting set problem for a weighted graph G if is provided with a t colouring of the vertices of G such that no copy of H in G is monochromatic.*

Proof. Suppose that we have such a t-colouring of G, and let $\mathcal{H}$ be the hypergraph defined by the vertices of G and the edges $e \in E$ corresponding to the copies of H in G. We will show that we are able to find a set of vertices of total weight at most $k(1 - \frac{1}{t})\tau(\mathcal{H})$ that covers $\mathcal{H}$.

Let V be the set of vertices in G which are contained in some copy H_i of H in G, and let $\mathcal{H}' = (V, E)$ be the hypergraph with vertex set V and the hyperedges corresponding to the copies of H in G. Let $g : V \to \mathbb{R}^+$ be a minimal fractional cover of $\mathcal{H}'$ and $f : E \to \mathbb{R}^+$ be a maximal fractional matching in $\mathcal{H}'$ with values $|g| = |f| = \tau^*$.

We consider two cases.

1. $g(v) > 0$ for every $v \in V$.
   According to the complementary slackness conditions,

$$w(V) = \sum_{v \in V} w(v) = \sum_{v \in V} \sum_{e \ni v} f(e) = \sum_{e \in E} f(e)|e \cap V|$$

$$\leq \sum_{e \in E} f(e)k = k \sum_{e \in E} f(e) = k\tau^*(H),$$

and thus

$$\tau^*(H) \geq \frac{w(V)}{k}. \tag{1}$$

On the other hand, the union of any $(t-1)$ colours of H is obviously a cover of H, so

$$\tau(H) \leq \frac{t-1}{t} w(V). \tag{2}$$

Let $S \subseteq V$ be the largest colour class in our colouring. We may then obtain a discrete $(1 - 1/t)k$ approximation of the fractional H-Hitting Set Problem by choosing the vertices $V \backslash S$.

2. There exists a $v \in V$ such that $g(v) = 0$.
   We argue by induction on the number of vertices in G. Consider an edge e corresponding to a copy of H containing v. Since $\sum_{v \in v} g(e) \geq 1$, there must be a vertex v' in e such that $g(v') > \frac{1}{k-1}$.
   Note that g restricted to $V \backslash v'$ is clearly a valid fractional covering of $G \backslash v'$ as every e in $G \backslash v'$ must clearly also be covered in G. Clearly, we also have a valid t-colouring of $G \backslash v'$. Thus, we may obtain a $(1 - 1/t)k$ approximation of the fractional H-Hitting Set Problem on $G \backslash v'$ by our induction hypothesis, which will have total weight no greater than $(1 - 1/t)k \sum_{v \in V \backslash v'} w(v)$. Let J be the set of vertices selected in this manner.
   Now adding v' to J yields a desired cover. Indeed clearly $J \cup \{v'\}$ is a cover, and since $g(v') \geq \frac{1}{k-1} \geq \frac{1}{(1-1/t)k}$, we have $|J \cup v'| \leq (1 - 1/t)k\tau^*$.

In other cases the statement of theorem follows as $\tau \geq \tau*$.

The proof of Theorem 4 gives us the following approximation algorithm:

Algorithm 2: ColorSimp

Data: On input (G, H, c)
Let V be the set of vertices of G that are in a copy of H in G. Let G' be the hypergraph defined on V with edges $e \in E$, the copies of H in G. Let $g : V \to \mathbb{R}^+$ be a minimal fractional cover of G'. **if** $g(v) > 0, \forall v \in V$ **then**
| Let $J = \arg\max_{S \subseteq V | c(s) = c(r) \forall s, r \in S} |S|$; **return** $V \setminus J$
else
| Choose $v \in V, g(v) = 0$;
| Choose $e \in E$ with $v \in e$;
| Let $v' = \arg\max_{u \in e}(g(v))$;
| Let G^* be the graph G without v' and any of its adjacent edges.
| **return** $v' \cup ColorSimp(G^*, H, c)$
end

Our approach for designing approximation algorithms for the H-hitting set problem is to find a set $\mathcal{K}$ of good graphs for H such that every $\mathcal{K}$-free graph G admits a colouring (that can be found efficiently). Then we can apply Phase I to the initial graph to simplify the graph to a $\mathcal{K}$-free graph, and then apply Theorem 4 to obtain a desired H-hitting set.

2.1   The Approximate-Easy Graphs

In this section, we will apply the method that was developed in Sect. 2 to establish that trees are approximate-easy. Our proof implies that a broader class of graphs are approximate-easy. In order to define this class, we need to introduce the notion of a semi-symmetric cut vertex.

A *rooted graph* is a graph G where one vertex v is distinguished as the root. We say that a rooted graph (H, u) is a subgraph of a rooted graph (G, v) if there is an edge-preserving injection from $V(H)$ to $V(G)$ that maps u to v.

Definition 3. *Let H be a graph consisting of m connected graph $F_1, \ldots, F_r$, all sharing a single vertex v, and otherwise having distinct vertices. We call v a semi-symmetric cut-vertex of H if there exists distinct $i, j \in [r]$ such that F_i is a subgraph of F_j as a v-rooted graph.*

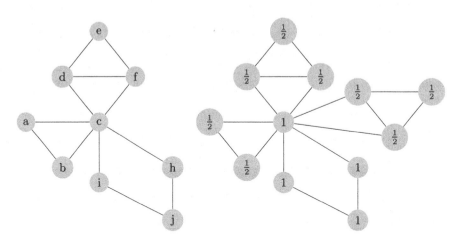

Fig. 1. On the left: c is a semi-symmetric cut-vertex in a graph H as the c-rooted subgraph induced by (c, d, e, f) contains the c-rooted subgraph induced by (c, a, b) as a subgraph. On the right: This graph is at least $k - \frac{1}{2}$ good for H where H is the graph displayed on the left side.

Theorem 5. *Every graph H containing a semi-symmetric cut-vertex is necessarily approximate easy. More precisely, there is an efficient $(|V(H)| - \frac{1}{2})$-factor approximation algorithm for the H-hitting set problem.*

Proof. Let v be a semi-symmetric cut-vertex in H, and let $F_1, \ldots, F_r$ be as in Definition 3, and without loss of generality let us assume that F_1 is a subgraph of F_2 as v-rooted graphs. Denote $k = |V(H)|$. Construct a new graph H' from H by attaching an additional copy of F_2 to v, say F_2'. We will show that H' is $\left(k - \frac{|F_1|-1}{2}\right)$ good for H. This can be easily verified by assigning a weight of $\frac{1}{2}$ to all vertices in $V(F_1) \cup V(F_2) \cup V(F_2')\backslash\{v\}$, and a weight of 1 to all other vertices in H. Note that every H-hitting set in H' either includes one of the vertices with weight 1, or at least two of the vertices with weight $\frac{1}{2}$. Since the total weight is $k - \frac{|F_1|-1}{2}$, we conclude that H' is a$(k - \frac{|F_1|-1}{2})$-good for H.

Next we run the algorithm in Sect. 2 with H' as the only good graph. We will arrive at an H'-free weighted graph G'. It remains to obtain a $(k - \frac{1}{2})$-factor approximation of the H-hitting set problem for G'.

We say that a copy of H in G' is centred at $u \in V(G')$ if u can correspond to the semi-symmetric cut-vertex v in this copy of H. We call $u \in V(G')$ central if some copy of H is centred at u. Similarly we say that a copy of F_2 is centred at u if u can correspond to v in F_2.

For each central vertex u in G', select an arbitrary copy H_u of H centred at u. Let S_u be the set of vertices inside H_u excluding u. Every copy of F_2 centred at u must intersect S_u, since otherwise we would be able to extend H_u to a copy of H' in the H'-free graph G'.

We now construct a directed graph D with vertices $V(G')$, and the directed edges (u, w) for any central vertex u, and every $w \in S_u$. Note that D has maximal out-degree $k - 1$.

Claim. Every directed graph of maximum out-degree m admits a proper $(2m + 1)$-colouring.

Proof. Since the sum of the in-degrees is equal to the sum of the out-degrees, such a graph must contain at least one vertex with total degree at most $2m$. We can remove this vertex, colour the rest of the graph inductively, and then colour this vertex with one of the available colours.

We may thus colour D with at most $2(k - 1) + 1 < 2k$ colours. Colour all vertices in G' accordingly. This is a valid $2k$-hypergraph colouring of the vertices in G' with hyperedges corresponding to the copies of H in G'. Every copy H_0 of H contains a central vertex u. Since there might be other copies of H centred at u, H_0 might not be the copy used to define S_u, however it still contains a copy of F_2 centred at t, and thus it has at least one element in S_u. This vertex is coloured differently than t, and thus H_0 is at least 2-coloured.

Applying Theorem 4, we obtain a $k - \frac{k}{2k} = k - \frac{1}{2}$ approximation of the hitting set problem for the copies of H in G'.

Corollary 1. *Every graph H with at least three vertices and at least one vertex of degree 1 is approximate-easy, and has an approximation factor of at most $k - \frac{1}{2}$.*

Proof. Let u be a vertex of degree 1, and let v be the vertex adjacent to u. If H has at least three vertices, v must be of minimum degree 2; v is necessarily a semi-symmetric cut vertex with the single edge uv as our choice of F_i, and any other adjacent component as F_j in Definition 3. We may apply the previous theorem.

Corollary 2. *Every tree T on at least 3 vertices is approximate-easy, and has an approximation factor of at most $k - \frac{1}{2}$.*

3 Hardness Results

In this section, we present our results regarding the hardness of the H-hitting set problem. In Theorem 6 below, we prove that unless H is an empty graph, the H-hitting set problem is NP-complete. We note in Theorem 7 that the same argument implies the NP-completeness of the *induced* H-hitting set problem for every H. Next in Sect. 3.2 we present the proof of Theorem 1 by using a simplified version of Guruswami-Lee's [7] argument to show that 2-vertex connected graphs are not approximate-easy. Finally, in Sect. 3.3, we prove Theorem 3, by adapting the proof of Theorem 1 to the induced setting to show that if H or its complement is 2-vertex connected, then H is not *induced* approximate-easy.

3.1 NP-Completeness

Theorem 6. *The H-hitting set problem is NP-complete for every connected graph H with at least two vertices.*

Theorem 6 follows immediately from Lemma 1 and Lemma 2 below.

Lemma 1. *The H-hitting set problem is NP-complete if H is a connected graph with minimum degree 2,*

Proof. The proof is by a reduction from the Vertex Cover Problem. Let G be an instance of the vertex cover problem.

It is well known that every connected graph can be uniquely decomposed into a tree of its maximal 2-connected components, called block-cut tree. Let J be a 2-connected subgraph corresponding to a leaf of the block-cut tree of H. Note that J contains at most one cut-vertex, and since the minimum degree of G is 2, there must be an edge e_0 in J such that neither of the endpoints of e_0 is a cut-vertex in H.

We construct a graph G' by "gluing" a copy H_e of H onto every edge e of G via the edge e_0: More precisely, we take the disjoint union of the two graphs and identify[2] the two edges e and e_0. We will show that solving the H-hitting problem on G' allows us to solve the vertex cover problem on the original graph G.

---

[2] We arbitrarily choose a start and an end point for both e_0 and e, and identify the starts together, and the ends together.

First note that every vertex cover S for G is an H-hitting set for G' as removing the vertices in S from G' eliminates all the edges in G, and moreover if u is a vertex in $V(G)\backslash S$, then since neither of the endpoints of e was a cut-vertex, u cannot belong to any copy of H in $G' - S$.

For the other direction, consider an H-hitting set T in G', and let $S = T \cap V(G)$. Let $E \subseteq E(G)$ be the set of the edges in G that are not covered by S. Note that for every $e \in E$, T must contain at least one vertex from H_e. Hence $|T| \geq |S| + |E|$, and the latter is obviously an upper bound on the size of a minimum vertex cover for G.

We conclude that the size of a minimum vertex cover in G is equal to the size of the smallest H-hitting set in G'.

Next in Lemma 2 we establish NP-completeness for the case where H contains a vertex of degree 1.

Lemma 2. *The H-hitting set problem is NP-complete if H is a connected graph with minimum degree 1.*

Proof. Again the proof is by a reduction from the vertex cover problem. Let G be an instance of the vertex cover problem. Let v_0 be a vertex of degree 1 in H, and let u_0 be the unique neighbour of v_0. Let $F = H - v_0$.

This time, we obtain a graph G' by gluing a copy of F on every *vertex* of G via the vertex u_0. More formally, for every vertex u of G, we add a disjoint copy F_u of F, and unify u and u_0.

Let S be a vertex cover for G. Removing S from G' turns it into a disjoint union of copies of F, which is H-free. Thus S is an H-hitting set for G'.

For the other direction, consider an H-hitting set T in G', and let $S = T \cap V(G)$. Let $R \subseteq V(G)$ be the set of the vertices in G that are involved in the remaining edges in $G - S$. Note that $S \cup R$ is a vertex cover for H. For every $u \in R$, T must contain at least one vertex from F_u. Hence $|T| \geq |S \cup R|$.

We conclude that the size of smallest H-hitting set in G' is equal to the size of a minimum vertex cover in G.

The proofs of Lemma 2 and 1 apply to the induced case as well. We conclude that Theorem 6 also holds for the induced H-hitting set problem.

Theorem 7. *The induced H-hitting set problem is NP-complete for every connected graph H with at least two vertices.*

3.2   Guruswami-Lee's Hardness Result: Theorem 1

In this section, we present Guruswami-Lee's hardness of approximation result albeit with minor modifications. The starting point is the hardness of the hypergraph vertex cover.

Theorem 8. *[10] Fix an integer $k > 2$, and let $\epsilon \in (0,1)$. Given a k-uniform hypergraph $\mathcal{H} = (V_\mathcal{H}, E_\mathcal{H})$, assuming the UGC and $P \neq NP$, there is no polynomial time algorithm that distinguishes the following cases.*

- **Completeness:** *There exist disjoint subsets $V_1, \ldots, V_k \subseteq V_{\mathcal{H}}$, each with $\frac{1-\epsilon}{k}$ fraction of vertices, such that each hyperedge has at most one vertex in each V_i. Note that in this case, every V_i together with the vertices in $V_0 := V_{\mathcal{H}} \setminus (V_1 \cup \ldots \cup V_k)$ form a vertex cover with $(\frac{1-\epsilon}{k} + \epsilon)$ fraction of vertices.*
- **Soundness:** *Every subset of $V_{\mathcal{H}}$ with a less than $(1 - \epsilon)$ fraction of vertices does not intersect at least one hyperedge. Equivalently, every subset C of ϵ-fraction of vertices wholly contains a hyperedge.*

In order to deduce a hardness result for the H-hitting set problem from Theorem 8, naturally one would think of replacing each hyperedge $e = (v_1, \ldots, v_k)$ of $\mathcal{H}$ with a copy of H. However, this can lead to a problem as one might create unintentional copies of H that come from a combination of different hyperedges. Thus a vertex cover for $\mathcal{H}$ might not necessarily correspond to an H-hitting set for the constructed graph. To overcome this problem, we will replace each vertex of $\mathcal{H}$ with a large "cloud" of vertices, and for each hyperedge $e = (v_1, \ldots, v_k)$ we randomly implant several copies of H on the clouds of these vertices.

Theorem 9. *Let H be a 2-vertex connected graph on k vertices, and let $\epsilon > 0$. Assuming the UGC, unless NP $\subseteq$ BPP, no polynomial time algorithm can distinguish between the following two cases for a graph G.*

- *Completeness: There is an H-hitting set with $\frac{1}{k} + \epsilon$ fraction of the vertices.*
- *Soundness: Every set with 2ϵ fraction of the vertices contains at least one copy of H.*

In particular, for every $\delta > 0$, no efficient algorithm can approximate the H-hitting set problem with an approximation factor of $k - \delta$.

Proof. Let $\mathcal{H}$ be the k-uniform hypergraph from Theorem 8. We will construct a polynomial size random graph G such that with probability at least $7/8$, it will satisfy the following property: approximating the H-hitting set problem on G would distinguish the two cases in Theorem 8. Since G is randomly constructed, we can only conclude the hardness result under the assumption that NP $\not\subseteq$ BPP.

Without loss of generality we will assume $V_H = [k]$. Given the k-uniform hypergraph $\mathcal{H}$, we put an arbitrary order on the vertices of every hyperedge $e = (v_1, \ldots, v_k)$ of $\mathcal{H}$.

We may assume that $n = |V_{\mathcal{H}}|$ is sufficiently large as a function of ϵ and k, as otherwise the hypergraph vertex cover problem on $\mathcal{H}$ could be solved efficiently. Let $B = B(n, \epsilon, k)$ be a sufficiently large number that polynomially depends on n to be determined later. Let $\lambda = \lambda(k, \epsilon)$ be a positive integer to be determined later as well. The random graph G is defined in the following manner:

- $V_G = V_{\mathcal{H}} \times [B]$. That is we replace every vertex v of the hypergraph $\mathcal{H}$ with B new vertices. We refer to **cloud**$(v) := \{v\} \times [B]$ as the *cloud* of v.
- For every edge $e = (v_1, \ldots, v_k) \in E_{\mathcal{H}}$, we plant λB copies of H in G. For $j = 1, \ldots, \lambda B$ repeat:

- Pick $(\ell_1, \ldots, \ell_k) \in [B]^k$ uniformly at random and plant a copy of H on $(v_1, \ell_1), \ldots, (v_k, \ell_k)$ by mapping the i-th vertex of H to (v_i, ℓ_i). We put a tag of $[e, j]$ on all the edges of this copy of H.

The above procedure produces a random graph G, together with a map $\psi : E_G \to E_H \times [\lambda B]$ representing the tag of each edge in G. Note that G can have multiple edges between two vertices. While we can replace these edges with a single edge without affecting the set of H-hitting sets, for the sake of the presentation, it will be convenient to keep them as multiple edges.

We refer to the planted copies of H in G as *intended* copies. Note that a copy of H in G is intended if and only if all the edges in the copy have the same tag. However G might also have other copies of H, which we refer to as *unintended*.

Completeness: Suppose that $\mathcal{H}$ satisfies the conditions of the completeness case of Theorem 8. Let $S = (V_0 \cup V_1) \times [B]$. Since $(V_0 \cup V_1)$ is a vertex cover in $\mathcal{H}$, S hits every intended copy of H. We will show that with probability at least $3/4$, there will be only few unintended copies of H that do not intersect S. Consequently, we can hit those copies by adding few extra vertices to S. Consider an unintended copy of H in G given by a map $\phi : [k] \to V_G$. Since this copy is unintended, there are $p > 1$ different tags $t_1, \ldots, t_p$ on its edges. Let $I_i \subseteq [k]$ be the set of the vertices of H that are incident to the edges with the tag t_i in that copy. Since H is 2-vertex connected, each I_i has at least two vertices that belong to some other I_j as well. This implies $|I_1| + \ldots + |I_p| \geq k + p$.

There are $(nB)^k$ choices for ϕ, and fixing p, there are at most $p^{|E(H)|}(\lambda B|E_{\mathcal{H}}|)^p$ choices for the tags on the edges of this copy of H. For a fixed ϕ and fixed tags, the probability that the corresponding tagged copy of H is in G is $B^{-(|I_1| + \ldots + |I_p|)} \leq B^{-k-p}$. We conclude that the expected number of unintended copies of H is at most

$$\sum_{p=2}^{|E(H)|} (nB)^k p^{|E(H)|} (\lambda B|E_{\mathcal{H}}|)^p B^{-k-p} \leq \lambda^{k^2} n^{k^2}.$$

Taking $B = \lambda^{k^2} n^{k^2}$, we see that the expected number of unintended copies of H is B, which is very small compared to $|V(G)| = nB$. Assuming $n > \frac{k}{\epsilon}$ and applying Markov's inequality, the probability that there are more than $4B \leq \frac{\epsilon}{k} nB = \frac{\epsilon}{k}|V(G)|$ unintended copies of H in G is at most $\frac{1}{4}$. Thus, with probability at least $\frac{3}{4}$, there is an H-hitting set in G of size at most

$$|S| + 4B \leq \left(\frac{1-\epsilon}{k} + \epsilon\right)|V_G| + \left(\frac{\epsilon}{k}\right)|V_G| = \left(\frac{1}{k} + \epsilon\right)|V_G|,$$

as desired.

Soundness: Next suppose that $\mathcal{H}$ satisfies the conditions of the soundness case of Theorem 8. First we show that with probability at least $7/8$, the random graph G satisfies the following property: For every edge $e = (v_1, \ldots, v_k) \in E_{\mathcal{H}}$, for any choice of subsets $A_i \subseteq \textbf{cloud}(v_i)$ with $|A_i| \geq \epsilon B$ for all $i \in [k]$, there is a copy of H on the induced subgraph of G on $A_1 \cup \ldots \cup A_k$. Indeed, for any choice of e and A_i's, the probability that none of the λB planted copies of H that are created by e fall into this set is at most

$$(1 - \epsilon^k)^\lambda \leq e^{-\lambda B \epsilon^k}.$$

Applying a union bound over e and A_i's, we can bound this probability by

$$|E_\mathcal{H}| 2^{kB} e^{-\lambda B \epsilon^k} \leq 1/8,$$

for $\lambda \geq k\epsilon^{-k}$.

Now consider a G that satisfies the above property, and let $D \subseteq V_G$ be a set with at least 2ϵ fraction of the vertices. Let $C \subseteq V_\mathcal{H}$ be the set of vertices $v \in \mathcal{H}$ such that $|\textbf{cloud}(v) \cap D| \geq \epsilon B$. Since $|D| \geq 2\epsilon|V(G)|$, we know that $|C| \geq \epsilon|V_\mathcal{H}|$, and thus there is a hyperedge e in C. Consequently, there is a copy of H in D.

3.3   Hitting Sets for Induced Subgraphs: Proof of Theorem 3

In this section we present the proof of Theorem 3 by proving an analogue of Theorem 9 for induced hitting sets.

Theorem 10. *Let H be a 2-vertex connected graph on k vertices. Assuming the UGC, unless* NP $\subseteq$ BPP, *no polynomial time algorithm can distinguish between the following two cases for a graph G.*

- *Completeness: There is an induced H-hitting set with $\frac{1}{k} + \epsilon$ fraction of the vertices.*
- *Soundness: Every set with 3ϵ fraction of the vertices contains at least one induced copy of H.*

In particular, for every $\delta > 0$, no efficient algorithm can approximate the induced H-hitting set problem with an approximation factor of $k - \delta$.

Proof. Create the random graph G precisely as in the proof of Theorem 9. Therefore, we can see that there are $|E_\mathcal{H}|\lambda B$ intended copies of H. However, some of these copies might not remain *induced* copies of H due to possible intersections with other intended copies. We call an intended copy of H in G *destroyed* if it is *not* an *induced* copy of H. Note that this happens exactly when another intended copy of H plants an edge between two vertices that are not supposed to be connected in this copy. As it is explained below, the proof follows by showing that with high probability the number of destroyed copies is small.

Completeness: In Theorem 9, it was proven that with probability at least $\frac{3}{4}$, there is an (not necessarily induced) H-hitting set of size at most $(\frac{1}{k} + \epsilon)|V(G)|$ in G. Since an H-hitting set is also an induced H-hitting set, the completeness follows.

Soundness: In this case, we need to show that the number of destroyed intended copies of H is small. For $e \in E_{\mathcal{H}}$ and $i \in \{1, 2, \ldots, \lambda B\}$, let $H_{e,i}$ denote the i-th intended copy of H in G arising from e. Note that for $H_{e,i}$ to be destroyed, there must be another pair (e', j) such that $e' \neq e$, and $H_{e,i}$ and $H_{e',j}$ intersect in at least two vertices. Note that $H_{e,i}$ cannot be destroyed by another $H_{e,j}$.

Let the random variable X denote the number of destroyed copies. From the above discussion, X is obviously bounded by the number of (e, i, e', j) such that $e \neq e'$ and $H_{e,i}$ and $H_{e',j}$ intersect in at least two vertices. Hence by linearity of expectation

$$\mathbb{E}[X] \leq \sum_{e \neq e'} \sum_{i,j} \Pr[|V(H_{e,i}) \cap V(H_{e',j})| \geq 2].$$

For fixed e, e', i, j, in order to have $|V(H_{e,i}) \cap V(H_{e',j})| \geq 2$, the hyperedges e and e' must intersect in at least two vertices $u, v \in V_{\mathcal{H}}$, and moreover $H_{e,i}$ and $H_{e',j}$ must have landed on the same vertices in **cloud**(u) and **cloud**(v). There are at most k^2 choices for u and v, and given u and v, the probability that these copies land on the same vertices on both clouds is exactly $1/B^2$. Hence by applying the union bound on all the possible choices of $u, v \in e \cap e'$, we have $\Pr[|V(H_{e,i}) \cap V(H_{e',j})| \geq 2] \leq k^2/B^2$. We conclude that

$$\mathbb{E}[X] \leq |E_{\mathcal{H}}|^2 (\lambda B)^2 \frac{k^2}{B^2} = |E_{\mathcal{H}}|^2 \lambda^2 k^2. \tag{3}$$

Now, using Markov's inequality, the probability that more than $10|E_{\mathcal{H}}|^2 \lambda^2 k^2$ intended copies are destroyed is at most $\frac{1}{10}$. Thus with probability at least $\frac{9}{10}$, the number of the vertices that are involved in destroyed copies of H is at most $k \times (10|E_{\mathcal{H}}|^2 \lambda^2 k^2) \leq \epsilon |V(G)|$. Now consider a subset of $V(G)$ of size at least $3\epsilon |V(G)|$. Then $2\epsilon |V(G)|$ of these vertices are not in any destroyed copies, and thus by the proof of Theorem 9, they contain an intended copy of H. This copy is induced as it is not part of any destroyed copy.

References

1. Aharoni, R., Holzman, R., Krivelevich, M.: On a theorem of Lovász on covers in r-partite hypergraphs. Combinatorica **16**(2), 149–174 (1996)
2. Brešar, B., Jakovac, M., Katrenič, J., Semanišin, G., Taranenko, A.: On the vertex k-path cover. Discrete Appl. Math. **161**(13–14), 1943–1949 (2013)
3. Brešar, B., Kardoš, F., Katrenič, J., Semanišin, G.: Minimum k-path vertex cover. Discrete Appl. Math. **159**(12), 1189–1195 (2011)
4. Camby, E., Cardinal, J., Chapelle, M., Fiorini, S., Joret, G.: A primal-dual 3-approximation algorithm for hitting 4-vertex paths. In: The 9th International Colloquium on Graph Theory and Combinatorics, p. 61 (2014)
5. Chudak, F.A., Goemans, M.X., Hochbaum, D.S., Williamson, D.P.: A primal-dual interpretation of two 2-approximation algorithms for the feedback vertex set problem in undirected graphs. Oper. Res. Lett. **22**(4–5), 111–118 (1998)
6. Fiorini, S., Joret, G., Schaudt, O.: Improved approximation algorithms for hitting 3-vertex paths. Math. Program. **182**(1), 355–367 (2019). https://doi.org/10.1007/s10107-019-01395-y

7. Guruswami, V., Lee, E.: Inapproximability of H-transversal/packing. SIAM J. Discrete Math. **31**(3), 1552–1571 (2017)
8. Kardoš, F., Katrenič, J., Schiermeyer, I.: On computing the minimum 3-path vertex cover and dissociation number of graphs. Theor. Comput. Sci. **412**(50), 7009–7017 (2011)
9. Katrenič, J.: A faster FPT algorithm for 3-path vertex cover. Inform. Process. Lett. **116**(4), 273–278 (2016)
10. Khot, S., Regev, O.: Vertex cover might be hard to approximate to within $2 - \epsilon$. J. Comput. Syst. Sci. **74**(3), 335–349 (2008)
11. Lee, E.: Partitioning a graph into small pieces with applications to path transversal. Math. Program. **177**(1), 1–19 (2018). https://doi.org/10.1007/s10107-018-1255-7
12. Tu, J., Zhou, W.: A factor 2 approximation algorithm for the vertex cover P_3 problem. Inform. Process. Lett. **111**(14), 683–686 (2011)

Isomorphic Unordered Labeled Trees up to Substitution Ciphering

Florian Ingels[(✉)] and Romain Azaïs

Laboratoire Reproduction et Développement des Plantes, Univ Lyon, ENS de Lyon, UCB Lyon 1, CNRS, INRAE, Inria, 69342 Lyon, France
{florian.ingels,romain.azais}@inria.fr

Abstract. Given two messages – as linear sequences of letters, it is immediate to determine whether one can be transformed into the other by simple substitution cipher of the letters. On the other hand, if the letters are carried as labels on nodes of topologically isomorphic unordered trees, determining if a substitution exists is referred to as marked tree isomorphism problem in the literature and has been show to be as hard as graph isomorphism. While the left-to-right direction provides the cipher of letters in the case of linear messages, if the messages are carried by unordered trees, the cipher is given by a tree isomorphism. The number of isomorphisms between two trees is roughly exponential in the size of the trees, which makes the problem of finding a cipher difficult by exhaustive search. This paper presents a method that aims to break the combinatorics of the isomorphisms search space. We show that in a linear time (in the size of the trees), we reduce the cardinality of this space by an exponential factor on average.

Keywords: Labeled unordered trees · Tree isomorphism · Substitution cipher

1   Introduction

A *simple substitution cipher* is a method of encryption that transforms a sequence of letters, replacing each letter from the original message by another letter, not necessarily taken from the same alphabet [7].

Assume you have at your disposal two messages of the same length, and you want to determine if there exists a substitution cipher that transforms one message onto the other. This question is easily solved, as the cipher is induced by the order of letters. One letter after the other, you can build the cipher by mapping them, until (i) either you arrive at the end of the message, and the answer is Yes, (ii) either you detect an inconsistency in the mapping and the answer is No. Actually, this procedure induces an equivalence relation on messages of the same length: two messages are equivalent (isomorphic) if and only if there is a cipher that transforms one message onto the other. See Fig. 1 for an illustration.

Supported by European Union H2020 project ROMI.

P. Flocchini and L. Moura (Eds.): IWOCA 2021, LNCS 12757, pp. 385–399, 2021.
https://doi.org/10.1007/978-3-030-79987-8_27

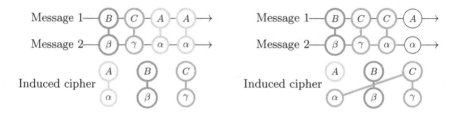

Fig. 1. Simple substitution cipher induced by the order of letters on two examples, one where the two messages are isomorphic (left), and one where there are not (right). In the latter, the last letter of both messages is ignored as an inconsistency is detected at the penultimate letter.

In this article, we are interested in the analogous problem of determining whether two messages are identical up to a substitution cipher, but instead of a linear sequence, the letters are placed as labels on nodes of unordered trees – i.e. for which the order among children of a same node is not relevant.

Instead of requiring that the two messages are of same length – as it was the case for sequences, we require that the two trees are *isomorphic*, i.e. they share the same topology. The reading order of letters is not induced by the sequence but by a *tree isomorphism*, that is a bijection between the nodes of both trees, that respect topology constraints. While the reading order is unique for sequences, for trees, the number of isomorphisms is given by a product of as many factorials as the number of nodes of the tree (see upcoming Eq. (1) and illustrative Fig. 4). Although this number depends highly on the topology, ignoring pathological cases, it is usually extremely large. To give an order of magnitude, for a million replicates of random recursive trees [14] of size 100, the average number of tree isomorphisms is 6.88×10^8 – with a median of 2.21×10^5. The *tree ciphering isomorphism problem* can then be precised as:

"Given two isomorphic unordered trees, is there any tree isomorphism that induces a substitution cipher of the labels of one tree onto the other?"

This question induces an equivalence relation on trees: two topologically isomorphic unordered trees with labels are equivalent if and only if there exists a tree isomorphism that induces a substitution cipher on the labels that transforms one tree onto the other – see Theorem 1. The problem is formally introduced in this paper in Sect. 2, while an example is provided now in Fig. 2.

Determining if two trees are *topologically* isomorphic can be achieved within linear time via the so-called AHU algorithm [1, Ex. 3.2]. Determining if two labeled trees are isomorphic under the definition above is, on the other hand, a difficult problem. It is an instance of labeled graph isomorphism – see [13] and [6] – that was introduced under the name *marked tree isomorphism* in [4, Section 6.4], where it has been proved *graph isomorphism complete*, i.e. as hard as graph isomorphism. The latter is still an open problem, where no proof of NP-completeness nor polynomial algorithm is known [10].

One classic family of algorithms trying to achieve graph isomorphism are *color refinement algorithms*, also known as Weisfeiler-Leman algorithms [12].

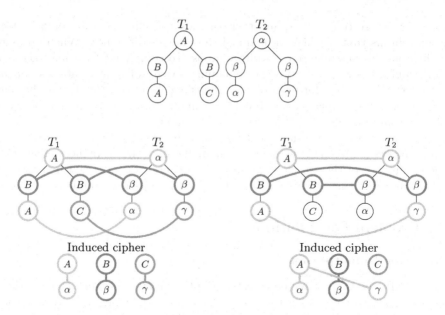

Fig. 2. Two messages encoded as labels on unordered trees T_1 and T_2 (top). T_1 and T_2 are topologically identical. There exist two tree isomorphisms between T_1 and T_2, one inducing a simple substitution cipher (below, left) and the other one that does not (below, right). In the latter, the full tree isomorphism is not parsed as an inconsistency is detected before. Overall, the two labeled trees T_1 and T_2 are isomorphic since at least one tree isomorphism leads to a substitution cipher.

Both graphs are colored according to some rules, and the color histograms are compared afterwards: if they diverge, the graphs are not isomorphic. However, this test is incomplete in the sense that there exist non-isomorphic graphs that are not distinguished by the coloring. The distinguishability of those algorithms is constantly improved – see [8] for recent results – but does not yet answer the problem for any graph. Actually, AHU algorithm for topological tree isomorphism can be interpreted as a color refinement algorithm.

To address the tree ciphering isomorphism problem, one strategy is to explore the space of tree isomorphisms and look for one that induces a ciphering, if it exists. As stated earlier and as discussed in Sect. 2, such a search space is factorially large. This paper does not seek to solve the tree ciphering isomorphism problem, but rather to break the combinatorial complexity of the search space.

In Sect. 3, we present an algorithm fulfilling this objective. Even if it uses AHU algorithm, our method does not involve a color refinement process. Actually, we adopt a strategy that is more related to constrained matching problems in bipartite graphs [5,9]. In details, since we are building two isomorphisms simultaneously – one on trees and the other on labels – that must be compatible, the general idea is to use the constraints of one to make deductions about the other, and vice versa. For instance, whenever two nodes must be mapped together, so are their labels, and therefore you can eliminate all potential tree isomorphisms that would have mapped those labels differently. When no more deductions are possible, our algorithm stops. To complete (if feasible) the two isomorphisms, and to explore the remaining space, different strategies can be considered, including, for example, backtracking. However, this is not the purpose of this paper which aims to break the combinatorial complexity of the space of tree isomorphisms compatible to substitution ciphering.

Finally, in Sect. 4, we show that our algorithm runs in linear time – at least experimentally. Moreover, we show on simulated data that it reduces on average the cardinality of the search space of an *exponential factor* – which shows the great interest of this approach especially considering its low computational cost.

2   Problem Formulation

2.1   Tree Isomorphisms

A (rooted) tree is a connected directed graph without cycle such that (i) there exists a special node called the root, which has no parent, and (ii) any node different from the root has exactly one parent. The parent of a node u is denoted by $\mathcal{P}(u)$, where its children are denoted as $\mathcal{C}(u)$. Trees are said to be unordered if the order among siblings is not significant. In a sequel, we use *tree* to designate a unordered rooted tree.

The *degree* of a node is defined as $\deg(u) = \#\mathcal{C}(u)$, and the degree of a tree is $\deg(T) = \max_{u \in T} \deg(u)$. The leaves $\mathcal{L}(T)$ of a tree T are all the nodes without any children. The depth $\mathcal{D}(u)$ of a node u is the length of the path between u and the root. The depth $\mathcal{D}(T)$ of T is the maximal depth among all nodes. For any node u of T, we define the *subtree* $T[u]$ rooted in u as the tree composed of u and all of its descendants.

Let T_1 and T_2 be two trees.

Definition 1. *A bijection $\varphi : T_1 \to T_2$ is a tree isomorphism if and only if, for any $u, v \in T_1$, if u is a child of v in T_1, then $\varphi(u)$ is a child of $\varphi(v)$ in T_2; in addition, roots must be mapped together.*

We can define $\mathrm{Isom}(T_1, T_2)$ as the set of all tree isomorphisms between T_1 and T_2. If this set is not empty, then T_1 and T_2 are *topologically isomorphic* and we denote $T_1 \equiv T_2$. It is well known that $\equiv$ is an equivalence relation over the set of trees [11, Chapter 4]. Figure 3 provides an example of tree isomorphism.

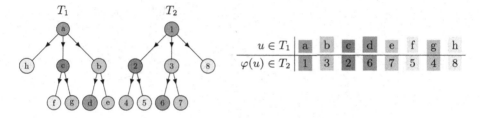

$u \in T_1$	a	b	c	d	e	f	g	h
$\varphi(u) \in T_2$	1	3	2	6	7	5	4	8

Fig. 3. Two topologically isomorphic trees T_1 and T_2 (left) and an example of tree isomorphism $\varphi \in \mathrm{Isom}(T_1, T_2)$ (right). Nodes are labeled and colored for ease of comprehension.

The class of equivalence of node $u \in T_i$ under $\equiv$ – denoted by $[u]$ – is the set of all nodes $v \in T_i$ such that $T_i[u] \equiv T_i[v]$. So-called AHU algorithm [1, Ex. 3.2] assigns in a bottom-up manner to each node u of both trees a color that represents $[u]$. The algorithm can thereby conclude in linear time whether two trees are isomorphic, if and only if their roots are identically colored.

Any tree isomorphism $\varphi : T_1 \to T_2$ maps $u \in T_1$ onto $v = \varphi(u) \in T_2$ only if $[u] = [v]$. Thus, all tree isomorphisms can be – recursively from the root – obtained by swapping nodes (i) of same equivalence class and (ii) children of a same node. Consequently, the number of tree isomorphisms between T_1 and T_2 depends only on the class of equivalence of T_1 (equivalently T_2), and will be denoted by $N_\equiv(T_1)$. For any tree T, we have

$$N_\equiv(T) = \prod_{u \in T} \prod_{q \in \{[v]: v \in \mathcal{C}(u)\}} (\#\{v \in \mathcal{C}(u) : [v] = q\})!. \quad (1)$$

Fig. 4. A tree T. Nodes susceptible to be swapped are boxed together, leading to $N_\equiv(T) = (2!)^3 = 8$.

An example is provided in Fig. 4.

2.2 Tree Cipherings

We now assume that each node of a tree carries a *label*. Let T be a tree and $u \in T$; we denote by $\overline{u}$ the label of node u. The alphabet of T, denoted by $\mathcal{A}(T)$, is defined as $\mathcal{A}(T) = \cup_{u \in T} \overline{u}$. We say that T is a labeled tree.

Let T_1 and T_2 be two topologically isomorphic labeled trees and $\varphi \in \mathrm{Isom}(T_1, T_2)$. φ naturally induces a binary relation R_φ over sets $\mathcal{A}(T_1)$ and $\mathcal{A}(T_2)$, defined as

$$\forall x \in \mathcal{A}(T_1), \forall y \in \mathcal{A}(T_2), x \ R_\varphi \ y \iff \exists u \in T_1, (x = \overline{u}) \wedge (y = \overline{\varphi(u)}).$$

Figure 5 illustrates this induced binary relation on an example.

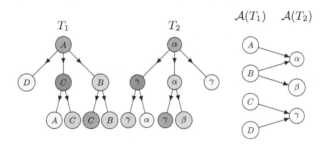

Fig. 5. Two topologically isomorphic labeled trees (left) and the induced binary relation (right). The tree isomorphism φ is displayed through node colors – cf. Fig. 3.

Such a relation R_φ is said to be a bijection if and only if for any $x \in \mathcal{A}(T_1)$, there exists a unique $y \in \mathcal{A}(T_2)$ so that $x \; R_\varphi \; y$, and conversely if for any $y \in \mathcal{A}(T_2)$, there exists a unique $x \in \mathcal{A}(T_1)$ so that $x \; R_\varphi \; y$. This is not the case of the relation induced by the example in Fig. 5, since C and D are both in relation to γ, and also B is in relation to both α and β.

When R_φ is a bijection, we can define a bijective function $f_\varphi : \mathcal{A}(T_1) \to \mathcal{A}(T_2)$ by $f_\varphi(x) = y \iff x \; R_\varphi \; y$. This function is called a *substitution cipher* (following the analogy developed in the introduction) and verifies $\forall u \in T_1, f_\varphi(\overline{u}) = \overline{\varphi(u)}$.

Definition 2. $\varphi \in Isom(T_1, T_2)$ *is said to be a* tree ciphering *if and only if R_φ is a bijection; in which case we denote* $T_1 \xrightarrow{\varphi} T_2$.

Let us denote by $\mathrm{Cipher}(T_1, T_2)$ the set of tree cipherings between T_1 and T_2. If $\mathrm{Cipher}(T_1, T_2)$ is not empty, then we write $T_1 \sim T_2$ and say that T_1 and T_2 are *isomorphic by substitution ciphering*, since the following results holds.

Theorem 1. $\sim$ *is an equivalence relation over the set of labeled trees.*

Proof. The proof is deferred to Appendix A.

Remark 1. It is possible to be more restrictive on the choices of substitution ciphers. Let $(G, \circ)$ be a subgroup of the bijections between $\mathcal{A}(T_1)$ and $\mathcal{A}(T_2)$. Then, if we replace "R_φ is a bijection" in Definition 2 by "$R_\varphi \in G$", the induced relation $\sim_G$ is also an equivalence relation. With $G = \{\mathrm{Id}\}$, $T_1 \sim_G T_2$ means $T_1 \equiv T_2$ plus equality of labels. It is actually the definition adopted for labeled tree isomorphism in [3, Section 5.1].

Determining if $T_1 \sim T_2$ implies to find $\varphi \in \mathrm{Isom}(T_1, T_2)$ such that φ is also in $\mathrm{Cipher}(T_1, T_2)$. Therefore, the cardinality of the search space is given by (1), and is potentially exponentially large compared to the size of the trees. In the sequel of the paper, we present an algorithm that aims to break this cardinality.

3   Breaking down the Combinatorial Complexity

Let be two labeled trees T_1 and T_2. To build a tree ciphering between T_1 and T_2 (if only it exists), a strategy is to ensure that $T_1 \equiv T_2$, and then explore $\text{Isom}(T_1, T_2)$, whose cardinality is given by (1). Since AHU algorithm [1, Ex. 3.2] solves the problem of determining whether $T_1 \equiv T_2$ in linear time, as well as assigning to each node u its equivalence class $[u]$ under $\equiv$, we use AHU as a preprocessing step.

In the case of linear messages, illustrated in Fig. 1, the isomorphism on labels is induced by the reading order, starting with the first letter. In our case, we know that the roots have to be mapped together and we start here. At each step of the algorithm, we will add elements to the two bijections we aim to build: φ for the nodes and f for the labels. We present in Subsection 3.1 how to update those bijections, with the EXTBIJ procedure.

Besides, the topological constraints imposed by tree isomorphism allow to sort the nodes of the trees and to group them by susceptibility to be mapped together. In Subsect. 3.2, we introduce two concepts, bags and collections, that reflects this grouping mechanism. The actual mapping of nodes is performed by the procedure MAPNODES, introduced in Subsect. 3.3.

Finally, the precise course of the algorithm is presented in Subsect. 3.4. Starting by grouping all the nodes together, we successively add topological filters to refine the groups of nodes. Whenever possible, if a filter allows us to deduce that two nodes should be mapped together, we do so, thus reducing the cardinality of the remaining possibilities. The last filter checks constraints on labels and allows a last phase of deductions, before concluding the algorithm – whose analysis is discussed in Sect. 4.

3.1   Extension of a Bijection

During the execution of the algorithm, we construct two mappings: φ for the nodes, and f for the labels. They start as empty mappings $\emptyset \mapsto \emptyset$, and will be updated through time. They must remain bijective at all times, and the rules for updating them are presented here.

A partial bijection ψ from E to F is an injective function from a subset S_ψ of E to F. Let $a \in E$ and $b \in F$; suppose we want to determine if the couple (a, b) is compatible with ψ – in the sense that it respects (or does not contradict) the partial bijection. First, if $a \in S_\psi$, then b must be equal to $\psi(a)$. Otherwise, if $a \notin S_\psi$, then b must not be in the image of ψ, i.e. $\forall s \in S_\psi, \psi(s) \neq b$. If those conditions are respected, then (a, b) is compatible with ψ; furthermore, if $a \notin S_\psi$, then we can extend ψ on $S_\psi \cup \{a\}$ by defining $\psi(a) = b$ so that ψ remains a partial bijection. Formally, for any $a \in E$ and $b \in F$, with ψ a partial bijection from E to F, we define

$$\text{EXTBIJ}(a, b, \psi) = \big(a \in S_\psi \implies \psi(a) = b\big) \wedge \big(a \notin S_\psi \implies \forall s \in S_\psi, \psi(s) \neq b\big);$$

so that $\text{EXTBIJ}(a, b, \psi)$ returns $\top$ if and only if the couple (a, b) is compatible with the partial bijection ψ. For the sake of brevity, we assume that the function

EXTBIJ also extends the partial bijection in the case $a \notin S_\psi$ by defining $\psi(a) = b$ – naturally only if the function returned $\top$.

EXTBIJ will be used in the sequel to update both partial bijections φ (from T_1 to T_2) and f (from $\mathcal{A}(T_1)$ to $\mathcal{A}(T_2)$). However, if one uses the restricted substitution ciphers presented in Remark 1, one must design a specific version of EXTBIJ to update f, accounting for the desired properties.

3.2   Bags and Collections

Remark that if two nodes $u \in T_1$ and $v \in T_2$ are mapped together via φ, then they must share a number of common features: (i) $\mathcal{D}(u) = \mathcal{D}(v)$, (ii) $[u] = [v]$, (iii) $f_\varphi(\overline{u}) = \overline{v}$, and (iv) $\varphi(\mathcal{P}(u)) = \mathcal{P}(v)$. Our goal is to gather together nodes that share such common features. For this purpose, we introduce the concepts of bags and collections.

We recall that a partition P of a set X is a set of non-empty subsets P_i of X such that every element $x \in X$ is in exactly one of these subsets P_i. Let P (resp. Q) be a partition of the nodes of T_1 (resp. T_2).

A bag B is a couple (P_i, Q_j) such that $P_i \in P, Q_j \in Q$ and $\#P_i = \#Q_j$ – this number is denoted by $\#B$. A bag contains nodes that share a number of common features, and are therefore candidates to be mapped together. If a bag is constructed such that P_i and Q_j each contain a single element, then those elements should be unambiguously mapped together – via the function MAPNODES that will be introduced in the next subsection. Formally, this rule is expressed as:

Deduction Rule 1. *While there exist bags $B = (P_i, Q_j)$ with $P_i = \{u\}$ and $Q_j = \{v\}$, call* MAPNODES(u, v, φ, f) *– and delete B.*

A collection C gathers several P_i's and Q_j's, that are candidates to form bags. Formally, $C : \mathbb{N} \to 2^P \times 2^Q$ with $C(n) = (\{P_i, i \in I\}, \{Q_j, j \in J\})$ – possibly $I = J = \emptyset$ – such that, denoting the components by $C_1(n)$ and $C_2(n)$,

(i) $\forall n, \#C_1(n) = \#C_2(n)$;
(ii) $\forall n, \forall P_i \in C_1(n), \#P_i = n$ and $\exists a \in \mathcal{A}(T_1), \forall u \in P_i, \overline{u} = a$;
(iii) $\forall n, \forall Q_j \in C_2(n), \#Q_j = n$ and $\exists b \in \mathcal{A}(T_2), \forall v \in Q_j, \overline{v} = b$.

We denote by $\#C(n)$ the common cardinality of (i); and $\overline{P_i}$ and $\overline{Q_j}$ the common labels of (ii) and (iii). Note that the number of n's such that $\#C(n) > 0$ is finite.

The elements of $C_i(n)$, since they share the same cardinality n, are candidates to form bags together. If $\#C(n) = 1$, we can form a bag with the two elements of $C_1(n)$ and $C_2(n)$:

Deduction Rule 2. *While there exist collections C and integers n for which $C(n) = (\{P_i\}, \{Q_j\})$; if* EXTBIJ$(\overline{P_i}, \overline{Q_j}, f)$, *create bag (P_i, Q_j) and delete $C(n)$ – otherwise stop and conclude that $T_1 \not\sim T_2$.*

As it will be described later on, each subset P_i or Q_j will belong to either a bag or a collection. Any node u will either be already mapped in φ, or attached to one bag or collection through the partitions. We denote by $p(u)$ the function that returns the bag or collection in which u belongs to, if any. We denote by $\mathbb{B}$ the set of all bags, and by $\mathbb{C}$ the set of all collections.

3.3   Mapping Nodes

We now present with Algorithm 1 the function MAPNODES that performs the mapping between nodes, while updating $\varphi, f, \mathbb{B}$ and $\mathbb{C}$. The latter two, $\mathbb{B}$ and $\mathbb{C}$, are considered to be "global" variables and are therefore not included in the pseudocode provided.

Once two nodes u and v are mapped, the topology constraints impose that $\mathcal{P}(u)$ and $\mathcal{P}(v)$ are mapped together, if not already the case, but also $\mathcal{C}(u)$ and $\mathcal{C}(v)$. These children are either (i) already mapped – and there is nothing to do, or (ii) in bags or collections potentially containing other nodes with which they can no longer be mapped – since their parents are not. In the latter case, it is then necessary to separate the children of u and v from these bags and collections. The procedure SPLITCHILDREN aims to do that, in the following manner. For each P_i (resp. Q_j) in the current partitions of nodes such that $P_u = P_i \cap \mathcal{C}(u) \neq \emptyset$ (resp. $Q_v = Q_j \cap \mathcal{C}(v) \neq \emptyset$):

- Either (P_i, Q_j) forms a bag, in which case we delete it and create instead two new bags formed by (P_u, Q_v) and $(P_i \setminus P_u, Q_j \setminus Q_v)$.
- Either there exists a collection C so that $P_i \in C_1(n)$ and $Q_j \in C_2(n)$ – with $n = \#P_i = \#Q_j$. In which case, we remove them from their set $C_i(n)$, and add instead P_u (resp. Q_v) to $C_1(q)$ (resp. $C_2(q)$) – with $q = \#P_u = \#Q_v$ – and $P_i \setminus P_u$ (resp. $Q_j \setminus Q_v$) to $C_1(n - q)$ (resp. $C_2(n - q)$). Note that this splitting operation changes the sets $C_i(\cdot)$ and therefore we need to apply Deduction Rule 2 to check whether some bags are to be created or not.

At any time, if MAPN-ODES returns $\bot$, then we can immediately conclude that $T_1 \not\sim T_2$ and stop. Similarly, if the procedure SPLITCHIL-DREN leads to the creation of a pathological object (e.g. a bag where $\#P_i \neq \#Q_j$), we can also conclude that $T_1 \not\sim T_2$ and stop. We can conclude that $T_1 \sim T_2$ only when all nodes have been mapped.

Algorithm 1: MAPNODES

Input: $u \in T_1, v \in T_2, \varphi, f$
if EXTBIJ($\overline{u}, \overline{v}, f$) **and** EXTBIJ($u, v, \varphi$) **then**
  Delete u from $p(u)$ and v from $p(v)$
  SPLITCHILDREN(u, v)
  if $\varphi(\mathcal{P}(u)) = \mathcal{P}(v)$ **then**
    Return $\top$
  else
    Return MAPNODES($\mathcal{P}(u), \mathcal{P}(v), \varphi, f$)
else
  Return $\bot$

3.4 The Algorithm

Let T_1 and T_2 be two labeled trees; we assume that $T_1 \equiv T_2$. Let $\varphi : \emptyset \mapsto \emptyset$ and $f : \emptyset \mapsto \emptyset$. We start with no collections and a single bag containing all nodes of T_1 and T_2. The general idea is to build a finer and finer partition of the nodes (by applying successive filters), and mapping nodes whenever possible to build the two isomorphisms considered – if they exist: φ and f.

Depth. We partition the only bag $B = (T_1, T_2)$, defining $T_i(d) = \{u \in T_i : \mathcal{D}(u) = d\}$ for $d = 0, \ldots, \mathcal{D}(T_i)$. We delete B from $\mathbb{B}$ and for each d, we create a new bag $(T_1(d), T_2(d))$. Then, apply Deduction Rule 1. Note that since SPLITCHILDREN modifies bags after mapping two nodes, the number of bags meeting the prerequisite of the mapping deduction rule can vary through the iterations. At this step, since the roots are the only nodes with depth of 0, they must be mapped together, and the deduction rule is then applied at least once.

Parents and Children Signature. For each bag $B = (S_1, S_2)$ in $\mathbb{B}$, we partition S_1 and S_2 by shared parent, i.e. we define $S_i(v) = \{u \in S_i : \mathcal{P}(u) = v\}$. For any such a parent v, we define its children signature $\sigma(v)$ as the multiset $\sigma(v) = \{[u] : u \in \mathcal{C}(v)\}$. Nodes from $S_1(v)$ and $S_2(v')$ should be mapped together only if $\sigma(v) = \sigma(v')$. We then group the nodes by signature – losing at the same time the parent information, but which will be recovered through the function MAPNODES – and define $S_i(s) = \cup_{\sigma(v)=s} S_i(v)$. We then create new bags $(S_1(s), S_2(s))$ for each such s, and finally delete B.

Once all bags have been partitioned, apply again Deduction Rule 1.

Equivalence Class Under $\equiv$. For each remaining bag $B = (S_1, S_2)$ in $\mathbb{B}$, we partition S_1 and S_2 by equivalence class under $\equiv$, i.e. we define $S_i(c) = \{u \in S_i : [u] = c\}$. We then create new bags $(S_1(c), S_2(c))$ for each such c, and finally delete B.

Once all bags have been partitioned, apply again Deduction Rule 1.

Labels. For each remaining bag $B = (S_1, S_2)$ in $\mathbb{B}$, we now look at the labels of nodes in S_1 and S_2. We define $S_i(a) = \{u \in S_i : \overline{u} = a\}$. Some of these labels may have been seen previously and may be already mapped in f, in which case we can form bags with the related sets $S_i(a)$. Formally, we apply the following deduction rule.

Deduction Rule 3. *While there exist two sets (of same cardinality) $S_1(a)$ and $S_2(b)$ with $f(a) = b$, create bag $(S_1(a), S_2(b))$. If only one of the two sets exists $(S_1(a)$ with $a \in D_f$ or $S_2(b)$ with $b \in I_f)$ but not its counterpart, we can conclude that $T_1 \not\sim T_2$ and stop.*

The remaining $S_i(a)$ are to be mapped together. However, since we do not know the mapping between their labels, we cannot yet regroup them in bags. We create instead a collection C that contains all those $S_i(a)$, and delete bag B.

Once all bags have been partitioned, either in new bags or in collections, apply Deduction Rule 2. Since this rule maps new labels between them, new bags may be created by virtue of Deduction Rule 3. Consequently, Deduction Rule 3 should be applied every time a bag is created by Deduction Rule 2 – including during the SPLITCHILDREN procedure. Finally, apply again Deduction Rule 1.

4 Analysis of the Algorithm

The analysis presented here is based on theoretical considerations and numerical simulations of labeled trees. For several given n and $\mathcal{A}$, we generated 500 couples (T_1, T_2) as follows. To create T_1, we generate a random recursive tree [14] of size n, and assign a label, randomly chosen from the alphabet $\mathcal{A}$, to each node. We build T_2 as a copy of T_1, before randomly shuffling the children of each node. In this case, $T_1 \sim T_2$. To get $T_1 \not\sim T_2$, we choose a node u of T_2 at random and replace its label by another one, drawn among $\mathcal{A}(T_1) \setminus \{\overline{u}\}$ – this is the most difficult case to determine if $T_1 \not\sim T_2$. The results are gathered in Figs. 6 and 8 and discussed later in the section. Remarkably, in terms of computation times and combinatorial complexity, they seem to mostly depend on n, and not $\#\mathcal{A}$.

4.1 The Algorithm Is Linear

In spite of an intricate back and forth structure between nodes, bags and collections (notably through deduction rules and the SPLITCHILDREN procedure), our algorithm is linear, in the following sense.

Proposition 1. *The number of calls to the function* MAPNODES *is bounded by the size of the trees.*

Proof. Each call to MAPNODES strictly reduces by one, in each tree, the number of nodes remaining to be mapped – and thus present among the bags and collections. As a result, MAPNODES cannot be called more times than the total number of nodes – including the recursive calls of MAPNODES on the parents.

It is important to note, however, that this does not guarantee the overall linearity of the algorithm. Indeed, the complexity of a call to MAPNODES depends on the number of deductions that will be made, notably though the SPLITCHILDREN procedure.

Nevertheless, it seems that this variation regarding the deductions is compensated globally, since experimentally, as shown in Fig. 6a, in the case $T_1 \sim T_2$, it appears quite clearly that the total computation time for the preprocessing phase is linear in the size of the trees. In the case $T_1 \not\sim T_2$, the algorithm allows to conclude negatively in a sublinear time on average – as shown in Fig. 6b.

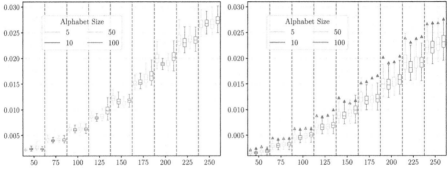

(a) Computation time when $T_1 \sim T_2$.   (b) Computation time when $T_1 \not\sim T_2$.

Fig. 6. Computation time (in s) for the execution of the algorithm of Subsect. 3.4, according to the size of the considered trees. The different sizes of alphabet are displayed with different colors. In Fig. 6a, the red triangles indicate the average value of the corresponding computation time in the case $T_1 \sim T_2$ (estimated from Fig. 6b)

4.2 The Algorithm Reduces the Complexity by an Exponential Factor on Average

At any moment during the execution of the algorithm, given $\mathbb{B}$ and $\mathbb{C}$, we can deduce the current size of the search space. Indeed, for each bag B, there are $(\#B)!$ ways to map the nodes between them (not all of them necessarily leading to a tree isomorphism); for a collection C and for given n, there are $(\#C(n))!$ ways to create bags, each giving $n!^{\#C(n)}$ possible mappings. The overall number of mappings associated to $C(n)$ is then given by $(n!)^{\#C(n)}(\#C(n))!$. Let us define the size of the current search space as

$$N(\mathbb{B}, \mathbb{C}) = \prod_{B \in \mathbb{B}} (\#B)! \prod_{C \in \mathbb{C}} \left(\prod_n (n!)^{\#C(n)} (\#C(n))! \right)$$

Applying the deduction rules does not reduce this number at first sight – since we transform into bags collections with $\#C(n) = 1$ and we map nodes when $\#B = 1$. On the other hand, each call to SPLITCHILDREN reduces this number. Indeed, for each bag or collection where a child of the mapped nodes appears, this object is divided into two parts, breaking the associated factorial:

- A bag with $(p+q)$ elements cut into two bags of size p and q reduces the size of the space by a factor of $\binom{p+q}{p}$.
- An element of $C(p+q)$ cut into two elements of size p and q induces that $\#C(n)$ decreases by 1, and both $\#C(p)$ and $\#C(q)$ increase by 1. Overall, the size of the search space is modified by a factor of $\binom{p+q}{p} \frac{\#C(p+q)}{(\#C(p)+1)(\#C(q)+1)}$.

Each filter during the execution of the algorithm, that consists in splitting each bag into several ones has also the same effect on the overall cardinality. We can measure the evolution of the size of the search space by looking at the log-ratio $r(\mathbb{B}, \mathbb{C})$, defined as follows – with $N_{\equiv}(T_1)$ as in (1):

$$r(\mathbb{B}, \mathbb{C}) = \log_{10} \frac{N(\mathbb{B}, \mathbb{C})}{N_{\equiv}(T_1)}$$

The search space is reduced if and only if $r(\mathbb{B}, \mathbb{C})$ is a negative number. It should be noted that we start the algorithm with a space size of $(\#T_1)!$, i.e. much more than $N_{\equiv}(T_1)$: the initial log-ratio is then positive. Note that despite having an initial search space bigger than $\mathrm{Isom}(T_1, T_2)$, the algorithm cannot build a bijection that is not a tree isomorphism. The first topological filters (depth, parents, equivalence class) bring the log-ratio close to 0 – as illustrated in Fig. 7 with 500 replicates of random trees of size 100 and an alphabet of size 5.

In more details, if we denote by $r_{\mathrm{final}}(\mathbb{B}, \mathbb{C})$ the log-ratio after the last filter on labels, Fig. 8 provides a closer look at the results, and we can see that apart from pathological exceptions obtained with small trees, the log-ratio is always a negative number, so the algorithm does reduce the search space.

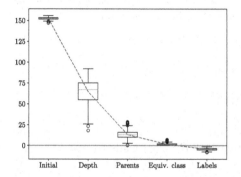

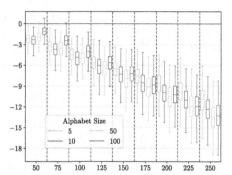

Fig. 7. Evolution of $r(\mathbb{B}, \mathbb{C})$ when $T_1 \sim T_2$.

Fig. 8. $r_{\mathrm{final}}(\mathbb{B}, \mathbb{C})$ when $T_1 \sim T_2$, according to the size of the considered trees. The different sizes of alphabet are displayed with different colors.

As a conclusion, we observe that the search space is reduced on average of an exponential factor and that this factor seems linear in the size of the tree. In other words, it seems that the larger the trees considered, the more exponentially the search space is reduced – which is a remarkable property and justifies the interest of our method, especially given its low computational cost.

Implementation. The algorithm presented in this paper has been implemented as a module of the Python library `treex` [2].

Acknowledgements. The authors would like to thank three anonymous reviewers for their valuable comments on the first version of this manuscript.

A   Proof of Theorem 1

We begin with some preliminary reminders. Let R be a relation over sets E and F. R is a bijection if and only if $\forall x \in E, \exists! y \in F, x\,R\,y$ and $\forall y \in E, \exists! x \in E, x\,R\,y$.

Let R be a relation over sets E and F; the converse relation R^{-1} over sets F and E is defined as $y\,R^{-1}\,x \iff x\,R\,y$. If R is a bijection, then so is R^{-1}.

Let R be a relation over sets E and F; and S a relation over sets F and G. The composition of R and S, denoted by $S \circ R$, is a relation over E and G, and defined as $x\,(S \circ R)\,z \iff \exists y \in F, (x\,R\,y) \wedge (y\,S\,z)$. If R and S are bijections, then so is $S \circ R$.

We now begin the proof. Let T_1, T_2 and T_3 be trees such that $T_1 \overset{\varphi}{\to} T_2$ and $T_2 \overset{\psi}{\to} T_3$. It should be clear that trivially, $T_1 \overset{\mathrm{Id}}{\to} T_1$. We aim to prove the following:

$$T_1 \xrightarrow{\psi \circ \varphi} T_3 \text{ and } T_2 \xrightarrow{\varphi^{-1}} T_1.$$

First of all, it is trivial that $\psi \circ \varphi \in \mathrm{Isom}(T_1, T_3)$. The proof then follows directly from the reminders above and the two following lemmas:

Lemma 1. $R_{\psi \circ \varphi} = R_\psi \circ R_\varphi$.

Proof. Let $x \in \mathcal{A}(T_1)$ and $z \in \mathcal{A}(T_3)$. It suffices to show

$$x\,R_{\psi \circ \varphi}\,z \iff \exists y \in \mathcal{A}(T_2), x\,R_\varphi\,y \wedge y\,R_\psi\,z.$$

$\implies$ There exists $u \in T_1$ so that $x = \overline{u}$ and $z = \overline{(\psi \circ \varphi)(u)}$. Let $v = \varphi(u)$ and $y = \overline{v}$; then $\overline{u}\,R_\varphi\,\overline{v}$, so $x\,R_\varphi\,y$; similarly $\overline{v}\,R_\psi\,\overline{\psi(v)}$ leads to $y\,R_\psi\,z$.

$\impliedby$ There exists $u \in T_1$ so that $\overline{u} = x$ and $y = \overline{\varphi(u)}$. Let $v = \varphi(u)$. As $y\,R_\psi\,\overline{\psi(v)}$, then $\overline{\psi(v)} = z$ and it follows $x\,R_{\psi \circ \varphi}\,z$.

Lemma 2. $R_\varphi{}^{-1} = R_{\varphi^{-1}}$.

Proof. Let $x \in \mathcal{A}(T_1)$ and $y \in \mathcal{A}(T_2)$. It suffices to show $x\,R_\varphi\,y \iff y\,R_{\varphi^{-1}}\,x$.

$\implies$ There exists $u \in T_1$ so that $x = \overline{u}$ and $y = \overline{\varphi(u)}$. Let $v = \varphi(u)$. Since $u = \varphi^{-1}(v), y\,R_{\varphi^{-1}}\,x$.

$\impliedby$ There exists $v \in T_2$ so that $\overline{v} = y$ and $x = \overline{\varphi^{-1}(v)}$. Let $u = \varphi^{-1}(v)$. Since $v = \varphi(u), x\,R_\varphi\,y$.

References

1. Aho, A.V., Hopcroft, J.E., Ullman, J.D.: The Design and Analysis of Computer Algorithms. Reading (1974)
2. Azaïs, R., Cerutti, G., Gemmerlé, D., Ingels, F.: Treex: a Python package for manipulating rooted trees. J. Open Source Softw. 4(38), 1351 (2019)

3. Azaïs, R., Ingels, F.: The weight function in the subtree kernel is decisive. J. Mach. Learn. Res. **21**, 1–36 (2020)
4. Booth, K.S., Colbourn, C.J.: Problems polynomially equivalent to graph isomorphism. University, Computer Science Department (1979)
5. Canzar, S., Elbassioni, K., Klau, G.W., Mestre, J.: On tree-constrained matchings and generalizations. Algorithmica **71**(1), 98–119 (2015)
6. Champin, P.-A., Solnon, C.: Measuring the similarity of labeled graphs. In: Ashley, K.D., Bridge, D.G. (eds.) ICCBR 2003. LNCS (LNAI), vol. 2689, pp. 80–95. Springer, Heidelberg (2003). https://doi.org/10.1007/3-540-45006-8_9
7. Gardner, M.: Codes, Ciphers and Secret Writing. Courier Corporation (1984)
8. Grohe, M., Schweitzer, P., Wiebking, D.: Deep Weisfeiler Leman. In: Proceedings of the 2021 ACM-SIAM Symposium on Discrete Algorithms (SODA), pp. 2600–2614. SIAM (2021)
9. Mastrolilli, M., Stamoulis, G.: Constrained matching problems in bipartite graphs. In: Mahjoub, A.R., Markakis, V., Milis, I., Paschos, V.T. (eds.) ISCO 2012. LNCS, vol. 7422, pp. 344–355. Springer, Heidelberg (2012). https://doi.org/10.1007/978-3-642-32147-4_31
10. Schöning, U.: Graph isomorphism is in the low hierarchy. In: Brandenburg, F.J., Vidal-Naquet, G., Wirsing, M. (eds.) STACS 1987. LNCS, vol. 247, pp. 114–124. Springer, Heidelberg (1987). https://doi.org/10.1007/BFb0039599
11. Valiente, G.: Algorithms on Trees and Graphs. Springer Science & Business Media, Heidelberg (2002). https://doi.org/10.1007/978-3-662-04921-1
12. Weisfeiler, B., Leman, A.: The reduction of a graph to canonical form and the algebra which appears therein. NTI Ser. **2**(9), 12–16 (1968)
13. Zemlyachenko, V.N., Korneenko, N.M., Tyshkevich, R.I.: Graph isomorphism problem. J. Sov. Math. **29**(4), 1426–1481 (1985)
14. Zhang, Y., Zhang, Y.: On the number of leaves in a random recursive tree. Braz. J. Probab. Stat. **29**, 897–908 (2015)

Intersecting Disks Using Two Congruent Disks

Byeonguk Kang[1]($\boxtimes$), Jongmin Choi[1]($\boxtimes$), and Hee-Kap Ahn[2]($\boxtimes$) (iD)

[1] Department of Computer Science and Engineering, Pohang University of Science and Technology, Pohang, Korea
{kbu417,icothos}@postech.ac.kr
[2] Department of Computer Science and Engineering, Graduate School of Artificial Intelligence, Pohang University of Science and Technology, Pohang, Korea
heekap@postech.ac.kr

Abstract. We consider the Euclidean 2-center problem for a set of n disks in the plane: find two smallest congruent disks such that every disk in the set intersects at least one of the two congruent disks. We present a deterministic algorithm for the problem that returns an optimal pair of congruent disks in $O(n^2 \log^3 n/\log \log n)$ time. We also present a randomized algorithm with $O(n^2 \log^2 n/\log \log n)$ expected time. These results improve the previously best deterministic and randomized algorithms, making a step closer to the optimal algorithms for the problem. We show that the same algorithms also work for the 2-piercing problem and the restricted 2-covering problem on disks.

Keywords: Euclidean 2-center · Disk covering · Parametric search

1 Introduction

The k-center problem for a set P of n points is to find k smallest congruent balls such that every point in P is contained in one of the k balls. For $k = 1$, the problem can be solved in $O(n)$ time for any fixed dimension [17]. For any fixed k it can be solved in $O(n^{O(\sqrt{k})})$ time for n points in the plane [15].

The special case of the problem for $k = 2$ in the plane, also known as the planar 2-center problem, has been studied extensively in 1990's. Its decision version was considered by Hershberger and Suri in 1991 [14]. Agarwal and Sharir gave an $O(n^2 \log^3 n)$-time algorithm [2] for the problem. A major breakthrough was made by Sharir who gave the first algorithm with near-linear time, $O(n \log^9 n)$, for the planar 2-center problem [19]. Since then, a fair amount of work has been done to improve the running time to the optimal. Eppstein gave an $O(n \log^2 n)$-time algorithm for the case that the optimal disks are *well-separated*, and showed

This research was supported by the Institute of Information & communications Technology Planning & Evaluation (IITP) grant funded by the Korea government (MSIT) (No. 2017-0-00905, Software Star Lab (Optimal Data Structure and Algorithmic Applications in Dynamic Geometric Environment)) and (No. 2019-0-01906, Artificial Intelligence Graduate School Program (POSTECH)).

© Springer Nature Switzerland AG 2021
P. Flocchini and L. Moura (Eds.): IWOCA 2021, LNCS 12757, pp. 400–413, 2021.
https://doi.org/10.1007/978-3-030-79987-8_28

$\Omega(n \log n)$-time lower bound for any deterministic algorithm for the problem [12]. Chan [6] gave an $O(n \log^2 n \log^2 \log n)$-time algorithm for the problem in 1999 using parametric search. Then there was little improvement for more than 20 years until Wang [22] gave an $O(n \log n \log \log n)$-time algorithm for the case of the optimal disks being close to each other in 2020. Shortly after this, Choi and Ahn [8] gave an $O(n \log n)$-time algorithm for the case, improving Wang's result. Finally, Cho and Oh [7] gave an $O(n \log n)$-time algorithm for the remaining well-separated case, improving Eppstein's $O(n \log^2 n)$-time algorithm. Thus, there is an $O(n \log n)$-time algorithm for the planar 2-center problem, matching the lower bound.

The planar k-center problem was also studied under the L_1 distance, and there are algorithms with running times $O(n)$ for $k = 2, 3$ [11,20], and $O(n \log n)$ for $k = 4$ [20]. The planar k-center problem under the L_1 distance for $k \geq 5$ can be solved in $O(n^{k-4} \log n)$ time [18].

There are fair amounts of work on some variations of the 2-center problem. For example, 2-center problems for points whose centers lie outside input obstacles [13], for weighted points [10], and for a convex polygon [21] in the plane, and 2-center problems for points in three dimensions [1].

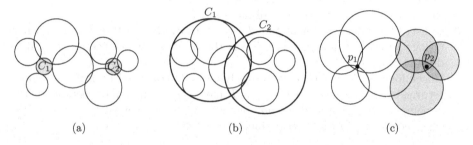

Fig. 1. (a) The disk 2-center problem on disks in the plane: every input disk intersects C_1 or C_2. (b) The restricted disk 2-cover problem on disks in the plane: every disk is fully contained in C_1 or C_2. (c) The 2-piercing problem on disks in the plane: every disk intersects p_1 or p_2.

Motivated by facility location problems for mobile demand points and geometric optimization for imprecise points, we consider a generalization of the 2-center problem in which given a set $\mathcal{D}$ of n disks of nonnegative radii in the plane, find two smallest congruent disks C_1 and C_2 satisfying $D \cap (C_1 \cup C_2) \neq \emptyset$ for every $D \in \mathcal{D}$. We call this problem the *2-center problem on disks*.

This problem was studied by Ahn et al. [3]. They gave a deterministic algorithm that returns an optimal pair of congruent disks in $O(n^2 \log^4 n \log \log n)$ time, and a randomized algorithm with $O(n^2 \log^3 n)$ expected time. They showed that their algorithms also work for the *restricted 2-cover problem* and the *2-piercing problem* on disks in the plane. In the restricted 2-cover problem on disks, every disk in $\mathcal{D}$ must be fully contained in one of two smallest congruent disks. In the 2-piercing problem on disks, every disk in $\mathcal{D}$ must be pierced by one of two optimal points. See Fig. 1 for an illustration of these problems.

Our Results. We present a deterministic algorithm with $O(n^2 \log^3 n / \log \log n)$ time and a randomized algorithm with $O(n^2 \log^2 n / \log \log n)$ expected time for the 2-center problem and the restricted 2-cover problem on n disks in the plane. Our algorithms improve the previously best known deterministic algorithm by $O(\log n \log^2 \log n)$ factor and randomized algorithm by $O(\log n \log \log n)$ factor. Our deterministic algorithm also works for the 2-piercing problem with $O(n^2 \log^2 n / \log \log n)$ time.

Our approach is different to the previous one in a few aspects. For a disk D and a real value $r \geq 0$, the disk inflated by r from D, denoted by $D(r)$, is centered at the center of D and its radius is the radius of D plus r. The algorithms by Ahn et al. [3] use an arrangement of the inflated disks of input disks and a segment tree in sequential decision algorithm, and apply Megiddo's parametric search [16] in optimization. Our algorithms use a point-line dual arrangement of centers of disks and a collection of search trees in a sequential decision algorithm and apply Cole's parametric search [9] in optimization.

The sequential decision algorithm by Ahn et al. works as follows.

1. Construct an arrangement $\mathcal{B}$ of the inflated disks such that each face of $\mathcal{B}$ is the intersection of some inflated disks.
2. Construct a path π traversing all faces in $\mathcal{B}$.
3. Construct a binary segment tree T over π that, given a face f in $\mathcal{B}$, returns $O(\log n)$ regions whose intersection is the intersections of the inflated disks disjoint from f.
4. Check for each face f (one center c_1 lying in f) traversed by π if the regions returned from the query with f on T have a nonempty intersection. If it is nonempty, the inflated disks not pierced by c_1 are pierced by a point (the other center c_2) in the intersection.

We use a dual arrangement $\mathcal{A}$ of the disk centers and construct a dual directed tree $T_\mathcal{E}$ of $\mathcal{A}$. Our sequential decision algorithm traverses the tree in directions of inserting inflated disks one by one and finds the centers. Our sequential decision algorithm works as follows.

1. Construct a point-line dual arrangement $\mathcal{A}$ of the disk centers such that each face of $\mathcal{A}$ represents the inflated disks whose centers lie in one side of a line in primal space.
2. Construct a directed tree $T_\mathcal{E}$ such that there is a one-to-one correspondence between the tree nodes and the faces of $\mathcal{A}$, and each edge is directed from a node to a neighboring node of lower level in $\mathcal{A}$.
3. Construct a collection T_t of t-ary search trees that, given a face f in $\mathcal{A}$, returns $O(\log n / \log \log n)$ regions whose common intersection is the intersection of the inflated disks represented by f.
4. Check for each face f while traversing $T_\mathcal{E}$ if the inflated disks represented by f have a nonempty intersection and the remaining inflated disks also have a nonempty intersection, using T_t and an insertion-only convex programming.

Our algorithm uses Cole's parametric search with an $O(n^2 \log^2 n/\log\log n)$-time sequential decision algorithm and an $O(\log n)$-time parallel decision algorithm using $O(n^2 \log^2 n/\log^2 \log n)$ processors, after $O(n^2 \log^3 n/\log\log n)$-time preprocessing. The improvement of the sequential decision algorithm comes from the insertion-only convex programming and the data structure $\mathcal{T}_t$. The parallel decision algorithm constructs $\mathcal{T}_t$ in the preprocessing phase and the convex programming runs in parallel. Putting them together using Cole's parametric search, we get an $O(n^2 \log^3 n/\log\log n)$-time algorithm. A randomized algorithm with $O(n^2 \log^2 n/\log\log n)$ expected time can be obtained by combining our sequential decision algorithm and Chan's optimization technique [5].

2 Preliminaries

We give some basic observations, definitions, and notations, some of which are from the previous work [3].

Observation 1 (Observation 1 in [3]). Let (C_1, C_2) be a pair of optimal covering disks. Let ℓ be the bisector of the segment connecting the centers of C_1 and C_2. Then, $C_i \cap D \neq \emptyset$ for every $D \in \mathcal{D}$ whose center lies on the same side of ℓ as the center of C_i, for $i = \{1, 2\}$.

We can obtain similar observations for the 2-center problem and the 2-piercing problem. For a line ℓ in the plane, let B_ℓ be a bipartition of $\mathcal{D}$ to $\mathcal{D}_\ell$ and $\mathcal{D}_\ell^c = \mathcal{D}\backslash\mathcal{D}_\ell$, where $\mathcal{D}_\ell$ is the set consisting of disks in $\mathcal{D}$ with centers lying strictly below ℓ. Based on Observation 1, B_ℓ defines a subproblem consisting of two 1-center problems such that the smallest radius for the subproblem is the larger one of the two radii from the 1-center problems.

For a real value $r \geq 0$, let $\mathcal{D}(r)$ be the set of the inflated disks $D(r)$ on disks $D \in \mathcal{D}$. Then the decision version of the 2-center problem on $\mathcal{D}$ with radius r reduces to the 2-piercing problem on $\mathcal{D}(r)$. Given a real value r, the decision 2-center problem on $\mathcal{D}$ is to determine whether $r \geq r^*$ or not, where r^* is the radius of the two smallest congruent disks of the 2-center problem on $\mathcal{D}$. In other words, it is to determine whether the 2-piercing problem on $\mathcal{D}(r)$ has a solution or not.

Cole's Parametric Search. For an optimization problem, let T_s denote the time complexity of a sequential decision algorithm, and let T_p denote the time complexity of a parallel decision algorithm using P processors. If these algorithms satisfy the requirement for applying *Cole's parametric search* [9], we can obtain an algorithm for the problem that runs in $O(PT_s + T_s(T_p + \log P))$ time.

Convex Programming. To determine whether the intersection of input convex sets is empty, we use convex programming. If convex programming has a feasible solution, then the intersection of input convex sets is nonempty.

Given compact convex subsets in the plane, each representing a constraint, and an objective function, a point that satisfies the constraints and minimizes the objective function value can be found using convex programming. There are two types of primitive operations in convex programming: finding the leftmost feasible point of two constraints, and determining whether a given point is contained in a constraint.

Lemma 1. *A problem with k convex constraints can be solved in $O(T_c + \log k)$ time using convex programming, with $O(k^2)$ processors, where T_c denotes the time per primitive operation.*

Proof. The leftmost point in the intersection of the constraints is either the leftmost point of a constraint or the leftmost point in the intersection of two constraints. We compute all leftmost points, each defined by a pair of constraints, in $O(T_c)$ time using $O(k^2)$ processors. Then we find the rightmost point over the leftmost points in $O(\log k)$ time using $O(k^2)$ processors. Finally, we check if the point is contained in every constraint in $O(T_c)$ time using $O(k)$ processors. □

Lemma 2. *Given a convex program with k constraints and the leftmost point v^* of the intersection of the constraints, the operation of adding a new constraint to the convex programming can be handled such that the leftmost point in the intersection of the $k + 1$ constraints can be found in $O(kT_c)$ time, where T_c denotes the time per primitive operation.*

Proof. The leftmost point in the intersection of the $k + 1$ constraints is either v^* or a point v' on the boundary of the newly added constraint. The point v' is the rightmost one among k points, each of which is the leftmost point in the intersection of the newly added constraint with one of the k constraints. The k leftmost points can be found in $O(kT_c)$ time. Then we check if v^* and v' are contained in every constraint in $O(kT_c)$ time. □

In the following, we consider the 2-piercing problem on inflated disks in $\mathcal{D}(r)$.

3   The 2-Piercing Problem on Disks

In this section, we present an algorithm for the 2-piercing problem on a set $\mathcal{E} = \{E_1, E_2, \ldots, E_n\}$ of n disks in the plane. For a disk set X, let $I(X)$ denote the intersection of the disks in X. If there is a bipartition B_ℓ such that both $I(\mathcal{E}_\ell)$ and $I(\mathcal{E}_\ell^c)$ are nonempty, the 2-piercing problem on $\mathcal{E}$ has the solution, where $\mathcal{E}_\ell$ is the set consisting of disks in $\mathcal{E}$ with centers lying strictly below ℓ, and $\mathcal{E}_\ell^c = \mathcal{E} \setminus \mathcal{E}_\ell$. For each bipartition B_ℓ, we perform the *emptiness test* which determines whether $I(\mathcal{E}_\ell) \neq \emptyset$ and $I(\mathcal{E}_\ell^c) \neq \emptyset$.

Dual Arrangement, Bipartitions, and Levels. We construct the dual arrangement $\mathcal{A}$ for the centers of the disks in $\mathcal{E}$ by the following point-line duality transform: For a point $p := (p_x, p_y)$ in the primal plane, its dual $\bar{p}$ is the line $\bar{p} := (y = p_x x - p_y)$ in the dual plane. Likewise, for a line $\ell := (y = \ell_x x + \ell_y)$ in the primal plane, its dual $\bar{\ell}$ is the point $\bar{\ell} := (\ell_x, -\ell_y)$ in the dual plane.

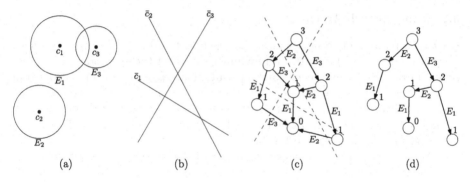

Fig. 2. (a) Three disks E_1, E_2, E_3 with centers at $c_1 = (-0.6, -0.08)$, $c_2 = (-1.9, -4.2)$, $c_3 = (1.7, -0.16)$ in the plane. (b) Three dual lines $\bar{c}_1: y = -0.6x + 0.08$, $\bar{c}_2: y = -1.9x + 4.2$, $\bar{c}_3: y = 1.7x + 0.16$ form the dual arrangement of the centers of disks in (a). (c) Dual graph $G_{\mathcal{E}}$ of the dual arrangement $\mathcal{A}$. The level of a node in $G_{\mathcal{E}}$ is the level of its corresponding face in $\mathcal{A}$. (d) Directed tree $T_{\mathcal{E}}$ from dual graph $G_{\mathcal{E}}$.

See Fig. 2(a,b). The duality transform preserves incidence ($p \in \ell$ if and only if $\bar{\ell} \in \bar{p}$) and order (p lies above ℓ if and only if $\bar{\ell}$ lies above $\bar{p}$) [4]. Thus, $\mathcal{A}$ is the arrangement induced by n lines in the dual plane, each of which is the dual of the center of an input disk. The level of a point in $\mathcal{A}$ is the number of lines in $\mathcal{A}$ lying on or below the point. For a face f of $\mathcal{A}$, let $\bar{\ell}$ be a point in f but not on the upper boundary chain of f. We define the level of f to be the level of $\bar{\ell}$. Let $\mathcal{E}_f$ denote the set of the disks in $\mathcal{E}$ such that the dual lines of their centers lie strictly above $\bar{\ell}$. Observe that $\mathcal{E}_f = \mathcal{E}_\ell$, and let $\mathcal{E}_f^c = \mathcal{E} \backslash \mathcal{E}_f$. Thus, they form the bipartition of the centers of input disks induced by ℓ in the primal plane.

3.1   Dual Directed Tree

Let $G_{\mathcal{E}}$ be a directed acyclic graph such that there is a one-to-one correspondence between the nodes of $G_{\mathcal{E}}$ and the faces in $\mathcal{A}$, and two nodes u, w of $G_{\mathcal{E}}$ are connected by a directed edge (u, w) from u to w if and only if the faces f_u and f_w corresponding to u and w, respectively, share a boundary edge and the level of f_u is larger than the level of f_w. There is a one-to-one correspondence between the bipartitions and the nodes of $G_{\mathcal{E}}$. For a node u in $G_{\mathcal{E}}$, let $\mathcal{E}_u = \mathcal{E}_f$ for face f of $\mathcal{A}$ corresponding to u, and let $\mathcal{E}_u^c = \mathcal{E} \backslash \mathcal{E}_u$. For each edge (u, w) of $G_{\mathcal{E}}$, $\mathcal{E}_w \backslash \mathcal{E}_u$ consists of exactly one disk, and (u, w) corresponds to the disk in $\mathcal{E}_w \backslash \mathcal{E}_u$. In Fig. 2(c), each directed edge is labeled with the disk corresponding to the edge.

Let v_r be the node of $G_{\mathcal{E}}$ that has no incoming edge. We construct from $G_{\mathcal{E}}$, a directed tree $T_{\mathcal{E}}$ rooted v_r that spans all vertices of $G_{\mathcal{E}}$, by choosing only one incoming edge for each node of $G_{\mathcal{E}}$. See Fig. 2(d). For two nodes u, w, let $p(u, w)$ denote the directed path from u to w in $T_{\mathcal{E}}$, if exists.

3.2   *t*-ary Search Trees

Let t be a parameter to be set later. For each leaf node v of $T_\mathcal{E}$, we construct a t-ary search tree $T_t(v)$ in bottom-up manner such that the leaf nodes of $T_t(v)$ are ordered from left to right, each corresponding to an edge in $p(v_r, v)$ in order from v_r to v, the leftmost t leaf nodes have the same parent node and the next t leaf nodes have the same parent node, and so on. This construction goes recursively to higher levels, and each nonleaf node has at most t child nodes. Thus, $T_t(v)$ has height $h = O(\log_t n)$. See Fig. 3.

The path $p(v_r, v)$ represents a sequence of disks, each corresponding to an edge of the path. The data structure $T_t(v)$ supports queries that given a path $p(v_r, w)$ for a node w in $p(v_r, v)$, returns $h = O(\log_t n)$ subpaths that together form $p(v_r, w)$, and h intersections of disks, each corresponding to a subpath.

We construct a collection of t-ary search trees, one for each leaf node of $T_\mathcal{E}$, avoiding duplications of nodes as follows. First, we apply depth-first search (DFS) at v_r of $T_\mathcal{E}$, which gives us an order of the edges of $T_\mathcal{E}$, traversed by DFS. These edges are the leaf nodes of the collection, ordered from left to right following the order by DFS. Then we construct t-ary trees, in the order of the leaf nodes of $T_\mathcal{E}$ visited by DFS. For two leaf nodes v, v' with v visited before v', let v_{split} denote the lowest common ancestor node of $p(v_r, v)$ and $p(v_r, v')$. Then the path $p(v_r, v_{\text{split}})$ is the longest common subpath of the paths. To avoid duplications, $T_t(v')$ simply maintains a pointer to the part of $T_t(v)$ corresponding to $p(v_r, v_{\text{split}})$ with respect to t value, instead of constructing the part again. Let $\mathcal{T}_t$ denote the collection of all t-ary search trees. See Fig. 3(b) for an illustration.

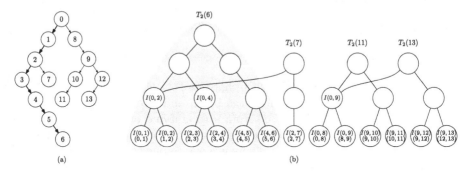

Fig. 3. (a) Dual directed tree $T_\mathcal{E}$. (b) $\mathcal{T}_2$ for all leaf nodes in $T_\mathcal{E}$. $T_2(6)$ is constructed on $p(0, 6)$ (thick path in (a)). Blank nodes in trees store no intersection of disks.

3.3   Intersections of Disks for Paths

For a path $p(u, w)$, let $I(u, w)$ denote the intersection of the disks corresponding to $p(u, w)$. Observe that $I(v_r, w) = I(\mathcal{E}_w)$. For a node ν in $T_t(v)$, let ν^- be the left sibling node of ν, ν^+ the node next (right) to ν at the same level, and $\mathrm{rc}(\nu)$ the rightmost child node of ν in $T_t(v)$. The leaf node ν of $T_t(v)$ corresponding to an edge e of $p(v_r, v)$ stores the intersection $I(\nu) = I(\nu^-) \cap D_e$ if ν^- is defined, and $I(\nu) = D_e$ otherwise, where D_e is the disk corresponding to e. A nonleaf node ν stores $I(\nu)$ if the subtree rooted at ν^+ is a perfect t-ary tree. We set $I(\nu) = I(\nu^-) \cap I(\mathrm{rc}(\nu))$ if ν^- is defined, and $I(\nu) = I(\mathrm{rc}(\nu))$ otherwise. See Fig. 3(b) for an illustration.

For a node ν, if $I(\nu)$ is stored at ν, $I(\nu) = I(w, w')$ for path $p(w, w')$ such that the edge of $p(w, w')$ incident to w corresponds to the leftmost leaf node in the subtree rooted at the parent node of ν, and the edge of $p(w, w')$ incident to w' corresponds to the rightmost leaf node in the subtree rooted at ν. Thus, $I(\nu) = \bigcap D_e$ for all edges e in $p(w, w')$.

Lemma 3. *Given a real value $r \geq 0$, we can construct T_t together with the intersections of disks stored at nodes in $O(tn^2 \log_t n)$ time using $O(tn^2 \log_t n)$ space.*

Proof. We construct $T_t(v)$'s for leaf nodes v of $T_{\mathcal{E}}$ in bottom up manner. Consider a path $p(v_r, v)$ and a node $\nu \in T_t(v)$. If ν^- is defined and the subtree rooted at ν^+ is a perfect t-ary tree, $I(\nu^-)$ and $I(\mathrm{rc}(\nu))$ are stored at ν^- and $\mathrm{rc}(\nu)$, respectively.

For a nonleaf node ν of $T_t(v)$ at height h, if $I(\nu)$ is stored at ν, the subtree rooted at ν^+ is a perfect t-ary tree with t^h leaf nodes. Thus, at height h, there are $O(n^2/t^h)$ nodes ν in T_t that store the intersections of disks with $|I(\nu)| = O(t^{h+1})$. Since the intersection $I(\nu)$ at height h can be computed in $O(t^{h+1})$ time, T_t together with the intersections stored at nodes can be constructed in $O(\sum_{h=0}^{\log_t n} t^{h+1}(n^2/t^h)) = O(tn^2 \log_t n)$ time and space. $\Box$

Lemma 4. *Given a node w in $T_{\mathcal{E}}$, we can find a set $\mathcal{W}$ of $O(\log_t n)$ nodes in T_t such that $\cap_{\nu \in \mathcal{W}} I(\nu) = I(\mathcal{E}_w)$.*

Proof. Note that $I(v_r, w) = I(\mathcal{E}_w)$. We consider $I(v_r, w)$ instead of $I(\mathcal{E}_w)$. Let v be a leaf node of $T_{\mathcal{E}}$ such that w is a node in $p(v_r, v)$. Let e_w be the edge incident to w in $p(v_r, w)$. Let $\pi(w)$ be the path in $T_t(v)$ from the root to the leaf node ν_w corresponding to e_w.

Then $I(v_r, w)$ is the intersection of $I(\nu_w)$ and $I(\nu_h^-)$ for all h ($0 \leq h \leq \log_t n$) if ν_h^- is defined, where ν_h^- is the left node of the node in $\pi(w)$ at height h. If a nonleaf node ν_h^- is defined but $I(\nu_h^-)$ is not stored in ν_h^-, $I(\nu_h^-) = I((\nu_h^-)^-) \cap I(\mathrm{rc}(\nu_h^-))$. If $I(\mathrm{rc}(\nu_h^-))$ is not stored at $\mathrm{rc}(\nu_h^-)$, $\mathrm{rc}(\nu_h^-)$ is the second to the last node at height $h - 1$, and thus $\mathrm{rc}(\nu_h^-)$ and ν_{h-1}^- are the same node, and $I(\nu_{h-1}^-)$ can be used for $I(\mathrm{rc}(\nu_h^-))$.

Therefore, we always get at most two intersections at each height of $T_t(v)$ such that their common intersection is $I(v_r, w)$, and we can get a set $\mathcal{W}$ of $O(\log_t n)$ nodes such that $\cap_{\nu \in \mathcal{W}} I(\nu) = I(v_r, w)$. $\Box$

3.4   Algorithm

We determine whether $I(\mathcal{E}_u)$ and $I(\mathcal{E}_u^c)$ nonempty using $\mathcal{T}_t$, for all nodes $u \in T_\mathcal{E}$. For each leaf node in $\mathcal{T}_t$ (path in $T_\mathcal{E}$), the path corresponding to the leaf node is decomposed to $O(\log_t n)$ subpaths such that each subpath is represented by a node in $\mathcal{T}_t$ while traversing $T_\mathcal{E}$. We determine whether a path whose corresponding intersection is empty by convex programming. While traversing $T_\mathcal{E}$, we add disks one by one, which is the special case of adding a constraint in Lemma 2. If there is a node $u \in T_\mathcal{E}$ such that both $I(\mathcal{E}_u)$ and $I(\mathcal{E}_u^c)$ are nonempty, then the 2-piercing problem has a solution. Using $\mathcal{T}_t$ (Lemmas 3 and 4), convex programming (Lemma 2) and setting $t = \log^\epsilon n$, we can solve the 2-piercing problem in $O(n^2 \log^2 n/\log\log n)$ time using $O(n^2 \log^{1+\epsilon} n)$ space for any constant $0 < \epsilon \le 1$.

Theorem 1. *Given a set of n disks in the plane, we can compute two points p_1 and p_2 such that every input disk contains p_1 or p_2 in $O(n^2 \log^2 n/\log\log n)$ time using $O(n^2 \log^{1+\epsilon} n)$ space for any constant $0 < \epsilon \le 1$.*

Proof. By Lemma 3, we can construct $\mathcal{T}_t$ in $O(tn^2 \log_t n)$ time using $O(tn^2 \log_t n)$ space. While traversing $T_\mathcal{E}$ from root v_r, we find the leftmost point of $I(\mathcal{E}_u)$ for each node u in $T_\mathcal{E}$. The intersection $I(\mathcal{E}_u) = I(v_r, u)$ is the intersection of $I(v_r, w)$ and $I(w, u)$, where w is the parent node of u in $T_\mathcal{E}$. When we traverse node u, the leftmost point of $I(v_r, w)$ is already computed, and we can get a set $\mathcal{W}$ consisting of $O(\log_t n)$ nodes such that the common intersection of $I(\nu)$'s for all $\nu \in \mathcal{W}$ is $I(v_r, w)$ by Lemma 4. Then, we can compute the leftmost point of $I(\mathcal{E}_u)$ using the leftmost point of $I(v_r, w)$ and $I(\nu)$'s for all $\nu \in \mathcal{W}$ by Lemma 2. Since the boundary of $I(\nu)$ consists of $O(n)$ circular arcs, the primitive operation in Lemma 2 takes time $T_c = O(\log n)$. Combining Lemma 2 and $T_c = O(\log n)$, we obtain an $O(n^2 \log^2 n/\log t)$-time algorithm. By setting $t = \log^\epsilon n$ for any constant ϵ with $0 < \epsilon \le 1$, the algorithm runs with $O(n^2 \log^2 n/\log\log n)$ time using $O(n^2 \log^{1+\epsilon} n)$ space. $\qquad\square$

4   The 2-Center Problem on Disks

Our algorithms use parametric search which requires a sequential decision algorithm and a parallel decision algorithm. Our sequential decision algorithm and parallel decision algorithm mainly consist of two parts: (1) Constructing the search trees for emptiness tests. (2) Applying the emptiness test for every bipartition.

4.1   Sequential Decision Algorithm

In this section, we consider the decision 2-center problem with a real value $r \ge 0$ on $\mathcal{D}$ in the plane. We can get a solution to the problem by applying an algorithm for the 2-piercing problem on inflated disks in $\mathcal{D}(r)$.

By solving the 2-piercing problem on the inflated disk in $\mathcal{D}(r)$, we can solve the decision 2-center problem with a given value r on $\mathcal{D}$.

Theorem 2. *Given a set of n disks in the plane and a real value $r \geq 0$, we can determine whether there are two congruent disks C_1 and C_2 of radius r such that every input disk intersects C_1 or C_2 in $O(n^2 \log^2 n / \log \log n)$ time using $O(n^2 \log^{1+\epsilon} n)$ space for any constant ϵ with $0 < \epsilon \leq 1$.*

4.2   Parallel Decision Algorithm

We describe a parallel decision algorithm. Recall that r^* is the radius of the two smallest congruent disks of the 2-center problem on $\mathcal{D}$. We first describe a sequential preprocessing algorithm for finding an interval $(r_a, r_{a+1}]$ such that $r_a < r^* \leq r_{a+1}$ and $\mathcal{T}_t$ has the same combinatorial structure for any $r \in (r_a, r_{a+1}]$, that is, for each intersection stored at nodes of $\mathcal{T}_t$, the circular arcs along the boundary are in the same order. Then we describe how to parallelize the emptiness test algorithm to apply Cole's parametric search.

Sequential Preprocessing Algorithm. The sequential preprocessing algorithm consists of the construction of $\mathcal{T}_t$ for all $r \geq 0$ and binary search to find the interval $(r_a, r_{a+1}]$. Let $F_i \subset \mathbb{R}^3$ be a frustum such that the bottom base of F_i is $D_i \in \mathcal{D}$ lying in the plane $z = 0$, and the intersection of F_i and the plane $z = r$ is $D_i(r)$, for $i = 1, ..., n$. The data structure $\mathcal{T}_t$ for all $r \geq 0$ can be constructed by storing the intersections of frustums at nodes of $\mathcal{T}_t$, instead of the intersections of disks.

Lemma 5. *Given a set of n disks in the plane, we can construct $\mathcal{T}_t$ for all $r \geq 0$ in $O(tn^2 \log_t^2 n \cdot \log t)$ time. The space complexity of $\mathcal{T}_t$ for all $r \geq 0$ is $O(tn^2 \log_t n)$.*

Proof. To construct $\mathcal{T}_t$ for all $r \geq 0$, we compute the intersections of frustums, instead of the intersections of disks. It takes $O(s \log s)$ time to compute the intersection of two intersections of frustums of $O(s)$ complexity. So, we can construct $\mathcal{T}_t$ for all $r \geq 0$ in $O(\sum_{h=0}^{\log_t n} t^{h+1} \log t^{h+1} n^2 / t^h) = O(tn^2 \log_t^2 n \cdot \log t)$ time. The space complexity of $\mathcal{T}_t$ for all $r \geq 0$ is the same as the space complexity of $\mathcal{T}_t$ for a fixed r. $\qquad\boxed{}$

By Lemma 5, there are $O(tn^2 \log_t n)$ curves in the intersections of the frustums stored in $\mathcal{T}_t$ for all $r \geq 0$. We make an array $R := (r_1, ..., r_{O(tn^2 \log_t n)})$ representing the z-values at which the combinatorial structure changes. See Fig. 4 for an illustration. We use binary search on R to find an interval $(r_a, r_{a+1}]$ such that $r_a < r^* \leq r_{a+1}$ and $\mathcal{T}_t$ has the same combinatorial structure for any $r \in (r_a, r_{a+1}]$. Each comparison of binary search determining whether $r_i \geq r^*$ can be done in $O(n^2 \log^2 n / \log \log n)$ time.

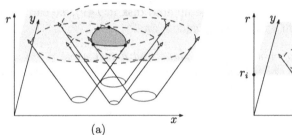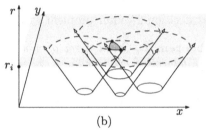

(a)   (b)

Fig. 4. There are four frustums. The light gray region represents the plane with a fixed r-value. The dark gray region represents the intersection of four inflated disks by the r-value of the light gray region. (a) The combinatorial structure consists of four arcs. (b) The combinatorial structure changes at r_i such that the structure in any r in $(r_{i-1}, r_i]$ consists of three arcs and the structure in any r in (r_i, ∞) consists of four arcs.

Lemma 6. *Given a set of n disks in the plane, we can find interval $(r_a, r_{a+1}]$ in $O(n^2 \log^3 n /\log \log n)$ time such that $r_a < r^* \leq r_{a+1}$ and $\mathcal{T}_t$ has the same combinatorial structure for any $r \in (r_a, r_{a+1}]$ and for $t = O(\log n)$.*

Proof. We first construct $\mathcal{T}_t$ for all $r \geq 0$ using Lemma 5. In each step of binary search, we find a median r in the array R and determine if $r \geq r^*$. Finding a median in an array can be done in $O(|R|)$. The size of R gets halved in each binary search step. It takes $O(n^2 \log^2 n /\log \log n)$ time to determine if $r \geq r^*$ by Theorem 2. Thus, in total it takes $O(n^2 \log^3 n /\log \log n + tn^2 \log_t^2 n \cdot \log t)$ time for finding the interval $(r_a, r_{a+1}]$ satisfying the property in the lemma.   □

Parallel Decision Algorithm. From the sequential preprocessing, we get $\mathcal{T}_t$ for an interval $(r_a, r_{a+1}]$ such that it has the same combinatorial structure for any $r \in (r_a, r_{a+1}]$ and $r^* \in (r_a, r_{a+1}]$. Using $\mathcal{T}_t$, we parallelize the process of determining $I(\mathcal{E}_u) = \emptyset$ and $I(\mathcal{E}_u^c) = \emptyset$ for all nodes $u \in T_{\mathcal{E}}$. Observe that each of $I(\mathcal{E}_u)$ and $I(\mathcal{E}_u^c)$ is represented by $O(\log_t n)$ intersections of disks. We can determine whether the common intersection of $O(\log_t n)$ intersections is empty by using parallel convex programming as in Lemma 1. The parallel decision algorithm takes $O(\log n)$ time using $O(n^2 \log^2 n /\log^2 \log n)$ processors.

Theorem 3. *Given a set of n disks in the plane and a real value $r \geq 0$, we can determine whether there are two congruent disks C_1 and C_2 of radius r such that every input disk intersects C_1 or C_2 in $O(\log n)$ time using $O(n^2 \log^2 n /\log^2 \log n)$ processors, after $O(n^2 \log^3 n /\log \log n)$-time preprocessing.*

Proof. By Lemma 6, we can construct $\mathcal{T}_t$ in $O(n^2 \log^3 n /\log \log n)$ time for the search the interval $(r_1, r_2]$ and $t = O(\log n)$. We determine whether the intersection of the selected intersections is empty or not, by parallel convex linear programming in Lemma 1. For $k = O(\log_t n)$ and $T_c = O(\log n)$, the parallel emptiness test takes $O(\log n + \log \log_t n)$ time using $O(n^2 \log_t^2 n)$ processors. Setting $t = O(\log n)$, we have $O(\log n)$ time using $O(n^2 \log^2 n / \log^2 \log n)$ processors, after $O(n^2 \log^3 n / \log \log n)$-time preprocessing.   ▣

4.3   Optimization Algorithms

Using Cole's parametric search [9], an optimal solution can be found in $O(PT_s + T_s(T_p + \log P))$ time. Here $T_s = O(n^2 \log^2 n / \log \log n)$, $T_p = O(\log n)$, $P = O(n^2 \log^2 n / \log^2 \log n)$, and therefore it can be done in $O(n^2 \log^3 n / \log \log n)$ time.

Theorem 4. *Given a set of n disks in the plane, we can compute two smallest congruent disks C_1 and C_2 such that every input disk intersects C_1 or C_2 in deterministic $O(n^2 \log^3 n / \log \log n)$ time.*

Proof. Our parallel algorithm satisfies the *bounded fan-in/fan-out* requirement of Cole's parametric search. Using Cole's parametric search, an optimal solution can be found in $O(PT_s + T_s(T_p + \log P))$ time. Here $T_s = O(n^2 \log^2 n / \log \log n)$, $T_p = O(\log n)$, $P = O(n^2 \log^2 n / \log^2 \log n)$, and therefore, the time complexity is $O(n^2 \log^3 n / \log \log n)$.   ▣

A randomized algorithm which runs in $O(n^2 \log^2 n / \log \log n)$ expected time can be done by combining our sequential decision algorithm and Chan's optimization technique [5].

Theorem 5. *Given a set of n disks in the plane, we can compute two smallest congruent disks C_1 and C_2 such that every input disk intersects C_1 or C_2 in $O(n^2 \log^2 n / \log \log n)$ expected time.*

Proof. Let $\mathcal{D}$ be a set of n disks in the plane, and let $\mathcal{D}_1, \ldots, \mathcal{D}_7$ be seven disjoint subsets of size at most $\lceil n/7 \rceil$ satisfying $\mathcal{D} = \bigcup_{i=1}^{7} \mathcal{D}_i$. Let $R(X)$ be the radius of two smallest congruent disks such that every disk in X intersects one of the two congruent disks. Observe that the two congruent disks are determined by at most six disks of X. Then $R(\mathcal{D}) = \max\{R(\mathcal{D} \setminus \mathcal{D}_1), R(\mathcal{D} \setminus \mathcal{D}_2), ..., R(\mathcal{D} \setminus \mathcal{D}_7)\}$. Therefore, by applying the Chan's optimization technique [5], we obtain a randomized algorithm that computes r^* in time linear to the running time of the sequential decision algorithm. By Theorem 2, we can conclude that this randomized algorithm takes $O(n^2 \log^2 n / \log \log n)$ expected time.   ▣

4.4   Restricted 2-Cover Problem

In the restricted 2-cover problem on disks in the plane, every input disk must be fully contained in one of the two optimal (smallest) congruent disks. The algorithm for the 2-center problem on disks in the plane also works for the restricted 2-cover problem. In the restricted 2-cover problem on disks, we define the *r-inflated disk* of a disk D as the disk with radius $r - r_D$ centered at the center of D, where r_D is the radius of disk D. Then we apply the decision algorithm for the 2-piercing problem on the r-inflated disks. The rest of the algorithm is the same as the algorithm for the 2-center problem on disks.

Corollary 1. *Given a set of n disks in the plane, we can compute two smallest congruent disks C_1 and C_2 such that every input disk is fully contained in either C_1 or C_2 in deterministic $O(n^2 \log^3 n / \log \log n)$ time.*

Corollary 2. *Given a set of n disks in the plane, we can compute two smallest congruent disks C_1 and C_2 such that every input disk is fully contained in either C_1 or C_2 in $O(n^2 \log^2 n / \log \log n)$ expected time.*

References

1. Agarwal, P.K., Avraham, R.B., Sharir, M.: The 2-center problem in three dimensions. In: Proceedings of the 26th Annual Symposium on Computational Geometry, pp. 87–96 (2010)
2. Agarwal, P.K., Sharir, M.: Planar geometric location problems. Algorithmica **11**(2), 185–195 (1994). https://doi.org/10.1007/BF01182774
3. Ahn, H.K., Kim, S.S., Knauer, C., Schlipf, L., Shin, C.S., Vigneron, A.: Covering and piercing disks with two centers. Comput. Geom. **46**(3), 253–262 (2013)
4. de Berg, M., Cheong, O., Kreveld, M.V., Overmars, M.: Computational Geometry Algorithms and Applications, 3rd edn. Springer, Berlin (2008). https://doi.org/10.1007/978-3-540-77974-2
5. Chan, T.M.: Geometric applications of a randomized optimization technique. Discrete Comput. Geom. **22**(4), 547–567 (1999)
6. Chan, T.M.: More planar two-center algorithms. Comput. Geom. **13**(3), 189–198 (1999)
7. Cho, K., Oh, E.: Optimal algorithm for the planar two-center problem. arXiv:2007.08784 (2020)
8. Choi, J., Ahn, H.K.: Efficient planar two-center algorithms. arXiv:2006.10365 (2020)
9. Cole, R.: Slowing down sorting networks to obtain faster sorting algorithms. J. ACM (JACM) **34**(1), 200–208 (1987)
10. Drezner, Z.: The planar two-center and two-median problems. Transp. Sci. **18**(4), 351–361 (1984)
11. Drezner, Z.: On the rectangular p-center problem. Naval Res. Logist. (NRL) **34**(2), 229–234 (1987)
12. Eppstein, D.: Faster construction of planar two-centers. In: Proceedings of the 8th Annual ACM-SIAM Symposium on Discrete Algorithms, pp. 131–138 (1997)
13. Halperin, D., Sharir, M., Goldberg, K.: The 2-center problem with obstacles. J. Algorithms **42**(1), 109–134 (2002)
14. Hershberger, J., Suri, S.: Finding tailored partitions. J. Algorithms **12**(3), 431–463 (1991)
15. Hwang, R.Z., Lee, R.C.T., Chang, R.C.: The slab dividing approach to solve the Euclidean p-center problem. Algorithmica **9**(1), 1–22 (1993)
16. Megiddo, N.: Applying parallel computation algorithms in the design of serial algorithms. J. ACM (JACM) **30**(4), 852–865 (1983)

17. Megiddo, N.: Linear programming in linear time when the dimension is fixed. J. ACM (JACM) **31**(1), 114–127 (1984)
18. Nussbaum, D.: Rectilinear p-piercing problems. In: Proceedings of the 10th International Symposium on Symbolic and Algebraic Computation, pp. 316–323 (1997)
19. Sharir, M.: A near-linear algorithm for the planar 2-center problem. Discrete Comput. Geom. **18**(2), 125–134 (1997)
20. Sharir, M., Welzl, E.: Rectilinear and polygonal p-piercing and p-center problems. In: Proceedings of the 12th Annual Symposium on Computational Geometry, pp. 122–132 (1996)
21. Shin, C.S., Kim, J.H., Kim, S.K., Chwa, K.Y.: Two-center problems for a convex polygon. In: Proceedings of the 6th European Symposium on Algorithms, pp. 199–210 (1998)
22. Wang, H.: On the planar two-center problem and circular hulls. In: Proceedings of the 36th International Symposium on Computational Geometry, vol. 164, pp. 68:1–68:14 (2020)

Disjoint Paths and Connected Subgraphs for H-Free Graphs

Walter Kern[1], Barnaby Martin[2], Daniël Paulusma[2], Siani Smith[2(✉)],
and Erik Jan van Leeuwen[3]

[1] Department of Applied Mathematics, University of Twente,
Twente, The Netherlands
w.kern@twente.nl
[2] Department of Computer Science, Durham University, Durham, UK
{barnaby.d.martin,daniel.paulusma,siani.smith}@durham.ac.uk
[3] Department of Information and Computing Sciences, Utrecht University,
Utrecht, The Netherlands
e.j.vanleeuwen@uu.nl

Abstract. The well-known DISJOINT PATHS problem is to decide if a
graph contains k pairwise disjoint paths, each connecting a different terminal pair from a set of k distinct pairs. We determine, with an exception
of two cases, the complexity of the DISJOINT PATHS problem for H-free
graphs. If k is fixed, we obtain the k-DISJOINT PATHS problem, which is
known to be polynomial-time solvable on the class of all graphs for every
$k \geq 1$. The latter does no longer hold if we need to connect vertices from
terminal sets instead of terminal pairs. We completely classify the complexity of k-DISJOINT CONNECTED SUBGRAPHS for H-free graphs, and
give the same almost-complete classification for DISJOINT CONNECTED
SUBGRAPHS for H-free graphs as for DISJOINT PATHS.

1 Introduction

A path from s to t in a graph G is an *s-t-path* of G, and s and t are called its
terminals. Two pairs (s_1, t_1) and (s_2, t_2) are *disjoint* if $\{s_1, t_1\} \cap \{s_2, t_2\} = \emptyset$. In
1980, Shiloach [19] gave a polynomial-time algorithm for testing if a graph with
disjoint terminal pairs (s_1, t_1) and (s_2, t_2) has vertex-disjoint paths P^1 and P^2
such that each P^i is an s_i-t_i path. This problem can be generalized as follows.

DISJOINT PATHS
> *Instance:* a graph G and pairwise disjoint terminal pairs $(s_1, t_1) \ldots, (s_k, t_k)$.
> *Question:* Does G have pairwise vertex-disjoint paths $P^1, \ldots, P^k$ such that P^i
> is an s_i-t_i path for $i \in \{1, \ldots, k\}$?

Karp [12] proved that DISJOINT PATHS is NP-complete. If k is fixed, that is, not
part of the input, then we denote the problem as k-DISJOINT PATHS. For every
$k \geq 1$, Robertson and Seymour proved the following celebrated result.

W. Kern—Recently passed away and we are grateful for his contribution.
D. Paulusma—Supported by the Leverhulme Trust (RPG-2016- 258).

© Springer Nature Switzerland AG 2021
P. Flocchini and L. Moura (Eds.): IWOCA 2021, LNCS 12757, pp. 414–427, 2021.
https://doi.org/10.1007/978-3-030-79987-8_29

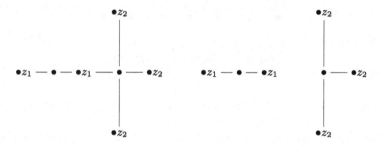

Fig. 1. An example of a yes-instance (G, Z_1, Z_2) of (2-)DISJOINT CONNECTED SUB-GRAPHS (left) together with a solution (right).

Theorem 1 ([18]). *For all $k \geq 2$, k-DISJOINT PATHS is polynomial-time solvable.*

The running time in Theorem 1 is cubic. This was later improved to quadratic time by Kawarabayashi, Kobayashi and Reed [13].

As DISJOINT PATHS is NP-complete, it is natural to consider special graph classes. The DISJOINT PATHS problem is known to be NP-complete even for graph of clique-width at most 6 [8], split graphs [9], interval graphs [15] and line graphs. The latter result can be obtained by a straightforward reduction (see, for example, [8,9]) from its edge variant, EDGE DISJOINT PATHS, proven to be NP-complete by Even, Itai and Shamir [5]. On the positive side, DISJOINT PATHS is polynomial-time solvable for cographs, or equivalently, P_4-free graphs [8].

We can generalize the DISJOINT PATHS problem by considering terminal sets Z_i instead of terminal pairs (s_i, t_i). We write $G[S]$ for the subgraph of a graph $G = (V, E)$ induced by $S \subseteq V$, where S is *connected* if $G[S]$ is connected.

DISJOINT CONNECTED SUBGRAPHS
    Instance: a graph G and pairwise disjoint terminal sets $Z_1, \ldots, Z_k$.
    Question: Does G have pairwise disjoint connected sets $S_1, \ldots, S_k$ such
            that $Z_i \subseteq S_i$ for $i \in \{1, \ldots, k\}$?

If k is fixed, then we write k-DISJOINT CONNECTED SUBGRAPHS. We refer to Fig. 1 for a simple example of an instance (G, Z_1, Z_2) of 2-DISJOINT CONNECTED SUBGRAPHS. Robertson and Seymour [18] proved in fact that k-DISJOINT CONNECTED SUBGRAPHS is cubic-time solvable as long as $|Z_1| + \ldots + |Z_k|$ is fixed (this result implies Theorem 1). Otherwise, van 't Hof et al. [22] proved that already 2-DISJOINT CONNECTED SUBGRAPHS is NP-complete even if $|Z_1| = 2$ (and $|Z_2|$ may have arbitrarily large size). The same authors also proved that 2-DISJOINT CONNECTED SUBGRAPHS is NP-complete for split graphs. Afterwards, Gray et al. [7] proved that 2-DISJOINT CONNECTED SUBGRAPHS is NP-complete for planar graphs. Hence, Theorem 1 cannot be extended to hold for k-DISJOINT CONNECTED SUBGRAPHS.

Fig. 2. The graph $H = 3P_1 + P_4$.

We note that in recent years a number of exact algorithms were designed for k-DISJOINT CONNECTED SUBGRAPHS. Cygan et al. [4] gave an $O^*(1.933^n)$-time algorithm for the case $k = 2$ (see [17,22] for faster exact algorithms for special graph classes). Telle and Villanger [20] improved this to time $O^*(1.7804^n)$. Recently, Agrawal et al. [1] gave an $O^*(1.88^n)$-time algorithm for the case $k = 3$. Moreover, the 2-DISJOINT CONNECTED SUBGRAPHS problem plays a crucial role in graph contractibility: a connected graph can be contracted to the 4-vertex path if and only if there exist two vertices u and v such that $(G - \{u, v\}, N(u), N(v))$ is a yes-instance of 2-DISJOINT CONNECTED SUBGRAPHS (see, e.g. [14,22]).

A class of graphs that is closed under vertex deletion is called *hereditary*. Such a graph class can be characterized by a unique set $\mathcal{F}$ of minimal forbidden induced subgraphs. Hereditary graphs enable a systematic study of the complexity of a graph problem under input restrictions: by starting with the case where $|\mathcal{F}| = 1$, we may already obtain more general methodology and a better understanding of the complexity of the problem. If $|\mathcal{F}| = 1$, say $\mathcal{F} = \{H\}$ for some graph H, then we obtain the class of H-*free* graphs, that is, the class of graphs that do not contain H as an induced subgraph (so, an H-free graph cannot be modified to H by vertex deletions only). In this paper, we start such a systematic study for DISJOINT PATHS and DISJOINT CONNECTED SUBGRAPHS, both for the case when k is part of the input and when k is fixed.

Our Results

By combining some of the aforementioned known results with a number of new results, we prove the following two theorems in Sects. 3 and 4, respectively. In particular, we generalize the polynomial-time result for DISJOINT PATHS on P_4-free graphs to hold even for DISJOINT CONNECTED SUBGRAPHS. See Fig. 2 for an example of a graph $H = sP_1 + P_4$; we refer to Sect. 2 for undefined terminology.

Theorem 2. *Let H be a graph. If $H \subseteq_i sP_1 + P_4$, then for every $k \geq 2$, k-DISJOINT CONNECTED SUBGRAPHS on H-free graphs is polynomial-time solvable; otherwise even 2-DISJOINT CONNECTED SUBGRAPHS is NP-complete.*

Theorem 3. *Let H be a graph not in $\{3P_1, 2P_1 + P_2, P_1 + P_3\}$. If $H \subseteq_i P_4$, then DISJOINT CONNECTED SUBGRAPHS is polynomial-time solvable for H-free graphs; otherwise even DISJOINT PATHS is NP-complete.*

Theorem 2 completely classifies, for every $k \geq 2$, the complexity of k-DISJOINT CONNECTED SUBGRAPHS on H-free graphs. Theorem 3 determines the complexity of DISJOINT PATHS and DISJOINT CONNECTED SUBGRAPHS on H-free graphs for every graph H except if $H \in \{3P_1, 2P_1 + P_2, P_1 + P_3\}$. In Sect. 5 we reduce the number of open cases from six to *three* by showing some equivalencies.

In Sect. 6 we give some directions for future work. In particular we prove that both problems are polynomial-time solvable for co-bipartite graphs, which form a subclass of the class of $3P_1$-free graphs and give exact algorithms for both problems based on Held-Karp type dynamic programming techniques [2, 10].

2 Preliminaries

We use $H \subseteq_i H'$ to indicate that H is an induced subgraph of H', that is, H can be obtained from H' by a sequence of vertex deletions. For two graphs G_1 and G_2 we write $G_1 + G_2$ for the *disjoint union* $(V(G_1) \cup V(G_2), E(G_1) \cup E(G_2))$. We denote the disjoint union of r copies of a graph G by rG. A graph is said to be a linear forest if it is a disjoint union of paths.

We denote the path and cycle on n vertices by P_n and C_n, respectively. The *girth* of a graph that is not a forest is the number of edges of a smallest induced cycle in it.

The *line graph* $L(G)$ of a graph G has vertex set $E(G)$ and there exists an edge between two vertices e and f in $L(G)$ if and only if e and f have a common end-vertex in G. The claw $K_{1,3}$ is the 4-vertex star. It is readily seen that every line graph is claw-free. Recall that a graph is H-free if it does not contain H as induced subgraph. For a set of graphs $\{H_1, \ldots, H_r\}$, we say that a graph G is $(H_1, \ldots, H_r)$-*free* if G is H_i-free for every $i \in \{1, \ldots, r\}$.

A *clique* is a set of pairwise adjacent vertices and an *independent set* is a set of pairwise non-adjacent vertices. A graph is *split* if its vertex set can be partitioned into two (possibly empty) sets, one of which is a clique and the other is an independent set. A graph is split if and only if it is (C_4, C_5, P_4)-free [6]. A graph is a *cograph* if it can be defined recursively as follows: any single vertex is a cograph, the disjoint union of two cographs is a cograph, and the join of two cographs G_1, G_2 is a cograph (the *join* adds all edges between the vertices of G_1 and G_2). A graph is a cograph if and only if it is P_4-free [3].

A graph $G = (V, E)$ is *multipartite*, or more specifically, r-*partite* if V can be partitioned into r (possibly empty) sets $V_1, \ldots, V_r$, such that there is an edge between two vertices u and v if and only if $u \in V_i$ and $v \in V_j$ for some i, j with $i \neq j$. If $r = 2$, we also say that G is *bipartite*. If there exist an edge between every vertex of V_i and every vertex of V_j for every $i \neq j$, then the multipartite graph G is *complete*.

The *complement* of a graph $G = (V, E)$ is the graph $\overline{G} = (V, \{uv \mid u, v \in V, u \neq v \text{ and } uv \notin E\})$. The complement of a bipartite graph is a *cobipartite* graph. A set $W \subseteq V$ is a *dominating set* of a graph G if every vertex of $V \backslash W$ has a neighbour in W, or equivalently, $N[W]$ (the closed neighbourhood of W) is equal to V. We say that W is a *connected dominating set* if W is a dominating set and $G[W]$ is connected.

3   The Proof of Theorem 2

We consider k-DISJOINT CONNECTED SUBGRAPHS for fixed k. First, we show a polynomial-time algorithm on H-free graphs when $H \subseteq_i sP_1 + P_4$ for some fixed $s \geq 0$. Then, we prove the hardness result.

For the algorithm, we need the following lemma for P_4-free graphs, or equivalently, cographs. This lemma is well known and follows immediately from the definition of a cograph: in the construction of a connected cograph G, the last operation must be a join, so there exists cographs G_1 and G_2, such that G obtained from adding an edge between every vertex of G_1 and every vertex of G_2. Hence, the spanning complete bipartite graph of G has non-empty partition classes $V(G_1)$ and $V(G_2)$.

Lemma 1. *Every connected P_4-free graph on at least two vertices has a spanning complete bipartite subgraph.*

Two instances of a problem Π are *equivalent* when one of them is a yes-instance of Π if and only if the other one is a yes-instance of Π. We note that if two adjacent vertices will always appear in the same set of every solution $(S_1, \ldots, S_k)$ for an instance $(G, Z_1, \ldots, Z_k)$, then we may contract the edge between them at the start of any algorithm. This takes linear time. Moreover, H-free graphs are readily seen (see e.g. [14]) to be closed under edge contraction if H is a linear forest. Hence, we can make the following observation.

Lemma 2. *For $k \geq 2$, from every instance of $(G, Z_1, \ldots, Z_k)$ of k-DISJOINT CONNECTED SUBGRAPHS we can obtain in polynomial time an equivalent instance $(G', Z'_1, \ldots, Z'_k)$ such that every Z'_i is an independent set. Moreover, if G is H-free for some linear forest H, then G' is also H-free.*

We can now prove the following lemma.

Lemma 3. *Let H be a graph. If $H \subseteq_i sP_1 + P_4$, then for every $k \geq 1$, k-DISJOINT CONNECTED SUBGRAPHS on H-free graphs is polynomial-time solvable.*

Proof. Let $H \subseteq_i sP_1 + P_4$ for some $s \geq 0$. Let $(G, Z_1, \ldots, Z_k)$ be an instance of k-DISJOINT CONNECTED SUBGRAPHS, where G is an H-free graph. By Lemma 2, we may assume without loss of generality that G is connected and moreover that $Z_1, \ldots, Z_k$ are all independent sets.

We first analyze the structure of a solution $(S_1, \ldots, S_k)$ (if it exists). For $i \in \{1, \ldots, k\}$, we may assume that S_i is inclusion-wise minimal, meaning there is no $S'_i \subset S_i$ that contains Z_i and is connected. Consider a graph $G[S_i]$. Either $G[S_i]$ is P_4-free or $G[S_i]$ contains an induced $rP_1 + P_4$ for some $0 \leq r \leq s - 1$. We will now show that in both cases, S_i is the (not necessarily disjoint) union of Z_i and a connected dominating set of $G[S_i]$ of constant size.

First suppose that $G[S_i]$ is P_4-free. As $G[S_i]$ is connected and Z_i is independent, we apply Lemma 1 to find that $S_i \backslash Z_i$ contains a vertex u that is adjacent to every vertex of Z_i. Hence, by minimality, $S_i = Z_i \cup \{u\}$ and $\{u\}$ is a connected dominating set of $G[S_i]$ of size 1.

Now suppose that $G[S_i]$ has an induced $rP_1 + P_4$ for some $r \geq 0$, where we choose r to be maximum. Note that $r \leq s - 1$. Let W be the vertex set of the induced $rP_1 + P_4$. Then, as r is maximum, W dominates $G[S_i]$. Note that $G[W]$ has $r + 1 \leq s$ connected components. Then, as $G[S_i]$ is connected and W is a dominating set of $G[S_i]$ of size $r + 4 \leq s + 3$, it follows from folklore arguments (see e.g. [21, Prop. 6.3.24]) that $G[S_i]$ has a connected dominating set W' of size at most $3s + 1$. Moreover, by minimality, $S_i = Z_i \cup W'$.

Hence, in both cases we find that S_i is the union of Z_i and a connected dominating set of $G[S_i]$ of size at most $t = 3s + 1$; note that t is a constant, as s is a constant.

Our algorithm now does as follows. We consider all options of choosing a connected dominating set of each $G[S_i]$, which from the above has size at most t. As soon as one of the guesses makes every Z_i connected, we stop and return the solution. The total number of options is $O(n^{tk})$, which is polynomial as k and t are fixed. Moreover, checking the connectivity condition can be done in polynomial time. Hence, the total running time of the algorithm is polynomial. □

The proof our next result is inspired by the aforementioned NP-completeness result of [22] for instances (G, Z_1, Z_2) where $|Z_1| = 2$ but G is a general graph.

Lemma 4. *The* 2-DISJOINT CONNECTED SUBGRAPHS *problem is* NP-*complete even on instances* (G, Z_1, Z_2) *where* $|Z_1| = 2$ *and* G *is a line graph.*

Proof. Note that the problem is in NP. We reduce from 3-SAT. Let $\phi = \phi(x_1, \ldots, x_n)$ be an instance of 3-SAT with clauses $C_1, \ldots, C_m$. We construct a corresponding graph $G = (V, E)$ as follows. We start with two disjoint paths P and $\bar{P}$ on vertices p_i, x_i, q_i and $\bar{p}_i, \bar{x}_i, \bar{q}_i$, respectively, where $x_i, \bar{x}_i$ correspond to the positive and negative literals in ϕ, respectively. To be more precise, we define:

$$P = p_1, x_1, q_1, p_2, x_2, q_2, \ldots, p_n, x_n, q_n, \text{ and } \overline{P} = \bar{p}_1, \bar{x}_1, \bar{q}_1, \ldots, \bar{p}_n, \bar{x}_n, \bar{q}_n,$$

We add the two edges $e = p_1\bar{p}_1$, and $f = q_n\bar{q}_n$. For $i = 1, \ldots, n-1$, we also add edges $q_i\bar{p}_{i+1}$ and $\bar{q}_ip_{i+1}$. We now replace each x_i by vertices $x_i^{j_1}, x_i^{j_2}, \ldots x_i^{j_r}$, where $j_1, \ldots, j_r$ are the indices of the clauses C_j that contain x_i. That is, we replace the subpath p_i, x_i, q_i of P by the path $p_i, x_i^{j_1}, x_i^{j_2}, \ldots x_i^{j_r}, q_i$. We do the same path replacement operation on $\bar{P}$ with respect to every $\bar{x}_i$. Finally, we add every clause C_j as a vertex and add an edge between C_j and x_i^j if and only if $x_i \in C_j$, and between C_j and $\bar{x}_i^j$ if and only if $\bar{x}_j \in C_j$. This completes the description of $G = (V, E)$. We refer to Fig. 3 for an illustration of our construction.

We now focus on the line graph $L = L(G)$ of G. Let $Z_1 = \{e, f\} \subseteq E = V(L)$ and let Z_2 consist of all vertices of L that correspond to edges in G that are incident to some C_j. Note that Z_1 and Z_2 are disjoint. Moreover, each clause C_j corresponds to a clique of size at most 3 in L, which we call the clause clique of C_j. We claim that ϕ is satisfiable if and only if the instance (L, Z_1, Z_2) of 2-DISJOINT CONNECTED SUBGRAPHS is a yes-instance.

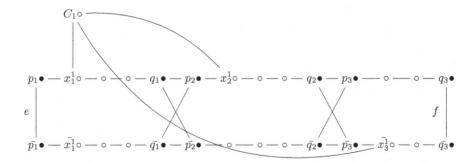

Fig. 3. The construction described with edges added for the clause $C_1 = (x_1 \vee x_2 \vee \bar{x}_3)$.

First suppose that ϕ is satisfiable. Let τ be a satisfying truth assignment for ϕ. In G, we let P^1 denote the unique path whose first edge is e and whose last edge is f and that passes through all $x_i^j \in V$ if $x_i = 0$ and through all $\bar{x}_i^j$ if $x_i = 1$. In L we let S_1 consist of all vertices of $L(P^1)$; note that $Z_1 = \{e, f\}$ is contained in S_1 and that S_1 is connected. We let P^2 denote the "complementary" path in G whose first edge is e and whose last edge is f but that passes through all x_i^j if and only if P^1 passes through all $\bar{x}_i^j$, and conversely ($i = 1, \ldots, n$). In L, we put all vertices of $L(P^2)$, except e and f, together with all vertices of Z_2 in S_2. As τ satisfies ϕ, some vertex of each clause clique is adjacent to a vertex of P^2. Hence, as P^2 is a path, S_2 is connected and we found a solution for (L, Z_1, Z_2).

Now suppose that (L, Z_1, Z_2) is a yes-instance of 2-DISJOINT CONNECTED SUBGRAPHS. Then $V(L)$ can be partitioned into two vertex-disjoint connected sets S_1 and S_2 such that $Z_1 \subseteq S_1$ and $Z_2 \subseteq S_2$. In particular, $L[S_1]$ contains a path P^1 from e to f. In fact, we may assume that $S_1 = V(P^1)$, as we can move every other vertex of S_1 (if they exist) to S_2 without disconnecting S_2.

Note that P^1 corresponds to a connected subgraph that contains the adjacent vertices p_1 and $\bar{p}_1$ as well as the adjacent vertices q_n and $\bar{q}_n$. Hence, we can modify P^1 into a path Q in G that starts in p_1 or $\bar{p}_1$ and that ends in q_n or $\bar{q}_n$. Note that Q contains no edge incident to a clause vertex C_j, as those edges correspond to vertices in L that belong to Z_2. Hence, by construction, Q "moves from left to right", that is, Q cannot pass through both some x_i^j and $\bar{x}_i^j$ (as then Q needs to pass through either x_i^j or $\bar{x}_i^j$ again implying that Q is not a path).

Moreover, if Q passes through some x_i^j, then Q must pass through all vertices x_i^{jh}. Similarly, if Q passes through some $\bar{x}_i^j$, then Q must pass through all vertices $\bar{x}_i^{jh}$. As Q connects the edges $p_1 \bar{p}_1$ and $q_n \bar{q}_n$, we conclude that Q must pass, for $i = 1, \ldots, n$, through either every x_i^{jh} or through every $\bar{x}_i^{jh}$. Thus we may define a truth assignment τ by setting

$$x_i = \begin{cases} 1 \text{ if } Q \text{ passes through all } \bar{x}_i^j \\ 0 \text{ if } Q \text{ passes through all } x_i^j. \end{cases}$$

We claim that τ satisfies ϕ. For contradiction, assume some clause C_j is not satisfied. Then Q passes through all its literals. However, then in S_2, the vertices of Z_2 that correspond to edges incident to C_j are not connected to other vertices of Z_2, a contradiction. This completes the proof of the lemma.   □

A straightforward modification of the reduction of Lemma 5 gives us Lemma 6. We can also obtain Lemma 6 by subdividing the graph G in the proof of Lemma 4 twice (to get a bipartite graph) or p times (to get a graph of girth at least p).

Lemma 5 ([22]). 2-Disjoint Connected Subgraphs *is* NP-*complete for split graphs, or equivalently, $(2P_2, C_4, C_5)$-free graphs.*

Lemma 6. 2-Disjoint Connected Subgraphs *is* NP-*complete for bipartite graphs and for graphs of girth at least p, for every integer $p \geq 3$.*

We are now ready to prove Theorem 2.

Theorem 2 (restated). *Let H be a graph. If $H \subseteq_i sP_1 + P_4$, then for every $k \geq 1$, k-Disjoint Connected Subgraphs on H-free graphs is polynomial-time solvable; otherwise even* 2-Disjoint Connected Subgraphs *is* NP-*complete.*

Proof. If H contains an induced cycle C_s for some $s \geq 3$, then we apply Lemma 6 by setting $p = s + 1$. Now assume that H contains no cycle, that is, H is a forest. If H has a vertex of degree at least 3, then H is a superclass of the class of claw-free graphs, which in turn contains all line graphs. Hence, we can apply Lemma 4. In the remaining case H is a linear forest. If H contains an induced $2P_2$, we apply Lemma 5. Otherwise H is an induced subgraph of $sP_1 + P_4$ for some $s \geq 0$ and we apply Lemma 3.   □

4   The Proof of Theorem 3

We first prove the following result, which generalizes the corresponding result of Disjoint Paths for P_4-free graphs due to Gurski and Wanke [8]. We show that we can use the same modification to a matching problem in a bipartite graph.

Lemma 7. Disjoint Connected Subgraphs *is polynomial-time solvable for P_4-free graphs.*

Proof. For some integer $k \geq 2$, let $(G, Z_1, \ldots, Z_k)$ be an instance of Disjoint Connected Subgraphs where G is a P_4-free graph. By Lemma 2 we may assume that every Z_i is an independent set. Now suppose that $(G, Z_1, \ldots, Z_k)$ has a solution $(S_1, \ldots, S_k)$. Then $G[S_i]$ is a connected P_4-free graph. Hence, by Lemma 1, $G[S_i]$ has a spanning complete bipartite graph on non-empty partition classes A_i and B_i. As every Z_i is an independent set, it follows that either $Z_i \subseteq A_i$ or $Z_i \subseteq B_i$. If $Z_i \subseteq A_i$, then every vertex of B_i is adjacent to every vertex of Z_i. Similarly, if $Z_i \subseteq B_i$, then every vertex of A_i is adjacent to every vertex of Z_i. We conclude that in every set S_i, there exists a vertex y_i such that $Z_i \cup \{y_i\}$ is connected.

The latter enables us to construct a bipartite graph $G' = (X \cup Y, E')$ where X contains vertices $x_1, \ldots, x_k$ corresponding to the set $Z_1, \ldots, Z_k$ and Y is the set of non-terminal vertices of G. We add an edge between $x_i \in X$ and $y \in Y$ if and only if y is adjacent to every vertex of Z_i. Then $(G, Z_1 \ldots Z_k)$ is a yes-instance of DISJOINT CONNECTED SUBGRAPHS if and only if G' contains a matching of size k. It remains to observe that we can find a maximum matching in polynomial time, for example, by using the Hopcroft-Karp algorithm for bipartite graphs [11].   □

The first lemma of a series of four is obtained by a straightforward reduction from the EDGE DISJOINT PATHS problem (see, e.g. [8,9]), which was proven to be NP-complete by Even, Itai and Shamir [5]. The second lemma follows from the observation that an edge subdivision of the graph G in an instance of DISJOINT PATHS results in an equivalent instance of DISJOINT PATHS; we apply this operation a sufficiently large number of times to obtain a graph of large girth. The third lemma is due to Heggernes et al. [9]. We modify their construction to prove the fourth lemma.

Lemma 8. DISJOINT PATHS *is* NP-*complete for line graphs.*

Lemma 9. *For every* $g \geq 3$, DISJOINT PATHS *is* NP-*complete for graphs of girth at least* g.

Lemma 10 ([9]). DISJOINT PATHS *is* NP-*complete for split graphs, or equivalently,* $(C_4, C_5, 2P_2)$-*free graphs.*

Lemma 11. DISJOINT PATHS *is* NP-*complete for* $(4P_1, P_1 + P_4)$-*free graphs.*

Proof. We reduce from DISJOINT PATHS on split graphs, which is NP-complete by Lemma 10. By inspection of this result (see [9, Theorem 3]), we note that the instances $(G, \{(s_1, t_1), \ldots, (s_k, t_k)\})$ have the following property: the split graph G has a split decomposition (C, I), where C is a clique, I an independent set, C and I are disjoint, and $C \cup I = V(G)$, such that $I = \{s_1, \ldots, s_k, t_1, \ldots, t_k\}$. Now let G' be obtained from G by, for each terminal s_i, adding edges to s_j and t_j for all $j \neq i$. Then consider the instance $(G', \{(s_1, t_1), \ldots, (s_k, t_k)\})$.

We note that $G'[C]$ is still a complete graph, while $G'[I]$ is a complete graph minus a matching. It is immediate that G' is $4P_1$-free. Moreover, any induced subgraph H of G' that is isomorphic to P_4 must contain at least two vertices of I and at least one vertex of C. If H contains two vertices of C, then as $G'[C]$ is a clique, H contains two non-adjacent vertices in I. Similarly, if H contains one vertex of C (and thus three vertices of I), then H contains two non-adjacent vertices in I. Since C is a clique in G' and every (other) vertex of I is adjacent in G' to any pair of non-adjacent vertices of I, it follows that G' is $P_1 + P_4$-free as well.

We claim that $(G, \{(s_1, t_1), \ldots, (s_k, t_k)\})$ is a yes-instance if and only if $(G', \{(s_1, t_1), \ldots, (s_k, t_k)\})$ is a yes-instance. This is because the edges that were added to G to obtain G' are only between terminal vertices of different pairs. These edges cannot be used by any solution of DISJOINT PATHS for $(G', \{(s_1, t_1), \ldots, (s_k, t_k)\})$, and thus the feasibility of the instance is not affected by the addition of these edges.   □

We are now ready to prove Theorem 3.

Theorem 3 (restated). *Let H be a graph not in $\{3P_1, 2P_1 + P_2, P_1 + P_3\}$. If $H \subseteq_i P_4$, then* DISJOINT CONNECTED SUBGRAPHS *is polynomial-time solvable for H-free graphs; otherwise even* DISJOINT PATHS *is* NP*-complete.*

Proof. First suppose that H contains a cycle C_r for some $r \geq 3$. Then DISJOINT PATHS is NP-complete for the class of H-free graphs, as DISJOINT PATHS is NP-complete on the subclass consisting of graphs of girth $r + 1$ by Lemma 9. Now suppose that H contains no cycle, that is, H is a forest. If H contains a vertex of degree at least 3, then the class of H-free graphs contains the class of claw-free graphs, which in turn contains the class of line graphs. Hence, we can apply Lemma 8. It remains to consider the case where H is a forest with no vertices of degree at least 3, that is, when H is a linear forest.

If H contains four connected components, then the class of H-free graphs contains the class of $4P_1$-free graphs, and we can use Lemma 11. If H contains an induced P_5 or two connected components that each have at least one edge, then H contains the class of $2P_2$-free graphs, and we can use Lemma 10. If H contains two connected components, one of which has at least four vertices, then H contains the class of $(P_1 + P_4)$-free graphs, and we can use Lemma 11 again. As $H \notin \{3P_1, 2P_1 + P_2, P_1 + P_3\}$, this means that in the remaining case H is an induced subgraph of P_4. In that case even DISJOINT CONNECTED SUBGRAPHS is polynomial-time solvable on H-free graphs, due to Lemma 7. □

5   Reducing the Number of Open Cases to Three

Theorem 3 shows that we have the same three open cases for DISJOINT PATHS and DISJOINT CONNECTED SUBGRAPHS, namely when $H \in \{3P_1, P_1 + P_3, 2P_1 + P_2\}$. We show that instead of six open cases, we have in fact only three.

Proposition 1. DISJOINT PATHS *and* DISJOINT CONNECTED SUBGRAPHS *are equivalent for $3P_1$-free graphs.*

Proof. Every instance of DISJOINT PATHS is an instance of DISJOINT CONNECTED SUBGRAPHS. Let $(G, Z_1, \ldots, Z_k)$ be an instance of DISJOINT CONNECTED SUBGRAPHS where G is a $3P_1$-free graph. By Lemma 2 we may assume that each Z_i is an independent set. Then, as G is $3P_1$-free, each Z_i has size at most 2. So we obtained an instance of DISJOINT PATHS. □

Proposition 2. DISJOINT PATHS *on $(P_1 + P_3)$-free graphs and* DISJOINT CONNECTED SUBGRAPHS *on $(P_1 + P_3)$-free graphs are polynomially equivalent to* DISJOINT PATHS *on $3P_1$-free graphs.*

Proof. We prove that we can solve DISJOINT CONNECTED SUBGRAPHS in polynomial time on $(P_1 + P_3)$-free graphs if we have a polynomial-time algorithm for DISJOINT PATHS on $3P_1$-free graphs. Showing this suffices to prove the theorem, as DISJOINT PATHS is a special case of DISJOINT CONNECTED SUBGRAPHS and $3P_1$-free graphs form a subclass of $(P_1 + P_3)$-free graphs.

Let $(G, Z_1, \ldots, Z_k)$ be an instance of DISJOINT CONNECTED SUBGRAPHS, where G is a $(P_1 + P_3)$-free graph. Olariu [16] proved that every connected $\overline{P_1 + P_3}$-free graph is either triangle-free or complete multipartite. Hence, the vertex set of G can be partitioned into sets $D_1, \ldots, D_p$ for some $p \geq 1$ such that

– every $G[D_i]$ is $3P_1$-free or the disjoint union of complete graphs, and
– for every i, j with $i \neq j$, every vertex of D_i is adjacent to every vertex of D_j.

Using this structural characterization, we first argue that we may assume that each Z_i has size 2, making the problem an instance of DISJOINT PATHS. Then we show that we can either solve the instance outright or can alter G to be $3P_1$-free.

First, we argue about the size of each Z_i. By Lemma 2 we may assume that every Z_i is an independent set and is thus contained in the same set D_j. If $G[D_j]$ is $3P_1$-free, then this implies that any Z_i that is contained in D_j has size 2. If $G[D_j]$ is a disjoint union of complete graphs, then each vertex of a Z_i that is contained in D_j belongs to a different connected component of D_j and $Z_i \cup \{v\}$ is connected for every vertex $v \notin D_j$. As at least one vertex $v \notin D_j$ is needed to make such a set Z_i connected, we may therefore assume that for a solution $(S_1, \ldots, S_k)$ (if it exists), $S_i = Z_i \cup \{v\}$ for some $v \notin D_j$. The latter implies that we may assume without loss of generality that every such Z_i has size 2 as well.

If $p = 1$, then each connected component of G is $3P_1$-free, and we are done. Hence, we assume that $p \geq 2$. In fact, since any two distinct sets D_i and D_j are complete to each other, the union of any two $3P_1$-free graphs induces a $3P_1$-free graph. Therefore we may assume without loss of generality that only $G[D_1]$ might be $3P_1$-free, whereas $G[D_2], \ldots, G[D_p]$ are disjoint unions of complete graphs.

Recall that $Z_i = \{s_i, t_i\}$ for every $i \in \{1, \ldots, k\}$ and we search for a solution $(P^1, \ldots, P^k)$ where each P^i is a path from s_i to t_i. First suppose s_i and t_i belong to D_1. Then P^i has length 2 or 3 and in the latter case, $V(P^i) \subseteq D_1$. Now suppose that s_i and t_i belong to D_h for some $h \in \{2, \ldots, k\}$. Then P^i has length exactly 2, and moreover, the middle (non-terminal) vertex of P^i does not belong to D_h.

We will now check if there is a solution $(P^1, \ldots, P^k)$ such that every P^i has length exactly 2. We call such a solution to be of *type 1*. In a solution of type 1, every $P^i = s_i u t_i$ for some non-terminal vertex u of G. If s_i and t_i belong to D_h for some $h \in \{2, \ldots, p\}$, then $u \in D_j$ for some $j \neq i$. If s_i and t_i belong to D_1, then $u \in D_j$ for some $j \neq 1$ but also $u \in D_1$ is possible, namely when u is adjacent to both s_i and t_i.

Verifying the existence of a type 1 solution is equivalent to finding a perfect matching in a bipartite graph $G' = A \cup B$ that is defined as follows. The set A consists of one vertex v_i for each pair $\{s_i, t_i\}$. The set B consists of all non-terminal vertices u of G. For $\{s_i, t_i\} \subseteq D_1$, there exists an edge between u and v_i in G' if and only if in G it holds that $u \in D_h$ for some $h \in \{2, \ldots, p\}$ or $u \in D_1$ and u is adjacent to both s_i and t_i. For $\{s_i, t_i\} \subseteq D_h$ with $h \in \{2, \ldots, p\}$, there exists an edge between u and v_i in G' if and only if in G it holds that $u \in D_j$

for some $j \in \{1, \ldots, p\}$ with $h \neq j$. We can find a perfect matching in G' in polynomial time by using the Hopcroft-Karp algorithm for bipartite graphs [11].

Suppose that we find that $(G, \{s_1, t_1\}, \ldots, \{s_k, t_k\})$ has no solution of type 1. As a solution can be assumed to be of type 1 if $G[D_1]$ is the disjoint union of complete graphs, we find that $G[D_1]$ is not of this form. Hence, $G[D_1]$ is $3P_1$-free. Recall that $G[D_j]$ is the disjoint union of complete graphs for $2 \leq i \leq p$. It remains to check if there is a solution that is of *type 2* meaning a solution $(P^1, \ldots, P^k)$ in which at least one P^i, whose vertices all belong to D_1, has length 3.

To find a type 2 solution (if it exists) we construct the following graph G^*. We let $V(G^*) = A_1 \cup A_2 \cup B_1 \cup B_2$, where

- A_1 consists of all terminal vertices from D_1;
- A_2 consists of all non-terminal vertices from D_1;
- B_1 consists of all terminal vertices from $D_2 \cup \cdots \cup D_p$; and
- B_2 consists of all non-terminal vertices from $D_2 \cup \cdots \cup D_p$.

Note that $V(G^*) = V(G)$. To obtain $E(G^*)$ from $E(G)$ we add some edges (if they do not exist in G already) and also delete some edges (if these existed in G):

(i) for each $\{s_i, t_i\} \subseteq B_1$, add all edges between s_i and vertices of B_2, and delete any edges between t_i and vertices of B_2;

(ii) add an edge between every two terminal vertices in B_1 that belong to different terminal pairs; and

(iii) add an edge between every two vertices of B_2.

We note that $G^*[D_1]$ is the same graph as $G[D_1]$ and thus $G^*[D_1]$ is $3P_1$-free. Moreover, $G^*[B_1 \cup B_2]$ is $3P_1$-free by part (i) of the construction. Hence, as there exists an edge between every vertex of $A_1 \cup A_2$ and every vertex of $B_1 \cup B_2$ in G and thus also in G^*, this means that G^* is $3P_1$-free. It remains to prove that $(G, \{s_1, t_1\}, \ldots, \{s_k, t_k\})$ and $(G^*, \{s_1, t_1\}, \ldots, \{s_k, t_k\})$ are equivalent instances.

First suppose that $(G, \{s_1, t_1\}, \ldots, \{s_k, t_k\})$ has a solution $(P^1, \ldots, P^k)$. Assume that the number of paths of length 3 in this solution is minimum over all solutions for $(G, \{s_1, t_1\}, \ldots, \{s_k, t_k\})$. We note that $(P^1, \ldots, P^k)$ is a solution for $(G^*, \{s_1, t_1\}, \ldots, \{s_k, t_k\})$ unless there exists some P^i that contains an edge of $E(G) \backslash E(G^*)$. Suppose this is indeed the case. As $G^*[D_1] = G[D_1]$ and every edge between a vertex of $A_1 \cup A_2$ and a vertex of $B_1 \cup B_2$ also exists in G^*, we find that the paths connecting terminals from pairs in D_1 are paths in G^*. Hence, s_i and t_i belong to D_h for some $h \in \{2, \ldots, p\}$ and thus $P^i = s_i u t_i$ where u is a vertex of D_j for some $j \in \{2, \ldots, p\}$ with $j \neq h$.

As we already found that $(G, \{s_1, t_1\}, \ldots, \{s_k, t_k\})$ has no type 1 solution, there is at least one $P^{i'}$ with length 3, so $P^{i'} = s_{i'} v v' t_{i'}$ is in $G[D_1]$. However, we can now obtain another solution for $(G, \{s_1, t_1\}, \ldots, \{s_k, t_k\})$ by changing P^i into $s_i v t_i$ and $P^{i'}$ into $s_{i'} u t_{i'}$, a contradiction, as the number of paths of length 3 in $(P^1, \ldots, P^k)$ was minimum. We conclude that every P^i only contains edges from $E(G) \cap E(G^*)$, and thus $(P^1, \ldots, P^k)$ is a solution for $(G^*, \{s_1, t_1\}, \ldots, \{s_k, t_k\})$.

Now suppose that $(G^*, \{s_1, t_1\}, \ldots, \{s_k, t_k\})$ has a solution $(P^1, \ldots, P^k)$. Consider a path P^i. First suppose that s_i and t_i both belong to B_1. Then we may assume without loss of generality that $P^i = s_i u t_i$ for some $u \in A_2$. As B_1 only contains terminals from pairs in $D_2 \cup \ldots \cup D_p$, the latter implies that P^i is a path in G as well. Now suppose that s_i and t_i both belong to A_1. Then we may assume without loss of generality that $P^i = s_i u t_i$ for some non-terminal vertex of $V(G) = V(G^*)$ or $P^i = s_i u u' t_i$ for two vertices u, u' in $A_2 \subseteq D_1$. Hence, P^i is a path in G as well. We conclude that $(P^1, \ldots, P^k)$ is a solution for $(G, \{s_1, t_1\}, \ldots, \{s_k, t_k\})$. This completes our proof. $\qquad\square$

6 Conclusions

We first gave a dichotomy for DISJOINT k-CONNECTED SUBGRAPHS in Theorem 2: for every k, the problem is polynomial-time solvable on H-free graphs if $H \subseteq_i sP_1 + P_4$ for some $s \geq 0$ and otherwise it is NP-complete even for $k = 2$. Two vertices u and v are a P_4-*suitable pair* if $(G - \{u, v\}, N(u), N(v))$ is a yes-instance of 2-DISJOINT CONNECTED SUBGRAPHS. Recall that a graph G can be contracted to P_4 if and only if G has a P_4-suitable pair. Deciding if a pair $\{u, v\}$ is a suitable pair is polynomial-time solvable for H-free graphs if H is an induced subgraph of $P_2 + P_4$, $P_1 + P_2 + P_3$, $P_1 + P_5$ or $sP_1 + P_4$ for some $s \geq 0$; otherwise it is NP-complete [14]. Hence, we conclude from our new result that the presence of the two vertices u and v that are connected to the sets $Z_1 = N(u)$ and $Z_2 = N(v)$, respectively, yield exactly three additional polynomial-time solvable cases.

We also classified, in Theorem 3, the complexity of DISJOINT PATHS and DISJOINT CONNECTED SUBGRAPHS for H-free graphs. Due to Propositions 1 and 2, there are three non-equivalent open cases left and we ask the following:

Open Problem 1. *Determine the computational complexity of* DISJOINT PATHS *on H-free graph for* $H \in \{3P_1, 2P_1 + P_2\}$ *and the computational complexity of* DISJOINT CONNECTED SUBGRAPHS *on H-free graphs for* $H = 2P_1 + P_2$.

The three open cases seem challenging. We were able to prove the following positive result for a subclass of $3P_1$-free graphs, namely cobipartite graphs, or equivalently, $(3P_1, C_5, \overline{C_7}, \overline{C_9}, \ldots)$-free graphs (proof omitted).

Theorem 4. DISJOINT PATHS *is polynomial-time solvable for cobipartite graphs.*

Finally, we briefly mention exact algorithms. Using Held-Karp type dynamic programming techniques [2,10], we can obtain exact algorithms for DISJOINT PATHS and DISJOINT CONNECTED SUBGRAPHS running in time $O(2^n n^2 m)$ and $O(3^n km)$, respectively (proofs omitted). Faster exact algorithms are known for k-DISJOINT CONNECTED SUBGRAPHS for $k = 2$ and $k = 3$ [1,4,20], but we are unaware if there exist faster algorithms for general graphs.

Open Problem 2. *Is there an exact algorithm for* DISJOINT PATHS *or* DISJOINT CONNECTED SUBGRAPHS *on general graphs where the exponential factor is* $(2 - \epsilon)^n$ *or* $(3 - \epsilon)^n$, *respectively, for some* $\epsilon > 0$?

References

1. Agrawal, A., Fomin, F.V., Lokshtanov, D., Saurabh, S., Tale, P.: Path contraction faster than 2^n. SIAM J. Discrete Math. **34**, 1302–1325 (2020)
2. Bellman, R.: Dynamic programming treatment of the travelling salesman problem. J. ACM **9**, 61–63 (1962)
3. Corneil, D.G., Lerchs, H., Burlingham, L.S.: Complement reducible graphs. Discrete Appl. Math. **3**, 163–174 (1981)
4. Cygan, M., Pilipczuk, M., Pilipczuk, M., Wojtaszczyk, J.O.: Solving the 2-disjoint connected subgraphs problem faster than 2^n. Algorithmica **70**, 195–207 (2014)
5. Even, S., Itai, A., Shamir, A.: On the complexity of timetable and multicommodity flow problems. SIAM J. Comput. **5**, 691–703 (1976)
6. Földes, S., Hammer, P.L.: Split graphs. Congressus Numerantium, XIX:311–315 (1977)
7. Gray, C., Kammer, F., Löffler, M., Silveira, R.I.: Removing local extrema from imprecise terrains. Comput. Geom. Theory Appl. **45**, 334–349 (2012)
8. Gurski, F., Wanke, E.: Vertex disjoint paths on clique-width bounded graphs. Theor. Comput. Sci. **359**, 188–199 (2006)
9. Heggernes, P., van 't Hof, P., van Leeuwen, E.J., Saei, R.: Finding disjoint paths in split graphs. Theor. Comput. Syst. **57**(1), 140–159 (2014). https://doi.org/10.1007/s00224-014-9580-6
10. Held, M., Karp, R.M.: A dynamic programming approach to sequencing problems. J. Soc. Ind. Appl. Math. **10**, 196–210 (1962)
11. Hopcroft, J.E., Karp, R.M.: An $n^{5/2}$ algorithm for maximum matchings in bipartite graphs. SIAM J. Comput. **2**, 225–231 (1973)
12. Karp, R.M.: On the complexity of combinatorial problems. Networks **5**, 45–68 (1975)
13. Kawarabayashi, K., Kobayashi, Y., Reed, B.A.: The disjoint paths problem in quadratic time. J. Comb. Theory Ser. B **102**, 424–435 (2012)
14. Kern, W., Paulusma, D.: Contracting to a longest path in H-free graphs. In: Proceedings of ISAAC 2020, LIPIcs, 181:22:1–22:18 (2020)
15. Natarajan, S., Sprague, A.P.: Disjoint paths in circular arc graphs. Nordic J. Comput. **3**, 256–270 (1996)
16. Olariu, S.: Paw-free graphs. Inf. Process. Lett. **28**, 53–54 (1988)
17. Paulusma, D., van Rooij, J.M.M.: On partitioning a graph into two connected subgraphs. Theor. Comput. Sci. **412**(48), 6761–6769 (2011)
18. Robertson, N., Seymour, P.D.: Graph minors. XIII. the disjoint paths problem. J. Comb. Theory Ser. B **63**, 65–110 (1995)
19. Shiloach, Y.: A polynomial solution to the undirected two paths problem. J. ACM **27**, 445–456 (1980)
20. Telle, J.A., Villanger, Y.: Connecting terminals and 2-disjoint connected subgraphs. In: Brandstädt, A., Jansen, K., Reischuk, R. (eds.) WG 2013. LNCS, vol. 8165, pp. 418–428. Springer, Heidelberg (2013). https://doi.org/10.1007/978-3-642-45043-3_36
21. van Leeuwen, E.J.: Optimization and Approximation on Systems of Geometric Objects. University of Amsterdam (2009)
22. van 't Hof, P., Paulusma, D., Woeginger, G.J.: Partitioning graphs into connected parts. Theor. Comput. Sci. **410**, 4834–4843 (2009)

Prophet Secretary for k-Knapsack and l-Matroid Intersection via Continuous Exchange Property

Soh Kumabe[1]($\boxtimes$) and Takanori Maehara[2]

[1] The University of Tokyo, Tokyo, Japan
soh_kumabe@mist.i.u-tokyo.ac.jp
[2] RIKEN, Wako, Japan

Abstract. We study the *k-knapsack and l-matroid constrained prophet secretary problem*, which is a combinatorial prophet secretary problem whose feasible domain is the intersection of k-knapsack constraints and l-matroid constraints. Here, the prices of the items and the structure of the matroids are deterministic and known in advance, and the values of the items are stochastic and their distributions are known in advance. We derive a constant-factor approximation algorithm for this problem. We adapt Ehsani et al. (2018)'s technique for the matroid constraint to the knapsack constraint via continuous relaxation. For this purpose, we prove an "exchange property" of the knapsack constraint.

Keywords: Online algorithm · Prophet secretary problem · Knapsack

1 Introduction

1.1 Background and Motivation

The *combinatorial prophet secretary problem* [10] is the following stochastic online optimization problem: Let E be a finite set of items, $\mathcal{D} \subseteq 2^E$ be a feasible domain, and $\vec{v} = (v_e)_{e \in E} \in \mathbb{R}_{\geq 0}^E$ be a random variable, representing the values of items, whose entries follow some known distributions independently. The items arrive one-by-one in uniform random order. When $e \in E$ arrives, we observe the realized value of v_e and irrevocably pick e or ignore e. The objective of the problem is to design a strategy that selects a feasible set $X \in \mathcal{D}$ with large objective value $v(X) = \sum_{j \in X} v_j$ in expectation, with respect to the random arrival order and the realization of $\vec{v}$.

A special case of the above problem, where $\mathcal{D}$ consists of the set of singletons, is called the prophet secretary problem, and was introduced as an "intersection problem" of the *prophet inequality problem* [15, 16, 18] and the *secretary problem* [9]. Here, in the former case, we know the distribution of values but we cannot assume the arrival order is random, and in the latter case, we can assume the arrival order is random but we do not know the distribution of the values. The above combinatorial prophet secretary problem was introduced by Ehsani et al. [10].

© Springer Nature Switzerland AG 2021
P. Flocchini and L. Moura (Eds.): IWOCA 2021, LNCS 12757, pp. 428–441, 2021.
https://doi.org/10.1007/978-3-030-79987-8_30

Table 1. Approximation ratio of the algorithm. (k, l)-th entry shows the approximation ratio of k-knapsack l-matroid constraint.

-	0	1	2	3	4	5	6	7
1	0.2134	0.1729	0.1419	0.1187	0.1014	0.0895	0.0800	0.0723
2	0.1441	0.1159	0.0982	0.0860	0.0764	0.0686	0.0622	0.0569
3	0.1056	0.0884	0.0756	0.0665	0.0605	0.0555	0.0512	0.0476
4	0.0818	0.0707	0.0621	0.0553	0.0500	0.0465	0.0435	0.0408
5	0.0662	0.0586	0.0525	0.0475	0.0434	0.0400	0.0377	0.0357
6	0.0554	0.0499	0.0454	0.0416	0.0384	0.0357	0.0333	0.0317
7	0.0476	0.0435	0.0400	0.0370	0.0345	0.0323	0.0303	0.0286

The prophet secretary problem attracts attention [2,6,7,10,11] because, by comparing the prophet problem and the secretary problem, we can see the "value" of knowing distributions and/or the random ordering. By definition, we can expect that the prophet secretary problem has a better approximation ratio than the corresponding prophet inequality and secretary problems. In fact, if we can select at most one item (i.e., original prophet secretary setting), the prophet secretary problem admits an approximation ratio of $1 - 1/e + 1/27 \approx 0.669$ [7], whereas the optimal approximation ratio in the prophet inequality problem is $1/2$ [16] and the optimal approximation ratio in the secretary problem is $1/e \approx 0.367$. Similarly, if the constraint is a matroid, the approximation ratio in the prophet secretary problem is $1 - 1/e \approx 0.638$ [10], whereas the optimal approximation ratio in the prophet inequality problem is $1/2$ [14] and the best-known approximation ratio in the secretary problem is $O(\log \log \mathrm{rank})$ in the secretary setting [12,17], where rank is the rank of the matroid. Note that the latter is conjectured to be $O(1)$ [4].

In this study, we consider the prophet secretary problem on knapsack-related constraints, say, k-knapsack and l-matroid intersection constraint. Here, the prices of the items are deterministic and known in advance, and the values of the items are stochastic and their distributions are known in advance. The knapsack constraint has not been studied in the prophet secretary setting, and it has a technical difficulty in applying the existing framework (see Sect. 1.4). This study aims to establish a new technique to overcome the difficulty and derive an algorithm with a constant-factor approximation ratio.

1.2   Our Contribution

In this study, we prove the following theorem.

Theorem 1. *There is an algorithm for the k-knapsack l-matroid prophet secretary problem that has the approximation ratio listed in Table 1 (see Theorem 2 for the formula to compute the table). Specifically, the approximation ratio is $\Omega(1/(k + l))$.*

The theorem is proved by setting a threshold $\epsilon > 0$ and splitting items into a "small" part whose weights are at most ϵ and a "large" part whose items have the weights greater than ϵ. Our main technical contribution is an improved approximation ratio for the small part (Lemma 7). If we consider only multiple knapsack constraint, this result is simplified as follows.

Corollary 1. *There is an algorithm for the k-knapsack prophet secretary problem that has the approximation ratio of* $(1 - \epsilon)(1 - e^{-k})/k$, *where ϵ is an upper bound of the weights of the items.*

1.3   Comparison with Existing Results and Discussion

First, we compare our result with the existing results on k-knapsack l-matroid setting. Feldman et al. [13] provided an *online contention resolutions scheme* with constant *selectability* for this setting, which means the approximation algorithm of the same factor under the prophet inequality setting. In their algorithm, combining constraints affects the approximation ratio multiplicatively and thus, approximation ratio is exponential with respect to k. Since the approximation ratio of our algorithm is $\Omega(1/(k + l))$, our result outperforms their result.

Next, we compare our result with the existing results on the single knapsack case ($k = 1$ and $l = 0$). The approximation ratio of Feldman et al. [13]'s algorithm is $3/2 - \sqrt{2} \approx 0.086$. Dütting et al. [8] provided an algorithm with the approximation ratio of $1/5 = 0.2$ under the prophet inequality setting. Babaioff et al. [3] provided an algorithm with the approximation ratio of $1/10e \approx 0.036$ under the secretary setting and Albers et al. [1] gave the current best approximation ratio 0.150. Thus, as expected, we see that the prophet secretary setting gives a better approximation ratio than the less information cases. However, if we adapt the proof in [8] to the secretary prophet setting, we can obtain a better approximation ratio as follows.

Proposition 1. *There is an algorithm for the knapsack constrained prophet secretary problem that has the approximation ratio of* 0.222.

Our result is inferior to their result because of the following reason. Their algorithm also set a threshold $\epsilon > 0$ and split the items into a small and a large parts. Their approximation ratio for the small part is $(1 - \epsilon)/(2 - \epsilon)$, which is better than our ratio of $(1 - \epsilon)(1 - e^{-1})$ in Corollary 1 when $\epsilon = 1/2$. For $\epsilon = 1/2$, the large part is equivalent to the original prophet secretary problem and current best approximation ratio is $1 - 1/e + 1/27$ [7]. Proposition 1 is obtained combining these bounds. See Sect. 4 for the proof. Note that our result outperforms their result if we can assume that all items have weight at most $1/3$. Furthermore, for the small part our approximation ratio $(1 - \epsilon)(1 - e^{-1})$ outperforms Albers et al.'s [1] one $1 - (1 - \epsilon)^{-1} \log(2 - \epsilon)$ on the secretary setting for any $\epsilon \leq 1/2$. Therefore, combining their result with the large part does not yields the better approximation ratio.

Finally, we compare our result with the existing results on the multiple knapsack case ($k \geq 2$). This constraint is a special case of the k-*sparse packing integer*

Algorithm 1 Ehsani et al.'s framework

1: Draw n random numbers from $U[0,1]$ and arrange them by $t_1 < \cdots < t_n$
2: $A = \emptyset$
3: **for** $j \in V$ **do**
4:     Observe v_j
5:     **if** $A \cup \{j\} \in \mathcal{D}$ and $v_j \geq \alpha(t_j) b_j(A)$ **then**
6:         $A \leftarrow A \cup \{j\}$
7:     **end if**
8: **end for**
9: **return** A

---

programming (PIP) constraint. Dütting et al. [8] proposed an algorithm with the approximation ratio of $1/8k$ if all the items have weights at most $1/2$. Our result (Corollary 1) gives $(1 - e^{-k})/(2k)$ approximation, which outperforms their result for all $k \geq 2$.

1.4   Proof Strategy

To obtain an algorithm for knapsack constraint, we extend Ehsani et al.'s technique in the matroid constraint case [10]. We first explain their technique. Then, we explain the difficulty of our problem and how to overcome the difficulty.

Prophet Secretary for Matroid Constraint. A *matroid* is a pair $(V, \mathcal{I})$ of a finite set V and a set family $\mathcal{I} \subseteq 2^V$ such that $\mathcal{I}$ contains the emptyset $\emptyset$, $\mathcal{I}$ is downward-closed, and for any $X, Y \in \mathcal{I}$ such that $|X| < |Y|$, there exists $i \in Y \setminus X$ such that $X \cup \{i\} \in \mathcal{I}$.

Their algorithm is presented in Algorithm 1. It assigns "arrival time" to each items, and it maintains an adaptive threshold $\alpha(t_j) b_j(A)$ based on the time and current solution. The algorithm includes the j-th item to the solution if its value v_j is greater than the threshold.

The threshold function consists of two parts: α and b. Here, b is defined by

$$b_j(A) = \mathbb{E}[R(A, \vec{v}) - R(A \cup \{j\}), \vec{v})], \tag{1}$$

where R denotes an auxiliary function defined by

$$R(A, \vec{v}) = \max \left\{ \sum_{j \in X \setminus A} v_j : X \in \mathcal{D}, A \subseteq X \right\}. \tag{2}$$

Intuitively, $b_j(A)$ is the "damage" of including j to a solution A. α is a monotone decreasing function that represents a discount factor.

Ehsani et al. used the following inequality to analyze the performance of their algorithm:

$$\sum_{j \in Y} (R(A, \vec{v}) - R(A \cup \{j\}, \vec{v})) \leq R(A, \vec{v}). \tag{3}$$

This inequality follows from Brualdi's lemma (Lemma 5). Using this inequality, they proved that the algorithm achieves an approximation ratio of $1 - e^{-1}$ by setting $\alpha(t) = 1 - e^{-t}$.

Difficulty with the Knapsack Constraint and How to Overcome the Difficulty. In general, if we obtain an inequality that has the same LHS of (3) and the RHS of (3) multiplied by a constant factor, we obtain another constant approximation ratio of the corresponding prophet secretary problem. Thus, we need to achieve such an inequality for the knapsack constraint.

In the matroid case, this inequality is derived from the exchange property (i.e., adding one item requires the removal of only one item). However, in the knapsack constraint case, there is a scenario that adding a small item requires us to exclude a large item. This makes it difficult to obtain such an inequality. We solve this difficulty by modifying the definition of R (and hence b) via a continuous relaxation of the problem.

In Sect. 2, we provide a precise condition of the auxiliary function R that allows us to achieve an approximation ratio of the prophet secretary problem. In Sect. 3, we prove that continuous relaxation of the knapsack constraint admits an exchange property that extends (3); this is the main technical contribution of this paper. Finally, in Sect. 4, we provide the proof of the main theorem by combining these results.

2   Auxiliary Function R via Continuous Relaxation

The following lemma states the requirement of the auxiliary function R.

Lemma 1. *Suppose that for each A and $\vec{v}$ there exists $\vec{x}(A, \vec{v}) \in [0,1]^{V \setminus A}$ such that*

1. *$x_j(A, \vec{v}) > 0$ implies $A \cup \{j\} \in \mathcal{F}$, where $\mathcal{F}$ is the set of feasible solutions to the original problem.*
2. *$\sum_j b_j(A, \vec{v}) x_j(A, \vec{v}) \leq \mu R(A, \vec{v})$, where $R(A, \vec{v}) = \sum_{j \in V \setminus A} v_j x_j(A, \vec{v})$ and $b_j(A, \vec{v}) = R(A, \vec{v}) - R(A \cup \{j\}, \vec{v})$.*
3. *$R(\emptyset, \vec{v}) = \vec{v}^{\top} \vec{x}(\emptyset, \vec{v})$ is a γ-approximation to the original problem.*

Then, Algorithm 1 with $b_j(A) = \mathbb{E}[b_j(A, \vec{v})]$ is a $\gamma(1 - e^{-\mu})/\mu$ approximation.

Proof. Let X be the solution obtained by the algorithm. We decompose the objective value as

$$\sum_{j \in X} v_j = \underbrace{\sum_{j \in X} \left(v_j - \alpha(t_j) b_j(A_{t_j})\right)}_{\text{utility}} + \underbrace{\sum_{j \in X} \alpha(t_j) b_j(A_{t_j})}_{\text{revenue}}, \tag{4}$$

where A_t represents the solution at the time immediately before t. We call the first and second terms on the right-hand side utility and revenue, respectively.

We first evaluate the revenue. Let $r(t) = \mathbb{E}\left[R(A_t, \vec{v})\right]$ and suppose $j \in X$. Then, for a sufficiently small $\epsilon > 0$, we have

$$Revenue(t_j + \epsilon) - Revenue(t_j) = \alpha(t_j)b_j(A_{t_j}) \tag{5}$$

$$= \alpha(t_j)(R(A_{t_j}, \vec{v}) - R(A_{t_j + \epsilon}, \vec{v})), \tag{6}$$

where $Revenue(t)$ represents the revenue before time t, i.e., $\sum_{t_j \leq t} \alpha(t_j)b_j(A_{t_j})$. By taking the integral and expectation, we have

$$\mathbb{E}\left[Revenue\right] = -\int_0^1 \alpha(t)r'(t)dt. \tag{7}$$

Next, we evaluate the utility. Because the algorithm picks element j only if $v_j \geq \alpha(t_j)b_j(A_{t_j})$, the utility is represented by

$$Utility_j = \left(v_j - \alpha(t_j)b_j(A_{t_j})\right)^+ \mathbb{1}[\{j\} \cup A_{t_j} \in \mathcal{F}] \tag{8}$$

$$\geq \left(v_j - \alpha(t_j)b_j(A_{t_j})\right)^+ x(A_{t_j}, \vec{v}) \tag{9}$$

$$\geq \left(v_j - \alpha(t_j)b_j(A_{t_j})\right) x(A_{t_j}, \vec{v}) \tag{10}$$

where $Utility_j$ represents the j-th summand of the utility and x^+ represents x if $x \geq 0$ and 0 otherwise. Here, in the first inequality, we used the first assumption of the lemma.

Before taking the expectation on $t = t_j \in [0, 1]$, we remove the dependence of A_t on t_j. In particular, we claim

$$\mathbb{E}\left[(v_j - \alpha(t)b_j(A_t)) x_j(A_t, \vec{v})|t_j = t\right] \geq \mathbb{E}\left[(v_j - \alpha(t)b_j(A_t)) x_j(A_t, \vec{v})\right]. \tag{11}$$

We prove this by the same argument as [10, Lemma 17]. If $t_j \geq t$ or $j \notin A_t$, it holds in equality because A_t does not depend on t_j. Otherwise, i.e., if $t_j < t$ and $j \in A_t$, the right hand side is zero. Therefore, the claim is proved.

Now, by taking the expectation on t and the summation over j, we have

$$\mathbb{E}\left[Utility\right] \geq \sum_{j \in V} \int_0^1 \mathbb{E}\left[(v_j - \alpha(t)b_j(A_t)) x_j(A_t, \vec{v})|t_j = t\right] dt \tag{12}$$

$$\geq \sum_{j \in V} \int_0^1 \mathbb{E}\left[(v_j - \alpha(t)b_j(A_t)) x_j(A_t, \vec{v})\right] dt \tag{13}$$

$$= \mathbb{E}\left[\int_0^1 \left(R(A_t, v) - \alpha(t)\sum_j b_j(A_t)x_j(A_t, \vec{v})\right) dt\right] \tag{14}$$

$$\geq \mathbb{E}\left[\int_0^1 (1 - \alpha(t)\mu) R(A_t, \vec{v})dt\right] \tag{15}$$

$$= \int_0^1 (1 - \alpha(t)\mu) r(t)dt. \tag{16}$$

Here, in the fourth line, we used the second assumption of the lemma.

By combining these inequalities, we obtain

$$\mathbb{E}\left[Utility\right] + \mathbb{E}\left[Revenue\right] \geq \int_0^1 \left(-\alpha(t)r'(t) + (1 - \alpha(t)\mu)r(t)\right) dt \tag{17}$$

$$= \alpha(0)r(0) - \alpha(1)r(1) + \int_0^1 \left(\alpha'(t) - \mu\alpha(t) + 1\right) r(t)dt. \tag{18}$$

We let $\alpha(t)$ as the solution of the differential equation

$$\alpha'(t) - \mu\alpha(t) + 1 = 0, \quad \alpha(1) = 0, \tag{19}$$

which eliminates the second and third terms in (18). The closed form of α is given by

$$\alpha(t) = \frac{1 - e^{\mu(t-1)}}{\mu}. \tag{20}$$

Thus, we obtain

$$Utility + Revenue \geq \alpha(0)r(0) = \frac{1 - e^{-\mu}}{\mu} \cdot \gamma\text{OPT}. \tag{21}$$

Here, OPT denotes the optimal value of the original problem and we used the third condition that $r(0)$ is a γ-approximation of the original problem.

3   Exchange Lemma for Knapsack Constraint

We want to apply Lemma 1 to the knapsack constraint to obtain our main theorem. The second condition of the lemma is shown via the exchange property of the constraints.

In this section, we prove an exchange lemma for the knapsack constraint. We first prove the following technical lemma, which is the fractional version of the inequality (3) for the knapsack case.

Lemma 2. Let $\vec{c} \in \mathbb{R}^n_{\geq 0}$, $\vec{y} \in \mathbb{R}^n_{\geq 0}$, and $C \in \mathbb{R}_{\geq 0}$ such that $C \geq c_j$ for all $j \in [n]$ and $C \geq \sum_{j=1}^n c_j y_j$. If $\rho \colon [0, C] \to \mathbb{R}_{\geq 0}$ is a monotone non-decreasing function, we have

$$\sum_{j=1}^n \int_0^{c_j} \rho(t)y_j dt \leq \int_0^C \rho(t)dt. \tag{22}$$

Proof. Without loss of generality, we assume $c_1 \leq c_2 \cdots \leq c_n$. Let $l \in [n]$ be the smallest index such that $y_l + \cdots + y_n \leq 1$. Then, we have

$$\sum_{j=1}^{n} \int_{0}^{c_j} \rho(t) y_j \, dt = \sum_{j=1}^{n} \int_{c_{j-1}}^{c_j} \rho(t)(y_j + \cdots + y_n) dt \tag{23}$$

$$= \underbrace{\sum_{j=1}^{l-1} \int_{c_{j-1}}^{c_j} \rho(t)(y_j + \cdots + y_n) dt}_{(a)} + \underbrace{\sum_{j=l}^{n} \int_{c_{j-1}}^{c_j} \rho(t)(y_j + \cdots + y_n) dt}_{(b)}, \tag{24}$$

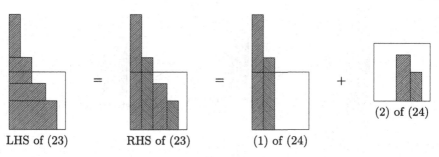

LHS of (23)   RHS of (23)   (1) of (24)   (2) of (24)

where $c_0 = 0$ for convention. We evaluate (a) and (b) separately. For (a), we have

$$(a) \leq \sum_{j=1}^{l-1} \int_{c_{j-1}}^{c_{j-1}+(c_j-c_{j-1})(y_j+\cdots+y_n)} \rho(t) dt \leq \int_{0}^{\sum_{j=1}^{l-1}(c_j-c_{j-1})(y_j+\cdots+y_n)} \rho(t) dt. \tag{25}$$

Here, the first inequality follows from the monotonicity of ρ; the second inequality follows from the fact that the adjacent integral ranges are overlapping. Further, aligning the ranges to the right to resolve the overlaps increases the integral because of the monotonicity of ρ. For integral (b),

$$(b) \leq \sum_{j=l}^{n} \int_{c_j-(c_j-c_{j-1})(y_j+\cdots+y_n)}^{c_j} \rho(t) dt \leq \int_{C-\sum_{j=l}^{n}(c_j-c_{j-1})(y_j+\cdots+y_n)}^{C} \rho(t) dt. \tag{26}$$

Here, the first inequality follows from the monotonicity of ρ; the second inequalities follows from the fact that adjacent integral ranges have gaps and filling such gaps by aligning to the right increases the integral because of the monotonicity of ρ.

The upper endpoint of the integral in (25) is at most the lower endpoint of the integral in (26) because $C \geq \sum_{j=1}^{n} c_j y_j = \sum_{j=1}^{n}(c_j - c_{j-1})(y_j + \cdots + y_n)$. Thus, by filling the gap, we obtain the result.

Remark 1. The above proof can be understood geometrically. We regard $\rho(t)$ as a density over the plane that depends on the x-coordinate; the right side is denser than the left side. The right-hand side of (22) is the weighted area of the rectangle of width C and height one, and the left-hand side of (22) is the weighted area of the shaded region in the first figure below. The proof deforms the shaded region to fit the rectangle for the right-hand side using the monotonicity of the density.

A *knapsack polytope* P is defined by

$$P = \{\vec{x} \in [0,1]^V : \vec{c}^\top \vec{x} \le 1\} \tag{27}$$

for some $\vec{c} \in [0,1]^V$. Now, we prove the exchange lemma for knapsack polytope.

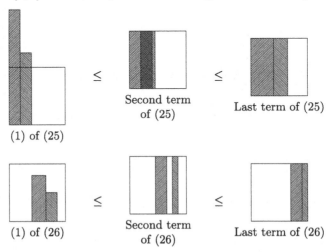

Lemma 3 (Fractional Exchange Lemma for Knapsack Constraint). *Let P be a knapsack polytope and $\vec{v} \in \mathbb{R}^V$ be a nonnegative vector. For $j \in V$, let $\vec{e}_j$ be the j-th unit vector. For all $\vec{x}, \vec{y} \in P$, there exists vectors $\vec{\chi}_j$ for each $j \in \text{supp}(y)$ such that the following conditions hold.*

1. $\vec{0} \le \vec{\chi}_j \le \vec{x}$,
2. $\vec{x} - \vec{\chi}_j + \vec{e}_j \in P$,
3. $\sum_{j \in V} y_j \vec{v}^\top \vec{\chi}_j \le \vec{v}^\top \vec{x}$.

Proof. Without loss of generality we assume that $v_1/c_1 \le \cdots \le v_m/c_m$. Let $t_p = \sum_{i=1}^p c_i x_i$. For convention, we define $t_0 = 0$. We define $\vec{\nu}(t)$ for $t \in (0, t_m]$ by

$$\vec{\nu}(t) = \frac{\vec{e}_p}{c_p}, \quad t \in (t_{p-1}, t_p], \tag{28}$$

If $\sum_{i=1}^j c_i x_i \le c_j$, we set $l_j = c_j$ and

$$\vec{\chi}_j = \int_{t=0}^{l_j} \vec{\nu}(t) dt. \tag{29}$$

Otherwise, we set $l_j = (1 - x_j)c_j$ and

$$\vec{\chi}_j = \int_{t=0}^{l_j} \vec{\nu}(t)dt + x_j\vec{e}_j. \tag{30}$$

For both cases, the first condition clearly holds because $\vec{\nu}(t) \geq 0$ (element-wise inequality), and the integral of $\vec{\nu}(t)$ is $\vec{x}$. The second condition holds because of the identity

$$\int_{t=0}^{l_j} \vec{c}^\top \vec{\nu}(t)dt = \int_{t=0}^{l_j} 1dt = l_j. \tag{31}$$

Now, we prove the third condition. Let Y' be the set of indices j in the "otherwise" case. Let $\rho(t) = \vec{v}^\top \vec{\nu}(t)$. Because the items are ordered by their efficiency, $\rho(t)$ is a monotone non-decreasing function. Therefore, by Lemma 2, we have

$$\sum_j y_j\vec{v}^\top \vec{\chi}_j = \sum_j \int_{t=0}^{l_j} y_j\rho(t)dt + \sum_{j\in Y'} \int_{t=t_{j-1}}^{t_j} y_j\rho(t)dt \tag{32}$$

$$\leq \int_{t=0}^{\sum_j y_j l_j} \rho(t)dt + \int_{t=\sum_j y_j l_j}^{\sum_j y_j c_j} \rho(t)dt = OPT. \tag{33}$$

The following lemma is a fractional version of Lemma 18 in [10].

Lemma 4 (Fractional exchange lemma for a matroid constraint). *Let $(V, \mathcal{I})$ be a matroid. Let $\vec{x}, \vec{y} \in [0,1]^V$ be vectors such that $\mathrm{supp}(\vec{x}), \mathrm{supp}(\vec{y}) \in \mathcal{I}$. Then, there exists vectors $\vec{\chi}_j$ for each $j \in \mathrm{supp}(y)$ such that*

1. $\vec{0} \leq \vec{\chi}_j \leq \vec{x}$,
2. $\mathrm{supp}(\vec{x} - \vec{\chi}_j + \vec{e}_j) \in \mathcal{I}$,
3. $\displaystyle\sum_{j\in V} y_j v^\top \vec{\chi}_j \leq \vec{v}^\top \vec{x}$.

The proof uses the following result, called *Brualdi's lemma*.

Lemma 5 (Brualdi [5]). *Let $(V, \mathcal{I})$ be a matroid. For any basis $A, B \in \mathcal{I}$, there exists a bijection π between $A \setminus B$ to $B \setminus A$ such that $A \setminus \{j\} \cup \{\pi(j)\} \in \mathcal{I}$.*

Proof (Proof of Lemma 4). By Brualdi's lemma, we obtain a bijection π from $\mathrm{supp}(\vec{y})$ to $\mathrm{supp}(\vec{x})$ such that $\mathrm{supp}(\vec{x}) \cup \{j\} \setminus \{\pi(j)\} \in \mathcal{I}$ for all $j \in \mathrm{supp}(\vec{y})$. Let $\vec{\chi}_j = x_{\pi(j)}\vec{e}_{\pi(j)}$. Then, we have

$$\sum_{j\in V} y_j v^\top \vec{\chi}_j \leq \sum_{j\in\mathrm{supp}(\vec{y})} v^\top \vec{\chi}_j = \sum_{j\in\mathrm{supp}(\vec{y})} v_{\pi(j)}x_{\pi(j)} = \vec{v}^\top \vec{x}. \tag{34}$$

We can extend these lemmas for the k-knapsack and l-matroid constraint case as follows.

Corollary 2 (Fractional exchange Lemma for k-Knapsack and l-Matroid Constraint). *Let $P_1, \ldots, P_k$ be knapsack polytopes, $(V, \mathcal{I}_1), \ldots, (V, \mathcal{I}_l)$ be matroids, and $\vec{v} \in \mathbb{R}^V$ be a nonnegative vector. For all $\vec{x}, \vec{y} \in P_1 \cap \cdots \cap P_k$ such that $\operatorname{supp}(\vec{x}), \operatorname{supp}(\vec{y}) \in \mathcal{I}_1 \cap \cdots \cap \mathcal{I}_l$, there exists vectors $\vec{\chi}_j$ for each $j \in \operatorname{supp}(y)$ such that the following conditions hold.*

1. $\vec{0} \le \vec{\chi}_j \le \vec{x}$,
2. $\vec{x} - \vec{\chi}_j + \vec{e}_j \in P_1 \cap \cdots \cap P_k$,
3. $\operatorname{supp}(\vec{x} - \vec{\chi}_j + \vec{e}_j) \in \mathcal{I}_1 \cap \cdots \cap \mathcal{I}_l$,
4. $\displaystyle\sum_{j \in V} y_j v^\top \vec{\chi}_j \le (k+l) \vec{v}^\top \vec{x}$.

Proof. We choose $\vec{\chi}_j^1, \ldots, \vec{\chi}_j^k$ for the knapsack constraints $1, \ldots, k$ using Lemma 3. Further, we choose $\vec{\chi}_j'^1, \ldots, \vec{\chi}_j'^l$ for the matroid constraints $1, \ldots, l$ using Lemma 4. Then, we define $\vec{\chi}_j = \vec{\chi}_j^1 \vee \cdots \vee \vec{\chi}_j^k \vee \vec{\chi}_j'^1 \vee \ldots \vec{\chi}_j'^l$, where $\vee$ denotes the element-wise maximum. Then, it is easy to see that these vectors satisfy the conditions. $\qquad\square$

4   Final Proof

Let $\epsilon > 0$ be a parameter determined later. Let V_L be the set of items such that for at least one knapsack constraint, the weight is greater than ϵ and $V_S = V \setminus V_L$. We first obtain an approximation ratio of the case when $V_S = \emptyset$ (*LargeCase*).

Lemma 6 (No Small Items). *If $V_S = \emptyset$, there exists an algorithm that has an approximation ratio of $(1 - e^{-t})/t$, where $t = k(\lceil 1/\epsilon \rceil - 1)$.*

Proof. First, we mention that, in this case, any feasible solution has the cardinality of at most t. Let $\mathcal{F}$ be the feasible domain. For all $A \subseteq V$ and $\vec{v}$, we define $\vec{x}(A, \vec{v})$ by the indicator vector of $\operatorname{argmax}_X \{ \sum_{j \in X} v_j : A \cup X \in \mathcal{F} \}$. Since such set X has the cardinality of at most t, we have

$$\sum_j b_j(A, \vec{v}) \vec{x}(A, \vec{v}) \le \sum_j R(A, \vec{v}) \vec{x}(A, \vec{v}) \le t R(A, \vec{v}). \tag{35}$$

Therefore, by Lemma 1, Algorithm 1 with an appropriate α has an approximation ratio $(1 - e^{-t})/t$. $\qquad\square$

Next, we obtain an approximation ratio of the case when $V_L = \emptyset$ (*SmallCase*).

Lemma 7 (No Large Items). *If $V_L = \emptyset$, there exists an algorithm that has an approximation ratio of $(1 - \epsilon)(1 - e^{-(k+l)})/(k+l)$.*

Proof. We introduce a polytope $P_p(A, \epsilon)$ for each item subset A and knapsack constraint p defined by

$$P_p(A, \epsilon) = \left\{ \vec{x} \in [0,1]^{V \setminus A} : \sum_{j \in V \setminus A} c_j^p x_j \le 1 - \epsilon - \sum_{j \in A} c_j^p \right\}. \tag{36}$$

We define $\vec{x}(A, \vec{v})$ the optimal solution to the problem

$$\text{maximize} \sum_{j \in V \setminus A} v_j x_j$$
$$\text{subject to } \vec{x} \in P_1(A, \epsilon) \cap \cdots \cap P_k(A, \epsilon),$$
$$\text{supp}(\vec{x}) \in \mathcal{I}_1 \cap \cdots \cap \mathcal{I}_l. \tag{37}$$

We apply Lemma 1 for this vector. By the definition of $P_p(A, \epsilon)$ and the assumption that there are no large items, this vector satisfies Condition 1 in Lemma 1. By Corollary 2, this vector satisfies Condition 2 with $\mu = (k + l)$ and Condition 3 with $\gamma = 1 - \epsilon$. Therefore, by Lemma 1, Algorithm 1 with an appropriate α provides an approximation ratio of $(1 - \epsilon)e^{-(k+l)}/(k + l)$.

By combining these two cases, we obtain the following theorem.

Theorem 2. *There exists a constant-factor approximation algorithm for the k-knapsack and l-matroid prophet secretary problem. The approximation ratio is given by*

$$\max_{p \in \mathbb{Z}_{>0}} \frac{\beta_L \beta_S}{\beta_L + \beta_S}, \tag{38}$$

where

$$\beta_L = \frac{1 - e^{-(p-1)k}}{(p-1)k}, \beta_S = \frac{(1 - 1/p)(1 - e^{-(k+l)})}{k + l}. \tag{39}$$

Proof. Let $0 < \epsilon \le 1$ be a parameter determined later. Let $\beta_L(\epsilon)$ and $\beta_S(\epsilon)$ be the approximation ratio in Lemma 6 and Lemma 7, respectively. With probability $\beta_S(\epsilon)/(\beta_L(\epsilon) + \beta_S(\epsilon))$, we run the algorithm for LargeCase by ignoring all elements whose size is less than or equals to ϵ; otherwise, we run the algorithm for SmallCase by ignoring all elements whose size is greater than ϵ. The expected value of this algorithm is

$$\text{ALG} = \frac{\beta_S(\epsilon)}{\beta_L(\epsilon) + \beta_S(\epsilon)}(\text{LargeCase}) + \frac{\beta_L(\epsilon)}{\beta_L(\epsilon) + \beta_S(\epsilon)}(\text{SmallCase}) \tag{40}$$

$$\ge \frac{\beta_S(\epsilon)}{\beta_L(\epsilon) + \beta_S(\epsilon)}\beta_L(\epsilon)(\text{Large part of OPT}) \tag{41}$$

$$+ \frac{\beta_L(\epsilon)}{\beta_L(\epsilon) + \beta_S(\epsilon)}\beta_S(\epsilon)(\text{Small part of OPT}) \tag{42}$$

$$= \frac{\beta_L(\epsilon)\beta_S(\epsilon)}{\beta_L(\epsilon) + \beta_S(\epsilon)}\text{OPT}, \tag{43}$$

where ALG denotes the value of the output of the algorithm. Without loss of generality, because of the form of $\beta_L(\epsilon)$ and $\beta_S(\epsilon)$, we can choose $\epsilon = 1/p$ for some integer p. Thus, the theorem is proved.

This theorem immediately proves our main theorem (Theorem 1) where Table 1 is obtained by a numerical calculation. Setting $\epsilon = 1/2$ yields the lower bound $\Omega(1/(k+l))$ of the approximation ratio. If there is no matroid contraint, we obtain a closed form of the approximation ratio as follows.

Corollary 3. *If there is no matroid constraint, i.e., $l = 0$, the optimal p is given by $p = 3$ if $k = 1$ and $p = 2$ otherwise. The corresponding approximation ratios are $(2e^2 - 2)/(7e^2 + 3e) \approx 0.213$ and $(1 - e^{-k})/3k$.*

Finally, for the self-completeness of the paper, we prove Proposition 1.

Proof (Proof of Proposition 1). The proof follows the similar strategy as the above. We let $\epsilon = 1/2$ and split the items into small and large. For the small case, we use the algorithm of [8], which has the approximation ratio of $1/3$. For the large case, we use the algorithm of [7], which has the approximation ratio of $1 - 1/e + 1/27$. By combining these results, we obtain $(28e - 27)/(111e - 81) \approx 0.222$.

References

1. Albers, S., Khan, A., Ladewig, L.: Improved online algorithms for knapsack and gap in the random order model. In: Approximation, Randomization, and Combinatorial Optimization. Algorithms and Techniques, pp. 22:1–22:23 (2019)
2. Azar, Y., Chiplunkar, A., Kaplan, H.: Prophet secretary: Surpassing the 1-1/e barrier. In: Proceedings of the 2018 ACM Conference on Economics and Computation, pp. 303–318. ACM (2018)
3. Babaioff, M., Immorlica, N., Kempe, D., Kleinberg, R.: A knapsack secretary problem with applications. In: Charikar, M., Jansen, K., Reingold, O., Rolim, J.D.P. (eds.) APPROX/RANDOM -2007. LNCS, vol. 4627, pp. 16–28. Springer, Heidelberg (2007). https://doi.org/10.1007/978-3-540-74208-1_2
4. Babaioff, M., Immorlica, N., Kleinberg, R.: Matroids, secretary problems, and online mechanisms. In: Proceedings of the Eighteenth Annual ACM-SIAM Symposium on Discrete Algorithms, pp. 434–443. Society for Industrial and Applied Mathematics (2007)
5. Brualdi, R.A.: Comments on bases in dependence structures. Bull. Aust. Math. Soc. **1**(2), 161–167 (1969)
6. Correa, J., Foncea, P., Hoeksma, R., Oosterwijk, T., Vredeveld, T.: Posted price mechanisms for a random stream of customers. In: Proceedings of the 2017 ACM Conference on Economics and Computation, pp. 169–186 (2017)
7. Correa, J., Saona, R., Ziliotto, B.: Prophet secretary through blind strategies. In: Proceedings of the Thirtieth Annual ACM-SIAM Symposium on Discrete Algorithms, pp. 1946–1961. SIAM (2019)
8. Dütting, P., Feldman, M., Kesselheim, T., Lucier, B.: Prophet inequalities made easy: stochastic optimization by pricing non-stochastic inputs. SIAM J. Comput. **49**(3), 540–582 (2020)
9. Dynkin, E.B.: The optimum choice of the instant for stopping a Markov process. Soviet Math. **4**, 627–629 (1963)
10. Ehsani, S., Hajiaghayi, M., Kesselheim, T., Singla, S.: Prophet secretary for combinatorial auctions and matroids. In: Proceedings of the Twenty-Ninth Annual ACM-SIAM Symposium on Discrete Algorithms, pp. 700–714. SIAM (2018)

11. Esfandiari, H., Hajiaghayi, M.T., Liaghat, V., Monemizadeh, M.: Prophet secretary. SIAM J. Discrete Math. **31**(3), 1685–1701 (2017)
12. Feldman, M., Svensson, O., Zenklusen, R.: A simple o (log log (rank))-competitive algorithm for the matroid secretary problem. In: Proceedings of the Twenty-sixth Annual ACM-SIAM Symposium on Discrete Algorithms, pp. 1189–1201 (2014)
13. Feldman, M., Svensson, O., Zenklusen, R.: Online contention resolution schemes. In: Proceedings of the Twenty-Seventh Annual ACM-SIAM Symposium on Discrete Algorithms, pp. 1014–1033. SIAM (2016)
14. Kleinberg, R., Matthew, S., Weinberg, M.: prophet inequalities. In: Proceedings of the Forty-Fourth Annual ACM Symposium on Theory of Computing, pp. 123–136. ACM (2012)
15. Krengel, U., Sucheston, L.: Semiamarts and finite values. Bull. Am. Math. Soc. **83**, 745–747 (1977)
16. Krengel, U., Sucheston, L.: On semiamarts, amarts, and processes with finite value. Probab. Banach Spaces **4**, 197–266 (1978)
17. Lachish, O.: O (log log rank) competitive ratio for the matroid secretary problem. In: 2014 IEEE 55th Annual Symposium on Foundations of Computer Science, pp. 326–335 (2014)
18. Samuel-Cahn, E.: Comparison of threshold stop rules and maximum for independent nonnegative random variables. Ann. Probab. **12**, 1213–1216 (1984)

Minimum Eccentricity Shortest Path Problem with Respect to Structural Parameters

Martin Kučera[ID] and Ondřej Suchý[(✉)][ID]

Department of Theoretical Computer Science, Faculty of Information Technology,
Czech Technical University in Prague, Prague, Czech Republic
martin@mkucera.cz, ondrej.suchy@fit.cvut.cz

Abstract. The MINIMUM ECCENTRICITY SHORTEST PATH PROBLEM
consists in finding a shortest path with minimum eccentricity in a given
undirected graph. The problem is known to be NP-complete and W[2]-
hard with respect to the desired eccentricity. We present fpt algorithms
for the problem parameterized by the modular width, distance to cluster
graph, the combination of distance to disjoint paths with the desired
eccentricity, and maximum leaf number.

Keywords: Graph theory · Minimum eccentricity shortest path ·
Parameterized complexity · Fixed-parameter tractable

1 Introduction

The MINIMUM ECCENTRICITY SHORTEST PATH (MESP) problem asks, given
an undirected graph and an integer k, to find a shortest path with eccentricity
at most k—a shortest path (between its endpoints) whose distance to all other
vertices in the graph is at most k. The shortest path achieving the minimum
k may be viewed as the "most accessible", and as such, may find applications
in communication networks, transportation planning, water resource manage-
ment, and fluid transportation [7]. Some large graphs constructed from reads
similarity networks of genomic data appear to have very long shortest paths
with low eccentricity [17]. Furthermore, MESP can be used to obtain the best
to date approximation for a minimum distortion embedding of a graph into the
line [7] which has applications in computer vision [16], computational biology
and chemistry [11,12]. The eccentricity of MESP is closely tied to the notion of
laminarity (minimum eccentricity of the graph's diameter) [1].

MESP was introduced by Dragan and Leitert [7] who showed that it is NP-
hard on general graphs and constructed a slice-wise polynomial (XP) algorithm,

This paper is eligible for the best student paper award.

M. Kučera–Work of this author was supported by the Student Summer Research Program 2020 of FIT CTU in Prague, and by grant SGS20/212/OHK3/3T/18.

O. Suchý–The author acknowledges the support of the OP VVV MEYS funded project
CZ.02.1.01/0.0/0.0/16_019/0000765 "Research Center for Informatics".

P. Flocchini and L. Moura (Eds.): IWOCA 2021, LNCS 12757, pp. 442–455, 2021.
https://doi.org/10.1007/978-3-030-79987-8_31

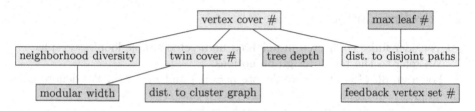

Fig. 1. Hasse diagram of the boundedness relation between structural parameters explored in this paper and related ones. An edge between a parameter **A** above and a parameter **B** below means that whenever **A** is bounded for some graph class, then so is **B**. The parameters for which MESP is FPT are in green (dark if the result is described in this paper, light if implied by those described). Yellow represents a parameter for which MESP is FPT in combination with the desired eccentricity. The complexity for parameters in gray is open. (Color figure online)

which finds a shortest path with eccentricity at most k in a graph with n vertices and m edges in $\mathcal{O}(n^{2k+2}m)$ time. They also presented a linear-time algorithm for trees. Additionally, they developed a 2-approximation, a 3-approximation, and an 8-approximation algorithm that runs in $\mathcal{O}(n^3)$ time, $\mathcal{O}(nm)$ time, and $\mathcal{O}(m)$ time, respectively. Birmelé et al. [1] further improved the 8-approximation to a 3-approximation, which still runs in linear time. Dragan and Leitert [8] showed that MESP can be solved in linear time for distance-hereditary graphs (generalizing the previous result for trees) and in polynomial time for chordal and dually chordal graphs. Later, they proved [9] that the problem is NP-hard even for bipartite subcubic planar graphs, and W[2]-hard with respect to the desired eccentricity for general graphs. Furthermore, they showed that in a graph with a shortest path of eccentricity k, a minimum k-dominating set can be found in $n^{\mathcal{O}(k)}$ time. A related problem of finding shortest isometric cycle was studied by Birmelé et al. [3]. Birmelé et al. [2] studied a generalization of MESP, where the task is to decompose a graph into subgraphs with bounded shortest-path eccentricity, the hub-laminar decomposition.

Our Contribution. We continue the research direction of MESP in structured graphs [8], focusing on parameters which can measure the amount of structure present in the graph. As treewidth, the most famous of such measures, remains out of reach, we focus on parameters providing even more structure (see [14] for an overview of some graph measures). We provide fpt algorithms for the problem with respect to the modular width, distance to cluster graph, distance to disjoint paths combined with the desired eccentricity, and maximum leaf number (see Fig. 1 for an overview of our results).

Outline. In Sect. 2, we provide necessary notations and formal definitions. In Sect. 3, we describe our parameterized algorithms. In Sect. 4, we discuss possible future work.

Proofs omitted in this extended abstract due to space restrictions can be found in the full version of the paper, preprint is available on ArXiv [13].

2 Preliminaries

We consider finite connected unweighted undirected simple loopless graphs.

We refer to Diestel [6] for graph notions.

For a graph $G = (V, E)$ we denote $n = |V|$ and $m = |E|$. We denote $G[S]$ the induced subgraph of G on vertices $S \subseteq V$ and $G \backslash S = G[V \backslash S]$.

We denote an ordered sequence of elements $\boldsymbol{s} = (s_1, \ldots, s_{|s|})$. For two sequences $\boldsymbol{s} = (s_1, \ldots, s_{|s|}), \boldsymbol{t} = (t_1, \ldots, t_{|t|})$ we denote their concatenation

$$\boldsymbol{s} \frown \boldsymbol{t} = (s_1, \ldots, s_{|s|}, t_1, \ldots, t_{|t|}).$$

A *path* is a sequence of vertices where every two consecutive vertices are adjacent. The first and last vertices of the path are called its *endpoints*. A *path between* u and v or *u-v-path* is a path with endpoints u and v. The length of a path P is the number of edges in it, i.e., $|P| - 1$. A u-v-path is *shortest* if it has the least length among all u-v-paths. The distance $d_G(u, v)$ between two vertices $u, v \in V$ is the length of the shortest u-v-path.

The distance between a vertex $u \in V$ and a set of vertices $S \subseteq V$ is $d_G(u, S) = \min_{s \in S} d_G(u, s)$. The eccentricity of a set $S \subseteq V$ is $\mathrm{ecc}_G(S) = \max_{u \in V} d_G(u, S)$. For a path P, we use P instead of $V(P)$ for its set of vertices, if there is no risk of confusion, e.g., $d_G(u, P) = d_G(u, V(P))$ and $\mathrm{ecc}_G(P) = \mathrm{ecc}_G(V(P))$.

For vertex $u \in V$ we denote $N_G(u) = \{v \mid \{u, v\} \in E\}$ the open neighborhood, $N_G[u] = N_G(u) \cup \{u\}$ the closed neighborhood, and $N_G^k[u] = \{v \in V \mid d_G(u, v) \le k\}$ the closed k-neighborhood of u.

In this paper, we focus on the following problem.

MINIMUM ECCENTRICITY SHORTEST PATH PROBLEM (MESP)
Input: An undirected graph G, desired eccentricity $k \in \mathbb{N}$.
Question: Is there a path P in G which is a shortest path between its endpoints with $\mathrm{ecc}_G(P) \le k$?

A parameterized problem Π is *fixed parameter tractable (FPT)* with respect to a parameter k if there is an algorithm solving any instance of Π with size n in $f(k) \cdot n^{O(1)}$ time for some computable function f. Such an algorithm is called a *parameterized* or an *fpt algorithm*. See Cygan et al. [5] for more information on parameterized algorithms.

In this paper, we present fpt algorithms for MESP with respect to the following structural parameters.

Definition 1 (Modular width, [10]). *Consider graphs that can be obtained from an algebraic expression that uses the following operations:*

(O1) *create an isolated vertex;*
(O2) *the disjoint union of 2 disjoint graphs (the disjoint union of graphs G_1 and G_2 is the graph $\big(V(G_1) \cup V(G_2), E(G_1) \cup E(G_2)\big)$);*
(O3) *the complete join of 2 disjoint graphs (the complete join of graphs G_1 and G_2 is the graph $\big(V(G_1) \cup V(G_2), E(G_1) \cup E(G_2) \cup \big\{\{v, w\} \mid v \in V(G_1), w \in V(G_2)\big\}\big)$);*

(O4) *the substitution with respect to some pattern graph T (for a graph T with vertices $t_1, \ldots, t_n$ and disjoint graphs $G_1, \ldots, G_n$ the substitution of the vertices of T by the graphs $G_1, \ldots, G_n$ is the graph with vertex set $\bigcup_{i=1}^n V(G_i)$ and edge set $\bigcup_{i=1}^n E(G_i) \cup \{\{u,v\} \mid u \in V(G_i), v \in V(G_j),$ and $\{t_i, t_j\} \in E(T)\}$).*

We define the width of an algebraic expression A as the maximum number of operands used by any occurrence of the operation (O4) in A. The modular-width of a graph G, denoted $\mathrm{mw}(G)$, can be defined as the least integer m such that G can be obtained from such an algebraic expression of width at most m.

Given a graph $G = (V, E)$, an algebraic expression of width $\mathrm{mw}(G)$ describing G can be constructed in $O(|V| + |E|)$ time [15].

Definition 2 (Distance to cluster graph). *For a graph $G = (V, E)$, a modulator to cluster graph is a vertex subset $X \subseteq V$ such that $G \backslash X$ is a vertex-disjoint union of cliques. The distance to cluster graph is the size of the smallest modulator to cluster graph.*

A modulator to cluster graph of a graph with distance to cluster graph p can be found in $\mathcal{O}(1.9102^p \cdot (n + m))$ time [4].

Definition 3 (Distance to disjoint paths). *For a graph $G = (V, E)$ a modulator to disjoint paths is a vertex subset $X \subseteq V$, such that $G \backslash X$ is a vertex-disjoint union of paths. The distance to disjoint paths is the size of the smallest modulator to disjoint paths.*

For completeness, we include the following result which is rather folklore.

Lemma 1 ($\star^1$). *The modulator to disjoint paths C of a graph G with distance to disjoint paths c can be found in $\mathcal{O}(4^c(n + m))$ time.*

Definition 4 (Maximum leaf number). *The maximum leaf number of a graph G is the maximum number of leaves in a spanning tree of G.*

The presented algorithms rely on the following lemma.

Lemma 2. *For any graph $G = (V, E)$, any set $S \subseteq V$, and any vertex $s \in V$, at most one permutation $\boldsymbol{\pi} = (\pi_1, \ldots, \pi_{|S|})$ of the vertices in S exists, such that there is a shortest path P with the following properties:*

1. *The first vertex on P is s,*
2. *P contains all vertices from S, and*
3. *the vertices from S appear on P in exactly the order given by $\boldsymbol{\pi}$.*

Moreover, given a precomputed distance matrix for G, the permutation $\boldsymbol{\pi}$ can be found in $\mathcal{O}(|S| \log |S|)$ time.

---

[1] Proofs of lemmas marked with ($\star$) can be found in the full version of the paper, preprint is available on ArXiv [13].

Proof. For the sake of deriving a contradiction, suppose that there are two different permutations π and π' satisfying the conditions, and let P, P' be the respective shortest paths. Let $i \in \{1, \ldots, |S| - 1\}$ be the first position such that $\pi_i \neq \pi'_i$ and let $j \in \{2, \ldots, |S|\}$ be the position of π_i in π' (clearly, $j > i$). Let P_1 be the subpath of P from s to π_i, and P'_2 the subpath of P' from π'_j to $\pi'_{|S|}$ (excluding the first vertex π'_j). Then, $P'' = P_1 \frown P'_2$ is a path which is strictly shorter than P' and has the same endpoints. This contradicts P' being a shortest path.

Sorting all vertices in S by increasing distance from the starting endpoint s yields our permutation π. It corresponds to some shortest path P if and only if

$$d_G(s, \pi_1) + \sum_{i=1}^{|S|-1} d_G(\pi_i, \pi_{i+1}) = d_G(s, \pi_{|S|}).$$

$\square$

3 Parameterized Algorithms

In this section, we present several fpt algorithms for MESP. In Subsect. 3.1, we present an algorithm parameterized by the modular width. In Subsect. 3.2 we define the CONSTRAINED SET COVER (CSC) problem. In Subsect. 3.3 we show an fpt algorithm for MESP parameterized by the distance to cluster graph which reduces MESP to CSC. In Subsect. 3.4, we present an fpt algorithm parameterized by the distance to disjoint paths and the desired eccentricity, combined. This algorithm also depends on the solution of the CSC problem. Subsect. 3.5 presents an algorithm parameterized by the maximum leaf number.

3.1 Modular Width

We present an fpt algorithm for MESP parameterized by the modular width.

Let $G = (V, E)$ be a graph with modular width w and A be the corresponding algebraic expression describing the graph. We take a look at the last operation applied in A. Operation (O1) is trivial and (O2) yields a disconnected graph, therefore we suppose the last operation is either (O3) or (O4).

If it is (O3) and G is a path (of length at most 3), then the whole path is trivially a shortest path with eccentricity 0. If G is not a path, then the minimum eccentricity shortest path is any single edge connecting the two original graphs with eccentricity 1.

If it is (O4), the pattern graph is $T = (V_T, E_T)$ with $V_T = \{v_1, \ldots, v_w\}$ and the substituted graphs are $G_1, \ldots, G_w$, then we suppose that $w \geq 3$ and T is not a clique (otherwise (O3) could be used as the last operation). We continue by showing that the structure of the pattern graph restricts the structure of any shortest path in the resulting graph significantly.

Lemma 3 ($\star$). *If the last operation in A is (O4), then there is a minimum eccentricity shortest path in G which contains at most one vertex from each G_i for $i \in \{1, \ldots, w\}$.*

Now, we show that with respect to eccentricity, all vertices in the same graph G_i are equivalent. I.e., a minimum eccentricity shortest path in G can be found by trying all shortest paths in T.

Lemma 4 ($\star$). *Let P be a shortest path in G and $p \in P \cap G_i$. We create a path P' by substituting p in P by any $p' \in G_i$. Then, $\mathrm{ecc}_G(P') = \mathrm{ecc}_G(P)$.*

Based on what we have shown, we can construct an algorithm to solve MESP. We handle separately the graphs created using (O1) or (O3) as the last operation. For (O4), we iterate through all possible shortest paths π in T. For each of them and each $i \in \{1, \ldots, |\pi|\}$ we let $p_i \in G_{\pi_i}$ arbitrarily, and let $P := (p_1, \ldots, p_{|\pi|})$. Then we check whether P is a shortest path with eccentricity at most k in G. By the above arguments, if there is a shortest path of eccentricity at most k, we will find one.

All shortest paths in a graph may be found by simply performing n DFS traversals (one starting in each vertex). Each forward step of the DFS represents a new shortest path; we skip edges that would break the shortest path property (this can easily be checked with a precomputed distance matrix).

By assuming a trivial upper bound 2^w on the number of shortest paths in T, we arrive at the following theorem.

Theorem 1. *There is an algorithm that solves MESP in $\mathcal{O}(2^w \cdot n^3)$ time, where w is the modular width of the input graph.*

3.2   Constrained Set Cover

In this subsection we define the CONSTRAINED SET COVER (CSC) problem. In the folllowing two subsections it will be used as a subroutine to solve MESP.

CONSTRAINED SET COVER
Input: A set $\mathcal{C} = C_1 \cup \cdots \cup C_m$ of candidates, a set $\mathcal{R} = \{r_1, \ldots, r_n\}$ of requirements to be satisfied, and a function $\Psi : \mathcal{C} \to 2^{\mathcal{R}}$ that determines for each candidate which requirements it satisfies.
Question: Is there a *constrained set cover*, that is, a set of candidates, exactly one from each set $s_1 \in C_1, \ldots, s_m \in C_m$ such that together they satisfy all the requirements, i.e., $\Psi(s_1) \cup \cdots \cup \Psi(s_m) = \mathcal{R}$?

Each candidate can be thought of as a set of (satisfied) requirements. Hence, if we drop the constraints $s_i \in C_i$, we get the ordinary SET COVER. In our definition several candidates can satisfy the same set of requirements.

In the next two subsections we use the following lemma.

Lemma 5 ($\star$). CONSTRAINED SET COVER *can be solved in $\mathcal{O}(2^{2|\mathcal{R}|}|\mathcal{R}| \cdot |\mathcal{C}|)$ time.*

3.3   Distance to Cluster Graph

In this subsection, we present an fpt algorithm for MESP parametrized by the distance to cluster graph. The trivial case where distance to cluster graph is 0 is omitted. Note that if G is a graph with a modulator to cluster graph U, then, for any edge $\{u, v\}$ in $G \backslash U$, u and v have the same neighborhood in $G \backslash U$.

The high-level idea of the algorithm is that we iteratively guess (by trying all possible combinations), for each vertex in the modulator to cluster U, whether it lies on the desired shortest path (we say it belongs to the set L), or it is at distance 1 or 2 from the shortest path (it belongs to the set $\mathcal{R}_1$ or $\mathcal{R}_2$, respectively), or at an even further distance. Then, we try to find a shortest path such that all vertices from L lie on it, and all vertices in $\mathcal{R}_1, \mathcal{R}_2$ have the respective distance from the path. Finding such a path is reduced to solving the CSC problem presented in Subsect. 3.2. Once we guess the correct combination of these sets, we actually construct the MESP.

First, we discuss some properties of graphs having the desired path.

Lemma 6 ($\star$). *Let G be a graph with modulator to cluster graph U and let P be a shortest path with $\mathrm{ecc}_G(P) = k$. Then, there exists a shortest path P' such that it contains at least one vertex from U and $\mathrm{ecc}_G(P') \leq k$.*

Definition 5. *Let G be a graph with a modulator to cluster graph U. Let P be a shortest path in G with $\mathrm{ecc}_G(P) \leq k$ and $U \cap P \neq \emptyset$. We denote $L^P = P \cap U$ and $\boldsymbol{\pi}^P = (\pi_1^P, \ldots, \pi_{|L^P|}^P)$ the permutation/order of vertices from L^P in which they appear on the path P. We denote $\mathcal{R}_i^P = \{u \in U \mid d_G(u, P) = i\}$ the set of vertices in U that are at distance i from P, for $i \in \{1, 2\}$.*

Let $V = V(G) \backslash U$. Since $G[V]$ is a disjoint union of cliques, and for every $i \in \{1, \ldots, |L^P| - 1\}$, all vertices that are between π_i^P and π_{i+1}^P on P are from V, we have $d_G(\pi_i^P, \pi_{i+1}^P) \leq 3$, as otherwise P would not be a shortest path.

Let $\boldsymbol{\pi} = (\pi_1, \ldots, \pi_{|\boldsymbol{\pi}|})$ be a candidate (guess) on the value of $\boldsymbol{\pi}^P$. Intuitively, if we had the correct values of $\boldsymbol{\pi} = \boldsymbol{\pi}^P$, we would only need to select the (at most two) vertices between each π_i, π_{i+1}.

To help us refer to those pairs π_i, π_{i+1} between which we still need to choose some vertices we denote $\boldsymbol{h}_\pi = (h_1, \ldots, h_\ell)$ the increasing sequence of all indices i such that $\{\pi_i, \pi_{i+1}\} \notin E$. For every $i \notin \boldsymbol{h}_\pi$, we have $\{\pi_i, \pi_{i+1}\} \in E(G)$ and, thus, there is no vertex between π_i and π_{i+1} on P. For every $h_i \in \boldsymbol{h}_\pi$: If $d_G(\pi_{h_i}, \pi_{h_i+1}) = 2$, then there is one vertex on P between π_{h_i} and π_{h_i+1}, and it is from V. If $d_G(\pi_{h_i}, \pi_{h_i+1}) = 3$, then there are two vertices from V on P between π_{h_i} and π_{h_i+1}.

Definition 6. *We define the set C_{h_i} of candidate vertices between π_{h_i} and π_{h_i+1} for each $h_i \in \boldsymbol{h}_\pi$.*

$$
C_{h_i} = \begin{cases} \left\{ (u, u) \in V^2 \mid \{\{\pi_{h_i}, u\}, \{u, \pi_{h_i+1}\}\} \subseteq E \right\} & \text{if } d_G(\pi_{h_i}, \pi_{h_i+1}) = 2 \\ \left\{ (u, v) \in V^2 \mid \{\{\pi_{h_i}, u\}, \{u, v\}, \{v, \pi_{h_i+1}\}\} \subseteq E \right\} & \text{if } d_G(\pi_{h_i}, \pi_{h_i+1}) = 3 \\ \emptyset & \text{otherwise} \end{cases}
$$

For $h_i \in \boldsymbol{h}_\pi$ with $d_G(\pi_{h_i}, \pi_{h_i+1}) = 2$, the set C_{h_i} contains pairs of the same vertices (u, u). To avoid adding some vertex into a path twice, we define a function μ which maps a pair of two elements to a sequence of length 1 or 2:

$$\mu(u, v) = \begin{cases} (u) & \text{if } u = v, \\ (u, v) & \text{if } u \neq v. \end{cases}$$

To solve MESP, we need to choose exactly one pair from each of $C_{h_1}, \ldots, C_{h_\ell}$. Later, we show that the problem of choosing these pairs is an instance of CSC.

First, we define a function $\delta^P : U \cup V \to \mathbb{N}$ that will help us prove that the path constructed from the CSC solution will have a small eccentricity:

$$\delta^P(u) = \min\left\{ d_G(u, L^P), \; d_G(u, \mathcal{R}_1^P) + 1, \; d_G(u, \mathcal{R}_2^P) + 2 \right\}.$$

Lemma 7 $(\star)$**.** *Function δ^P is a good estimate of the distance from P:*

1. *$\delta^P(u) = d_G(u, P)$ for every $u \in U$, and*
2. *$\delta^P(u) = d_G\big(u, P\backslash(N_G[u] \cap V)\big)$ for every $u \in V$.*

Now we show how to choose optimal vertices from each C_i by solving CSC.

Lemma 8 $(\star)$**.** *Suppose that P is a shortest path in G with $\text{ecc}_G(P) \leq k$, both endpoints of P are in U, and we have the corresponding values of $L^P, \pi^P, \mathcal{R}_1^P, \mathcal{R}_2^P$ as described in Definition 5. Let $\boldsymbol{h}_{\pi^P} = (h_1, \ldots, h_\ell)$ and $(s_{h_1}, \ldots, s_{h_\ell})$ be a solution of the CSC instance with requirements $\mathcal{R}^P = \mathcal{R}_1^P \cup \mathcal{R}_2^P$, sets of candidates $\mathcal{C} = C_{h_1} \cup \cdots \cup C_{h_\ell}$, and function $\Psi(u, v) = N_G(u) \cup N_G(v) \cup \left((N_G^2[u] \cup N_G^2[v]) \cap \mathcal{R}_2^P \right)$. Then*

$$P' = (\pi_1^P, \ldots, \pi_{h_1}^P) \frown \mu(s_{h_1}) \frown (\pi_{h_1+1}^P, \ldots, \pi_{h_2}^P)$$
$$\cdots \frown \mu(s_{h_i}) \frown (\pi_{h_i+1}^P, \ldots, \pi_{h_{i+1}}^P) \frown \mu(s_{h_{i+1}})$$
$$\cdots \frown (\pi_{h_{\ell-1}+1}^P, \ldots, \pi_{h_\ell}^P) \frown \mu(s_{h_\ell}) \frown (\pi_{h_\ell+1}^P, \ldots, \pi_{|L^P|}^P)$$

is a shortest path and $\text{ecc}_G(P') \leq \max\{2, k\}$.

Clearly, if $k \geq 2$, then we can use Lemma 8 to construct a shortest path with eccentricity at most k. Now, we discuss the case when $k = 1$.

Observation 1 $(\star)$**.** *If $\text{ecc}_G(P) = 1$, then for every $u \in U \cup V$ we have $\delta^P(u) \leq 2$.*

Corollary 1 $(\star)$**.** *If $\text{ecc}_G(P) = 1$, then a path P' with $\text{ecc}_G(P') \leq 1$ can be constructed similarly as in Lemma 8 with the following modification. For each candidate set C_i which contains some pair $(x, y) \in V^2$ such that there is a neighbor $z \in V$ of x with $\delta^P(z) = 2$, remove every $(u, v) \in V^2$ such that z is not a neighbor of u from C_i.*

We have shown how to construct a shortest path with eccentricity at most k by solving the CSC problem, even if $k = 1$. Finally, we observe that such a path can be constructed even if one or both of its endpoints are in V.

Lemma 9 ($\star$). *If P has an endpoint $s \in V$, its neighbor $t \in P$ might also be in V. Let $P = (s, t, \ldots)$. We may obtain a path P' with $\mathrm{ecc}_G(P') \le k$ by removing $\Psi(s, s)$ (and $\Psi(t, t)$ if $t \in V$) from $\mathcal{R}^P$, finding $s_{h_1}, \ldots, s_{h_\ell}$ by solving the CSC, and prepending s (and t if $t \in V$) to P'.*

MESP can be solved by trying all possible combinations of $(L, s, \mathcal{R})$: $L \subseteq U, s \in L, \mathcal{R} = \mathcal{R}_1 \cup \mathcal{R}_2 \subseteq (U \backslash L)$. For each combination, do:

1. Find a permutation $\boldsymbol{\pi}$ of L, such that $\pi_1 = s$ and $\sum_{i=1}^{|L|-1} d_G(\pi_i, \pi_{i+1}) = d_G(\pi_1, \pi_{|L|})$. If it does not exist, continue with the next combination.
2. For each $h_i \in \boldsymbol{h}_\pi$: create set C_{h_i} according to Definition 6 and Corollary 1.
3. Solve the CSC instance as described in Lemma 8.
4. If the CSC instance has a solution, construct path P' as in Lemma 8.
5. Check if $\mathrm{ecc}_G(P') \le k$. If yes, return P'. If not, try the same after prepending and/or appending all combinations of single vertices and of pairs of vertices to P' (see Lemma 9).

Note that by Lemma 2, there is at most one such permutation $\boldsymbol{\pi}$ in step 1.

Theorem 2 ($\star$). *In a graph with distance to cluster graph p, MESP can be solved in $\mathcal{O}(2^{4p} p \cdot n^6)$ time.*

3.4   Distance to Disjoint Paths

In this subsection, we present an fpt algorithm for MESP parameterized by the distance to disjoint paths and the desired eccentricity, combined.

The high-level idea of the algorithm is similar to that in Subsect. 3.3. We iteratively guess (by trying all possible combinations), for each vertex in the modulator to disjoint paths C, what is the distance to the desired shortest path. Then, we try to find a shortest path which satisfies all the guessed distance requirements by solving an instance of the CSC problem. We argue that if these requirements are guessed correctly, the resulting path will indeed be the desired MESP.

We start by discussing some properties of graphs in which a shortest path $P = (p_1, \ldots, p_{|P|})$ with $\mathrm{ecc}_G(P) \le k$ does exist. Assume that P is such a path, fixed for the next few lemmas and definitions.

Definition 7. *Let $\widehat{C}^P = C \cup \{p_1, p_{|P|}\}$. Let $L^P = P \cap \widehat{C}^P$. We denote $\boldsymbol{\pi}^P = (\pi_1^P, \ldots, \pi_{|L^P|}^P)$ the permutation/order of vertices from L^P on the path P. We define function $\delta^P(v) = d_G(v, P)$ for every $v \in V$.*

Let $\widehat{C}, L$ be candidates for $\widehat{C}^P, L^P$, respectively. Similarly as in Subsect. 3.3, the permutation $\boldsymbol{\pi} = (\pi_1, \ldots, \pi_{|L|})$ of the vertices in L is unique (if it exists), and can be found in polynomial time. For each consecutive pair of vertices $\pi_i, \pi_{i+1} \in L$, there may be multiple shortest paths connecting them, such that they do not contain any other vertices from $\widehat{C}$. Exactly one of these shortest paths is contained in P for each pair. We say $\bar{\sigma}$ is a *candidate segment* if it is a sequence of vertices on some shortest path from π_i to π_{i+1} excluding the endpoints π_i, π_{i+1} and $\bar{\sigma} \cap \widehat{C} = \emptyset$. We define $\mathcal{S}(\pi_i, \pi_{i+1})$ as a set of all candidate segments $\bar{\sigma}$ between π_i and π_{i+1}. We denote $\widetilde{\mathcal{S}} = \bigcup_{i=1}^{|L|-1} \mathcal{S}(\pi_i, \pi_{i+1})$ the set of all candidate segments in G. We say that a candidate segment $\sigma \in \widetilde{\mathcal{S}}$ is a *necessary segment* if it must be part of any shortest path P' such that $\widehat{C} = \widehat{C}^{P'}$, $L = L^{P'}$, $\pi = \pi^{P'}$, and $\text{ecc}_G(P') \leq k$.

Intuitively, if we had the correct values of π, we would only need to select one segment out of each $\mathcal{S}(\pi_i, \pi_{i+1})$ for $i \in \{1, \ldots, |L|-1\}$, in order to construct the path P. To do so, we need the following function, which estimates the distance from a vertex to the path P.

Definition 8 (estimate distance to P). *For a graph $G = (V, E)$, a set of vertices $\widehat{C} \subseteq V$ and a function $\delta : \widehat{C} \to \mathbb{N}$ we define $d_G^{\delta} : V \times 2^{\widehat{C}} \to \mathbb{N}$ as*

$$d_G^{\delta}(v, S) = \min_{s \in S} d_G(v, s) + \delta(s).$$

Observation 2 ($\star$). *If $\widehat{C} = \widehat{C}^P$ and $\delta = \delta^P|_{\widehat{C}}$ (that is, the restriction of δ^P to $\widehat{C}$) for some shortest path P in G, then for every $v \in V$ we have $d_G(v, P) \leq d_G^{\delta}(v, \widehat{C})$. In particular, if $d_G^{\delta}(v, \widehat{C}) \leq k$, then $d_G(v, P) \leq k$.*

If we had the correct values for the permutation π of vertices from $\widehat{C}$ that are on P, we would still have to take care of those vertices $v \in V$ with $d_G^e(v, \widehat{C}) > k$, in order to solve MESP. In particular, we would have to choose a segment from each $\mathcal{S}(\pi_i, \pi_{i+1})$ in a way that for every vertex v with $d_G^e(v, \widehat{C}) > k$, there would be some chosen segment at distance at most k from v. We say that a candidate segment $\bar{\sigma} \in \widetilde{\mathcal{S}}$ satisfies $v \in V \setminus \widehat{C}$ if $d_G(v, \bar{\sigma}) \leq k < d_G^{\delta}(v, \widehat{C})$.

We continue by showing that the number of vertices v with $d_G^{\delta}(v, \widehat{C}) > k$ which *do not* lie on P is bounded by the size of L.

Lemma 10 ($\star$). *Let $\bar{\sigma} \in \widetilde{\mathcal{S}}$ be a candidate segment and $D = \{v \in V \setminus P \mid \bar{\sigma}$ satisfies $v\}$. Then $|D| \leq 2$.*

Corollary 2. *Let P be a shortest path in G with $\text{ecc}_G(P) \leq k$. Let $U = \{v \in V \setminus P \mid d_G^{\delta^P}(v, \widehat{C}^P) > k\}$. There are $|L^P| - 1$ segments on P, therefore $|U| \leq 2(|L^P| - 1)$.*

We have shown that there are not many vertices $v \notin P$ with $d_G^{\delta^P}(v, \widehat{C}^P) > k$. Now, we show that all such vertices actually have $d_G^{\delta^P}(v, \widehat{C}^P) = k + 1$.

Lemma 11 ($\star$). *Let P be a shortest path in G with $\text{ecc}_G(P) \leq k$. Let $v \in V$ be such that $d_G^{\delta^P}(v, \widehat{C}^P) \geq k + 1$. Let $u \in P$ be the nearest vertex to v on P. Then either $u = v$, or $d_G(u, v) = k$ and $d_G(u, \widehat{C}^P) = 1$.*

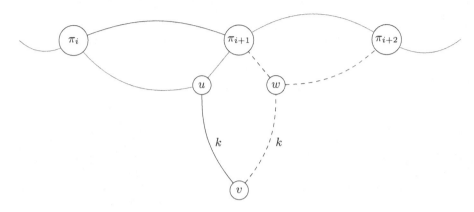

Fig. 2. Example of a situation from Lemma 11. Path P is red, candidate segments are blue. The segment containing vertex u satisfies v. If all the dashed parts are present in G, then the segment containing vertex w also satisfies v. (Color figure online)

Let $v \in V \backslash P$ be such that $d_G^{\delta^P}(v, \widehat{C}^P) = k+1$ and $S \subseteq \widetilde{S}$ be a set of candidate segments that satisfy v. As shown in Fig. 2, there may be multiple such segments in S. However, if v itself lies on some segment, then only segments in the same $\mathcal{S}(\pi_i, \pi_{i+1})$ may satisfy v.

Observation 3 ($\star$). *Let P be a shortest path in G with $ecc_G(P) \leq k$. Let $\bar{\sigma} \in \mathcal{S}(\pi_i^P, \pi_{i+1}^P)$ be a candidate segment that contains some vertex u such that $d_G^{\delta^P}(u, \widehat{C}^P) = k+1$. Let $u' \in P$ be the nearest vertex to u on P. Then u' lies on a segment $\sigma \in \mathcal{S}(\pi_i^P, \pi_{i+1}^P)$.*

We already know from Lemma 11 that if a candidate segment contains some vertex v with $d_G^\delta(v, \widehat{C}) > k + 1$, then it is a necessary segment. Now, we show another sufficient condition for a candidate segment to be a necessary segment.

Lemma 12 ($\star$). *Let $\bar{\sigma} \in \mathcal{S}(\pi_i, \pi_{i+1})$ be a candidate segment that contains some vertices $u, v \in \bar{\sigma}$ such that $u \neq v$ and $d_G^\delta(u, \widehat{C}) = d_G^\delta(v, \widehat{C}) = k+1$. Then, $\bar{\sigma}$ is a necessary segment.*

Let us summarize what we have shown so far. If we had the correct values for the permutation $\boldsymbol{\pi}$ of vertices from $\widehat{C}$ that are on P, and of δ, we would only need to select one segment out of each $\mathcal{S}(\pi_i, \pi_{i+1})$ to find a shortest path with eccentricity at most k. There are some vertices $u \in V$ such that $d_G^\delta(u, \widehat{C}) \leq k$ and for these vertices, the distance to the resulting path will be at most k, no matter which segments we choose.

A segment which contains some vertex v with $d_G^\delta(v, \widehat{C}) > k + 1$ is a necessary segment. A segment which contains two vertices $u \neq v$ with $d_G^\delta(u, \widehat{C}) = d_G^\delta(v, \widehat{C}) = k+1$ is a necessary segment as well. For the remaining segments, we know that for every $u \in V$ with $d_G^\delta(u, \widehat{C}) > k$, the shortest path with eccentricity at most k needs to contain some $\sigma_u \in \widetilde{S}$ such that $d_G(u, \sigma_u) \leq k$. Furthermore, for every $v \in \widehat{C}$ with $d_G(v, L) > \delta(v)$, the path needs to contain some $\sigma_v \in \widetilde{S}$ such that $d_G(v, \sigma_v) \leq \delta(v)$.

Clearly, the problem of selecting one segment out of each set of candidate segments is an instance of CSC: the sets of candidates are $\mathcal{C} = \mathcal{S}(\pi_1, \pi_2) \cup \cdots \cup \mathcal{S}(\pi_{|L|-1}, \pi_{|L|})$, the requirements are $\mathcal{R} = \{v \in V \backslash \widehat{C} \mid d_G^\delta(v, \widehat{C}) > k\} \cup \{v \in \widehat{C} \backslash L \mid d_G(v, L) > \delta(v)\}$, and the function $\Psi(\bar{\sigma}) = \{v \in V \backslash \widehat{C} \mid \bar{\sigma} \text{ satisfies } v\} \cup \{v \in \widehat{C} \backslash L \mid d_G(v, \sigma) \leq \delta(v)\}$.

We know that the number of vertices outside of P that the segments can satisfy is bounded by the size of L. Furthermore, we know that if a segment contains at least two vertices that need to be satisfied, then it is a necessary segment. Lastly, we know that if a segment from some $\mathcal{S}(\pi_i, \pi_{i+1})$ contains one vertex v with $d_G^\delta(v, \widehat{C}) = k + 1$, then only segments from the same $\mathcal{S}(\pi_i, \pi_{i+1})$ may satisfy v. Thus, all segments in $\mathcal{S}(\pi_i, \pi_{i+1})$ that do not satisfy v may be disregarded. By this, we ensure that v will be satisfied no matter which segment is chosen, and v does not need to be added to the requirements $\mathcal{R}$. Hence, the requirements $\mathcal{R}$ do not need to contain any vertices from P, and the size of $\mathcal{R}$ is bounded by the size of C.

In the following lemma, we show that we do not need to explicitly check whether a segment contains some vertex v with $d_G^\delta(v, \widehat{C}) > k + 1$ to decide that it is a necessary segment. This will simplify our algorithm a bit.

Lemma 13 ($\star$). *If a segment $\sigma \in \widetilde{S}$ contains a vertex u such that $d_G^\delta(u, \widehat{C}) > k + 1$, then it must also contain two vertices v, v' with $d_G^\delta(v, \widehat{C}) = d_G^\delta(v', \widehat{C}) = k + 1$.*

Finally, we propose an algorithm that solves MESP. It finds the correct values for $p_1, \widehat{C}, L$, and $\delta: \widehat{C} \to \{0, \ldots, k\}$ by trying all possible combinations. For each combination, it performs the following steps.

1. Find a permutation π of L, such that $\pi_1 = p_1$ and $\sum_{i=1}^{|L|-1} d_G(\pi_i, \pi_{i+1}) = d_G(\pi_1, \pi_{|L|})$. If it does not exist, continue with the next combination.
2. For each π_i, π_{i+1}, check all candidate segments in $\mathcal{S}(\pi_i, \pi_{i+1})$.
   (a) If there are any segments containing a vertex u with $d_G^\delta(u, \widehat{C}) = k + 1$, then we may disregard all candidate segments which do not satisfy u.
   (b) After disregarding these segments, if there is only one candidate segment left, it is a necessary segment. If there is no candidate segment left, then no solution exists.
3. If there is a vertex v such that $d_G^\delta(v, \widehat{C}) > k + 1$, and it does not lie on a segment that we have marked as a necessary segment, then no solution exists.
4. Construct the set U of vertices v that are not contained in any segment and have $d_G^\delta(v, \widehat{C}) = k + 1$. If $|U| > 2(|L| - 1)$, then no solution exists.
5. Choose the rest of the segments from all candidate segments (except those disregarded in step 1) by solving the CSC instance, with requirements $u \in \widehat{C} \backslash L$ whose distance to the parts of P selected so far is greater than $\delta(u)$, and all of U.

6. If the CSC instance has a solution, construct a path from π and from the chosen candidate segments. If the resulting path has eccentricity at most k, return it.

Note that by Lemma 2, there is at most one such permutation π in step 1. We arrive at the following theorem.

Theorem 3 ($\star$). *For a graph with distance to disjoint paths c, MESP can be solved in $\mathcal{O}(2^{5c}k^c c \cdot n^4)$.*

3.5 Maximum Leaf Number

Theorem 4 ($\star$). *There is an algorithm that solves MESP in $\mathcal{O}(2^\ell \cdot n^3)$ time, where ℓ is the maximum leaf number of the input graph.*

4 Future Directions

We have shown that MESP is fixed-parameter tractable with respect to several structural parameters. This partially answers an open question of Dragan and Leitert [8] on classes where the problem is polynomial time solvable, as this is the case whenever we limit ourselves to a class where one of the studied parameters is a constant.

The natural next steps in the research of parameterized complexity of MESP would be to investigate the existence of fpt algorithms with respect to the distance to disjoint paths alone, and with respect to other structural parameters, such as tree depth or feedback vertex set number (see Fig. 1).

References

1. Birmelé, É., de Montgolfier, F., Planche, L.: Minimum eccentricity shortest path problem: an approximation algorithm and relation with the k-laminarity problem. In: Chan, T.-H.H., Li, M., Wang, L. (eds.) COCOA 2016. LNCS, vol. 10043, pp. 216–229. Springer, Cham (2016). https://doi.org/10.1007/978-3-319-48749-6_16
2. Birmelé, E., de Montgolfier, F., Planche, L., Viennot, L.: Decomposing a graph into shortest paths with bounded eccentricity. In: ISAAC 2017. LIPIcs, vol. 92, pp. 15:1–15:13. Dagstuhl (2017)
3. Birmelé, E., de Montgolfier, F., Planche, L., Viennot, L.: Decomposing a graph into shortest paths with bounded eccentricity. Discrete Appl. Math. **284**, 353–374 (2020)
4. Boral, A., Cygan, M., Kociumaka, T., Pilipczuk, M.: A fast branching algorithm for cluster vertex deletion. Theory Comput. Syst. **58**(2), 357–376 (2016)
5. Cygan, M., et al.: Parameterized Algorithms. Springer, Cham (2015)
6. Diestel, R.: Graph Theory, 5th Edition, Graduate texts in mathematics, vol. 173. Springer, Cham (2016). https://doi.org/10.1007/978-3-319-31940-7
7. Dragan, F.F., Leitert, A.: On the minimum eccentricity shortest path problem. In: Dehne, F., Sack, J.-R., Stege, U. (eds.) WADS 2015. LNCS, vol. 9214, pp. 276–288. Springer, Cham (2015). https://doi.org/10.1007/978-3-319-21840-3_23

8. Dragan, F.F., Leitert, A.: Minimum eccentricity shortest paths in some structured graph classes. J. Graph Algorithms Appl. **20**(2), 299–322 (2016)
9. Dragan, F.F., Leitert, A.: On the minimum eccentricity shortest path problem. Theor. Comput. Sci. **694**, 66–78 (2017)
10. Gajarský, J., Lampis, M., Ordyniak, S.: Parameterized algorithms for modular-width. In: Gutin, G., Szeider, S. (eds.) IPEC 2013. LNCS, vol. 8246, pp. 163–176. Springer, Cham (2013). https://doi.org/10.1007/978-3-319-03898-8_15
11. Indyk, P.: Algorithmic applications of low-distortion geometric embeddings. In: FOCS 2001, pp. 10–33. IEEE Computer Society (2001)
12. Indyk, P., Matousek, J.: Low-distortion embeddings of finite metric spaces. In: Handbook of Discrete and Computational Geometry, pp. 177–196. CRC Press, Boca Raton (2004)
13. Kučera, M., Suchý, O.: Minimum eccentricity shortest path problem with respect to structural parameters (2021). https://arxiv.org/abs/2008.07898
14. Sorge, M., Weller, M.: The graph parameter hierarchy (2016). https://manyu.pro/assets/parameter-hierarchy.pdf
15. Tedder, M., Corneil, D., Habib, M., Paul, C.: Simpler linear-time modular decomposition via recursive factorizing permutations. In: Aceto, L., Damgård, I., Goldberg, L.A., Halldórsson, M.M., Ingólfsdóttir, A., Walukiewicz, I. (eds.) ICALP 2008. LNCS, vol. 5125, pp. 634–645. Springer, Heidelberg (2008). https://doi.org/10.1007/978-3-540-70575-8_52
16. Tenenbaum, J.B., Silva, V.D., Langford, J.C.: A global geometric framework for nonlinear dimensionality reduction. Science **290**(5500), 2319–2323 (2000)
17. Völkel, F., Bapteste, E., Habib, M., Lopez, P., Vigliotti, C.: Read networks and k-laminar graphs. CoRR abs/1603.01179 (2016)

Non-preemptive Tree Packing

Stefan Lendl[1], Gerhard Woeginger[2], and Lasse Wulf[3(✉)]

[1] Department of Operations and Information Systems, University of Graz, Graz, Austria
stefan.lendl@uni-graz.at
[2] Department of Computer Science, RWTH Aachen, Aachen, Germany
woeginger@algo.rwth-aachen.de
[3] Institute of Discrete Mathematics, Graz University of Technology, Graz, Austria
wulf@math.tugraz.at

Abstract. An instance of the non-preemptive tree packing problem consists of an undirected graph $G = (V, E)$ together with a weight $w(e)$ for every edge $e \in E$. The goal is to activate every edge e for some time interval of length $w(e)$, such that the activated edges keep G connected for the longest possible overall time.

We derive a variety of results on this problem. The problem is strongly NP-hard even on graphs of treewidth 2, and it does not allow a polynomial time approximation scheme (unless P = NP). Furthermore, we discuss the performance of a simple greedy algorithm, and we construct and analyze a number of parameterized and exact algorithms.

1 Introduction

The Tree Packing Problem of Nash-Williams. For a given undirected connected graph $G = (V, E)$ and a weight function $w : E \to \mathbb{N}$, Nash-Williams [6] considered the following optimization problem:

$$\text{maximize} \quad \sum \{x_T : T \text{ is a spanning tree}\} \tag{1}$$

$$\text{such that} \quad \sum \{x_T : e \in T\} \leq w(e) \qquad \text{for every } e \in E \tag{2}$$

$$x \geq 0, \text{ integral} \tag{3}$$

Nash-Williams [6] derives a min-max relation for problem (1)–(3) that connects it to certain cut conditions. Building on these results, Cunningham [3] constructs a polynomial time algorithm for the problem by reducing it to a polynomial number of maximum flow problems. Barahona [2] presents another algorithm with a better time complexity.

Problem (1)–(3) may also be interpreted as a scheduling problem: Every edge $e \in E$ is a resource that can be activated for a total of $w(e)$ time units. The objective now is to activate the edges in such a way that the graph remains connected for the longest possible overall time. Under this interpretation, the variable x_T indicates the number of time units in which the spanning tree T is used in the activation schedule. Figure 1 contains a simple illustrating example

© Springer Nature Switzerland AG 2021
P. Flocchini and L. Moura (Eds.): IWOCA 2021, LNCS 12757, pp. 456–468, 2021.
https://doi.org/10.1007/978-3-030-79987-8_32

on the three-vertex cycle. There are three spanning trees (that each consist of two edges), and an optimal schedule uses each of these spanning trees for exactly one time unit. It is easy to see that in every optimal schedule, one of the three edges will be active during the first and the third time slot and will be inactive during the second time slot; in the language of scheduling, we say that the execution of that edge is preempted at time 1 and afterwards resumed at time 2. (In the schedule shown in Fig. 1, edge e_3 is the preempted edge.)

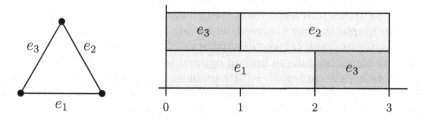

Fig. 1. The three edges in the graph on the left hand side have weights $w(e_1) = w(e_2) = w(e_3) = 2$. The schedule on the right hand side keeps the graph connected for a total of three time units.

The Non-preemptive Version of Tree Packing. We consider a non-preemptive variant of the above tree packing problem, where the execution of edges must not be preempted: Every edge e is activated at some time point $\tau(e)$ chosen by the scheduler, and then remains active without interruption during the full time interval $[\tau(e), \tau(e) + w(e)]$. The objective is again to activate the edges in such a way that the graph remains connected for the longest possible overall time. The resulting combinatorial optimization problem is called *non-preemptive tree packing* (N-TREEPACK for short), and the optimal objective value for a graph $G = (V, E)$ with edge weights $w : E \to \mathbb{N}_0$ will be denoted $\mathsf{ntp}(G, w)$.

In the example in Fig. 1, every reasonable non-preemptive schedule will either activate two of the edges at time 0, or activate the three edges respectively at times $0, 1, 2$. As there is no way of keeping the graph connected for more than two time units, the optimal objective value is $\mathsf{ntp}(G, w) = 2$.

Contributions of this Paper. We analyze the computational complexity and the approximability of non-preemptive tree packing, and we also provide some parameterized and exact algorithms for it. The complexity results are devastating:

– N-TREEPACK is strongly NP-hard, even on complete bipartite graphs $K_{2,n}$.
– N-TREEPACK is strongly NP-hard, even on graphs of bandwidth 2.

Since complete bipartite graphs $K_{2,n}$ are series-parallel and since graphs of bandwidth 2 are outerplanar, our results yield strong NP-hardness for essentially all natural subclasses of graphs with treewidth 2. As edges of zero-weight are irrelevant for the objective value of N-TREEPACK, NP-hardness immediately propagates from graphs to supergraphs; hence our results also imply strong NP-hardness for all the standard families of specially structured graphs, like planar graphs, bipartite graphs, interval graphs, cographs, etc. The only notable exception are the trees and the cactus graphs. Furthermore, we analyze the complexity of cases with small objective values: Deciding whether $\mathsf{ntp}(G, w) \geq \beta$ can be done in polynomial time for $\beta = 3$ and is NP-hard for $\beta = 7$; the intermediate cases with $\beta \in \{4, 5, 6\}$ remain open.

With respect to polynomial time approximation, we introduce a simple greedy algorithm which has a worst case guarantee of $n - 1$ on n-vertex graphs. On cactus graphs it always succeeds in finding an optimal solution, whereas for every non-cactus graph G there exist edge weights w so that on the input (G, w) the greedy algorithm fails to find an optimal solution. We show by means of a gap-reduction that (unless $\mathrm{P} = \mathrm{NP}$) problem N-TREEPACK does not allow a polynomial time approximation algorithm with worst case ratio strictly better than $7/6$; this of course excludes the existence of a PTAS.

Finally, we derive a number of FPT-results in the area of parameterized complexity. The special case of N-TREEPACK where both the treewidth and the maximum edge weight are bounded by a constant k allows an FPT-algorithm whose running time is linear in $|E|$. (The case where only the treewidth is bounded and the case where only the maximum edge weight is bounded are both para-NP-hard, and hence unlikely to belong to FPT.) Furthermore we design an exact algorithm for N-TREEPACK whose (exponential) running time is bounded by $|E|! \cdot \mathrm{poly}(|E|)$.

Organization of the Paper. Section 2 provides central definitions and summarizes the notation. Section 3 contains the NP-hardness results for specially structured graph classes. Sections 4 and 5 contain the negative and positive results for small objective values. Section 6 discusses the greedy algorithm, and Sect. 7 states some parameterized and exact algorithms for N-TREEPACK.

2   Preliminaries

We write $\mathbb{N}_0 = \mathbb{N} \cup \{0\}$ for the set of nonnegative integers. For $a \leq b$, the term $[a, b]$ denotes the *time slot* starting at a and ending at b. For $t \geq 1$, the time slot $[t - 1, t]$ is often called the t-th time slot or time slot t. Every graph $G = (V, E)$ in this paper is simple, undirected and without loops. We write $V(G) = V$ and $E(G) = E$. For $V' \subseteq V$, the *edge cut* $\delta(V')$ is the set of edges with one endpoint in V' and one endpoint in $V - V'$; for $v \in V$, we write $\delta(v) = \delta(\{v\})$. We denote by $G[V']$ the *induced subgraph* on V'. By removing the vertex v from G, we obtain the graph $G - v = G[V - \{v\}]$. Similarly, for $E' \subseteq E$ and $e \in E$ we have $G - E' = (V, E - E')$ and $G - e = G - \{e\}$. For all other graph-theoretic concepts used in the paper, we refer the reader to the text book by West [8].

An instance for problem N-TREEPACK is a weighted graph (G, w), where $G = (V, E)$ and $w : E \to \mathbb{N}_0$. A *schedule* for instance (G, w) is a map $\sigma :$

$E \to \mathbb{N}_0$, that maps each edge e to its activation time $\sigma(e)$. For a schedule σ and an edge e, the *activity interval of* e is $[\sigma(e), \sigma(e) + w(e)]$. For $t \geq 1$, we let $E_t^\sigma = \{e \in E : \sigma(e) + 1 \leq t \leq \sigma(e) + w(e)\}$ denote the set of edges that are active in the t-th time slot, and we let $G_t^\sigma = (V, E_t^\sigma)$ denote the graph on vertex set V with all the edges that are active in the t-th time slot. Finally, we define the *objective value* $\mathsf{ntp}(\sigma)$ of schedule σ as the number of time slots $[t-1, t]$ for which G_t^σ is connected. When the schedule σ is clear from the context, we often simply write E_t and G_t instead of E_t^σ and G_t^σ. The following statement can be shown by a simple exchange argument.

Lemma 1. *For every instance* (G, w) *of* N-TREEPACK *with* $\mathsf{ntp}(G, w) = T$, *there exists an optimal schedule* σ *for which the graphs* $G_1^\sigma, \ldots, G_T^\sigma$ *(in the first T time slots) are trees.* □

3   NP-Hardness

In this section, we establish the NP-hardness of problem N-TREEPACK for certain families of highly restricted graphs. All proofs are done by reductions from the strongly NP-hard 3-PARTITION problem; see Garey and Johnson [5].

Problem 3-PARTITION:
Instance: Positive integers $q_1, \ldots, q_{3n}$ with sum $\sum_{i=1}^{3n} q_i = nQ$ that satisfy $Q/4 < q_i < Q/2$ for all i.
Question: Is there a partition of these $3n$ numbers into n into triplets, such that in every triplet the three numbers sum up to Q?

In the following, we show that N-TREEPACK is strongly NP-hard both for series-parallel graphs (Theorem 1) and for outerplanar graphs (Theorem 2).

Theorem 1. *Problem* N-TREEPACK *is strongly NP-hard, even on complete bipartite graphs* $K_{2,\ell}$.

Proof. Let $q_1, \ldots, q_{3n}$ be an instance of 3-PARTITION as defined above. Let $\beta = nQ + n - 1$. We construct an instance of N-TREEPACK on the complete bipartite graph $K_{2,4n-1}$ with bipartition $\{a, b\}$ and $\{x_1, \ldots, x_{n-1}, y_1, \ldots, y_{3n}\}$. For $i \in \{1, \ldots, n-1\}$, we set $w(\{x_i, a\}) = i(Q+1)$ and $w(\{x_i, b\}) = (n-i)(Q+1)$. For $i \in \{1, \ldots, 3n\}$, we set $w(\{y_i, a\}) = q_i$ and $w(\{y_i, b\}) = \beta$. We claim that the constructed instance of N-TREEPACK possesses a schedule with objective value β, if and only if the underlying 3-PARTITION instance has answer YES.

(Only if) Assume that for the constructed N-TREEPACK instance there exists a schedule σ with objective value β. Note that for every $i \in \{1, \ldots, n-1\}$, we have $w(\{x_i, a\}) + w(\{x_i, b\}) = \beta + 1$. Hence the sum of all the edge weights in the graph is $(n-1)(\beta+1) + 3n\beta + (\beta - n + 1) = 4\beta n$. As every spanning tree for $K_{2,4n-1}$ has $4n$ edges, each of the connected graphs $G_1, \ldots, G_\beta$ must have exactly $4n$ edges (and must actually be a spanning tree). Furthermore, the activity interval of every edge is contained in $[0, \beta]$. Now consider some vertex x_i with $i \in \{1, \ldots, n-1\}$. Since $w(\{x_i, a\}) + w(\{x_i, b\}) = \beta + 1$, there is exactly one $k \in \{1, \ldots, \beta\}$ so that the edge set E_k contains both edges incident to vertex x_i. It is easy to see that either $k = i(Q + 1)$ or $k = (n - i)(Q + 1)$ holds, and we say that this value k is associated with vertex x_i. If some value k is associated with two distinct vertices x_i and x_j, then G_k contains a cycle; a contradiction. We conclude that each of the values $k \in \{Q + 1, 2(Q + 1), \ldots, (n - 1)(Q + 1)\}$ is associated with exactly one of the vertices $x_1, \ldots, x_{n-1}$. This means that for the corresponding time slots $[k - 1, k]$, vertex a is connected to vertex b via the two edges that are incident to the associated vertex x_i. The remaining $\beta - n + 1$ time slots form n (maximal) intervals each of length Q. It is easily verified that during each such interval exactly three vertices y_i, y_j, y_ℓ ensure the connection between a and b, and that the weights of the three edges $\{y_i, a\}, \{y_j, a\}, \{y_\ell, a\}$ satisfy $q_i + q_j + q_\ell = Q$. Hence the corresponding triplets form a solution for the 3-PARTITION instance.

(If) Now assume that the 3-PARTITION instance has a solution. For $1 \le i \le n - 1$, we activate edge $\{x_i, a\}$ at time 0 and edge $\{x_i, b\}$ at time $i(Q + 1) - 1$. For $1 \le i \le n$, we activate edge $\{y_i, b\}$ at time 0; finally, the edges $\{y_i, a\}$ are grouped into triplets according to the solution of the 3-PARTITION instance and scheduled as indicated in the proof of the (only if) part.   □

Theorem 2. *Problem* N-TREEPACK *is strongly NP-hard, even on graphs of bandwidth* 2.

Proof. Let $q_1, \ldots, q_{3n}$ be an instance of 3-PARTITION as defined above. We construct an instance of N-TREEPACK as follows. The graph G has $4n + 1$ vertices $u_0, \ldots, u_n$ and $v_1, \ldots, v_{3n}$. We will sometimes denote vertex u_k also by the name v_{k-n}, for $1 \le k \le n + 1$; in particular we use $u_n = v_0$ and $u_{n-1} = v_{-1}$. Furthermore we define $\beta = (2n - 1)Q$.

- For $k = 0, \ldots, n - 1$, the edge $\{u_k, u_{k+1}\}$ receives weight $w(\{u_k, u_{k+1}\}) = 2(n - k - 1)Q$.
- For $k = 0, \ldots, n - 2$, the edge $\{u_k, u_{k+2}\}$ receives weight $w(\{u_k, u_{k+2}\}) = 2(k + 1)Q$.
- For $k = 1, \ldots, 3n$, the edge $\{v_{k-1}, v_k\}$ has weight $w(\{v_{k-1}, v_k\}) = q_k$.
- For $k = -1, \ldots, 3n - 2$, the edge $\{v_k, v_{k+2}\}$ has weight $w(\{v_k, v_{k+2}\}) = \beta$.

In the ordering $u_0, u_1, \ldots, u_n, v_1, v_2, \ldots, v_{3n}$, every edge either connects two adjacent vertices or two vertices at distance 2. Hence, the constructed graph $G = (V, E)$ has bandwidth 2.

We will study schedules σ of objective value β. As the sum of all edge weights in G equals $\beta(|V| - 1)$, each of the graphs $G_1^\sigma, \ldots G_\beta^\sigma$ is a spanning tree and the activity interval of every edge is contained in $[0, \beta]$. We will discuss the behavior of σ on the induced subgraph $H_i = G[\{u_0, \ldots, u_i\}]$ for $1 \le i \le n$. Since the only connections between H_i and the rest of the graph are via the two vertices u_{i-1} and u_i, during any time slot $[t - 1, t]$ with $1 \le t \le \beta$, graph H_i will consist of one or two connected components under schedule σ. If there is a single connected component, we say that H_i is *fully-connected* during the t-th time slot. If there are two connected components (one containing vertex u_{i-1}, the other one containing vertex u_i), we say that H_i is *semi-connected* during the t-th time slot. Note that if H_i is semi-connected during the t-th time slot, then the edge $\{u_{i-1}, u_{i+1}\}$ must be active during that slot, as there are no other edges that would be able to connect the component containing u_{i-1} to the rest of the graph.

The graph H_1 consists of the vertices u_0 and u_1 and of the edge $\{u_0, u_1\}$ of weight $(2n - 2)Q$. Suppose for the sake of contradiction that H_1 is neither fully-connected during the first time slot nor during the β-th time slot. Then the edge $\{u_0, u_2\}$ (of value $2Q$) must be contained both in E_1 and in E_β, which is impossible. By symmetry, we will henceforth assume that under schedule σ the graph G_1 is fully-connected during the β-th time slot. This implies that $\{u_0, u_1\}$ is active during $[Q, \beta]$ and that $\{u_0, u_2\}$ is active during $[0, 2Q]$. For graph H_i (with $1 \le i \le n$) one can show by induction that H_i is semi-connected during the time intervals $[0, Q], [2Q, 3Q], \ldots, [(2i - 2)Q, (2i - 1)Q]$ and fully-connected at all other moments in $[0, \beta]$. The induction uses the following facts and observations on the two edges $\{u_{i-2}, u_i\}$ and $\{u_{i-1}, u_i\}$ that are in H_i but not in H_{i-1}:

- Graph H_i is semi-connected during the first time slot: By the inductive hypothesis we have H_{i-1} semi-connected during the first time slot. If H_i would be fully-connected during the first time slot, we would get $\sigma(\{u_{i-2}, u_i\}) = \sigma(\{u_{i-1}, u_i\}) = 0$. Since all involved edges have weight $w(e) > Q$, this yields a cycle at time $t = Q + 1$ as the desired contradiction.
- Since graph H_{i-1} is semi-connected at time 0, the edge $\{u_{i-2}, u_i\}$ must be active at time 0 and hence must be active during $[0, (2i - 2)Q]$.
- Graph H_i is fully-connected during the β-th time slot: Otherwise, H_i is fully-connected neither during the first time slot nor during the β-th time slot. Then the edge $\{u_{i-1}, u_{i+1}\}$ would have to be active for β time units.
- Since H_i is fully-connected during the β-th time slot, the edge $\{u_{i-1}, u_i\}$ must be active during the β-th time slot, and hence must be active during the interval $[(2i - 1)Q, \beta]$.

The induction yields for $i = n$ that the induced subgraph H_n is semi-connected during the time intervals $[0, Q], [2Q, 3Q], \ldots, [(2n - 2)Q, (2n - 1)Q]$ (that is, all the intervals of length Q that start at an even multiple of Q) and fully-connected during the time intervals $[Q, 2Q], [3Q, 4Q], \ldots, [(2n - 3)Q, (2n - 2)Q]$ (that is, all the intervals of length Q that start at an odd multiple of Q).

Next, consider the subgraph G' that is induced by the $3n + 2$ vertices $v_{-1} = u_{n-1}$, $v_0 = u_n$ and $v_1, \ldots, v_{3n}$. As the edges $\{v_k, v_{k+2}\}$ with $k = -1, \ldots, 3n - 2$ all have value β, there is an active path P_0 through the vertices with even index during the full interval $[0, \beta]$ and there is an active path P_1 through the vertices with odd index during $[0, \beta]$. By the above discussion, graph H_n connects these two paths P_0 and P_1 to each other during the time intervals $[Q, 2Q], [3Q, Q4], \ldots, [(2n - 3)Q, (2n - 2)Q]$. The only way for connecting P_0 and P_1 to each other during the remaining time intervals $[0, Q], [2Q, 3Q], \ldots, [(2n - 2)Q, (2n - 1)Q]$ is by using the edges $\{v_{k-1}, v_k\}$ with $k = 1, \ldots, 3n$ of weight q_k. As this groups the numbers $q_1, \ldots, q_{3n}$ into n groups with sum Q, we get a solution for the instance of 3-PARTITION.

Vice versa, if the 3-PARTITION instance has a solution, then we build a schedule σ of objective value β: For $k = 0, \ldots, n-1$, we activate edge $\{u_k, u_{k+1}\}$ at time $(2k+1)Q$. For $k = 0, \ldots, n-2$, we activate edge $\{u_k, u_{k+2}\}$ at time 0. For $k = -1, \ldots, 3n-2$, we activate edge $\{v_k, v_{k+2}\}$ at time 0. Finally, the edges $\{v_{k-1}, v_k\}$ are grouped into triplets and scheduled as described in the other direction of the proof.   □

4   A Negative Result for Objective Value Seven

In this section, we show that it is NP-hard to decide whether there exists a schedule of objective value at least 7. The reduction is from the following version of the Hamilton cycle problem; see Akiyama, Nishizeki and Saito [1]

Problem HAMILTON-3-REG
Instance: A bipartite, 3-regular graph H'.
Question: Does H' possess a Hamilton cycle?

The reduction is done in two steps. The first step transforms an instance H' of HAMILTON-3-REG into a new 4-regular graph H with the properties described in Lemma 2. As this transformation is done by routine arguments, we omit the details. The second step then transforms the 4-regular graph H from Lemma 2 into a corresponding instance of problem N-TREEPACK.

Lemma 2. *There is a polynomial time algorithm that takes an instance H' of* HAMILTON-3-REG *as input and outputs a 4-regular bipartite graph H together with a so-called special edge $\{u, z\} \in E(H)$, so that the following holds:*

(i) If H' is a YES-instance of HAMILTON-3-REG, *then the new graph H contains a Hamilton cycle that traverses the special edge $\{u, z\}$.*

(ii) If H' is a NO-instance of HAMILTON-3-REG, *then the new graph H has no Hamilton path starting in vertex u.*   □

Now let H be a 4-regular bipartite graph as described in Lemma 2. Let $U = \{u_1, \ldots, u_k\}$ and $Z = \{z_1, \ldots, z_k\}$ denote the two parts in the bipartition of H, and let $\{u_k, z_k\}$ be its special edge. We create an instance (G, w) of N-TREEPACK from H. The graph G has the vertex set

$$V(G) = \{x, y\} \cup U \cup Z \cup \bigcup_{i=1}^{k} \{v_{i1}, \ldots, v_{i4}\} \cup \bigcup_{i=1}^{k} \{v'_{i1}, v'_{i2}\}.$$

Furthermore, the graph G has the following edges and edge weights:

- For every edge $e \in E(H)$, the graph G also contains e. We set $w(e) = 2$, if $e \neq \{u_k, z_k\}$ and $w(\{u_k, z_k\}) = 1$.
- For every $i = 1, \ldots, k$, the induced subgraph $L_i = G[\{x, u_i, v_{i1}, v_{i2}, v_{i3}, v_{i4}\}]$ is called the *i-th gadget of type L* and has edges and edge weights as depicted in Fig. 2.

– For every $i = 1, \ldots, k$, the induced subgraph $R_i = G[\{x, z_i, v'_{i1}, v'_{i2}\}]$ is called
  the i-th gadget of type R and has edges and edge weights as depicted in Fig. 2.
– Finally, the two edges $\{x, y\}$ and $\{y, z_k\}$ have $w(\{x, y\}) = w(\{y, z_k\}) = 4$.

Assume that (G, w) allows some schedule σ of objective value 7. During any time
slot $[t - 1, t]$ with $1 \leq t \leq 7$, the i-th gadget of type L will consist of either one
or two connected components under schedule σ. If there is a single connected
component, we say that L_i is *fully-connected* during the t-th time slot. If there
are two connected components (one containing x, and one containing u_i), we say
that L_i is *semi-connected* during the t-th time slot. In the same matter, during
the t-th time slot, gadget R_i is either fully-connected or semi-connected with x
and z_i in different components. Likewise, the induced subgraph $G[\{x, y, z_k\}]$ is
either fully-connected, or semi-connected with x and z_k in different components.

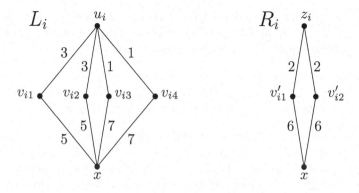

Fig. 2. Gadget of type L (on the left) and gadget of type R (on the right).

Lemma 3. *Let (G, w) be the instance described above. Every schedule of objective value 7 for (G, w) satisfies the following.*

(i)   *A gadget of type R is fully-connected during time slots 2 and 6, and semi-connected during each of the remaining five time slots.*

(ii) *For each $i = 1, \ldots, k$, there are $t_1, t_2 \in \{1, 2, 4, 6, 7\}$ with $t_1 \neq t_2$, such that the gadget L_i is fully-connected during time slots $3, 5, t_1, t_2$, and semi-connected during each of the remaining three time slots.*

(iii) *The induced subgraph $G[\{x, y, z_k\}]$ is fully-connected during time slot 4, and semi-connected during each of the remaining six time slots.* □

Lemma 4. *Let the graph H and the special edge $\{u, z\}$ be as described in Lemma 2, and let (G, w) be the corresponding N-TREEPACK instance. If H contains a Hamilton cycle which uses the special edge, then $\mathsf{ntp}(G, w) \geq 7$.*

Proof. Let $e_0 = \{u, z\} = \{u_k, z_k\}$. There is a Hamilton cycle C using e_0. Then $H - E(C)$ is 2-regular, hence there exist pairwise disjoint matchings $M_1, \ldots, M_4$

such that $M_1 \,\dot\cup\, M_2 = E(C)$ and $M_3 \,\dot\cup\, M_4 = E_H - E(C)$ and $e_0 \in M_1$. We describe a schedule σ with objective value 7:

- All gadgets of type R are fully-connected during time slots 2 and 6, and semi-connected otherwise.
- All gadgets of type L are fully-connected during time slots 1, 3, 5, and 7, and semi-connected otherwise.
- The subgraph $G[\{x, y, z_k\}]$ is fully-connected during time slot 4, and semi-connected otherwise.
- All edges $e \in M_1 - \{e_0\}$ have activity interval $[2, 4]$. The edge e_0 has activity interval $[2, 3]$.
- All edges $e \in M_2$ have activity interval $[3, 5]$.
- All edges $e \in M_3$ have activity interval $[0, 2]$.
- All edges $e \in M_4$ have activity interval $[5, 7]$.

It is easy to see that the active edges in H form a matching of H during the time slots 1, 2, 3, 5, 6, and 7, and form a Hamilton path of H during the time slot 4. All of the graphs $G_1^\sigma, \ldots, G_7^\sigma$ are connected, hence $\mathsf{ntp}(\sigma) = 7$.   □

Lemma 5. *Let the graph H and the special edge $\{u, z\}$ be as described in Lemma 2, and let (G, w) be the corresponding* N-TREEPACK *instance. If* $\mathsf{ntp}(G, w) \geq 7$, *then H contains a Hamilton path starting at vertex u.*

Proof. So assume there exists a schedule σ of objective value 7. For a vertex $v \in U \cup Z$, and $t \in \{1, \ldots, 7\}$, let $d_t(v) = |\delta(v) \cap E(H) \cap E_t|$ denote the number of incident edges of v, which are both in $E(H)$ and active during the t-th time slot. The strategy of the proof will be to repeatedly deduce some conditions for $d_t(v)$. Let $e_0 = \{u, z\} = \{u_k, z_k\}$ be the special edge.

First, recall Lemma 3. Consider vertex z_i for some $i \in \{1, \ldots, k\}$. We know that for each $t \in \{1, 3, 5, 7\}$ in the t-th time slot both the gadget R_i and $G[\{x, y, z_k\}]$ are semi-connected. But of course at least one edge incident to z_i must be active during the t-th time slot. Hence $d_1(z_i), d_3(z_i), d_5(z_i), d_7(z_i) \geq 1$. Note that the four edges in $\delta(z_i) \cap E(H)$ each have weight at most 2 (in the case $i \neq k$ we have four times weight 2, and for $i = k$ we have three times weight 2, and $w(e_0) = 1$). For the sake of contradiction, assume $d_1(z_i) > 1$. Then from the four edges in $\delta(z_i) \cap E(H)$ at least two are scheduled at time 0. This is a contradiction to $d_3(z_i), d_5(z_i), d_7(z_i) \geq 1$. Hence $d_1(z_i) = 1$. By the same argument, $d_7(z_i) = 1$. A similar argument shows that $e_0 \notin E_4$.

Next, consider the graph G_1 of active edges in the first time slot. We know that $d_1(z_i) = 1$ for all $i = 1, \ldots, k$. Hence the induced subgraph $G_1[U \cup Z]$ is acyclic, has k edges, and therefore has k connected components. But because G_1 is connected, and because all the gadgets of type R and $G[\{x, y, z_k\}]$ are semi-connected during time slot 1, this implies that every single gadget of type L is actually fully-connected during time slot 1. The same argument holds for time slot 7. In total, together with Lemma 3, we have that a gadget of type L is fully-connected during the t-th time slot, if and only if $t \in \{1, 3, 5, 7\}$. This in turn implies that for all $i = 1, \ldots, k$, one has $d_2(u_i), d_4(u_i), d_6(u_i) \geq 1$. The two facts $d_2(u_i) \geq 1$ and $d_6(u_i) \geq 1$ together imply $d_4(u_i) \leq 2$. Likewise, for all $i = 1, \ldots, k$, the two facts $d_1(z_i) = 1$ and $d_7(z_i) = 1$ together imply $d_4(z_i) \leq 2$.

Finally, we claim $d_4(u_k) = 1$. In fact, $d_4(u_k) = 0$ is impossible, because gadgets of type L are semi-connected during time slot 4. For the sake of contradiction, assume $d_4(u_k) > 1$. We know that $d_2(u_k) \geq 1, d_6(u_k) \geq 1$ and $e_0 \notin E_4$ and $w(e_0) = 1$. So our assumption $d_4(u_k) > 1$ is only possible if $e_0 \in E_2$ or $e_0 \in E_6$. But for the vertex z_k, we also know $d_1(z_k), d_3(z_k), d_5(z_k), d_7(z_k) \geq 1$. This is a contradiction to $e_0 \in E_2 \cup E_6$, hence our assumption was wrong and $d_4(u_k) = 1$. In summary, during time slot 4, all gadgets of type L and R are semi-connected. We also have for all $i = 1, \ldots, k$ that $d_4(u_i) \leq 2$ and $d_4(z_i) \leq 2$. Also, $d_4(u_k) = 1$. However, G_4 is connected. These facts together imply that the induced subgraph $G_4[U \cup Z]$ is a Hamilton path in H starting at $u_k = u$.   □

By combining Lemmas 2, 4 and 5 we get the following summarizing theorem.

Theorem 3. *For* N-TREEPACK *it is strongly NP-hard to decide whether there exists a schedule of objective value at least 7.*   □

All edge weights in the above reduction are in the set $\{1, \ldots, 7\}$. A minor modification yields the following corollary.

Corollary 1. *Problem* N-TREEPACK *is strongly NP-hard, even if all edge weights are in* $\{1, \ldots, 6\}$.   □

As it is NP-hard to distinguish between N-TREEPACK instances with optimal objective value 6 and N-TREEPACK instances with optimal objective value 7, we also get the following in approximability result.

Corollary 2. *Unless* $P = NP$, *there is no polynomial time approximation algorithm for* N-TREEPACK *with worst case guarantee better than* $7/6$.   □

5   A Positive Result for Objective Value Three

In Sect. 4, we have established the NP-hardness of deciding whether there exists a schedule of objective value at least 7. As a complementary result, we now show that it can be decided in polynomial time whether there is a schedule of objective value at least 3.

Theorem 4. *For an instance of* N-TREEPACK *on a graph with m edges, it can be decided in* $\mathcal{O}(m^3)$ *time whether* $\mathsf{ntp}(G, w) \geq 3$.

Proof. Let $G = (V, E)$. We partition the edge set E into set W_1 (edges of weight 1), set W_2 (edges of weight 2), and set $W_{\geq 3}$ (edges of weight at least 3). In a schedule of objective value 3, we may activate all edges in $W_{\geq 3}$ at time 0. We interpret an edge in W_2 as a pair of two edges of weight 1: one of these two edges is scheduled during the middle time slot $[1, 2]$; the other edge can either be scheduled during slot $[0, 1]$ or during slot $[2, 3]$. Edges in W_1 are scheduled during one of the three slots $[0, 1], [1, 2], [2, 3]$.

By Lemma 1, we may assume that in a feasible schedule of length 3 the graphs (V, E_t) with $t = 1, 2, 3$ are trees. Let H_3 be the graph that results from G after contracting the edge set $W_{\geq 3}$, and let H_{23} be the graph that results from G after contracting the edge set $W_2 \cup W_{\geq 3}$. We introduce three matroids:

- The matroid $\mathcal{F}_1$ has the ground set $W_1 \cup W_2$. A set $F \subseteq W_1 \cup W_2$ is independent in $\mathcal{F}_1$, if and only if F is acyclic in graph H_3.
- The matroid $\mathcal{F}_2$ has the ground set W_1. A set $F \subseteq W_1$ is independent in $\mathcal{F}_2$, if and only if F is acyclic in graph H_{23}.
- The matroid $\mathcal{F}_3$ has ground set $W_1 \cup W_2$, and coincides with $\mathcal{F}_1$.

It is easily seen that $\mathsf{ntp}(G, w) \geq 3$ holds, if and only if there exist three pairwise disjoint subsets S_1, S_2, S_3 of E, such that S_t forms a base of the matroid $\mathcal{F}_t$ for $t = 1, 2, 3$. This can be checked in $\mathcal{O}(m^3)$ time by using Edmond's matroid partitioning algorithm [4]. $\qquad\square$

By a similar (but simpler) argument we can also decide in polynomial time whether $\mathsf{ntp}(G, w) \geq 2$. Deciding whether $\mathsf{ntp}(G, w) \geq 1$ is trivial. The complexity of deciding whether $\mathsf{ntp}(G, w) \geq \beta$ remains open for $\beta \in \{4, 5, 6\}$.

6 The Greedy Algorithm

We introduce a greedy algorithm that maintains connectivity by always activating edges of the largest possible weight. Formally, we let $F_t \subseteq E_t$ denote the set of edges whose activity intervals end at time t. By $U_t = E - (E_1 \cup E_2 \cup \cdots \cup E_t)$ we denote the set of edges that have not been used and activated before time t.

Now the Greedy algorithm starts its work by initializing $E_0 := \emptyset$, $F_0 := \emptyset$, and $U_0 := E$. For $t \geq 0$, the set E_{t+1} for time slot $[t, t+1]$ is computed as follows. If the graph $(V, E_t - F_t)$ is a tree, we set $E_{t+1} := E_t$. If the graph $(V, E_t - F_t)$ is a forest with c components, we turn it into a tree by adding a maximum weight subset $A \subseteq U_t$ of cardinality $c - 1$; then we set $E_{t+1} := (E_t - F_t) \cup A$. In case no such set A exists, the Greedy algorithm terminates. (The set A can be computed for instance by applying Kruskal's algorithm for maximum spanning trees; ties are broken arbitrarily.)

Theorem 5. *For every graph $G = (V, E)$ on n vertices and for every $w : E \to \mathbb{N}_0$, the Greedy algorithm computes a schedule of length at least $\mathsf{ntp}(G, w)/(n - 1)$. Furthermore, there exist instances on which the schedule computed by the Greedy algorithm is a factor $\lfloor n/2 \rfloor$ below the optimal objective value.*

Proof. For the positive result, we consider the time slot $[T, T+1]$ at which Greedy terminates. Then the graph $(V, (E_T - F_T) \cup U_T)$ is not connected. We consider the vertex set $C \subseteq V$ of one of the components of that graph, and the corresponding edge cut $\delta(C)$. Then the weight $w(\delta(C)) = \sum_{e \in \delta(C)} w(e)$ yields a trivial upper bound for the optimal objective value:

$$\mathsf{ntp}(G, w) \leq w(\delta(C)) \tag{4}$$

Since every edge set E_j with $1 \leq j \leq T$ induces a tree, we have $|E_j| = n-1$ and hence $|E_j \cap \delta(C)| \leq n-1$. As all edges in the cut $\delta(C)$ have been activated and run to completion before the time slot $[T, T+1]$, we conclude $T \geq w(\delta(C))/(n-1)$, which together with (4) yields the desired approximation guarantee.

For the negative result, we consider the complete graph $K_n = (V, E)$ on n vertices with weights $w(e) = 1$ for all $e \in E$. A folklore result (see for instance Palmer [7]) says that the maximum number of edge-disjoint spanning trees that can be packed into K_n is $\lfloor n/2 \rfloor$. This implies $\mathsf{ntp}(K_n, w) = \lfloor n/2 \rfloor$. On the other hand, if the Greedy algorithm at time 0 activates the $n-1$ edges in the edge cut $\delta(v)$ for some $v \in V$, the objective value of the resulting schedule equals 1.   □

Theorem 6. *For every connected graph $G = (V, E)$, the following two statements are equivalent.*

(i) G is a cactus graph.

(ii) For every choice $w : E \to \mathbb{N}_0$ of edge weights, the Greedy algorithm solves the N-TREEPACK instance (G, w) to optimality.   □

7   Parameterized Complexity

In this section, we show that problem N-TREEPACK is fixed parameter tractable with respect to various parameters. Note that Theorems 1 and 3 imply the NP-hardness of problem N-TREEPACK, even if either the treewidth or the edge weights are bounded by a constant.

Theorem 7. *If both the treewidth and the maximum edge weight of the input graph $G = (V, E)$ are bounded, problem N-TREEPACK allows for an FPT-algorithm whose running time is linear in $|E|$.*   □

Theorem 8. *On input graphs $G = (V, E)$, problem N-TREEPACK is solvable in exponential time $\mathcal{O}(|E|^2 \cdot |E|!)$.*   □

Theorem 9. *Problem N-TREEPACK is fixed parameter tractable with respect to the size k of a feedback edge set. There is a kernel with $\mathcal{O}(k)$ vertices and edges.*   □

The last theorem of this section shows that problem N-TREEPACK is tractable on instances that in a certain sense are close to the preemptive tree packing problem (1)–(3).

Theorem 10. *Let (G, w) be an instance of N-TREEPACK on m edges, so that $m - k$ edges have weight 1 and the remaining k edges have weight at most k. Then an optimal solution can be found in $\mathcal{O}(k^{2k}m^3)$ time.*   □

Acknowledgements. Stefan Lendl and Lasse Wulf acknowledge the support of the Austrian Science Fund (FWF): W1230. Gerhard Woeginger acknowledges support by the DFG RTG 2236 "UnRAVeL".

References

1. Akiyama, T., Nishizeki, T., Saito, N.: NP-completeness of the Hamiltonian cycle problem for bipartite graphs. J. Inf. Process. **3**, 73–76 (1980)
2. Barahona, F.: Packing spanning trees. Math. Oper. Res. **20**, 104–115 (1995)
3. Cunningham, W.H.: Optimal attack and reinforcement of a network. J. ACM **32**, 549–561 (1985)
4. Edmonds, J.: Minimum partition of a matroid into independent subsets. J. Res. Natl. Bur. Stand. **69**, 67–72 (1965)
5. Garey, M.R., Johnson, D.S.: Computers and Intractability: A Guide to the Theory of NP-Completeness. Freeman W. H, New York (1979)
6. Nash-Williams, C.S.J.A.: Edge-disjoint spanning trees of finite graphs. J. Lond. Math. Soc. **36**, 445–450 (1961)
7. Palmer, E.M.: On the spanning tree packing number of a graph: a survey. Discret. Math. **230**, 13–21 (2001)
8. West, D.B.: Introduction to Graph Theory. Pearson College Div (2000)

Card-Based Cryptographic Protocols for Three-Input Functions Using Private Operations

Yoshifumi Manabe$^{(\boxtimes)}$ (ID) and Hibiki Ono

Kogakuin University, Shinjuku, Tokyo 163–8677, Japan
manabe@cc.kogakuin.ac.jp

Abstract. This paper shows card-based cryptographic protocols to calculate several Boolean functions using private operations under the semi-honest model. Private operations, introduced by Nakai et al. are the most powerful operations for card-based protocols. We showed that copy, logical AND, and logical XOR can be calculated with the minimum number of cards using three private operations, private random bisection cuts, private reverse cuts, and private reveals. This paper shows that by using these private operations, all of the following Boolean functions can be calculated without additional cards other than the input cards: (1) Any three input Boolean functions, (2) Half adder and full adder, and (3) Any n-input symmetric Boolean functions. The numbers of cards used in these protocols are smaller than the ones without private operations.

Keywords: Card-based cryptographic protocols · Multi-party secure computation · Boolean functions · Half adder · Symmetric functions

1 Introduction

Card-based cryptographic protocols [14,30] have been proposed in which physical cards are used instead of computers to securely calculate values. They can be used when computers cannot be used or users cannot trust software in the computers. They can also be used to teach the foundation of cryptography [4,26]. den Boer [2] first showed a five-card protocol to securely calculate logical AND of two inputs. Since then, many protocols have been proposed to calculate logical AND, logical XOR, and copy primitives to compute general Boolean functions [1,5,8,13,15,18,31,32,36,37,39,42,45,55,58] and specific computations such as a class of Boolean functions [22,27,29,38,46,51], computation using garbled circuits [52], simulation of universal computation such as Turing machines [7,16], millionaires' problem [23,34,40], voting [28,35,59,60], random permutation [9,11,12,33], grouping [10], ranking [56], lottery [53], proof of knowledge of a puzzle solution [3,6,20,24,25,43,44,47–49], and so on. This paper considers the calculation of Boolean functions under the semi-honest model.

The protocols are executed by two players, Alice and Bob. Though this paper and many other papers assume semi-honest model, malicious actions or mistakes

© Springer Nature Switzerland AG 2021
P. Flocchini and L. Moura (Eds.): IWOCA 2021, LNCS 12757, pp. 469–484, 2021.
https://doi.org/10.1007/978-3-030-79987-8_33

might occur in real cases. Preventing or detecting such actions were considered [17,21,42,57].

There are several types of protocols regards to the inputs and outputs of the computations. The first type is committed inputs, where the inputs are given as committed values. The players do not know the input values. The other type is non-committed inputs, where players give their own private inputs to the protocol. Protocols with committed inputs are desirable since they can be used for non-committed inputs: each player can give his own private input value as a committed value. Some protocols output their computation results as committed values. The players do not know the output values. The other type of protocols output the result as a non-committed value, that is, the final result is obtained by opening some or all cards. Protocols with committed outputs are desirable since the committed output result can be used as an input to another computation. If further computation is unnecessary, the players just open the committed outputs and obtain the result. Thus, this paper discusses protocols with committed inputs and committed outputs.

This paper assumes the standard two-type card model, in which one bit data is represented by two cards. The detail is shown in Sect. 2.

Operations that a player executes where the other players cannot see are called private operations. They are considered to be executed under the table, in the back, and so on. They were first introduced by Nakai et al. to solve millionaires' problem [34]. Using private operations, committed-input and committed-output logical AND, logical XOR, and copy protocols can be achieved with four cards, that is, without additional cards other than the input (output) cards, with finite steps, and without non-uniform shuffles [42]. The AND protocol in [31] without private operations uses six cards. It is proved to be impossible to achieve finite-runtime AND with four cards by the model without private operations [13,15]. As for the number of cards used for copy protocols, six was the minimum for finite-runtime copy [31] without private operations. It is proved to be impossible to achieve a copy with four cards by the model without private operations [13]. Thus private operations are effective in card-based protocols.

Another type of private operations, we call private input operations, were introduced to calculate Boolean functions with non-committed inputs [19,54]. A player uses the private input operations to input his own private values to the protocol. The operations were used also in millionaires' problem [40], voting [59], and so on. Since this paper considers committed inputs, private input operations are not used.

Though the private operations are powerful, it is shown that we can calculate any n-input Boolean functions with four additional cards [42]. Thus the research question is obtaining the class of Boolean functions that can be calculated without additional cards using the private operations. This paper shows new card-based protocols using private operations to calculate (1) any three input Boolean functions and (2) half adder and full adder, and (3) any n-input symmetric Boolean functions. All of these protocols need no additional cards other than the input cards. Thus these protocols are optimal regards to the number of cards. In [37,38] two additional cards were necessary to calculate these functions without private operations.

In Sect. 2 basic definitions are shown. Section 3 shows the private operations introduced by [42]. Section 4 shows the sub-protocols shown in [42] that are used in this paper. Section 5 shows protocols to calculate three input Boolean functions. Section 6 shows protocols to calculate half and full adder, and n-input symmetric Boolean functions. Section 7 concludes the paper.

2 Preliminaries

This section gives the notations and basic definitions of card-based protocols. This paper is based on the standard two-type card model. In the two-type card model, there are two kinds of marks, ♣ and ♡ . Cards of the same marks cannot be distinguished. In addition, the back of both types of cards is ? . It is impossible to determine the mark in the back of a given card with ? .

One bit of data is represented by two cards as follows: ♣♡ = 0 and ♡♣ = 1.

One pair of cards that represents one bit $x \in \{0, 1\}$, whose face is down, is called a commitment of x, and denoted as $commit(x)$. It is written as $\underbrace{?\ ?}_{x}$.

Note that when these two cards are swapped, $commit(\overline{x})$ can be obtained. Thus, NOT can be calculated without private operations.

A linearly ordered cards is called a sequence of cards. A sequence of cards S whose length is n is denoted as $S = s_1, s_2, \ldots, s_n$, where s_i is i-th card of the sequence. $S = \underbrace{?}_{s_1}\ \underbrace{?}_{s_2}\ \underbrace{?}_{s_3}\ \ldots, \underbrace{?}_{s_n}$. A sequence whose length is even is called an even sequence. $S_1 || S_2$ is a concatenation of sequence S_1 and S_2.

All protocols are executed by multiple players. Throughout this paper, all players are semi-honest, that is, they obey the rule of the protocols, but try to obtain information x of $commit(x)$. There is no collusion among players executing one protocol together. No player wants any other player to obtain information of committed values.

The space complexity of card-based protocols is evaluated by the number of cards. The time complexity of card-based protocols using private operations is evaluated by the number of rounds [41]. The first round is (possibly parallel) local executions by each player using the cards initially given to each player, from the initial state to the instant when no further local execution is possible without receiving cards from another player. The local executions in each round include sending cards to some other players but do not include receiving cards. The $i(> 1)$-th round begins with receiving all the cards sent during $(i - 1)$-th round. Each player executes local executions using the received cards and the cards left to the player at the end of the $(i - 1)$-th round. Each player executes until no further local execution is possible without receiving cards from another player. The number of rounds of a protocol is the maximum number of rounds necessary to output the result among all possible inputs and all possible choices of the random values.

3   Private Operations

We show three private operations introduced in [42], private random bisection cuts, private reverse cuts, and private reveals.

Primitive 1 *(Private random bisection cut). A private random bisection cut is the following operation on an even sequence $S_0 = s_1, s_2, \ldots, s_{2m}$. A player selects a random bit $b \in \{0, 1\}$ and outputs*

$$S_1 = \begin{cases} S_0 & if\, b = 0 \\ s_{m+1}, s_{m+2}, \ldots, s_{2m}, s_1, s_2, \ldots, s_m & if\, b = 1 \end{cases}$$

The player executes this operation in a place where the other players cannot see. The player does not disclose the bit b.   □

Note that the protocols in this paper uses the operation only when $m = 1$ and $S_0 = commit(x)$. Given $S_0 = \boxed{?}\boxed{?}$, The player's output $S_1 = \boxed{?}\boxed{?}$, which is $\underbrace{}_{x}$ $\underbrace{}_{x \oplus b}$

$\boxed{?}\boxed{?}$ or $\boxed{?}\boxed{?}$.
$\underbrace{}_{x}$ $\underbrace{}_{\overline{x}}$

Note that a private random bisection cut is the same as the random bisection cut [31], but the operation is executed in a hidden place.

Primitive 2 *(Private reverse cut, Private reverse selection). A private reverse cut is the following operation on an even sequence $S_2 = s_1, s_2, \ldots, s_{2m}$ and a bit $b \in \{0, 1\}$. A player outputs*

$$S_3 = \begin{cases} S_2 & if\, b = 0 \\ s_{m+1}, s_{m+2}, \ldots, s_{2m}, s_1, s_2, \ldots, s_m & if\, b = 1 \end{cases}$$

The player executes this operation in a place where the other players cannot see. The player does not disclose b.

Note that in many protocols below, selecting left m cards is executed after a private reverse cut. The sequence of these two operations is called a private reverse selection. A private reverse selection is the following procedure on an even sequence $S_2 = s_1, s_2, \ldots, s_{2m}$ and a bit $b \in \{0, 1\}$. A player outputs

$$S_3 = \begin{cases} s_1, s_2, \ldots s_m & if\, b = 0 \\ s_{m+1}, s_{m+2}, \ldots, s_{2m} & if\, b = 1 \end{cases}$$ □

The difference between the private random bisection cut and the private reverse cut is that b is not newly selected by the player.

Next, we define a private reveal.

Primitive 3 *(Private reveal). A player privately opens a given committed bit. The player does not disclose the obtained value.*   □

Using the obtained value, the player privately sets a sequence of cards.

Consider the case when Alice executes a private random bisection cut on $commit(x)$ and Bob executes a private reveal on the bit. Since the committed bit is randomized by the bit b selected by Alice, the opened bit is $x \oplus b$. Bob obtains no information about x if b is randomly selected and not disclosed by Alice. Bob must not disclose the obtained value. If Bob discloses the obtained value to Alice, Alice knows the value of the committed bit.

4   Protocols for XOR, AND, Copy, and Other Boolean Functions

This section shows the sub-protocols presented in [41, 42] used in this paper's protocols. The correctness proof is shown in [41, 42].

4.1   XOR Protocol

Protocol 1 *(XOR protocol) [41]*
   Input: $commit(x)$ and $commit(y)$.
   Output: $commit(x \oplus y)$.

1. *Alice executes a private random bisection cut on input $S_0 = commit(x)$ and $S_0' = commit(y)$ using the same random bit b. Let the output be $S_1 = commit(x')$ and $S_1' = commit(y')$, respectively. Note that $x' = x \oplus b$ and $y' = y \oplus b$. Alice sends S_1 and S_1' to Bob.*
2. *Bob executes a private reveal on $S_1 = commit(x')$. Bob executes a private reverse cut on S_1' using x'. Let the result be S_2. Bob outputs S_2.*   □

The protocol is two rounds. Note that the protocol uses no cards other than the input cards.

4.2   And Protocol

Protocol 2 *(AND protocol) [42]*
   Input: $commit(x)$ and $commit(y)$.
   Output: $commit(x \wedge y)$.

1. *Alice executes a private random bisection cut on $commit(x)$. Let the output be $commit(x')$. Alice hands $commit(x')$ and $commit(y)$ to Bob.*
2. *Bob executes a private reveal on $commit(x')$. Bob sets*

$$S_2 = \begin{cases} commit(y) \| commit(0) & if x' = 1 \\ commit(0) \| commit(y) & if x' = 0 \end{cases}$$

and hands S_2 to Alice.

3. *Alice executes a private reverse selection on S_2 using the bit b generated in the private random bisection cut. Let the obtained sequence be S_3. Alice outputs S_3.*   □

In Step 2, the cards of $commit(x')$ are re-used to set $commit(0)$. Thus the protocol uses no cards other than the input cards. The protocol is three rounds.

4.3   COPY Protocol

Protocol 3 *(COPY protocol) [42]*
  Input: commit(x).
  Output: m copies of commit(x).

1. *Alice executes a private random bisection cut on commit(x). Let the output be commit(x'). Alice hands commit(x') to Bob.*
2. *Bob executes a private reveal on commit(x'). Bob makes m copies of x'. Bob faces down these cards. Bob hands these cards, m copies of commit(x'), to Alice.*
3. *Alice executes a private reverse cut to each copy of commit(x') using the bit b Alice generated in the private random bisection cut. Alice outputs these copies.*   □

The protocol is three rounds. Note that the protocol does not need additional cards other than $2m$ output cards.

4.4   Any Two-Input Boolean Functions

Though the previous subsection showed AND and XOR, any two-input Boolean functions can also be calculated by a similar protocol by three rounds and four cards [42]. Any two-input Boolean function $f(x, y)$ can be written as

$$f(x, y) = \begin{cases} f(1, y) \text{ if } x = 1 \\ f(0, y) \text{ if } x = 0 \end{cases}$$

where $f(1, y)$ and $f(0, y)$ are y, $\overline{y}$, 0, or 1.
  First consider the case when both of $f(1, y)$ and $f(0, y)$ are 0 or 1. ($f(1, y)$, $f(0, y)) = (0, 0)$ (or $(1, 1)$) means that $f(x, y) = 0$ (or $f(x, y) = 1$), thus we do not need to calculate f. ($f(1, y), f(0, y)) = (1, 0)$ (or $(0, 1)$) means the $f(x, y) = x$ (or $f(x, y) = \overline{x}$), thus we do not need to calculate f by a two player protocol.
  Next consider the case when both of $(f(1, y), f(0, y))$ are y (or $\overline{y}$). This case is when $f(x, y) = y$ (or $f(x, y) = \overline{y}$), thus we do not need to calculate f by a two player protocol.
  Next case is when $(f(1, y), f(0, y))$ is $(y, \overline{y})$ or $(\overline{y}, y)$. $(f(1, y), f(0, y)) = (\overline{y}, y)$ is $x \oplus y$ (XOR). $(f(1, y), f(0, y)) = (y, \overline{y})$ is $\overline{x \oplus y}$, thus this function can be calculated as follows: use XOR protocol and NOT is taken to the output. Thus, this function can also be calculated.

The remaining case is when one of $(f(1, y), f(0, y))$ is y or $\overline{y}$ and the other is 0 or 1. We can use the AND protocol and Bob sets

$$S_2 = \begin{cases} commit(f(1, y))||commit(f(0, y)) \text{ if } x' = 1 \\ commit(f(0, y))||commit(f(1, y)) \text{ if } x' = 0 \end{cases}$$

using one $commit(y)$ in the second step of the protocol.

Thus, any two-input Boolean function can be calculated without additional cards.

4.5 Parallel Computations

The above two-input Boolean function calculations can be executed in parallel [42]. Consider the case when $commit(x)$ and $commit(y_i)(i = 1, 2, \ldots, n)$ are given and $commit(f_i(x, y_i))(i = 1, 2, \ldots, n)$ need to be calculated. They can be executed in parallel. Alice executes a private random bisection cut on $commit(x)$ and hands $commit(x')$ and $commit(y_i)(i = 1, 2, \ldots, n)$ to Bob. Bob sets $S_2^i(i = 1, 2, \ldots, n)$ using x' and $commit(y_i)$ according to f_i. Alice executes a private reverse cut or a private reverse selection on each of $S_2^i(i = 1, 2, \ldots, n)$ using the bit b selected at the private random bisection cut. By the procedure, $commit(f_i(x, y_i))$ $(i = 1, 2, \ldots, n)$ are simultaneously obtained.

Note that if f_i is calculated by an AND-type protocol, two new cards are necessary and the two cards of $commit(x')$ can be used. Thus, when at most one f_i is executed by an AND-type protocol and all the others are executed by XOR-type protocols, they can be executed in parallel without additional cards.

4.6 Preserving an Input

In the above protocols to calculate Boolean functions, the input commitment values are lost. If the input is not lost, the input commitment can be used as an input to another calculation. Thus input preserving calculation is discussed [37,42].

In the XOR protocol, $commit(x')$ is no more necessary after Bob sets S_2. Thus, Bob can send back $commit(x')$ to Alice. Then, Alice can recover $commit(x)$ using the private reverse cut. In this modified protocol, the output is $commit(x \oplus y)$ and $commit(x)$ without additional cards.

An input preserving calculation without increasing the number of cards can be executed for AND type protocols just like [37]. When we execute the AND type protocol, two cards are selected by Alice at the final step. The remaining two cards are used to recover an input value. The unused two cards' value is

$$\begin{cases} f(0, y) \text{ if } x = 1 \\ f(1, y) \text{ if } x = 0 \end{cases}$$

thus the output is $commit((\overline{x} \wedge f(1, y)) \oplus (x \wedge f(0, y)))$.

Note that the function f satisfies that one of $(f(0, y), f(1, y))$ is y or $\overline{y}$ and the other is 0 or 1. Otherwise, we do not need to calculate f by the AND type two player protocol.

The output $f(x, y)$ can be represented as $(x \wedge f(1, y)) \oplus (\overline{x} \wedge f(0, y))$. Execute the above input preserving XOR protocol for these two output values so that the input $f(x, y)$ is preserved. The output of XOR protocol is $(\overline{x} \wedge f(1, y)) \oplus (x \wedge f(0, y)) \oplus (x \wedge f(1, y)) \oplus (\overline{x} \wedge f(0, y)) = f(1, y) \oplus f(0, y)$. Since one of $(f(0, y), f(1, y))$ is y or $\overline{y}$ and the other is 0 or 1, the output is y or $\overline{y}$ (depending on f). Thus, input y can be recovered without additional cards. Thus, preserving an input can be realized by 4 cards, which is the minimum. In [37], two additional cards are necessary.

4.7 n-input Boolean Functions

Since any 2-input Boolean function, NOT, and COPY can be executed, any n-input Boolean function can be calculated by the combination of the above protocols.

Using the technique in [37] and above input preserving Boolean function calculations, any n-input Boolean function can be calculated with $2n + 4$ cards as follows [42].

Any Boolean function $f(x_1, x_2, \ldots, x_n)$ can be represented as follows:
$f(x_1, x_2, \ldots, x_n) = (\overline{x_1} \wedge \overline{x_2} \wedge \cdots \wedge \overline{x_n} \wedge f(0, 0, \ldots, 0)) \oplus (x_1 \wedge \overline{x_2} \wedge \cdots \wedge \overline{x_n} \wedge f(1, 0, \ldots, 0)) \oplus (\overline{x_1} \wedge x_2 \wedge \cdots \wedge \overline{x_n} \wedge f(0, 1, \ldots, 0)) \oplus \cdots \oplus (x_1 \wedge x_2 \wedge \cdots \wedge x_n \wedge f(1, 1, \ldots, 1))$.

Since the terms with $f(i_1, i_2, \ldots, i_n) = 0$ can be removed, this function f can be written as $f = \bigoplus_{i=1}^{k} v_1^i \wedge v_2^i \wedge \cdots \wedge v_n^i$, where $v_j^i = x_j$ or $\overline{x_j}$. Let us write $T_i = v_1^i \wedge v_2^i \wedge \cdots \wedge v_n^i$. The number of terms $k(< 2^n)$ depends on f.

Protocol 4 *(Protocol for any Boolean function) [42]*
  Input: commit$(x_i)(i = 1, 2, \ldots, n)$.
  Output: commit$(f(x_1, x_2, \ldots, x_n))$.
  The additional four cards (two pairs of cards) p_1 and p_2 are used as follows.
  p_1: the intermediate value to calculate f is stored.
  p_2: the intermediate value to calculate T_i is stored.
Execute the following steps for $i = 1, \ldots, k$.

1. *Copy v_1^i from the input x_1 to p_2.*
2. *For $j = 2, \ldots, n$, execute the following procedure: Apply the input-preserving AND protocol to p_2 and input x_j (If AND is taken between $\overline{x_j}$, first execute NOT to the input, then apply the AND protocol, and return the input to x_j again.)*
  At the end of this step, T_i is obtained at p_2.
3. *If $i = 1$, move p_2 to p_1. If $i > 1$, apply the XOR protocol between p_1 and p_2. The result is stored to p_1.*

At the end of the protocol, $f(x_1, x_2, \ldots x_n)$ is obtained at p_1.   □

5   Protocols for Three-Input Boolean Functions

This section shows protocols for three-input Boolean functions. [38] has shown that any three-input Boolean functions can be calculated with at most eight cards. We show that these functions can be calculated with six cards using private operations, that is, no additional cards are necessary other than the input cards.

The arguments to show the protocols with six cards are just the same as the one in [38]. The main difference is that AND-type functions can be calculated by four cards using the private operations.

There are $2^{2^3} = 256$ different functions with three inputs. However, some of these functions are equivalent by replacing variables and taking negations. NPN-classification [50] was considered to reduce the number of different functions considering the equivalence class of functions. The rules of NPN-classification are as follows.

1. Negation of input variables (Example: $x_i \leftrightarrow \overline{x_i}$).
2. Permutations of input variables (Example: $x_i \leftrightarrow x_j$).
3. Negation of the output ($f \leftrightarrow \overline{f}$).

For example, consider $f_1(x_1, x_2, x_3) = (x_1 \wedge x_2) \vee x_3$. Several functions in the same equivalence class that includes f_1 are: $f_2 = (\overline{x_1} \wedge \overline{x_2}) \vee x_3$, $f_3 = (\overline{x_1} \wedge \overline{x_3}) \vee x_2$, $f_4 = \overline{f_3}$, and so on.

Input negation and output negation can be executed by card-based protocols without increasing the number of cards. They are executed by just swapping input cards or output cards. Permutations of input variables can also be executed without increasing the number of cards. They can be achieved by just changing the positions of the input values. Therefore, all functions in the same NPN equivalence class can be calculated with the same number of cards.

Theorem 1. *Any three input Boolean functions can be securely calculated without additional cards other than the input cards when we use private operations.*

Proof. When the number of inputs is 3, there are the following 14 NPN-representative functions [50]. (Note that x, y, and z are used to represent input variables.)

1. $NPN_1 = 1$
2. $NPN_2 = x$
3. $NPN_3 = x \vee y$
4. $NPN_4 = x \oplus y$
5. $NPN_5 = x \wedge y \wedge z$
6. $NPN_6 = (x \wedge y \wedge z) \vee (\overline{x} \wedge \overline{y} \wedge \overline{z})$
7. $NPN_7 = (x \wedge y) \vee (x \wedge z)$
8. $NPN_8 = (x \wedge y) \vee (\overline{x} \wedge \overline{y} \wedge z)$
9. $NPN_9 = (x \wedge y \wedge \overline{z}) \vee (x \wedge \overline{y} \wedge z) \vee (\overline{x} \wedge y \wedge z)$
10. $NPN_{10} = (x \wedge \overline{y} \wedge \overline{z}) \vee (\overline{x} \wedge y \wedge \overline{z}) \vee (\overline{x} \wedge \overline{y} \wedge z) \vee (x \wedge y \wedge z) = x \oplus y \oplus z.$
11. $NPN_{11} = (x \wedge y) \vee (x \wedge z) \vee (y \wedge z)$

12. $NPN_{12} = (x \wedge \overline{z}) \vee (y \wedge z)$
13. $NPN_{13} = (x \wedge y \wedge z) \vee (x \wedge \overline{y} \wedge \overline{z})$
14. $NPN_{14} = (x \wedge y) \vee (x \wedge z) \vee (\overline{x} \wedge \overline{y} \wedge \overline{z})$

Among these 14 functions, NPN_1 - NPN_4 depend on less than three inputs. Since any two-variable function can be calculated without additional cards, these functions can be calculated with at most six cards.

We show a calculation protocol for each of the remaining functions.

For NPN_5, $x \wedge y$ can be calculated without additional cards. Then $x \wedge y \wedge z$ can be calculated without additional cards other than the input cards, $x \wedge y$ and z.

NPN_7 can be represented as $NPN_7 = x \wedge (y \vee z)$, thus this function can also be calculated without additional cards.

NPN_{10} can be calculated as $(x \oplus y) \oplus z$ without additional cards.

NPN_{13} can be represented as $NPN_{13} = x \wedge (\overline{y \oplus z})$, thus this function can also be calculated without additional cards.

NPN_{14} can be represented as $NPN_{14} = \overline{x} \oplus (y \vee z)$, thus this function can also be calculated without additional cards.

NPN_6 can be represented as $NPN_6 = (\overline{x \oplus y}) \wedge (\overline{x \oplus z})$. First, calculate $x \oplus y$ and $x \oplus z$ in parallel, where a private random bisection cut is executed to x. Then NOT is applied to each result. Next, calculate AND to these results.

NPN_8 can be represented as $NPN_8 = (\overline{x \oplus y}) \wedge (y \vee z)$. First, calculate $x \oplus y$ and $y \vee z$ in parallel, where a private random bisection cut is executed to y. Then NOT is applied to the first result. Next, calculate AND to these results.

NPN_9 can be represented as $NPN_9 = (\overline{x \oplus y \oplus z}) \wedge (x \vee z)$. First, calculate $x \oplus y$ with preserving input x. Next, calculate $(x \oplus y) \oplus z$ and $x \vee z$ in parallel, where a private random bisection cut is executed to z. Then NOT is applied to the first result. Next, calculate AND to these results.

NPN_{12} can be calculated as follows. First, calculate $x \wedge \overline{z}$ with preserving input z. Next, calculate $y \wedge z$. Then, calculate OR to these results.

NPN_{11} can be represented as

$$NPN_{11} = \begin{cases} z & \text{if } x \oplus y = 1 \\ x & \text{if } x \oplus y = 0 \end{cases}$$

This function can be calculated as follows. First, calculate $x \oplus y$ with preserving input x. Thus, x, z, and $x \oplus y$ are obtained. Then, modify the AND-type protocol as follows.

1. Alice executes private random bisection cut on $x \oplus y$. The obtained value is $x \oplus y \oplus b$, where b is the random value.
2. Bob executes private reveal on $x \oplus y \oplus b$. Bob sets

$$S_2 = \begin{cases} commit(z)||commit(x) \text{ if } x \oplus y \oplus b = 1 \\ commit(x)||commit(z) \text{ if } x \oplus y \oplus b = 0 \end{cases}$$

3. Alice executes a private reverse selection on S_2 using the bit b generated in the private random bisection cut. Let the obtained sequence be S_3. Alice outputs S_3.

The output is $commit(z)$ if ($x \oplus y \oplus b = 1$ and $b = 0$) or ($x \oplus y \oplus b = 0$ and $b = 1$). The case equals to $x \oplus y = 1$. The output is $commit(x)$ if ($x \oplus y \oplus b = 1$ and $b = 1$) or ($x \oplus y \oplus b = 0$ and $b = 0$). The case equals to $x \oplus y = 0$. Thus the result is correct. Therefore, NPN_{11} can also be calculated without additional cards. □

6   Half Adder and Full Adder, and Symmetric Functions

This section first shows a realization of half adder and full adder.

The input and output of the secure half adder are as follows:

- Input: $commit(x)$ and $commit(y)$
- Output: $S = commit(x \oplus y)$ and $C = commit(x \wedge y)$

The half adder is realized by the following steps, whose idea is just the same as the one in [37].

1. Execute XOR protocol with preserving input x. Thus x and $x \oplus y$ are obtained.
2. Obtain $\overline{x \oplus y}$ by swapping the two cards of $x \oplus y$.
3. Execute AND protocol to x and $\overline{x \oplus y}$ with preserving input $\overline{x \oplus y}$. Thus $\overline{x \oplus y}$ and $x \wedge \overline{(x \oplus y)} = x \wedge y$ are obtained.
4. Obtain $x \oplus y$ by swapping the two cards of $\overline{x \oplus y}$.

No additional cards are necessary other than the four input cards. The protocol in [37] needs two additional cards, thus the number of cards is reduced by our protocol.

The input and output of the secure full adder are as follows:

- Input: $commit(x), commit(y)$, and $commit(C_I)$
- Output: $S = commit(x \oplus y \oplus C_I)$, $C_O = commit((x \wedge y) \vee (x \wedge C_I) \vee (y \wedge C_I))$

Since the half adder can be calculated without additional cards, the full adder can also be calculated without additional cards by the following protocol.

1. Add x and y using the half adder. The outputs are $x \oplus y$ and $x \wedge y$.
2. Add C_I to the result $x \oplus y$ using the half adder. The outputs are $x \oplus y \oplus C_I$ and $C_I \wedge (x \oplus y)$.
3. Execute OR protocol to $C_I \wedge (x \oplus y)$ and $x \wedge y$. Since $(C_I \wedge (x \oplus y)) \vee (x \wedge y) = (x \wedge y) \vee (x \wedge C_I) \vee (y \wedge C_I)$, the carry C_O is obtained.

Using the half adder and full adder, calculation of symmetric function can be done by the technique in [37]. n-input symmetric function $f(x_1, x_2, \ldots, x_n)$ depends only on the number of variables such that $x_i = 1$. Let $Y = \sum_{i=1}^{n} x_i$. Then the function f can be written as $f(x_1, x_2, \ldots, x_n) = g(Y)$. When Y is given by a binary representation, $Y = y_k y_{k-1} \cdots y_1$, g can be written as $g(y_1, y_2, \ldots, y_k)$, where $k = \lfloor \log n \rfloor + 1$.

Given input $x_1, x_2, \ldots, x_n$, first obtain the sum of these inputs using the half adder and full adder protocols without additional cards. The sum is obtained as $y_1, y_2, \ldots, y_k$. Then, calculate g using y_is. When $n \leq 7$, $k \leq 3$, thus any three input Boolean function g can be calculated without additional cards. When $n \geq 8$, Y is represented with $k = \lfloor \log n \rfloor + 1$ bits. Since $n - k \geq 4$, at least 8 input cards are unused after y_is are calculated. Any Boolean function can be calculated with four additional cards, thus g can be calculated without additional cards other than the input cards.

Theorem 2. *Any symmetric Boolean function can be securely calculated without additional cards other than the input cards when we use private operations.*

7   Conclusion

This paper showed card-based cryptographic protocols to calculate three input Boolean functions, half adder, full adder, and symmetric functions using private operations. One of the important open problems is obtaining another class of Boolean functions that can be calculated without additional cards using private operations.

Acknowledgement. The authors would like to thank anonymous referees for their careful reading of our manuscript and their many insightful comments and suggestions.

References

1. Abe, Y., Hayashi, Y.I., Mizuki, T., Sone, H.: Five-card and computations in committed format using only uniform cyclic shuffles. New Gener. Comput. **39**(1), 97–114 (2021)
2. den Boer, B.: More efficient match-making and satisfiability *The Five Card Trick*. In: Quisquater, J.-J., Vandewalle, J. (eds.) EUROCRYPT 1989. LNCS, vol. 434, pp. 208–217. Springer, Heidelberg (1990). https://doi.org/10.1007/3-540-46885-4_23
3. Bultel, X., et al.: Physical zero-knowledge proof for Makaro. In: Izumi, T., Kuznetsov, P. (eds.) SSS 2018. LNCS, vol. 11201, pp. 111–125. Springer, Cham (2018). https://doi.org/10.1007/978-3-030-03232-6_8
4. Cheung, E., Hawthorne, C., Lee, P.: Cs 758 project: secure computation with playing cards (2013). http://cdchawthorne.com/writings/secure_playing_cards.pdf
5. Crépeau, C., Kilian, J.: Discreet solitary games. In: Stinson, D.R. (ed.) CRYPTO 1993. LNCS, vol. 773, pp. 319–330. Springer, Heidelberg (1994). https://doi.org/10.1007/3-540-48329-2_27
6. Dumas, J.-G., Lafourcade, P., Miyahara, D., Mizuki, T., Sasaki, T., Sone, H.: Interactive physical zero-knowledge proof for Norinori. In: Du, D.-Z., Duan, Z., Tian, C. (eds.) COCOON 2019. LNCS, vol. 11653, pp. 166–177. Springer, Cham (2019). https://doi.org/10.1007/978-3-030-26176-4_14
7. Dvořák, P., Koucký, M.: Barrington plays cards: the complexity of card-based protocols. arXiv preprint arXiv:2010.08445 (2020)

8. Francis, D., Aljunid, S.R., Nishida, T., Hayashi, Y., Mizuki, T., Sone, H.: Necessary and sufficient numbers of cards for securely computing two-bit output functions. In: Phan, R.C.-W., Yung, M. (eds.) Mycrypt 2016. LNCS, vol. 10311, pp. 193–211. Springer, Cham (2017). https://doi.org/10.1007/978-3-319-61273-7_10

9. Hashimoto, Y., Nuida, K., Shinagawa, K., Inamura, M., Hanaoka, G.: Toward finite-runtime card-based protocol for generating hidden random permutation without fixed points. IEICE Trans. Fund. Electron. Commun. Comput. Sci. **101-A**(9), 1503–1511 (2018)

10. Hashimoto, Y., Shinagawa, K., Nuida, K., Inamura, M., Hanaoka, G.: Secure grouping protocol using a deck of cards. IEICE Trans. Fund. Electron. Commun. Comput. Sci. **101**(9), 1512–1524 (2018)

11. Ibaraki, T., Manabe, Y.: A more efficient card-based protocol for generating a random permutation without fixed points. In: Proceedings of 3rd International Conference on Mathematics and Computers in Sciences and in Industry (MCSI 2016), pp. 252–257 (2016)

12. Ishikawa, R., Chida, E., Mizuki, T.: Efficient card-based protocols for generating a hidden random permutation without fixed points. In: Calude, C.S., Dinneen, M.J. (eds.) UCNC 2015. LNCS, vol. 9252, pp. 215–226. Springer, Cham (2015). https://doi.org/10.1007/978-3-319-21819-9_16

13. Kastner, J., Koch, A., Walzer, S., Miyahara, D., Hayashi, Y., Mizuki, T., Sone, H.: The minimum number of cards in practical card-based protocols. In: Takagi, T., Peyrin, T. (eds.) ASIACRYPT 2017, Part III. LNCS, vol. 10626, pp. 126–155. Springer, Cham (2017). https://doi.org/10.1007/978-3-319-70700-6_5

14. Koch, A.: The landscape of optimal card-based protocols. IACR Cryptology ePrint Archive, Report 2018/951 (2018)

15. Koch, A., Schrempp, M., Kirsten, M.: Card-based cryptography meets formal verification. New Gener. Comput. **39**(1), 115–158 (2021)

16. Koch, A., Walzer, S.: Private function evaluation with cards. Cryptology ePrint Archive, Report 2018/1113 (2018). https://eprint.iacr.org/2018/1113

17. Koch, A., Walzer, S.: Foundations for actively secure card-based cryptography. In: Proceedings of 10th International Conference on Fun with Algorithms (FUN 2020). Schloss Dagstuhl-Leibniz-Zentrum für Informatik (2020)

18. Koch, A., Walzer, S., Härtel, K.: Card-Based cryptographic protocols using a minimal number of cards. In: Iwata, T., Cheon, J.H. (eds.) ASIACRYPT 2015. LNCS, vol. 9452, pp. 783–807. Springer, Heidelberg (2015). https://doi.org/10.1007/978-3-662-48797-6_32

19. Kurosawa, K., Shinozaki, T.: Compact card protocol. In: Proceedings of 2017 Symposium on Cryptography and Information Security (SCIS 2017), pp. 1A2-6 (2017). (in Japanese)

20. Lafourcade, P., Miyahara, D., Mizuki, T., Sasaki, T., Sone, H.: A physical ZKP for Slitherlink: how to perform physical topology-preserving computation. In: Heng, S.-H., Lopez, J. (eds.) ISPEC 2019. LNCS, vol. 11879, pp. 135–151. Springer, Cham (2019). https://doi.org/10.1007/978-3-030-34339-2_8

21. Manabe, Y., Ono, H.: Secure card-based cryptographic protocols using private operations against malicious players. In: Maimut, D., Oprina, A.-G., Sauveron, D. (eds.) SecITC 2020. LNCS, vol. 12596, pp. 55–70. Springer, Cham (2021). https://doi.org/10.1007/978-3-030-69255-1_5

22. Marcedone, A., Wen, Z., Shi, E.: Secure dating with four or fewer cards. IACR Cryptology ePrint Archive, Report 2015/1031 (2015)

23. Miyahara, D., Hayashi, Y.i., Mizuki, T., Sone, H.: Practical card-based implementations of Yao's millionaire protocol. Theor. Comput. Sci. **803**, 207–221 (2020)

24. Miyahara, D., et al.: Card-based ZKP protocols for Takuzu and Juosan. In: Proceedings of 10th International Conference on Fun with Algorithms (FUN 2020). Schloss Dagstuhl-Leibniz-Zentrum für Informatik (2020)

25. Miyahara, D., Sasaki, T., Mizuki, T., Sone, H.: Card-based physical zero-knowledge proof for Kakuro. IEICE Trans. Fund. Electron. Commun. Comput. Sci. **102**(9), 1072–1078 (2019)

26. Mizuki, T.: Applications of card-based cryptography to education. IEICE Techinical Report ISEC2016-53, pp. 13–17 (2016). (in Japanese)

27. Mizuki, T.: Card-based protocols for securely computing the conjunction of multiple variables. Theor. Comput. Sci. **622**, 34–44 (2016)

28. Mizuki, T., Asiedu, I.K., Sone, H.: Voting with a logarithmic number of cards. In: Mauri, G., Dennunzio, A., Manzoni, L., Porreca, A.E. (eds.) UCNC 2013. LNCS, vol. 7956, pp. 162–173. Springer, Heidelberg (2013). https://doi.org/10.1007/978-3-642-39074-6_16

29. Mizuki, T., Kumamoto, M., Sone, H.: The five-card trick can be done with four cards. In: Wang, X., Sako, K. (eds.) ASIACRYPT 2012. LNCS, vol. 7658, pp. 598–606. Springer, Heidelberg (2012). https://doi.org/10.1007/978-3-642-34961-4_36

30. Mizuki, T., Shizuya, H.: Computational model of card-based cryptographic protocols and its applications. IEICE Trans. Fund. Electron. Commun. Comput. Sci. **100**(1), 3–11 (2017)

31. Mizuki, T., Sone, H.: Six-card secure AND and four-card secure XOR. In: Deng, X., Hopcroft, J.E., Xue, J. (eds.) FAW 2009. LNCS, vol. 5598, pp. 358–369. Springer, Heidelberg (2009). https://doi.org/10.1007/978-3-642-02270-8_36

32. Mizuki, T., Uchiike, F., Sone, H.: Securely computing XOR with 10 cards. Australas. J. Comb. **36**, 279–293 (2006)

33. Murata, S., Miyahara, D., Mizuki, T., Sone, H.: Efficient generation of a card-based uniformly distributed random derangement. In: Uehara, R., Hong, S.-H., Nandy, S.C. (eds.) WALCOM 2021. LNCS, vol. 12635, pp. 78–89. Springer, Cham (2021). https://doi.org/10.1007/978-3-030-68211-8_7

34. Nakai, T., Misawa, Y., Tokushige, Y., Iwamoto, M., Ohta, K.: How to solve millionaires' problem with two kinds of cards. New Gener. Comput. **39**(1), 73–96 (2021)

35. Nakai, T., Shirouchi, S., Iwamoto, M., Ohta, K.: Four cards are sufficient for a card-based three-input voting protocol utilizing private permutations. In: Shikata, J. (ed.) ICITS 2017. LNCS, vol. 10681, pp. 153–165. Springer, Cham (2017). https://doi.org/10.1007/978-3-319-72089-0_9

36. Niemi, V., Renvall, A.: Secure multiparty computations without computers. Theor. Comput. Sci. **191**(1), 173–183 (1998)

37. Nishida, T., Hayashi, Y., Mizuki, T., Sone, H.: Card-based protocols for any Boolean function. In: Jain, R., Jain, S., Stephan, F. (eds.) TAMC 2015. LNCS, vol. 9076, pp. 110–121. Springer, Cham (2015). https://doi.org/10.1007/978-3-319-17142-5_11

38. Nishida, T., Hayashi, Y., Mizuki, T., Sone, H.: Securely computing three-input functions with eight cards. IEICE Trans. Fund. Electron. Commun. Comput. Sci. **98**(6), 1145–1152 (2015)

39. Nishimura, A., Nishida, T., Hayashi, Y., Mizuki, T., Sone, H.: Card-based protocols using unequal division shuffles. Soft. Comput. **22**(2), 361–371 (2018)

40. Ono, H., Manabe, Y.: Efficient card-based cryptographic protocols for the millionaires' problem using private input operations. In: Proceedings of 13th Asia Joint Conference on Information Security (AsiaJCIS 2018), pp. 23–28 (2018)

41. Ono, H., Manabe, Y.: Card-based cryptographic protocols with the minimum number of rounds using private operations. In: Pérez-Solà, C., Navarro-Arribas, G., Biryukov, A., Garcia-Alfaro, J. (eds.) DPM/CBT -2019. LNCS, vol. 11737, pp. 156–173. Springer, Cham (2019). https://doi.org/10.1007/978-3-030-31500-9_10

42. Ono, H., Manabe, Y.: Card-based cryptographic logical computations using private operations. New Gener. Comput. **39**(1), 19–40 (2021)

43. Robert, L., Miyahara, D., Lafourcade, P., Mizuki, T.: Physical zero-knowledge proof for Suguru puzzle. In: Devismes, S., Mittal, N. (eds.) SSS 2020. LNCS, vol. 12514, pp. 235–247. Springer, Cham (2020). https://doi.org/10.1007/978-3-030-64348-5_19

44. Robert, L., Miyahara, D., Lafourcade, P., Mizuki, T.: Interactive physical ZKP for connectivity: applications to nurikabe and hitori. In: Proceedings of 17th International Conference on Computability in Europe (CiE 2021). LNCS (2021)

45. Ruangwises, S., Itoh, T.: AND protocols using only uniform shuffles. In: van Bevern, R., Kucherov, G. (eds.) CSR 2019. LNCS, vol. 11532, pp. 349–358. Springer, Cham (2019). https://doi.org/10.1007/978-3-030-19955-5_30

46. Ruangwises, S., Itoh, T.: Securely computing the n-variable equality function with $2n$ cards. In: Chen, J., Feng, Q., Xu, J. (eds.) TAMC 2020. LNCS, vol. 12337, pp. 25–36. Springer, Cham (2020). https://doi.org/10.1007/978-3-030-59267-7_3

47. Ruangwises, S., Itoh, T.: Physical zero-knowledge proof for number link puzzle and k vertex-disjoint paths problem. New Gener. Comput. **39**(1), 3–17 (2021)

48. Ruangwises, S., Itoh, T.: Physical zero-knowledge proof for ripple effect. In: Uehara, R., Hong, S.-H., Nandy, S.C. (eds.) WALCOM 2021. LNCS, vol. 12635, pp. 296–307. Springer, Cham (2021). https://doi.org/10.1007/978-3-030-68211-8_24

49. Sasaki, T., Miyahara, D., Mizuki, T., Sone, H.: Efficient card-based zero-knowledge proof for Sudoku. Theor. Comput. Sci. **839**, 135–142 (2020)

50. Sasao, T., Butler, J.T.: Progress in Applications of Boolean Functions. Morgan and Claypool Publishers, San Franciso (2010)

51. Shinagawa, K., Mizuki, T.: The six-card trick: secure computation of three-input equality. In: Lee, K. (ed.) ICISC 2018. LNCS, vol. 11396, pp. 123–131. Springer, Cham (2019). https://doi.org/10.1007/978-3-030-12146-4_8

52. Shinagawa, K., Nuida, K.: A single shuffle is enough for secure card-based computation of any Boolean circuit. Discrete App. Math. **289**, 248–261 (2021)

53. Shinoda, Y., Miyahara, D., Shinagawa, K., Mizuki, T., Sone, H.: Card-based covert lottery. In: Maimut, D., Oprina, A.-G., Sauveron, D. (eds.) SecITC 2020. LNCS, vol. 12596, pp. 257–270. Springer, Cham (2021). https://doi.org/10.1007/978-3-030-69255-1_17

54. Shirouchi, S., Nakai, T., Iwamoto, M., Ohta, K.: Efficient card-based cryptographic protocols for logic gates utilizing private permutations. In: Proceedings of 2017 Symposium on Cryptography and Information Security (SCIS 2017), p. 1A2-2 (2017). (in Japanese)

55. Stiglic, A.: Computations with a deck of cards. Theor. Comput. Sci. **259**(1), 671–678 (2001)

56. Takashima, K., et al.: Card-based protocols for secure ranking computations. Theor. Comput. Sci. **845**, 122–135 (2020)

57. Takashima, K., Miyahara, D., Mizuki, T., Sone, H.: Actively revealing card attack on card-based protocols. Nat. Comput. 1–14 (2021)

58. Toyoda, K., Miyahara, D., Mizuki, T., Sone, H.: Six-card finite-runtime XOR protocol with only random cut. In: Proceedings of the 7th ACM Workshop on ASIA Public-Key Cryptography, pp. 2–8 (2020)

59. Watanabe, Y., Kuroki, Y., Suzuki, S., Koga, Y., Iwamoto, M., Ohta, K.: Card-based majority voting protocols with three inputs using three cards. In: Proceedings of 2018 International Symposium on Information Theory and its Applications (ISITA), pp. 218–222. IEEE (2018)

60. Yasunaga, K.: Practical card-based protocol for three-input majority. IEICE Transactions on Fundamentals of Electronics, Commun. Comput. Sci. **E103.A**(11), 1296–1298 (2020). https://doi.org/10.1587/transfun.2020EAL2025

Königsberg Sightseeing: Eulerian Walks in Temporal Graphs

Andrea Marino[1(✉)] and Ana Silva[2(✉)]

[1] Dipartimento di Sistemi, Informatica, Applicazioni,
Università degli Studi di Firenze, Firenze, Italy
andrea.marino@unifi.it

[2] Departamento de Matemática, Universidade Federal do Ceará,
Fortaleza, CE, Brazil
anasilva@mat.ufc.br

Abstract. An Eulerian walk (or Eulerian trail) is a walk (resp. trail) that visits every edge of a graph G at least (resp. exactly) once. This notion was first discussed by Leonhard Euler while solving the famous Seven Bridges of Königsberg problem in 1736. But what if Euler had to take a bus? In a temporal graph (G, λ), with $\lambda : E(G) \to 2^{[\tau]}$, an edge $e \in E(G)$ is available only at the times specified by $\lambda(e) \subseteq [\tau]$, in the same way the connections of the public transportation network of a city or of sightseeing tours are available only at scheduled times. In this scenario, even though several translations of Eulerian trails and walks are possible in temporal terms, only a very particular variation has been exploited in the literature, specifically for infinite dynamic networks (Orlin, 1984). In this paper, we deal with temporal walks, local trails, and trails, respectively referring to edge traversal with no constraints, constrained to not repeating the same edge in a single timestamp, and constrained to never repeating the same edge throughout the entire traversal. We show that, if the edges are always available, then deciding whether (G, λ) has a temporal walk or trail is polynomial, while deciding whether it has a local trail is NP-complete even if it has lifetime 2. In contrast, in the general case, solving any of these problems is NP-complete, even under very strict hypotheses.

1 Introduction

An Eulerian walk (or Eulerian trail) is a walk (resp. trail) that visits every edge of a graph G at least (resp. exactly) once. The Eulerian trail notion was first discussed by Leonhard Euler while solving the famous Seven Bridges of Königsberg problem in 1736, where one wanted to pass by all the bridges over

Partially supported by MIUR under PRIN Project n. 20174LF3T8 AHeAD (Efficient Algorithms for HArnessing Networked Data), University of Florence under Project GRANTED (GRaph Algorithms for Networked TEmporal Data), CNPq 303803/2020-7 and 437841/2018-9, FUNCAP/CNPq PNE-0112-00061.01.00/16, and STIC-AMSUD 360/2019 - 88881.197438/2018-01.

P. Flocchini and L. Moura (Eds.): IWOCA 2021, LNCS 12757, pp. 485–500, 2021.
https://doi.org/10.1007/978-3-030-79987-8_34

the river Preger without going twice over the same bridge. Imagine now a similar problem, where you have a set of tourist sights linked by possible routes. If the routes themselves are also of interest, a sightseeing tourism company might want to plan visits on different days that cover all the routes. One could do that with no constraints at all (thus performing a walk), or with the very strict constraint of never repeating a route (thus getting a trail), or constraining oneself to at least not repeating the same route on the same day (thus getting what we called a local trail). If we further assume that some routes might not be always accessible, we then get distinct problems defined on temporal graphs.

In a temporal graph (G, λ), with $\lambda : E(G) \to 2^{[\tau]}$, an edge $e \in E(G)$ is available only at the times specified by $\lambda(e) \subseteq [\tau]$, in the same way the connections of the public transportation network of a city or of sightseeing tours are available only at scheduled times. In this scenario, paths and walks are valid only if they traverse a sequence of adjacent edges $e_1, \ldots, e_k$ at non-decreasing times $t_1 \leq \ldots \leq t_k$, respectively, with $t_i \in \lambda(e_i)$ for every $i \in [k]$ (similarly, one can consider strictly increasing sequences, i.e. with $t_1 < \ldots < t_k$).

Several translations of Eulerian trails and walks are possible in temporal terms, depending on the constraints we consider. In particular, we study the following variations. Below, all the walks and trails are implicitly considered to be temporal, as defined in the previous paragraph.

Problem 1. Given a temporal graph (G, λ), we consider the following problems:

- EULERIAN WALK: deciding whether (G, λ) has an Eulerian walk, i.e. a walk traversing each edge of G at least once.
- EULERIAN LOCAL TRAIL: deciding whether (G, λ) has an Eulerian local trail, i.e. a walk traversing each edge of G at least once, and at most once in each timestamp.
- EULERIAN TRAIL: deciding whether (G, λ) has an Eulerian trail, i.e. a walk traversing each edge of G exactly once.

We also consider the related problems where the walks/trails are closed (first vertex equal to the last one), respectively referring to them as EULERIAN CLOSED WALK, EULERIAN LOCAL TOUR, and EULERIAN TOUR. Finally, for all of the above problems, we add the prefix STRICT to refer to the variation in which walks must be strictly increasing sequences of edges. Observe that, when $\tau = 1$, then both EULERIAN TRAIL and EULERIAN LOCAL TRAIL degenerate into the original formulation of the Seven Bridges of Königsberg problem. This is why we think they appear to be more natural adaptations of the static version of the problem.

The research on temporal graphs has attracted a lot of attention in the past decade (we refer the reader to the surveys [17, 19] and the seminal paper [16]), and temporal graphs appeared also under different names, e.g. as time-varying graphs [5], as evolving networks [3], and as link streams [17]. Even after all the received interest, surprisingly enough the above problems have received very little attention, as we discuss in what follows.

One of the concepts closest to ours is the one defined by Orlin in [21], where he gives a polynomial algorithm to check the existence of an Eulerian closed walk (i.e. a *tour*) in dynamic graphs. However, the dynamic graph model is quite different from the temporal graph model used in this paper, as pointed out in [19]. Indeed, looking at the corresponding time-expanded graph related to [21], temporal edges can go back in time and the graph is infinite. Nevertheless, the results presented there seemed to point towards the polinomiality of the problems investigated here, as observed in [19]: "the results proved for it [the dynamic graph model] are resounding and possibly give some first indications of what to expect when adding to combinatorial optimization problems a time dimension". We found however that this is not the case, as we will show that even EULERIAN WALK turns out to be much harder on temporal graphs. Taking inspiration in [21], we also define a dynamic-based temporal graph as a temporal graph whose edges are always available, and we analyze the complexity of the above problems on these particular instances.

In this paper we prove the following results. These are summarized in Table 1, which also reports some recent results that will be discussed shortly.

Theorem 1. *Given a temporal graph* (G, λ) *with lifetime* τ,

1. EULERIAN WALK *is* NP-*complete, even if either each snapshot of* (G, λ) *is a forest of constant size, or each edge appears at most 3 times. Also, it is polynomial-time solvable if* (G, λ) *is dynamic-based, and is in* XP *when parameterized by* τ.
2. EULERIAN LOCAL TRAIL *is* NP-*complete for each* $\tau \geq 2$, *even if* (G, λ) *is dynamic-based.*
3. EULERIAN TRAIL *is* NP-*complete for each* $\tau \geq 2$. *It is polynomial if* (G, λ) *is dynamic-based.*

Same applies to tours, i.e. EULERIAN CLOSED WALK, EULERIAN LOCAL TOUR, *and* EULERIAN TOUR.

Theorem 1 gives a complete taxonomy of our problems, also focusing on the possibility of getting polynomial algorithms when we have a small lifetime τ. In particular, for EULERIAN TRAIL and EULERIAN LOCAL TRAIL, since they become polynomial when $\tau = 1$, the bound for τ is optimal, giving us a complete dichotomy with respect to the lifetime of (G, λ), excluding the possibility of any FPT algorithm with parameter τ unless P = NP. In contrast, EULERIAN WALK is easily solvable for every fixed τ, showing that walks are easier than trails even on the temporal context.

EULERIAN WALK is related to the TEXP problem [20], which consists of, given a temporal graph (G, λ), finding a temporal walk that visits all vertices in G (possibly, more than once) whose arrival time is minimum. In [20], they prove that TEXP is NP-complete and even not approximable unless P = NP; this is in stark contrast with the static version of the problem, which can be trivially solved in linear time. A lot of research has been devoted to temporal exploration, e.g. bounding the arrival time of such walks in special instances [9, 10] and extending

Table 1. Our results concerning Problem 1. For general temporal graphs (first row) and for dynamic-based temporal graphs (second row). [†] corresponds to deciding whether G is connected. [*] corresponds to deciding whether G has an Eulerian trail. All the results in the first row, when $\tau = \Omega(m)$, with $m = |E(G)|$, extend to STRICT EULERIAN WALK, STRICT EULERIAN LOCAL TRAIL, STRICT EULERIAN TRAIL (if $\tau < m$, then the answer is trivial).

	WALK	LOCAL TRAIL	TRAIL
GENERAL	NP-c	NP-c for $\tau \geq 2$	NP-c for $\tau \geq 2$
	XP by τ		NP-c if $tw(G) = 2$ [4]
			FPT by $k + imw(G, \lambda)$ [4]
DYNAMIC-BASED	Poly[†]	NP-c for $\tau \geq 2$	Poly[*]

previous results in the case of non-strictly increasing paths [11]. In [1], they proved that TEXP is NP-complete even when restricted to temporal stars in which each edge appears at most k times, for all fixed $k \geq 6$. On the other hand, they showed that, if each edge appears at most $k = 3$ times, then the problem is polynomial-time solvable on temporal stars. Observe that in a star, passing by all the leaves translates also into passing by all the edges. Therefore their result implies already NP-completeness for EULERIAN WALK with the same constraints as before. Our proof complements the latter result as we prove that EULERIAN WALK is NP-complete on general temporal graphs even when every edge appears at most 3 times.

Up to our knowledge, there is only one other paper that investigates problems similar to EULERIAN TRAIL and EULERIAN LOCAL TRAIL, that will appear in the same volume as this paper and focuses on the EULERIAN TRAIL variation, giving independent and broadly different results [4]. In particular, they prove that EULERIAN TRAIL is NP-complete even if each edge appears at most k times, for every fixed $k \geq 3$. Observe that our result improves that to $k = 2$, since we prove it is NP-complete even if the lifetime is 2. Nevertheless, even though their reduction produces a temporal graph with unbounded lifetime, it also gives a base graph with very simple structure (a set of triangles intersecting in a single vertex, which is a graph with treewidth 2). In addition, they also introduce a parameter for temporal graphs, that they called *interval-membership width* (denoted by $imw(G, \lambda)$), and provide an FPT algorithm parametrized by k plus $imw(G, \lambda)$, as also reported in Table 1.

It is important to remark that none of the variations we considered immediately implies any of the others. We will show indeed that the property of being Eulerian for the static base graph G is in general a necessary but not sufficient condition for the existence of an Eulerian trail, becoming sufficient only if we restrict to dynamic-based temporal graphs. In the case of Eulerian local trail, we will see that this property is not even necessary. In addition, if only strictly increasing temporal walks/trails are allowed, then our reductions for the first row of Table 1 can be easily modified, thus giving NP-completeness results also in this case. Observe that in this case a necessary condition for a positive answer is $\tau \geq |E(G)|$; this is why we do not have the same bounds for τ as in Theorem 1.

Corollary 1. *(i)* STRICT EULERIAN WALK, *(ii)* STRICT EULERIAN LOCAL TRAIL, *(iii)* STRICT EULERIAN TRAIL *are* NP-*complete in a general temporal graph* (G, λ) *with lifetime* $\tau = \Omega(|E(G)|)$.

Also in the case of dynamic-based temporal graphs (second row of Table 1), the polynomiality is preserved for strictly increasing Eulerian walks and Eulerian trails and we leave open the question whether STRICT EULERIAN LOCAL TRAIL is still NP-complete on dynamic-based temporal graphs.

Finally, as a byproduct of our reductions we get the following result about static graphs, which can be of independent interest.

Corollary 2. *Given a graph* G, *deciding whether the edges of* G *can be covered with two trails is* NP-*complete.*

Further Related Work. When considering dynamic-based temporal graphs, as edges are assumed to be always available during the lifetime τ, we could relate our problems to several other problems on static graphs. A closely related one would be the Chinese Postman problem, where the edges of the graph have positive weights and one wants to find an Eulerian closed walk on G with minimum weight; in other words, one wants to add copies of existing edges in order to obtain an Eulerian graph of minimum sum weight. Even if we regard the Chinese Postman problem where the weights are all equal to 1, this is very different from our approach since for us, repetition of a long common trail in different snapshots does not make the solution worse, while it would considering the Chinese Postman problem. It is easy to see though that the solution for the Chinese Postman would give us an upper bound for the amount of time spent on an Eulerian local tour of a dynamic-based graph, as we could start a new trail on a new snapshot whenever an edge repetition was detected. The Chinese Postman problem is largely known to be polynomial [15], and some variations that take time into consideration have been investigated, mostly from the practical point of view (see e.g. [6,22,23]), but none of which is equivalent to our problem.

The problem of trying to obtain an Eulerian subgraph (as opposed to a supergraph, as was the case in the previous paragraph) has also been studied. In [7], the authors study a family of problems where the goal is to make a static graph Eulerian by a minimum number of deletions. They completely classify the parameterized complexity of various versions of the problem: vertex or edge deletions, undirected or directed graphs, with or without the requirement of connectivity. Also in [12], the parameterized complexity of the following Euler subgraph problems is studied: (i) Largest Euler Subgraph: for a given graph G and integer parameter k, does G contain an induced Eulerian subgraph with at least k vertices?; and (ii) Longest Circuit: for a given graph G and integer parameter k, does G contain an Eulerian subgraph with at least k edges?

EULERIAN LOCAL TRAIL on dynamic-based graphs is actually more closely related to the problem of covering the edges of a graph with the minimum number of (not necessarily disjoint) trails, whereas the aforementioned problems are more concerned with either minimizing edge repetitions or maximizing the subgraph covered by a single trail. Even if the trail cover problem can be so naturally defined and involve such a basic structure as trail, up to our knowledge it has not been previously investigated yet. Note that EULERIAN LOCAL TRAIL is slightly different from trail cover, since we also require that together the trails form a walk. In any case, a small modification of our proof of Theorem 1.2 implies that deciding whether the edges of a graph can be covered with at most two trails is NP-complete (Corollary 2). Interestingly enough, the vertex version of this problem, namely the path cover problem, has been largely investigated (see e.g. [2,14,18]).

Preliminaries. A static graph G has an Eulerian tour (trail) if and only if G has at most one non-trivial component and all the vertices have even degree (at most two vertices have odd degree). A graph is called Eulerian if it has an Eulerian tour. We use standard notation for graphs and we use and extend the notation in [19]. A temporal graph is a graph together with a function on the edges saying when each edge is active; more formally, a *temporal graph* is a pair (G, λ), where $\lambda : E(G) \to 2^{\mathbb{N} - \{0\}}$. Here, we consider only finite temporal graphs, i.e., graphs such that $\max \bigcup_{e \in E(G)} \lambda(e)$ is defined. This value is called the *lifetime of* (G, λ) and denoted by τ. Given $i \in [\tau]$, we define the *snapshot* G_i as being the subgraph of G containing exactly the edges active in time i; more formally, $V(G_i) = V(G)$ and $E(G_i) = \{e \in E(G) \mid i \in \lambda(e)\}$.

Given vertices v_0, v_k in a graph G, a v_0, v_k-*walk* in G is an alternating sequence $(v_0, e_1, v_1, \ldots, e_k, v_k)$ of vertices and edges such that e_i goes from v_{i-1} to v_i for $i \in \{1, \ldots, k\}$. We define a walk in a temporal graph similarly, except that a walk cannot go back in time. More formally, given a temporal graph (G, λ) and a v_0, v_k-walk $W = (v_0, e_1, v_1, \ldots, e_k, v_k)$, we say that W is a *temporal v_0, v_k-walk* if $\lambda(e_1) \le \lambda(e_2) \le \ldots \le \lambda(e_k)$. It is *closed* if it starts and finishes on the same vertex of G, i.e., if $v_0 = v_k$.

We say that a temporal walk W is a *local trail* if there are no two occurrences of the same edge of G in the same snapshot, i.e., if W restricted to G_i is a trail in G for every $i \in [\tau]$. We say that W is a *trail* if there are no two occurrences of the same edge of G in W. A closed (local) trail is also called a *(local) tour*. Finally, a temporal walk W is called *Eulerian* if at least one copy of each edge of G appears at least once in W. Observe that, by definition, an Eulerian trail visits every edge exactly once.

A *dynamic-based graph* is a temporal graph (G, λ) where the edges are always available, i.e. $\lambda(e) = [\tau]$ for each $e \in E(G)$.[1] We denote a dynamic-based graph simply by $(G, [\tau])$ where τ is the lifetime of the temporal graph.

---

[1] This is the reason why we use the term dynamic-based, as they are similar to the dynamic networks used in [21] when studying Eulerian trails, except that edges cannot go back in time and the lifetime is finite.

2 Eulerian Walk

In this section we focus on EULERIAN WALK, i.e. deciding if there is a temporal walk passing by each edge at least once, proving the results in Item 1 in Theorem 1, summarized in the first column of Table 1.

In particular, a preliminary result concerns the case where the lifetime τ is bounded. It consists basically of checking whether there is a choice of connected components $H_1, \ldots, H_\tau$, one for each timestamp i, that together cover all the edges of G and is such that H_i intersects H_{i+1}, for each $i \in [\tau - 1]$.

Lemma 1. *Given a temporal graph (G, λ) with fixed lifetime τ, solving EULE-RIAN WALK on (G, λ) can be done in time $O((n + m) \cdot n^{\tau-1})$, where $n = |V(G)|$ and $m = |E(G)|$.*

Proof. Let $G_1, \cdots, G_\tau$ be the snapshots of G; note first that if $E(G_i)$ is empty, then this snapshot can be suppressed. Our problem reduces to deciding whether there is a choice of connected components $H_1, \ldots, H_\tau$, one for each timestamp i, that together cover all the edges of G and is such that H_i intersects H_{i+1}, for each $i \in [\tau - 1]$. As for each $i \in [\tau]$, there are at most n nodes in the intersections, there are at most $O(n^{\tau-1})$ choices. For each choice the test can be done in $O(\tau(n + m))$, obtaining $O(\tau(n + m)n^{\tau-1})$ running time, which is $O((n + m)n^{\tau-1})$.

In the following, we show that when τ is unbounded, deciding whether (G, λ) admits an Eulerian walk is NP-complete by reducing from 3-SAT. This is best possible because of the above lemma.

Theorem 2. *Given a temporal graph (G, λ), deciding whether (G, λ) admits an Eulerian walk is NP-complete, even if either each snapshot of (G, λ) is a forest of constant size, or each edge appears in at most 3 snapshots.*

Proof. We make a reduction from 3-SAT. Let ϕ be a 3-CNF formula on variables $\{x_1, \cdots, x_n\}$ and clauses $\{c_1, \cdots, c_m\}$, and construct G as follows. For each clause c_i, add vertices $\{a_i, b_i\}$ to G and edge $a_i b_i$. Now consider a variable x_i, and let $c_{i_1}, \cdots, c_{i_p}$ be the clauses containing x_i positively, and $c_{j_1}, \cdots, c_{j_q}$ be the clauses containing x_i negatively. Add two new vertices $x_i, \overline{x}_i$ to G, and edges $\{x_i a_{i_k} \mid k \in [p]\} \cup \{\overline{x}_i a_{j_k} \mid k \in [q]\}$; denote the spanning subgraph of G formed by these edges by H_i, and let H_i' be equal to H_i together with edges $\{a_i b_i \mid i \in \{i_1, \cdots, i_p, j_1, \cdots, j_q\}\}$. We can suppose that $\{i_1, \cdots, i_p\} \cap \{j_1, \cdots, j_q\} = \emptyset$ as otherwise the clauses in the intersection would always be trivially valid; thus we get that H_i, H_i' are forests. Finally, add a new vertex T and make it adjacent to every vertex in $\{x_i, \overline{x}_i \mid i \in [n]\}$.

We now describe the snapshots of (G, λ). See Fig. 1 to follow the construction. We first build 2 consecutive snapshots in (G, λ) related to x_i, for each $i \in [n]$. The first one is equal to H_i', and the second one contains exactly the edges $\{Tx_i, T\overline{x}_i, Tx_{i+1}, T\overline{x}_{i+1}\}$ if $i < n$, and if $i = n$, then the second snapshot is equal to $G - \{a_j b_j \mid j \in [m]\}$; this can be done because this subgraph is connected.

Denote by S_i^1, S_i^2 the first and second snapshot of x_i, for each $i \in [n]$. Put these snapshots consecutively in timestamps 1 through $2n$, in the order of the indexing of the variables. For now, observe that only the last snapshot might not be a forest; this will be fixed later. We now prove that ϕ is a satisfiable formula if and only if (G, λ) admits an Eulerian walk.

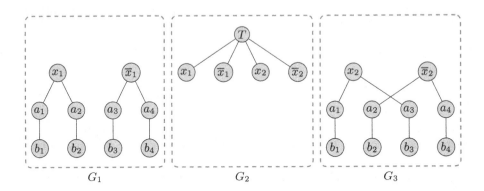

Fig. 1. First three snapshots of the construction. For simplicity, we represent only the non-trivial components of each snapshot. In this example, we have c_1 containing $(x_1 \vee x_2)$, c_2 containing $(x_1 \vee \overline{x}_2)$, c_3 containing $(\overline{x}_1 \vee x_2)$, and c_4 containing $(\overline{x}_1 \vee \overline{x}_2)$.

Because of space constraints, the proof of equivalence is not presented, but we believe that the construction itself is already convincing. Also, the reader can verify that the number of appearances of each edge is at most 3. As for the constraint on the structure of the graph, it can be done by repeating the same pattern used for the first $2n - 1$ snapshots, and the fact that 3-SAT is NP-complete even if each variable appears at most three times [8].   □

If instead we are considering strictly increasing walks, a small modification of our construction will also imply NP-completeness, hence proving Corollary 1(i). Indeed, it suffices to relate each variable x_i to a window big enough to ensure we will be able to visit all the edges of the considered component. Because each variable appears at most three times, one can see that it is enough that the edges are available for a period of 12 timestamps. So, our previous snapshot S_i^1 remains available for 12 consecutive timestamps, after which we will make S_i^2 available for 2 timestamps. Because the spare time can never be used to go from the component containing x_i to the component containing $\overline{x}_i$, the same arguments used in Theorem 2 still hold.

Now, if we consider a dynamic-based graph (G, λ), since all the edges are active throughout its lifetime, we clearly have that there exists an Eulerian walk if and only if G is connected, as highlighted by the following Lemma.

Lemma 2. EULERIAN WALK *is polynomial for dynamic-based temporal graphs.*

By Lemma 1, Theorem 2, and Lemma 2, we obtain Item 1 of Theorem 1. Finally, note that if one is interested in closed walks instead, not only our NP-completeness reduction can be adapted in order to ensure that we can always go back to the initial vertex, but also the complexity results still hold.

3   Eulerian Local Tours and Trails

In this section we focus on Item 2 of Theorem 1. In the whole section, we will focus on dynamic-based temporal graphs as the hardness results for general temporal graphs are implied by the ones we prove for this restricted class. After the preliminary result in Lemma 3, we focus on proving the hardness result for the problem of deciding whether $(G, [2])$ has an Eulerian local tour, explaining the construction behind our reduction from NAE 3-SAT, whose correctness is proved in Theorem 3. We also argue that, if G is a cubic graph, then being Hamiltonian is a necessary but not sufficient condition for $(G, [2])$ to admit an Eulerian local tour, arguing the need of an *ad hoc* reduction for our problem. As the reduction in Theorem 3 focuses on solving EULERIAN LOCAL TOUR for $\tau = 2$, in Corollary 3 we extend this result to each fixed τ and to trails, thus completing the proof of Item 2 of Theorem 1. The following lemma helps us in our proof.

Lemma 3. *Let G be a graph. If $(G, [2])$ has an Eulerian local tour T, then T restricted to timestamp i must pass by all vertices of odd degree in G, for each $i \in [2]$.*

Proof. For each $i \in [2]$, denote by T_i the trail in G equal to T restricted to timestamp i, and suppose, by contradiction, that $u \in V(G)$ is a vertex with odd degree not contained in T_1. Because T is a temporal tour, observe that T_1 is a trail in G starting at some s and finishing at some t, and T_2 is a trail in G starting at t and finishing at s, with possibly $s = t$. This means that the subgraph of G formed by the edges of T_2 is such that every $x \in V(G) \setminus \{s, t\}$ has even degree. This is a contradiction because, since no edge incident to u is visited in T_1, we get that all the edges incident to u must be visited in T_2, i.e., u would have odd degree in T_2. The same argument holds in case u is not in T_2, and the lemma follows.   □

A simple consequence of the above lemma is that, as previously said, if G is cubic, then G must be Hamiltonian in order for $(G, [2])$ to have an Eulerian local tour. Since deciding whether a cubic graph is Hamiltonian is NP-complete [13], this hints towards the NP-completeness of the problem. However, since the other way around is not necessarily true (see e.g. the graph in Fig. 2), we need an explicit reduction. Indeed, the construction in Fig. 2 shows us that we might need an arbitrarily large lifetime in order to be able to visit all the edges of $(G, [\tau])$ even if G is a 2-connected outerplanar cubic graph (which is trivially Hamiltonian).

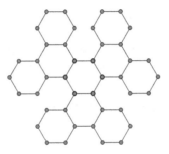

Fig. 2. Example of outerplanar graph G such that $(G, [2])$ does not have an Eulerian local tour.

In the following we explain the construction behind our reduction from NAE 3-SAT. Let ϕ be a CNF formula on variables $\{x_1, \cdots, x_n\}$ and clauses $\{c_1, \cdots, c_m\}$. We start by presenting a meta-construction, in the sense that part of the constructed graph will be presented for now as black boxes and the actual construction is done later, as depicted in Fig. 3. The meta part concerns the clauses; so for now, denote by C_i the black box related to clause c_i. Without going into details, C_i will contain exactly one entry vertex for each of its literal, with some additional vertices, that will be presented later. So, given a literal ℓ contained in c_i, denote by $I_i(\ell)$ the entry vertex for ℓ in C_i. All defined three vertices are distinct.

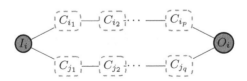

Fig. 3. Edge gadget with clause black boxes.

Now, for each variable x_i, let $c_{i_1}, \cdots, c_{i_p}$ be the clauses containing x_i positively and $c_{j_1}, \cdots, c_{j_q}$ containing x_i negatively. Add two new vertices, I_i and O_i (these will be the entry and exit vertices for the variable gadget), and add the following edges (these compose the paths shown in Fig. 3):

$$E_i = \{I_i I_{i_1}(x_i), I_i I_{j_1}(\overline{x}_i), I_{i_p}(x_i)O_i, I_{j_q}(\overline{x}_i)O_i\}$$
$$\cup \{I_{i_h}(x_i)I_{i_{h+1}}(x_i) \mid h \in [p-1]\}$$
$$\cup \{I_{j_h}(\overline{x}_i)I_{j_{h+1}}(\overline{x}_i) \mid h \in [q-1]\}$$

The paths will function as a switch, telling us whether the variable is true or false within the considered snapshot; we then denote by P_i the set of edges in the path $(I_i, I_{i_1}(x_i), \cdots, I_{i_p}(x_i), O_i)$, and by $\overline{P}_i$ the set of edges in the path $(I_i, I_{j_1}(\overline{x}_i), \cdots, I_{j_q}(\overline{x}_i), O_i)$. Now, to link the variable gadgets and to construct the clause gadgets, we will need a gadget that will function as an edge that must appear in the trail performed in G_1 and the one performed in G_2. For this, we use Lemma 3 applied to the gadget in Fig. 4a; when adding such a gadget between a pair u, v, we simply say that we are adding the *forced edge uv*.

Now, to link the variable gadgets, we add three new vertices s_1, s_2, t and the following forced edges.

$$E' = \{s_i t \mid i \in [2]\} \cup \{tI_1, O_n t\} \cup \{O_i I_{i+1} \mid i \in [n]\}.$$

The new vertices simply help us assume where the trail starts and finishes. Now, let T be an Eulerian local tour of $(G, [2])$ and denote by T_i the trail in G defined by T restricted to G_i, for $i \in [2]$. It is fairly easy to see (and we will prove it shortly) that if we can ensure that T_1 uses P_i if and only if T_2 uses $\overline{P}_i$, then we can prove equivalence with NAE 3-SAT. In other words, the clause gadget must be so that, for every clause c_j containing x_i (or equivalently $\overline{x}_i$), we get that either both edges incident to $I_j(x_i)$ in P_i (or equivalently $I_j(\overline{x}_i)$ in $\overline{P}_i$) are used, or none of them is used. Such a gadget is presented in Fig. 4b, where the red edges are forced.

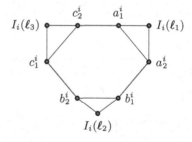

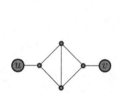

(a) Gadget related to a forced edge uv.

(b) Gadget related to clause c_i. Red edges represent forced edges.

Fig. 4. Gadgets for the reduction in Theorem 3.

Theorem 3. *Let G be a graph with degree at most 4. Then* EULERIAN LOCAL TOUR *is NP-complete on* $(G, [2])$.

Proof. Let ϕ and G be as previously stated. First, consider a truth NAE assignment f to ϕ. We construct $T_1, T_2 \subseteq E(G)$ and prove that they form an Eulerian local tour of G. Start by putting P_i in T_1 and $\overline{P}_i$ in T_2 if x_i is true, and the other way around if x_i is false. From now on, whenever we add a forced edge to T_1 and T_2, we are actually adding the trails depicted in Fig. 5.

Now, add E' to both T_1 and T_2, and consider c_i with literals ℓ_1, ℓ_2, ℓ_3. Suppose, without loss of generality, that ℓ_1 is true and ℓ_2 is false. We then add to T_1 the trail depicted in Fig. 6a, and to T_2 the one depicted in Fig. 6b. Observe that

(a) Trail added to T_1 when forced edge uv is added to T_1.

(b) Trail added to T_2 when forced edge uv is added to T_2.

Fig. 5. Trails related to forced edges.

all internal edges of C_i are covered. Also note that the value of ℓ_3 is irrelevant (the choice remains the same, let it be true or false). We know that the remaining edges are also covered by $T_1 \cup T_2$ by construction. Finally, notice that both T_1 and T_2 touch all odd-degree vertices in a way that every vertex (including the even-degree ones) has even degree in T_1 and in T_2, except s_1, s_2 which have degree exactly 1. Also note that they form a connected graph; indeed they are formed by the cycle passing through the variable gadgets and t, together with some pending trails passing by the clause gadgets. Therefore, we can find an s_1, s_2-trail passing by all edges of T_1, and an s_2, s_1-trail passing by all edges of T_2, thus getting our Eulerian local tour.

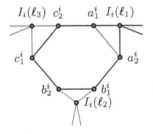

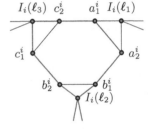

(a) Added to T_1 when ℓ_1 is true.

(b) Added to T_2 when ℓ_2 is false.

Fig. 6. Trails in C_i related to a given NAE assignment.

In order to prove that if Eulerian local tour of $(G, [2])$, then ϕ has an NAE satisfying assignment, it suffices to show that for every x_i, either P_i or $\overline{P}_i$ is contained in T_k for each $k \in [2]$.   □

Observe that if we add two new vertices of degree one adjacent to vertex t, then we get a reduction to the problem of deciding whether the edges of G can be covered by two trails, proving Corollary 2. The following corollary concludes the proof of Item 2 in Theorem 1.

Corollary 3. EULERIAN LOCAL TOUR *and* EULERIAN LOCAL TRAIL *are* NP-*complete on temporal graphs with lifetime* τ *for every fixed* $\tau \geq 2$. *This also holds on dynamic-based graphs.*

Proof. We first make a reduction from EULERIAN LOCAL TOUR on $(G, [2])$ to EULERIAN LOCAL TRAIL on $(G', [\tau])$. Given $(G, [2])$, let G' be obtained from G by adding a star on $\tau + 1$ vertices and identifying one of its leaves with a vertex $s \in V(G)$. We argue that $(G, [2])$ has an Eulerian local tour starting and finishing in s if and only if $(G', [\tau])$ has an Eulerian local trail. The lemma follows because we can then obtain a Turing reduction by building a distinct instance for each $s \in V(G)$. Denote the vertices of the initial star by $u, v_1, \cdots, v_{\tau+1}$, where u is the central vertex, and v_2 is the vertex where G is pending. Let T be an Eulerian local tour of $(G, [2])$ starting and finishing in s, and T_1, T_2 be the trails in G defined by T. Build an Eulerian local trail of $(G', [\tau])$ by visiting $v_1 u$, $u v_2$ and T_1 in G'_1, then performing T_2 and visiting $v_2 u$ and $u v_3$ in G'_2, and finish visiting the remaining edges of the star in the obvious way.

Now, let T be an Eulerian local trail of $(G', [\tau])$, and denote by T_i the trail in G' defined by T restricted to G'_i, for each $i \in [\tau]$. Observe that because we have $\tau + 1$ cut edges, we get that each T_i contains at most 2 of them, and in case it contains exactly 2, say $v_1 u, u v_2$, then T_{i+1} either does not contain any cut edge, or must intersect T_i in $v_1 u, u v_2$. This means that the best we can do in order to finish by time τ is to visit exactly two of them in the first snapshot, and exactly one more in each of the subsequent snapshots. We can therefore suppose, without loss of generality that T_i contains $v_i u, v_{i+1} u$ for each $i \in [\tau]$. Note that this implies that every edge of $(G, [2])$ must be visited in T_1 and T_2, with T_1 starting in v_2 and T_2 finishing in v_2, as we wanted to prove.

Finally, note that $(G', [\tau])$ constructed above has an Eulerian local trail if and only if $(G', \tau + 1)$ has an Eulerian local tour. This completes our proof. □

Finally, in order to prove Corollary 1.(ii), which considers strictly increasing local trails, we can make a modification similar to the one made for walks, meaning that the lifetime of (G, λ) will be sliced into windows, where each window allows only for the edges of a given variable to appear. Observe that this transformation results in a non-dynamic-based temporal graph.

4   Eulerian Tours and Trails

We finally focus on EULERIAN TRAIL and EULERIAN TOUR, proving that in the general case they are both NP-complete, hence, proving Item 3 in Theorem 1. To this aim, we make an adaptation of the construction in Theorem 3. Observe that here the base graph needs to be Eulerian as otherwise the answer to EULERIAN TRAIL is trivially NO. This also implies that the problem restricted to dynamic-based graphs is trivial: if the base graph is Eulerian, then the answer to EULERIAN TOUR is YES; otherwise, then the answer is NO. The trick now is to take advantage of the function λ in order to enforce the edges.

Theorem 4. EULERIAN TOUR *and* EULERIAN TRAIL *are* NP-*complete, even on temporal graphs with fixed lifetime* $\tau \geq 2$.

Proof. We first prove the case $\tau = 2$. For this, we simply replace the gadget to enforce an edge uv in the construction of Sect. 3 by two paths of length 2, P_{uv}^1 and P_{uv}^2, where the edges in P_{uv}^i are active only in snapshot G_i, for each $i \in [2]$. Because the arguments used in Sect. 3 depended only on the fact of uv be indeed an enforced edge, we can apply the same arguments here. The only difference is that the trails in G_1 and G_2 now cannot intersect, which indeed is the case since the intersection between T_1 and T_2 in Sect. 3 is exactly the set of forced edges, and since here each appearance of a forced edge uv is actually related either to P_{uv}^1 or to P_{uv}^2.

Now, in order to prove the NP-completeness for higher values of τ, we can simply add new vertices $v_3, \cdots, v_\tau$ and edges $\{s_1 v_3\} \cup \{v_i v_{i+1} \mid i \in \{3, \cdots, \tau\}\}$. This gives us that EULERIAN TRAIL is NP-complete on (G, λ) with lifetime τ for every fixed $\tau \geq 2$. And if we want a closed trail, it suffices to identify v_τ with s_1, if $\tau \geq 4$, and if $\tau = 3$, we add a new vertex v_4 and edges $v_3 v_4, v_4 s_1$ active in snapshot G_3. This concludes our proof. □

Again, in order to prove Corollary 1.(iii), a modification similar to the one made for walks works. We give a more formal argument below. One can observe that similar arguments can be applied to prove Corollary 1.(ii), as previously claimed.

Proof (Sketch of the proof of Corollary 1.(iii)). As previously said, we will slice the lifetime of (G, λ) into windows, each window allowing only for the edges of a given variable to appear. For this, first observe that, for each clause c_j, a pass in our previous T_1 used 10 edges inside of C_j (already considering that a forced edge is being replaced by two paths on 2 edges), and that a pass in our previous T_2 used 11 edges of C_j. This means that if we allow the edges of a variable x_i appearing in c_j to live long enough, we will be able to visit C_j in a strictly increasing way. For simplicity, consider again that each variable appears at most 3 times in ϕ. We assign to each x_i two time windows, one for the first passing, one for the second, each of size 40 (could be 39, but we choose that for roundness). Thus, variable x_1 will take windows $\{1, \cdots, 40\}$ and $\{40n + 1, \cdots, 40(n+1)\}$, with the edges of $P_1 \cup \overline{P}_1 \cup \{O_1 I_2\}$ being active in the following times: the first edge of P_1 and of $\overline{P}_1$ are active in time 1 and $40n + 1$, the last edges of P_1 and of $\overline{P}_1$ are active in time 38 and $40n + 38$, forced edge O_1, I_2 is active in times $\{39, 40, 40n + 39, 40(n+1)\}$, and the remaining edges are active in the period $\{2, \cdots, 37\} \cup \{40n + 2, \cdots, 40n + 37\}$. Similarly the window of x_i will be $\{40(i-1), \cdots, 40i\} \cup \{40(n+i-1), \cdots, 40(n+i)\}$. Note that we can link O_n to I_1 directly, making them active during $\{40n - 1, 40n\}$ (this is the end of the first window of x_n). Finally, the edges inside a clause c_j will be active during the windows of the corresponding literals. One can verify that the key Property (II) in the proof of Theorem 3 holds, and NP-completeness follows. □

References

1. Eleni, C.A., Mertzios, G.B., Spirakis, P.G.: The temporal explorer who returns to the base. In: 11th International Conference on Algorithms and Complexity - CIAC 2019, Rome, Italy, pp. 13–24 (2019)
2. Arumugam, S., Hamid, I., Abraham, V.M.: Decomposition of graphs into paths and cycles. J. Discrete Math. (2013)
3. Borgnat, P., Fleury, E., Guillaume, J-P., Magnien, C., Robardet, C., Scherrer, A.: Evolving networks. In: Mining Massive Data Sets for Security, pp. 198–203 (2007)
4. Bumpus, B.M., Meeks, K.: Edge exploration of temporal graphs. In: Proceedings of Combinatorial Algorithms - 32nd International Workshop, IWOCA 2021, Ottawa, Canada, 5–7July 2020 (2021). To appear
5. Casteigts, A., Flocchini, P., Quattrociocchi, W., Santoro, N.: Time-varying graphs and dynamic networks. Int. J. Parallel Emerg. Distrib. Syst. $27(5)$, 387–408 (2012)
6. Kayacı Çodur, M., Yılmaz, M.: A time-dependent hierarchical Chinese postman problem. Central Euro. J. Oper. Res. $28(1)$, 337–366 (2018). https://doi.org/10.1007/s10100-018-0598-8
7. Cygan, M., Marx, D., Pilipczuk, M., Pilipczuk, M., Schlotter, I.: Parameterized complexity of Eulerian deletion problems. Algorithmica $68(1)$, 41–61 (2014)
8. Dahlhaus, E., Johnson, D.S., Papadimitriou, C.H., Seymour, P.D., Yannakakis, M.: The complexity of multiterminal cuts. SIAM J. Comput. $23(4)$, 864–894 (1994)
9. Erlebach, T., Hoffmann, M., Kammer, F.: In: 42nd International Colloquium on Automata, Languages, and Programming - ICALP 2015, Kyoto, Japan, volume 9134 of Lecture Notes in Computer Science, pp. 444–455. Springer (2015)
10. Erlebach, T., Spooner, J.T.: Faster exploration of degree-bounded temporal graphs. In: 43rd International Symposium on Mathematical Foundations of Computer Science - MFCS 2018. Schloss Dagstuhl-Leibniz-Zentrum fuer Informatik (2018)
11. Erlebach, T., Spooner, J.T.: Non-strict temporal exploration. In: Richa, A.W., Scheideler, C. (eds.) SIROCCO 2020. LNCS, vol. 12156, pp. 129–145. Springer, Cham (2020). https://doi.org/10.1007/978-3-030-54921-3_8
12. Fomin, F.V., Golovach, P.A.: Long circuits and large Euler subgraphs. SIAM J. Discrete Math. $28(2)$, 878–892 (2014)
13. Garey, M.R., Johnson, D.S., Tarjan, R.E.: The planar Hamiltonian circuit problem is NP-complete. SIAM J. Comput. $5(4)$, 704–714 (1976)
14. Gómez, R., Wakabayashi, Y.: Covering a graph with nontrivial vertex-disjoint paths: existence and optimization. In: International Workshop on Graph-Theoretic Concepts in Computer Science - WG 2018, Cottbus, Germany, 27–29 June, pp. 228–238. Springer (2018)
15. Guan, M.: Graphic programming using odd or even points. Acta Math. Sin. (in Chinese), 10, 263–266 (1960). Translated in Chinese Mathematics 1. American Mathematical Society, 273–277
16. Kempe, D., Kleinberg, J., Kumar, A.: Connectivity and inference problems for temporal networks. In: 32nd annual ACM Symposium on Theory of Computing - STOC 2000, Portland, Oregon, 21–23 May 2000
17. Latapy, M., Viard, T., Magnien, C.: Stream graphs and link streams for the modeling of interactions over time. Soc. Netw. Anal. Mining $8(1)$, 1–29 (2018). https://doi.org/10.1007/s13278-018-0537-7
18. Manuel, P.: Revisiting path-type covering and partitioning problems. arXiv preprint arXiv:1807.10613 (2018)

19. Michail, O.: An introduction to temporal graphs: an algorithmic perspective. Internet Math. **12**(4), 239–280 (2016)
20. Michail, O., Spirakis, P.G.: Traveling salesman problems in temporal graphs. Theor. Comput. Sci. **63**(4), 1–23 (2016)
21. Orlin, J.B.: Some problems on dynamic/periodic graphs. In: Progress in Combinatorial Optimization, pp. 273–293. Elsevier (1984)
22. Sun, J., Tan, G., Honglei, Q.: Dynamic programming algorithm for the time dependent Chinese postman problem. J. Inf. Comput. Sci. **8**, 833–841 (2011)
23. Wang, H.-F., Wen, Y.-P.: Time-constrained Chinese postman problems. Comput. Math. Appl. **44**(3–4), 375–387 (2002)

Reconfiguring Simple s, t Hamiltonian Paths in Rectangular Grid Graphs

Rahnuma Islam Nishat[✉], Venkatesh Srinivasan, and Sue Whitesides

Department of Computer Science, University of Victoria, Victoria, BC, Canada
{rnishat,srinivas,sue}@uvic.ca

Abstract. We study the following reconfiguration problem: given two s, t Hamiltonian paths connecting diagonally opposite corners s and t of a rectangular grid graph G, can we transform one to the other using only *local* operations in the grid cells? In this work, we introduce the notion of *simple* s, t Hamiltonian paths, and give an algorithm to reconfigure such paths of G in $O(|G|)$ time using local operations in unit grid cells. We achieve our algorithmic result by proving a combinatorial *structure theorem* for simple s, t Hamiltonian paths in rectangular grid graphs.

1 Introduction

An $m \times n$ *rectangular grid graph* G is a subgraph of the infinite integer grid embedded on m rows and n columns; the outer boundary of G is a rectangle $\mathcal{R}$, and the inner faces of G are 1×1 grid cells, so G has mn vertices. An s, t *Hamiltonian path* P of G is a Hamiltonian path of G with endpoints at the top left and bottom right vertices s and t of $\mathcal{R}$. See Fig. 2(a) for an example. The *reconfiguration of s, t Hamiltonian paths* in grid graphs is concerned with transforming one s, t Hamiltonian path into another such Hamiltonian path of the same graph using some *operation* that preserves Hamiltonicity in each intermediate step of the transformation. Here, we give a reconfiguration algorithm for a class of s, t Hamiltonian paths that we call 'simple' paths. Moreover, we use an operation (namely, *pairs of switch operations* to be defined in Sect. 4) that is *local* to the grid graph, not the path. See Fig. 1.

Each internal node v of G lies on an *internal subpath* of P, namely the subpath joining the first boundary vertices v_s and v_t met when travelling along P from v toward s and toward t, respectively. This internal subpath must make at least two bends (turns) if v_s and v_t lie on the same side of $\mathcal{R}$, at least one bend if they lie on adjacent sides, and does not need to bend if v_s and v_t lie directly opposite each other. An s, t Hamiltonian path P is *simple* if each such internal subpath has the minimum possible number of bends. Thus each internal node of G either lies on a bend-free internal subpath between directly opposite boundary nodes,

© Springer Nature Switzerland AG 2021
P. Flocchini and L. Moura (Eds.): IWOCA 2021, LNCS 12757, pp. 501–515, 2021.
https://doi.org/10.1007/978-3-030-79987-8_35

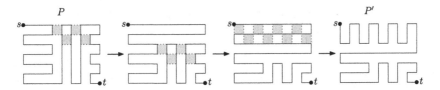

Fig. 1. Reconfiguring simple s, t Hamiltonian path P to another such path P' using pairs of switch operations.

or lies on a one-bend internal subpath between nodes on adjacent sides, or lies on a two-bend internal subpath between nodes on the same side. We call such paths P simple because they cannot wind and twist inside $\mathcal{R}$: internal subpaths can travel between any pair of sides of $\mathcal{R}$, yet can make only the minimum possible number of bends to do so. The Hamiltonian path in Fig. 2(a) is simple. See Fig. 11 for an example of a 'twisty' and 'windy' s, t Hamiltonian path.

Although linear time reconfiguration algorithms have been designed for "1-complex" Hamiltonian cycles in rectangular grid graphs [13] (i.e., Hamiltonian cycles where each internal node of the grid is connected to the boundary by a straight line segment along the cycle), similar techniques do not carry over to s, t Hamiltonian paths. Whereas all the internal paths in a cycle must start and end on adjacent nodes on the boundary, paths allow internal subpaths to have endpoints that are far apart on the boundary. In this paper we study the structure and reconfiguration of simple s, t Hamiltonian paths as an important step towards understanding the structure of general s, t Hamiltonian paths and designing reconfiguration algorithms for them.

Hamiltonian paths and cycles in grid graphs have attracted interest in part due to their many applications (e.g., in robot navigation [5], 3D printing [11], and polymer science [16]). They have the potential to reduce turn costs and travel time and to increase navigation accuracy (e.g., [1,4,21]). We believe the general study of reconfiguration of paths in grids is both interesting on its own and also useful for exploring the space of route possibilities in grid-like environments such as warehouses and 3D printer platforms.

Our Contributions. Our work is the first to study reconfiguration of paths in grid graphs, while previous work on paths focused mainly on existence, enumeration, and generation, not reconfiguration. In particular, (1) we establish the structure of simple s, t Hamiltonian paths in rectangular grid graphs; (2) we introduce a *zip* operation, which uses a *switch* in a grid cell as the atomic local operation, to reconfigure simple s, t Hamiltonian paths; our zip operation is comprised of pairs of switch operations that preserve Hamiltonicity and its accompanying data structure facilitate running time analysis and possible implementation of our algorithm and other reconfiguration algorithms; (3) using the structure theorem, we give an algorithm to reconfigure any simple s, t Hamiltonian path to any other such path in time linear in the size $|G|$ of the grid graph G.

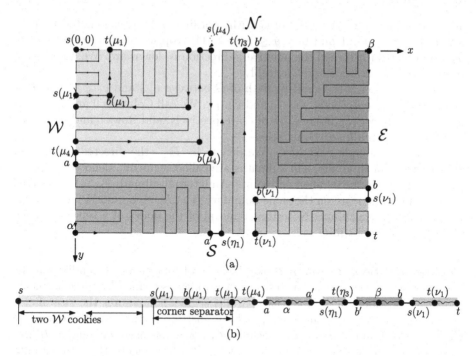

Fig. 2. (a) Grid layout of a simple path $P = P_{s,t}$. (b) The combinatorial layout of P. (See Sects. 2 and 3 for explanation of notation.)

Related Work. Itai *et al.* [6] gave necessary and sufficient conditions for the existence of a Hamiltonian path between any pair of vertices in a rectangular grid graph. The existence problem for classes of non-rectangular grid graphs was studied in [3,9,20]. Ruskey and Sawada studied existence of 'bent' Hamiltonian cycles in grid graphs in d-dimension, $d \geq 2$, where each edge in a pair of successive edges on the path lies in a different dimension [18]. Combinatorial aspects of Hamiltonian paths in grid graphs such as enumeration [8,17] and generating functions [2] have been explored.

Reconfiguration problems have attracted attention for some time [7,15]. There has been some recent work on reconfiguration of Hamiltonian cycles in grid graphs. Takaoka [19] has shown that for some unembedded graph classes, deciding whether there is a sequence of "switch" operations between two given Hamiltonian cycles is a PSPACE-complete problem. Nishat and Whitesides studied reconfiguration of Hamiltonian cycles of "bend complexity 1" in grid graphs without holes [12–14].

2   Terminology and Basics

Throughout this paper, a *simple path* means a simple s, t Hamiltonian path of G; P visits each node of G exactly once and uses only edges in G. A *cell* of G is an internal face of G. A vertex of G with coordinates (x, y) is denoted by $v_{x,y}$, where $0 \leq x \leq n - 1$ and $0 \leq y \leq m - 1$. The top left corner vertex s of G has coordinates $(0, 0)$, and the positive y-direction is downward. We use the two terms *node* and *vertex* interchangeably.

Column x of G is the shortest path of G between $v_{x,0}$ and $v_{x,m-1}$, and *Row* y is the shortest path between $v_{0,y}$ and $v_{n-1,y}$. We call Columns 0 and $n - 1$ the *west* ($\mathcal{W}$) and *east* ($\mathcal{E}$) boundaries of G, respectively, and Rows 0 and $m - 1$ the *north* ($\mathcal{N}$) and *south* ($\mathcal{S}$) boundaries.

Let P be a simple path of G. We denote by $P_{u,w}$ the directed subpath of P from vertex u to w. Straight subpaths are called segments, denoted $seg[u, v]$, where u and v are the segment endpoints. An internal subpath $P_{u,v}$ of P is called a *cookie* if both u, v are on the same boundary (i.e., $\mathcal{N}$, $\mathcal{S}$, $\mathcal{E}$, and $\mathcal{W}$); otherwise, $P_{u,v}$ is called a *separator*. (Note that removal of the nodes of a separator from G separates s from t in G.)

Cookies and Separators. A cookie has one of four types, depending on the boundary where the cookie has its *base*. A cookie c is formed by three segments of P; the common length of the two parallel segments measures the *size* of c.

Since separators of P have endpoints on distinct boundaries, there are two kinds, as shown in Fig. 2: a *corner separator* μ_i or ν_i has one bend, and a *straight separator* η_i has no bends. Traveling along $P_{s,t}$, we denote the i-th straight separator we meet by η_i where $0 \leq i \leq k$, and its endpoints by $s(\eta_i)$ and $t(\eta_i)$, where $s(\eta_i)$ is the first endpoint met. We say a corner separator *cuts off* a corner (s or t). Traveling along $P_{s,t}$, we denote the i-th corner separator cutting off s by μ_i, where $0 \leq i \leq j$. We denote its internal bend by $b(\mu_i)$, and its endpoints by $s(\mu_i)$ and $t(\mu_i)$, where $s(\mu_i)$ is the first endpoint met. Similarly, we denote the i-th corner separator cutting off t by ν_i; endpoint $s(\nu_i)$ is met before $t(\nu_i)$, with internal bend at $b(\nu_i)$, where $0 \leq i \leq \ell$. A corner separator that has one of its endpoint connected to s or t by a segment of P is called a *corner cookie*. We have j corner separators μ_i cutting off s, and k straight separators, and ℓ corner separators ν_i cutting off t. In Fig. 2, $j = 4$, $k = 3$ and $\ell = 1$. (We will see that only s and t can be cut off.)

Runs of Cookies. A *run of cookies* is a subpath of P consisting of cookies of the same type, spaced one unit apart and joined by the single boundary edges between them, possibly extended at either end by an edge joining a cookie endpoint to an adjacent boundary vertex. A run of cookies is denoted $Run[u, v]$, where u and v belong to the same boundary and delimit the range of boundary vertices covered; $Run[u, v]$ may consist of a single boundary edge (u, v). In Fig. 2, $Run[s, s(\mu_1)]$ is a run of $\mathcal{W}$ cookies. To describe the path structure, we define three types of runs, depending on the cookie sizes along the run: the sizes may remain the same, or be non-increasing (denoted $Run^{\geq}[u, v]$) or non-decreasing ($Run^{\leq}[u, v]$). Runs are assumed to have cookies of the same size unless specified otherwise. In Fig. 2, $Run^{\geq}[a, \alpha]$ is non-increasing; $Run[s, s(\mu_1)]$, which has same size cookies, could also be viewed as non-increasing or non-decreasing.

Canonical Paths. A *canonical path* is a simple path P with no bends at internal vertices. If m is odd, P can be $\mathcal{E}$-$\mathcal{W}$ and fill rows of G one by one; if n is odd, P can be $\mathcal{N}$-$\mathcal{S}$ and fill columns (see Fig. 3(d)). There are no other types.

Assumption. Let α and β denote the bottom left and top right corner vertices of G. Without loss of generality, *we assume the input simple path $P_{s,t}$ visits α before β.* The target simple path for the reconfiguration as well as intermediate configurations may visit β before α. In the rest of this section and the next, P denotes the input simple path.

Blocks. As shown in Fig. 3, corner separators μ_j and ν_1 define rectangular subgraphs, called *blocks*, of G, denoted $\mathcal{R}_s$ and $\mathcal{R}_t$. Straight separators η_1 and η_k determine a block $\mathcal{R}_{mid}$. The remaining nodes of G determine two blocks: $\mathcal{R}_\alpha$, with corners at α, a (upper left) and a' (lower right), and $\mathcal{R}_\beta$, with corners at β, and b' (upper left) and b (lower right) as shown. Blocks $\mathcal{R}_s$ and $\mathcal{R}_t$ vanish when $j=0$ and $\ell=0$, respectively. For now the goal is to define the blocks. Later we will observe that $P_{s,t}$ connects the blocks with "links" as shown in the figure.

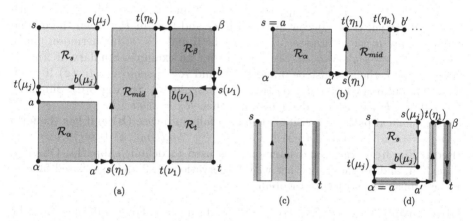

Fig. 3. (a) Blocks of $P_{s,t}$; the *links* between blocks are darkened. (b)–(d) Some special cases. See Sect. 3 for details. (c) An $\mathcal{N}$-$\mathcal{S}$ canonical path.

We make two observations (the second is a consequence of the first) that prepare for the structure theorem in the next section.

Observation 1. The subpath $P_{s,\alpha}$ of $P_{s,t}$ must cover all the $\mathcal{W}$ vertices as otherwise P fails to be Hamiltonian or non-crossing. For the same reason, no corner separators cut off α or β, and $P_{s,t}$ must visit β before visiting any other $\mathcal{E}$ vertices. It follows that the corner separators μ_i cutting off s must occur in $P_{s,t}$ before α, and that the corner separators ν_i cutting off t occur in $P_{s,t}$ after β. It also follows that all straight separators η_i occur between α and β, and that the number $k \geq 1$ of them must be odd. □

Observation 2. $P_{s,t}$ breaks G into at most five blocks ($\mathcal{R}_s$, $\mathcal{R}_\alpha$, $\mathcal{R}_{mid}$, $\mathcal{R}_\beta$, and $\mathcal{R}_t$) as shown in Fig. 3. $P_{s,t}$ must join blocks with *link* edges, one on each boundary, directed as shown in the figure. In certain cases, detailed in Sect. 3, blocks (and their links) may vanish, or blocks may shrink to segments. □

3   Structure of a Simple Path *P*

We regard P in its directed form $P_{s,t}$ as composed of an *initial* subpath $P_{s,s(\eta_1)}$, followed by a *middle* subpath $P_{s(\eta_1),t(\eta_k)}$, and a *final* subpath $P_{t(\eta_k),t}$. Except for three special cases **(a)–(c)** listed below, the initial subpath first covers all the vertices in $\mathcal{R}_s$, next takes an edge $(t(\mu_j),a)$, denoted $link(t(\mu_j),a)$, to $\mathcal{R}_\alpha$. It then covers $\mathcal{R}_\alpha$ and then takes an edge $link(a',s(\eta_1))$ to $s(\eta_1)$. The possible forms for the final subpath are the same as for the initial subpath $P_{t,s}$, which is $P_{s,t}$ in reverse.

Three Special Cases: **(a)** If α is not adjacent to $s(\eta_1)$ in G and $j = 0$, then $s = a$ because $\mathcal{R}_s$ disappears and the initial subpath begins at $\mathcal{R}_\alpha$, where each side of $\mathcal{R}_\alpha$ has length at least 1. See Fig. 3(b). Similarly, if β is not adjacent to $t(\eta_k)$ and $\ell = 0$, then $\mathcal{R}_t$ disappears and $b = t$ and each side of $\mathcal{R}_\beta$ has length at least 1. **(b)** If α is adjacent to $s(\eta_1)$ in G and thus in $P_{s,t}$, then there is no room for any μ_i so as in case (a), j must be 0 and $a = s$; furthermore, unlike case (a), $\mathcal{R}_\alpha = seg[s = a, \alpha]$. See Fig. 3(c) for an example. Similary, if $t(\eta_k)$ is adjacent to β, then ℓ must be 0 and $b = t$, and $\mathcal{R}_\beta = seg[\beta, b = t]$. **(c)** If $j > 0$ and $P_{s,t}$ contains edge $(t(\mu_j), \alpha)$ then $\mathcal{R}_\alpha = seg[\alpha, a']$. See Fig. 3(d). Similarly, if $\ell > 0$ and $P_{s,t}$ contains edge $(\beta, s(\nu_1))$, then $\mathcal{R}_\beta = seg[b', \beta]$.

A canonical path that visits α before β falls into case **(b)** and has the form
$\| \ seg[s = a, \alpha] \ \| \ link(\alpha, s(\eta_1)) \ \| \ P_{s(\eta_1),t(\eta_k)} \ \| \ link(t(\eta_k), \beta) \ \| \ seg[\beta, b = t] \ \|$.
(Recall Fig. 3(c).) Here, and later on, $\|$ is used to delimit subpaths. Observations 3, 4, and 5 below, together with Observations 1, 2 and associated figures, will establish our structure theorem.

Observation 3 (about $\mathcal{R}_s$). For $j = 0$, $\mathcal{R}_s$ disappears. For $j > 0$, there must be j endpoints of μ_i on $\mathcal{W}$ and j on $\mathcal{N}$. Endpoint $t(\mu_j)$ must lie on $\mathcal{W}$, not $\mathcal{N}$, and it must be the corner separator endpoint nearest to α on $\mathcal{W}$. The endpoints must alternate on $\mathcal{W}$ as shown in Figs. 4(a), (d). Thus for j odd, the top endpoint on $\mathcal{W}$ is $t(\mu_1)$, so $s(\mu_1)$ is on $\mathcal{N}$; for $j > 0$ even, $s(\mu_1)$ is on $\mathcal{W}$. Either way, $P_{s,s(\mu_1)} = Run[s, s(\mu_1)]$.

$P_{s(\mu_1),t(\mu_j)}$ makes round trips (see Fig. 4(e)) between $\mathcal{W}$ and $\mathcal{N}$ via the μ_i. These trips have the form $\|\mu_i \| Run[t(\mu_i), s(\mu_{i+1})] \ \|\mu_{i+1}\|$ and alternate leaving from $\mathcal{N}$ or from $\mathcal{W}$, as the return leg of one trip is the outgoing leg of the next. Furthermore, $t(\mu_{i+1})$ must be adjacent to $s(\mu_i)$ to ensure the Hamiltonicity of $P_{s,t}$. Between $t(\mu_i)$ and $s(\mu_{i+1})$, the path must have the form $Run[t(\mu_i), s(\mu_{i+1})]$, which may be just a single edge $(t(\mu_i), s(\mu_{i+1}))$.

Thus for $j > 0$, $P_{s,t(\mu_j)} = \|Run[s, s(\mu_1)] \ \| \ \mu_1 \ \| \ Run[t(\mu_1), s(\mu_2)] \ \| \ \mu_2 \ \| \cdots \| \ \mu_i \ \| \ Run[t(\mu_i), s(\mu_{i+1})] \ \| \ \mu_{i+1} \ \| \cdots \| \ \mu_{j-1} \ \| \ Run[t(\mu_{j-1}), s(\mu_j)] \ \| \ \mu_j\|$. □

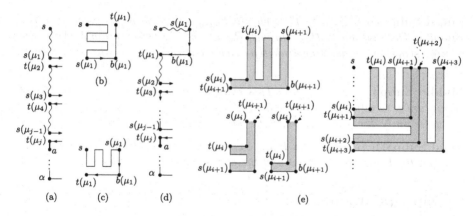

Fig. 4. Endpoint ordering on $\mathcal{W}$ and form of $P_{s,s(\mu_1)}$ for: (a), (b) even $j > 0$; (c), (d) j odd; (e) round trips $P_{s(\mu_i),t(\mu_{i+1})}$ start on $\mathcal{W}$ (red) or on $\mathcal{N}$ (blue). (Color figure online)

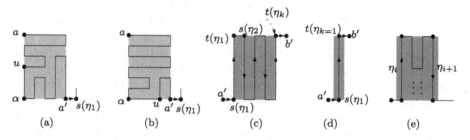

Fig. 5. (a) Block $\mathcal{R}_\alpha$ where $|seg[u,\alpha]| > 1$ and (b) where $|seg[\alpha,u]| > 1$; (c) $\mathcal{R}_{mid}$ for $k > 1$; (d) $\mathcal{R}_{mid}$ is a segment when $k = 1$. (e) $\mathcal{R}_{mid}$ cannot have cookies.

Observation 4 (about $\mathcal{R}_\alpha$). In special case **(b)**, in which $s = a$ and α is adjacent to $s(\eta_1)$, we have that $\alpha = a'$ and so $P_{a,a'} = seg[s = a, \alpha]$ on $\mathcal{W}$. In special case **(c)**, in which $P_{s,t}$ contains edge $(t(\mu_j), a = \alpha)$, we have that $P_{a,a'} = seg[a = \alpha, a']$ on $\mathcal{S}$. Otherwise, two segments of P meet at α, one on $\mathcal{W}$, one on $\mathcal{S}$; one has unit length and the other is strictly longer. See Fig. 5(a),(b). It follows from Observation 1 that either $P_{a,a'} = \|Run^{\geq}[a,u]\| seg[u,\alpha]\| Run^{\leq}[\alpha,a']\|$, with u at least two units from α on $\mathcal{W}$, or $P_{a,a'} = \|Run^{\geq}[a,\alpha]\| seg[\alpha,u]\| Run^{\leq}[u,a']\|$, with u at least two units from α on $\mathcal{S}$. $\square$

Observation 5 (about $\mathcal{R}_{mid}$). (See Fig. 5(c), (d), (e).) Block $\mathcal{R}_{mid}$ has no cookies. Extending run notation, $P_{s(\eta_1),t(\eta_k)} = \|\eta_1 \| Run[t(\eta_1), t(\eta_k)]\|$. $\square$

The next theorem, based on Observations 1–5 and associated figures, establishes the structure of P by giving the forms that the subpaths of $P_{s,t}$ may take inside the blocks. See Fig. 3.

Theorem 1 (Structure of Simple Paths). *Let $P_{s,t}$ be a simple path with k straight separators and j and ℓ corner separators cutting off s and t, respectively.*

Initial Subpath $[P_{s,s(\eta_1)}]$. *The subpath* $P_{s,t(\mu_j)}$ *through* $\mathcal{R}_s$ *is given in Obser-vation 3. The subsubpath* $P_{a,a'}$ *through* $\mathcal{R}_\alpha$ *is given in Observation 4. Appending* $link(a', s(\eta_1))$ *and inserting* $link(t(\mu_j), a)$ *(if needed), gives the structure for* $P_{s,s(\eta_1)}$.

Middle Subpath $[P_{s(\eta_1),t(\eta_k)}]$. *This consists of an odd number* k *of straight separators* η_i *in adjacent grid columns, joined by edges* $(t(\eta_i), s(\eta_{i+1}))$, $1 \leq i < k$, *which lie on* N *for odd* i *and on* S *for even* i. *See Observation 5.*

Final Subpath $[P_{t(\eta_k),t}]$. *As* $P_{t,t(\eta_k)}$ *is the initial subpath of* $P_{t,s}$ *(the reverse of* $P_{s,t}$*), the forms for the final and the initial subpaths are the same.*

4   Zip Operation

In this section we define the *zip* operation that can be applied on an s, t Hamiltonian path P, not necessarily simple. We will use zips in our reconfiguration algorithm in the next section. We first define some terminology, including a *switch* operation that is used as the atomic local operation of *zip*.

A *vertical track* tr_x^v is the subgraph of G induced by Columns x and $x + 1$, and a *horizontal track* tr_y^h is the subgraph induced by Rows y and $y + 1$. A *cycle-path cover* $\mathbb{P}$ of G is a set of cycles and paths that collectively cover all the vertices of G. Let f be a cell of G with face cycle a, b, c, d, a in G such that edges (a, d) and (c, b) are the only edges of f that are in $\mathbb{P}$ (see Fig. 6). A *switch* operation on $\mathbb{P}$ in f replaces (a, d) and (c, b) with (a, b) and (c, d) and thus produces another cycle-path cover $\mathbb{P}'$. We call f a *switchable* cell of G in $\mathbb{P}$ (or in P for short, if $\mathbb{P} = \{P\}$). Note that our switch operation is applied locally to a cell of an *embedded* grid graph, and hence a single switch does not preserve Hamiltonicity. We, therefore, always apply switches in pairs such that the second switch operation in a pair patches the two parts created by the first switch and returns an s, t Hamiltonian path. In contrast, the switch operation defined by Lignos [10] and Takaoka [19] is applied to a four cycle of the graph, where the graph is not necessarily planar, and may not be embedded; and their switch operations do not need to be paired. See Fig. 6(c) for an example of the switch operation defined in [10,19].

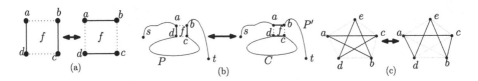

Fig. 6. (a) A *switch* in a cell f. (b) A switch in a switchable cell of an *embedded* s, t Hamiltonian path P in cycle-path cover $\mathbb{P} = \{P\}$ yields new cover $\mathbb{P}' = \{P', C\}$. (c) A single switch (i.e., edges (a, b) and (d, c) is replaced by edges (a, d) and (b, c)) as defined in [10,19] on a non-planar graph preserves Hamiltonicity.

Observation 6. A switch operation in a grid cell that is switchable for an s,t Hamiltonian path P of G gives a cycle-path cover of G: $\mathbb{P}' = \{C, P'\}$ where C is a cycle and P' is a path with ends s and t. A switch in a switchable cell f of $\mathbb{P}'$, where each of C and P' contains one edge of f, gives an s,t Hamiltonian path.

A *zipline* $l_z^{q_1,q_2}$ (the superscript may be omitted for short) is a straight directed line in G from node q_1 to node q_2; l_z lies in a row or a column of G, and two tracks (horizontal or vertical) of G contain l_z. We designate one of those tracks as the *cookie track* tr, and the other track the *side track* tr'. A switchable cell f of tr or tr' is called a *perpendicular switchable cell* if the two edges that f contributes to P are perpendicular to tr (and tr'); we denote the first of those edges met while walking along l_z from q_1 to q_2 by $e_1(f)$, the second edge by $e_2(f)$, and we say that f is *between q_1 and q_2* if $e_1(f)$ and $e_2(f)$ occur between q_1 and q_2. Note that not all the switchable cells of tr and tr' between q_1 and q_2 are necessarily perpendicular switchable cells.

The *zip set* S of zipline l_z is the set of perpendicular switchable cells from tr and tr' constructed as follows: walking along l_z from q_1 to q_2, we look for the first perpendicular switchable cell in the cookie track tr; if we find such a switchable cell, say f_1, in tr between q_1 and q_2, we look for the next perpendicular switchable cell in the side track tr' between $e_2(f_1)$ and vertex q_2; if we find such a cell (call it f_1') in tr' then we add both f_1 and f_1' to S; otherwise we do not add f_1 to S as it cannot be paired with a switchable cell from the side track. If we find f_1' before reaching q_2, we repeat the above process for cells between $e_2(f_1')$ and q_2 to find pairs (f_2, f_2'), (f_3, f_3') and so on. At the end, S is either empty or consists of pairs of perpendicular switchable cells, one from tr and one from tr'. Moreover, no two cells of S share an edge, and each pair f_i and f_i' is met consecutively on l_z between f_{i-1}' and f_{i+1}. We now define the *zip* operation; see Fig. 7.

Definition 1 (Zip). *Let $l_z^{q_1,q_2}$ be a zipline of G with cookie track tr and side track tr', and let $S = \{f_1, f_1', f_2, f_2', \ldots, f_p, f_p'\}$, where $|S| \geq 0$ is even, be the zip set of l_z. The* zip *operation $Z = zip(l_z^{q_1,q_2}, tr, tr')$ applies switch to all the cells of S in the following order: $f_1', f_1, f_2', f_2, \ldots, f_p', f_p$.*

We note here that the zip operation may not preserve simplicity when applied on a simple path. However, as Lemma 1 shows, each zip and indeed each pair of its switches, preserves s,t Hamiltonicity.

Let $S = \{f_1, f_1', f_2, f_2', \ldots, f_p, f_p'\}$ be the zip set of the zip operation $Z = zip(l_z^{q_1,q_2}, tr, tr')$, where $|S| > 0$. S is *intra-pair overlapping* if for each pair of cells f_i, f_i' of S, the edges appear in the following *path order* on $P_{s,t}$: $e_1(f_1)$, $e_1(f_1')$, $e_2(f_1)$, $e_2(f_1')$. Note that the path order of those four edges is different from their *zipline order*, i.e., the order in which they are met on l_z. Zip set S is *inter-pair overlapping* when S is intra-pair overlapping, and the first edge $e_1(f_{i+1})$ of the $(i+1)^{st}$ pair is between $e_2(f_i)$ and $e_2(f_i')$ on P. In other words, the path order of the edges of the cells of S is $e_1(f_1)$, $e_1(f_1')$, $e_2(f_1)$, $e_1(f_2)$, $e_2(f_1')$, $e_1(f_2')$, $e_2(f_2)$, $e_1(f_3)$, $e_2(f_2')$, $\ldots$, $e_2(f_p')$.

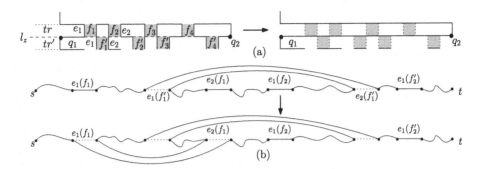

Fig. 7. (a) An example of a zip operation on an s, t Hamiltonian path P where l_z is horizontal. (b) The cycle-path cover $\mathbb{P}$ obtained by a switch in f_1', and the new s, t Hamiltonian path P' obtained from the cycle-path cover by a switch in f_1.

Lemma 1. *Let $Z = zip(l_z^{q_1, q_2}, tr, tr')$ be a zip operation on an s, t Hamiltonian path P with the zip set $S = \{f_i, f_i'$ for $1 \leq i \leq p\}$, $|S| > 0$. If S is inter-pair overlapping, then after switching the cells f_1' and f_1 we obtain an s, t Hamiltonian path P' such that $S' = S - \{f_1, f_1'\}$ is the zip set of $Z' = zip(l_z^{q, q_2}, tr, tr')$, where q is the endpoint of $e_2(f_1')$ on l_z, and S' is inter-pair overlapping.*

Proof. We first prove that P' is an s, t Hamiltonian path. Since S is intra-pair overlapping, the path order of the edges of f_1 and f_1' is $e_1(f_1)$, $e_1(f_1')$, $e_2(f_1)$, $e_2(f_1')$. By Observation 6, a switch in f_1' must produce a cycle-path cover, say $\mathbb{P}$, of G containing exactly one cycle and one path from s to t; see Fig. 7(b). Now edge $e_1(f_1)$ is on the path and the edge $e_2(f_1)$ is on the cycle in $\mathbb{P}$. Therefore, the switch in f_1 in $\mathbb{P}$ gives an s, t Hamiltonian path by Observation 6. Since by definition of zip set, f_2 is the first perpendicular switchable cell encountered in cookie track tr when walking from $e_2(f_1')$ to q_2, $S' = S - \{f_1, f_1'\}$ is the zip set of Z'. By the inter-pair overlapping property of S, only one edge $e_1(f_2)$ of a cell in $S - \{f_1, f_1'\}$ is on the subpath of P from s to the second endpoint of $e_2(f_1')$. Since P and P' only differ in that subpath, the path order of the edges of the cells of S' on P' is the same as their path order on P. Therefore, S' is inter-pair overlapping. $\qquad\square$

Using Lemma 1, we can prove the following lemma by induction. To achieve the stated time complexity, a suitable data structure is used to store the path edges that allows $O(1)$ time retrieval and update.

Lemma 2. *A zip operation on an s, t Hamiltonian path P of G returns another s, t Hamiltonian path P' in $O(\max\{m, n\})$ time, where P and P' differ only at the cells belonging to the zip set.*

5   Reconfiguring Simple Paths

In this section, we give an algorithm to reconfigure any simple path P to another simple path P'. The algorithm reconfigures P and P' to canonical paths $\mathbb{P}$ and $\mathbb{P}'$, respectively; it then reconfigures $\mathbb{P}$ to $\mathbb{P}'$ in case they are not the same path (i.e., one of $\mathbb{P}$ and $\mathbb{P}'$ is the $\mathcal{N}$-$\mathcal{S}$ canonical path and the other is the $\mathcal{E}$-$\mathcal{W}$ canonical path); and finally the algorithm reverses the steps taken from P' to $\mathbb{P}'$.

Reconfiguring P to $\mathbb{P}$. We give an algorithm that we call RECONFIGSIMP to reconfigure any simple path $P = P_{s,t}$ to a canonical path $\mathbb{P}$, where $\mathbb{P}$ might be either $\mathcal{N}$-$\mathcal{S}$ or $\mathcal{E}$-$\mathcal{W}$. The algorithm runs in three steps: (a) reconfigure the initial subpath of $P_{s,t}$ such that either $P_{s,\alpha}$ is a segment $seg[s, \alpha]$ and $\alpha = a'$, or $Run[s, \alpha]$ is a run of $\mathcal{W}$ cookies of size $x(a')$ and $P_{\alpha,a'}$ is a segment $seg[\alpha, a']$; (b) reconfigure the final subpath (that is, the initial subpath of $P_{t,s}$) similarly to the previous step; (c) if the path resulting from Step (b) contains both the segments $seg[s, \alpha]$ and $seg[\beta, t]$ then we have the $\mathcal{N}$-$\mathcal{S}$ canonical path and the algorithm terminates; otherwise, we have at least one run of $\mathcal{E}$ or $\mathcal{W}$ cookies and we reconfigure $P_{s,t}$ to obtain the $\mathcal{E}$-$\mathcal{W}$ canonical path.

Step (a). The algorithm reconfigures the initial subpath of P by calling two procedures, first RECON_R_s and then RECON_s_to_a'.

Procedure. RECON_R_s reconfigures R_s by dissolving an even number of corner separators cutting off s. If the number j is even, then all the corner separators are dissolved; otherwise, all but μ_j are dissolved.

If $j > 0$ is even, we apply zips on a horizontal zipline $l_z^{q_1,q_2}$, where q_1 and q_2 are on the $\mathcal{W}$ boundary and on Column $x(s(\mu_j))$, respectively, for all the zip operations. The zipline l_z is first placed on Row 1. Track tr_0^h above l_z is the cookie track and track tr_1^h below l_z is the side track. The zip operation then grows a $\mathcal{W}$ cookie of size $x(q_2)$ in the cookie track. We then move the zipline two rows down and apply a similar zip. In this way the zipline is *swept downward* until it is on Row $y(t(\mu_j))$, where the final zip is applied to obtain a simple path P' with no corner separators cutting off s. We call this procedure a SWEEPDOWN; see Fig. 8.

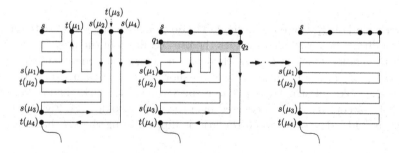

Fig. 8. Intermediate steps of a SWEEPDOWN procedure on $\mathcal{R}_s$ of P.

If j is odd, we apply a procedure SWEEPRIGHT similar to SWEEPDOWN. Zips are applied on a vertical zipline $l_z^{q_1,q_2}$, where q_1 and q_2 are on the $\mathcal{N}$ boundary and on Row $y(t(\mu_j))$, respectively, and the cookie track is on the left and the side track is on the right of l_z. The zipline starts from Column 1 and finishes on Column $x(s(\mu_j)) - 1$, sweeping two columns eastward after every zip except the last one.

Let P' be the path returned by RECON_R_s. The following lemma shows that P' is a simple path with at most one corner separator cutting off s.

Lemma 3. P' *is a simple path with no corner separator if j is even, and P' has a single corner separator that coincides with μ_j of P if j is odd.*

Proof. By Lemma 2, each of the zip operations in the SWEEPDOWN procedure gives an s, t Hamiltonian path, although it might not necessarily be simple. First assume that j is even. To prove that the P' is a simple s, t Hamiltonian path, we show that each zip operation satisfies the following pre- and post-conditions.

Pre-condition: P_{q_1, q_2} occupies only the rows on or below the cookie track.

Post-condition: Cookie track tr contains a $\mathcal{W}$ cookie of size $x(q_2)$.

Since the cookie track of the first zip Z_1 is incident to the $\mathcal{N}$ boundary, the input P satisfies the pre-condition. Let P_1 be the path obtained by applying Z_1 to P. Every switch in tr' creates an island in tr and the next switch in tr connects that island to the $\mathcal{W}$ cookie, thus increasing the size of the $\mathcal{W}$ cookie. The first switch of Z_1 is illustrated in Fig. 9.

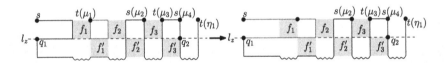

Fig. 9. The 1×1 island, depicted by red line, in the cookie track created by the switch in f_1' is attached to the $\mathcal{W}$ cookie by the switch in f_1. The W cookie after the switch is depicted by red line. (Color figure online)

At the end of Z_1, there is one single $\mathcal{W}$ cookie of size $x(q_2)$ in P_1 satisfying the post-condition. Then P_1 also satisfies the pre-condition of the next zip operation where the cookie track is tr_2^h, as there are no edges perpendicular to the side track tr_1^h of Z_1 in P_1. In this way, we can show that after each zip operation we have a $\mathcal{W}$ cookie of size $x(q_2)$ in the respective cookie track. Therefore, at the end of the sweep, $\mathcal{R}_s$ of P' is covered by a run of $\mathcal{W}$ cookies; see Fig. 8. The case when j is odd is the same when the grid is flipped along the s, t diagonal. $\square$

Procedure. RECON_s_to_a' applies zips on P' to obtain another simple path P'', where the initial subpath of P'' has either a run of $\mathcal{W}$ cookies or no cookies at all. If there is an odd number of columns, including the $\mathcal{W}$ boundary, to the left of η_1, we apply a *SweepLeft* procedure similar to the *SweepRight* procedure above. The zips are applied on a vertical zipline $l_z^{q_1, q_2}$, where q_1 and q_2 are on the $\mathcal{S}$ boundary and $\mathcal{N}$ boundary, respectively, and the cookie track is on the right

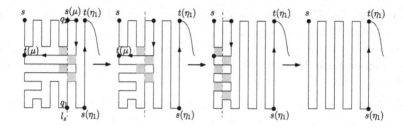

Fig. 10. The intermediate paths in the SWEEPLEFT procedure in P'.

and the side track is on the left of l_z. The zipline starts from Column $x(s(\eta_1)) - 2$ and finishes on Column 1, sweeping two columns westward after every zip except the last one. See Fig. 10. Otherwise, when there is an even number of columns covered by the initial subpath of P, the number of rows must be odd [2]. We apply a *SweepDown* procedure from Row 1 to Row $m - 2$, where q_1 is on the $\mathcal{W}$ boundary and q_2 is on η_1 for all the ziplines. The following lemma proves the correctness of procedure RECON_s_to_a'.

Lemma 4. *The initial subpath of P'' has either a run of $\mathcal{W}$ cookies or no cookies at all.*

Proof. If $x(s(\eta_1))$ of P' is even, it is easy to show that the number of corner separators cutting off s cannot be odd. Therefore, j for P' must be 0 by Lemma 3. Since the initial subpath covers a rectangular region from corner s to its diagonally opposite corner a', if the number of columns is even then the number of rows must be odd [2]. Then the SWEEPDOWN procedure grows a run of $\mathcal{W}$ cookies in a similar way as in the proof of Lemma 3. If $x(s(\eta_1))$ is odd, then there are two cases to consider based on the existence of a corner separator cutting off s in P'. In both cases, SWEEPLEFT works similarly to SWEEPDOWN if we flip the $P_{s,s(\eta_1)}$ subpath about the diagonal. □

Since Step (b) is very similar to Step (a), we now describe Step (c).

Step (c). Let the path obtained after Steps (a) and (b) be $\mathcal{P}$. If there exists any $\mathcal{E}$ or $\mathcal{W}$ cookies in $\mathcal{P}$, then SWEEPDOWN is applied on the whole path, where the starting point q_1 of each zipline is on the $\mathcal{W}$ boundary and the ending point q_2 is on the $\mathcal{E}$ boundary. The following theorem follows from Lemmas 3 and 4.

Theorem 2. *Algorithm* RECONFIGSIMP *reconfigures a simple path in a rectangular grid graph G to a canonical path of G in $O(|G|)$ time.*

Reconfiguring $\mathbb{P}$ to $\mathbb{P}'$: This step is similar to Step (c) of Algorithm RECONFIGSIMP; if $\mathbb{P}$ is $\mathcal{N}$-$\mathcal{S}$ then we grow horizontal straight separators by sweeping downward; otherwise, we grow vertical separators by sweeping eastward. We now have the following theorem.

Theorem 3. *Let $\mathbb{P}$ and $\mathbb{P}'$ be two different canonical paths of G. Then $\mathbb{P}$ can be reconfigured to $\mathbb{P}'$ using zips in $O(|G|)$ time.*

Main Algorithmic Result: We now state the central algorithmic result of our paper in the following theorem.

Theorem 4. *Let P and P' be two simple paths of a rectangular grid graph G. Then P can be reconfigured to P' using zips in $O(|G|)$ time.*

6 Conclusion

We have opened the exploration of reconfiguration of families of Hamiltonian paths in grid graphs. We have established the structure of any *simple* s, t Hamiltonian path in a rectangular grid graph G and given an $O(|G|)$ algorithm to reconfigure any such path to any other using *zip* operations. It would be interesting to find new families of s, t Hamiltonian paths, as shown in Fig. 11, and local operations that can reconfigure them.

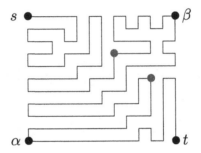

Fig. 11. An s, t Hamiltonian path, where the red vertices are connected to a boundary by at least 4 segments on the path. (Color figure online)

The problems about reconfiguring Hamiltonian paths with arbitrary end vertices remain open. Another future direction of research could be reconfiguration of Hamiltonian paths in d dimension, where $d \geq 3$.

References

1. Arkin, E.M., Bender, M.A., Demaine, E.D., Fekete, S.P., Mitchell, J.S.B., Sethia, S.: Optimal covering tours with turn costs. In: Proceedings of the Twelfth Annual ACM-SIAM Symposium on Discrete Algorithms, SODA 2001, pp. 138–147. Society for Industrial and Applied Mathematics, Philadelphia (2001)
2. Collins, K.L., Krompart, L.B.: The number of Hamiltonian paths in a rectangular grid. Discrete Math. **169**(1–3), 29–38 (1997)
3. Everett, H.: Hamiltonian paths in nonrectangular grid graphs. Master's thesis, University of Saskatchewan, Canada (1986)

4. Fellows, M., et al.: Milling a graph with turn costs: a parameterized complexity perspective. In: Thilikos, D.M. (ed.) WG 2010. LNCS, vol. 6410, pp. 123–134. Springer, Heidelberg (2010). https://doi.org/10.1007/978-3-642-16926-7_13

5. Gorbenko, A., Popov, V., Sheka, A.: Localization on discrete grid graphs. In: He, X., Hua, E., Lin, Y., Liu, X. (eds.) CICA 2011. LNEE, vol. 107, pp. 971–978. Springer, Dordrecht (2012). https://doi.org/10.1007/978-94-007-1839-5

6. Itai, A., Papadimitriou, C.H., Szwarcfiter, J.L.: Hamilton paths in grid graphs. SIAM J. Comput. **11**(4), 676–686 (1982)

7. Ito, T., et al.: On the complexity of reconfiguration problems. Theoret. Comput. Sci. **412**(12), 1054–1065 (2011)

8. Jacobsen, J.L.: Exact enumeration of Hamiltonian circuits, walks and chains in two and three dimensions. J. Phys. A Math. Gen. **40**, 14667–14678 (2007)

9. Keshavarz-Kohjerdi, F., Bagheri, A.: Hamiltonian paths in L-shaped grid graphs. Theoret. Comput. Sci. **621**, 37–56 (2016)

10. Lignos, I.: Reconfigurations of combinatorial problems: graph colouring and Hamiltonian cycle. Ph.D. thesis, Durham University (2017)

11. Muller, P., Hascoet, J.Y., Mognol, P.: Toolpaths for additive manufacturing of functionally graded materials (FGM) parts. Rapid Prototyping J. **20**(6), 511–522 (2014)

12. Nishat, R.I.: Reconfiguration of Hamiltonian cycles and paths in grid graphs. Ph.D. thesis, University of Victoria, Canada (2020)

13. Nishat, R.I., Whitesides, S.: Bend complexity and Hamiltonian cycles in grid graphs. In: Cao, Y., Chen, J. (eds.) COCOON 2017. LNCS, vol. 10392, pp. 445–456. Springer, Cham (2017). https://doi.org/10.1007/978-3-319-62389-4_37

14. Nishat, R.I., Whitesides, S.: Reconfiguring Hamiltonian cycles in L-shaped grid graphs. In: Sau, I., Thilikos, D.M. (eds.) WG 2019. LNCS, vol. 11789, pp. 325–337. Springer, Cham (2019). https://doi.org/10.1007/978-3-030-30786-8_25

15. Nishimura, N.: Introduction to reconfiguration. Algorithms **11**(4), 52 (2018)

16. Bodroza Pantić, O., Pantić, B., Pantić, I., Bodroza Solarov, M.: Enumeration of Hamiltonian cycles in some grid graphs. MATCH Commun. Math. Comput. Chem. **70**, 181–204 (2013)

17. Pettersson, V.: Enumerating Hamiltonian cycles. Electron. J. Comb. **21**(4), P4.7 (2014)

18. Ruskey, F., Sawada, J.: Bent Hamilton cycles in d-dimensional grid graphs. Electron. J. Comb. **10**(1), R1 (2003)

19. Takaoka, A.: Complexity of Hamiltonian cycle reconfiguration. Algorithms **11**(9), 140 (2018)

20. Umans, C., Lenhart, W.: Hamiltonian cycles in solid grid graphs. In: 38th Annual Symposium on Foundations of Computer Science, FOCS 1997, pp. 496–505 (1997)

21. Winter, S.: Modeling costs of turns in route planning. GeoInformatica **6**(4), 345–361 (2002)

New Approximations and Hardness Results for Submodular Partitioning Problems

Richard Santiago[✉]

ETH Zürich, Zürich, Switzerland
rtorres@ethz.ch

Abstract. We consider the following class of submodular k-multiway partitioning problems: (Sub-k-MP) min $\sum_{i=1}^{k} f(S_i) : S_1 \uplus S_2 \uplus \cdots \uplus S_k = V$ and $S_i \neq \emptyset$ for all $i \in [k]$. Here f is a non-negative submodular function, and $\uplus$ denotes the union of disjoint sets. Hence the goal is to partition V into k non-empty sets $S_1, S_2, \ldots, S_k$ such that $\sum_{i=1}^{k} f(S_i)$ is minimized. These problems were introduced by Zhao et al. partly motivated by applications to network reliability analysis, VLSI design, hypergraph cut, and other partitioning problems.

In this work we revisit this class of problems and shed some light on their hardness of approximation in the value oracle model. We provide new unconditional hardness results for Sub-k-MP in the special settings where the function f is either monotone or symmetric. We then extend Sub-k-MP to a larger class of partitioning problems, where the functions $f_i(S_i)$ can be different, and there is a more general partitioning constraint $S_1 \uplus S_2 \uplus \cdots \uplus S_k \in \mathcal{F}$ for some family $\mathcal{F} \subseteq 2^V$ of feasible sets. We provide a black box reduction that allows us to leverage several existing results from the literature; leading to new approximations for this class of problems.

1 Introduction

Submodularity is a property of set functions equivalent to the notion of diminishing returns. We say that a set function $f : 2^V \to \mathbb{R}$ is *submodular* if for any two sets $A \subseteq B \subseteq V$ and an element $v \notin B$, the corresponding marginal gains satisfy $f(A \cup \{v\}) - f(A) \geq f(B \cup \{v\}) - f(B)$. Submodular functions are a classical object in combinatorial optimization and operations research [20]. They arise naturally in many contexts such as set covering problems, cuts in graphs, and facility location problems. In recent years, they have found a wide range of applications in different computer science areas.

Since a submodular function is defined over an exponentially large domain, as is typical in the field we assume access to a *value oracle* that returns $f(S)$ for a given set S. A great variety of submodular maximization and minimization problems under a wide range of constraints have been considered in the literature. In this work, we are primarily interested in the following class of problems:

SUBMODULAR k-MULTIWAY PARTITIONING (SUB-k-MP): Given a non-negative submodular function $f : 2^V \to \mathbb{R}_+$, the goal is to partition V into

© Springer Nature Switzerland AG 2021
P. Flocchini and L. Moura (Eds.): IWOCA 2021, LNCS 12757, pp. 516–530, 2021.
https://doi.org/10.1007/978-3-030-79987-8_36

k non-empty sets $S_1, S_2, \ldots, S_k$ such that $\sum_{i=1}^{k} f(S_i)$ is minimized. That is,

$$(\text{SUB-}k\text{-MP}) \quad \min \sum_{i=1}^{k} f(S_i) : S_1 \uplus S_2 \uplus \cdots \uplus S_k = V \text{ and } S_i \neq \emptyset \text{ for all } i \in [k],$$

where we use $\uplus$ to denote the union of disjoint sets.

Special important cases occur when in addition the function f is either monotone or symmetric. We refer to those as MON-SUB-k-MP and SYM-SUB-k-MP respectively. Recall that a set function f is monotone if $f(A) \leq f(B)$ whenever $A \subseteq B \subseteq V$, and symmetric if $f(S) = f(V \setminus S)$ for any $S \subseteq V$.

In the absence of the non-emptyness constraints $S_i \neq \emptyset$, the problem is trivial since the partition $(V, \emptyset, \ldots, \emptyset)$ is always optimal by submodularity. However, although at first glance the non-emptyness constraints may seem inconspicuous, they lead to interesting models and questions in terms of tractability. We discuss this in more detail next.

These problems were introduced by Zhao, Nagamochi, and Ibaraki in [32] partly motivated by applications to hypergraph cut and partition problems. They mention how SUB-k-MP arises naturally in settings like network reliability analysis [31] and VLSI design [7]. They also discuss how this class captures several important problems as special cases. For instance, the well-studied GRAPH-k-CUT problem in graphs where the goal is to remove a subset of edges of minimum weight such that the remaining graph has at least k connected components. This problem is a special case of SYM-SUB-k-MP, where f corresponds to a cut function in a graph and hence it is symmetric and submodular. Another example is the more general HYPERGRAPH-k-CUT problem on hypergraphs, where the goal is to remove a subset of hyperedges of minimum weight such that the remaining hypergraph has at least k connected components. This problem is a special case of SUB-k-MP (see [4, 32] for further details).

The above class of submodular partitioning problems, however, is not as well understood. No hardness of approximation for these problems seems to be known under the standard $P \neq NP$ assumption. In fact, it is not known whether these problems are in P for fixed values of $k > 4$ (even in the simpler monotone and symmetric cases). We discuss this in more detail in Sect. 1.1.

One goal of this work is to revisit these problems and shed some light onto their hardness of approximation in the value oracle model. These hardness results are, thus, information theoretic. That is, limits on the approximability of a problem when only polynomially many queries to the value oracle are allowed. We provide new hardness results for SYM-SUB-k-MP and MON-SUB-k-MP.

A second goal is to extend SUB-k-MP to a more general class of problems and initiate the study of its tractability. This seems natural given that SUB-k-MP already captures fundamental problems such as GRAPH-k-CUT and HYPERGRAPH-k-CUT, and in addition, its complexity is not as well understood. We consider the class of problems given by

$$k\text{-WAY MA-Min}(\mathcal{F}) \quad \min \sum_{i=1}^{k} f_i(S_i) : S_1 \uplus \cdots \uplus S_k \in \mathcal{F} \text{ and } S_i \neq \emptyset \; \forall i \in [k],$$

where the functions f_i are all non-negative submodular and potentially different, and the family $\mathcal{F} \subseteq 2^V$ can be any collection of subsets of V. We denote this class by k-WAY MULTI-AGENT MINIMIZATION (k-WAY MA-Min). This is partially motivated by the work of Santiago and Shepherd [26] which considers the following class of multi-agent submodular minimization problems:

$$\text{MA-Min}(\mathcal{F}) \quad \min \sum_{i=1}^{k} f_i(S_i) : S_1 \uplus \cdots \uplus S_k \in \mathcal{F}.$$

We study the connections between these multi-agent problems and their k-WAY versions. In particular, we show that in many cases the approximation guarantees for MA-Min$(\mathcal{F})$ can be extended to the corresponding k-WAY versions at a small additional loss.

1.1   Related Work

Zhao et al. [32] show that a simple greedy splitting algorithm achieves a $(2-2/k)$-approximation for both MON-SUB-k-MP and SYM-SUB-k-MP (Queyranne [24] announced the same result for symmetric submodular functions), and a $(k-1)$-approximation for the more general SUB-k-MP. All these approximations hold for arbitrary (i.e., not necessarily fixed) values of k. Okumoto et al. [23] showed that SUB-k-MP is polytime solvable for $k = 3$, and Guiñez and Queyranne [14] showed that the symmetric version SYM-SUB-k-MP is polytime solvable for $k = 4$. We next discuss in more detail the cases where k is fixed (i.e., not part of the input) and when k is part of the input.

For fixed values of k, the GRAPH-k-CUT problem can be solved in polynomial time [13, 18]. In recent work Chandrasekaran et al. [3] gave a randomized polytime algorithm for HYPERGRAPH-k-CUT, whose complexity had remained an intriguing open problem even for fixed values of k. In subsequent work Chandrasekaran and Chekuri [2] gave a deterministic polytime algorithm. For the more general submodular multiway partitioning problems, Chekuri and Ene [4] gave a $(1.5 - 1/k)$-approximation for SYM-SUB-k-MP and a 2-approximation for SUB-k-MP. The latter was improved to $2 - 2/k$ by Ene et al. [9].

When k is part of the input GRAPH-k-CUT is NP-Hard [13]. Hence all the above problems are also NP-Hard (see full version for details about the monotone case). Moreover, the symmetric and general version are also APX-Hard since they generalize GRAPH-k-CUT. We note that GRAPH-k-CUT was claimed to be APX-Hard by Papadimitriou (see [28]), although a formal proof never appeared in the literature until the recent work of Manurangsi [22]. The latter gave conditional hardness by showing that assuming the Small Set Expansion Hypothesis, it is NP-hard to approximate GRAPH-k-CUT to within a $2 - \epsilon$ factor of the optimum for every constant $\epsilon > 0$. Chekuri and Li [6] give a simple reduction showing that an α-approximation for HYPERGRAPH-k-CUT implies an $O(\alpha^2)$-approximation for DENSEST-k-SUBGRAPH. This gives conditional hardness of approximation for HYPERGRAPH-k-CUT since the best known approximation for DENSEST-k-SUBGRAPH is $O(n^{1/4+\epsilon})$ [1], and Manurangsi [21] shows that assuming the Exponential Time Hypothesis there is no polynomial-time algorithm with an approximation factor of $n^{1/(\log \log n)^c}$ for some constant $c > 0$. The DENSEST-k-SUBGRAPH problem is believed to not admit an efficient constant factor approximation assuming $P \neq NP$. Since SUB-k-MP generalizes HYPERGRAPH-k-CUT, the above gives conditional hardness on SUB-k-MP.

The hardness of approximation for the class of submodular multiway partitioning problems, however, is not as well understood. No hardness of approximation for these problems seems to be known under the $P \neq NP$ assumption or under the value oracle model. In fact, it is not even known whether MON-SUB-k-MP, SYM-SUB-k-MP, or SUB-k-MP, are in P for fixed low values of k (read $k > 3$ for the monotone and general versions, and $k > 4$ for the symmetric version).

For the class of multi-agent minimization problems MA-Min($\mathcal{F}$), the special case where $\mathcal{F} = \{V\}$ is known as Minimum Submodular Cost Allocation and has been previously studied [5,8,15,30]. The works of Goel et al. [11] and Santiago and Shepherd [26] studied these problems under more general families. For a comprehensive review of multi-agent submodular optimization problems see [25].

We are not aware of previous work for the k-WAY MA-Min($\mathcal{F}$) class of problems besides the special case of SUB-k-MP. That is, the case with $\mathcal{F} = \{V\}$ and $f_i = f$ for all i.

1.2   Our Contributions

The contributions of this work are three-fold: new hardness results, a black box reduction, and new applications. We discuss each of these three blocks next.

In this work we initiate the study of the hardness of approximation in the value oracle model for different variants of SUB-k-MP. We provide the first unconditional hardness of approximation results for SYM-SUB-k-MP and MON-SUB-k-MP in the value oracle model. For the latter problem we are not aware of any previous (even conditional) hardness result. For SYM-SUB-k-MP our bound matches the (conditional) inapproximability factor of $2 - \epsilon$ from the work of [22]. See Sect. 2 for proof details and further discussion.

Theorem 1. *Given any $\epsilon > 0$, any algorithm achieving a $(2 - \epsilon)$-approximation for the* SYM-SUB-k-MP *problem when k is part of the input, requires exponentially many queries in the value oracle model.*

Theorem 2. *Given any $\epsilon > 0$, any algorithm achieving a $(\frac{4}{3} - \epsilon)$-approximation for the* MON-SUB-k-MP *problem when k is part of the input, requires exponentially many queries in the value oracle model.*

Our main algorithmic result is a black box procedure which, at a small additional loss, turns a solution for the multi-agent problem (i.e., MA-Min($\mathcal{F}$)) into a solution for the k-WAY version (i.e., k-WAY MA-Min($\mathcal{F}$)). We do this in the case where the objective functions f_i are non-negative and monotone, and for families $\mathcal{F}$ that are upwards closed (i.e., if $S \in \mathcal{F}$ and $T \supseteq S$ then $T \in \mathcal{F}$). The latter is a mild assumption given that the functions are monotone. Our guarantees are tight up to a small constant additive term.

Table 1. Comparison of some of our results with previous work.

	Mon-Sub-k-MP		Sym-Sub-k-MP		Mon-Sub-k-MP for vertex covers	
	Approx	Hardness	Approx	Hardness	Approx	Hardness
Known	$2 - 2/k$ [32]	–	$2 - 2/k$ [32]	$2 - \epsilon$ (conditional) [22]	–	$2 - \epsilon$ [11]
This work	2 (faster)	$4/3 - \epsilon$	–	$2 - \epsilon$ (unconditional)	3	–

Theorem 3. *Let $\mathcal{F}$ be an upwards closed family. Then an $\alpha(n, k)$-approximation for monotone MA-Min($\mathcal{F}$) implies an $(\alpha(n, k) + 1)$-approximation for monotone k-WAY MA-Min($\mathcal{F}$). In addition, there are instances where achieving an $(\alpha(n, k) + \frac{1}{3} - \epsilon)$-approximation requires exponentially many queries in the value oracle model for any $\epsilon > 0$.*

We remark that improving the above additive term would lead to an improvement of the best current approximation factor for Mon-Sub-k-MP (i.e., the setting where $\mathcal{F} = \{V\}$ and $f_i = f$ for all i). The multi-agent version of this problem: $\min \sum_{i=1}^{k} f(S_i) : S_1 \uplus \cdots \uplus S_k = V$, has a trivial 1-approximation by taking the partition $V, \emptyset, \ldots, \emptyset$, and hence by Theorem 3 we obtain a 2-approximation for the corresponding k-WAY version, i.e., for Mon-Sub-k-MP. This matches asymptotically the best known approximation of $2 - 2/k$ for this problem given in [32]. It is not known however whether this is tight.

The above black box result leads to interesting applications (see Sect. 3 for full details). For instance, the problem k-WAY MA-Min($\mathcal{F}$) with monotone functions f_i and where $\mathcal{F}$ corresponds to the family of vertex covers of a graph, admits a tight $O(\log n)$-approximation. Moreover, in the case where all the functions f_i are the same, this becomes a 3-approximation. To the best of our knowledge this is the current best approximation for this problem. The special case $k = 1$ corresponds to the submodular vertex cover problem studied in [11], where a hardness of $2 - \epsilon$ is shown in the value oracle model.

Corollary 1. *There is a 3-approximation for the problem $\min \sum_{i \in [k]} f(S_i) : S_1 \uplus \cdots \uplus S_k \in \mathcal{F}$ and $S_i \neq \emptyset$ for all $i \in [k]$, where f is non-negative monotone submodular and $\mathcal{F}$ is the family of vertex covers of a graph.*

Another direct consequence of Theorem 3 is providing a very simple 2-approximation for Mon-Sub-k-MP. The argument in fact shows that one very specific partition achieves the desired bound. In addition to simple, the procedure to build such a partition is also fast. Indeed, the running time of the $(2 - 2/k)$-approximation algorithm provided by Zhao et al. [32] is $kn^3 EO$, where EO denotes the time that a call to the value oracle takes. On the other hand, the running time of this procedure is $O(n \, EO + n \log n)$ and hence almost linear.

Corollary 2. *There is a 2-approximation algorithm for Mon-Sub-k-MP running in time $O(n \, EO + n \log n)$, where EO denotes the time that a value oracle call takes.*

We compare some of our results with previous work in Table 1.

2 Hardness Results in the Value Oracle Model

In this section we provide the first unconditional hardness of approximation results in the value oracle model for SYM-SUB-k-MP and MON-SUB-k-MP. In addition, for MON-SUB-k-MP we are not aware of any (even conditional) hardness of approximation result previous to this work.

Our results are based on the technique of building two functions that are hard to distinguish with high probability for any (even randomized) algorithm. This was first used in the work of Goemans et al. [12], and has since then been used in several subsequent works [10,11,16,27,29].

2.1 A 2-Factor Inapproximability Oracle Hardness for SYM-SUB-k-MP

The current best known (conditional) hardness of approximation for SYM-SUB-k-MP follows from the result of Manurangsi [22], where it is shown that assuming the Small Set Expansion Hypothesis, it is NP-hard to approximate GRAPH-k-CUT to within a $2 - \epsilon$ factor of the optimum for every constant $\epsilon > 0$. Since SYM-SUB-k-MP generalizes GRAPH-k-CUT, the same conditional hardness of approximation automatically applies. In this section we prove an unconditional lower bound hardness for SYM-SUB-k-MP in the value oracle model (Theorem 1). To the best of our knowledge this is the first result of this kind for this problem.

To prove the desired result, we first build two indistinguishable functions as follows. Let $|V| = n$ be an even number, R be a random set of size $\frac{n}{2}$, and $\bar{R}$ denote its complement. Define parameters $\epsilon^2 = \frac{1}{n}\omega(\ln n)$ and $\beta = \frac{n}{4}(1 + \epsilon)$, such that β is an integer. Consider the functions

$$f_1(S) = \min\left\{|S|, \frac{n}{2}\right\} - \frac{|S|}{2} \text{ and } f_2(S) = \min\left\{|S|, \frac{n}{2}, \beta + |S \cap R|, \beta + |S \cap \bar{R}|\right\} - \frac{|S|}{2}.$$

These functions were already used in the work of Svitkina and Fleischer [29] to prove polynomial hardness of approximation for the submodular sparsest cut and submodular balanced cut problems. They show that the above two functions are non-negative symmetric submodular and hard to distinguish. That is, any (even randomized) algorithm that makes a polynomial number of oracle queries has probability at most $n^{-\omega(1)}$ of distinguishing the functions f_1 and f_2.

We use this to show hardness of approximation for SYM-SUB-k-MP as follows.

Claim. Consider the SYM-SUB-k-MP problem with $k = \frac{n}{2} + 1$ and inputs f_1 and f_2. Then, if the input is f_1 any feasible solution has objective value at least $\frac{n}{2}$, while if the input is f_2 then the optimal value is at most $\frac{n}{4}(1 + \epsilon)$.

Proof. Since we have n elements that must be split into $\frac{n}{2} + 1$ non-empty sets, no more than $\frac{n}{2}$ items can be assigned to any given set. That is, for any feasible solution $S_1, S_2, \ldots, S_k$ we must have that $|S_i| \leq \frac{n}{2}$. It then follows that when the input is f_1, any feasible solution $S_1, S_2, \ldots, S_k$ has objective value exactly $\sum_{i=1}^{k} f_1(S_i) = \sum_{i=1}^{k} \left(|S_i| - \frac{|S_i|}{2} \right) = \frac{n}{2}$.

On the other hand, when the input is f_2, a feasible solution is given by taking $S_1 = \bar{R}, S_2 = \{r_1\}, S_3 = \{r_2\}, \ldots, S_k = \{r_{n/2}\}$, where $R = \{r_1, r_2, \ldots, r_{n/2}\}$. This has objective value

$$\sum_{i=1}^{k} f_2(S_i) = f_2(\bar{R}) + \sum_{i=2}^{k} f_2(r_{i-1}) = \left(\beta - \frac{n}{4} \right) + \sum_{i=2}^{k} \left(1 - \frac{1}{2} \right)$$

$$= \beta - \frac{n}{4} + \frac{n}{4} = \beta = \frac{n}{4}(1 + \epsilon).$$

$\square$

From the above result it follows that the gap between the optimal solutions for SYM-SUB-k-MP when the inputs are f_1 and f_2 is at least

$$\frac{OPT_1}{OPT_2} \geq \frac{\frac{n}{2}}{\frac{n}{4}(1 + \epsilon)} = \frac{2}{1 + \epsilon}.$$

Since $\epsilon = o(1)$ this gap can be arbitrarily close to 2 for large values of n. Given that f_1 and f_2 are hard to distinguish, this now leads to Theorem 1.

Proof [Theorem 1]. Assume there is an algorithm that makes polynomially many queries to the value oracle and that achieves a $(2 - \delta)$-approximation for SYM-SUB-k-MP for some constant $\delta > 0$. Let the functions f_1 and f_2 be as defined above, and choose n and the parameter $\epsilon(n)$ so that $(1 + \epsilon(n))(2 - \delta) < 2$, i.e., so that $\epsilon(n) < \delta/(2 - \delta)$. Since $\epsilon(n) = o(1)$ and δ is a constant, this is always possible.

Consider the output of the algorithm when the input is f_2. By Claim 2.1 in this case the optimal solution is at most $\frac{n}{4}(1 + \epsilon)$, and hence the algorithm finds a feasible solution $(S_1, S_2, \ldots, S_k)$ such that $\sum_{i \in [k]} f_2(S_i) \leq (2 - \delta)(1 + \epsilon)\frac{n}{4} < \frac{n}{2}$, where the last inequality follows from the choice of ϵ. However, there is no feasible solution $(S_1', S_2', \ldots, S_k')$ such that $\sum_{i \in [k]} f_1(S_i') < \frac{n}{2}$, since by Claim 2.1 any feasible solution for f_1 has value at least $n/2$. That means that if the input is the function f_1 the algorithm outputs a different answer, thus distinguishing between f_1 and f_2. A contradiction. $\square$

2.2 A 4/3-Inapproximability Oracle Hardness for MON-SUB-k-MP

In this section we prove an unconditional lower bound hardness of approximation for MON-SUB-k-MP in the value oracle model (Theorem 2). To the best of our knowledge, this is the first hardness of approximation result (either conditional or unconditional) for MON-SUB-k-MP. As discussed in the full version of this work, the conditional hardness of approximation for GRAPH-k-CUT does not extend to MON-SUB-k-MP, since the objective function in that case must take negative values.

The argument is similar to the one from Sect. 2.1. Consider the functions

$$f_3(S) = \min\left\{|S|, \frac{n}{2}\right\} \quad \text{and} \quad f_4(S) = \min\left\{|S|, \frac{n}{2}, \beta + |S \cap R|, \beta + |S \cap \bar{R}|\right\},$$

where all the parameters are as defined in Sect. 2.1. Note that $f_3 = f_1 + g$ and $f_4 = f_2 + g$, where $g(S) = |S|/2$. Since both f_1 and f_2 are submodular, and g is modular, it follows that f_3 and f_4 are also submodular. Moreover, it is straightforward to check that both f_3 and f_4 are also non-negative and monotone.

Since f_1 and f_2 are hard to distinguish, and $f_3(S) \neq f_4(S)$ if and only if $f_1(S) \neq f_2(S)$, it follows that f_3 and f_4 are also hard to distinguish.

The following result shows the gap between the optimal solutions of the corresponding problems.

Claim. Consider the MON-SUB-k-MP problem with $k = \frac{n}{2} + 1$ and inputs f_3 and f_4. Then, if the input is f_3 any feasible solution has objective value at least n, while if the input is f_4 then the optimal value is at most $\frac{3+\epsilon}{4}n$.

Proof. The argument is very similar to that of Claim 2.1. Since any feasible solution $S_1, S_2, \ldots, S_k$ must satisfy $|S_i| \leq \frac{n}{2}$ for all i, it then follows that when the input is f_3, any feasible solution $S_1, S_2, \ldots, S_k$ has objective value exactly $\sum_{i=1}^{k} f_3(S_i) = \sum_{i=1}^{k} |S_i| = n$.

On the other hand, when the input is f_4, a feasible solution is given by $S_1 = \bar{R}, S_2 = \{r_1\}, S_3 = \{r_2\}, \ldots, S_k = \{r_{n/2}\}$, where $R = \{r_1, r_2, \ldots, r_{n/2}\}$. This has objective value $\sum_{i=1}^{k} f_4(S_i) = \frac{n}{4}(1 + \epsilon) + \frac{n}{2} = \frac{3+\epsilon}{4}n$. $\qquad\square$

It follows that the gap between the optimal solutions for MON-SUB-k-MP when the inputs are f_3 and f_4 is at least $OPT_3/OPT_4 \geq 4/(3+\epsilon)$. Since $\epsilon = o(1)$ this gap can be arbitrarily close to $4/3$ for large values of n. Given that f_3 and f_4 are hard to distinguish, this now leads to Theorem 2.

Proof [Theorem 2]. Assume there is an algorithm that makes polynomially many queries to the value oracle and that achieves a $(4/3 - \delta)$-approximation for MON-SUB-k-MP for some constant $\delta > 0$. Let the functions f_3 and f_4 be as defined above, and choose n and the parameter $\epsilon(n)$ so that $(\frac{3+\epsilon}{4})(\frac{4}{3} - \delta) < 1$, i.e., so that $\epsilon < \frac{9\delta}{4-3\delta}$. Since $\epsilon(n) = o(1)$ and δ is a constant, this can always be done.

Consider the output of the algorithm when the input is f_4. By Claim 2.2 in this case the optimal solution is at most $\frac{3+\epsilon}{4}n$, and hence the algorithm finds a feasible solution $(S_1, S_2, \ldots, S_k)$ such that $\sum_{i \in [k]} f_4(S_i) \leq (\frac{4}{3} - \delta)\frac{3+\epsilon}{4}n < n$, where the last inequality follows from the choice of ϵ. However, there is no feasible solution $(S_1', S_2', \ldots, S_k')$ such that $\sum_{i \in [k]} f_3(S_i') < n$, since by Claim 2.2 any feasible solution for f_3 has value at least n. That means that if the input is the function f_3 the algorithm outputs a different answer, thus distinguishing between f_3 and f_4. A contradiction. $\qquad\square$

3   From Multi-agent Minimization to the k-WAY versions

In this section we show that if the functions f_i are monotone, then a feasible solution to the MA-Min($\mathcal{F}$) problem can be turned into a feasible solution to the corresponding k-WAY version (i.e., k-WAY MA-Min($\mathcal{F}$)) at almost no additional loss. Moreover, our argument is completely black box with respect to how the approximation for the MA-Min instance is obtained (i.e., it could be via a greedy algorithm, a continuous relaxation, or any other kind of approach). We show the following.

Theorem 3. *Let $\mathcal{F}$ be an upwards closed family. Then an $\alpha(n, k)$-approximation for monotone MA-Min($\mathcal{F}$) implies an $(\alpha(n, k) + 1)$-approximation for monotone k-WAY MA-Min($\mathcal{F}$). In addition, there are instances where achieving an $(\alpha(n, k) + \frac{1}{3} - \epsilon)$-approximation requires exponentially many queries in the value oracle model for any $\epsilon > 0$.*

Proof. Denote by OPT the value of the optimal solution to the k-WAY problem and by $\overline{OPT}$ the value of the optimal solution to MA-Min($\mathcal{F}$). Then it is clear that $\overline{OPT} \leq OPT$ since any feasible solution for the k-WAY version is also feasible for MA-Min($\mathcal{F}$).

Let $G = ([k] \uplus V, E)$ denote the complete bipartite graph where the weight of an edge (i, v) is given by $f_i(v)$. Let M be a minimum $[k]$-saturating matching in G, that is a minimum cost matching such that every node in $[k]$ gets assigned at least one element. Since the edges have non-negative weights it is clear that $|M| = k$, i.e., each node $i \in [k]$ gets assigned exactly one element from V. Denote the edges of the matching by $M = \{(1, u_1), (2, u_2), \ldots, (k, u_k)\}$, and let $U := \{u_1, u_2, \ldots, u_k\}$ be the elements in V that M is incident to.

We then have that the cost of M is at most OPT. Indeed, if $(S_1^*, S_2^*, \ldots, S_k^*)$ is an optimal solution to the k-WAY instance, we can remove elements from the sets S_i^* arbitrarily until each of the sets consists of exactly one element. By monotonicity, removing elements can only decrease the objective value of the solution. Moreover, since now each set consists of exactly one element, this is a feasible $[k]$-saturating matching, and hence its cost is at least the cost of M. That is, $\sum_{i \in [k]} f_i(u_i) \leq OPT$.

Let $(S_1, S_2, \ldots, S_k)$ be an α-approximation for the MA-Min($\mathcal{F}$) instance. Then we have $\sum_{i \in [k]} f_i(S_i) \leq \alpha \cdot \overline{OPT} \leq \alpha \cdot OPT$. We combine this solution with the matching M by defining new sets $S_i' := (S_i \setminus U) \uplus \{u_i\}$ for each $i \in [k]$. It is clear that this is now a feasible solution to the k-WAY problem since all the sets S_i' are non-empty and pairwise disjoint, and their union $\cup_{i \in [k]} S_i' = U \cup \left(\cup_{i \in [k]} S_i\right)$ belongs to $\mathcal{F}$ since $\cup_{i \in [k]} S_i \in \mathcal{F}$ and $\mathcal{F}$ is upwards closed. Moreover, the cost of the new solution is given by

$$\sum_{i \in [k]} f_i(S_i') = \sum_{i \in [k]} f_i(S_i \setminus U + u_i) \leq \sum_{i \in [k]} f_i(S_i \setminus U) + \sum_{i \in [k]} f_i(u_i)$$

$$\leq \sum_{i \in [k]} f_i(S_i) + \sum_{i \in [k]} f_i(u_i) \leq \alpha \cdot OPT + OPT = (\alpha + 1) \cdot OPT,$$

where the first inequality follows from subadditivity (since the functions are non-negative submodular) and the second inequality from monotonicity.

For the inapproximability result part, consider the family $\mathcal{F} = \{V\}$ and the setting where all the functions f_i are the same. Then the corresponding MA-Min($\mathcal{F}$) problem has a trivial 1-approximation, while the k-WAY version corresponds to MON-SUB-k-MP. The latter, by Theorem 2, cannot be approximated in the value oracle model to a factor of $(\frac{4}{3} - \epsilon)$ for any $\epsilon > 0$ without making exponentially many queries. It follows that for these instances, the MA-Min version has an exact solution while the k-WAY versions have an inapproximability lower bound of $4/3 - \epsilon = 1 + 1/3 - \epsilon$. Hence, there are instances where for any $\epsilon > 0$, achieving an $(\alpha(n, k) + \frac{1}{3} - \epsilon)$-approximation requires exponentially many queries in the value oracle model. This completes the proof.   □

For proving or improving the result from Theorem 3, one could be tempted to first compute an optimal (or approximate) solution $(S_1, \ldots, S_k)$ to the MA-Min($\mathcal{F}$) problem, and then find an allocation of some of the elements of $F :=$ $\biguplus_{i \in [k]} S_i$ among the agents that did not get any item. However, this approach can lead to a large additional loss, since a set $F \in \mathcal{F}$ could be optimal for the MA-Min problem but highly suboptimal for the k-WAY version. The following example shows this, even for the case of modular functions.

Let $T \subsetneq V$ be an arbitrary set of size $2(k - 1)$. Let $f_1(S) = |S|$ and $f_i(S) = w(S)$ for all $i \geq 2$ where $w : V \to \mathbb{R}_+$ is the modular function given by $w(v) = 1 + \epsilon$ for $v \notin T$ and $w(v) = M$ for $v \in T$, for some large value M. Moreover, let $\mathcal{F} = \{S : |S| \geq 2(k - 1)\}$. Then a feasible (and optimal) solution to the MA-Min($\mathcal{F}$) problem is given by the allocation $(T, \emptyset, \ldots, \emptyset)$ with objective value $f_1(T) = |T| = 2(k - 1)$. However, any splitting of some of the items of T among the other $k - 1$ agents leads to a solution of cost at least $M(k - 1) + (k - 1) = (M + 1)(k - 1)$. On the other hand, an optimal solution for the k-WAY version is given by any partition of the form $(S_1, \{v_2\}, \{v_3\}, \ldots, \{v_k\})$ where $S_1 \subseteq V$ is any set of $k - 1$ elements, and $\{v_2, v_3, \ldots, v_k\} \subseteq V \setminus T$. This leads to a solution of cost $(k - 1) + (1 + \epsilon)(k - 1) = (2 + \epsilon)(k - 1)$. Thus having a gap in terms of objective value of at least $\frac{(M+1)(k-1)}{(2+\epsilon)(k-1)} = \frac{M+1}{2+\epsilon}$.

Theorem 3 allows us to extend several results from monotone multi-agent minimization to the k-WAY versions. We discuss some of these consequences next. An interesting application is obtained using the $O(\log n)$-approximation from [30].

Corollary 3. *There is a tight $O(\log n)$-approximation for the allocation problem*

$$\min \sum_{i=1}^{k} f_i(S_i) : S_1 \uplus S_2 \uplus \cdots \uplus S_k = V \text{ and } S_i \neq \emptyset \text{ for all } i \in [k],$$

where all the functions f_i are non-negative monotone submodular.

Proof. The approximation factor follows from Theorem 3 and the tight $O(\log n)$-approximation [30] for the corresponding MA-Min instance. To see why this is tight assume that an (asymptotically) better approximation factor of $o(\log n)$ is possible. Then given a MA-Min instance we can reduce it to an instance of the k-WAY version by adding a set $D := \{d_1, d_2, \ldots, d_k\}$ of k dummy elements to the ground set. That is, consider an instance of the k-WAY version with $V' := V \cup D$, $\mathcal{F}' = \{V'\}$, and $f_i'(S) := f_i(S \cap V)$ for each $i \in [k]$ and $S \subseteq V'$. Then by assumption we have a $o(\log(n + k))$-approximation for this problem, and hence we also have the same approximation factor for the original MA-Min instance. But this contradicts the lower bound of $\Omega(\log n)$ for the MA-Min problem. $\square$

More generally, we can obtain approximation guarantees for families with a bounded blocker. Given a family $\mathcal{F}$, there is an associated blocking clutter $\mathcal{B}(\mathcal{F})$ which consists of the minimal sets B such that $B \cap F \neq \emptyset$ for each $F \in \mathcal{F}$. We refer to $\mathcal{B}(\mathcal{F})$ as the blocker of $\mathcal{F}$. We say that $\mathcal{B}(\mathcal{F})$ is β-bounded if $|B| \leq \beta$ for all $B \in \mathcal{B}(\mathcal{F})$. Families such as $\mathcal{F} = \{V\}$ or vertex covers in a graph, are examples of families with a bounded blocker. Indeed, the family $\mathcal{F} = \{V\}$ has a 1-bounded blocker, since $\mathcal{B}(\mathcal{F}) = \{\{v_1\}, \{v_2\}, \ldots, \{v_n\}\}$. The family $\mathcal{F}$ of vertex covers of a graph G has a 2-bounded blocker, since $\mathcal{B}(\mathcal{F}) = \{\{u, v\} : (u, v) \text{ is an edge in } G\}$. Recall that a set $S \subseteq V$ is a vertex cover in a graph G if every edge in G is incident on a vertex in S.

It is shown in [26] that families with a β-bounded blocker admit a $O(\beta \log n)$-approximation for the multi-agent monotone minimization problem. This, combined with Theorem 3, implies a $O(\beta \log n)$-approximation for the k-WAY versions. In particular, this leads to a tight $O(\log n)$-approximation for the k-WAY MA-Min$(\mathcal{F})$ problem with monotone functions f_i and where $\mathcal{F}$ corresponds to the family of vertex covers of a graph. The tightness follows from Corollary 3 and the fact that vertex covers generalize the family $\mathcal{F} = \{V\}$.

3.1   The Special Case Where all the Functions f_i are the Same

Theorem 3 also leads to interesting consequences in the special case where $f_i = f$ for all i. In that setting, it is easy to see that the single-agent and multi-agent versions are equivalent. That is,

$$\min f(S) : S \in \mathcal{F} \quad = \quad \min \sum_{i \in [k]} f(S_i) : S_1 \uplus S_2 \uplus \cdots \uplus S_k \in \mathcal{F}, \qquad (1)$$

and moreover $F \in \mathcal{F}$ is an optimal solution to the single-agent problem if and only if the trivial partition $(F, \emptyset, \ldots, \emptyset)$ is an optimal solution to the multi-agent version. Again this just follows from submodularity and non-negativity since then $f(T) \leq f(S) + f(T - S)$ for any $S \subseteq T \subseteq V$. That is, partitioning the elements of a set can only increase the value of the solution. This leads to the following result.

Corollary 4. *Let $\mathcal{F}$ be any upwards closed family, and assume there is an $\alpha(n)$-approximation for the single-agent monotone minimization problem: $\min f(S)$: $S \in \mathcal{F}$. If $f_i = f$ for all i, then there is an $\alpha(n)$-approximation for monotone MA-Min$(\mathcal{F})$, and hence an $(\alpha(n) + 1)$-approximation for monotone k-WAY MA-Min$(\mathcal{F})$.*

Algorithm 1: Simpler and faster algorithm for MON-SUB-k-MP

Input: A ground set $V = \{v_1, v_2, \ldots, v_n\}$, and a set function $f : 2^V \to \mathbb{R}$ with oracle access.

Sort and rename the elements so that $f(v_1) \leq f(v_2) \leq \ldots \leq f(v_n)$.

$S_1 \leftarrow \{v_1\}$, $S_2 \leftarrow \{v_2\}$, $\ldots$, $S_{k-1} \leftarrow \{v_{k-1}\}$, $S_k \leftarrow V \setminus \{v_1, v_2, \ldots, v_{k-1}\}$.

Output: $(S_1, S_2, \ldots, S_k)$

Proof. By Eq. (1), an $\alpha(n)$-approximation for the single-agent monotone minimization problem implies an $\alpha(n)$-approximation for monotone MA-Min($\mathcal{F}$) in the setting where $f_i = f$ for all i. Now the result for the corresponding k-WAY versions immediately follows from Theorem 3. □

We remark that by taking $\mathcal{F} = \{V\}$ the above corollary leads to a 2-approximation for MON-SUB-k-MP, which matches asymptotically the currently best known. Hence improving the plus one additive term would lead to an improvement on the approximation factor of the latter problem.

Corollary 4 leads to new results. For instance, using the 2-approximation from [11,16] for single-agent monotone minimization over families of vertex covers, we immediately get a 3-approximation for the corresponding monotone k-WAY MA-Min($\mathcal{F}$) problem over the same type of families. This now proves Corollary 1. More generally, given the β-approximation results ([17,19]) for minimizing a monotone submodular function over families with a β-bounded blocker, we have the following.

Corollary 5. *Let $\mathcal{F}$ be an upwards closed family with a β-bounded blocker. Then there is a $(\beta + 1)$-approximation algorithm for the problem $\min \sum_{i \in [k]} f(S_i)$: $S_1 \uplus \cdots \uplus S_k \in \mathcal{F}$ and $S_i \neq \emptyset$ for all $i \in [k]$, where f is a non-negative monotone submodular function.*

3.2 A Simpler and Faster 2-Approximation for MON-SUB-k-MP

Another direct consequence of Theorem 3 is to provide a very simple and fast 2-approximation for MON-SUB-k-MP. We describe the procedure in Algorithm 1. The running time of the $(2 - 2/k)$-approximation algorithm provided by Zhao et al. [32] is $kn^3 EO$, where EO denotes the time that a call to the value oracle takes. On the other hand, the running time of our procedure is $O(n\,EO + n\log n)$ and hence almost linear. All we need to do is first make n oracle calls to evaluate $f(v)$ for each $v \in V$, and then sort the elements so that $f(v_1) \leq f(v_2) \leq \ldots \leq f(v_n)$ (which requires $O(n\log n)$ time).

Corollary 2. *Algorithm 1 is a 2-approximation for MON-SUB-k-MP running in time $O(n\,EO + n\log n)$, where EO denotes the time that a value oracle call takes.*

4   Conclusion and Open Problems

We revisited the class of Submodular k-Multiway Partitioning problems (SUB-k-MP). We proved new unconditional inapproximability results for the monotone and symmetric cases of SUB-k-MP in the value oracle model.

We introduced and explored a new class of submodular partitioning problems which generalizes SUB-k-MP. We showed that several results from multi-agent submodular minimization can be extended to their k-WAY counterparts at a small additional loss. Thus obtaining several new results for this class of problems.

Many interesting open questions remain, perhaps the most important being about the approximation hardness of SUB-k-MP. It remains completely open whether these problems are polytime solvable for fixed values of $k > 4$. In addition, given the conditional hardness of approximation for HYPERGRAPH-k-CUT based on DENSEST-k-SUBGRAPH, we believe it may be possible to prove strong unconditional hardness of approximation results for SUB-k-MP in the value oracle model when k is part of the input.

It also remains open whether the 2-approximation for MON-SUB-k-MP is tight. And more generally, to close the gap between the upper bound and lower bound in Theorem 3.

Acknowledgements. The author thanks Bruce Shepherd for valuable discussions and suggestions that motivated some of this work.

References

1. Bhaskara, A., Charikar, M., Chlamtac, E., Feige, U., Vijayaraghavan, A.: Detecting high log-densities: an $o(n^{1/4})$ approximation for densest k-subgraph. In: Proceedings of the Forty-second ACM Symposium on Theory of Computing, pp. 201–210. ACM (2010)
2. Chandrasekaran, K., Chekuri, C.: Hypergraph k-cut for fixed k in deterministic polynomial time. In: 2020 IEEE 61st Annual Symposium on Foundations of Computer Science (FOCS), pp. 810–821 (2020)
3. Chandrasekaran, K., Xu, C., Yu, X.: Hypergraph k-cut in randomized polynomial time. Math. Program. 1–29 (2019)
4. Chekuri, C., Ene, A.: Approximation algorithms for submodular multiway partition. In: 2011 IEEE 52nd Annual Symposium on Foundations of Computer Science (FOCS), pp. 807–816. IEEE (2011)
5. Chekuri, C., Ene, A.: Submodular cost allocation problem and applications. In: International Colloquium on Automata, Languages, and Programming, pp. 354–366. Springer (2011). Extended version: arXiv preprint arXiv:1105.2040
6. Chekuri, C., Li, S.: On the hardness of approximating the k-way hypergraph cut problem. Theory Comput. **16**(1), 1–8 (2020)
7. Chopra, S., Owen, J.H.: A note on formulations for the a-partition problem on hypergraphs. Discrete Appl. Math. **90**(1–3), 115–133 (1999)
8. Ene, A., Vondrák, J.: Hardness of submodular cost allocation: lattice matching and a simplex coloring conjecture. In: Approximation, Randomization, and Combinatorial Optimization. Algorithms and Techniques (APPROX/RANDOM 2014), vol. 28, pp. 144–159 (2014)

9. Ene, A., Vondrák, J., Wu, Y.: Local distribution and the symmetry gap: approximability of multiway partitioning problems. In: Proceedings of the Twenty-Fourth Annual ACM-SIAM Symposium on Discrete Algorithms, pp. 306–325. SIAM (2013)

10. Feige, U., Mirrokni, V.S., Vondrak, J.: Maximizing non-monotone submodular functions. SIAM J. Comput. **40**(4), 1133–1153 (2011)

11. Goel, G., Karande, C., Tripathi, P., Wang, L.: Approximability of combinatorial problems with multi-agent submodular cost functions. In: 50th Annual IEEE Symposium on Foundations of Computer Science, FOCS 2009, pp. 755–764. IEEE (2009)

12. Goemans, M.X., Harvey, N.J., Iwata, S., Mirrokni, V.: Approximating submodular functions everywhere. In: Proceedings of the Twentieth Annual ACM-SIAM Symposium on Discrete Algorithms, pp. 535–544. Society for Industrial and Applied Mathematics (2009)

13. Goldschmidt, O., Hochbaum, D.S.: A polynomial algorithm for the k-cut problem for fixed k. Math. Oper. Res. **19**(1), 24–37 (1994)

14. Guinez, F., Queyranne, M.: The size-constrained submodular k-partition problem (2012). https://docs.google.com/viewer?a=v&pid=sites&srcid=ZGVmYXVsdGR vbWFpbnxmbGF2aW9ndWluZXpob21lcGFnZXxneneDo0NDVlMThkMDg4ZWRl OGI1

15. Hayrapetyan, A., Swamy, C., Tardos, É.: Network design for information networks. In: Proceedings of the Sixteenth Annual ACM-SIAM Symposium on Discrete Algorithms, pp. 933–942. Society for Industrial and Applied Mathematics (2005)

16. Iwata, S., Nagano, K.: Submodular function minimization under covering constraints. In: 2009 50th Annual IEEE Symposium on Foundations of Computer Science, FOCS 2009, pp. 671–680. IEEE (2009)

17. Iyer, R., Jegelka, S., Bilmes, J.: Monotone closure of relaxed constraints in submodular optimization: Connections between minimization and maximization: Extended version (2014)

18. Karger, D.R., Stein, C.: A new approach to the minimum cut problem. J. ACM (JACM) **43**(4), 601–640 (1996)

19. Koufogiannakis, C., Young, N.E.: Greedy δ-approximation algorithm for covering with arbitrary constraints and submodular cost. Algorithmica **66**(1), 113–152 (2013)

20. Lovász, L.: Submodular functions and convexity. In: Bachem, A., Korte, B., Grotschel, M. (eds.) Mathematical Programming The State of the Art, pp. 235–257. Springer, Heidelberg (1983). https://doi.org/10.1007/978-3-642-68874-4_10

21. Manurangsi, P.: Almost-polynomial ratio eth-hardness of approximating densest k-subgraph. In: Proceedings of the 49th Annual ACM SIGACT Symposium on Theory of Computing, pp. 954–961 (2017)

22. Manurangsi, P.: Inapproximability of maximum biclique problems, minimum k-cut and densest at-least-k-subgraph from the small set expansion hypothesis. Algorithms **11**(1), 10 (2018)

23. Okumoto, K., Fukunaga, T., Nagamochi, H.: Divide-and-conquer algorithms for partitioning hypergraphs and submodular systems. Algorithmica **62**(3–4), 787–806 (2012)

24. Queyranne, M.: On optimum size-constrained set partitions. Presented at 3rd Combinatorial Optimization Workshop, AUSSOIS 1999, France (1999). http://www.iasi.cnr.it/iasi/aussois99/queyranne.html

25. Santiago, R.: Multi-Agent submodular optimization: variations and generalizations. Ph.D. thesis, McGill University (2019)

26. Santiago, R., Shepherd, F.B.: Multi-agent submodular optimization. In: Approximation, Randomization, and Combinatorial Optimization. Algorithms and Techniques (APPROX/RANDOM 2018), vol. 116, pp. 23:1–23:20 (2018)

27. Santiago, R., Shepherd, F.B.: Multivariate submodular optimization. In: Proceedings of the 36th International Conference on Machine Learning (ICML). pp. 5599–5609 (2019), http://proceedings.mlr.press/v97/santiago19a.html

28. Saran, H., Vazirani, V.V.: Finding k cuts within twice the optimal. SIAM J. Comput. **24**(1), 101–108 (1995)

29. Svitkina, Z., Fleischer, L.: Submodular approximation: sampling-based algorithms and lower bounds. SIAM J. Comput. **40**(6), 1715–1737 (2011)

30. Svitkina, Z., Tardos, É.: Facility location with hierarchical facility costs. ACM Trans. Algorithms (TALG) **6**(2), 37 (2010)

31. Zhao, L.: Approximation algorithms for partition and design problems in networks. Ph.D. Thesis (2002)

32. Zhao, L., Nagamochi, H., Ibaraki, T.: Greedy splitting algorithms for approximating multiway partition problems. Math. Program. **102**(1), 167–183 (2005)

An FPT Algorithm for Matching Cut
and d-Cut

N. R. Aravind and Roopam Saxena[✉]

Department of Computer Science and Engineering, IIT Hyderabad, Hyderabad, India
{aravind,cs18resch11004}@iith.ac.in

Abstract. For a positive integer d, the d-CUT is the problem of deciding if an undirected graph $G = (V, E)$ has a nontrivial bipartition (A, B) of V such that every vertex in A (resp. B) has at most d neighbors in B (resp. A). When $d = 1$, this is the MATCHING CUT problem. Gomes and Sau [9] gave the first fixed-parameter tractable algorithm for d-CUT parameterized by the maximum number of the crossing edges in the cut (i.e., the size of edge cut). However, their paper does not provide an explicit bound on the running time, as it indirectly relies on an MSOL formulation and Courcelle's Theorem [5]. Motivated by this, we design and present an FPT algorithm for the MATCHING CUT (and more generally for d-CUT) for general graphs with running time $2^{O(k \log k)} n^{O(1)}$ where k is the maximum size of the edge cut. This is the first FPT algorithm for the MATCHING CUT (and d-CUT) with an explicit dependence on this parameter. We also observe that MATCHING CUT cannot be solved in $2^{o(k)} n^{O(1)}$ unless ETH fails.

Keywords: Matching cut · Fixed-parameter tractable · Algorithms

1 Introduction

For a graph $G = (V, E)$, (A, B) is a partition of G if $A \cup B = V$ and $A \cap B = \emptyset$. Further, if both $A \neq \emptyset$ and $B \neq \emptyset$ then (A, B) is called a *cut*. The set of all the edges with one endpoint in A and another in B denoted by $E(A, B)$ is called an *edge cut* and the size of the edge cut is defined as $|E(A, B)|$. A matching is an edge set $M \subseteq E$ such that no two edges $e_i, e_j \in M$ share any endpoint. A cut (A, B) is a *matching cut* if every vertex in A (resp. B) has at most 1 neighbor in B (resp A). Equivalently, a cut (A, B) is a matching cut if the edge cut $E(A, B)$ is a matching. Note that a matching cut can be empty, and a matching whose removal disconnects a graph is not necessarily a matching cut. The MATCHING CUT is the problem of deciding if a given undirected graph G has a matching cut or not.

Recently Gomes and Sau [9] considered a generalization of matching cut and called it d-cut. For a positive integer $d \geq 1$, a cut (A, B) is a d-cut if every vertex in A (resp. B) has at most d neighbors in B (resp A). They named d-CUT the problem of deciding if a given graph G has a d-cut or not. They showed that for

© Springer Nature Switzerland AG 2021
P. Flocchini and L. Moura (Eds.): IWOCA 2021, LNCS 12757, pp. 531–543, 2021.
https://doi.org/10.1007/978-3-030-79987-8_37

every $d \geq 1$, d-CUT is NP-hard for regular graphs even when restricted to $(2d + 2)$-regular graphs [9]. They considered various structural parameters to study d-CUT and provided FPT results. They also showed fixed-parameter tractability of d-CUT when parameterized by the maximum size of the edge cut, using results provided by Marx, O'Sullivan and Razgon [16]. However, they did not provide an explicit bound on the running time as the treewidth reduction technique of [16] relies on MSOL formulation and Courcelle's Theorem [5] to show fixed-parameter tractability. Marx et al. [16] also observed that their method may actually increase the treewidth of the graph, however the treewidth will remain $f(k)$ for some function f. This motivated us to investigate an FPT algorithm for d-CUT parameterized by the maximum size of the edge cut where we can explicitly bound the dependence on the parameter. In this paper, we will discuss an FPT algorithm for d-CUT. Note that when $d = 1$, we can refer to the problem as MATCHING CUT.

Let us now formally define d-CUT in the context of parameterized complexity with maximum size of the edge cut as parameter.

---

k-d-CUT:
Input: An instance $I = (G, k, d)$. Where graph $G = (V, E)$, $|V| = n$ and $k, d \in \mathbb{N}$.
Parameter: k.
Output: yes if G contains a d-cut (A, B) such that $|E(A, B)| \leq k$, *no* otherwise.

1.1   Previous Work

The matching cut problem has been extensively studied. It was first introduced by Graham [10]. Chvátal [4] proved matching cut to be NP-Complete for graphs with maximum degree 4. Bonsma [2] proved matching cut to be NP-complete for planar graphs with maximum degree 4 and with girth 5. Kratsch and Le [13] provided an exact algorithm with running time $O^*(1.4143^n)$[1] and also provided a single exponential algorithm parameterized by the vertex cover number. Komusiewicz, Kratsch and Le [12] further improved the running time of the branching based exact algorithm to $O^*(1.3803^n)$ and also provided a SAT based $O^*(1.3071^n)$-time randomized algorithm. They also provided a single exponential algorithm parameterized by distance to cluster and distance to co-cluster. Aravind, Kalyanasundaram and Kare [1] provided fixed-parameter tractable algorithms for various structural parameters, including treewidth. Recently hardness and polynomial time results have also been obtained for various structural assumptions in [11,14,15].

---

[1] We use O^* notation which suppresses polynomial factors.

1.2 Our Contribution

Our main contribution is the following theorem.

Theorem 1. *k-d-CUT can be solved in time $2^{O(k \log k)} n^{O(1)}$.*

We designed a dynamic programming based algorithm for the proof of the above theorem, which we will discuss in Sect. 3. Cygan, Komosa, Lokshtanov, Pilipczuk, Pilipczuk, Saurabh and Wahlström [7] provided a compact tree decomposition with bounded adhesion along with a framework to design FPT algorithms and showed its application to Minimum Bisection and other problems. We used this framework and tree decomposition along with k-d-CUT specific calculations and proofs to design an algorithm for k-d-CUT.

We also observe the parameterized lower bound for MATCHING CUT.

Theorem 2. *Unless ETH fails, the problem to decide if a given n vertex graph has a matching cut with edge cut size at most k cannot be solved in $2^{o(k)} n^{O(1)}$.*

Proof. Kratsch and Le (Sect. 3.3 in [13]) have shown that for an n vertex graph, MATCHING CUT cannot be solved in $2^{o(n)}$ assuming exponential time hypothesis (ETH). For every matching cut, the maximum size of the edge cut k is linearly bounded by the number of vertices in the graph. Thus, for an n vertex graph, MATCHING CUT cannot be solved in $2^{o(k)} n^{O(1)}$ time unless ETH fails.   □

2 Preliminaries

2.1 Multiset Notations

Considering a set U as universe, a multiset is a 2-tuple $P = (U, m_P)$ where multiplicity function $m_P : U \to \mathbb{Z}_{\geq 0}$ is a mapping from U to non negative integers such that for an element $e \in U$, the value $m_P(e)$ is the multiplicity of e in P that is the number of occurrence of e in P. Cardinality of a multiset P is the sum of multiplicity of all the distinct elements of P. We write $e \in P$ if $m_P(e) \geq 1$. P is considered *empty* and denoted by $P = \emptyset$ iff $\forall e \in U$, $m_P(e) = 0$. For two multiset A and B on universe U, let m_A and m_B be their respective multiplicity functions. We will use the following operations on multisets for our purposes.

Equality: A is equal to B, denoted by $A = B$, if $\forall e \in U$, $m_A(e) = m_B(e)$. Otherwise, we say that A and B are distinct.

Inclusion: A is included in B, denoted by $A \subseteq B$, if $\forall e \in U$, $m_A(e) \leq m_B(e)$.

Sum Union: P is a sum union of A and B, denoted by $P = A \uplus B$, if $\forall e \in U$, $m_P(e) = m_A(e) + m_B(e)$, where m_P is the multiplicity function for P.

Throughout this paper, if the context is clear, for any multiset X we will use m_X to denote the multiplicity function of X.

2.2   Graph Notations

All the graphs we consider are simple, undirected and connected. $G = (V, E)$ is a graph with vertex set V and edge set E. $E(G)$ denotes the set of edges of graph G, and $V(G)$ denotes the set of vertices of G. For $E' \subseteq E$, $V(E')$ denotes the set of all vertices of G with at least one edge in E' incident on it. We use $G' \subseteq G$ to denote that G' is a subgraph of G. For a vertex set $V' \subseteq V$, $G[V']$ denotes the induced subgraph of G on vertex set V'. For an edge set $E' \subseteq E$, $G[E']$ denotes the subgraph of G on edge set E' i.e., $G[E'] = (V(E'), E')$.

For disjoint vertex sets $A \subseteq V$ and $B \subseteq V$, $E_G(A, B)$ denotes the set of all the edges of G with one endpoint in A and another in B. For a subgraph $G' \subseteq G$, $E_{G'}(A, B)$ denotes the set of edges $E_G(A, B) \cap E(G')$. For a vertex $v \in V$, we use $N_G(v)$ to denote the set of all adjacent vertices of v in G, if the context of the graph is clear we will simply use $N(v)$.

Partition of a Graph: For a graph $G = (V, E)$, (A, B) is a partition of G if $A \cup B = V$ and $A \cap B = \emptyset$. We call a partition (A, B) trivial if either $A = \emptyset$ or $B = \emptyset$.

Cut: For a graph G, a partition (A, B) is a cut if both $A \neq \emptyset$ and $B \neq \emptyset$.

d-Cut: A cut (A, B) is a d-cut if every vertex in A has at most d neighbors in B, vice versa every vertex in B has at most d neighbors in A.

d-Matching: For a graph $G = (V, E)$, an edge set $M \subseteq E$ is called a d-matching if every vertex $v \in V$ has at most d edges in M incident on it. Observe that a cut (A, B) is d-cut iff $E(A, B)$ is a d-matching.

2.3   Fixed-Parameter Tractability

We refer the reader to [6, 8] for basic background on parameterized complexity. We also recall some basic definitions. A *parameterized problem* is a language $L \subseteq \Sigma^* \times \mathbb{N}$ where Σ is a fixed and finite alphabet. For an instance $I = (x, k) \in \Sigma^* \times \mathbb{N}$, k is called the parameter. A parameterized problem is called *fixed-parameter tractable* (FPT) if there exists an algorithm $\mathcal{A}$ (called a *fixed-parameter algorithm*), a computable function $f : \mathbb{N} \to \mathbb{N}$, and a constant c such that, the algorithm $\mathcal{A}$ correctly decides whether $(x, k) \in L$ in time bounded by $f(k).|(x, k)|^c$.

2.4   Tree Decomposition

A tree decomposition of a graph G is a pair (T, β) where T is a tree and β (called a bag) is a mapping that assigns to every $t \in V(T)$ a set $\beta(t) \subseteq V(G)$, such that the following holds:

1. For every $e \in E(G)$, there exists a $t \in V(T)$ such that $V(e) \subseteq \beta(t)$;
2. For $v \in V(G)$, let $\beta^{-1}(v)$ be the set of all vertices $t \in V(T)$ such that $v \in \beta(t)$, then $T[\beta^{-1}(v)]$ is a connected nonempty subgraph of T.

If the tree T in decomposition (T, β) is rooted at some node r, we call it a *rooted tree decomposition*.

Definition 1 (Adhesion in a Tree Decomposition). *In a tree decomposition (T, β), For an edge $e \in E(T)$ where $e = t_1 t_2$, a set $\sigma(e) = \beta(t_1) \cap \beta(t_2)$ is called an adhesion of e. For a rooted tree decomposition (T, β), adhesion of a node $t \in V(T)$, denoted by $\sigma(t)$, is $\sigma(t, t')$ where t' is parent node of t in T. The adhesion of a root node r is $\emptyset$.*

Some Functions for Convenience: For a rooted tree decomposition (T, β) at some node r, for $s, t \in V(t)$ we say that s is a *descendent* of t, if t lies on the unique path from s to r. This implies that a node is a descendant of itself.

$$\gamma(t) = \bigcup_{c: \ descendant \ of \ t} \beta(c), \quad \alpha(t) = \gamma(t) \setminus \sigma(t), \quad G_t = G[\gamma(t)] - E(G[\sigma(t)]).$$

Definition 2 (Compact Tree Decomposition [7]). *A rooted tree decomposition (T, β) of G is compact if for every non root-node $t \in V(T) : G[\alpha(t)]$ is connected and $N(\alpha(t)) = \sigma(t)$.*

Definition 3 (Separation [7]). *A pair (A, B) of vertex subsets in a graph G is a separation if $A \cup B = V(G)$ and there is no edge with one endpoint in $A \setminus B$ and the other in $B \setminus A$; the order of the separation (A, B) is $|A \cap B|$.*

In [7] the edge cut (A, B) is defined as a pair $A, B \subseteq V(G)$ such that $A \cup B = V(G)$ and $A \cap B = \emptyset$, which we refer to as partition (A, B). And the order of the edge cut (A, B) is defined as $|E(A, B)|$. These terminologies are required for following definition.

Definition 4 (Unbreakability [7]). *Let G be a graph, A vertex subset $X \subseteq V(G)$ is (q,k)-unbreakable if every separation(A,B) of order at most k satisfies $|A \cap X| \leq q$ or $|B \cap X| \leq q$. A vertex subset $Y \subseteq V(G)$ is (q,k)-edge-unbreakable if every edge cut (A,B) of order at most k satisfies $|A \cap Y| \leq q$ or $|B \cap Y| \leq q$.*

Observe that every set that is (q,k)-unbreakable is also (q,k)-edge-unbreakable.

Theorem 3 ([7]). *Given an n-vertex graph G and an integer k, one can in time $2^{O(k \log k)} n^{O(1)}$ compute a rooted compact tree decomposition (T, β) of G such that*

1. *every adhesion of (T, β) is of size at most k;*
2. *every bag of (T, β) is (i, i)-unbreakable for every $1 \leq i \leq k$.*

$\square$

Note that since every bag of the output decomposition (T, β) of Theorem 3 is (k, k)-unbreakable, it is also (k, k)-edge-unbreakable. Further, the construction provided for the proof of theorem 3 in [7] maintained that the number of edges in decomposition is always upper bounded by $|V(G)|$ and hence $|V(T)| \leq |V(G)| + 1$.

3   An FPT Algorithm for k-d-CUT (Proof of Theorem 1)

A disconnected graph G trivially has a d-cut of size 0 and thus, (G, k, d) is always a yes instance of k-d-CUT for every $k, d \in \mathbb{N}$. Thus, it remains to find if (G, k, d) is a yes instance when the graph G is connected. From here onward, we will always assume that the input graph G is simple and connected. Further, we can also assume that $d < k$, otherwise the problem is equivalent to deciding if G has a min-cut of size at most k, which is polynomial time solvable [17].

We will start by invoking Theorem 3 on input n vertex graph G with parameter k. This gives us a rooted compact tree decomposition (T, β) of G where every bag is (k, k)-edge-unbreakable and every node $t \in V(T)$ has adhesion of size at most k. Consider the following definition.

Definition 5 (*Matched Candidate Set of a Vertex Set* $Q \subseteq V$). *For a vertex set $Q \subseteq V$ and $d \in \mathbb{N}$, we call a multiset $P = (V, m_P)$ a d-matched candidate set of Q if following holds.*

– $\forall v \in Q$, $m_P(v) \leq d$,
– $\forall v \in V \setminus Q$, $m_P(v) = 0$,
– $|P| \leq k$.

Note that if $Q = \emptyset$, then the empty multiset is the only d-matched candidate set of Q.

Proposition 1 ($\star^2$). *Given $Q \subseteq V$, if $|Q| \leq k$, then there are at most $2^{O(k \log k)}$ distinct d-matched candidate sets of Q and in time $2^{O(k \log k)} n^{O(1)}$ we can list all of them.*

We perform bottom up dynamic programming on (T, β). For every vertex $t \in V(T)$, every set $S \subseteq \sigma(t)$, $\bar{S} = \sigma(t) \setminus S$, every d-matched candidate set $P = (V, m_P)$ of $\sigma(t)$ and $n_e \in \{0, 1\}$ we compute an integer $M[t, S, P, n_e] \in \{0, 1, 2, \ldots, k, \infty\}$ with the following properties.

(1) If $M[t, S, P, n_e] \leq k$, then there exists a partition (A, B) of G_t such that the following holds.
   – If $n_e = 1$ then both A and B are non empty, otherwise either A or B is empty,
   – $A \cap \sigma(t) \in \{S, \bar{S}\}$,
   – $E_{G_t}(A, B)$ forms a d-matching,
   – every vertex v in $\sigma(t)$ has at most $m_P(v)$ neighbors in the other side of the partition in G_t i.e., $\forall v \in \sigma(t)$, $|N_{G[E_{G_t}(A,B)]}(v)| \leq m_P(v)$,
   – $|E_{G_t}(A, B)| \leq M[t, S, P, n_e]$.
(2) For every partition (A, B) of the entire graph G that satisfies the following conditions:
   – $A \cap \sigma(t) \in \{S, \bar{S}\}$,
   – $E_G(A, B)$ forms a d-matching,

---

[2] Due to space constraint, the proofs of statements marked with a $\star$ have been omitted.

- every vertex v in $\sigma(t)$ has at most $m_P(v)$ neighbors in the other side of the partition in G_t i.e., $\forall v \in \sigma(t)$, $|N_{G[E_{G_t}(A,B)]}(v)| \leq m_P(v)$,
- $|E_G(A,B)| \leq k$.

It holds that $|E_{G_t}(A,B)| \geq M[t,S,P,1]$ if both $V(G_t) \cap A$ and $V(G_t) \cap B$ are non empty, otherwise $|E_{G_t}(A,B)| \geq M[t,S,P,0]$ if either $V(G_t) \cap A$ or $V(G_t) \cap B$ is empty.

Note that $|E_{G_t}(A,B)| \geq \infty$ imply that such partition (A,B) doesn't exist. Let us now formally prove that table $M[.]$ is sufficient for our purpose.

Lemma 1. *(G,k,d) is a yes instance of k-d-CUT if and only if $M[r,\emptyset,\emptyset,1] \leq k$ where r is the root of T.*

Proof. For the first direction, a non trivial partition (A,B) for G_r whose existence is asserted by property 1 for $M[r,\emptyset,\emptyset,1]$ is a d-cut of G with $|E_G(A,B)| \leq k$, as $G_r = G$.

For the other direction, let (A,B) be a d-cut of G such that $|E_G(A,B)| \leq k$. Since $\sigma(r) = \emptyset$, (A,B) satisfies all the prerequisites of property (2) for $t = r$, $S = \emptyset$ and $P = \emptyset$. Further, $V(G_r) \cap A$ and $V(G_r) \cap B$ are both non empty as (A,B) is a non trivial partition. Thus, $k \geq |E_{G_r}(A,B)| \geq M[r,\emptyset,\emptyset,1]$. This finishes the proof. $\qquad\square$

Proposition 2 ($\star$). *For every t, S, and P, if either S or $\bar{S}$ is empty, then $M[t,S,P,0] = 0$ satisfies properties (1) and (2).*

Proposition 3 ($\star$). *For every t, S, and P, if both S and $\bar{S}$ are non empty, then $M[t,S,P,0] = \infty$.*

Proposition 4 ($\star$). *For every t, S, and P, if both S and $\bar{S}$ are non empty, then $M[t,S,P,1] \geq 1$.*

To prove Theorem 1, it would suffice to compute the $M[.]$ table for every node $t \in V(T)$ in time $2^{O(k \log k)} n^{O(1)}$. Further, as the number of nodes in T is bounded by $O(n)$, it would suffice if we show that for a fixed $t \in V(T)$, the entries $M[t,.,.,.]$ can be computed in $2^{O(k \log k)} n^{O(1)}$.

For every $t \in V(t)$, the number of distinct d-matched candidate sets of $\sigma(t)$ is bounded by $2^{O(k \log k)}$ and we can obtain them in time $2^{O(k \log k)} n^{O(1)}$ due to $|\sigma(t)| \leq k$ and Proposition 1. Thus, the number of cells $M[t,.,.,.]$ at every vertex $t \in V(T)$ are bounded by $2^{O(k \log k)}$. Given entries $M[c,.,.,.]$ for every children c of a node $t \in V(T)$, if we can show that a single cell $M[t,S,P,n_e]$ can be computed in time $2^{O(k \log k)} n^{O(1)}$, then we can bound the time required to compute all the entries $M[t,.,.,.]$ to $2^{O(k \log k)} \cdot 2^{O(k \log k)} n^{O(1)}$ which is essentially $2^{O(k \log k)} n^{O(1)}$. Thus, we focus on the calculation of a single cell $M[t,S,P,n_e]$.

3.1   Calculating the Value of $M[t,S,P,n_e]$

In this section we will discuss the calculation of a single cell $M[t,S,P,n_e]$ for the given $t \in V(t)$, $S \subseteq \sigma(t)$ and P such that P is a d-matched candidate set of

$\sigma(t)$. We use Propositions 2 and 3 to set $M[t, S, P, 0] \in \{0, \infty\}$. Thus, we move on to the calculation of $M[t, S, P, 1]$. Let $Z(t)$ be the set of all the children of t in T. From here on we will assume that entries $M[c, ., ., .]$ are calculated for every $c \in Z(t)$. Note that if t is a leaf vertex then $Z(t)$ is empty.

Intuitively, in this step of dynamic programming we will focus on to partitioning $\beta(t)$ and use entries $M[c, ., ., .]$ as black boxes to find the best way to partition subgraphs G_c. Within this framework we can think of every edge $e \in E(G_t[\beta(t)])$ as a subgraph of G_t and to find the best way to partition subgraph $G[e]$ we construct a table $M_E[e, S', P']$ for every edge $e = uv \in E(G_t[\beta(t)])$ where $S' \subseteq \{u, v\}$, and P' is a 1-matched candidate set of $\{u, v\}$. We are taking a 1-matched candidate set of $V(e)$, as there is only 1 edge in $G[e]$.

We assign following values to $M_E[e, S', P']$.

- $M_E[e, \emptyset, P'] = M_E[e, \{u, v\}, P'] = 0$ for every 1-matched candidate set P' of $V(e)$;
- $M_E[e, \{u\}, P'] = M_E[e, \{v\}, P'] = 1$ for 1-matched candidate set P' of $V(e)$ such that $m_{P'}(u) = m_{P'}(v) = 1$;
- $M_E[e, \{u\}, P'] = M_E[e, \{v\}, P'] = \infty$ for every 1-matched candidate set P' of $V(e)$ such that $m_{P'}(u) = 0$ or $m_{P'}(v) = 0$.

Intuitively, if both u and v fall into the same side of the partition then $M_E[e, S', P']$ costs 0. Otherwise, if u and v fall into different side of the partition and both are allowed to have a neighbor in the other side of the partition in $G[e]$ as per P' then $M_E[e, S', P']$ costs 1 and if at least one of u or v is not allowed to have a neighbor in the other side of the partition in $G[e]$ as per P' then $M_E[e, S', P']$ costs ∞. Every 1-matched candidate set of $V(e)$ can be considered as a subset of $V(e)$. Thus, number of cells $M_E[.]$ is bounded by $n^{O(1)}$ and we can calculate them as per above assignment in time $n^{O(1)}$.

Definition 6 (S-Compatible Set of a Bag). *For $S \subseteq \sigma(t)$ and $\bar{S} = \sigma(t) \setminus S$, a set $A_s \subseteq \beta(t)$ is called an S-Compatible set of $\beta(t)$ if*

- $|A_s| \leq k$,
- $A_s \cap \sigma(t) \in \{S, \bar{S}\}$,
- A_s *is non-empty proper subset of $\beta(t)$ i.e., $A_s \neq \emptyset$ and $A_s \neq \beta(t)$.*

For an S-compatible set A_s of $\beta(t)$, let $S_c = A_s \cap \sigma(c)$ and $\bar{S}_c = \sigma(c) \setminus S_c$ for every $c \in Z(t)$. And let $S_e = A_s \cap V(e)$ and $\bar{S}_e = V(e) \setminus S_e$ for every $e \in E(G_t[\beta(t)])$.

We define $Z_{ab}^{A_s} = \{c \mid c \in Z(t) \wedge (S_c \neq \emptyset) \wedge (\bar{S}_c \neq \emptyset)\}$ and call it the set of *broken children* of t with respect to A_s, and also define $E_{ab}^{A_s} = \{e \mid e \in E(G_t[\beta(t)]) \wedge (S_e \neq \emptyset) \wedge (\bar{S}_e \neq \emptyset)\}$ and call it the set of *broken edges* of $E(G_t[\beta(t)])$ with respect to A_s. Given A_s, we can find $Z_{ab}^{A_s}$ and $E_{ab}^{A_s}$ in time $n^{O(1)}$.

Definition 7 (P-Compatible Family). *For t, P and A_s such that $t \in V(T)$, P is a d-matched candidate set of $\sigma(t)$ and A_s is an S-compatible set of $\beta(t)$. A family $F_{P|A_s} = \{P_c \mid c \in Z_{ab}^{A_s}\} \cup \{P_e \mid e \in E_{ab}^{A_s}\}$ is called an A_s-restricted P-compatible family of t if the following holds:*
let $P_z = \biguplus_{P_v \in F_{P|A_s}} P_v,$

– *for each $c \in Z_{ab}^{A_s}$, P_c is a d-matched candidate set of $\sigma(c)$,*
– *for each $e \in E_{ab}^{A_s}$, P_e is a 1-matched candidate set of $V(e)$,*
– $|P_z| \leq 2k$,
– $\forall v \in V$, $m_{P_z}(v) \leq d$,
– $\forall v \in \sigma(t)$, $m_{P_z}(v) \leq m_P(v)$.

The intuition behind the A_s-restricted P-compatible family is as follows. We will use entries $M[c,.,.,.]$ and $M_E[.,.,.]$ for our calculation, to access a particular entry, we will guess P_c and P_e which are consistent with P and maintain the property of d-matching.

We say that two A_s-restricted P-compatible families $F_{P|A_s} = \{P_c| \, c \in Z_{ab}^{A_s}\} \cup \{P_e| \, e \in E_{ab}^{A_s}\}$ and $F'_{P|A_s} = \{P'_c| \, c \in Z_{ab}^{A_s}\} \cup \{P'_e| \, e \in E_{ab}^{A_s}\}$ are equal iff $\forall c \in Z_{ab}^{A_s}$, $P_c = P'_c$ and $\forall e \in E_{ab}^{A_s}$, $P_e = P'_e$. We say that $F_{P|A_s}$ and $F'_{P|A_s}$ are distinct iff they are not equal.

Proposition 5 ($\star$). *For an S-compatible set A_s of $\beta(t)$, if $|Z_{ab}^{A_s}| + |E_{ab}^{A_s}| \leq k$, then there are at most $2^{O(k \log k)}$ distinct A_s-restricted P-compatible families of t and in time $2^{O(k \log k)} n^{O(1)}$ we can list all of them.*

Assuming that $M[c,.,.,.]$ table is calculated for every $c \in Z(t)$ and $M_E[e,.,.]$ are available as per above assignment for every $e \in E(G_t[\beta(t)])$. For an S-compatible set A_s of $\beta(t)$ and A_s-restricted P-compatible family $F_{P|A_s} = \{P_c| \, c \in Z_{ab}^{A_s}\} \cup \{P_e| \, e \in E_{ab}^{A_s}\}$ of t, we define the cost of A_s and $F_{P|A_s}$ for t as follows.

$$cs(t, A_s, F_{P|A_s}) = \sum_{c \in Z_{ab}^{A_s}} M[c, A_s \cap \sigma(c), P_c, 1] + \sum_{e \in E_{ab}^{A_s}} M_E[e, A_s \cap V(e), P_e]. \quad (1)$$

Proposition 6 ($\star$). *For every A_s-restricted P-compatible family $F_{P|A_s}$ of t, $cs(t, A_s, F_{P|A_s}) \geq |Z_{ab}^{A_s}| + |E_{ab}^{A_s}|$.*

Proposition 7 ($\star$). *Assuming $M[c,.,.,.]$ and $M_E[e,.,.]$ tables are calculated for every $c \in Z(t)$ and every $e \in E(G_t[\beta(t)])$. Given an S-compatible set A_s of $\beta(t)$ and an A_s-restricted P-compatible family $F_{P|A_s}$ of t, $cs(t, A_s, F_{P|A_s})$ can be calculated in time $2^{O(k \log k)} n^{O(1)}$.*

We define minimum cost of an S-compatible set A_s of $\beta(t)$ and d-matched candidate set P of $\sigma(t)$ as follows:

$$mcs(t, A_s, P) = min\{cs(t, A_s, F_{P|A_s})| \, F_{P|A_s} \text{ is } A_s\text{-re.}P\text{-com. family of t}\}. \quad (2)$$

Lemma 2 ($\star$). *Assuming the values $M[c,.,.,.]$ satisfies properties (1) and (2) for every $c \in Z(t)$ then $mcs(t, A_s, P)$ satisfies the following properties.*

(a) *If $mcs(t, A_s, P) \leq k$, then there exists a partition (A, B) of G_t, such that:*
   – $A \cap \beta(t) = A_s$,

  – $|E_{G_t}(A, B)| \leq mcs(t, A_s, P)$,
  – $E_{G_t}(A, B)$ *forms a d-matching*,
  – $\forall v \in \sigma(t),\ |N_{G[E_{G_t}(A,B)]}(v)| \leq m_P(v)$.
(b) *For every partition* (A, B) *of the entire graph* G *that satisfy following conditions:*
  – $A \cap \beta(t) = A_s$,
  – $|E_G(A, B)| \leq k$,
  – $E_G(A, B)$ *forms a d-matching*,
  – $\forall v \in \sigma(t),\ |N_{G[E_{G_t}(A,B)]}(v)| \leq m_P(v)$.
  It holds that $|E_{G_t}(A, B)| \geq mcs(t, A_s, P)$.

We note that if $mcs(t, A_s, P) > k$ then there doesn't exist a partition (A, B) of G satisfying conditions of property (b).

Lemma 3. *For an S-compatible set* A_s *of* $\beta(t)$, *in time* $2^{O(k \log k)} n^{O(1)}$ *we can either decide that* $mcs(t, A_s, P) > k$ *or calculate* $mcs(t, A_s, P)$.

Proof. Given A_s, we check if $|Z_{ab}^{A_s}| + |E_{ab}^{A_s}| \leq k$, if not, then using Proposition 6 we conclude that $mcs(t, A_s, P) > k$. Else, if $|Z_{ab}^{A_s}| + |E_{ab}^{A_s}| \leq k$, then we use Proposition 5 to get all the $2^{O(k \log k)}$ distinct A_s- restricted P-compatible families of t in time $2^{O(k \log k)} n^{O(1)}$ and calculate $mcs(t, A_s, P)$ as per Eq. 2. We need to calculate $cs(t, A_s, F_{P|A_s})$ for $2^{O(k \log k)}$ distinct A_s-restricted P-compatible families, which we can accomplish by invoking Proposition 7, $2^{O(k \log k)}$ times. Thus, we conclude that calculation of $mcs(t, A_s, P)$ would take time $2^{O(k \log k)} n^{O(1)}$.   □

We now move on to give an assignment to $M[t, S, P, 1]$. Consider the following assignments.

$$MIN_c = \min\{\min\{M[c, \emptyset, P_c, 1] \mid P_c \text{ is a } d\text{- mat. can. set of } \sigma(c) \text{ such that} \quad (3)$$
$$\forall v \in \sigma(t), m_{P_c}(v) \leq m_P(v)\} \mid c \in Z(t)\}.$$

If t is a leaf vertex and $Z(t)$ is empty, then we set $MIN_c = \infty$.

$$MIN_{\beta(t)} = \min\{mcs(t, A_s, P) \mid A_s \text{ is an } S \text{ -compatible set of } \beta(t)\}. \quad (4)$$

Consider the following assignment of $M[t, S, P, 1]$.

1. Case: $S = \emptyset$ or $S = \sigma(t)$.

$$M[t, S, P, 1] = min\{MIN_c, MIN_{\beta(t)}\}. \quad (5)$$

2. Case: $S \neq \emptyset$ and $S \neq \sigma(t)$.

$$M[t, S, P, 1] = MIN_{\beta(t)}. \quad (6)$$

In Eqs. (5) and (6) if the right hand side exceed k then we set $M[t, S, P, 1] = \infty$.

Lemma 4 ($\star$). *Assuming values $M[c,.,.,.]$ satisfy properties (1) and (2) for every $c \in Z(t)$ and $mcs(t, A_s, P)$ satisfies properties (a) and (b) for every S-compatible set A_s of $\beta(t)$, assignment of $M[t, S, P, 1]$ as per Eq. (5) and (6) satisfies properties (1) and (2).*

Calculation of MIN_c is straightforward and requires to iterate over every $M[c,.,.,.]$ for every c. As number of cells $M[c,.,.,.]$ at each c are bounded by $2^{O(k \log k)}$ and $|Z(t)|$ can be at most $O(n)$, we can calculate MIN_c in $2^{O(k \log k)} n^{O(1)}$.

To calculate $MIN_{\beta(t)}$ a simple brute force approach of guessing all the S-compatible sets of $\beta(t)$ will not work, as it will exceed the running time budget that we have. However, as it is required to calculate $MIN_{\beta(t)}$ only if $MIN_{\beta(t)} \leq k$, we can restrict our search space.

To this end, let us assume that $MIN_{\beta(t)} \leq k$, and let us fix a minimizing argument A_s^*, then A_s^* is the S-compatible set such that $MIN_{\beta(t)} = mcs(t, A_s^*, P) \leq k$. In such a scenario due to Proposition 6 we have that $|Z_{ab}^{A_s^*}| + |E_{ab}^{A_s^*}| \leq k$. Recalling $S_c = A_s^* \cap \sigma(c)$ and $\bar{S}_c = \sigma(c) \setminus S_c$ for every $c \in Z(t)$. And $S_e = A_s^* \cap V(e)$ and $\bar{S}_e = V(e) \setminus S_e$ for every $e \in E(G_t[\beta(t)])$.

Lemma 5 ([3]). *Given a set U of size n, and integers $0 \leq a, b \leq n$, one can in $2^{O(min(a,b) \log(a+b))} n \log n$ time construct a family $\mathcal{F}$ of at most $2^{O(min(a,b) \log(a+b))} \log n$ subsets of U, such that following holds: for any sets $A, B \subseteq U$, $A \cap B = \emptyset$, $|A| \leq a$, $|B| \leq b$, there exists a set $S \in \mathcal{F}$ with $A \subseteq S$ and $B \cap S = \emptyset$.*

Let $B^* = (\cup_{c \in Z_{ab}^{A_s^*}} \bar{S}_c) \bigcup (\cup_{c \in E_{ab}^{A_s^*}} \bar{S}_e)$. Due to $|Z_{ab}^{A_s^*}| + |E_{ab}^{A_s^*}| \leq k$ and $|\sigma(c)| \leq k$, we can observe that $|B^*| \leq k^2$. Invoking Lemma 5 for the universe $\beta(t)$ and integers $k, k^2 + k$, we obtain a family $\mathcal{F}$ of subsets of $\beta(t)$ such that there exists a set $A_g \in \mathcal{F}$ such that $A_g \supseteq A_s^*$ and $A_g \cap (B^* \cup (\sigma(t) \setminus A_s^*)) = \emptyset$. We call such a set A_g a *good set*. Further, the size of $\mathcal{F}$ is bounded by $2^{O(k \log k)} \log n$.

We now construct an auxiliary graph H on the vertex set $\beta(t)$ such that an edge $uv \in E[H]$ if and only if one of the following holds:

1. $u, v \in \sigma(t)$;
2. there exists a $c \in Z(t)$ such that $u, v \in \sigma(c)$;
3. $uv \in E(G_t[\beta(t)])$.

Observe that $\sigma(t)$ forms a clique in H, similarly every $\sigma(c)$ forms a clique in H and $G_t[\beta(t)]$ is a subgraph of H. For $X \subseteq \beta(t)$, we call a connected component C_s of $H[X]$ an *S-compatible component* if $V(C_s)$ is an S-compatible set of $\beta(t)$.

Proposition 8 ($\star$). *If A_g is a good set, then there exists an S-compatible component C_s in the subgraph $H[A_g]$ such that $mcs(t, A_s^*, P) = mcs(t, V(C_s), P)$.*

Proposition 8 allow us to efficiently calculate $MIN_{\beta(t)}$. We need to iterate over every $A_g \in \mathcal{F}$ and for each S-compatible component C_s in $H[A_g]$ we need to invoke Lemma 3 so that we can either calculate $mcs(t, V(C_s), P)$ or decide if $mcs(t, V(C_s), P) > k$. If $mcs(t, V(C_s), P) > k$ then we assume it to be ∞. We take the minimum value $mcs(t, V(C_s), P)$ encountered among all the S-compatible component C_s in $H[A_g]$ over all the choices $A_g \in \mathcal{F}$ and assign it to $MIN_{\beta(t)}$. Correctness of this procedure comes due to the minimality of $mcs(t, A_s^*, P)$ among all the S-compatible sets of $\beta(t)$ and due to Proposition 8. If we don't encounter any S-compatible component during this process then we can conclude that the assumption $MIN_{\beta(t)} \le k$ doesn't hold and we set $MIN_{\beta(t)} = \infty$.

As the size of $\mathcal{F}$ is bounded by $2^{O(k \log k)} \log n$ and we can obtain it using Lemma 5 in time $2^{O(k \log k)} n \log n$. And for every $A_g \in \mathcal{F}$, $H[A_g]$ can contain at most n S-compatible components and we can find all of them in time $n^{O(1)}$ by using standard graph traversal methods. Thus, we need to invoke Lemma 3 for at most $2^{O(k \log k)} n^{O(1)}$ S-compatible components(sets), and each invocation takes $2^{O(k \log k)} n^{O(1)}$, thus, calculation of $MIN_{\beta(t)}$ takes time $2^{O(k \log k)} n^{O(1)}$. Recalling that calculation of MIN_c takes $2^{O(k \log k)} n^{O(1)}$. This conclude that a single cell $M[t, S, P, 1]$ can be calculated in time $2^{O(k \log k)} n^{O(1)}$. Further, we use Propositions 2 and 3 to set values of $M[t, S, P, 0]$. This concludes that a single cell $M[t, S, P, n_e]$ can be calculated in time $2^{O(k \log k)} n^{O(1)}$. Recalling Lemma 1, this suffices to conclude the proof of Theorem 1.

4   Conclusion

In this paper, we discussed a $2^{O(k \log k)} n^{O(1)}$ time fixed-parameter tractable algorithm for d-CUT where k is the maximum size of the edge cut. We also observed that MATCHING CUT cannot be solved in $2^{o(k)} n^{O(1)}$ unless ETH fails. It will be an interesting problem to reduce the gap between lower and upper bound for MATCHING CUT.

Acknowledgements. We thank Fahad Panolan for useful discussions, in particular his suggestion of the tree decomposition that we used in the paper.

References

1. Aravind, N.R., Kalyanasundaram, S., Kare, A.S.: On structural parameterizations of the matching cut problem. In: Gao, X., Du, H., Han, M. (eds.) COCOA 2017. LNCS, vol. 10628, pp. 475–482. Springer, Cham (2017). https://doi.org/10.1007/978-3-319-71147-8_34
2. Bonsma, P.S.: The complexity of the matching-cut problem for planar graphs and other graph classes. J. Graph Theor. **62**(2), 109–126 (2009). https://doi.org/10.1002/jgt.20390
3. Chitnis, R.H., Cygan, M., Hajiaghayi, M., Pilipczuk, M., Pilipczuk, M.: Designing FPT algorithms for cut problems using randomized contractions. In: 53rd Annual IEEE Symposium on Foundations of Computer Science, FOCS 2012, New Brunswick, NJ, USA, 20–23 October 2012, pp. 460–469. IEEE Computer Society (2012). https://doi.org/10.1109/FOCS.2012.29
4. Chvátal, V.: Recognizing decomposable graphs. J. Graph Theor. **8**(1), 51–53 (1984)

5. Courcelle, B.: The monadic second-order logic of graphs. i. Recognizable sets of finite graphs. Inf. Comput. **85**(1), 12–75 (1990)
6. Cygan, M., et al.: Parameterized Algorithms. Springer, Cham (2015). https://doi.org/10.1007/978-3-319-21275-3
7. Cygan, M., et al.: Randomized contractions meet lean decompositions. ACM Trans. Algorithm. **17**(1), 6:1–6:30 (2021)
8. Downey, R.G., Fellows, M.R.: Fundamentals of Parameterized Complexity. TCS. Springer, London (2013). https://doi.org/10.1007/978-1-4471-5559-1
9. Gomes, G.C.M., Sau, I.: Finding cuts of bounded degree: complexity, FPT and exact algorithms, and kernelization. Algorithmica (2021). https://doi.org/10.1007/s00453-021-00798-8
10. Graham, R.L.: On primitive graphs and optimal vertex assignments. Ann. New York Acad. Sci **175**, 170–186 (1970)
11. Hsieh, S.-Y., Le, H.-O., Le, V.B., Peng, S.-L.: Matching cut in graphs with large minimum degree. In: Du, D.-Z., Duan, Z., Tian, C. (eds.) COCOON 2019. LNCS, vol. 11653, pp. 301–312. Springer, Cham (2019). https://doi.org/10.1007/978-3-030-26176-4_25
12. Komusiewicz, C., Kratsch, D., Le, V.B.: Matching cut: Kernelization, single-exponential time FPT, and exact exponential algorithms. Discret. Appl. Math. **283**, 44–58 (2020). https://doi.org/10.1016/j.dam.2019.12.010
13. Kratsch, D., Le, V.B.: Algorithms solving the matching cut problem. Theor. Comput. Sci. **609**, 328–335 (2016). https://doi.org/10.1016/j.tcs.2015.10.016
14. Le, H., Le, V.B.: On the complexity of matching cut in graphs of fixed diameter. In: Hong, S. (ed.) 27th International Symposium on Algorithms and Computation, ISAAC 2016. LIPIcs, Sydney, Australia, 12–14 December 2016, vol. 64, pp. 50:1–50:12. Schloss Dagstuhl - Leibniz-Zentrum für Informatik (2016). https://doi.org/10.4230/LIPIcs.ISAAC.2016.50
15. Le, H., Le, V.B.: A complexity dichotomy for matching cut in (bipartite) graphs of fixed diameter. Theor. Comput. Sci. **770**, 69–78 (2019). https://doi.org/10.1016/j.tcs.2018.10.029
16. Marx, D., O'Sullivan, B., Razgon, I.: Treewidth reduction for constrained separation and bipartization problems. In: Marion, J., Schwentick, T. (eds.) 27th International Symposium on Theoretical Aspects of Computer Science, STACS 2010. LIPIcs, Nancy, France, 4–6 March 2010, vol. 5, pp. 561–572. Schloss Dagstuhl - Leibniz-Zentrum für Informatik (2010). https://doi.org/10.4230/LIPIcs.STACS.2010.2485
17. Stoer, M., Wagner, F.: A simple min-cut algorithm. J. ACM **44**(4), 585–591 (1997). https://doi.org/10.1145/263867.263872

Backtrack Search for Parallelisms
of Projective Spaces

Svetlana Topalova[iD] and Stela Zhelezova[(⊠)][iD]

Institute of Mathematics and Informatics, Bulgarian Academy of Sciences,
Sofia, Bulgaria
{svetlana,stela}@math.bas.bg

Abstract. A *spread* in $PG(n, q)$ is a set of lines which partition the
point set. A partition of the set of lines by spreads is called a *parallelism*.
The numerous relations and applications of parallelisms determine a significant interest in the methods for their construction. We consider two
different backtrack search algorithms which can be used for that purpose. The first one implies search on a set of spreads, and the second
- on the lines of the projective space. The authors have used them for
the classification of parallelisms invariant under definite automorphism
groups. The present paper concerns the applicability of each of the two
algorithms to cases with different peculiarities, and some ways to modify
them for usage on parallel computers. Suitable examples are given.

Keywords: Finite projective space · Parallelism · Resolution of a
combinatorial design

1 Introduction

Let $PG(n, q)$ be the n-dimensional projective space over the finite field $\mathbb{F}_q$. If n
is odd, it is possible to find a set of lines such that each point is in exactly one
of these lines. Such a set is called a spread. Two spreads are isomorphic if an
automorphism of $PG(n, q)$ maps one to the other. A parallelism is a partition
of the set of all lines of the projective space to spreads. Two parallelisms are
isomorphic if there is an automorphism of $PG(n, q)$ which maps the spreads
of one parallelism to spreads of the other. Background material on projective
spaces, spreads and parallelisms, can be found, for instance, in [13] or [23].

General constructions of parallelisms of $PG(n, 2)$ are presented in [1] and
[31], of $PG(2^n - 1, q)$ in [5], and of $PG(3, q)$ in [7,9,12,15]. All parallelisms of
$PG(3, 2)$ and $PG(3, 3)$ are known [2,13]. For larger projective spaces, however,
the classification problem is open. That is why computer-aided constructions
of parallelisms with certain predefined features (usually given automorphism
groups as in [3,16,18,20,22,24,25]) contribute significantly to the study of the
properties and possible applications of parallelisms.

The research of both authors is partially supported by the Bulgarian National Science
Fund under Contract No KP-06-N32/2-2019.

P. Flocchini and L. Moura (Eds.): IWOCA 2021, LNCS 12757, pp. 544–557, 2021.
https://doi.org/10.1007/978-3-030-79987-8_38

Parallelisms are related to translation planes [23], network coding [8], error-correcting codes [11], and cryptography [21]. Of major importance is their relation to 2-designs.

Let $V = \{P_i\}_{i=1}^{v}$ be a finite set of *points*, and $\mathcal{B} = \{B_j\}_{j=1}^{b}$ a finite collection of k-element subsets of V, called *blocks*. $D = (V, \mathcal{B})$ is a *2-design* with parameters 2-(v,k,λ) if any 2-subset of V is contained in exactly λ blocks of $\mathcal{B}$. A *parallel class* is a set of blocks such that each point is in exactly one block. A *resolution* of the design is a partition of the collection of blocks to parallel classes.

The point-line incidence in $PG(n, q)$ defines a 2-design [23, 2.35-2.36] which is called the point-line design. There is a one-to-one correspondence between parallelisms and the resolutions of this design. That is why constructing parallelisms is sometimes considered as identical to the more general problem of constructing resolutions. Point-line $PG(n, q)$ designs, however, are Steiner 2-designs with very rich automorphism groups and plenty of resolutions. This imposes some specific problems. We have already constructed some of the parallelisms in $PG(3, 4)$ [3,4,24,25,27] and $PG(3, 5)$ [26,28,30]. In the present paper we would like to share our experience in doing this by backtrack search algorithms. The algorithms described here can also be used for the construction of resolutions of designs, and in particular, of Steiner 2-designs, but we only discuss on their performance for the case of parallelisms (resolutions of point-line designs) for which case we already have quite a lot of experimental data.

Backtrack search is exponential. It can only be successful with problems in which relatively small parameters are concerned, many restrictions are set and specific mathematical properties of the constructed objects can be taken into account by the algorithm. The basic types of backtrack search algorithms for classification of combinatorial structures are described by Kaski and Östergård in [14]. Our approach is based on the algorithm known as *orderly generation* [14] proposed by Faradžev [10] and Read [19].

Before us a computer-aided search for parallelisms has been used by Stinson and Vanstone [22], Fuji-Hara [9], Prince [16–18], Sarmiento [20], Braun [6], and recently by Betten [2]. Their algorithms are not universal (the same holds for the algorithms that we have used so far). They are based on theoretical results implying the specifics of the respective parameters, properties and restrictions. Moreover, compared to the theoretical considerations, the computer-aided work in some papers (for instance [18,20]) is considered minor and trivial and almost not described. It is therefore very difficult to compare the algorithms that are implemented, and this is **not** our aim here. There is something common in all these works, however. In all of them the authors first consider the spreads of the projective space, determine which of them can possibly take part in the parallelisms which have to be constructed, and construct these spreads. After that they use the possible spreads to obtain parallelisms.

We have followed the same procedure, but to partially check the correctness of our computer-aided results, we have independently used two basically different algorithms for the construction of parallelisms. They are described in Sect. 2 of the present paper together with the general framework in which they can be applied. The main difference between them is that in Algorithm 1 the backtrack search is on the spreads that have been constructed in advance, while in Algorithm 2 it is on the lines of $PG(n, q)$. Algorithm 2 takes into consideration the restrictions which are imposed on the parallelisms, but it practically does not use the results of the preliminary investigation on the spreads of the projective space. We have not found such an approach described in the works of other authors.

Our aim is to show that Algorithm 2 also has its advantages in some cases. Section 3 contains observations on the performance of the two algorithms with respect to the particularity of the considered problem and suitable examples. Possible ways to modify these algorithms for usage on parallel computers are considered in Sect. 4, and concluding remarks can be found in Sect. 5.

2   Two Construction Algorithms

2.1   Framework

We will start with a brief description of the framework in which each of these algorithms can be used. Before applying one of the algorithms, we have found generators of the automorphism group of the point-line 2-design and have sorted its blocks (they correspond to the lines of the projective space) with respect to a defined lexicographic order. Denote by v the number of points, and by b the number of lines. Let k be the number of points in one line, and let r be the number of lines containing a definite point. For some purposes it might be more convenient to treat the lines as sets of points. This is illustrated in Fig. 1 by an example for $PG(3, 2)$.

The point-line incidence matrix of $PG(3, 2)$ is an incidence matrix of a 2-(15, 3, 1) design

P\B	1	2	3	4	5	6	7	8	9	10	11	12	13	14	15	16	17	18	19	20	21	22	23	24	25	26	27	28	29	30	31	32	33	34	35
1	1	1	1	1	1	1	1	0	0	0	0	0	0	0	0	0	0	0	0	0	0	0	0	0	0	0	0	0	0	0	0	0	0	0	0
2	1	0	0	0	0	0	0	1	1	1	1	1	1	0	0	0	0	0	0	0	0	0	0	0	0	0	0	0	0	0	0	0	0	0	0
3	1	0	0	0	0	0	0	0	0	0	0	0	0	1	1	1	1	1	1	0	0	0	0	0	0	0	0	0	0	0	0	0	0	0	0
4	0	1	0	0	0	0	0	1	0	0	0	0	0	1	0	0	0	0	0	1	1	1	1	0	0	0	0	0	0	0	0	0	0	0	0
5	0	1	0	0	0	0	0	0	1	0	0	0	0	0	1	0	0	0	0	0	0	0	0	1	1	1	1	0	0	0	0	0	0	0	0
6	0	0	1	0	0	0	0	1	0	0	0	0	0	0	1	0	0	0	0	0	0	0	0	0	0	0	0	1	1	1	1	0	0	0	0
7	0	0	1	0	0	0	0	0	1	0	0	0	0	1	0	0	0	0	0	0	0	0	0	0	0	0	0	0	0	0	0	1	1	1	1
8	0	0	0	1	0	0	0	0	0	1	0	0	0	0	0	1	0	0	0	1	0	0	0	1	0	0	0	1	0	0	0	1	0	0	0
9	0	0	0	1	0	0	0	0	0	0	1	0	0	0	0	0	1	0	0	0	1	0	0	0	1	0	0	0	1	0	0	0	1	0	0
10	0	0	0	0	1	0	0	0	0	1	0	0	0	0	0	0	1	0	0	0	0	1	0	0	0	1	0	0	0	1	0	0	0	1	0
11	0	0	0	0	1	0	0	0	0	0	1	0	0	0	0	1	0	0	0	0	0	0	1	0	0	0	1	0	0	0	1	0	0	0	1
12	0	0	0	0	0	1	0	0	0	0	0	1	0	0	0	0	0	1	0	1	0	0	0	0	1	0	0	0	0	1	0	0	0	0	1
13	0	0	0	0	0	1	0	0	0	0	0	0	1	0	0	0	0	0	1	0	1	0	0	1	0	0	0	0	0	0	1	0	0	1	0
14	0	0	0	0	0	0	1	0	0	0	0	1	0	0	0	0	0	0	1	0	0	1	0	0	0	0	1	1	0	0	0	0	1	0	0
15	0	0	0	0	0	0	1	0	0	0	0	0	1	0	0	0	0	1	0	0	0	0	1	0	0	1	0	0	1	0	0	1	0	0	0

The lines by their points

b_1	b_2	b_3	b_4	b_5	b_6	b_7	b_8	b_9	b_{10}	b_{11}	b_{12}	b_{13}	b_{14}	b_{15}	b_{16}	b_{17}	b_{18}	b_{19}	b_{20}	b_{21}	b_{22}	b_{23}	b_{24}	b_{25}	b_{26}	b_{27}	b_{28}	b_{29}	b_{30}	b_{31}	b_{32}	b_{33}	b_{34}	b_{35}
1	1	1	1	1	1	1	2	2	2	2	2	2	3	3	3	3	3	3	4	4	4	4	5	5	5	5	6	6	6	6	7	7	7	7
2	4	6	8	10	12	14	4	5	8	9	12	13	4	5	8	9	12	13	8	9	10	11	8	9	10	11	8	9	10	11	8	9	10	11
3	5	7	9	11	13	15	6	7	10	11	14	15	7	6	11	10	15	14	12	13	14	15	13	12	15	14	14	15	12	13	15	14	13	12

Fig. 1. The point-line incidence of $PG(3, 2)$

A parallelism contains r spreads which can (for the sake of the construction) be considered as spread 1, spread 2, ... spread r. The lines incident with the i-th point should be in different spreads. That is why without loss of generality we can choose a spread for each line containing the first point. This can be seen on the parallelism of PG(3, 2) from Fig. 2.a where it is assumed that line b_1 is in the first spread, b_2 in the second, etc. This way we have one predefined line in each spread.

a) A parallelism of $PG(3,2)$ has 7 spreads of 5 lines each

b_1 b_{20} b_{26} b_{31} b_{33}	b_2 b_{10} b_{19} b_{29} b_{35}	b_3 b_{11} b_{18} b_{22} b_{24}	b_4 b_{12} b_{15} b_{23} b_{34}	b_5 b_{13} b_{14} b_{25} b_{28}	b_6 b_8 b_{17} b_{27} b_{32}	b_7 b_9 b_{16} b_{21} b_{30}
1 4 5 6 7	1 2 3 6 7	1 2 3 4 5	1 2 3 4 7	1 2 3 5 6	1 2 3 5 7	1 2 3 4 6
2 8 10 11 9	4 8 13 9 11	6 9 12 10 8	8 12 5 11 10	10 13 4 9 8	12 4 9 11 8	14 5 8 9 10
3 12 15 13 14	5 10 14 15 12	7 11 15 14 13	9 14 6 15 13	11 15 7 12 14	13 6 10 14 15	15 7 11 13 12

b) number of lines to consider at each step

b_1 b_{20} b_{26} b_{31} b_{33}	b_2 b_{10} b_{19} b_{29} b_{35}	b_3 b_{11} b_{18} b_{22} b_{24}	b_4 b_{12} b_{15} b_{23} b_{34}	b_5 b_{13} b_{14} b_{25} b_{28}	b_6 b_8 b_{17} b_{27} b_{32}	b_7 b_9 b_{16} b_{21} b_{30}
1 4 4 4 4	1 6 6 3 3	1 5 5 3 3	1 4 4 2 2	1 3 3 2 2	1 2 2 1 1	1 1 1 1 1

c) number of lines to put at each step

b_1 b_{20} b_{26} b_{31} b_{33}	b_2 b_{10} b_{19} b_{29} b_{35}	b_3 b_{11} b_{18} b_{22} b_{24}	b_4 b_{12} b_{15} b_{23} b_{34}	b_5 b_{13} b_{14} b_{25} b_{28}	b_6 b_8 b_{17} b_{27} b_{32}	b_7 b_9 b_{16} b_{21} b_{30}
1 4 2 1 1	1 4 2 1 1	1 3 1 1 1	1 4 2 1 1	1 3 1 1 1	1 2 2 1 1	1 1 1 1 1

Fig. 2. A parallelism of $PG(3,2)$

We next consider the restrictions which should hold for the constructed parallelisms (automorphism groups, special type of the spreads, etc.) and decide at which stage to take them in account. Then we apply Algorithm 1 or Algorithm 2 to construct all the desired parallelisms.

2.2   Algorithm 1

Algorithm 1, Part 1. In Algorithm 1 we first have to obtain, sort in a convenient way, and save all the spreads that can possibly take part in a parallelism which we want to construct. We use backtrack search illustrated by the following code segment, where `SpreadConstruct(2)` is called to obtain all the spreads containing a given first line.

`SpreadConstruct(int Line)` starts with finding `Point` - the first point that is not contained in any of the (`Line-1`) lines which have already been added to the spread. `FirstLine[Point]` is the first line (the lines are sorted in lexicographic order) which is incident with `Point`, and `LastLine[Point]` is the last one. `Possible(Line, li)` checks if the line `li` can be part of the current spread, `Put(Line, li)` adds the line `li` to the spread and `Take(Line)` removes it.

If `Line` is the last line in the spread, the spread is ready and `WriteSpread()` is called. If more lines have to be added, `SpreadConstruct (Line+1)` is called to choose the next line. The steps in the construction of the eight spreads of PG(3, 2) containing line b_1, are presented in Fig. 3.

```
void SpreadConstruct(int Line)
{
  int Point = FirstMissingPoint(Line);
  for(int li=FirstLine[Point]; li<=LastLine[Point]; li++)
  {
    if(Possible(Line, li))
    {
      Put(Line, li);
      if(Line==v/k) WriteSpread();
      else SpreadConstruct(Line+1);
      Take(Line);
    }
  }
}
```

The candidates for the first spread of the constructed parallelisms are further tested for isomorphism, and only nonisomorphic ones remain. In the example for $PG(3,2)$ (Fig. 3) only one of the spreads containing b_1 remains.

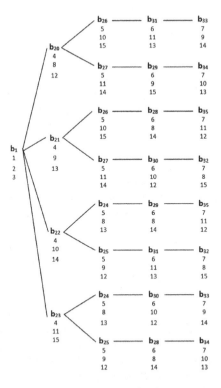

Fig. 3. The eight spreads of $PG(3,2)$ containing line b_1

Algorithm 1, Part 2. After all the possible spreads have been constructed, sorted and saved in a convenient way, the main part of Algorithm 1 starts, namely the parallelisms are constructed by a backtrack search on the possible spreads. The search is illustrated in the following code segment.

```
void ParallConstr(int SpreadNum)
{
  for(int i=1; i<=NumberOfVariants[SpreadNum]; i++)
  {
    if(SpreadOK(SpreadNum, i))
    {
      PutSpread(SpreadNum, i);
      if(SpreadNum==r) WriteParallelism();
      else ParallConstr(SpreadNum+1);
      TakeSpread(SpreadNum);
    }
  }
}
```

ParallConstr(1) is called to construct the parallelisms. The function considers all spread variants for the SpreadNum-th spread. SpreadOK(SpreadNum, i) is called to check if the i-th spread variant can extend the obtained until this moment partial parallelism. If this is the case, the function PutSpread(SpreadNum, i) adds it to the current solution.

If all the r spreads are already in the partial parallelism, Write Parallelism() is called to write the obtained solution, otherwise ParallConstr (SpreadNum+1) attempts to add the next spread.

2.3   Algorithm 2

Without loss of generality we assume that the first line in the i-th spread is b_i (the grey lines in Fig. 2.a). This algorithm constructs the parallelisms by adding the other spread lines in a consequent way that is illustrated by the code segment presented below, where the parallelisms are constructed by MakeParall(2, 1).

The first missing point in spread SpreadNum is denoted by Point and found by FirstMissingPoint(Line, SpreadNum). The lines that we try to add to the current partial solution are between FirstLine[Point] and LastLine[Point]. Possible(Line, li, SpreadNum) checks if the current partial spread can be extended with the line in consideration. If this is the case, Put(Line, li, SpreadNum) adds it to the current solution.

We continue to add lines until all points are covered (i.e. the number of lines is v/k) and when this happens, the function MakeParall(2, SpreadNum+1) starts adding the next spread. If there are already r spreads in the partial parallelism, WriteParallelism() is called.

```
void MakeParall(int Line, int SpreadNum)
{
  int Point = FirstMissingPoint(Line, SpreadNum);
  for(int li = FirstLine[Point]; li<=LastLine[Point]; li++)
  {
    if(Possible(Line, li, SpreadNum))
    {
      Put(Line, li, SpreadNum);
      if(Line==v/k)
      {
        if(SpreadNum==r) WriteParallelism();
        else MakeParall(2, SpreadNum+1);
      }
      else MakeParall(Line+1, SpreadNum);
      Take(Line, SpreadNum);
    }
  }
}
```

2.4   Restrictions

If the construction of all parallelisms of $PG(n, q)$ is infeasible, families of parallelisms with definite properties are usually constructed. The imposed restrictions can make the classification feasible because they narrow the search space.

The restrictions might be on the type of the spreads, or on the whole parallelisms. If the spreads of the projective space have been classified, then we might want to construct only parallelisms which have spreads belonging to a given isomorphism class, or spreads with some geometric properties. Such restrictions are usually easy to take in consideration in the functions Possible(int Line, int li) and Possible(int Line, int li, int SpreadNum) of Algorithms 1 and 2 respectively. Depending on the desired property, a check for it might be applied either for all, or only for some values of Line.

Out of the restrictions on the whole parallelisms, most popular are the predefined automorphism groups. Such a group is a subgroup of the automorphism group of the projective space that preserves the parallelism. Consider parallelisms invariant under a subgroup G_p of prime order p. The orbit of a spread under G_p can be either of length 1, or of length p. In the first case it is made of several whole line orbits, and in the second of lines from different line orbits under G_p. This is illustrated in Fig. 4 by an example of a parallelism of $PG(3, 2)$ with an assumed automorphism of order 3. It is not difficult to see that in this case parallelisms must consist of one fixed spread and two spread orbits of length 3. The number of possible spreads with a predefined line is 8 (Fig. 3), but with respect to the assumed automorphism, there are 2 possibilities for the fixed spread and 3 for a spread (with one predefined line) with a spread orbit of length 3.

The predefined automorphism group G_3 acts on the points as:
$(1)(2)(3)(4, 8, 12)(5, 9, 13)(6, 10, 14)(7, 11, 15)$,
and on the lines as: $(b_1)(b_2, b_4, b_6)$ (b_3, b_5, b_7) (b_8, b_{10}, b_{12}) (b_9, b_{11}, b_{13}) (b_{14}, b_{16}, b_{18}) (b_{15}, b_{17}, b_{19}) (b_{20}) (b_{21}, b_{24}, b_{25}) (b_{22}, b_{28}, b_{30}) (b_{23}, b_{32}, b_{35}) (b_{26}, b_{33}, b_{31}) (b_{27}, b_{29}, b_{34}).
The three spread orbits under G_3 are:

An orbit of length 1: the spread is made of line orbits $\{b_1\}$, $\{b_{20}\}$ and $\{b_{26}, b_{31}, b_{33}\}$

b_1	b_{20}	b_{26}	b_{31}	b_{33}
1	4	5	6	7
2	8	10	11	9
3	12	15	13	14

An orbit of length 3: the spreads contain one line from each of the line orbits $\{b_2, b_4, b_6\}$, $\{b_8, b_{10}, b_{12}\}$, $\{b_{15}, b_{17}, b_{19}\}$, $\{b_{23}, b_{32}, b_{35}\}$, and $\{b_{27}, b_{29}, b_{34}\}$

b_2	b_{10}	b_{19}	b_{29}	b_{35}		b_4	b_{12}	b_{15}	b_{23}	b_{34}		b_6	b_8	b_{17}	b_{27}	b_{32}
1	2	3	6	7		1	2	3	4	7		1	2	3	5	7
4	8	13	9	11		8	12	5	11	10		12	4	9	11	8
5	10	14	15	12		9	14	6	15	13		13	6	10	14	15

An orbit of length 3: the spreads contain one line from each of the line orbits $\{b_3, b_5, b_7\}$, $\{b_9, b_{11}, b_{13}\}$, $\{b_{14}, b_{16}, b_{18}\}$, $\{b_{21}, b_{24}, b_{25}\}$, $\{b_{22}, b_{28}, b_{30}\}$

b_3	b_{11}	b_{18}	b_{22}	b_{24}		b_5	b_{13}	b_{14}	b_{25}	b_{28}		b_7	b_9	b_{16}	b_{21}	b_{30}
1	2	3	4	5		1	2	3	5	6		1	2	3	4	6
6	9	12	10	8		10	13	4	9	8		14	5	8	9	10
7	11	15	14	13		11	15	7	12	14		15	7	11	13	12

Fig. 4. A parallelism of $PG(3, 2)$ with a predefined automorphism of order 3

Parallelisms of $PG(3, 2)$ are the smallest possible example. When larger parameters are concerned and automorphism groups are assumed, there usually are many spreads that are neither made of whole line orbits, nor contain lines from different line orbits of the same length. So these spreads are excluded from the search set for Part 2 of Algorithm 1.

2.5   Isomorphism Test

We use the function SmallerExists(int SprNum) which checks if there exists an automorphism of the projective space which maps the current partial solution of SprNum spreads to a lexicographically smaller one, and returns *true* if so. A call to SmallerExists (int SprNum) is included in the function SpreadOK(int SpreadNum, int var) of Algorithm 1 for some values of SpreadNum and in the function Possible(int Line, int li, int SpreadNum) of Algorithm 2 for some values of SpreadNum if Line==v/k.

We have to mention here that the main difference between our approach and the *orderly generation* algorithm [14] is that we do not call SmallerExists for each value of SpreadNum. We apply the isomorphism test only to some of the partial solutions and to all complete parallelisms that are obtained. Different choices of the set of values of SpreadNum for which the test is applied, might lead to considerable differences in the computation time.

If we construct parallelisms with a predefined automorphism group G_p, the check in SmallerExists(int SprNum) covers only the automorphisms from the normalizer of G_p in the automorphism group of the projective space. If G_p is not a Sylow subgroup, such a normalizer-based test might not reject all isomorphic solutions, and it will be necessary to apply another type of isomorphism test on the obtained parallelisms [29].

3   Comparison

Algorithm 1 is intuitively expected to be the better one, but our experience shows that the computation time needed by the two algorithms to construct the parallelisms has been quite similar in most of the problems that we have considered so far. The main reason is that by backtrack search on the lines, the number of lines which have to be considered at each step is big for the lines of the first several spreads of the parallelism, but decreases for the next spreads. This is illustrated in Fig. 2.b. The number of lines which meet the requirements decreases too (Fig. 2.c). This significantly helps Algorithm 2 to compete with Algorithm 1. Moreover, Algorithm 1 avoids the repeated backtracking that Algorithm 2 does for the construction of the spreads within the parallelism, but Part 2 of Algorithm 1 usually performs a search on a set of much bigger cardinality than the line set.

Example 1. Parallelisms of $PG(3,4)$ with automorphism groups of order 7 are constructed in [24]. Algorithm 2 performs a search on the 357 lines of the projective space, while Part 2 of Algorithm 1 searches on the possible spreads that have been constructed in advance. There are 22860 spreads containing a given line and having an orbit of length 7. Three such orbits are needed to obtain a parallelism. The computation times of both algorithms are comparable.

Algorithm 2 usually performs worse when there are restrictions on the type of the spreads because usually a considerable number of lines have to be added to the spread before the nonexistence of the corresponding spread property can be established, and therefore many not applicable partial solutions are not rejected.

Example 2. Subregular parallelisms of $PG(3,4)$ with a given automorphism group of order 2 are constructed in [4]. There are 691968 possible spreads with orbits of length 2, but only 213760 of them are subregular, and Part 2 of Algorithm 1 performs a search on them. In this case Algorithm 2 can establish that a spread is not subregular when at least 13 of all its 17 lines are added. Algorithm 2 needs 5 times more time to find the parallelisms.

Algorithm 1 is difficult to use if the number of the spreads in the search set is too big and there is not enough memory for all of them. The usage of disk memory is possible (we have done it for the construction in [27]), but the computation time rises significantly. And if the number of possible spreads is extremely big, Algorithm 1 is practically unusable, and Algorithm 2 has to be used instead.

Example 3. Parallelisms of PG(3, 5) with a cyclic automorphism group of order 8 are considered in [30]. There are too many parallelisms, and therefore we impose some restrictions on the partial parallelisms consisting of spread orbits of length less than 8, namely we extend to parallelisms only 19 partial solutions with rich automorphism groups. The extension implies the addition of three spread orbits of length 8. There are, however, 14227090 possibilities for each spread with an orbit of length 8, and this is too much for Algorithm 1. That is why we do it using Algorithm 2.

4   Parallel Versions

There are plenty of ways to create a parallel version of an algorithm. We established that a very simple communication-free MPI based implementation of the search for parallelisms on a parallel computer might be quite useful in many cases. The parallel implementations of the two algorithms might look like that:

Algorithm 1, Part 2:

```
int aa=0;
void ParallConstr(int SpreadNum)
{
  for(int i=1; i<=NumberOfVariants[SpreadNum]; i++)
  {
    if(SpreadOK(SpreadNum, i))
    {
      if(SpreadNum==N)
      {
        aa++;
        if((aa%prnum)!=mynum) continue;
      }
      PutSpread(SpreadNum, i);
      if(SpreadNum==r) WriteParallelism();
      else ParallConstr(SpreadNum+1);
      TakeSpread(SpreadNum);
    }
  }
}
```

Algorithm 2:

```
int aa=0;
void MakeParall(int Line, int SpreadNum)
{
  int Point = FirstMissingPoint(Line, SpreadNum);
  for(int li = FirstLine[Point]; li<=LastLine[Point]; li++)
  {
    if(Possible(Line, li, SpreadNum))
    {
      if(SpreadNum==N&&Line==L)
      {
        aa++;
        if((aa%prnum)!=mynum) continue;
      }
      Put(Line, li, SpreadNum);
      if(Line==v/k)
      {
        if(SpreadNum==r) WriteParallelism();
        else MakeParall(2, SpreadNum+1);
      }
      else MakeParall(Line+1, SpreadNum);
      Take(Line, SpreadNum);
    }
  }
}
```

We denote by **prnum** the number of processes, and by **mynum** the number of this process. Let N and L be positive integer constants not greater than the number of spreads in the parallelism and the number of lines in a spread respectively. And let **aa** be a global variable that counts the number of partial solutions of a definite size. Splitting the work among the processes can be done by allowing each process to extend only those partial solutions of the definite size whose number modulo **prnum** equals **mynum**.

It is of major importance to choose a suitable size of the partial solutions after which the job is split to the different processes. On the one hand, we want that the time to obtain these partial solutions will be relatively short, so that the time when all processes do the same job will be as small as possible. On the other hand, we want the number of the partial solutions of the chosen size to be relatively big, so that each process will extend many of them, because in that case the differences between the running times of the processes will be negligible.

If we use Algorithm 1 we can only choose partial solutions containing a definite number of whole spreads (**SpreadNum==N**), while by Algorithm 2 we may choose to split the job after a partial solution containing several whole spreads and several lines of the next spread (**SpreadNum==N&&Line==L**). That is why in some cases Algorithm 2 might be more suitable for a successful parallel implementation in the way described above.

5 Conclusion

Each problem which needs computer-aided search for parallelisms, has specifics that have to be carefully considered before organizing the search. We have tried to point out some cases when one of the two algorithms described here is less applicable than the other one. In most of the cases, however, both algorithms can be successfully used and the running times are comparable. This is particularly useful for checking the correctness of the results in two different ways.

Acknowledgements. For testing the algorithms from Sect. 4 the authors acknowledge the provided access to the e-infrastructure of the NCHDC – part of the Bulgarian National Roadmap on RIs, with the financial support by the Grant No D01-221/03.12.2018.

References

1. Baker, R.D.: Partitioning the planes of $AG_{2m}(2)$ into 2-designs. Discrete Math. **15**, 205–211 (1976). https://doi.org/10.1016/0012-365X(76)90025-X
2. Betten, A.: The packings of PG(3,3). Des. Codes Cryptogr. **79** (3), 583–595 (2016). https://doi.org/10.1007/s10623-015-0074-6.
3. Betten, A., Topalova, S., Zhelezova, S.: Parallelisms of $PG(3, 4)$ invariant under cyclic groups of order 4. In: Ćirić, M., Droste, M., Pin, J.É. (eds.) CAI 2019. LNCS, vol. 11545, pp. 88–99. Springer, Cham (2019). https://doi.org/10.1007/978-3-030-21363-3_8
4. Betten, A., Topalova, S., Zhelezova, S.: New uniform subregular parallelisms of PG(3,4) invariant under an automorphism of order 2. Cybern. Inf. Technol. **20**(6), 18–27 (2020). https://doi.org/10.2478/cait-2020-0057
5. Beutelspacher, A.: On parallelisms in finite projective spaces. Geom. Dedicata. **3**(1), 35–40 (1974). https://doi.org/10.1007/BF00181359
6. Braun, M.: Construction of a point-cyclic resolution in $PG(9, 2)$. Innov. Incid. Geom. Algebr. Topol. Comb. **3**(1), 33–50 (2006). https://doi.org/10.2140/iig.2006.3.33
7. Denniston, R. H. F.: Some packings of projective spaces. Atti Accad. Naz. Lincei Rend. Cl. Sci. Fis. Mat. Natur. **52**(8), 36–40 (1972)
8. Etzion, T., Silberstein, N.: Codes and designs related to lifted MRD codes. IEEE Trans. Inform. Theory **59**(2), 1004–1017 (2013). https://doi.org/10.1109/ISIT.2011.6033969
9. Fuji-Hara, R.: Mutually 2-orthogonal resolutions of finite projective space. Ars Combin. **21**, 163–166 (1986)
10. Faradžev, I. A.: Constructive enumeration of combinatorial objects. In Problèmes Combinatoires et Théorie des Graphes, (Université d'Orsay, July 9–13, 1977). Colloq. Internat. du C.N.R.S., vol. 260, pp. 131–135, CNRS, Paris (1978)
11. Gruner, A., Huber, M.: New combinatorial construction techniques for low-density parity-check codes and systematic repeat-accumulate codes. IEEE Trans. Commun. **60**(9), 2387–2395 (2012). https://doi.org/10.1109/TCOMM.2012.070912.110164
12. Johnson, N.L.: Some new classes of finite parallelisms, Note Mat. **20**(2), 77–88 (2000). https://doi.org/10.1285/i15900932v20n2p77

13. Johnson, N.L.: Combinatorics of Spreads and Parallelisms. Pure and Applied MathematicsPure and Applied Mathematics, Chapman & Hall. CRC Press, Boca Raton (2010)
14. Kaski, P., Östergård, P.: Classification algorithms for codes and designs. Algorithms and Computation in Mathematics, vol. 15. Springer, Heidelberg (2006). https://doi.org/10.1007/3-540-28991-7
15. Penttila, T., Williams, B.: Regular packings of $PG(3,q)$. Eur. J. Comb. **19**(6), 713–720 (1998)
16. Prince, A.R.: Parallelisms of $PG(3,3)$ invariant under a collineation of order 5. In: Johnson, N.L. (ed.) Mostly Finite Geometries. Lecture Notes in Pure and Applied Mathematics, vol. 190, pp. 383–390. Marcel Dekker, New York (1997)
17. Prince, A.R.: Uniform parallelisms of PG(3,3). In: Hirschfeld, J., Magliveras, S., Resmini, M. (eds.) Geometry, Combinatorial Designs and Related Structures, London Mathematical Society Lecture Note Series, vol. 245, pp. 193–200. Cambridge University Press, Cambridge (1997). https://doi.org/10.1017/CBO9780511526114.017
18. Prince, A.R.: The cyclic parallelisms of $PG(3,5)$. Eur. J. Comb. **19**(5), 613–616 (1998)
19. Read, R.C.: Every one a winner; or, how to avoid isomorphism search when cataloguing combinatorial configurations. Ann. Discrete Math. **2**, 107–120 (1978). https://doi.org/10.1016/S0167-5060(08)70325-X
20. Sarmiento, J.: Resolutions of $PG(5,2)$ with point-cyclic automorphism group. J. Comb. Des. **8**(1), 2–14 (2000). https://doi.org/10.1002/(SICI)1520-6610(2000)8:1⟨2::AID-JCD2⟩3.0.CO;2-H
21. Stinson, D.R.: Combinatorial Designs: Constructions and Analysis. Springer, New York (2004). https://doi.org/10.1007/b97564
22. Stinson, D.R., Vanstone, S.A.: Orthogonal packings in $PG(5,2)$. Aequationes Math. **31**(1), 159–168 (1986). https://doi.org/10.1007/BF02188184
23. Storme, L.: Finite Geometry. In: Colbourn, C., Dinitz, J. (eds.) Handbook of Combinatorial Designs. 2nd edn. Rosen, K. (eds.) Discrete mathematics and its applications, pp. 702–729. CRC Press, Boca Raton (2007)
24. Topalova, S., Zhelezova, S.: On transitive parallelisms of $PG(3,4)$. Appl. Algebra Engrg. Comm. Comput. **24**(3–4), 159–164 (2013). https://doi.org/10.1007/s00200-013-0194-z
25. Topalova, S., Zhelezova, S.: On point-transitive and transitive deficiency one parallelisms of $\boldsymbol{PG}(3,4)$. Designs, Codes Cryptogr. **75**(1), 9–19 (2013). https://doi.org/10.1007/s10623-013-9887-3
26. Topalova, S., Zhelezova, S.: New Regular Parallelisms of $PG(3,5)$. J. Comb. Des. **24**, 473–482 (2016). https://doi.org/10.1002/jcd.21526
27. Topalova, S., Zhelezova, S.: New parallelisms of $PG(3,4)$. Electron. Notes Discrete Math. **57**, 193–198 (2017). https://doi.org/10.1016/j.endm.2017.02.032
28. Topalova, S., Zhelezova, S.: Types of spreads and duality of the parallelisms of $PG(3,5)$ with automorphisms of order 13. Des. Codes Cryptogr. **87** (2–3), 495–507 (2019). https://doi.org/10.1007/s10623-018-0558-2
29. Topalova, S., Zhelezova, S.: Isomorphism and invariants of parallelisms of projective spaces. In: Bigatti, A., Carette, J., Davenport, J., Joswig, M., De Wolff, T. (eds.), Mathematical Software - ICMS 2020, Lecture Notes in Computer Science, vol. 12097, pp. 162–172, Springer, Cham (2020). https://doi.org/10.1007/978-3-030-52200-1_16

30. Topalova, S., Zhelezova, S.: Some parallelisms of $PG(3,5)$ involving a definite type of spread. In: 2020 Algebraic and Combinatorial Coding Theory (ACCT), pp. 135–139 (2020). https://doi.org/10.1109/ACCT51235.2020.9383404
31. Zaicev, G., Zinoviev, V., Semakov, N.: Interrelation of Preparata and Hamming codes and extension of Hamming codes to new double-error-correcting codes. In: Proceedings of Second International Symposium on Information Theory, (Armenia, USSR, 1971), Budapest, Academiai Kiado, pp. 257–263 (1973)

Approximating Multistage Matching Problems

Markus Chimani[iD], Niklas Troost[(✉)][iD], and Tilo Wiedera[iD]

Theoretical Computer Science, Osnabrück University, Osnabrück, Germany
{markus.chimani,niklas.troost,tilo.wiedera}@uos.de

Abstract. In *multistage perfect matching* problems, we are given a sequence of graphs on the same vertex set and are asked to find a sequence of perfect matchings, corresponding to the sequence of graphs, such that consecutive matchings are as similar as possible. More precisely, we aim to maximize the intersections, or minimize the unions between consecutive matchings.

We show that these problems are NP-hard even in very restricted scenarios. As our main contribution, we present the first non-trivial approximation algorithms for these problems: On the one hand, we devise a tight approximation on graph sequences of length two (2-stage graphs). On the other hand, we propose several general methods to deduce multistage approximations from blackbox approximations on 2-stage graphs.

Keywords: Temporal graphs · Approximation algorithms · Perfect matchings

1 Introduction

The study of graphs that evolve over time emerges naturally in several applications. As such, it is a well-known subject in graph theory [1–10,12–16,19,22]. While there are many possible approaches to model these problems (cf. the discussion of related work), the paradigm of *multistage graphs* has attracted quite some attention in recent years [2–5,13–15]. In this setting, we are given a sequence of separate, but related graphs (*stages*). A typical goal is to find a sequence of solutions for each individual graph such that the change in the solutions between consecutive graphs is minimized. Since multistage graph problems usually turn out to be NP-hard, one often resorts to FPT- or approximation algorithms. To the best of our knowledge, all approximation results in this setting discuss combined objective functions that reflect a trade-off between the quality of each individual solution and the cost of the change over time (cf., e.g., [3,14]). However, this is a drawback if one requires each stage's solution to attain a certain quality guarantee (e.g., optimality). Trying to ensure this by adjusting the trade-off weights in the above approximation algorithms leads to approximation ratios that no longer effectively bound the cost of change. Here, we discuss a multistage graph problem where each individual solution is necessarily optimal,

© Springer Nature Switzerland AG 2021
P. Flocchini and L. Moura (Eds.): IWOCA 2021, LNCS 12757, pp. 558–570, 2021.
https://doi.org/10.1007/978-3-030-79987-8_39

but we can still obtain an approximation ratio on the cost of the change over time.

A classical example are *multistage matching* problems, i.e., natural multistage generalizations of traditional matching problems (e.g., perfect matching, maximum weight matching, etc.). This is particularly interesting as optimality for a single stage would be obtainable in polynomial time, but all known multistage variants are NP-hard already for two stages. There are several known approximation algorithms for multistage matching problems [3]; however, they all follow the trade-off paradigm.

In this paper, we are concerned with maintaining a perfect matching on a multistage graph, such that the changes between consecutive matchings are minimized. After showcasing the complexities of our problems (Sect. 2), we will devise efficient approximation algorithms (Sect. 3).

Definitions and Preliminaries. Let $G = (V, E)$ be an undirected graph. For a set $W \subseteq V$ of vertices, let $\delta(W) := \{uv \in E \mid u \in W, v \in V \setminus W\}$ denote the set of its cut edges. For a singleton $\{v\}$, we may write $\delta(v)$ instead of $\delta(\{v\})$. A set $M \subseteq E$ of edges is a *matching* if every vertex is incident to at most one edge of M; it is a *perfect* matching if $|\delta(v) \cap M| = 1$ for every $v \in V$. A k-*cycle* (k-*path*) is a cycle (path, respectively) consisting of exactly k edges. The parity of a k-cycle is the parity of k. For a set $F \subseteq E$ of edges, let $V(F) := \{v \in V \mid \delta(v) \cap F \neq \varnothing\}$ denote its incident vertices.

For $x \in \mathbb{N}$, we define $[x] := \{1, ..., x\}$ and $[\![x]\!] := \{0\} \cup [x]$. A *temporal graph* (or τ-*stage graph*) is a tuple $\mathcal{G} = (V, E_1, ..., E_\tau)$ consisting of a vertex set V and multiple edge sets E_i, one for each $i \in [\tau]$. The graph $G_i := (V(E_i), E_i)$ is the *i*th *stage* of $\mathcal{G}$. We define $n_i := |V(E_i)|$, and $n := |V|$. A temporal graph is *spanning* if $V(E_i) = V$ for each $i \in [\tau]$.

Let $\mu := \max_{i \in [\tau-1]} |E_i \cap E_{i+1}|$ denote the maximum number of edges that are common between two adjacent stages. Let $E_\cap := \bigcap_{i \in [\tau]} E_i$ and $E_\cup := \bigcup_{i \in [\tau]} E_i$. The graph $G_\cup := (V(E_\cup), E_\cup)$ is the *union graph* of $\mathcal{G}$. A *multistage perfect matching* in $\mathcal{G}$ is a sequence $\mathcal{M} := (M_i)_{i \in [\tau]}$ such that for each $i \in [\tau]$, M_i is a perfect matching in G_i.

All problems considered in this paper (MIM, MUM, Min-MPM, Max-MPM; see below) are of the following form: Given a temporal graph $\mathcal{G}$, we ask for a multistage perfect matching $\mathcal{M}$ optimizing some objective function. In their respective decision variants, the input furthermore consists of some value κ and we ask whether there is an $\mathcal{M}$ with objective value at most (minimization problems) or at least (maximization problems) κ.

Definition 1 (MIM and τ-IM). *Given a temporal graph $\mathcal{G}$, the multistage intersection matching problem (MIM) asks for a multistage perfect matching $\mathcal{M}$ of $\mathcal{G}$ with maximum profit $p(\mathcal{M}) := \sum_{i \in [\tau-1]} |M_i \cap M_{i+1}|$. For fixed τ, we denote the problem by τ-IM.*

We also consider the natural inverse objective, i.e., minimizing the unions. While the problems differ in the precise objective function, an optimal solution of MIM is optimal for MUM as well, and vice versa.

Definition 2 (MUM and τ-UM). *Given a temporal graph $\mathcal{G}$, the multistage union matching problem (MUM) asks for a multistage perfect matching $\mathcal{M}$ of $\mathcal{G}$ with minimum cost $c(\mathcal{M}) := \sum_{i \in [\tau-1]} |M_i \cup M_{i+1}|$. For fixed τ, we denote the problem by τ-UM.*

Consider either MIM or MUM. Given a temporal graph $\mathcal{G}$, we denote with opt the optimal solution value and with apx the objective value achieved by a given algorithm with input $\mathcal{G}$. The *approximation ratio* of an approximation algorithm for MIM (MUM) is the infimum (supremum, respectively) of apx/opt over all instances.

Related Work. The classical *dynamic graph* setting often considers small modifications, e.g., single edge insertions/deletions [12,22]. There, one is given a graph with a sequence of modifications and asked for a feasible solution after each modification. A natural approach to tackle matchings in such graphs is to make local changes to the previous solutions [7–9,21].

A more general way of modeling changes is that of *temporal graphs*, introduced by Kempe et al. [16] and used herein. Typically, each vertex and edge is assigned a set of time intervals that specify when it is present. This allows an arbitrary number of changes to occur at the same time. Algorithms for this setting usually require a more global perspective and many approaches do not rely solely on local changes. In fact, many temporal (matching) problems turn out to be hard, even w.r.t. approximation and fixed-parameter-tractability [1,6,10,18,19].

One particular flavor of temporal graph problems is concerned with obtaining a sequence of solutions—one for each stage—while optimizing a global quantity. These problems are often referred to as *multistage problems* and gained much attention in recent years [2–5,13–15], including in the realm of matchings: e.g., the authors of [15] show W[1]-hardness for finding the largest edge set that induces a matching in each stage.

In the literature we find the problem Max-MPM, where the graph is augmented with time-dependent edge weights, and we want to maximize the value of each individual perfect matching (subject to the given edge costs) *plus* the total profit [3]. MIM is the special case where all edge costs are zero, i.e., we only care about the multistage properties of the solution, as long as each stage is perfectly matched. There is also the inverse optimization problem Min-MPM, where we minimize the value of each perfect matching *plus* the number of matching edges newly introduced in each stage. We have APX-hardness for Max-MPM and Min-MPM [3,14] (for Min-MPM one may assume a complete graph at each stage, possibly including edges of infinite weight). The latter remains APX-hard even for spanning 2-stage graphs with bipartite union graph and no edge weights (i.e., we only minimize the number of edge swaps) [3]. For uniform edge weights 0, the objective of Min-MPM is to minimize $\sum_{i \in [\tau-1]} |M_{i+1} \setminus M_i|$; similar but slightly different to MUM (equal up to additive $\sum_{i \in [\tau-1]} n_i/2$). For Min-MPM on metric spanning 2- or 3-stage graphs, the authors of [3] show 3-approximations. They also propose a $(1/2)$-approximation for Max-MPM on spanning temporal graphs with any number of stages, which is unfortunately wrong (see Appendix A of the arXiv version [11] of this paper for a detailed discussion).

When restricting Max-MPM and Min-MPM to uniform edge weights 0, optimal solutions for MIM, MUM, Max-MPM, and Min-MPM are identical; thus MIM and MUM are NP-hard. However, the APX-hardness of Min-MPM does not imply APX-hardness of MUM as the objective functions slightly differ. Furthermore, the APX-hardness reduction to Max-MPM inherently requires non-uniform edge weights and does not translate to MIM. To the best of our knowledge, there are no non-trivial approximation algorithms for any of these problems on more than three stages.

Our Contribution. We start with showing in Sect. 2 that the problems are NP-hard even in much more restricted scenarios than previously known, and that (a lower bound for) the integrality gap of the natural linear program for 2-IM is close to the approximation ratio we will subsequently devise. This hints that stronger approximation ratios may be hard to obtain, at least using LP techniques.

As our main contribution, we propose several approximation algorithms for the multistage problems MIM and MUM, as well as for their stage-restricted variants, see Fig. 1. In particular, in Sect. 3.1, we show a $(1/\sqrt{2\mu})$-approximation for 2-IM and that this analysis is tight. Then, in Sect. 3.2, we show that any approximation of 2-IM can be used to derive two different approximation algorithms for MIM, whose approximation ratios are a priori incomparable. In Sect. 3.3, we further show how to use all these algorithms to approximate MUM (and 2-UM). We also observe that it is infeasible to use an arbitrary MUM algorithm to approximate MIM. In particular, we propose the seemingly first approximation algorithms for MIM and MUM on arbitrarily many stages. We stress that our goal is to always guarantee a perfect matching in each stage; the approximation ratio deals purely with optimizing the transition costs. Recall that approximation algorithms optimizing a weighted sum between intra- and interstage costs cannot guarantee such solutions in general.

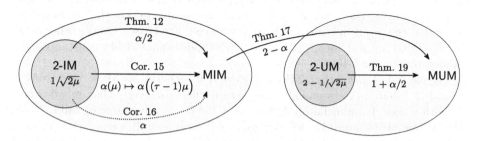

Fig. 1. Relations of our approximation results. An arc from problem A to B labeled $f(\alpha)$ denotes the existence of an $f(\alpha)$-approximation for B, given an α-approximation for A. In Corollary 16, α has to be constant. In Corollary 15, $\alpha(\cdot)$ is a function of μ. The ratio of 2-IM is by Theorem 8; combining this with Theorem 17 yields the ratio for 2-UM.

Preprocessing and Observations. Given a graph $G = (V, E)$, a single edge e is *allowed* if there exists a perfect matching M in G with $e \in M$ and *forbidden* otherwise. A graph is *matching-covered* if all its edges are allowed (cf. [17] for a concise characterization of matching-covered graphs). Forbidden edges can easily be found in polynomial time; see e.g. [20] for an efficient algorithm. A simple preprocessing for MIM and MUM is to remove the forbidden edges in each stage, as they will never be part of a multistage matching. Thereby, we obtain an equivalent *reduced* temporal graph, i.e., a temporal graph whose stages are matching-covered. If any stage in the reduced temporal graph contains no edges (but vertices), the instance is infeasible. In the following, we thus assume w.l.o.g. that the given temporal graph is reduced and *feasible*, i.e., in each stage there exists some perfect matching.

Observation 3. *Let $\mathcal{G}$ be a reduced 2-stage graph. For any $e \in E_\cap$, there is a perfect matching in each stage that includes e. Thus, there is a multistage perfect matching with profit at least 1 if $E_\cap \neq \varnothing$.*

Observation 4. *For any multistage perfect matching $(M_i)_{i \in [\tau]}$, it holds for each $i \in [\tau - 1]$ that $\max(n_i/2, n_{i+1}/2) \leq |M_i \cup M_{i+1}| = c(M_i, M_{i+1}) \leq 2\max(n_i/2, n_{i+1}/2)$. Thus, computing any multistage perfect matching is an immediate 2-approximation for MUM.*

Observation 5. *Consider the following algorithm: Enumerate every possible sequence $(F_i)_{i \in [\tau-1]}$ such that $F_i \subseteq E_i \cap E_{i+1}$ for each $i \in [\tau - 1]$; then check for each $i \in [\tau]$ whether there is a perfect matching M_i in G_i such that $F_{i-1} \cup F_i \subseteq M_i$, where $F_0 = F_\tau = \varnothing$. Thus, MIM and MUM are in FPT w.r.t. parameter $\sum_{i \in [\tau-1]} |E_i \cap E_{i+1}|$ (or similarly $\tau \cdot \mu$).*

2   Setting the Ground

Before we present our main contribution, the approximation algorithms, we motivate the intrinsic complexities of the considered problems. On the one hand, we show that the problem is already hard in very restricted cases. On the other hand, we show that natural linear programming methods cannot yield a constant-factor approximation for 2-IM.

While it is known that 2-IM is NP-hard in general, we show that 2-IM is already NP-hard in the seemingly simple case where each vertex has only degree 2 in both stages. It immediately follows that the decision variants of MIM, 2-UM, MUM, Min-MPM, and Max-MPM remain NP-hard as well, even if restricted to this set of temporal graphs. The proof of the following theorem is in Appendix B of [11].

Theorem 6. *Deciding 2-IM is NP-hard on spanning temporal graphs whose union graph is bipartite, even if both stages consist only of disjoint even cycles and $E_\cap$ is a collection of disjoint 2-paths.*

Algorithm 1: Approximation of 2-IM

---

1 set (M_1, M_2) to $(\varnothing, \varnothing)$

2 **for** $i = 1, 2, \dots$ **do**

3   set edge weights of G_1 to $\mathbb{1}\left(e \in E_\cap \setminus \bigcup_{j \in [i-1]} M_1^{(j)}\right)$

4   compute a maximum weight perfect matching $M_1^{(i)}$ on G_1

5   set edge weights of G_2 to $\mathbb{1}\left(e \in M_1^{(i)}\right)$

6   compute a maximum weight perfect matching $M_2^{(i)}$ on G_2

7   **if** $|M_1^{(i)} \cap M_2^{(i)}| \geq |M_1 \cap M_2|$ **then** set (M_1, M_2) to $(M_1^{(i)}, M_2^{(i)})$ **if**
     $E_\cap \subseteq \bigcup_{j \in [i]} M_1^{(j)}$ **then return** (M_1, M_2)

Linear Programs (LPs)—as relaxations of integer linear programs (ILPs)—are often used to provide dual bounds in the approximation context. Here, we consider the natural LP-formulation of 2-IM and show that the integrality gap (i.e., the ratio between the optimal objective value of the ILP and the optimal objective value of its relaxation) is at least $\sqrt{\mu}$, even already for spanning instances with a bipartite union graph. Up to a small constant factor, this equals the (inverse) approximation ratio guaranteed by Algorithm 1, which we will propose in Sect. 3. This serves as a hint that overcoming the approximation dependency $\sqrt{\mu}$ for 2-IM may be hard. A proof of the following theorem is in Appendix C of [11].

Theorem 7. *The natural LP for 2-IM has at least an integrality gap of $\sqrt{\mu}$, independent of the number μ of edges in the intersection.*

3 Approximation

We start with the special case of 2-IM, before extending the result to the multi-stage MIM scenario. Then we will transform the algorithms for use with 2-UM and MUM.

3.1 Approximating 2-IM

We first describe Algorithm 1, which is an approximation for 2-IM. Although its ratio is not constant but grows with the rate of $\sqrt{\mu}$, Theorem 7 hints that better approximations may be hard to obtain. Algorithm 1 roughly works as follows: Given a 2-stage graph $\mathcal{G}$, we iterate the following procedure on G_1 until every edge of $E_\cap$ has been in at least one perfect matching: Compute a perfect matching M_1 in G_1 that uses the maximum number of edges of $E_\cap$ that have not been used in any previous iteration; then compute a perfect matching M_2 in G_2 that optimizes the profit with respect to M_1. While doing so, keep track of the maximal occurring profit. Note that by choosing weights appropriately, we can construct a perfect matching that contains the maximum number of edges of some prescribed edge set in polynomial time [17].

We show:

Theorem 8. *Algorithm 1 is a tight $(1/\sqrt{2\mu})$-approximation for 2-IM.*

We prove this via two Lemmata; the bad instance of Lemma 9 in conjunction with the approximation guarantee (Lemma 10) establishes tightness.

Lemma 9 (Bad instance). *The approximation ratio of Algorithm 1 is at most $(1/\sqrt{2\mu})$. (Proof in Appendix D of [11].)*

Lemma 10 (Guarantee). *The approximation ratio of Algorithm 1 is at least $(1/\sqrt{2\mu})$.*

Proof. Let $\mathcal{G}$ be a feasible and reduced 2-stage graph with non-empty $E_\cap$. Clearly, our algorithm achieves apx ≥ 1 as described in Observation 3. Let k denote the number of iterations. For any $i \in [k]$, let $(M_1^{(i)}, M_2^{(i)})$ denote the 2-stage perfect matching computed in the ith iteration. The algorithm picks at least one new edge of $E_\cap$ per iteration into $M_1^{(i)}$ and hence terminates. Let (M_1^*, M_2^*) denote an optimal 2-stage perfect matching and $M_\cap^* := M_1^* \cap M_2^*$ its intersection. Let $R_i := (M_1^{(i)} \cap E_\cap) \setminus \bigcup_{j \in [i-1]} R_j$ denote the set of edges in $M_1^{(i)} \cap E_\cap$ that are not contained in $M_1^{(j)}$ for any previous iteration $j < i$ and let $r_i := |R_i|$. Note that in iteration i, the algorithm first searches for a perfect matching $M_1^{(i)}$ in G_1 that maximizes the cardinality r_i of its intersection with $E_\cap \setminus \bigcup_{j \in [i-1]} R_j$. We define $R_i^* := (M_1^{(i)} \cap M_\cap^*) \setminus \bigcup_{j \in [i-1]} R_j^*$ and $r_i^* := |R_i^*|$ equivalently to R_i, but w.r.t. $M_\cap^*$ (cf. Fig. 2). Observe that $R_i \cap M_\cap^* = R_i^*$.

Let $q := \sqrt{2\mu}$. For every $i \in [k]$ the algorithm chooses $M_2^{(i)}$ such that $|M_1^{(i)} \cap M_2^{(i)}|$ is maximized. Since we may choose $M_2^{(i)} = M_2^*$, it follows that apx $\geq \max_{i \in [k]} r_i^*$. Thus, if $\max_{i \in [k]} r_i^* \geq \mathsf{opt}/q$, we have a $(1/q)$-approximation. In case $\mathsf{opt} \leq q$, any solution with profit at least 1 yields a $(1/q)$-approximation. We show that we are in one of these cases.

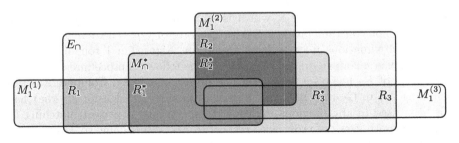

Fig. 2. Visualization of the relationships between $E_\cap, M_\cap^*, M_1^{(i)}, R_i$ and R_i^* for $i \in [3]$.

Algorithm 2: General multistage approximation

Input: Temporal graph $\mathcal{G}$, 2-stage perfect matching algorithm $\mathcal{A}$

1 create path $P := \{e_1, ..., e_{\tau-1}\}$
2 **foreach** $i \in [\tau - 1]$ **do**
3 $\quad$ set (S_i, T_{i+1}) to $\mathcal{A}(V, E_i, E_{i+1})$ $\qquad$ // approximate 2-stage graphs
4 $\quad$ set weight of e_i to $w_i := |S_i \cap T_{i+1}|$

5 compute maximum weight matching M_P in (P, w)
6 set $(M_i)_{i \in [\tau-1]}$ to $(S_i)_{i \in [\tau-1]}$ and M_τ to T_τ $\qquad$ // set initial solution
7 **foreach** $i \in [\tau - 1]$ **do** $\qquad\qquad$ // modify solution according to M_P
8 $\quad$ **if** $e_i \in M_P$ **then** set M_i to S_i and M_{i+1} to T_{i+1}

9 **return** $(M_1, ..., M_\tau)$

Let $\bar{q} := \lceil q \rceil$. Assume that $\mathsf{opt} > q$ (thus $\mathsf{opt} \geq \bar{q}$) and simultaneously $r_i^* < \mathsf{opt}/q$ for all $i \in [k]$. Since we distribute $M_\cap^*$ over the disjoint sets $\{R_i^* \mid i \in [k]\}$, each containing less than opt/q edges, we know that $k > q$ (thus $k \geq \bar{q}$). In iteration i, M_1^* has weight $|(M_1^* \cap E_\cap) \setminus \bigcup_{j \in [i-1]} R_j| \geq |M_\cap^* \setminus \bigcup_{j \in [i-1]} R_j| = |M_\cap^* \setminus \bigcup_{j \in [i-1]} R_j^*|$. Hence, the latter term is a lower bound on r_i, that we estimate as follows: $r_i \geq |M_\cap^* \setminus \bigcup_{j \in [i-1]} R_j^*| = \mathsf{opt} - \sum_{j \in [i-1]} r_j^* \geq \mathsf{opt} - \sum_{j \in [i-1]} \mathsf{opt}/q = \mathsf{opt} \cdot (1 - (i-1)/q)$. The above assumptions give a contradiction:

$$\mu = \left| \bigcup_{i \in [k]} R_i \right| \geq \sum_{i \in [\bar{q}]} r_i \geq \mathsf{opt} \cdot \sum_{i \in [\bar{q}]} \left(1 - \tfrac{i-1}{q}\right) \geq \bar{q} \cdot \sum_{i \in [\bar{q}]} \left(1 - \tfrac{i-1}{q}\right)$$

$$= \bar{q}\left(\bar{q} - \sum_{i \in [\bar{q}-1]} \tfrac{i}{q}\right) = \bar{q}\left(\bar{q} - \tfrac{(\bar{q}-1)\bar{q}}{2q}\right) = \bar{q}^2\left(1 - \tfrac{\bar{q}-1}{2q}\right) > \bar{q}^2\left(1 - \tfrac{q}{2q}\right) \geq \mu. \qquad \square$$

3.2 Approximating MIM

Let us extend the above result to an arbitrary number of stages. We show that we can use *any* 2-IM approximation algorithm (in particular also Algorithm 1) as a black box to obtain an approximation algorithm for MIM, while only halving the approximation ratio: Algorithm 2 uses an edge weighted path (P, w) on τ vertices as an auxiliary graph. We set the weight of the edge between the ith and $(i + 1)$th vertex to an approximate solution for the 2-IM instance that arises from the ith and $(i + 1)$th stage of the MIM instance. A maximum weight matching M_P in (P, w) induces a feasible solution for the MIM problem: If an edge $(j, j + 1)$ is in M_P, we use the corresponding solutions for the jth and $(j + 1)$th stage; for stages without incident edge in M_P, we select an arbitrary solution. Since no vertex is incident to more than one edge in M_P, there are no conflicts.

Observation 11. *For $F \subseteq E(P)$, denote $w(F) := \sum_{e \in F} w(e)$. Let e_i denote the ith edge of P. For $b \in [2]$, the matchings $M_b := \{e_i \in E(P) \mid i = b \mod 2\}$ are disjoint and their union is exactly $E(P)$. Thus, any maximum weight matching M_P in P achieves $2 \cdot w(M_P) \geq w(E(P))$.*

Theorem 12. *For a 2-IM α-approximation, Algorithm 2 $\frac{\alpha}{2}$-approximates MIM.*

Proof. Let $\mathcal{G} = (V, E_1, ..., E_\tau)$ be the given temporal graph. For any $i \in [\tau - 1]$, (S_i, T_{i+1}) is the output of the 2-IM α-approximation $\mathcal{A}(V, E_i, E_{i+1})$; let $w_i :=$ $|S_i \cap T_{i+1}|$. Let $\mathcal{M}^* := (M_1^*, ..., M_\tau^*)$ denote a multistage perfect matching whose profit $p(\mathcal{M}^*)$ is maximum. Since $\mathcal{A}$ is an α-approximation for 2-IM, we know that $|M_i^* \cap M_{i+1}^*| \leq w_i/\alpha$ for every $i \in [\tau - 1]$. Thus $p(\mathcal{M}^*) \leq (1/\alpha)\sum_{i \in [\tau-1]} w_i$. Algorithm 2 computes a maximum weight matching M_P in (P, w) and constructs a multistage solution $\mathcal{M}$. By Observation 11, we obtain $p(\mathcal{M}^*) \leq$ $(1/\alpha)\sum_{i \in [\tau-1]} w_i = (1/\alpha) \cdot w(E(P)) \leq (2/\alpha) \cdot w(M_P) \leq (2/\alpha) \cdot p(\mathcal{M})$.

We compute a maximum weight matching in a path in linear time using straightforward dynamic programming. Hence, assuming running time f for $\mathcal{A}$, Algorithm 2 requires $\mathcal{O}\left(\sum_{i \in [\tau-1]} |f(G_i, G_{i+1})|\right)$ steps.

Corollary 13. *Algorithm 1 in Algorithm 2 yields a $(1/\sqrt{8\mu})$-approximation for MIM.*

There is another way to approximate MIM via an approximation for 2-IM, which neither dominates nor is dominated by the above method:

Theorem 14. *There is an S-reduction from MIM to 2-IM, i.e., given any MIM instance $\mathcal{G}$, we can find a corresponding 2-IM instance $\mathcal{G}'$ in polynomial time such that any solution for $\mathcal{G}$ bijectively corresponds to a solution for $\mathcal{G}'$ with the same profit. Furthermore, $|E(G_1') \cap E(G_2')| = \sum_{i \in [\tau-1]} |E(G_i) \cap E(G_{i+1})|$.*

Proof. We will construct a 2-stage graph $\mathcal{G}'$ whose first stage G_1' consists of (subdivided) disjoint copies of G_i for odd i; conversely its second stage G_2' consists of (subdivided) disjoint copies of G_i for even i. More precisely, consider the following construction: Let $b(i) := 2 - (i \bmod 2)$. For each $i \in [\tau]$, we create a copy of G_i in $G_{b(i)}'$ where each edge $e \in E(G_i)$ is replaced by a 7-path p_i^e. We label the 3rd (5th) edge along p_i^e (disregarding its orientation) with e_i^- (e_i^+, respectively). To finally obtain $\mathcal{G}'$, for each $i \in [\tau-1]$ and $e \in E(G_i) \cap E(G_{i+1})$, we identify the vertices of e_i^+ with those of e_{i+1}^- (disregarding the edges' orientations); thereby precisely the edges e_i^+ and e_{i+1}^- become an edge common to both stages. No other edges are shared between both stages. This completes the construction of $\mathcal{G}'$ and we have $|E(G_1') \cap E(G_2')| = \sum_{i \in [\tau-1]} |E(G_i) \cap E(G_{i+1})|$.

Assume $\mathcal{M}' := (M_1', M_2')$ is a solution for $\mathcal{G}'$. Clearly, each path p_i^e in $G_{b(i)}'$ is matched alternatingly and hence either all or none of e_i^-, e_i^+, the first, and the last edge of p_i^e are in $M_{b(i)}'$. We derive a corresponding solution $\mathcal{M}$ for $\mathcal{G}$: For every $i \in [\tau]$ and $e \in E(G_i)$, we add e to M_i if and only if $e_i^- \in M_{b(i)}'$. Suppose that M_i is not a perfect matching for G_i, i.e., there exists a vertex v in G_i that is not incident to exactly one edge in M_i. Then also the copy of v in the copy of G_i in $G_{b(i)}'$ is not incident to exactly one edge of $M_{b(i)}'$, contradicting the feasibility of $\mathcal{M}'$.

Consider the profit achieved by $\mathcal{M}$: Every edge in $M_1' \cap M_2'$ corresponds to a different identification $\langle e_i^+, e_{i+1}^- \rangle$. We have $e \in M_i \cap M_{i+1}$ if and only if $e_i^- \in M_{b(i)}'$, $e_{i+1}^- \in M_{b(i+1)}'$, and $e_i^+ = e_{i+1}^-$. It follows that this holds if and only if $e_i^+ \in M_{b(i)}' \cap M_{b(i+1)}'$ and hence the profit of $\mathcal{M}$ is equal to that of $\mathcal{M}'$. The inverse direction proceeds in the same manner.

Since the new $\mu' := |E(G_1') \cap E(G_2')|$ is largest w.r.t. the original μ if $|E(G_i) \cap E(G_{i+1})|$ is constant for all i, we obtain:

Corollary 15. *For any 2-IM $\alpha(\mu)$-approximation where $\alpha(\mu)$ is a (typically decreasing) function of μ, there is an $\alpha\big((\tau-1)\mu\big)$-approximation for MIM. Using Algorithm 1, this yields a ratio of $1/\sqrt{2(\tau-1)\mu}$; for 3-IM and 4-IM this is tighter than Theorem 12.*

Assume the approximation ratio for 2-IM would not depend on μ. Then the above would yield a surprisingly strong result:

Corollary 16. *Any 2-IM α-approximation with constant α results in an α-approximation of MIM. If MIM is APX-hard, so is 2-IM.*

3.3   Approximating MUM

Consider the MUM-problem which minimizes the cost. As noted in Observation 4, a 2-approximation is easily accomplished. However, by exploiting the previous results for MIM, we obtain better approximations.

Theorem 17. *Any α-approximation of MIM is a $(2-\alpha)$-approximation of MUM.*

Proof. Recall that an optimal solution of MIM constitutes an optimal solution of MUM. As before, we denote the heuristic sequence of matchings by $(M_i)_{i \in [\tau]}$ and the optimal one by $(M_i^*)_{i \in [\tau]}$. Let $\xi := \sum_{i \in [\tau-1]} (n_i + n_{i+1})/2$. Consider the solutions' values w.r.t. MUM:

$$\frac{\mathsf{apx}_\cup}{\mathsf{opt}_\cup} = \frac{\sum_{i\in[\tau-1]} c(M_i, M_{i+1})}{\sum_{i\in[\tau-1]} c(M_i^*, M_{i+1}^*)} = \frac{\xi - \sum_{i\in[\tau-1]}|M_i \cap M_{i+1}|}{\xi - \sum_{i\in[\tau-1]}|M_i^* \cap M_{i+1}^*|} \leq \frac{\xi - \alpha \cdot \mathsf{opt}_\cap}{\xi - \mathsf{opt}_\cap} =: f.$$

As $0 < \alpha < 1$, f is monotonously increasing in $\mathsf{opt}_\cap$ if $0 \leq \mathsf{opt}_\cap < \xi$. Thus, since $\mathsf{opt}_\cap \leq \sum_{i\in[\tau-1]} \min(n_i, n_{i+1})/2 \leq \sum_{i\in[\tau-1]}(n_i + n_{i+1})/4 = \xi/2$, it follows that $\mathsf{apx}_\cup/\mathsf{opt}_\cup \leq (\xi - \alpha \cdot \xi/2)/(\xi - \xi/2) = 2 - \alpha$.

Corollary 18. *Let $r := \min\{8, 2(\tau-1)\}$. We have a $\big(2 - 1/\sqrt{r \cdot \mu}\big)$-approximation for MUM.*

Note that a similar reduction from MIM to MUM is not achieved as easily: Consider any $(1+\varepsilon)$-approximation for MUM. Choose an even integer $k \geq 6$ such that $k/(k-1) \leq 1 + \varepsilon$; consider a spanning 2-stage instance where each stage is a k-cycle and $E_\cap$ consists of a single edge e. The optimal 2-stage perfect matching $\mathcal{M}^*$ that contains e in both stages has profit $p(\mathcal{M}^*) = 1$ and cost $c(\mathcal{M}^*) = 2 \cdot k/2 - 1 = k - 1$. A 2-stage perfect matching $\mathcal{M}$ that does not contain e still satisfies $c(\mathcal{M}) = k$ and as such is an $(1+\varepsilon)$-approximation for MUM. However, its profit $p(\mathcal{M}) = 0$ does not provide any approximation of $p(\mathcal{M}^*) = 1$.

As for MIM, we aim to extend a given approximation for 2-UM to a general approximation for MUM. Unfortunately, we cannot use Theorems 14 and 17 for this, as an approximation for 2-UM does not generally constitute one for 2-IM (and MIM). On the positive side, a similar approach as used in the proof of Theorem 12 also works for minimization.

Theorem 19. *Any α-approximation $\mathcal{A}$ for 2-UM results in a $(1 + \alpha/2)$-approximation for MUM by using $\mathcal{A}$ in Algorithm 2.*

Proof. As before, let $(M_i^*)_{i\in[\tau]}$ denote an optimal solution for MUM. For each $i \in [\tau - 1]$, (S_i, T_i) denotes the output of $\mathcal{A}(V, E_i, E_{i+1})$. For $L \subseteq [\tau - 1]$, let $\xi(L) := \sum_{i\in L}(n_i + n_{i+1})/2$ and $\sigma(L) := \sum_{i\in L}|S_i \cup T_i|$. Note that $w_i := \xi(\{i\}) - \sigma(\{i\})$ equals the weight of e_i. We define $I := \{i \in [\tau-1] \mid e_i \in M_P\}$ as the set of indices corresponding to M_P and $J := [\tau - 1] \setminus I$ as its complement. By Observation 11, we have $w\big(E(P)\big) \leq 2 \cdot w(M_P)$, thus

$$\xi(I) - \sigma(I) + \xi(J) - \sigma(J) = w\big(E(P)\big) \leq 2 \cdot w(M_P) = 2\big(\xi(I) - \sigma(I)\big)$$
$$\Rightarrow \sigma(I) + \xi(J) \leq \xi(I) + \sigma(J) \Rightarrow 2\big(\sigma(I) + \xi(J)\big) \leq \xi(I \cup J) + \sigma(I \cup J).$$

The trivial upper bound ξ suffices to bound the algorithm's solution value:

$$\mathsf{apx} = \sigma(I) + \sum_{j\in J}|M_j \cup M_{j+1}| \leq \sigma(I) + \xi(J) \leq \tfrac{1}{2}\big(\xi(I \cup J) + \sigma(I \cup J)\big).$$

Since $\sigma(I \cup J)$ α-approximates the sum of all 2-UM instances' solution values, we have $\sigma(I\cup J) \leq \alpha \cdot \mathsf{opt}$. For each transition, any solution satisfies $(n_i + n_{i+1})/4 \leq |M_i \cup M_{i+1}|$ and hence $\xi(I \cup J) \leq 2 \cdot \mathsf{opt}$. Finally, we obtain the claimed ratio: $\mathsf{apx} \leq 1/2 \cdot \big(2 \cdot \mathsf{opt} + \alpha \cdot \mathsf{opt}\big) = (1 + \alpha/2) \cdot \mathsf{opt}$. $\square$

4   Conclusion

In this paper we presented the first approximation algorithm for 2-IM, having a tight approximation ratio of $1/\sqrt{2\mu}$. It remains open if a constant factor approximation for 2-IM is possible; however, we showed that this would imply a constant factor approximation for MIM. We further showed two ways in which MIM and MUM can be approximated by using any algorithm that approximates 2-IM, thereby also presenting the first approximation algorithms for multistage matching problems with an arbitrary number of stages. We are confident that our techniques are applicable to a broader set of related problems as well.

References

1. Akrida, E.C., Mertzios, G.B., Spirakis, P.G., Zamaraev, V.: Temporal vertex cover with a sliding time window. J. Comput. Syst. Sci. **107**, 108–123 (2020). https://doi.org/10.1016/j.jcss.2019.08.002
2. Bampis, E., Escoffier, B., Kononov, A.: LP-based algorithms for multistage minimization problems. arXiv (2019). https://arxiv.org/abs/org/abs/1909.10354
3. Bampis, E., Escoffier, B., Lampis, M., Paschos, V.T.: Multistage matchings. In: 16th Scandinavian Symposium and Workshops on Algorithm Theory (SWAT 2018), vol. 101, pp. 7:1–7:13 (2018). https://doi.org/10.4230/LIPIcs.SWAT.2018.7

4. Bampis, E., Escoffier, B., Schewior, K., Teiller, A.: Online multistage subset maximization problems. In: 27th Annual European Symposium on Algorithms (ESA 2019), vol. 144, pp. 11:1–11:14 (2019). https://doi.org/10.4230/LIPIcs.ESA.2019.11
5. Bampis, E., Escoffier, B., Teiller, A.: Multistage knapsack. In: 44th International Symposium on Mathematical Foundations of Computer Science (MFCS 2019), vol. 138, pp. 22:1–22:14 (2019). https://doi.org/10.4230/LIPIcs.MFCS.2019.22
6. Baste, J., Bui-Xuan, B.M., Roux, A.: Temporal matching. Theor. Comput. Sci. **806**, 184–196 (2020). https://doi.org/10.1016/j.tcs.2019.03.026
7. Bernstein, A., Stein, C.: Fully dynamic matching in bipartite graphs. In: Halldórsson, M.M., Iwama, K., Kobayashi, N., Speckmann, B. (eds.) ICALP 2015. LNCS, vol. 9134, pp. 167–179. Springer, Heidelberg (2015). https://doi.org/10.1007/978-3-662-47672-7_14
8. Bhattacharya, S., Henzinger, M., Italiano, G.F.: Deterministic fully dynamic data structures for vertex cover and matching. SIAM J. Comput. **47**(3), 859–887 (2018). https://doi.org/10.1137/140998925
9. Bosek, B., Leniowski, D., Sankowski, P., Zych, A.: Online bipartite matching in offline time. In: 2014 IEEE 55th Annual Symposium on Foundations of Computer Science, pp. 384–393 (2014). https://doi.org/10.1109/FOCS.2014.48
10. Casteigts, A.: A Journey through Dynamic Networks (with Excursions). Habilitation, Université de Bordeaux (2018). https://tel.archives-ouvertes.fr/tel-01883384
11. Chimani, M., Troost, N., Wiedera, T.: Approximating multistage matching problems. arXiv (2020). https://arxiv.org/abs/2002.06887
12. Eppstein, D.: Offline algorithms for dynamic minimum spanning tree problems. In: Dehne, F., Sack, J.-R., Santoro, N. (eds.) WADS 1991. LNCS, vol. 519, pp. 392–399. Springer, Heidelberg (1991). https://doi.org/10.1007/BFb0028278
13. Fluschnik, T., Niedermeier, R., Rohm, V., Zschoche, P.: Multistage vertex cover. In: 14th International Symposium on Parameterized and Exact Computation (IPEC 2019), vol. 148, pp. 14:1–14:14 (2019). https://doi.org/10.4230/LIPIcs.IPEC.2019.14
14. Gupta, A., Talwar, K., Wieder, U.: Changing bases: multistage optimization for matroids and matchings. In: Esparza, J., Fraigniaud, P., Husfeldt, T., Koutsoupias, E. (eds.) ICALP 2014. LNCS, vol. 8572, pp. 563–575. Springer, Heidelberg (2014). https://doi.org/10.1007/978-3-662-43948-7_47
15. Heeger, K., Himmel, A.S., Kammer, F., Niedermeier, R., Renken, M., Sajenko, A.: Multistage graph problems on a global budget. Theor. Comput. Sci. **868**, 46–64 (2021). https://doi.org/10.1016/j.tcs.2021.04.002
16. Kempe, D., Kleinberg, J.M., Kumar, A.: Connectivity and inference problems for temporal networks. In: Proceedings of the Thirty-Second Annual ACM Symposium on Theory of Computing (STOC 2000), pp. 504–513 (2000). https://doi.org/10.1145/335305.335364
17. Lovász, L., Plummer, M.: Matching Theory. American Mathematical Society (1986)
18. Mertzios, G.B., Molter, H., Niedermeier, R., Zamaraev, V., Zschoche, P.: Computing maximum matchings in temporal graphs. In: 37th International Symposium on Theoretical Aspects of Computer Science (STACS 2020), pp. 27:1–27:14 (2020). https://doi.org/10.4230/LIPIcs.STACS.2020.27
19. Michail, O., Spirakis, P.G.: Traveling salesman problems in temporal graphs. In: Csuhaj-Varjú, E., Dietzfelbinger, M., Ésik, Z. (eds.) MFCS 2014. LNCS, vol. 8635, pp. 553–564. Springer, Heidelberg (2014). https://doi.org/10.1007/978-3-662-44465-8_47

20. Rabin, M.O., Vazirani, V.V.: Maximum matchings in general graphs through randomization. J. Algorithms. **10**, 557–567 (1989). https://doi.org/10.1016/0196-6774(89)90005-9
21. Sankowski, P.: Faster dynamic matchings and vertex connectivity. In: Proceedings of the Eighteenth Annual ACM-SIAM Symposium on Discrete Algorithms (SODA 2007), pp. 118–126 (2007)
22. Thorup, M.: Near-optimal fully-dynamic graph connectivity. In: Proceedings of the Thirty-Second Annual ACM Symposium on Theory of Computing (STOC 2000), pp. 343–350 (2000). https://doi.org/10.1145/335305.335345

Heuristically Enhanced IPO Algorithms for Covering Array Generation

Michael Wagner$^{(\boxtimes)}$, Ludwig Kampel, and Dimitris E. Simos

SBA Research, 1040 Vienna, Austria
{mwagner,lkampel,dsimos}@sba-research.org

Abstract. The construction of covering arrays (CAs) with a small number of rows is a difficult optimization problem. CAs generated by greedy methods are often far from optimal, while many metaheuristics and search techniques become inefficient once larger instances are concerned. In this work, we propose to incorporate improvement heuristics directly into the constructing process of widely used in-parameter-order (IPO) algorithms for CA generation. We discuss how this approach can significantly reduce the search space of the heuristics and implement some of the discussed concepts in the SIPO algorithm, which enhances greedy IPO algorithms with Simulated Annealing. Using SIPO, we improved the best known upper bound on the number of rows of binary CAs of strength 6 for 43 different instances.

Keywords: Covering arrays · Heuristics · Optimization · Simulated annealing

1 Introduction

Due to the reliance of modern society on technology and software, efficient testing of software systems for failures and unintended behavior is crucial. At the same time many systems are too large to be tested exhaustively. Combinatorial testing (CT) is a testing methodology that makes it possible to test large systems within reasonable time while maintaining certain coverage guarantees, for an introduction see [14]. In past works, CT was successfully applied to find faults in systems with more than 2000 parameters [10].

Covering arrays (CAs) are the combinatorial design underlying the test sets used in CT and can be considered generalizations of orthogonal arrays. A (uniform) CA, denoted as $CA(N; t, k, v)$, is an array with N rows and k columns with values arising from an alphabet of cardinality v. Further, the defining property of a CA is that in every possible selection of t columns, every t-tuple $\{0, 1, ..., v - 1\}^t$ appears in at least one row. The parameter t is known as the *strength* of the CA, while we further refer to a selection of t columns as *column configuration*. Figure 1 gives an example of a $CA(6; 2, 10, 2)$. In this binary 6×10 array, any possible selection of 2 columns contains all binary 2-tuples, $(0, 0), (0, 1), (1, 0)$ and $(1, 1)$, in at least one row.

© Springer Nature Switzerland AG 2021
P. Flocchini and L. Moura (Eds.): IWOCA 2021, LNCS 12757, pp. 571–586, 2021.
https://doi.org/10.1007/978-3-030-79987-8_40

This defining property of CAs can further be expressed using the notion of *t-way interactions*. A *t*-way interaction is defined as a set of t pairs $\{(p_0, v_0), ..., (p_{t-1}, v_{t-1})\}$, representing a column index $0 \leq p_i < k$ and a corresponding value $0 \leq v_i < v$ for this column. A *t*-way interaction is considered *covered* if there exists at least one row in the array that contains all t column/value pairs of the *t*-way interaction. Hence, a CA is an array where every *t*-way interaction is covered. For example, in the first two columns, the CA in Fig. 1 covers the 2-way interaction $\{(1,0), (2,0)\}$ in the first two rows, $\{(1,0), (2,1)\}$ in row 3, $\{(1,1), (2,0)\}$ in row 4 and $\{(1,1), (2,1)\}$ in the last two rows. We call the data structure that is used to store the coverage information of all *t*-way interactions *coverage map*, which is discussed in detail in [13].

We refer to the problem of generating a CA$(N; t, k, v)$ with a given strength t, number of columns k and alphabet of cardinality v as *CA instance*. A CA with the minimal number of rows possible is considered *optimal*. The number of rows N of an optimal CA is called *covering array number* (CAN). Generating optimal CAs is tightly coupled with other hard optimization problems [12], therefore the precise value of CAN is only known for a small number of CA instances. For the majority of

$$\begin{pmatrix} 0 & 0 & 0 & 0 & 0 & 0 & 0 & 0 & 0 & 0 \\ 0 & 0 & 1 & 1 & 1 & 1 & 1 & 0 & 1 & 0 \\ 0 & 1 & 0 & 1 & 1 & 0 & 0 & 1 & 1 & 1 \\ 1 & 0 & 1 & 0 & 0 & 1 & 0 & 1 & 1 & 1 \\ 1 & 1 & 0 & 1 & 0 & 1 & 1 & 1 & 0 & 0 \\ 1 & 1 & 1 & 0 & 1 & 0 & 1 & 0 & 0 & 1 \end{pmatrix}$$

Fig. 1. CA$(6; 2, 10, 2)$

CA instances, only lower and upper bounds on CAN are known. The minimization of rows is also of interest in practical applications such as combinatorial testing, where CAs with a small number of rows directly translate to small test sets, effectively reducing the resources needed for testing.

Further, CAs can be generalized to mixed-level alphabets, which allows for different alphabet sizes in the columns. These structures are called mixed-level covering arrays (MCAs) and are denoted as MCA$(N; t, k, (v_1, ..., v_k))$. For simplicity, throughout the majority of this work, we focus on the case of uniform CAs, but note that generalization of all proposed concepts to MCAs is straight forward.

This work discusses how metaheuristics and search techniques can be integrated directly into the construction process of In-Parameter-Order algorithms and is structured as follows. In Sect. 2, we review various past works for the generation of CAs. Section 3 proposes a concept of enhancing the well studied in-parameter-order strategy with other heuristic methods. Parts of this concept are then implemented in the SIPO algorithm presented in Sect. 4, which is evaluated in Sect. 5. Last, in Sect. 6 we present improvements to the best known upper bounds on CAN and discuss future work in the conclusion in Sect. 7.

2   Related Work

Over the past 30 years, many different methods to generate CAs have been developed and evaluated, for a detailed survey see [21]. This includes exact approaches, mathematical and recursive constructions, greedy algorithms and metaheuristics.

In this section we want to give a brief overview about the different approaches and discuss their advantages and limitations.

Exact methods are mainly used to determine the CAN for a CA instance. Various techniques were applied to this extent, including SAT encodings and constraint models [9]. While such approaches will theoretically yield optimal solutions, they generally only terminate within reasonable time when applied to very small instances.

Mathematical constructions can yield excellent results, however they often are only applicable to a limited range of CA instances, e.g. the so called *Bush construction* [3], which constructs orthogonal arrays of index unity over prime fields. Cyclic constructions for CAs are based on cyclic shifts of one or more cyclotomic vectors, therefore generating CAs where the number of rows is a multiple of the number of columns. Last, the application of permutation vectors and covering perfect hash families (CPHF) [6] in particular have been used to improve many best known upper bounds on CAN for CAs of alphabet $v > 2$. A CA is obtained by inserting permutation vectors into the CPHF. Therefore a CPHF can be considered a compact version of a CA and due to its smaller size and the resulting reduction in search space, optimization techniques such as metaheuristics can efficiently be applied to generate CAs that would usually be too difficult.

Many different metaheuristic algorithms have been designed for the generation of CAs, including population-based metaheuristics, such as genetic algorithms [20] and particle swarm optimization [1]. Out of all these approaches, single-solution metaheuristics such as Simulated Annealing (SA) has often proven to produce the best results, achieving many improvements of to the best known upper bounds on CAN. Especially the works of Torres-Jimenez et al. [22] and [24] stand out, improving the best known upper bounds on CAN for many binary CA instances. The search-based software testing tool (SBSTT) algorithm proposed in [22] starts with an initial CA and extends it one column at a time. The entries in the new columns are greedily set to the values that maximize the number of covered t-way interactions. Thereafter, SA is used to reduce the number of uncovered t-way interactions to zero, where, if necessary, rows are added in order to help the Simulated Annealing algorithm to achieve this goal. This is continued until either the desired number of columns is reached, or a given number of rows is exceeded.

Further, various post-optimization methods were applied to reduce the number of rows of initial CAs, often generated by greedy methods. In [23], a metaheuristic post-optimization (MPO) algorithm using Simulated Annealing was proposed and applied to the NIST repository of covering arrays [18]. The MPO algorithm iteratively removes the row with the most redundant entries, using SA to cover any uncovered t-way interactions that can occur due to row removals.

Last, greedy algorithms are widely used in combinatorial test generation tools, due to their fast execution time and flexibility in regards to CAs with mixed alphabet sizes as well as constraint handling. Many greedy approaches build a CA one row at a time, adding rows until all t-way interactions are covered.

Prominent examples of such algorithms are AETG [4,11] and the deterministic density algorithm DDA [2].

2.1 Revisiting In-Parameter-Order Algorithms

Another greedy construction method is the In-Parameter-Order (IPO) strategy. The characteristic of this strategy is that a $v^t \times t$ array covering all possible t-way combinations for the first t columns, is extended one column at a time until a CA with k columns is constructed. The addition of a new column is referred to as *horizontal extension*. To ensure that all t-way interactions are covered, every horizontal extension step is succeeded by a *vertical extension* step, where all uncovered t-way interactions are added to the array, appending new rows if necessary. An overview of the IPO strategy is given in Algorithm 1.

Algorithm 1. IPO Strategy

$Array \leftarrow$ cross-product of first t columns
for $i \leftarrow t, \ldots, k$ **do**
    HorizontalExtension(i)
    if there are uncovered t-way interactions **then**
        VerticalExtension(i)
    end if
end for

In other words, the IPO strategy splits the CA generation process into smaller sub problems, in which an existing CA_i with i columns is extended to a CA_{i+1} with $i + 1$ columns. During such extension steps, values assigned to CA_i are considered fixed and are not changed anymore. When a new row is added to the array during a vertical extension step, initially all its values are set to so called *star values*. Star values represent entries for which no value has been set during the generation process thus far. In greedy IPO algorithms, once a star value is changed to an explicit value, it will no longer be considered for further optimization. Figure 2 gives an example of the different sections where star values can occur. First, the blue section corresponds to the horizontal extension, i.e.

a	b	c	d	e
0	0	0	0	
0	1	1	1	
1	0	1	0	
1	1	0	1	
*	0	*	1	
*	1	*	0	

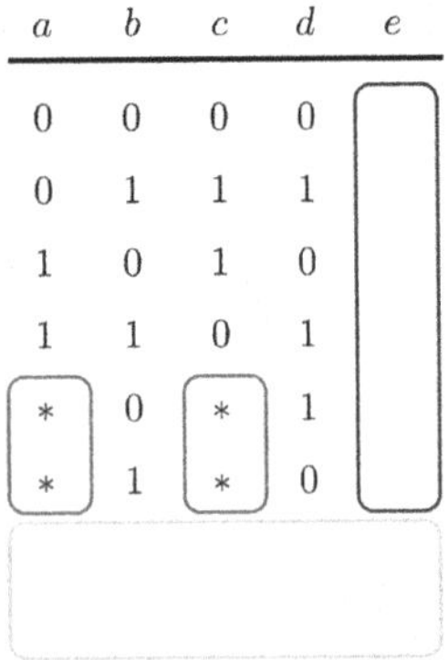

Fig. 2. Star values (Color figure online)

the newly added column, which values are initialized as star values. Second, the red sections contain star values that were created in previous extension steps, while the green section represents new rows that might be added during the current vertical extension.

The initial IPO algorithm, proposed by Lei and Tai in [16], was exclusively designed for pairwise testing, but the concept was generalized to arbitrary strengths in [15]. Further, two additional strategies for the greedy selection of values during the horizontal extension were proposed in [8]. During the horizontal extension of the original IPOG algorithm the rows are iterated from top to bottom and for each row, the value maximizing the number of newly covered t-way interactions is selected. IPOG-F further optimizes the order in which rows are assigned values, so that at any point of time during the horizontal extension, the row and value that maximize the number of covered t-way interactions is greedily selected. Various later works contributed in further improving IPO algorithms, e.g. the work in [7] improves the efficiency of the vertical extension by applying a graph coloring algorithm. In [26], the IPO strategy is used to generate binary CAs with quantum-inspired evolutionary algorithms. Last, [13] introduces many algorithmic optimizations that improve the performance of IPOG algorithms while constructing identical CAs. Two examples of algorithms that implement these optimizations are FIPOG and FIPOG-F, which can be considered a fast version of the IPOG and IPOG-F algorithms described above. Both algorithms are available in the combinatorial test generation tool CAgen [25].

3   Using Heuristics to Enhance IPO Algorithms

The IPO strategy manages to drastically reduce the size of the search space by fixing the values of any entries upon first assignment. At the same time, this restriction often leads to solutions that might be far from optimal. For this reason we propose to incorporate improvement heuristics directly into the IPO generation process. A pseudocode for this concept is given in Algorithm 2. A greedy IPO algorithm is simply extended by applying an improvement heuristic after each greedy extension step. While the concept seems very straight forward, various design choices have to be considered, which will have a large impact on the performance of the algorithm.

Algorithm 2. Heuristically enhanced IPO

Require: Greedy IPO algorithm, Improvement Heuristic
    Array ← cross-product of first t columns
    for $i \leftarrow t, \ldots, k$ **do**
        GreedyHorizontalExtension(i)
        EnhanceHorizontal()
        if there are uncovered t-way interactions **then**
            GreedyVerticalExtension()
            EnhanceVertical()
        end if
    end for

First, in addition to the v different values an entry is allowed to have, in our proposed approach star values need to be considered as well. Setting a value that does not contribute to any missing t-way interaction to a star value can help the algorithm by allowing it to use the entry for optimization during later extension steps. This can be achieved by considering star values as part of the objective function of the improvement algorithm.

Next we need to determine which parts of the array will be optimized at once, i.e. which entries of the array will be considered as modifiable by the improvement heuristic. For example, a heuristic for enhancing the horizontal extension might only act on the newly added column or in addition consider any star values in the array that were left over from previous vertical extensions. On the one hand, enhancing the new column together with old star values allows the algorithm to optimize on a more global scale, which can result in solutions of higher quality. On the other hand, this also increases the size of the search space substantially. The size of the search space during the i_{th} extension step is $(v+1)^m$ where m refers to the number of modifiable entries and $(v+1)$ to the number of different values an entry can take including a star value. If only entries in the new column are considered, m is equal to the number of rows N_{i-1} of the CA constructed during the last extension step, while if we also consider old star values, the number of remaining star values in the array needs to be added to this exponent on top of N_{i-1}. This difference becomes even more significant if we consider a global optimization, where every value in the previous array is considered as modifiable. In this case the search space increases to $(v+1)^{(i \cdot N_{i-1})}$. Further, an extension algorithm following the IPO strategy has the advantage that it only needs to ensure coverage of all t-way interactions in the $\binom{i}{t-1}$ column configurations that contain the newly added column, while a global approach might destroy previously covered t-way interactions and therefore has to consider all $\binom{i}{t}$ column configurations. Due to these limitations, in the past metaheuristics often yielded good results only for smaller instances, but proved inefficient when applied to large instances.

In the remainder of the work, we use Simulated Annealing to enhance the horizontal extension of greedy FIPOG and FIPOG-F algorithms, to show that applying improvement heuristics can improve the size of generated CAs significantly. Further, our experiments demonstrate that the search space reduction of this IPO strategy allows us to apply Simulated Annealing to larger instances than previous algorithms using the same metaheuristic, while even improving multiple best known upper bounds for binary CAs of strength $t = 6$.

4   SIPO: Enhancing IPO Algorithms with Simulated Annealing

As a proof of concept, to enhance the horizontal extension of greedy FIPOG algorithms we considered the use of Simulated Annealing, which in the past has proven to be a potent heuristic when applied to the problem of CA generation, e.g. [22] and [24]. The pseudocode is outlined in Algorithm 3. To this extent we made use of a basic One-Flip neighborhood. We distinguish between two different *enhance types*: NewColumn and FullHorizontal. With the NewColumn configuration, only entries that are part of the newly added column are considered modifiable, while FullHorizontal also considered star values generated during previous extension steps as modifiable. In each iteration, a new candidate move m is generated by randomly selecting a modifiable entry and setting it to a new random value. Next, the objective value s attained by applying m to the array is calculated. As objective functions, expressed as a minimization problem, we selected the number of missing t-way interactions minus the number of star values in the array. This provides an incentive for the algorithm to create star values if an entry does not contribute to the number of covered t-way interactions in the array. Therefore, the objective value of a move is given as the difference in the number of covered t-way interactions before and after applying the move. In addition, whenever a move would add a star value, the objective value is reduced by one, while if the entry changed away from a star value, s is incremented. The objective function is given in Eq. 1. Normalization of the objective function is difficult, as the number of star values in an optimal solution and therefore the best objective value for an extension step is unknown. Hence, we will leave the use of normalization as well as a comparison of the effects of it on different instances for future work.

A move will always be accepted if its objective value s is smaller or equal to 0. Otherwise, it will be accepted according to the Metropolis criterion with a probability of $e^{-s/T}$, where T refers to the current temperature. If a move is accepted, the respective modifiable entry is set to the new value and the coverage map needs to be updated accordingly. At the end of each iteration, the temperature is updated based on a geometric cooling schedule, such that $T_{i+1} = \alpha * T_i$, with $0 < \alpha < 1$. The algorithm terminates once the current temperature falls below a certain threshold, further referred to as final temperature T_f. In this work, the value alpha was calculated at the beginning of an SA run, based on a targeted number of iterations. We made the number of iterations at each extension step dependent on the instance size as well as a *base* number of iterations, more precisely *iterations* $= base \cdot (t - 1) \cdot i$, where i is the current number of columns.

$$s(m) = C_{before}(m) - C_{after}(m) + D(m) \tag{1}$$

$C_{before/after}(m) =$ number of covered t-way interactions before/after move m

$$D(m) = \begin{cases} -1 & \text{if } v_i \to * & \text{(from value to star value)} \\ 0 & \text{if } v_i \to v_j & \text{(from value to value)} \\ 1 & \text{if } * \to v_j & \text{(from star value to value)} \end{cases}$$

Finally, we want to note that our proposed algorithm can easily be generalized to generate CAs over higher or mixed alphabets as well as test sets with constraints with slight modifications. However, for the sake of compact presentation this work focuses on the generation of binary CAs.

Algorithm 3. SIPO - Horizontal

Require: SA Parameters, Termination criterion, EnhanceType
  procedure ENHANCEHORIZONTAL(Array, CoverageMap, new column i)
      ModifiableEntries $\leftarrow \emptyset$
      for $row \leftarrow 0, \ldots, N$ **do**
         ModifiableEntries $\leftarrow$ ModifiableEntries $\cup$ new ModifiableEntry(row, i, v)
      end for
      if EnhanceType is FullHorizontal **then**
         for every entry e_{jl} in the Array\column i **do**
            if e_{jl} is star value **then**
               ModifiableEntries $\leftarrow$ ModifiableEntries $\cup$ new ModifiableEntry(j, l, v)
            end if
         end for
      end if
      while Termination Criterion not met **do**
         Move $m \leftarrow$ Generate a random value for a random ModifiableEntry
         Calculate objective value s for move m
         if $s \leq 0$ or acceptance criteron met **then**
            Apply m to the array
            Update CoverageMap
         end if
      end while
  end procedure

---

5 SIPO Parameter Tuning and Algorithm Evaluation

To determine suitable values for the initial temperature T_i and the final temperature T_f used in SIPO, we generated binary covering arrays of strength $t = 4$ using the FIPOG algorithm as greedy construction method and using $base = 10000$ to calculate the number of iterations per extension step. To evaluate the performance of different configurations over multiple different instances we generated 10 CAs for each instance with $5 \leq k \leq 64$ columns and recorded the minimum and average number of rows of the generated CAs for each instance and configuration. Last, for comparison of the average performance of configurations over multiple CA instances, we consider the sum of average and minimum rows respectively over all considered instances.

In various previous works for CA generation that feature objective functions based on the number of covered t-way interactions, the final temperature was set to $T_f = 10^{-10}$, e.g. in [24]. However, in our experiments we noticed that the algorithm failed to improve the best found solution after a relatively small number of iterations. This suggests that, using this final temperature, the algorithm gets stuck in local minima too quickly. In our experiments the average temperature at which the last improvement to the best found solution occurred was greater than 0.1. Thus, we set T_f to 0.1, which had the effect that the last improvements were achieved far later during an SA run, indicating that the algorithm was able to use a larger percentage of iterations effectively, see Fig. 3a. Table 1 further

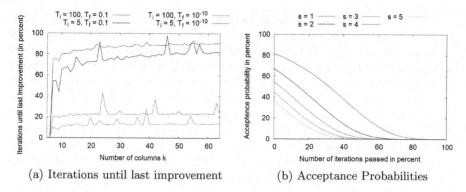

(a) Iterations until last improvement   (b) Acceptance Probabilities

Fig. 3. Parameter tuning: The number of iterations until the last improvement (in percent) is depicted for all extension steps in Fig. 3a. Figure 3b shows the acceptance probabilities of scores $s > 0$ over the course of an SA run with $T_i = 5$ and $T_f = 0.1$.

confirms that, given the same number of iterations, all configurations using a final temperature of $T_f = 0.1$ found better CAs on average than configurations using $T_f = 10^{-10}$.

We further experimented with various different initial temperatures and present the results for $T_i = 5$ and $T_i = 100$. In our experiments, the value for the initial temperature had no significant impact on solution quality. However, as Table 1 shows, the experiments with an initial temperature of $T_i = 5$ yielded slightly better results than those with $T_i = 100$, so for the remainder of this work, we set T_i to 5. We note that, since in these experiments the SA algorithm is limited to the same number of iterations for all tested SIPO configurations, their run times are relatively similar and therefore omitted from this comparison. The acceptance probability for different objective values $s > 0$ for this configuration is shown in Fig. 3b. It nicely shows that the last 20% of an SA run are mostly a local search around the previously best found solution. We believe this is a reasonable compromise between exploration and exploitation.

Table 1. Results of parameter tests averaged over all extension steps for the instance $CA(N; 4, 64, 2)$.

Configuration			Since improvement			Sum of rows	
Type	T_i	T_f	avg	min	T_{min}	avg	min
FullHorizontal	100	0.1	84.17%	93.30%	0.21	4318.1	4247
FullHorizontal	100	10^{-10}	22.73%	43.07%	0.13	4346.3	4265
FullHorizontal	5	0.1	74.22%	97.71%	0.22	4304.2	4230
FullHorizontal	5	10^{-10}	13.21%	27.57%	0.16	4340.5	4262
NewColumn	100	0.1	68.54%	79.52%	0.87	4344.8	4270
NewColumn	5	0.1	45.40%	62.58%	0.84	4350.1	4276
FIPOG-F	–	–	–	–	–	4734	4734
FIPOG	–	–	–	–	–	4862	4862

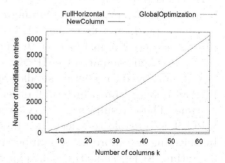

Fig. 4. Number of modifiable entries

Additionally, we compare the performance of SIPO when using the two different enhance types NewColumn and FullHorizontal. Table 1 shows that, while the configuration using FullHorizontal produced smaller arrays on average than configurations using NewColumn, the difference between the two enhance types is surprisingly small. This suggests that the reduction in search space might not have as much of a negative impact on the quality of obtained solutions as one might expect, further supporting the idea of reducing the search space in order to help the metaheuristic algorithms. To highlight how much of a reduction in search space the two different enhance types offer, Fig. 4 depicts the average number of modifiable entries that the SIPO configurations had for each instance. Further, the figure shows the number of modifiable entries that are subject to optimization when applying a global optimization method, assuming it generates CAs with the same number of rows as FullHorizontal. Recall that, as described in Sect. 3, the entire search space for the optimization of binary CAs is 3^m in all cases, where o is the number of modifiable entries.

Last, we compare the performance of SIPO and FIPOG-F by comparing the sum of the number of rows of the generated CAs over all considered instances. As mentioned introductory, FIPOG-F adds an additional level of optimization when compared to FIPOG by optimizing the order in which rows are assigned values. The results in Table 1 show, that all SIPO configurations were able to improve the total number of rows over all instances by a significantly larger margin than FIPOG-F.

To evaluate the run time and verify the effectiveness of this approach for higher alphabets and mixed-level covering arrays, we generated MCAs for four different models coming from real world applications, also used in [19] and [25]. We compared two configurations of the SIPO algorithm with the FIPOG-F implementation of the combinatorial test generation tool CAgen [25] and report the minimum and average number of rows of the generated MCAs, as well as the average run time over 10 runs in Table 2. Configuration SIPO 1k (col) used 1000 iterations as base and enhance type NewColumn, while SIPO 10k (full) used 10000 iterations as base and enhance type FullHorizontal. Both configurations use FIPOG as underlying greedy algorithm, therefore we also list the results of FIPOG, as implemented in CAgen, in the table, to show the improvement in terms of size achieved by the Simulated Annealing algorithm.

Our experiments show that SIPO 10k (full) generates the smallest MCAs in the vast majority of instances, reducing the number of rows by over 15% in some instances when compared to FIPOG. Further, in three different instances, SIPO 1k (col) managed to generate smaller MCAs in less time than the well established FIPOG-F algorithm, while in all other instances it performs better in either size or time. These results seem very promising, especially considering that using SIPO it is possible to select how much time should be spent on optimization by selecting an appropriate number of iterations for the Simulated Annealing algorithm. At the same time, the experiments showed that currently the number of iterations does not scale sufficiently with the strength of the instance, when

Table 2. Results of generated MCAs for testing real world applications. Bolded entries mark results where SIPO performs as good or better than FIPOG-F in terms of size, while underlined entries mark instances where the SIPO configuration terminated faster than FIPOG-F.

	t	FIPOG		FIPOG-F		SIPO 1k (col)			SIPO 10k (full)		
		size	time(ms)	size	time(ms)	min	avg	time(ms)	min	avg	time(ms)
wireless $5^9,4^5,3^7,2^3$	2	45	0	45	2	**40**	**42.8**	47.9	**40**	**43.1**	414.9
	3	315	6	309	158.5	**277**	**293.9**	437.5	**262**	**269.7**	3253.6
	4	1841	167.6	1768	9007.3	**1729**	**1734.4**	_4237.7_	**1627**	**1650.7**	31334.9
	5	11064	3777.9	10124	716572.7	10852	10869.6	_78811.5_	**9101**	**9136**	_287098.2_
	6	57633	80838.9	54170	41443197.3	57540	57550.8	770063.2	54686	54760.9	_3863834.1_
flex $5^2,3^4,2^{23}$	2	26	0	25	1	**25**	**25**	76.3	**25**	**25**	626.6
	3	91	2	100	28	**94**	**98.3**	863	**91**	**96.4**	6050.9
	4	347	45.9	341	682.5	**316**	**321.3**	7300	**314**	**318.1**	69670.2
	5	1164	743.4	1121	20349.5	**1045**	**1053.7**	64721.4	**1026**	**1031.3**	440585.7
	6	3527	9498.1	3398	532310.7	**3176**	**3192**	924954.1	**3093**	**3102.9**	5169165.1
make $6,5,4^2,3^4,2^{14}$	2	25	0	24	0	**24**	**24**	29.4	**24**	**24**	255.6
	3	102	0	107	12	**97**	**101.7**	181.9	**93**	**98.5**	1417
	4	387	10	381	223.5	**371**	**375.5**	926.8	**355**	**358.1**	8711.2
	5	1320	104.8	1260	4260.7	1270	1273	_4250.8_	**1160**	**1190.2**	39183.1
	6	4183	840.6	3941	69391.1	3993	4004.9	_16036.5_	3945	3954.5	160523.7
nanoxml $6,4,3^6,2^{11}$	2	30	0	30	0	**30**	**30**	38	**30**	**30**	349.9
	3	138	1.1	142	26.1	**123**	**125.1**	344.8	**120**	**122**	2506.6
	4	607	24	588	661.1	**542**	**546.2**	2008.3	**513**	**519.4**	18448
	5	2208	322.7	2167	17419.2	**2066**	**2082.8**	_11330.3_	**1954**	**1980.3**	88530.9
	6	7153	3189.2	6993	374719	**6871**	**6888.8**	_125152.3_	**6789**	**6806**	697260.3

applied to instances of higher alphabet. Improving this scaling is subject to future work.

6   New Upper Bounds on CAN

As mentioned in the introduction, generating CAs that are close to optimal is a very difficult combinatorial optimization problem in the general case. However, such highly optimized CAs can provide tighter upper bounds on CAN and will help to improve our understanding of this class of combinatorial design. Currently there is no distinguished optimization strategy that performs *best* for covering array generation in all instances. Therefore, a variety of strategies collectively constitutes the state of the art, which is recorded in [5]. While many different construction methods for CAs have been proposed over the past years, it is remarkable that the last improvement to upper bounds of binary CAs was achieved by SBSTT [22], which was published over 5 years ago. This once again indicates how difficult it is to push the limits of the best known upper bounds even further.

To show the effectiveness of our concept, we conducted experiments to generate binary CAs of strength $t = 6$ and compare them to the current state of the art. We chose these instances, because the previous best known upper bounds on CAN for up to $k = 69$ columns were found using metaheuristic, simulated annealing based, approaches and the instances seemed large enough for our search space reduction to be effective. For smaller instances with $18 \leq k \leq 37$ columns SBSTT was used to set the previously best known upper bounds on CAN, while for larger instances, metaheuristic post optimization (MPO) [23] found the smallest CAs to date. This makes these instances very interesting, as SBSTT is similar to the SIPO algorithm, but acts on the entire array instead of only on selected parts, while MPO also improves IPOG-F generated CAs using Simulated Annealing, but by means of post optimization.

Over the course of our experiments, we were able to improve the currently best known upper bounds on CAN for 43 instances. For all experiments we selected FIPOG-F as greedy construction heuristic. For the Simulated Annealing, an initial temperature of $T_i = 5$ and a final temperature of $T_f = 0.1$ were used. We experimented with four different configurations, with a base number of iterations between 1000 and 1000000. Table 3 contains the results for all CA instances where SIPO managed to improve the currently best known upper bound. The constructed CAs were verified using the combinatorial analysis tool CAmetrics [17] and are available online[1]. Further, Fig. 5 depicts the results for all binary $t = 6$ CA instances for up to $k = 73$ columns.

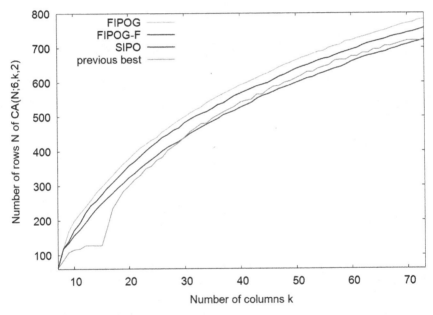

Fig. 5. New upper bounds on CAN$(6, k, 2)$ for $30 \leq k \leq 72$.

As expected, for small instances SBSTT produces significantly smaller CAs due to the global optimization, e.g. for $k = 18$ columns SBSTT found a CA with

---

[1] https://matris.sba-research.org/data/sipo.

Table 3. New upper bounds on CAN for binary arrays of strength $t = 6$ constructed with the SIPO algorithm. The values for the previous state of the art are taken from [5].

k	FIPOG-F	SIPO	prev best	Improvement	prev method
30	483	**440**	441	1	SBSTT
31	493	**452**	457	5	SBSTT
32	502	**461**	468	7	SBSTT
33	509	**470**	480	10	SBSTT
34	515	**479**	482	3	SBSTT
35	524	**487**	496	9	SBSTT
36	535	**495**	505	10	SBSTT
37	540	**506**	514	8	SBSTT
38	551	**512**	522	10	MPO
39	561	**518**	530	12	MPO
40	568	**527**	542	15	MPO
41	575	**534**	547	13	MPO
42	582	**541**	550	9	MPO
43	589	**553**	565	12	MPO
44	594	**561**	565	4	MPO
45	599	**566**	578	12	MPO
46	608	**573**	588	15	MPO
47	617	**581**	590	9	MPO
48	623	**586**	600	14	MPO
49	631	**592**	604	12	MPO
50	639	**599**	612	13	MPO
51	645	**605**	620	15	MPO
52	650	**612**	630	18	MPO
53	656	**615**	630	15	MPO
54	663	**623**	640	17	MPO
55	668	**628**	645	17	MPO
56	671	**634**	650	16	MPO
57	678	**640**	663	23	MPO
58	684	**644**	665	21	MPO
59	689	**651**	665	14	MPO
60	694	**660**	675	15	MPO
61	699	**666**	675	9	MPO
62	707	**672**	685	13	MPO
63	713	**677**	685	8	MPO
64	717	**680**	695	15	MPO
65	721	**685**	695	10	MPO
66	725	**688**	705	17	MPO
67	731	**695**	705	10	MPO
68	735	**699**	710	11	MPO
69	738	**705**	715	10	MPO
70	743	**709**	720	11	Cyclic
71	746	**714**	720	6	Cyclic
72	751	**718**	720	2	Cyclic

$N = 260$ rows, while the smallest CA found by SIPO had 295 rows. At the same time, for instances with $k \geq 30$ columns, SIPO managed to generate smaller CAs than SBSTT. While the MPO algorithm managed to significantly reduce the number of rows of IPOG-F generated CAs, the SIPO algorithm generated smaller CAs for all instances, in one case even improving the previously best known upper bound by 23 rows.

7   Conclusion and Future Work

In this work, we introduced the concept of heuristically enhanced IPO algorithms, which aims to make metaheuristics and heuristic search algorithms applicable to the generation of covering arrays for larger instances. We used Simulated Annealing to implement and compare some of the proposed concepts, discussed parameter choices and evaluated different configurations of the SIPO algorithm. By applying SIPO to the problem of generating binary CAs of strength 6, we achieved improvements to the best known upper bound on CAN for 43 instances.

In the future we want to continue our work on heuristically enhanced IPO algorithms. First, we want to analyze different configurations of the SIPO algorithm further in depth and potentially integrate constraint handling as well. In addition, we want to include the vertical extension of the IPO strategy into the optimization process. We believe this could improve the solution quality substantially. Last but not least, we plan to use different heuristics, consider hybrid approaches that are able to switch between global and local optimization and make use of parallelization and high performance computing to optimize even larger instances.

Acknowledgement. This research was carried out partly in the context of the Austrian COMET K1 program and publicly funded by the Austrian Research Promotion Agency (FFG) and the Vienna Business Agency (WAW).

References

1. Ahmed, B.S., Zamli, K.Z., Lim, C.P.: Application of particle swarm optimization to uniform and variable strength covering array construction. Appl. Soft Comput. **12**(4), 1330–1347 (2012)
2. Bryce, R.C., Colbourn, C.J.: A density-based greedy algorithm for higher strength covering arrays. Softw. Test. Verif. Reliab. **19**(1), 37–53 (2009)
3. Bush, K.A., et al.: Orthogonal arrays of index unity. Ann. Math. Stat. **23**(3), 426–434 (1952)
4. Cohen, D.M., Dalal, S.R., Fredman, M.L., Patton, G.C.: The AETG system: an approach to testing based on combinatorial design. IEEE Trans. Softw. Eng. **23**(7), 437–444 (1997)
5. Colbourn, C.J.: Covering Array Tables for t = 2, 3, 4, 5, 6. http://www.public. asu.edu/~ccolbou/src/tabby/catable.html. Accessed 28 July 2020
6. Colbourn, C.J., Lanus, E., Sarkar, K.: Asymptotic and constructive methods for covering perfect hash families and covering arrays. Des. Codes Cryptogr. **86**(4), 907–937 (2018)

7. Duan, F., Lei, Y., Yu, L., Kacker, R.N., Kuhn, D.R.: Improving IPOG's vertical growth based on a graph coloring scheme. In: 2015 IEEE Eighth International Conference on Software Testing, Verification and Validation Workshops (ICSTW), pp. 1–8 (2015)
8. Forbes, M., Lawrence, J., Lei, Y., Kacker, R.N., Kuhn, D.R.: Refining the in-parameter-order strategy for constructing covering arrays. J. Res. Natl. Inst. Stand. Technol. **113**(5), 287 (2008)
9. Hnich, B., Prestwich, S.D., Selensky, E., Smith, B.M.: Constraint models for the covering test problem. Constraints **11**(2), 199–219 (2006)
10. Jarman, D., et al.: Applying combinatorial testing to large-scale data processing at adobe. In: 2019 IEEE International Conference on Software Testing, Verification and Validation Workshops (ICSTW), pp. 190–193 (2019)
11. Kampel, L., Leithner, M., Simos, D.E.: Sliced AETG: a memory-efficient variant of the AETG covering array generation algorithm. Optim. Lett. **14**(6), 1543–1556 (2020)
12. Kampel, L., Simos, D.E.: A survey on the state of the art of complexity problems for covering arrays. Theor. Comput. Sci. **800**, 107–124 (2019)
13. Kleine, K., Simos, D.E.: An efficient design and implementation of the in-parameter-order algorithm. Math. Comput. Sci. **12**(1), 51–67 (2018)
14. Kuhn, D., Kacker, R., Lei, Y.: Introduction to Combinatorial Testing. Chapman & Hall/CRC. Innovations in Software Engineering and Software Development Series. Taylor & Francis (2013)
15. Lei, Y., Kacker, R., Kuhn, D.R., Okun, V., Lawrence, J.: IPOG: a general strategy for t-way software testing. In: 14th Annual IEEE International Conference and Workshops on the Engineering of Computer-Based Systems (ECBS 2007), pp. 549–556 (March 2007)
16. Lei, Y., Tai, K.C.: In-parameter-order: a test generation strategy for pairwise testing. In: Proceedings Third IEEE International High-Assurance Systems Engineering Symposium (Cat. No. 98EX231), pp. 254–261 (November 1998)
17. Leithner, M., Kleine, K., Simos, D.E.: CAmetrics: a tool for advanced combinatorial analysis and measurement of test sets. In: 2018 IEEE International Conference on Software Testing, Verification and Validation Workshops (ICSTW), pp. 318–327 (2018)
18. Forbes, M., Lawrence, J., Lei, Y., Kacker, R., Kuhn, D.: NIST repository of CAs. https://math.nist.gov/coveringarrays/ipof/ipof-results.html
19. Petke, J., Cohen, M.B., Harman, M., Yoo, S.: Practical combinatorial interaction testing: empirical findings on efficiency and early fault detection. IEEE Trans. Softw. Eng. **41**(9), 901–924 (2015)
20. Shiba, T., Tsuchiya, T., Kikuno, T.: Using artificial life techniques to generate test cases for combinatorial testing. In: Proceedings of the 28th Annual International Computer Software and Applications Conference, 2004, COMPSAC 2004, vol. 1, pp. 72–77 (2004)
21. Torres-Jimenez, J., Izquierdo-Marquez, I., Avila-George, H.: Methods to construct uniform covering arrays. IEEE Access **7**, 42774–42797 (2019)
22. Torres-Jimenez, J., Avila-George, H.: Search-based software engineering to construct binary test-suites. In: Mejia, J., Muñoz, M., Rocha, Á., Calvo-Manzano, J. (eds.) Trends and Applications in Software Engineering. AISC, vol. 405, pp. 201–212. Springer, Cham (2016). https://doi.org/10.1007/978-3-319-26285-7_17
23. Torres-Jimenez, J., Rodriguez-Cristerna, A.: Metaheuristic post-optimization of the NIST repository of covering arrays. CAAI Trans. Intell. Technol. **2**(1), 31–38 (2017)

24. Torres-Jimenez, J., Rodriguez-Tello, E.: New bounds for binary covering arrays using simulated annealing. Inf. Sci. **185**(1), 137–152 (2012)
25. Wagner, M., Kleine, K., Simos, D.E., Kuhn, R., Kacker, R.: CAgen: a fast combinatorial test generation tool with support for constraints and higher-index arrays. In: 2020 IEEE International Conference on Software Testing, Verification and Validation Workshops (ICSTW), pp. 191–200 (2020)
26. Wagner, M., Kampel, L., Simos, D.E.: IPO-Q: a quantum-inspired approach to the IPO strategy used in CA generation. In: Slamanig, D., Tsigaridas, E., Zafeirakopoulos, Z. (eds.) MACIS 2019. LNCS, vol. 11989, pp. 313–323. Springer, Cham (2020). https://doi.org/10.1007/978-3-030-43120-4_24

Author Index